STRONGLY COUPLED
PLASMA PHYSICS

North-Holland
Delta Series

NORTH-HOLLAND
AMSTERDAM • OXFORD • NEW YORK • TOKYO

Strongly Coupled Plasma Physics

Proceedings of the Yamada Conference XXIV on
Strongly Coupled Plasma Physics
Lake Yamanaka, Japan,
August 29 – September 2, 1989

Edited by

Setsuo Ichimaru
Department of Physics
University of Tokyo
Japan

1990

NORTH-HOLLAND
AMSTERDAM • OXFORD • NEW YORK • TOKYO

© Elsevier Science Publishers B.V./Yamada Science Foundation, 1990

All rights reserved. No part of this publication may be reproduced, stored in a retrieval system, or transmitted, in any form or by any means, electronic, mechanical, photocopying, recording or otherwise, without the prior written permission of the copyright holders.

Special regulations for readers in the U.S.A. - This publication has been registered with the Copyright Clearance Center Inc. (CCC), Salem, Massachusetts. Information can be obtained from the CCC about conditions under which photocopies of parts of this publication may be made in the U.S.A. All other copyright questions, including photocopying outside of the U.S.A., should be referred to the Publisher, unless otherwise specified.

No responsibility is assumed by the Publisher for any injury and/or damage to persons or property as a matter of products liability, negligence or otherwise, or from any use or operation of any methods, products, instructions or ideas contained in the material herein.

ISBN: 0 444 883630

Published by:

North-Holland
Elsevier Science Publishers B.V.
P.O. Box 211
1000 AE Amsterdam
The Netherlands

Sole distributors for the U.S.A. and Canada:

Elsevier Science Publishing Company, Inc.
655 Avenue of the Americas
New York, N.Y. 10010
U.S.A.

Library of Congress Cataloging-in-Publication Data

```
Yamada Conference on Strongly Coupled Plasma Physics (1989 : Lake
  Yamanaka, Japan)
    Strongly coupled plasma physics : proceedings of the XXIV Yamada
  Conference on Strongly Coupled Plasma Physics, Lake Yamanaka, Japan,
  August 29-September 2, 1989 / edited by Setsuo Ichimaru.
       p.   cm. -- (North-Holland delta series)
    Sponsored by the Yamada Science Foundation.
    ISBN 0-444-88363-0 (U.S.)
    1. Plasma (Ionized gases)--Congresses.  2. Plasma density-
  -Congresses.   I. Ichimaru, Setsuo, 1935-    . II. Yamada Kagaku
  Shinkō Zaidan.  III. Title.  IV. Series.
  QC717.6.Y36  1989
  530.4'4--dc20                                              90-6727
                                                                CIP
```

Printed in The Netherlands

PREFACE

The Twenty-Fourth YAMADA CONFERENCE on Strongly Coupled Plasma Physics was held from August 29 to September 2, 1989, at Hotel Mount Fuji near Lake Yamanaka on the outskirts of Tokyo. Participants consisted of 66 scientists from within Japan and 50 scientists representing 13 countries from abroad.

Charged particles in dense matter exhibit strong correlations due to the exchange and Coulomb interactions, and thus make a strongly coupled plasma. Examples in laboratory and astrophysical settings include: solid and liquid metals, semiconductors, charged particles in lower dimensions such as those trapped in interfacial states of condensed matter, dense multi-ionic systems such as superionic conductors, laser-cooled ions in the magnetic traps, inertial-confinement-fusion plasmas, shock-compressed ultrahigh-pressure metals, interiors of giant planets and brown dwarfs, stellar interiors, white dwarf interiors and the crustal matter of a neutron star. The Conference was devoted to elucidation of various physical processes involved in such dense materials. The subject areas covered in the Conference ranged over plasma physics, atomic and molecular physics, condensed matter physics and astrophysics.

The main program of the Conference consisted of three invited lectures, forty invited reviews and forty-eight contributed papers. The present volume compiles almost all of those papers, and its Chapter headings reflect the titles of the Conference sessions.

At the conclusion of this editorial task for the Proceedings, I wish to express my heartfelt thanks to all the participants who contributed in many ways to make the Conference a useful forum to promote scientific progress and to cultivate international cooperation and friendship. Special thanks are due to Professor Naoki Itoh, who oversaw every detail in the preparation and organization of the Conference as the Secretary of the Organizing Committee, to Dr. Hiroshi Iyetomi, who helped me in the editorial work for the Proceedings, to Dr. Shigenori Tanaka, who took care of financial transactions and bookkeeping for the Conference, and to Dr. Aiichiro Nakano, who kept record of the meetings of the Preparatory and Organizing Committees. We acknowledge the considerable assistance in the editorial processes provided by Drs. M.A.J.R. Eligh, Ms. Mary Carpenter, and Mr. Yasuo Nishikawa of Elsevier Science Publishers B.V.

The Conference was supported totally by the Yamada Science Foundation. The majority of the work reported by the Japanese participants was supported in part by the Ministry of Education, Science and Culture of Japan through Grant-in-Aid for Scientific Research, No. 62302058, on Dynamics of Strongly Coupled Plasmas. All of these are gratefully acknowledged.

Setsuo Ichimaru
Santa Barbara, California

LIST OF COMMITTEES

International Advisory Board

N. Ashcroft (Ithaca)
J.-P. Hansen (Lyon)
T. O'Neil (San Diego)
E. Schatzman (Meudon)
M. Tosi (Trieste)

H. DeWitt (Livermore)
W. Kohn (Santa Barbara)
D. Pines (Urbana)
K. Singwi (Evanston)
H. Van Horn (Rochester)

Organizing Committee

R. Abe (Tokyo)
Y. Furutani (Okayama)
S. Ichimaru, *Chairman* (Tokyo)
K. Kawasaki (Fukuoka)
K. Nishihara (Osaka)
S. Takeno (Kyoto)
A. Ueda (Kyoto)
M. Watabe (Hiroshima)

T. Ando (Tokyo)
Y. Hiwatari (Kanazawa)
N. Itoh (Tokyo)
Y. Nagaoka (Nagoya)
K. Nomoto (Tokyo)
M. Tanaka (Sendai)
Y. Wada (Tokyo)
F. Yonezawa (Yokohama)

OPENING ADDRESS

Ladies and Gentlemen:

On behalf of the Organizing Committee for the Twenty-Fourth Yamada Conference on Strongly Coupled Plasma Physics, I wish to begin the Opening Address by recalling how this Conference was conceived, prepared, and organized.

First of all, it is my privilege to express our heartiest thanks to the Yamada Science Foundation and Professor Yasusada Yamada for their total support of the Conference. As I shall explain shortly, Professor Yamada was essential in the realization of this Conference from the very beginning of its preparation; he supported the idea of having this sort of international conference in Japan and put forward such a proposal to the Board of Directors, of which he himself is a Member.

Internationally, a proposal for a scientific meeting on strongly coupled plasma physics in Japan this year arose at the closing of the Strongly Coupled Plasma Physics Meeting held three years ago in the summer of 1986 at the University of California, Santa Cruz. After my return to Japan from that meeting, I initiated an effort in such a direction, along with some of my colleagues here, by submitting a proposal for a three-year Research Grant to the Ministry of Education, Science and Culture, for the specific purpose of preparing for an eventual realization of such an international conference in Japan; we were encouraged by successful funding of this proposal. Soon after the approval of the Grant, the International Advisory Board[1] and the local Preparation Committee were formed to consolidate the organizational structure for the Conference. Subsequently, we contacted Professor Yamada and submitted the proposal for the Conference to the Yamada Science Foundation with his support. When the proposal was approved officially by the Board of Directors of the Foundation in May 1988, the current Organizing Committee[1] was formed. We are, therefore, very grateful for the support and

---

[1] see page vii

encouragement extended to us from various sources; we are pleased that the conference is now a reality.

Scientifically, the subject area covered by this Conference is fairly wide, ranging over plasma physics, atomic and molecular physics, condensed matter physics, and astrophysics. Some people may remark that the area to be covered is too broad, but in this age of acute specialization, we believe that this sort of conference has its place in promoting cross fertilization between various disciplines in science. It is the hope of all the Members of the Organizing Committee that the week spent together at the foot of Mount Fuji by Lake Yamanaka will enhance scientific activities, mutual friendship, and the spirit of international cooperation.

Thank you.

Setsuo Ichimaru

WELCOME ADDRESS

Ladies and Gentlemen:

On behalf of the Yamada Science Foundation, I would like to extend our hearty welcome to all of you who are participating in the Twenty-Fourth Yamada Conference on Strongly Coupled Plasma Physics, especially to those who have come a long way to Japan from abroad.

The Yamada Science Foundation develops its activities by giving support to creative projects in the field of basic natural science, providing travel funds for scientists, organizing conferences, promoting international collaboration projects and so forth. Among these, the organization of Yamada Conferences, usually held two or three times a year, is one of the most important activities. As the guiding principles to promote these activities, the Board of Directors of the Foundation puts emphasis on the following three symbolic letter 'I's. The first 'I' represents 'International', the second 'I' means 'Interdisciplinary' and the third and the most important 'I' symbolizes 'Innovative'.

In this context, the present Conference is particularly suited for the scope of the Foundation: It is international, having many guests from all over the world; it treats an interdisciplinary field traversing astrophysics, plasma physics, superconductivity, physics of glass states etc., and above all, its subject is one of the most creative and updated areas. For instance, if I understand correctly, the topic is directly related to the basic mechanism of high-T_c superconductivity. I believe that this Conference would fulfill the purpose of our Foundation through most active participation of all of you.

Another important aspect of this Conference is to provide opportunities to develop new friendships among the participants (particularly among young scientists who are not acquainted with each other) not only through hot scientific discussions but also through heartwarming chattering drinking Yamanashi wine; I expect your most active participation in this aspect too.

Finally, I would like to express our thanks to the Organizing Committee members for their efforts towards the successful performance of the Conference, choosing such a nice, scenic and wine producing district. I hope that all of you will enjoy the discussions and also relax at the foot of Mount Fuji, a symbol of the beauty of Japan.

Thank you.

Yasusada Yamada
Member, Board of Directors
Yamada Science Foundation

YAMADA SCIENCE FOUNDATION AND THE SCOPE OF YAMADA CONFERENCES

The Yamada Science Foundation was established in February 1977 in Osaka through the generosity of Mr. Kiro Yamada. Mr. Yamada was President of Rohto Pharmaceutical Company, Limited, a well-known manufacturer of medicines in Japan. He recognized that creative, unconstrained, basic research is indispensable for the future welfare and prosperity of mankind and he has been deeply concerned with its promotion. Therefore funds for this Foundation were donated from his private holdings.

The principal activity of the Yamada Science Foundation is to offer financial assistance to creative research in the basic natural sciences, particularly in interdisciplinary domains that bridge established fields. Projects which promote international cooperation are also favored. By assisting in the exchange of visiting scientists and encouraging international meetings, this Foundation intends to greatly further the progress of science in the global environment.

In this context the Yamada Science Foundation sponsors international Yamada Conferences once or twice a year in Japan. Subjects to be selected by the Foundation should be most timely and stimulating. These conferences are expected to be of the highest international standard so as to significantly foster advances in their respective fields.

EXECUTIVE MEMBERS OF YAMADA SCIENCE FOUNDATION

Officers:
Board of Directors
Leo ESAKI
Kenichi FUKUI
Osamu HAYAISHI
Noburô KAMIYA
Takeo NAGAMIYA, Director General
Shuntaro OGAWA, Standing Director
Syûzô SEKI
Tomoji SUZUKI
Yasusada YAMADA

Auditors:
Shigekiyo MUKAI
Jin-ichi TAKAMURA

Advisors:
Shiro AKABORI

TABLE OF CONTENTS

Preface	v
List of Committees	vii
Group Photograph	viii
Opening Address	ix
Welcome Address	xi
Yamada Science Foundation and the Scope of Yamada Conferences	xiii
Executive Members of Yamada Science Foundation	xiv

Chapter I: ASTROPHYSICS

Phase Transitions in Dense Astrophysical Plasmas
    H.M. Van Horn   3

Plasma Thermodynamics and the Evolution of Brown Dwarfs and Planets
    W.B. Hubbard   21

Discovery of Low Mass Objects in Taurus
    W.J. Forrest, Z. Ninkov, J.D. Garnett, M.F. Skrutskie and M. Shure   33

Topics in X-Ray Astronomy from Observations with Ginga
    K. Makishima   43

Proton Abundance in Hot Neutron Star Matter
    T. Takatsuka and J. Hiura   55

Thermonuclear Reaction Rates of Dense Carbon-Oxygen Mixtures in White Dwarfs
    S. Ogata, H. Iyetomi and S. Ichimaru   59

Chapter II: COMPUTER SIMULATIONS OF QUANTUM AND CLASSICAL MANY-BODY SYSTEMS

Quantum Monte Carlo Simulation of Hydrogen Plasmas
    J. Theilhaber and B.J. Alder   65

Quantum Simulation of Superconductivity
    M. Imada   81

Dynamic Simulation of Mixed Quantum-Classical Systems
    R.K. Kalia, P. Vashishta, S.W. de Leeuw and J. Harris   93

Monte Carlo Simulation Study of Dense Plasmas:
Freezing, Transport and Nuclear Reaction
    S. Ichimaru and S. Ogata   101

Static and Dynamic Properties of Confined, Cold Ion Plasmas:
MD Simulations
    J.P. Schiffer   113

Molecular Dynamics Study of Rapidly Quenched OCP
    M. Tanaka   125

Monte-Carlo Simulations for the Surface Properties of
the Strongly Coupled One-Component Plasma
    A. Ishida, T. Shirakawa, M. Hasegawa and M. Watabe   129

Chapter III: GLASS AND FREEZING TRANSITIONS

Freezing of Coulomb Liquids
    M.P. Tosi   135

Density Functional Theory of Quantum Wigner
Crystallization
    G. Pastore and G. Senatore   145

Molecular Dynamics Studies of Glassy States: Supercooled
Liquids and Amorphized Solids
    S. Yip   149

Stochastic Dynamics of Atoms near a Glass Transition Point
    Y. Hitawari and T. Odagaki   163

Molecular-Dynamics Study of Binary Alloys: Dynamical
Correlations of the Supercooled Liquids near the Glass
Transition of Binary Soft-Sphere Mixtures
    H. Miyagawa and Y. Hiwatari   167

Effect of the Quantum Electrons to Formation of a
Crystalline Order in Alkali Metals
    S. Nagano and S. Ohnishi   171

Chapter IV: STRONG-COUPLING THEORIES AND EXPERIMENTS IN SPECIFIC GEOMETRIES

Observation of Correlations in Finite, Strongly Coupled Ion Plasmas
    J.J. Bollinger, S.L. Gilbert, D.J. Heinzen, W.M. Itano and D.J. Wineland   177

Theory of Strongly-Correlated Pure Ion Plasma in Penning Traps
    D.H.E. Dubin and T.M. O'Neil   189

Surface Properties of the Coulomb Liquids: From the Classical One-Component Plasma to Liquid Metals
    M. Hasegawa and M. Watabe   201

Classical Charged Particle Systems with Interfaces
    H. Totsuji   213

Surface Correlations in Classical Finite Coulomb Systems
    Ph. Choquard   225

Pattern Formation Processes in Binary Mixtures with Surfactants
    T. Kawakatsu and K. Kawasaki   237

Chapter V: CHARGED PARTICLES IN LOWER DIMENSIONS AND/OR IN MAGNETIC FIELDS

Excitations in Conducting Polymers
    Y. Wada   243

Doping Disorder and Band Structures in Conjugated Polymers
    K. Harigaya, Y. Wada and K. Fesser   255

Strongly Coupled One-Dimensional System and the Polymer
    X. Sun, C. Wu, R. Fu, D.L. Lin and T.F. George   259

Many-Body Effects in Quantum Wells
    T. Ando   263

Strongly Correlated Two-Dimensional Electrons Formed on Dielectric Materials
    K. Kajita   275

Two-Dimensional Coulomb Systems: Solvable Models at $\Gamma = 2$
    B. Jancovici   285

Approximate Thermodynamic Functions for the Two-Dimensional Two-Component Coulomb Gas
    M. Lavaud and S. Brochot   297

Strongly Coupled 2D OCP in a Magnetic Field
    A. Isihara   301

Collisional Relaxation of a Strongly Magnetized Pure Electron Plasma (Theory and Experiment)
    T.M. O'Neil, P.G. Hjorth, B. Beck, J. Fajans and J.H. Malmberg   313

Long-Time Tails of Time Correlation Functions for an Ionic Mixture in a Magnetic Field and the Validity of Magnetohydrodynamics
    L.G. Suttorp   325

Chapter VI: QUANTUM ELECTRON LIQUIDS IN STRONG COUPLING

Density Functional Theory of Superconductors Regarded as Two-Component Plasmas
    W. Kohn   331

Green's Function and Dynamic Correlations of Electrons in Metals
    A. Nakano and S. Ichimaru   337

Frequency-Dependent Local-Field Factor $G(k,\omega)$ for a Two-Dimensional Electron Gas
    K.S. Singwi   349

Variational Theory of Electron Liquid
    Y. Takada   357

Landau Interaction Function and Effective Mass of an Electron Liquid
    H. Yasuhara and Y. Ousaka   369

RPA, Vertex Correction and Superconductivity in Two-Dimensional Models
    K. Yonemitsu   373

Absence of Exponential Screening in Quantum Mechanical Plasmas
    A. Alastuey and Ph.A. Martin   377

Chapter VII: METALLIC SYSTEMS

Nature of Phonons, Isotope Effect, and Superconductivity in $Ba_{1-x}K_xBiO_3$
    M.H. Degani, R.K. Kalia and P. Vashishta   385

Microscopic Derivation of Landau-Ginzburg Free Energy for an Ion-Electron Two-Component Plasma
    K. Ebina and M. Kaburagi   397

Thermodynamic Properties of a Liquid Metal Using a Soft-Sphere Reference System
    M. Hasegawa, I. Kondo, M. Watabe and W.H. Young   401

Electron-Ion Strong Coupling Effects in Dense Hydrogen Plasmas I. Equation of State and Electric Conductivity
    S. Tanaka, X.-Z. Yan and S. Ichimaru   405

Density Functional Approach to Particle Correlations and Electronic Structure in Dense Plasmas
    C. Dharma-wardana   409

Effect of the Electron-Ion Correlation Potentials on Thermodynamic Functions in Dense H and He Plasmas
    F. Perrot, Y. Furutani and C. Dharma-wardana   421

Energy Loss of Charged Particles in Liquid and Amorphous Metals
    F. Yoshida   425

Studies of a Strongly Coupled Plasma Produced in a Capillary Discharge
    J.F. Benage, Jr., L.A. Jones, R.J. Trainor, Jr., W.R. Shanahan, R.L. Shepherd and D.P. Nothwang   429

The Measurement of Transport Properties in Strongly Coupled Plasmas
    R.L. Shepherd, D.R. Kania, L.A. Jones, D.H. Schneider and R.E. Stewart   433

Electrical Resistivity of Strongly Coupled Plasmas in
Intense Fields
    R. Cauble, F.J. Rogers and W. Rozmus   439

Generation of a Strongly Coupled Plasma with Electron
Temperature around 4.2 K in Cryogenic Helium Gases
    K. Minami, K. Kato, A. Sugawara and T. Nomura   445

Measurement of the Dynamic Form Factor at Low
Frequencies for a Plasma with $\Gamma = .06$
    A.W. DeSilva and Y.Q. Zhang   449

Chapter VIII: METAL-INSULATOR TRANSITION

Thermodynamic and Structural Properties of Fluid Metals
in the Metal-Insulator Transition Range
    F. Hensel   455

Theoretical Study of Atomic and Electronic Structures in
Microclusters of Potassium and Mercury
    F. Yonezawa and S. Sakamoto   467

Ionization Effects in a Model Fluid
    J.P. Hernandez   479

The Insulator-Metal Transition in Dense Plasmas
    H. Hess   483

Pressure Ionization in Fluid Hydrogen
    G. Charbier and D. Saumon   495

Thermodynamics and Transport in Dense Partially Ionized
Plasmas
    W.-D. Kraeft, M.K. Kilimann and D. Kremp   507

Chapter IX: ATOMIC AND MOLECULAR STATES AND RADIATION

Generlized Schrodinger Equations for Shifts, Widths, and Wave
Functions of Atomic and Molecular States in Dense Matter
    M.D. Girardeau   521

Dynamics of Electric Fields in Strongly Coupled Plasmas
    J.W. Dufty and L. Zogaib   533

Electron-Ion Strong Coupling Effects in Dense Hydrogen
Plasmas II. Electric Levels of Impurity Ions
    S. Tanaka and S. Ichimaru   545

Equation of State and Opacity of Dense Plasmas
    F.J. Rogers and C.A. Iglesias   549

Some Interpretation of Experimental Values of DC
Electrical Conductivity and Spectral Line Shape
    M.M. Popovic, Y. Vitel and A.A. Mihajlov   561

Experimental Study of Optical Properties of Strongly
Coupled Plasmas
    V.E. Fortov, V.E. Bespalov, M.I. Kulish and S.I. Kuz   571

Many-Electron Effects on Dynamic Processes in Dense Matter
    S.M. Younger   583

Chapter X: SHOCK-COMPRESSED PLASMAS AND INERTIAL-CONFINEMENT-FUSION PLASMAS

Laser Produced Optically-Thin Strongly Coupled Plasmas
    A.N. Mostovych, K.J. Kearney and J.A. Stamper   589

Ion Beam-Plasma Interaction: A Standard Model Approach
    C. Deutsch   601

Particle Simulations on Static and Dynamic Properties of
Two Component Hot Dense Plasmas
    H. Furukawa, K. Nishihara, M. Kawaguchi, H. Sakagami
    T. Hiramatsu and H. Yasui   613

Optical Observation of Laser-Compressed Material
    Y. Sakagami, T. Nomura and H. Yoshida   617

Mechanism of Fuel Compression in ICF and Property of
Compressed Fuel Plasma
    K. Niu   621

Charge Neutralization During Propagation of Intense Light Ion
Beam for ICF Driver
    T. Kaneda and K. Niu   625

Chapter XI: DENSE MULTI-IONIC SYSTEMS

Dynamics and Mechanism of Diffusion in Superionic Conductors
    Y. Kaneko and A. Ueda   631

Properties of Strongly Coupled Multi-Ionic Plasmas
    H.E. DeWitt, W.L. Slattery and G.S. Stringfellow   635

Linear and Electronic Transport in Strongly Coupled
Binary Ionic Mixtures
    D. Leger and C. Deutsch   649

Statistical-Mechanical Effects on Cold Nuclear Fusion
in Metal Hydrides
    S. Ichimaru, S. Ogata, A. Nakano, H. Iyetomi and
    T. Tajima   653

Chapter XII: STRONG-COUPLING THEORIES AND EXPERIMENTS IN GENERAL

Critical Compressibility Factory of Lattice Gas
    R. Abe   659

Structural Phase Transitions in Dense Hydrogen
    H. Nagara   663

Plasma Contributions to the Cohesive Energy of Charge
Stabilised Colloidal Systems
    E. Canessa, M.J. Grimson and M. Silbert   675

A Two-Dimensional Polymer Chain with Short-Range
Interactions
    M. Takasu, J. Takashima and Y. Hiwatari   679

New Empirical Bridge Functions of Integral Equation:
Application to the Binary Supercooled Liquids of the Twelfth
Inverse Power Potential
    S. Kambayashi and Y. Hiwatari   683

Extended Mean Density Approximation for Structure Factors
of Fluids
    M. Itoh, O. Honda and K. Nakayama   687

Integral Equation Approach for Charged Colloidal Dispersions
    M. Fushiki   691

Density Functional Theory and Langevin-Diffusion Equation
    T. Munakata 695

Author Index 699
Subject Index 701

Chapter I:
Astrophysics

PHASE TRANSITIONS IN DENSE ASTROPHYSICAL PLASMAS

H. M. VAN HORN

Department of Physics and Astronomy, C. E. Kenneth Mees Observatory, and Laboratory for Laser Energetics, University of Rochester, Rochester, NY 14627-0011

The realization that dense plasmas may freeze, forming crystallized cores in white dwarfs and crusts in neutron stars, initiated an exploration of phase transitions in dense stars. In the intervening 20 years, several other types of phase transitions have been studied. In this review, I shall summarize the advances and identify areas that seem to me particularly interesting and important from the point of view of astrophysics. These include the following: (1) The transition to a glassy state has been proposed as an alternative to crystallization in white dwarfs. (2) The freezing and possible phase separation of binary mixtures, which had been suggested as a possible energy source in very cool C/O white dwarfs, seems unlikely to occur, according to recent calculations. (3) Fe/H phase separation, first suggested as a possibility to account for the solar neutrino problem, almost certainly does not occur in the Sun. It may be a significant effect in the evolution of low mass stars and "brown dwarfs," however. (4) H/He phase separation almost certainly does occur in giant planets and provides a significant energy source in these bodies. (5) It has been suggested that the metallization of H may occur via a "plasma phase transition". Some recent theoretical work supports this idea, but experimantal tests are still necessary.

1. INTRODUCTION

The idea that the plasma in the interior of a dense star can exist in more than one phase was first introduced to astrophysics about 30 years ago. Initially, this concept seems to have been regarded mainly as a technical device to permit calculations of the properties of matter at high densities. Thus, Kirzhnits[1], Abrikosov[2], and Salpeter[3] recognized that a crystalline lattice is the lowest energy state of fully ionized matter at $T = 0$, and they used this fact in computing the equation of state of dense astrophysical plasmas.

The pioneering calculations of Brush, Sahlin, and Teller[4] were the first to explore seriously the fluid/solid phase transition in dense matter. They performed Monte Carlo calculations of the Coulomb interaction energy of the so-called "one-component plasma" (OCP), a distribution of classical, point ions in a uniform, neutralizing background. They found that for sufficiently large values of the Coulomb coupling parameter $\Gamma = (Ze)^2/ak_BT$, where a is the radius of the Wigner-Seitz cell containing a single ion, the system undergoes a spontaneous transition from a fluid phase to a solid phase. This discovery stimulated immediate applications to dense stars. Mestel and Ruderman[5] used the existence of the high-temperature solid phase to compute the thermal properties of matter in cooling white dwarfs, and Van Horn[6] showed that white dwarf matter freezes while the star may still be hot enough to be observable. The discovery of pulsars and their interpretation as rotating neutron stars provided another application, as it was realized immediately that the surface layers of these stars would freeze into solid crusts[7].

Since this early work, our understanding of phase transitions in dense, astrophysical plasmas has advanced considerably. The purpose of this review is to summarize those advances, concentrating primarily on the astrophysical applications.

2. SOLIDIFICATION OF DENSE ASTROPHYSICAL PLASMAS

2.1. Crystallization of white dwarfs

Following the seminal investigation by Brush *et al.*[4], Hansen[8] and his coworkers[9] have made the OCP one of the best-studied models in modern statistical physics. It has now become possible to compute tens of millions of configurations for systems containing thousands of particles, so that thermodynamic averages can be calculated with precision, accurate interparticle correlation functions can be obtained, and various transport processes can be evaluated. The most recent and accurate calculations for the OCP of which I am aware are those by Slattery, Doolen, and DeWitt[10,11], who find that the OCP freezes into a *bcc* lattice at $\Gamma = 178 \pm 1$. If crystallization of the plasma is prevented entirely, the OCP can undergo a transition to a "Coulomb glass" at still higher values of Γ[12] (*cf*. §2.2. below).

It is important to recognize that a real dense plasma, such as that found in the core of a white dwarf or the crust of a neutron star, differs from an OCP in one very significant way. The OCP by definition has rigid, uniform background without any physical properties. In a real dense plasma, however, the neutralizing background in which the ions move consists of electrons with very definite physical properties. In particular, the Fermi energy E_F of the electrons completely dominates the pressure of the matter, while the ions dominate the thermal properties. The latter are strongly affected by the Coulomb energy E_{Coul}, particularly at low temperatures, where $E_{Coul} \gg k_B T$ or equivalently $\Gamma \gg 1$. At sufficiently high densities, where $E_F \gg E_{Coul}$, the electron density is very nearly uniform, so that the OCP provides a good approximation for the properties of the ionic component of the plasma. This is the case in the deep interior of a white dwarf. At lower densities, however, the polarization of the electron background by the ions must be taken into account, and this complicates the calculation of the phase diagram of dense plasmas considerably[13].

The first calculations of the evolution of white dwarfs which included the full effects of the fluid/solid phase transition were carried out by Lamb[14,15]. For a 1 $M_\odot$, pure ^{12}C white dwarf model, they found that core crystallization begins at an age $\sim 10^9$ years, when the star has cooled to a luminosity $L \approx 1.6 \times 10^{-3} L_\odot$ and an effective temperature $T_{\text{eff}} \approx 13,000$ K. This is indeed sufficiently hot and bright to permit direct observational study. Real white dwarfs are observed to have luminosities as faint as $\approx 10^{-4.5} L_\odot$ and temperatures as low as ~ 4500 K. However, they are believed to have cores consisting of a mixture of C and O, rather than pure C, and they generally have masses closer to $0.6 M_\odot$ than to $1.0 M_\odot$. Nevertheless, more recent calculations[16] continue to indicate that crystallizing white dwarfs are in principle observable.

Crystallization releases the latent heat of fusion associated with the formation of the solid lattice. This is a significant fraction of the thermal energy of the plasma and initially slows the cooling of the star appreciably. Eventually, however, the core temperature T of the white dwarf falls below the Debye temperature Θ_D of the lattice. The heat capacity subsequently falls as $(T/\Theta_D)^3$, and the star begins to cool increasingly rapidly. Both of these effects are evident in the theoretical white dwarf "luminosity function," the number density of stars (per pc^{-3}) per unit interval in $\log(L/L_\odot)$. The difficulty of determining the observational luminosity function at very low luminosities has prevented a sufficiently accurate determination of this quantity to test these theoretical predictions with present data, but we expect that data from the Hubble Space Telescope will make this possible in the near future. (See also §III below).

2.2. Transition to a glassy state

Ichimaru et al.[12] have used an improved HNC scheme to study the OCP at large values of Γ. They found that the second peak in pair correlation function $g(r)$ broadened near $\Gamma \approx 200$ and subsequent split in two for $\Gamma \geq 300$. As such splitting is common to glassy substances, they therefore concluded that the supercooled, metastable state is indeed a "Coulomb glass." They estimated the transition time to the stable, crystalline phase to be $\sim 10^5$ years and thus suggested that astrophysical dense matter with $171 \lesssim \Gamma \lesssim 210$, may be in such an amorphous, glassy state, rather than a crystalline state. More recently, Ogata and Ichimaru[17,18] have investigated the dynamic evolution of the microstructure in supercooled OCPs using Monte Carlo simulations. They studied four cases of rapid quenches from an equilibrium fluid at $\Gamma = 160$ to $\Gamma = 200, 300$, or 400. Except for the quench to $\Gamma = 200$, which formed a supercooled fluid rather than glass, all of these cases relaxed to metastable states (Coulomb glasses) with internal energies lying distinctly below that of the fluid, but above the bcc lattice energy.

These results are consistent with experience in terrestrial laboratories, where the preparation of a metastable state requires either (i) "splat cooling," which provides such a rapid quench that the system has no chance to reach the state of lowest internal energy or else (ii) very gradual cooling through the fluid/solid phase transition without external disturbances.

Neither of these situations obtains in white dwarfs, however. The white dwarf cooling timescale near the onset of freezing is billions of years, amply long to achieve microscopic equilibrium and avoid the metastable glassy state. Further, white dwarfs are sufficiently "noisy" that it seems unlikely that a metastable state could remain undisturbed. Surface convection zones, which are present in all cool white dwarfs, provide a noise source intrinsic to the star. In addition, all DA white dwarfs believed to pass through the pulsationally unstable ZZ Ceti phase, where the star undergoes global oscillations. Thus, the conditions necessary for the formation of a metastable, glassy state do not appear to occur in a white dwarf, and I therefore expect them to freeze into a crystalline lattice rather than a glass.

3. FREEZING OF C/O WHITE DWARFS AND THE AGE OF THE GALAXY

Until quite recently, the generally accepted view was that our Galaxy had been formed in a rapid inital collapse[19]. In this picture, the ages of all the components of the Galaxy must be the same as the ages of the oldest globular clusters, $(15 \pm 3) \times 10^9$ years $\equiv 15 \pm 3$ Gyr[20]. Recently, however, methods based on nucleocosmochronology and on the cooling ages of the white dwarfs have been introduced, both of which yield ages for the galactic disk that are much younger than this. For example, Malaney and Fowler[21] have obtained a galactic age $t_G \lesssim 12$ Gyr from both Th/Nd and Eu/Ba ratios.

The other new method of obtaining the age of the galactic disk, introduced by Winget et al.[22], combines the observed deficiency of white dwarfs having luminosities $L < 10^{-4.5} L_\odot$[23] with white dwarf cooling theory. The result gives $t_G \approx 9$ Gyr, consistent with the results of nucleocosmochronology but in conflict with the ages of the globular clusters. A more recent and completely independent calculation by Iben and Laughlin[24] obtains a disk age ~ 9 Gyr, in agreement with the results of Winget et al.[22].

Can carbon-oxygen phase separation explain this discrepancy? Until recently, it had been thought that separation of the C/O plasma in a white dwarf into C-rich and O-rich phases upon freezing, as first proposed Stevenson[25], might account for the difference between the white dwarf cooling age and the ages of the globular clusters. As shown by Mochkovitch[26], the sinking of the denser, O-rich solid releases substantial gravitational energy, slowing the cooling of a white dwarf and lengthening its age. If this were to occur, the disk of the Galaxy could be much older than the current estimate given by the "white dwarf chronometer."

Lengthening the age of a cool white dwarf also increases the luminosity function at these faint magnitudes, however. Garcia-Berro et al.[27] have recently computed the effect of complete phase separation upon the luminosity function, and they find a large and potentially observable effect. Thus, if phase separation were to occur, it could eventually be subjected to observational test.

It now seems very unlikely that C/O phase separation will take place, however. Barrat et al.[28] have recomputed the phase diagram for a C/O mixture and have found it to be of the spindle type rather than the eutectic type suggested by Stevenson. They conclude that significant phase separation does not occur, and that the maximum increase in the white dwarf ages associated with the freezing of this binary plasma is about 0.5 Gyr. Still more recently, Ichimaru, Iyetomi, and Ogata[29] have also recomputed the C/O phase diagram and found it to be of an "azeotropic" form. Like Barrat et al., they have concluded that significant phase separation does not occur and that the effect on white dwarf ages is minimal. These two studies essentially close the book on this effect.

The most attractive possibility for resolving the difference between the ages obtained from white dwarf cooling and nucleocosmochronology, on the one hand, and from cluster ages, on the other, seems to me to be to abandon the hypothesis that the disk of the Galaxy is the same age as the halo. Indeed, according to Norris and Green[30], "There seems no compelling reason to believe that the Galactic disk in the solar neighborhood has any major stellar component as old as the disk globular clusters." They favor a more gradual formation process and point out that the pressure-supported collapse models of Larson[31] are in best accord with observations. In Larson's model 6, after ~ 2 Gyr, the disk is confined to within ~ 5 kpc of center. Only after a further several Gyr does the disk form at the solar distance from the center. They regard this as "the most natural explanation of the apparent relative youth of the disk in the solar neighborhood." Larson[32], too, advocates this solution.

In the end, the definitive determination of the age of the galactic disk will almost certainly require the Hubble Space Telescope. In preparation for the launch of this instrument, Tamanaha et al.[33] have recently undertaken a calculation of the contribution to the white dwarf luminosity function of the stars which completed their evolution during the formation of the galactic halo. Such a possibility was first discussed by Larson[34] who postulated "bimodal star formation," with a "high-mass" star formation mode occurring preferentially in the early history of the galactic disk.

Tamanaha et al.[33] point out that after ~ 15 Gyr, a 0.8 $M_\odot$ white dwarf will have cooled to $L \sim 10^{-6} L_\odot$, implying $M_V \sim +20$, just below the current limit of detection. Though there exist increasing uncertainties in the microphysics at lower luminosities, these authors have made a first effort to explore this domain. Their most interesting finding is that the most extreme models do allow the entire halo dark matter to consist of white dwarfs, with ages $t_{halo} = 12$ to 13 Gyr. Younger halos cannot supply all the dark matter. This is a testable result, and when the HST is launched, it will surely be one of the priorities for observational investigation.

4. THE POSSIBILITY OF Fe/H PHASE SEPARATION IN LOW MASS STARS

The cumulative result from the ^{37}Cl experiment which has been running in the Homestake Mine for the past two decades yields a solar neutrino production rate of 2.0 ± 0.3 SNU[35]. For comparison, the most recent value from the standard solr model is $7.9(1 \pm 0.33)$ SNU[36]. This discrepancy has stimulated numerous theoretical efforts to find a solution[36,37].

One suggestion for the resolution of this problem was the hypothesis that the center of the Sun may be enriched in heavy elements relative to the abundances detected at the solar surface. A solar model with a small, inert Fe core, formed by draining Fe out of the surrounding H-burning region, would have a significantly different thermal structure from the standard model, perhaps yielding a substantially lower neutrino emission rate.

Motivated by this idea, Pollock and Alder[38] were led to consider the possibility that Fe may undergo phase separation from the H-rich plasma at the high pressures in the solar interior. They performed a hypernetted chain (HNC) calculation of the thermodynamic properties of an Fe/H plasma under conditions like those in Sun ($T = 1.5 \times 10^7$ K, $P = 10^5$ Mbar, Fe concentration $= 2.5 \times 10^{-5}$ ionic mole fraction), using a Debye-Hückel model for the interactions. From the resulting Gibbs free energy G, they computed various simplified Fe/H plasma phase diagrams, which they used to study phase separation in this system. Their results suggested that Fe might indeed undergo phase separation under these conditions, but they emphasized that more accurate calculations would be needed to confirm or refute this suggestion.

This prospect led to several subsequent efforts to carry out more accurate theoretical calculations of the phase diagram of Fe/H plasmas. The most recent of which I am aware is that of Iyetomi and Ichimaru[39], who have used more accurate calculations for electron-screened ion plasmas in the HNC approximation. They find that screening is a substantial effect, but that it is not enough to produce Fe/H phase separation. They find the critical point for demixing to be given by $T_{\text{crit}} \approx 5.5 \times 10^6$ K, $x_{\text{crit}} \approx 1.7 \times 10^{-2}$ at $P = 10^5$ Mbar, where $x = N_{\text{Fe}}/(N_{\text{H}} + N_{\text{Fe}})$. As the temperature at the critical point is only about 1/3 the value at the center of the Sun, Iyetomi and Ichimaru conclude that phase separation does not occur in Sun and thus cannot resolve the solar neutrino problem.

The conclusion that the Sun is too hot for Fe/H phase separation does not exclude other interesting astrophysical possibilities, however. For example, this may occur in very low-mass main sequence stars, which have significantly lower internal temperatures than does the Sun. This possibility has not yet been explored, but preliminary estimates lend credibility to this idea. In their detailed study of the evolution of very low mass stars, D'Antona and Mazzitelli[40,41] found $T_c \approx 4 \times 10^6$ K and $\rho_c \approx 600$ g cm^{-3} at the center of a 0.1 $M_\odot$ H-burning main sequence star. This temperature is well below the critical temperature calculated by Iyetomi and Ichimaru[39], and as the density is about four times greater than that at the center of the Sun, it seems quite likely that Fe/H phase separation will occur in such low-mass stars. Further calculations, under conditions appropriate to such objects, are clearly essential.

"Brown dwarf" stars, which become degenerate and cease gravitational contraction before they become hot enough to ignite H-burning, are still lower in temperature than low-mass main sequence stars. They are thus even more promising candidates for Fe/H phase separation[42]. For example, in a 0.05 $M_\odot \equiv 50\ M_J$ brown dwarf, where M_J is the mass of the planet Jupiter, D'Antona and Mazzitelli[40] find a central temperature $T_c \approx 2 \times 10^6$ K and a central density $\rho_c \approx 450$ g cm^{-3} at an age of about 2×10^9 years. In this object, the temperature has already begun to drop, and – as the temperature already appears to be below the critical temperature for Fe/H phase separation – it seems inescapable that phase separation must be taking place.

5. H/He PHASE SEPARATION IN GIANT PLANETS

The discovery that Jupiter radiates significantly more energy than it receives from the Sun triggered several theoretical efforts to find an explanation. Smoluchowski[41] seems to have been the first to discuss H/He phase equilibrium in Jupiter and to relate phase separation to the excess luminosity observed. Salpeter[14] subsequently pointed out that H/He phase separation can occur even in the presence of the vigorous convection present throughout the interior of Jupiter, with the denser He "raining out" to form a He-enriched core. The associated release of gravitational energy provides a substantial energy source, which greatly lengthens the cooling time. Salpeter also pointed out the possibility of H/He phase separation as an energy source in low-mass main sequence stars and brown dwarfs, but to my knowledge, no detailed evolutionary calculations have yet been carried out incorporating this energy source.

Stevenson and Salpeter[45] have carried out a comprehensive study of the H/He phase diagram, using Stevenson's[46] detailed computation of the Gibbs free energy for this system. For a mixture with a solar H/He ratio (10 % He by number), these calculations predict a miscibility gap in fluid He/metallic H mixtures at Mbar pressures for $T \lesssim 10^4$ K. MacFarlane[47], however, finds from a Thomas-Fermi-Dirac calculation that the critical temperatures are much lower than those obtained by Stevenson and Salpeter, so the issue is currently unresolved.

Stevenson and Salpeter[48] next applied these phase diagrams to explore the effect of H/He phase separation on the evolution of the giant planets. They noted that the success of the homogeneous evolutionary calculations for Jupiter[49] suggests that H/He phase separation has not yet begun in this planet, or else has begun only recently ($\lesssim$ 1 Gyr ago). This in turn implies that the critical temperature for the postulated molecular-to-metallic H phase transition (the "plasma phase transition" $\equiv$ PPT; $cf.$ §VI below) cannot exceed $\sim$ 20,000 K. If there is no first-order PPT, the phase diagrams predict H/He phase separation, with the denser He droplets "raining out," when the internal temperature of the planet falls sufficiently low. For a reasonable choice of parameters, this occurs within the domain of the giant planets. Curiously, the core temperatures may actually increase during phase separation, even though the surface temperature decreases monotonically with increasing age. Alternatively, if the PPT does occur, phase separation may proceed by the formation of He-poor "bubbles" of metallic H, which rise buoyantly and enrich the overlying layers. In either case, H/He phase separation releases sufficient gravitational energy to increase the cooling time of a giant planet by as much as a factor of four or five.

Saturn, like Jupiter, radiates more energy than it receives from the Sun. The Voyager 1 flyby showed this factor to be 1.78 ± 0.009[50]. Unlike the case of Jupiter, however, homogeneous evolutionary models for Saturn reach its current luminosity after only $\sim$ 2 Gyr, rather than the 4.5 Gyr age of the solar system. Thus, H/He phase separation with the accompanying release of gravitational energy appears to be essential to explain the observed excess luminosity[48].

Hubbard and Stevenson[51] have estimated the gravitational energy release from H/He phase separation in Saturn to be $\sim (1.5 \pm 0.7) \times 10^{12}$ erg $(g - He)^{-1}$. From this they estimate the time since the onset of He differentiation to be $\sim 2 \times 10^9$ years. Again, detailed evolutionary calculations, including the effects of phase separation, have not yet been done.

6. THE METALLIZATION OF H

Since Wigner and Huntington[52] first pointed out that H must become a monatomic metal at high densities, there has been considerable interest in the "metallization" of this element. A thorough review of the literature on these topics is beyond the scope of this paper, but significant advances have occurred recently, which I must at least mention. A good summary of the theoretical and experimental situation at high pressures has recently been provided by Mao and Hemley[53].

6.1. The Metallization of Hydrogen at Low Temperatures

Immediately following Wigner and Huntington's[52] prediction of the existence of metallic H, Wildt[54] and Critchfield[55] pointed out that this would have consequences for the internal structures of the giant planets. A rather thorough early discussion of the phase diagram of H and the transition from molecular H_2 to metallic H was given by de Marcus[56], who located the transition pressure between 1.93 and 3.5 Mbar, and who applied the results to construct detailed models of Jupiter and Saturn.

In the 1970s, various experimental investigations of the properties of compressed H became possible, at pressures approaching those anticipated for the transition to the metallic phase. Shock tube experiments had extended up to pressures of 760 kbar[57,58]. Some claims had been made that metallization of H had been actually achieved experimentally[45,48,53], but these results are generally discounted because of questions about the reliability of the experiments. A second major group of high pressure experiments has utilized the diamond anvil cell[59]. Ross[60] has recently summarized both static, diamond anvil-cell measurements and shock measurements on dense hydrogen, and he estimates the presumed first-order molecular-to-metallic phase transition to occur between 3.1 to 3.6 Mbar. Since then, the diamond anvil cell experiments have achieved pressures in excess of 200 GPa ($\equiv$ 2 Mbar) and have produced quite exciting new results, which I summarize below[61].

The issue for the physics of compressed H is the exact nature of the mechanism by which the transition to the metallic phase occurs. Does it happen as a first-order phase transition or through band-overlap[62]?

Friedli and Ashcroft[62] carried out one of the first detailed band-structure calculations for compressed molecular H_2, assuming it to be in the Pa_3 phase. Their calculations were done for a static lattice, but they recognized the importance of the zero-point motion of the very light protons. They found the band gap to vanish at $r_s = 1.48$, where the highest valence band crosses the lowest conduction band, and corresponds to a second-order metal-insulator transition. Chakravarty et al.[62] subsequently performed a density-functional calculation of the equation of state for static lattices of both molecular and metallic H. They conclude that remnant molecular pairing is preferred in the band-overlap metallic state; i.e. the metallic state is molecular. In their calculations, the metal-insulator transition occurs at $\sim$ 2 Mbar, while complete dissociation occurs only at very high densities ($r_s \sim 1.1$). A completely different approach was taken by Cepperley and Alder64, who did quantum Monte Carlo calculations of the energy of light elements at $T = 0$. Application of their method to hydrogen yields a molecular H_2 to monatomic metallic H transition at 2.8 Mbar.

More recently, Min, Jansen, and Freeman[65] have computed the band structure for solid molecular H_2 in the Pa_3 lattice structure, with results very similar to those obtained by Freidli and Ashcroft[63]. Min et al. find two pressure-induced transitions. (1) At 1.7 ± 0.2 Mbar they obtain an insulator-to-metal transition within solid molecular H_2, which occurs by a combination of band-overlap and bond-length relaxation. This is a second-order electronic transition that occurs without a corresponding structural rearrangement. (2) At 4 ± 1 Mbar Min et al. find a first-order structural phase transition to the monatomic metallic hcp phase.

Experimental support for some of these ideas has been obtained quite recently in very high pressure diamond anvil cell experiments. Hemley and Mao[65] have discovered that solid H_2 undergoes a structural phase transition at 145 GPa, as evidenced by an abrupt discontinuity in the intramolecular vibron frequency. They initially interpreted this as the orientational ordering transition which is expected to occur at some pressure. Ashcroft[67] has pointed out that the issue of whether or not rotation of the H_2 molecule is hindered is critical to understanding the nature of the high pressure transitions. If the H_2 rotation is stopped, it cannot prevent band overlap, and an abrupt transition to the metallic state occurs. The absence of such an abrupt onset of metallic properties thus implies that hindered rotation is *not* the explanation for the Carnegie experiments. Conversely, rotation of the H_2 molecule *does prevent* band overlap.

More recently, Barbee et al.[68] have computed the total energy for several different lattice structures, using a plane-wave basis extending up to 36 Ry. They included the important proton zero-point motion in the quasiharmonic approximation. For $P < 50$ Gpa, they found the enthalpies of the Pa_3 and the $m - hcp$ structures to be indistinguishable within their limits of error. For $P > 80 \pm 15$ GPa, however, the $m - hcp$ structure is more stable, in agreement with experiments[69,70]. The H_2 molecules are already oriented (rotationally hindered) in this pressure range, according to their calculations. There is no phase transition associated with this change, because the molecules can orient parallel with the hexagonal axis of this structure without changing the symmetry. Thus, they believe that the phase transition observed by Hemley and Mao[66] at ~ 150 GPa is *not* associated with molecular orientation. For $P > 380 \pm 50$ Gpa, Barbee et al. find a highly anisotropic, filamentary, primitive-hexagonal structure to be the most stable. The transition to the monatomic bcc phase occurs at 860 ± 100 Gpa. They also note that a metal-insulator transition is expected near 200 GPa in the $m-hcp$ phase, but their method of calculation is unable to predict this accurately.

The most recent experiments by Mao and Hemley[53] may finally have detected hydrogen metallization. In a set of seven experiments, they found the intensity of Raman scattering of the molecular vibron of H_2 to appear first near 180 GPa, suggesting the onset of electronic transitions near the laser excitation energy (2.54 eV). At higher pressure, the resonance frequency shifts to longer wavelengths, implying a redshift of the electronic transitions. The authors emphasize that the change of properties with pressure is not discontinuous. Above ~ 150 GPa, light transmission through the part of the hydrogen sample at the highest pressure within the cell gradually decreased, suggesting that the electronic excitation threshold had decreased from the zero-pressure value of 10.9 eV into the visible region. Reflectivity measurements also show an increase above 150 GPa. The lack of a sharp absorption edge may indicate the closure of an indirect gap; this depends upon the exact crystal structure, which is not yet known experimentally for this pahse. Mao and Hemley believe that the maximum P achieved in this study was $>$ 250 GPa. There is some evidence that dissociation may already have occurred occurred at the highest pressures, but the results are not yet conclusive.

Thus, theory and experiment seem to be converging, and our understanding of the pressure-induced transitions of hydrogen has improved dramatically in just the past few years.

6.2. The "Plasma Phase Transition" in Dense H

Considerations of the transition to the fully ionized, dense H plasma at elevated temperature have been motivated in part by an interest in computing the opacity of matter at high temperatures and densities[71] and in part by interest in the nature of the phase diagram[72]. The opacity-motivated research, biased by the knowledge that thermal ionization is a continuous process, has focused on rather ad hoc modifications of the Saha equation which lead to "pressure-ionization" through "lowering of the ionization potential"[73]. Conversely, research on the phase diagram of H at high pressures and temperatures has been stimulated by the availability of shock-wave experiments capable of reaching the regime where pressure-ionization becomes important[57].

In their detailed study of the H/He phase diagram, Stevenson and Salpeter[45] developed a simple model for the metal-insulator transition in H and pointed out that it is likely to remain a first-order transition to 10^3 K and possibly to $\lesssim 10^4$ K. If this is true, the coexistence curve must end in a "second critical point" (the first critical point marking the termination of the gas-liquid coexistence curve), because ionization is a continuous process at lower pressures. The model calculation of Stevenson and Salpeter assumes a first-order molecular H_2-to-monatomic metallic H phase transition and predicts the second critical point to lie at $T_c = 3500$ K, $P_c \approx 3$ Mbar.

The next step was taken by Robnik and Kundt[74], who computed a model for the free energy of a mixture of protons, electrons, and H atoms, but with no H_2 molecules. They ignored electron energy band effects (which Stevenson[46] had included) and the interactions between the H atoms and the charged particles. With these approximations, they found the second critical point to lie at $T_c = 19,000$ K, $P_c = 24$ GPa $\equiv 0.24$ Mbar. They also pointed out the connection between this "plasma phase transition" (PPT) and the molecular-metallic hydrogen phase transition. Ebeling and Richert[75] (1985) subsequently performed an independent calculation of the coexistence curve and second critical point. They also found a dramatic lowering of the pressure along the neutral/ionized coexistence curve for $T > 2000$ K, with the location of the second critical point given by $T_c = 16,500$ K, $P_c = 22.8$ GPa, $\rho_c = 0.13$g cm^{-3}. Marley and Hubbard[76,77] next computed the PPT by equating the chemical potential computed for the molecular phase to that of the metallic phase. For molecular H_2, they used a model free energy fitted to the shock-wave experiments of Nellis *et al.*[57], while for the metallic phase they used the free energy given by Hubbard and DeWitt[13]. Their method of calculation does not yield a critical point. Interestingly, like the calculations by Robnik and Kundt[74] and by Ebeling and Richert[75], Marley and Hubbard find $dT/dP > 0$ along the coexistence curve, implying $\Delta S < 0$ at the presumed first-order phase transition. Thus, the "latent heat" associated with this transition is negative.

The most recent contribution to this discussion is a new calculation by Saumon and Chabrier[78]. They have computed a new equation of state for fluid hydrogen, based on sophisticated models for the free energies of the neutral and ionized states. For the low-density, low-temperature, neutral state, the model free energy is computed from hard-sphere fluid perturbation theory for a mixture of H_2 molecules and H atoms. The model treats the internal states of the atoms and molecules using the "occupation probability" formalism recently developed for opacity calculations by Hummer and Mihalas[79]. For the high-temperature or high-density, fully ionized phase ($kT \gtrsim 1$ Ry or $\rho \gtrsim 2$ g cm^{-3}, corresponding to $r_s \lesssim 1$), the plasma is treated as a linearly screened ionic fluid plus a partially degenerate electron liquid, using an accurate fit to the exchange and correlation energy[80].

In the region of partial ionization, Saumon and Chabrier's two separate calculations are combined into a single, unified model for the free energy. In this regime, interactions between neutral and charged species are included through a polarization potential. The resulting free energy is minimized to obtain the relative abundances of H_2, H, H^+, and e^-, with the requirement of electrical neutrality imposed. This calculation *predicts*, rather than *assumes* a first-order plasma phase transition. The second critical point is found to lie at $T_c = 15,000$ K, $P_c = 0.646$ Mbar $= 64.6$ GPa, and $\rho_c = 0.36$g cm^{-3}. The latent heat is again found to be negative, there is a substantial density discontinuity across the phase boundary, and molecular dissociation occurs simultaneously with pressure-ionization. In contrast to some previous work, molecules are found to be the dominant species in the neutral phase at high density.

While it is the most detailed calculation of the plasma phase transition to date, Saumon and Chabrier's result still employs some potentially important approximations. For example, they neglect band structure effects in the dense fluid. Thus, we still need experiments to tell us the real facts. Fortunately, the new calculations keep the possibility of experimental test within reach in the near future.

7. SUMMARY AND CONCLUSIONS

In the past few decades, phase transitions have assumed an increasingly important role in astrophysics. We now believe that the centers of the coolest white dwarfs and the surface layers of neutron stars freeze solid as they cool, although they remain fully ionized. The apparent lack of C/O phase separation during crystallization of the cores of white dwarfs has led us to assign an age $\sim$ 9 Gyr to the disk of our Galaxy and is helping to bring about a new view of the process by which the Galaxy formed. Phase separation of Fe from H, originally conceived as a possible solution to the solar neutrino problem, seems not to occur in the Sun but almost certainly occurs in low-mass stars and brown dwarfs. H/He phase separation seems essential to explain the excess IR luminosity of Saturn. Quite recently, striking new evidence has been obtained in support of the band-overlap model for the metallization of hydrogen, which may have consequences for models of the interiors of the giant planets, and new theoretical calculations support the existence of the plasma phase transition in hydrogen at high densities.

Many interesting problems remain to be attacked. For example, I have not even discussed several other types of phase transitions, including the possible formation of a "pion condensate" in the dense interior of a neutron star[80], the possibility of formation of a neutron crystal in the centers of massive neutron stars[81], or the suggested existence of phase transitions in dense neutron star matter, from nuclei to "bubbles" and from "bubbles" to uniform matter[82].

In addition, we need to carry out detailed evolutionary calculations for low-mass stars and giant planets, which include the full effects of Fe/H and H/He phase separation. The H/He phase diagram must first be recomputed with sufficient accuracy to give accurate values for the critical point for H/He phase separation, however. We also need to explore the possibility of Fe/C phase separation as an energy source in cooling white dwarfs[44].

Metallic hydrogen remains a stimulating challenge to experimentalists and theorists alike. Has the monatomic metllic state finally been formed[53]? Does metallic H solidify at all at low temperatures[83]? Is metallic H a superconductor, either in the solid phase, if it exists[84,85], or in the liquid phase[86,87]? Does the plasma phase transition exist at high temperatures?

Clearly there will be enough challenges to keep us all happily occupied for years to come.

ACKNOWLEDGEMENTS

I am grateful to Neil Ashcroft, Gilles Chabrier, Bill Forrest, David Stevenson, Hugh DeWitt, Marvin Ross, Didier Saumon, and David Young for numerous helpful discussions. This work has been supported in part by the National Science Foundation under grants PHY 88-08146 and AST 88-20322 and in part by the National Aeronautics and Space Administration under grant NAGW- 1476, all through the University of Rochester.

REFERENCES
1. Kirzhnits, D. A., Soviet Phys. − J. E. T. P., 11, 365 (1960).

2. Abrikosov, A. A., Soviet Phys. − J. E. T. P., 12, 1254 (1961).

3. Salpeter, E. E., Ap. J., 134, 669 (1961).

4. Brush, S. G., Sahlin, H. L., and Teller, E., J. Chem. Phys., 45, 2102 (1966).

5. Mestel, L., and Ruderman, M. A., M. N. R. A. S., 136, 27 (1967).

6. Van Horn, H. M., Ap. J., 151, 227 (1968).

7. Ruderman, M. A., Nature, 218, 1129 (1968).

8. Hansen, J. P., Phys. Rev. A, 8, 3096 (1973).

9. Hansen, J. P., in J. de Phys, Coll. C2, 41, C2-43 (1979).

10. Slattery, W. L., Doolen, G. D., and DeWitt, H. E., Phys. Rev A, 21, 2087 (1980).

11. Slattery, W. L., Doolen, G. D., and DeWitt, H. E., Phys. Rev. A, 26, 2255 (1982).

12. Ichimaru, S., Iyetomi, H., Mitake, S., and Itoh, N., Ap. J., 265, L83 (1983).

13. Hubbard, W. B., and DeWitt, H. E., Ap. J., 290, 388 (1985).

14. Lamb, D. Q., Ph.D. thesis, University of Rochester. 1974.

15. Lamb, D. Q., and Van Horn, H. M., Ap. J., 200, 306 (1975).

16. Wood, M. A., Ph.D. thesis, University of Texas (Austin) 1989.

17. Ogata, S., and Ichimaru, S., Phys. Rev. A, 39, 1333 (1989a).

18. Ogata, S., and Ichimaru, S., J. Phys. Soc. Jap. 58, 356 (1989b).

19. Eggen, O. J., Lynden-Bell, D., and Sandage, A., Ap. J., 136, 748 (1962).

20. Iben, I., and Renzini, A., Phys. Rep., 105, 330 (1984).

21. Malaney, R. A. and Fowler, W. A., M. N. R. A. S., 237, 67 (1989).

22. Winget, D. E., Hansen, C. J., Liebert, J. W., Van Horn, H. M., Fontaine, G., Nather, R. E., Kepler, S. O., and Lamb, D. Q., Ap. J., 315, L77 (1987).

23. Liebert, J., Dahn, C. C., and Monet, D. G., Ap. J., 332, 891 (1988).

24. Iben, I., and Laughlin, G., Ap. J., 341, 312 (1989).

25. Stevenson, D. J., J. de Phys., Coll. C2, 41, C2-61 (1980).

26. Mochkovitch, R., Astron. Ap., 122, 212 (1983).

27. Garcia-Berro, E., Hernanz, M., Mochkovitch, R., and Isern, J., Astron. Ap., 193, 141 (1988).

28. Barrat, J. L., Hansen, J. P., and Mochkovitch, R., Astron. Ap., 199, L15 (1988).

29. Ichimaru, S., Iyetomi, H., and Ogata, S., Ap. J., 334, L17 (1988).

30. Norris, J. E., and Green, E. M., Ap. J., 337, 272 (1989).

31. Larson, R. B., M. N. R. A. S., 176, 31 (1976).

32. Larson, R. B., in Frontiers of Stellar Evolution, ed. D. Lambert, in press (1989).

33. Tamanaha, F., Silk, J., Wood, M., Winget, D., and Mochkovitch, R., preprint, 1989.

34. Larson, R. B., M. N. R. A. S., 218, 409 (1986).

35. Rowley, J. K., Cleveland, B. T., and Davis, R., Jr., in AIP Conference Proceedings No. 126: Solar Neutrinos and Neutrino Astronomy, ed. M. L. Cherry, K. Lande, and W. A. Fowler (American Institute of Physics: New York, 1985), p. 1.

36. Bahcall, J. N., and Ulrich, R. K., Revs. Mod. Phys., 60, 297 (1988).

37. Wolfenstein, L., and Beier, E. W., Physics Today, 42, 28 (1989).

38. Pollock, E. L., and Alder, B. J., Nature, 275, 41 (1978).

39. Iyetomi, H., and Ichimaru, S., Phys. Rev. A, 34, 3203 (1986).

40. D'Antona, F., and Mazzitelli, I., Ap. J., 296, 502 (1985).

41. Dorman, B., Nelson, L. A., and Chau, W. Y., Ap. J., 342, 1003 (1989).

42. Stevenson, D. J., in Astrophysics of Brown Dwarfs, ed. M. C. Kafatos, R. S. Harrington, and S. P. Maran (Cambridge University Press: Cambridge, 1986), p. 218.

43. Smoluchowski, R., Nature, 215, 691 (1967).

44. Salpeter, E. E., Ap. J., 181, L83 (1973).

45. Stevenson, D. J., and Salpeter, E. E., Ap. J. Suppl., 35, 221 (1977a).

46. Stevenson, D. J., Phys. Rev. B, 12, 3999 (1975).

47. Macfarlane, J. J., Ap. J., 280, 339 (1984).

48. Stevenson, D. J., and Salpeter, E. E., Ap. J. Suppl., 35, 239 (1977b).

49. Graboske, H. C., Pollack, J. B., Grossman, A. S., and Olness, R. J., Ap. J., 199, 265 (1975).

50. Hanel, R. A., Conrath, B. J., Kunde, V. G., Pearl, J. C., and Pirraglia, J. A., Icarus, 53, 262 (1983).

51. Hubbard, W. B., and Stevenson, D. J., in Saturn, ed. T. Gehrels and M. S. Matthews (University of Arizona Press: Tucson, 1984), p. 47.

52. Wigner, E. P., and Huntington, H. B., J. Chem. Phys., 3, 764 (1935).

53. Mao, H. K., and Hemley, R. J., Science, 244, 1462 (1989).

54. Wildt, R., Ap. J., 87, 508 (1938).

55. Critchfield, C. L., Ap. J., 96, 1 (1942).

56. deMarcus, W. C., Astron. J., 63, 2 (1959).

57. Nellis, W. J., Mitchell, A. C., van Thiel, M., Devine, G. J., and Trainor, R. J., J. Chem Phys., 79, 1480 (1983).

58. Duvall, G. E., and Graham, R. A., Revs. Mod. Phys., 49, 523 (1977).

59. Jayaraman, A., Revs. Mpd. Phys., 55, 65 (1983).

60. Ross, M., in High Pressure Chemistry and Biochemistry, ed. R. van (1987).

61. Ashcroft, N. W., Nature, 340, 345 (1989).

62. Chakravarty, S., Rose, J. H., Wood, D. M., and Ashcroft, N. W., Phys. Rev. B, 24, 1624 (1981).

63. Friedli, C., and Ashcroft, N. W., Phys. Rev. B, 16, 662 (1977).

64. Cepperley, D., and Alder, B., Science, 231, 555 (1986).

65. Min, B. I., Jansen, H. J. F., and Freeman, A. J., Phys. Rev. B, 33, 6383 (1986).

66. Hemley, R. J., and Mao, H. K., Phys. Rev. Letters, 61, 857 (1988).

67. Ashcroft, N. W., personal communication, 1988.

68. Barbee, T. W., III, García, A., Cohen, M. L., and Martins, J. L., Phys. Rev. Letters, 62, 1150 (1989b).

69. Hazen, R. M., Mao, H. K., Finger, L. W., and Hemley, R. J., Phys. Rev. B, 36, 3944 (1987).

70. Mao, H. K., Jephcoat, A. P., Hemley, R. J., Finger, L. W., Zha, C. S., Hazen, R. M., and Cox, D. E., Science, 239, 1131 (1988).

71. Cox, A. N., in Stellar Structure, ed. L. H. Aller and D. B. McLaughlin (University of Chicago Press: Chicago, 1965), p. 195.

72. Hess, H., and Ebeling, W., in Strongly Coupled Plasma Physics, ed. F. J. Rogers and H. E. DeWitt (Plenum Press: New York, 1987), p. 185.

73. Stewart, J. C., and Pyatt, K. D., Ap. J., 144, 1203 (1966).

74. Robnik, M., and Kundt, W., Astron. Ap., 120, 227 (1983).

75. Ebeling, W., and Richert, W., Phys. Lett., 108A, 80 (1985).

76. Marley, M. S., and Hubbard, W. B., Bull. Am. Astron. Soc., 18, 780 (1986).

77. Hubbard, W. B., and Marley, M. S. 1987, in Strongly Coupled Plasma Physics, ed. F. J. Rogers and H. E. DeWitt (Plenum Press: New York, 1987), p. 185.

78. Saumon, D., and Chabrier, G., Phys. Rev. Letters, 62, 2397 (1989).

79. Hummer, D. G., and Mihalas, D., Ap. J., 331, 794 (1988).

78. Ichimaru, S., Iyetomi, H., and Tanaka, S., Phys. Rep., 149, 93 (1987).

81. Shapiro, S. L., and Teukolsky, S. A., Black Holes, White Dwarfs, and Neutron Stars (John Wiley & Sons: New York, 1983).

82. Lamb, D. Q., Lattimer, J. M., Pethick, C. J., and Ravenhall, D. G., Nucl. Phys., A411, 449 (1983).

83. Oliva, J., and Ashcroft, N. W., Phys. Rev. B, 23, 6399 (1981).

84. Ashcroft, N. W., Phys. Rev. Letters, 21, 1748 (1968).

85. Barbee, T. W., III, García, A., and Cohen, M. L., Nature, 340, 369 (1989a).

86. Ascroft, N. W., in Modern Trends in the Theory of Condensed Matter, eds., A. Pekalski and J. Przystway (Springer-Verlag: Berlin, 1979).

87. Jaffe, J., and Ashcroft, N. W., Phys. Rev. B, 23, 6176 (1981).

PLASMA THERMODYNAMICS AND THE EVOLUTION OF BROWN DWARFS AND PLANETS

W.B. HUBBARD

Lunar and Planetary Laboratory, University of Arizona, Tucson, Arizona 85721, U.S.A.

To what extent can the observed properties of brown dwarfs (BD's) and giant planets (GP's) be used to test the hydrogen phase diagram at high pressure? In a partial answer to this question, we summarize recent theoretical work on the evolution of BD's and GP's. The baseline models assume that (a) interior temperature-pressure trajectories do *not* cross any first-order phase boundaries; (b) there is no demixing of helium; (c) atmospheric opacities are determined both by gases and by cloud particles. We then investigate how these models would be affected by the existence of a plasma phase transition. Uncertainty in the atmospheric boundary condition is quite pronounced, and affects the evolution of BD's in a way similar to that of plasma phase transitions. Baseline models have reasonable agreement with recent astronomical observations of BD's, suggesting that effects of hydrogen phase transitions are not yet discernible in the data. We evaluate predictions of numbers of BD's based on standard hypotheses about atmospheric cloud formation and the hydrogen phase diagram, and find that, for a Salpeter birth function, large numbers of low-luminosity BD's (with luminosity on the order of a few millionths of a solar luminosity) should now exist in the galaxy, at a density of about one per cubic parsec. Slower cooling of such BD's, as might be produced by proposed plasma phase transitions with an entropy increase from neutral to metallic phase on the order of 0.3 k_B per proton, would make them slightly more detectable. But a plausible change in the atmospheric boundary condition, e.g. from an effective low-opacity to a high-opacity atmosphere, has a similar but much larger effect, in that it can raise the object's interior temperature profile by a factor up to about 2 for a given surface gravity and effective temperature.

In Jupiter and Saturn, thermal contributions to the pressure-density relations are not negligible although the metallic-hydrogen layers are electron-degenerate. There is a tradeoff between increased interior temperature profiles as would be produced by a phase transition with higher entropy in the ionized phase, and the hydrogen abundance of the planet as a whole. Such considerations may be used to set limits on the size of the entropy jump in these planets.

1. INTRODUCTION

Brown dwarfs and giant planets evolve in a similar manner: a fully convective, electron-degenerate fluid interior cools to space at a rate which is governed by the radiative properties of the atmosphere. Such objects' predicted luminosity L and effective temperature T_e as a function of age can then be calculated; these quantities depend on interior thermodynamics as well as atmospheric properties. The evolutionary trajectories of these objects carry them through the hydrogen phase diagram in the vicinity of the proposed transition from molecular to metallic hydrogen, and also near various proposed critical points in the hydrogen phase diagram. Demixing of a two-component plasma (hydrogen and helium) must also be considered for the cooler objects, principally giant planets.

In principle, observations of the thermal evolution of strongly-coupled objects could be used to empirically probe the hydrogen phase diagram. However, the comparison of observations and theory is not straightforward, and is unfortunately obscured by effects which are unrelated to the phase diagram.

2. EVOLUTION OF BROWN DWARFS

2.1 Observable Characteristics

BD's are customarily defined as hydrogen-rich objects with mass M in the range $0.01 M_S < M < 0.08 M_S$, where M_S is a solar mass. The upper limit is defined by the minimum mass for equilibrium nuclear burning of hydrogen on the main sequence[1], while the lower limit is somewhat arbitrary and actually grades into GP's (the Jovian mass is $0.001 M_S$). If BD's form like stars, the lower limit is set by fragmentation theory. Objects with masses slightly larger than $0.08 M_S$ are very low mass stars (VLM's).

Although only a few *bona fide* BD candidates are known to exist[2], their predicted numbers in the Galaxy are very large[3]. Despite the small masses of these objects, their physics is very complex, and thus their predicted properties depend upon a number of poorly-understood areas of physics, including the properties of the high-pressure hydrogen phase diagram, upon which we focus here.

Apart from spectral characteristics of the emitted radiation, the main information that we have about a BD is its luminosity L and effective temperature T_e. These are related to the BD's radius R by

$$L = 4\pi R^2 \sigma T_e^4, \qquad (1)$$

where σ is the Stefan-Boltzmann constant.

Because BD's are not held in a state of thermal equilibrium by nuclear fusion reactions, all of these quantities are functions of time. The dependence on time t is investigated by equating L to the time derivative of the sum of the total internal energy and the gravitational binding energy of the BD. At any given point in the evolution, the value of R and all interior variables are obtained by integrating the equation of hydrostatic equilibrium. Integrating equation (1) after it has been cast in this form then gives the functions $L = L(M,t)$, $T_e = T_e(M,T)$. Very roughly, for BD's which are older than about 10^8 years, one has

$$\log_{10}(L/L_S) = -1.5 - 1.3 \log_{10}(t_9) + 2.5 \log_{10}(M/M_S), \qquad (2)$$

where L_S is the luminosity of the sun and t_9 is the object's age in 10^9 years. A similar equation applies for $T_e(t,M_S)$. Integration of eq. (1) to obtain eq. (2) requires a relationship between M, R, T_e, and the distribution of temperature T and pressure P in the object's interior. This calculation is greatly simplified by the well-known result that BD's and VLM's transport their interior heat by efficient convection throughout most of their interior[1], and thus the interior relation between T and P can be taken to be isentropic. The position of the isentrope on the T,P plane is critically dependent upon atmospheric properties in the region where the heat transfer changes from convection to radiation, as was elucidated by Hayashi and Nakano in one of the first studies of BD's[4].

2.2 Hydrogen Phase Diagram and BD Isentropes

Figure 1 shows a high-pressure phase diagram for hydrogen, along with three isentropes pertaining to BD's and GP's. In the high-temperature region (T > $10^{4.5}$ K), hydrogen transforms in a continuous manner from atomic to ionized form, and the electron gas develops significant degeneracy at the higher pressures. All species (atoms, protons, electrons) are weakly coupled. At very low temperatures (T < 10^3 K), all species are strongly coupled, and solid molecular hydrogen is believed to transform via a first-order phase transition to a metallic solid or liquid[5]. In the intermediate temperature regime, which is of interest for BD's and GP's, there exists the possibility that the envelope isentropes may cross first-order phase boundaries between coexisting liquid phases.

Four possible phase transitions between neutral and ionized liquid phases of hydrogen are illustrated in Fig. 1. The two nearly-vertical solid lines labeled MH correspond to two alternative phase boundaries calculated by Marley and Hubbard[6]. The light solid line labeled SC corresponds to the phase boundary for a plasma phase transition calculated by Saumon and Chabrier[7]. Similar plasma phase transitions were predicted by Robnik and Kundt (RK)[8], and by Ebeling and Richert[9] (not shown).

In order to conveniently label isentropes in BD's, we adopt the following nomenclature. An interior isentrope is uniquely labeled by the chemical composition (which we take to be solar) and by the parameter T_{10}, the temperature in K at 10 bars pressure. Note that in many BD's (and GP's) the actual temperature at 10 bars will not equal T_{10}, either because the atmosphere is optically thin at that pressure, or because the atmosphere at that pressure is at a different specific entropy than the deep interior because a phase boundary is crossed. In all cases, the value of T_{10} specifies the entropy of the deep, metallic-hydrogen portion of the object. In Fig. 1, the isentropes shown correspond to (top to bottom) BD's with T_{10} = 3000 K, 1000 K, and Jupiter at T_1 (temperature at 1 bar) = 165 K. The curves in the molecular/atomic hydrogen zone are calculated using the theory of Marley and Hubbard, corresponding to the right-hand phase boundary between molecular and metallic hydrogen fluids. For the BD curves, we also show points along the isentrope which lie at 0.99, 0.98, ..., of the total BD mass. The Jupiter curve terminates at the center of Jupiter's metallic-hydrogen zone.

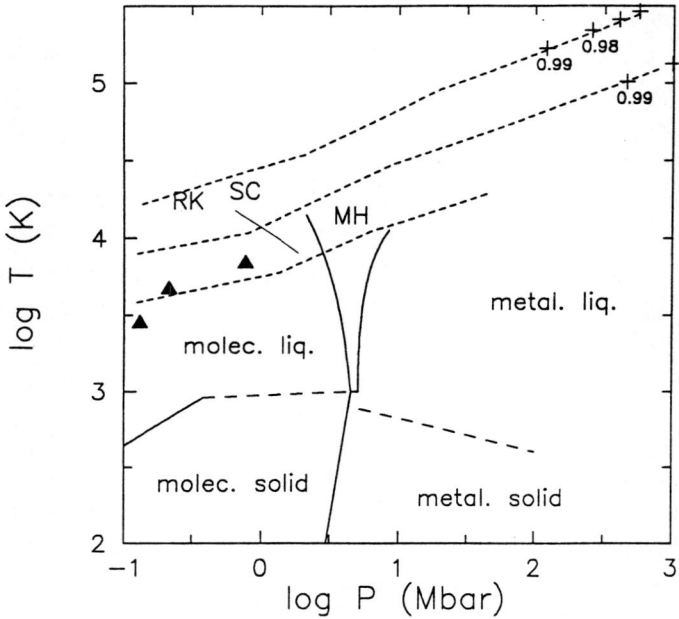

FIGURE 1
Hydrogen phase diagram, along with isentropes for two BD's (upper short-dashed curves), and Jupiter (lower short-dashed curve). Solid lines show possible phase boundaries (dashed where boundaries are more uncertain). Triangles show experimental shock-compression data points[10].

The isentropes shown in Fig. 1 do not take into account the possibility of a phase transition being crossed. The effect of such a phase boundary was first discussed by Stevenson and Salpeter[11], who showed that in a fully-convective BD or GP, the temperature remains constant across the boundary rather than the entropy. Thus if the entropy jump ΔS for a transition from the neutral (low-pressure) phase to the ionized (high-pressure) phase is positive, then value of T_{10} must be higher than would be the case for a continuous isentrope. Similarly, if $\Delta S < 0$, T_{10} will be lowered. In the case of an isentrope whose temperatures always lie higher than any critical point (as is the case for the highest isentrope in Fig. 1), the value of T_{10} is unchanged.

2.3 T_e vs. T_{10} relations

As discussed by Lunine et al.[3], the surface condition for a BD with a given interior isentrope specified by T_{10} can be represented by a relation

$$T_{10} = T_{10}(T_e, g), \qquad (3)$$

where g is the surface gravity of the BD. This relation absorbs all of the detailed physics of radiative and convective heat transport in the atmosphere, cloud formation, various opacity sources, and, as we shall discuss, interior plasma phase transitions. Equation (3) is coupled with eq. (1) to obtain a luminosity vs. time relation such as eq. (2). Our principal point in this section is that the variations in relation (3) due to different atmospheric opacities and cloud formation are much more pronounced than the variations attributable to a plasma phase transition.

Figures 2 and 3 show T_e vs. T_{10} relations for various gravities, background gas opacities, and cloud models. Note that the sensitivity to cloud models begins for $T_e < 2500$ K, while the sensitivity to overall opacity is substantial throughout the interval plotted. The gravity range corresponds to the approximate range of surface gravities for BD's, with the coolest and most massive BD's having $g = 3 \times 10^5$ cm/s^2.

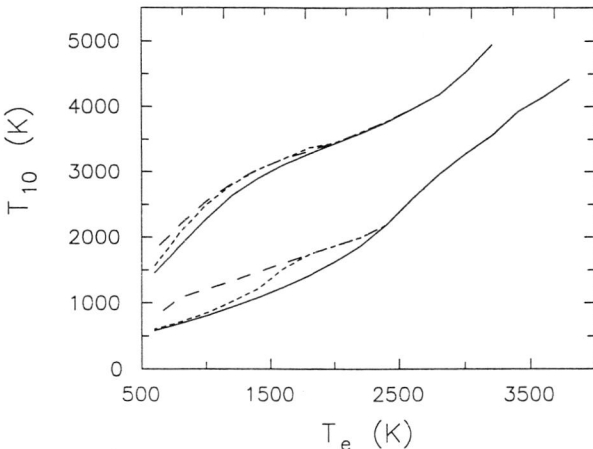

FIGURE 2

T_e vs. T_{10} for $g = 3 \times 10^3$ cm/s^2 (upper curves), and for $g = 3 \times 10^5$ cm/s^2 (lower curves). Short dashes show effect of condensation clouds in the BD atmosphere; long dashes show effect of frozen-in clouds; solid curves are for atmospheres without clouds. This model is for Tsuji high opacity case[12]; see reference 3 for more details.

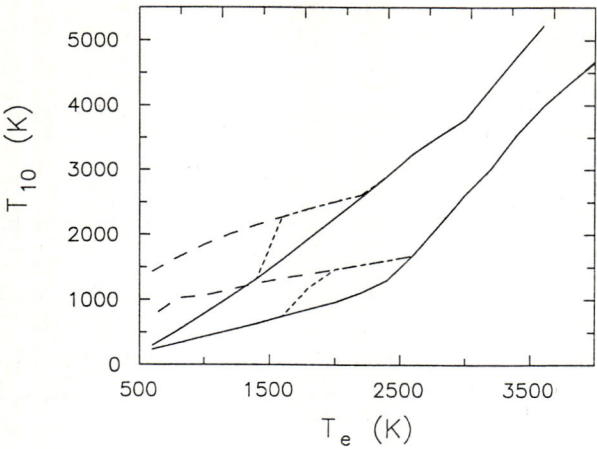

FIGURE 3
Same as Fig. 2, but for the Tsuji low opacity case.

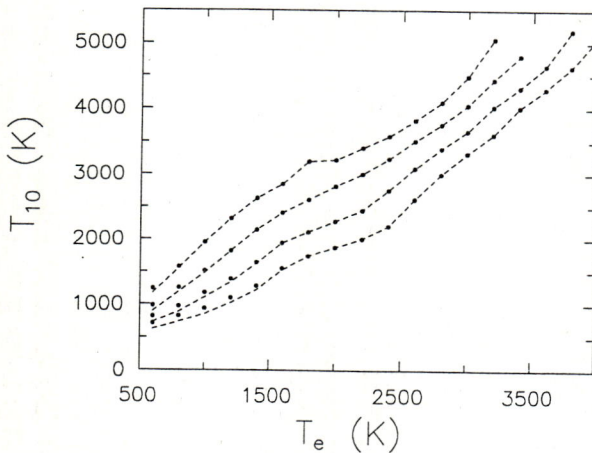

FIGURE 4
T_e vs. T_{10} for $g = 10^4$, 3×10^4, 10^5, and 3×10^5 cm/s^2 (top to bottom), for condensation clouds and Tsuji high opacity (short dashes). Dots show the same curves with the effect of a plasma phase transition included.

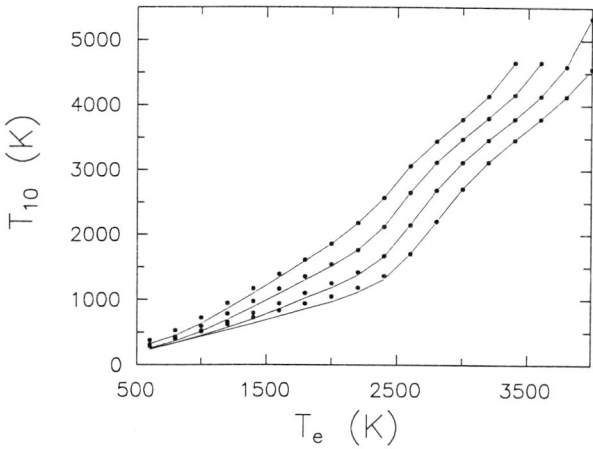

FIGURE 5

T_e vs. T_{10} for g = 10^4, 3×10^4, 10^5, and 3×10^5 cm/s^2 (top to bottom), for no clouds and Tsuji low opacity (solid lines). Dots show the same curves with the effect of a plasma phase transition included.

The models shown in Figs. 2 and 3 were calculated assuming that no phase boundaries were crossed in the BD's deep interior. Next, in order to evaluate the effect of crossing a plasma phase transition, we adopted the results of the thermodynamic calculations of Saumon and Chabrier[7] as being representative of the general class of models with such a transition. Their phase transition, which is shown as a line in Fig. 1, terminates at a critical point at T = 15000 K and P = 0.646 Mbar, where ΔS = 0. At the lowest temperature investigated, T = 8000 K, the entropy jump increases to ΔS = 0.3 k_B per proton (k_B = Boltzmann's constant). We incorporated this phase transition in our models in an approximate way, by calculating the appropriate shift in T_{10} for all models in which the isentrope intersected the phase boundary. The results are shown in Figs. 4 and 5.

2.4 Effect of Plasma Phase Transition on Brown Dwarf Evolution

As is clear from an examination of Figs. 2-5, the jump in entropy associated with a plasma phase transition plays a significant role only in the coolest brown dwarfs, those in which the interior temperature profile is cool enough to intersect the phase boundary. Our model atmospheres show that the effect begins to appear for T_e < 1500 K for the case of high Tsuji opacities, and for T_e < 2500 K for low Tsuji opacities. But the effect is small, of third order compared with the much larger first-order effects of the background gas opacity and the second-order effects of grain opacity (clouds).

Stevenson[13] speculated that effects of grain opacity might play a dominant role in brown dwarf evolution for $T_e < 1500$ K, and that the onset of cloud opacity might even be so sudden that the sign of the slope of the T_e vs. T_{10} relation might change within a limited interval of T_e. This would have major consequences on the evolution of cool brown dwarfs, creating a forbidden zone in T_e (somewhat analogous to a phase transition!) and having a substantial effect on the population of BD's in this region. However, although we find that cloud condensation has a substantial effect on the T_e vs. T_{10} relation, the relation never develops enough of a kink to create a forbidden zone.

Lunine et al.[3] have calculated N, the predicted number of BD's in a cubic parsec in the Galaxy per unit interval in 0.1 log (L/L_S). This function was calculated assuming that the birth-rate function n, the number of BD's created per cubic parsec in a given mass and time interval, is given by an extrapolation of the Salpeter[14] birth-rate function for stars. The resulting function depends to a moderate degree on the T_e vs. T_{10} relation for the BD's, but not to a testable extent at present. For either of the extreme models: the Tsuji low-opacity model with no clouds, or the Tsuji high-opacity model with clouds, we predict on the order of one BD per cubic parsec in the solar vicinity at the present age of the Galaxy, for BD luminosities in the range $-6 \geq \log_{10}(L/L_S) \geq -5$. The number of such objects is a convolution over a spectrum of masses and ages, and thus effects due to opacity fluctuations tend to be smeared out. The number of BD's in this luminosity range is slightly higher for the Tsuji high-opacity case, compared with the Tsuji low-opacity case, but the difference is not large. The effect of a plasma phase transition would not be noticeable in the statistics. At higher luminosities, where putative BD's such as Giclas 29-38B have been found[15] ($\log_{10}(L/L_S) = -4.4$), the statistics are more sensitive to the overall opacity, and at the reported effective temperature $T_e = 1200$ K, the adiabats could intersect the phase boundary. The plasma phase transition would indeed prolong the lifetime of the BD because of the modification of the T_e vs. T_{10} relation as well as the latent heat release, but the effect appears to be small compared with the much larger uncertainties in the atmospheric boundary coundition due to cloud formation and background opacity. In this calculation we have not included the possible effects of hydrogen-helium phase separation at the phase boundary, because the size or even the sign of this effect has not yet been evaluated.

3. STRUCTURE OF JUPITER

Stevenson and Salpeter[11] examined the effect of a possible entropy discontinuity ΔS on the temperature, density, and pressure structure of the Jovian interior. In principle Jupiter is a better subject for study of such a discontinuity than BD's because it is much cooler (see Fig. 1) and therefore more likely to have temperatures below the critical temperature of a plasma phase transition. Also, more is known about Jupiter's interior structure, composition, and heat flow. To date, there is no compelling evidence from any of these for a plasma phase transition in Jupiter.

Assuming $\Delta S = 0$, a model of Jupiter's interior[16] gives the following values at the base of the metallic-hydrogen/helium envelope: P = 45 Mbar, T = 19300 K, mass density ρ = 4.15 g/cm^3. This model has solar composition with modest enhancement of elements heavier than H and He, consistent with the observed composition of the atmosphere. Stevenson and Salpeter considered models with extreme values of ΔS = 1.0 k_B/proton and ΔS = -1.0 k_B/proton, leading to a factor of two uncertainty in the temperature at the base of the envelope. Since the thermal component of the pressure there is about 15% of the total pressure, such excursions have a significant effect on the deduced interior composition. More detailed thermodynamic models constructed since the work of Stevenson and Salpeter have tended to find smaller values of ΔS. For example, Hubbard and Marley[16] find ΔS = -0.6 k_B/proton in Jupiter for their right-hand phase curve shown in Fig. 1; the SC theory gives instead ΔS = 0.3 k_B/proton. Thus, the uncertainty has possibly been reduced.

Figure 6 shows a pressure-density profile and mass distribution in a recent model of Jupiter[16]. The dots correspond to the right-hand scale, and show the fractional mass m/M as a function of mass density ρ, where m is the mass enclosed within a sphere at that value of the density. The solid curve corresponds to the left-hand scale, and is a model for the metallic-hydrogen region of Jupiter which fits the planet's gravitational harmonics. The crosses are calculated points along a theoretical isentrope, using a helium mass fraction Y = 0.22. The lower crosses are calculated with ΔS = 0, while the upper crosses (which are almost indistinguishable) are calculated with ΔS = 0.3 k_B/proton, in accordance with the SC model. The observed value of Y in Jupiter's atmosphere[17] is 0.18 ± 0.04, so any of these isentropes could be brought into agreement with the solid line and with the atmospheric helium abundance by increasing the density slightly, as would be the case if the metallic-hydrogen envelope contained a moderate (and plausible) enhancement of elements heavier than helium. Fig. 6 suggests that ΔS = -0.6 k_B/proton would also fit the pressure-density model without any adjustment of chemical composition. We conclude that at the predicted level of ΔS, Jupiter does not provide a sensitive test of the hydrogen phase diagram. A more sensitive test would be provided by a prediction of the hydrogen-helium phase separation which would accompany a plasma phase transition, but no quantitative predictions of the behavior of such a binary mixture have been made. (Stevenson[18] has predicted liquid-liquid immiscibility in mixtures of metallic hydrogen and helium, but this is a different phase transition than the plasma phase transition under discussion here.)

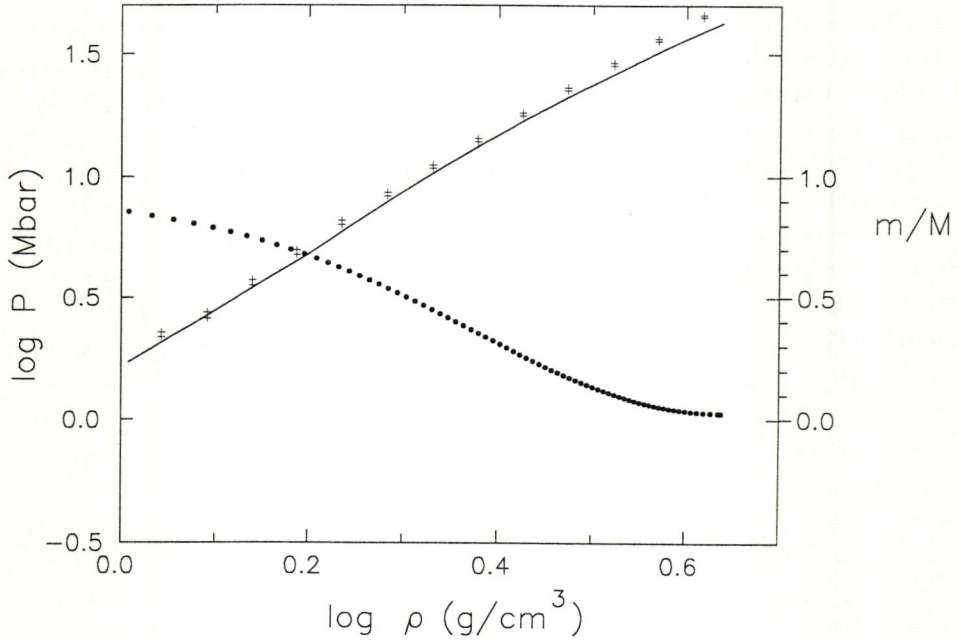

FIGURE 6
A model pressure-density profile for Jupiter (solid curve), compared with isentropes with and without a plasma phase transition (crosses). Dots show fractional mass at each point along the model profile.

4. CONCLUSIONS

Several studies have predicted a first-order phase transition between a largely neutral, strongly-coupled hydrogen fluid and a coexisting partially ionized hydrogen fluid. The phase boundary, if it exists, must terminate at a critical point, at a temperature ~ 10^4 K, and at a pressure in the range ~ 0.5

to 5 Mbar. BD's with effective temperatures cooler than 1500 K (high opacity models), or 2500 K (low opacity models) would cross this phase boundary. However, for predicted values of the entropy jump, the effect on BD evolution is not large and is likely to be obscured by more pronounced effects having to do with the atmospheric boundary condition. A more important question is the effect of the phase transition on partitioning of hydrogen, helium, and other species between the neutral atmosphere and the pressure-ionized interior of a BD.

GP's such as Jupiter and Saturn are certain to intersect the predicted phase boundary because of their cooler interior temperatures. As in BD's, the effect of the entropy jump is minor compared with other uncertainties. For example, if $\Delta S = 0.3$ k_B/proton, the temperature at the base of Jupiter's metallic-hydrogen zone (P = 45 Mbar) increases from 19300 K to 22000 K. On the other hand, partitioning of elements between the molecular-hydrogen zone and the metallic-hydrogen zone could be significant, and could have observable consequences. If there is no first-order phase boundary between these zones, but only a gradual change of state, then chemical differences between Jupiter and Saturn may be attributable to solely to liquid-liquid immiscibility in the metallic phase, as predicted by Stevenson[18], or to differences in the processes of formation of these bodies.

Thus, further astrophysical tests of the hydrogen phase diagram would seem to require theoretical studies of the phase behavior of multicomponent systems in the vicinity of the proposed critical point.

ACKNOWLEDGMENT

I thank Mark Marley, Jonathan Lunine, and Adam Burrows for comments on this paper. Figures 2 and 3 were reproduced from The Astrophysical Journal 338 (1989) 314 [reference 3].

REFERENCES

1) A. Burrows, W. B. Hubbard, and J. I. Lunine, Astrophys. J. in print (1989).

2) J. L. Greenstein, Comments Astrophys. 13 (1989) 303.

3) J. I. Lunine, W. B. Hubbard, A. Burrows, Y.-P. Wang, and K. Garlow, Astrophys. J. 338 (1989) 314.

4) C. Hayashi and T. Nakano, Prog. Theor. Phys. 30 (1963) 460.

5) D. Ceperley, Quantum Monte Carlo simulations of systems at high pressure, in: Simple Molecular Systems at Very High Density, eds. A. Poulian, P. Loubeyre, and N. Boccara (Plenum, New York, 1989) pp. 477-489.

6) M. S. Marley and W. B. Hubbard, Icarus 73 (1988) 536.

7) D. Saumon and G. Chabrier, Phys. Rev. Lett. 62 (1989) 2397.

8) M. Robnik and W. Kundt, Astron. Astrophys. 120 (1983) 227.

9) W. Ebeling and W. Richert, Phys. Lett. 108A (1985) 80.

10) W. J. Nellis, M. Ross, A. C. Mitchell, M. van Thiel, D. A. Young, F. H. Ree, and R. J. Trainor, Phys. Rev. A 27 (1983) 608.

11) D. J. Stevenson and E. E. Salpeter, Interior models of Jupiter, in Jupiter, ed. T. Gehrels (U. of Ariz. Press, Tucson, 1976) pp. 85-112.

12) T. Tsuji, Publ. Astron. Soc. Japan 3 (1971) 553.

13) D. J. Stevenson, High mass planets and low mass stars, in Astrophysics of Brown Dwarfs, eds. M. C. Kafatos, R. S. Harrington, and S. P. Maran (Cambridge Univ. Press, Cambridge, 1986) pp. 218-232.

14) E. E. Salpeter, Astrophys. J. 121 (1955) 161.

15) B. Zuckerman and E. E. Becklin, Nature 330 (1987) 138.

16) W. B. Hubbard and M. S. Marley, Icarus 78 (1989) 102.

17) D. Gautier and T. Owen, The composition of outer planet atmospheres, in Origin and Evolution of Planetary and Satellite Atmospheres, eds. S. K. Atreya, J. B. Pollack, and M. S. Matthews (Univ. of Ariz. Press, Tucson, 1989) pp. 487-512.

18) D. J. Stevenson, Phys. Rev. 12 (1975) 3999.

Strongly Coupled Plasma Physics
S. Ichimaru (Editor)
© Elsevier Science Publishers B.V. / Yamada Science Foundation, 1990

DISCOVERY OF LOW MASS OBJECTS IN TAURUS

*W. J. FORREST, *Z. NINKOV, *J. D. GARNETT, †M. F. SKRUTSKIE, +M. SHURE

*Department of Physics & Astronomy, University of Rochester, Rochester, NY 14627-0011 USA
†517G, LGRTB, University of Massachusetts, Amherst, MA 01003, USA
+Institute for Astronomy, University of Hawaii, 2680 Woodlawn Drive, Honolulu, HI 96822 USA

An infrared (2.2μm, K-band) search of small regions (25″ square) near 26 members of the Taurus star-forming association has revealed 20 dim ($K = 13$-16 mag) stellar objects near 13 of them. Of these 20 objects, 9 are exceptionally red. It is argued that these 9 are probably also Taurus members. From the luminosities ($0.8 - 4 \times 10^{-3} L_\odot$) and ages (estimated at 10^6 years), masses can be determined by reference to theoretical low-mass cooling curves. The masses are in the range 0.006-0.015 $M_\odot$, i.e. low-mass brown dwarfs. Proper motion studies of 7 of the objects visible on the POSS plates conducted by Burton Jones establish that many of the faint infrared objects really are physically associated with the Taurus-Auriga dark clouds.

I. INTRODUCTION

The theory of self-gravitating gaseous bodies predicts that below a mass of about 0.08 $M_\odot$, the central temperatures will never be high enough to initiate stable nuclear burning. These sub-stellar objects, which subsist on gravitational energy and cool eternally, have been dubbed brown dwarfs[1]. To date, there are no unambiguous indentifications of a bona-fide brown dwarf, though there are several candidates which are either low mass stars or high mass brown dwarfs, depending on their precise ages. It is of great interest to know if brown dwarfs do exist in the galaxy and, if so, how many there are and what are their masses. A sufficient population of brown dwarfs could represent a large mass which is dim visually. This would have enormous implications for the "missing mass" problem in our galaxy and perhaps others.

Previous searches for brown dwarfs have concentrated on objects ranging in age from $\sim 10^8$ to 10^{10} years. At these advanced ages, brown dwarfs are cool and dim, especially low mass brown dwarfs. Therefore such searches require high sensitivity to cool objects and will be most sensitive to higher mass brown dwarfs. At very young ages, however, even relatively low mass brown dwarfs of 0.01 $M_\odot$, will be relatively luminous. Furthermore, because of the nature of the Hayashi evolutionary tracks, dominated by convection internally and H^- and molecular radiative opacity at the surface, the effective temperatures will be relatively high, around 2500 -3500 K in the early stages of evolution. For instance, a 0.01 $M_\odot$ brown dwarf 10^6 years old will have a luminosity of $\sim 1.6 - 1.8 \times 10^{-3} L_\odot$ and an effective temperature

$\sim 2300 - 2600\text{K}^{2,3}$.

Based on these ideas, we searched in the vicinities of 26 T-Tauri stars in the Taurus star-forming association. The stars were selected to be relatively free of circumstellar material, i.e. so called "naked" or "weak" T-Tauri stars. A total of 20 dim (apparent K magnitude 13-16) objects were discovered near 13 of the T-Tauri stars. Of these 20, at least 9 were exceptionally red. We argue that it is extremely unlikely that the red objects could be foreground or background stars and are therefore probably also members of Taurus. Assuming ages equal to the average age of Taurus members (10^6 years) and using the above theoretical evolutionary curves, we estimate masses of $0.006 - 0.015 M_\odot$ based on the observed luminosities.

The key question of true Taurus membership has received strong support from the proper motion studies of Burton Jones[4]. He finds a strong similarity between the proper motions of the dim infrared objects found here and known Taurus-Auriga cloud members. These motions are distinct from those of distant background stars. This constitutes strong evidence that the faint infrared objects are physically associated with the Taurus-Auriga dark clouds. We conclude that the bulk of the red stars really are Taurus members and therefore low mass brown dwarfs as calculated above.

II. OBSERVATIONS

Infrared observations of 26 T-Tauri stars in Taurus were made at the NASA 3.0m Infrared Telescope Facility on Mauna Kea on 6 nights in September, 1988 using the Rochester 58×62 InSb infrared camera. Our procedure was to place the T-Tauri star in the center of the 25" square field-of-view of the camera. Then the telescope was nodded to presumed blank sky, usually 100-200" S, to acquire a background frame. The telescope was then nodded back to the T-Tauri star and 2 exposures of 33 seconds each in the photometric K-band ($\lambda_o = 2.23\mu$m, $\Delta\lambda = 0.41\mu$m) were acquired. The images were inspected by subtracting the background frame and displaying them on a video monitor with display levels set to reveal the dimmest possible object. A stellar object near the T-Tauri star was seen as a bright (positive) image, while a stellar object in the sky frame was seen as a dark (negative) image. We estimate a limiting magnitude of circa 16 for definite detection of positive stars and somewhat brighter for negative stars.

If a positive star was seen, measurements in the H ($\lambda_o = 1.65\mu$m, $\Delta\lambda = 0.32\mu$m) and J ($\lambda_o \cong 1.25\mu$m, $\Delta\lambda \cong 0.21\mu$m) bands were also acquired. In addition, if the original telescope position placed the newly discovered object in a bad part of the array, either near the edge or near the crack in the southeast corner, the telescope was moved to re-position the source more optimally. In most cases, a negative star, i.e. one present in the distant sky beam, was not followed up with J and H photometry unless a positive star was also present. On some occasions, our sky beams were positioned only 24" from the main beam. In these cases, negative stars were measured in all bands, based on the presumption that they might be associated with the known T-Tauri star.

A typical 2.23μm image is shown in figure 1. The $K \sim 13$ mag stellar object $8''$ S of LkCa 4 was first seen. Moving the telescope for better viewing revealed the second $K \sim 14$ mag stellar object $13''$ W of LkCa 4. In this case, a $75''$ square region surrounding LkCa 4 was also surveyed, to a limit of $K = 15 - 16$ mag. No further stellar objects were found.

The images were processed as follows to derive photometric magnitudes. First, as discussed by Forrest et al.[5], the output of the array is inherently non-linear. Data, both source and background images, were therefore first linearized as described in Forrest et al.[5]. After background subtraction, the pixel-to-pixel nonuniformities were eliminated through division by a flat field. We considered candidate flat fields (linearized) from the blue morning or evening sky and the inside of the dome illuminated by five $1000W$ incandescent light bulbs. There was very little difference between any of the flat fields taken throughout the run. We tested the photometric performance of the flat fields by imaging standard stars at 25 grid positions on the array. Using a combination of dome and blue sky flat fields, the best photometric performance was a standard deviation of 4.5% of the average. As discussed in Forrest et al.[5], we attribute a large component of the variation to pincushion distortion in our reimaging optics. After correcting the flat field for this effect, the variation improved to 2.6% at K. The uncertainty due to flat fielding will be somewhat higher at J and H.

Standard stars were observed throughout each night for calibration purposes. Measurements in a synthetic octagonal aperture $4.6''$ in diameter indicated quite stable values in the J, H, and K bands. The companions were often very faint, ranging down to a J magnitude of 17. For accurate photometry, a method of estimating the total signal from the star without contribution from nearby sky signal was developed. An octagonal aperture $3.8''$ in diameter was centered on the companion. The outer annulus $0.42''$ (one pixel) wide was used to estimate the sky level. This surface brightness level (counts per pixel) was subtracted from the inner octagon, $2.9''$ in diameter, to give the total star signal. This method of differential photometry is dependent on the seeing, which will smear true star signal into the sky annulus. Measurements on the standard stars indicated a $\pm 2\%$ uncertainty at H and K and $\pm 5\%$ uncertainty at J due to this effect.

We estimate a net uncertainty $\pm 10\%$ in the K, H, J fluxes and colors. To this must be added uncertainty due to detector noise for the fainter sources. Tests of the accuracy of the photometry and colors were generally favorable. For a red standard, we used Giclas 77-31[6] with $K = 7.84$, $J - H = 0.58$ and $H - K = 0.32$. Calibrating this in the same manner as the companion stars gave $K = 7.90$, $H - K = 0.26$ and $J - H = 0.56$, well within our quoted 10% uncertainties. For dim standards we observed the Hyades white dwarfs H 27 and VR 16 from Zuckerman and Becklin (ZB)[7]. For H 27 we got $J = 14.68 \pm 0.15, H = 14.71 \pm 0.1, K = 14.72 \pm .14$ while ZB quote $J = 14.70, K = 14.90$. The J magnitude is satisfactory but the K magnitudes differ by 0.18, which is somewhat larger than our estimated uncertainty of

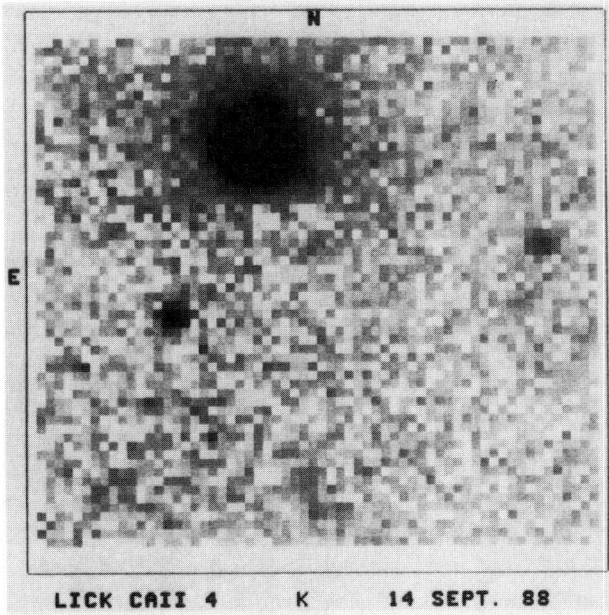

Figure 1 2.2μm (K) image of a 24″ × 26″ region near the T Tauri star LkCa 4 obtained on 14 September 1988. North is up and East is to the left. The T Tauri star is the bright object near the top. The dim object to the right (west) can also be seen on the POSS red plate no. 1454 from 23 October 1955. Its proper motion (2.7″ /century E and 2.2″ /century S (Jones, 1989)) is consistent with that of LkCa 4 and other Taurus members but inconsistent with a distant background star. It is therefore a highly probable Taurus member. Its luminosity derived from the measured infrared fluxes is $2 \times 10^{-3} L_\odot$. Its mass is inferred to be $\simeq 0.01 M_\odot$ based on an estimated age of 10^6 years. The brighter object to the left and below LkCa 4 is too close to be seen on the POSS plates and, therefore, its proper motion is unavailable at this time. Its red colors indicate probable Taurus membership. If this is confirmed spectroscopically, its luminosity would be $4 \times 10^{-3} L_\odot$ and its inferred mass about $0.015 M_\odot$.

0.14 mag. This may indicate a slight problem in our photometry, or it could indicate that H 27 has an infrared excess similar to the white dwarf LB 1497 studied by ZB. For VR 16, we found the white dwarf had a dim companion 4.4″ to the W (as well as a star comparable in brightness to the white dwarf 6″ W in our 100″ S sky beam). For the white dwarf we derive $J = 14.58, H = 14.61$ and $K = 14.74$, somewhat brighter than ZB's $J = 14.64, K = 14.85$. The 4.4″ W companion was $J = 16.14, H = 15.97, K = 15.86$. ZB used 12″ diameter aperture, which should have included the 4″ W companion. However, the sum of the white dwarf and the 4″ W companion is brighter than ZB's measurements by 0.3-0.4 magnitudes, which is well outside the uncertainties. (We conclude that ZB probably missed this companion.) If the 4″ W companion is also in the Hyades, its absolute K magnitude would be $M_K = 12.6$, which would be the dimmest known main sequence star. Since it is not outstandingly red, we suggest it may be a cool white dwarf. A typical background star would be a K1 main sequence star reddened by $A_K = 0.22$. This appears incompatible with the observed colors of the 4″ W companion.

We conclude there are no systematic inaccuracies in our J, H, K photometry of dim, red objects exceeding 0.1 in the magnitudes or colors. The photometric results for our objects are given in Table 1. In Table 2 we list the T-Tauri stars lacking nearby objects.

The plate scale, orientation, and optics distortion of our camera were determined by measuring the star pairs γ Ari and γ Del and the stars θ^1 A, B, C, and D in the Trapezium. The average plate scale at K was 0.42″/pixel and the columns of the array were tilted 0.3° from north. By modeling the pincushion distortion, an accuracy of 0.5% in plate scale and rotation was achieved over the whole array. The position of the companions relative to the T-Tauri stars were measured by calculating the centroid in the same aperature used for photometry. The resulting offsets, appropriate for the 1988.8 epoch are given in Table 1.

III. DISCUSSION

We don't expect that all of the 20 objects discovered here are members of the Taurus star-forming association. To predict the number of background stars expected, a star count model based on the method of Elias[8,9] was developed. The star densities given by Elias[8] were used. These were supplemented for main sequence spectral types later than K3 using the V luminosity function of D'Antona and Mazzitelli[10]. This was transformed to a K luminosity function using the relationship $M_K = 0.521 M_V + 0.80$, which was found to fit the range M_V = 4.4 to 17. This relationship is derived from Liebert and Probst's[11] fig. 2 for low mass stars and Allen[12] for higher mass stars. A further extension of the Elias model was the inclusion of a screen of extinction $A_K = 0.2$ mag. at 140 pc to account for the average extinction through Taurus of $A_V \simeq 2$. The model was normalized to the K star count data of Elias[9] in Taurus by variation of the radial scale length α^7. The resulting value of $\alpha = 3$ kpc gave a predicted value of 10.7 stars per square degree for $K < 7.5$, in adequate agreement with the actual counts of 11. The cumulative number of stars brighter than $K = 15.5$ from this model is 3033 stars/square degree. For the observed counts, we have taken only the companions seen at the initial telescope setting, with the T-Tauri star centered on the array. (We do not include additional stars revealed when the telescope was moved or stars appearing in our sky beams). We found 9 such objects brighter than $K = 15.5$ in our search of 26, $(25'')^2$ fields. The model predicts 3.7, a factor of 2.4 lower. Therefore, we anticipate that somewhat less than half of our objects may be field stars, while the rest are possible Taurus members, based on the star counts alone.

To select the likely Taurus members, we consider the infrared colors $J - H$ and $H - K$. The colors from table 1 are plotted in figure 2. Also shown are the colors of main sequence stars, the colors generated by interstellar extinction of $A_V = 2$ mag., and the colors of 2000-3500 K black bodies. The star count model predicts that the average field star will be a K1 main sequence star 1.3 kpc distant with $A_K = 0.32$ (corresponding to $A_V = 3.2$ mag.). Such a star would have colors $J - H = 0.75$ and $H - K = 0.33$. Nine of our objects have red colors, which differ significantly from these predicted colors, suggestive of low mass objects in Taurus. We suggest these objects are likely Taurus members, based on their colors and the large number found.

Red giant stars have colors similar to our objects, but a red giant would have to be $\sim$ 200 kpc away to appear at $K = 15$ magnitude. Reddened late M dwarfs would have colors similar to our companions. However, from the star count model described above we expect only 1 M-dwarf in our survey. Reddening in Taurus which exceeded the 0.2 mag. A_K (2 mag. A_V), which we have already included, could cause a G to K background star to appear as red as our objects. However detailed star counts in Taurus by Cernicharo and Bachiller[13] show a maximum of $A_V = 3.4$ at the LkCa 5 position and considerably less for the other objects. Therefore, this is not plausible either, since a value of $A_V > 4$ in Taurus would be necessary to account for the observed colors.

A very strong test of Taurus membership is proper motion. Members of the Taurus star forming region have typical proper motions of 2.3″/century in P.A. 164° east of north[14]. A typical background star 1.3 kpc distant will have very small proper motion, $\sim$ .26″/century. Burton Jones[4] has inspected the POSS plates from the mid 1950's and finds about half of our objects can be seen on the red plate. The others are either not visible or too close to the primary to be seen. For 7 of the objects with proper motion data, he finds a strong similarity between the proper motions of the dim infrared objects found here and known Taurus-Auriga cloud members. These motions are distinct from those of distant background stars. This constitutes strong evidence that the faint infrared objects are physically associated with the Taurus-Auriga dark clouds. We conclude that a large fraction of our 11 most likely candidates, selected from their infrared colors, actually are members of the Taurus star forming association.

In order to estimate the masses of the dim, red Taurus members, we compare the observed

Table 1. Stellar Objects Discovered Near 13 T Tauri Stars

Object	T Tauri Star	HBC #[c]	Offset from T-Tauri(arcsec)[b]		K	H-K (mag)	J-H	Proper Motion, Notes[a]
a. Red Objects								
1	LkCa 4	370	3.82E	8.28S	13.44	0.45	1.1	too close
2	LkCa 4	370	13.52W	4.75S	14.25	0.41	1.0	YES
3	FP Tau	26	(9.4W)	(7.9S)	15.96	0.50	0.8	NO, background p.m.
4	LkCa 7	379	13.39W	4.34S	15.25	0.34	0.46	YES
5	LkCa 7	379	3.37E	17.08S	15.81	0.0	0.7	YES
6	LkCa 7	379	1.64E	29.68S	14.38	0.4	0.6	not visible, tel. offset
7	LkCa 5	371	12.33E	7.81S	14.54	0.64	1.07	not visible
8	LkCa 8	385	12.65E	0.39N	14.10	0.39	–	MAYBE
9	LkCa 8	385	11.03E	10.10N	13.24	0.43	–	NO, high p.m.
10	NTTS 35120 NE	353	19.35E	4.48S	15.26	0.33	1.2	not visible, low lat. (−10°)
14	V836 Tau	429	(12.6W)	(6.1S)	16.3	0.9	–	only J, H measured
15	LkCa 3	386	9.94W	11.40N	15.9(H)	–	0.7	too close, v. bright obj. 13″ SSE
16	NTTS 45251+3016	427	9.04E	0.23S	15.24	0.08	1.3	MAYBE
17	NTTS 41559+1716	376	4.50E	9.97N	14.14	0.02	0.76	
b. Others								
11	NTTS 35120 NE	353	11.75E	8.91N	14.14	0.05	0.43	not red
12	NTTS 35120 NE	353	21.84E	17.14N	12.38	–	–	tel. offset
13	NTTS 35120 NE	353	9.27W	13.18N	13.73	–	–	tel.offset
18	NTTS 40234+2143	362	3.27W	4.94S	14.55	−0.01	0.27	too close
19	LkCa 21		(7.7 W)	(6.8N)	(16.4)	(0.1)	–	through clouds
20	V819 Tau		1.24E	10.44S	12.04	–	–	

Meaning Key:

[a] YES: proper motion indicates Taurus membership.
NO: proper motion more characteristic of a field star.
MAYBE: proper motion indeterminate between field stars and a Taurus members.
too close: Object too close to T-Tauri star to be seen on the POSS red plate.
not visible: object not visible on POSS red plate.
through clouds: observations made through thin clouds, photometry highly uncertain.
tel. offset: offset position derived using telescope motion, somewhat more uncertainty.

[b] Offset uncertainty ±0.15 arcsec. Offsets in parenthesis and those derived using telescope motion somewhat more uncertain.
[c] Herbig, Bell (1988) Catalog number[16]

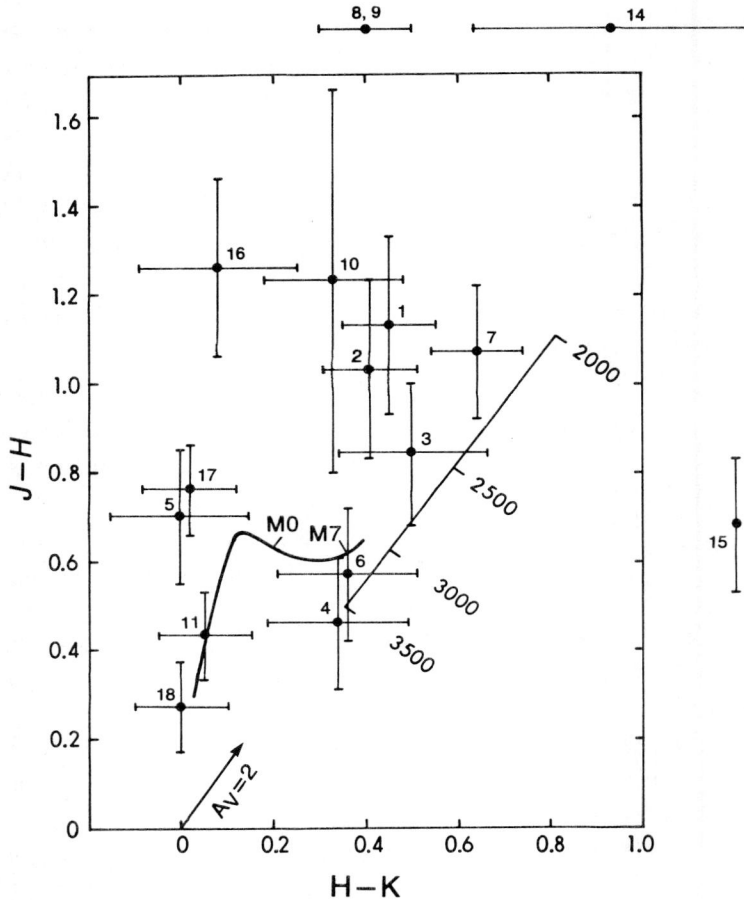

Figure 2 Near infrared color-color plot of the objects (data points with error bars) found in our Taurus survey. The data are from table 1. 10,000 K stars such as Vega (α Lyr) define the zero point of H-K and J-H colors. Also shown is a vector corresponding to $A_V=2$ mag of interstellar reddening in H-K and J-H. The curve with a J-H peak near H-K = 0.15 represents the average colors of main sequence stars from Probst and Liebert (1983 ApJ, **274**, 245) with the approximate positions of M0 and M7 spectral types indicated. The straight line with tick marks shows the infrared colors of black bodies with temperatures of 2000 to 3500 K as indicated. The points outside the box represent stars with only H-K (top) or J-H (right side) measurements. Exceptionally red objects, toward the top right in this figure, are suggested to be possible low mass brown dwarfs in the Taurus association.

Table 2. 13 T Tauri Stars Lacking Nearby Objects

V410 Tau, V827 Tau, HDE 283572, LkCa 1, V826 Tau,
SAO 76428, LkCa 16, Hubble 4, NTTS 42417+1744,
NTTS 42950+1757 (may be very faint object 10"N)
NTTS 4283+1700, NTTS 34903+2431, NTTS 43220+1815

luminosities to those predicted by the theoretical evolutionary models of Nelson et al.[2] and Lunine et al.[3]. Using the distance to Taurus of 140 pc, the luminosities are $0.8 - 4 \times 10^{-3} L_\odot$. For an age of 10^6 years, which is the average age of the previously known Taurus members, the inferred masses are $0.006 - 0.015 M_\odot$, i.e. low mass brown dwarfs. The ages would have to exceed 10^8 years for these objects to be stars ($M > 0.08 M_\odot$). This is counter to experience in Taurus, where the average age is only 10^6 years and star formation apparently began in earnest only 10^7 years ago[15,16]. Furthermore, the lifetime of star-forming molecular clouds is believed to be only $10^7 - 10^8$ years.

IV. CONCLUSIONS

We conclude that the Taurus clouds harbor a considerable number of low-mass objects, typically 0.01 $M_\odot$ or 10 $M_{Jupiter}$, which could be described as low mass brown dwarfs or high mass planets. Detailed studies of the low mass objects identified here are necessary to better understand this new type of object.

Spectroscopic studies can lead to an understanding of the atmospheric structure. This will allow determination of the effective temperatures which are needed to place them on the Hertzsprung-Russell diagram and to compare them with theoretical evolutionary models of such objects. It may also be possible to determine the surface gravities, which will give a direct estimate of the masses. For the objects which can't be seen on the POSS plates for proper motion studies, the spectra will be needed to definitely establish Taurus membership. K-band spectra will reveal the CO absorption ($> 2.3\mu$m) and H_2O absorption (1.9μm and 2.7μm) if the object is actually a 2000-3000K substellar object on its Hayashi convective track, rather than a reddened G-K background star.

Further survey observations are necessary to establish how widespread this phenomenon is. The present survey examined approximately 0.002 square degrees in Taurus. Most of our strongest brown dwarf candidates were found in the vicinity of the Lynds 1495 dark cloud, which comprises approximately 2 square degrees on the sky. If the low mass brown dwarfs are spread uniformly over this cloud, the total number would be about 10^4. If the population of brown dwarfs follows the molecular column density in the Taurus-Auriga clouds, traced by CO, the number could be as high as 10^6. In either case, they would greatly outnumber the ~ 200 known stars in all of Taurus-Auriga, discovered during the course of over 40 years of surveying in this region. If the numbers of low mass brown dwarfs is as great as

10^4 or 10^6 in the Taurus clouds, they could represent a heretofore unknown but dynamically important mass component of our galaxy.

REFERENCES

1. Tarter, J. C. 1986 in "Astrophysics of Brown Dwarfs" ed. Kafatos, Harrington and Maran (Cambridge U. Press, Cambridge) p. 121.

2. Nelson, L. A., Rappaport, S. A. and Joss, P. C. 1986 ApJ, 311, 226.

3. Lunine, J. I., Hubbard, W. B., Burrows, A. S., Wang, Y.-P., and Garlow, K. 1989 ApJ, 338, 314.

4. Jones, B. F. 1989 private communication.

5. Forrest, W. J., Pipher, J. L., Ninkov, Z. and Garnett, J. D. 1989 (Proceedings of Third NASA Ames Detector Conference, Craig R. McCreight, Ed.).

6. Elias, J. H., Frogel, J. A., Matthews, K. and Neugebauer, G. 1982 AJ, 87, 1029.

7. Zuckerman, B. and Becklin, EE 1987 ApJ (Letters), 319, L 99 (ZB).

8. Elias, J. H. 1978a Ap J, 223, 859.

9. Elias, J. H. 1978b Ap J, 224, 857.

10. D'Antona, F. and Mazzitelli, I. 1986 Astron. Ap., 162, 80.

11. Liebert, J. and Probst, R. G. 1987 Am. Rev. Astron. Ap, 25, 473.

12. Allen, C. W. 1973 "Astrophysical Quantities" (The Athlone Press, London).

13. Cernicharo, J. and Bachiller, R. 1984 A & A Supp, 58, 327

14. Jones, B. F. and Herbig, G. H. 1979, AJ, 84, 1872.

15. Hartmann, L. W., Soderblom, D. R., and Stauffer, J. R. 1987 AJ, 93, 907.

16. Herbig, G. H. and Bell, K. R., 1988 "Third Catalog of Emission-Line Stars of the Orion Population", Lick Obs. Bull. No. 1111 (U. Calif.)

TOPICS IN X-RAY ASTRONOMY FROM OBSERVATIONS WITH GINGA

Kazuo MAKISHIMA

Department of Physics, University of Tokyo, 7-3-1 Hongo
Bunkyo-ku, Tokyo, Japan 112

We review highlights from cosmic X-ray observations made with the GINGA
observatory, with a particular emphasis upon astrophysical plasma phenomena.
We first review observational aspects of astrophysical thin hot plasmas,
which are found in various environment with a wide range of scales and
densities. X-ray observations provide wealth of new information, including
magnetic field strengths, chemical abundance, and depth of gravitational
potential. We next discuss denser plasmas produced by mass accretion onto
compact gravitating objects. X-ray emission from these objects is highly
variable, and often exhibits self-similar time variability which may be a
manifestation of plasma turbulence. Finally we report recent progress on our
understanding of the quantum cyclotron resonance under ultra-strong magnetic
fields of neutron stars.

1. INTRODUCTION

Launched on February 5 1987 by the Institute of Space and Astronautical Science (ISAS), GINGA[1] (which means "galaxy" in Japanese) is a 3rd high-energy astronomy satellite of Japan, following HAKUCHO and TENMA.

As illustrated in Fig.1, there are three experiments on board GINGA: LAC (Large Area Proportional Counter; a Japan-UK collaboration), ASM (X-ray All Sky Monitor), and GBD (γ-ray Burst Detector; a Japan-US joint project). Domestic and overseas institutions involved in the GINGA project are listed in Table 1.

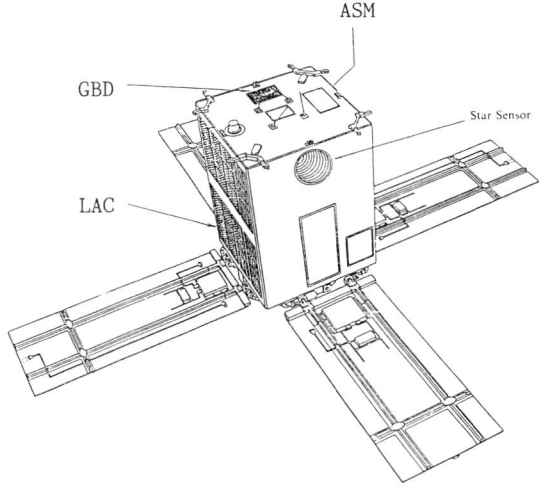

Fig.1 The GINGA spaceraft.

Table 1. Organizations involved in the GINGA project.

Japan	: ISAS, University of Tokyo, Nagoya University, Osaka University, Riken, Osaka City University
UK (LAC)	: Leicester University, Rutherford-Appleton Laboratory
US (GBD)	: Los Alamos National Laboratory

With the 4000 cm^2 total effective area and an extremely low background level, the LAC has the highest sensitivity (down to $\sim 5 \times 10^{-12}$ erg/cm^2/s) to the 2-35 keV cosmic X-rays ever achieved. These instrumentations, as well as many scientific results, have been published in Publ. Astron. Soc. Japan, **41**, No.3 as a special issue (hereafter referred to as [PASJ]).

GINGA observations aim at both galactic and extragalactic objects. The galactic targets include nearby stellar coronae, accreting white dwarfs, neutron-star binaries, black hole candidates, young supernova remnants (SNRs), isolated young pulsars, hot diffuse matter in the Milky Way, and the Galactic center. The extragalactic targets include nearby normal galaxies, hot coronae around galaxies, active galactic nuclei, clusters of galaxies, hot intergalactic medium, and the enigmatic cosmic diffuse X-ray background.

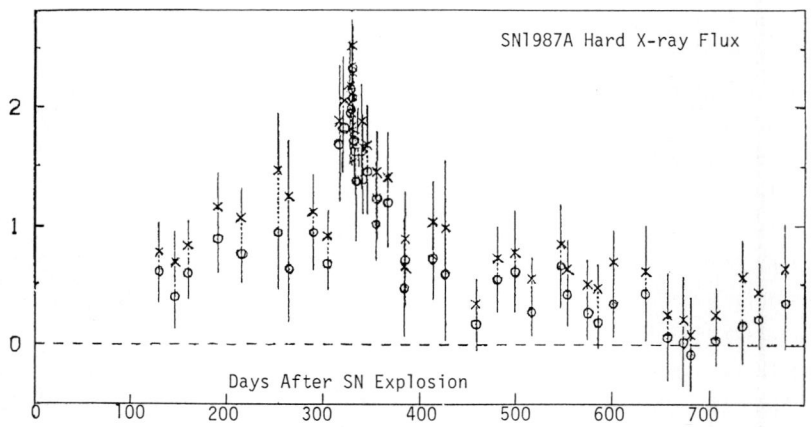

Fig.2 : The 16-28 keV X-ray lightcurve of SN1987A.

GINGA had not been in the orbit for more than 3 weeks until the supernova SN1987A appeared in the Large Magellanic Cloud. Since then we repeatedly monitored the SN1987A region. The first detection of hard X-rays from SN1987A was made in August 1987[2], with GINGA and with the Soviet Mir-Kvant Mission[3]. The observed hard X-rays are thought to be Compton-degraded nuclear gamma rays from Co-57, produced in the explosive nucleosynthesis. The observed hard X-ray lightcurve, Fig.2, has given constraints to the mixing of heavy elements in the expanding debris of the exploded star[4]. The reported 0.5 ms optical pulsations[5], however, have not yet been confirmed with the GINGA data, mainly hampered by the limited time resolution available (0.98 ms). Another set of important GINGA results are the detections of more than a dozen QSOs (quasars) and the determinations of their 2-20 keV spectra[6]. These pioneering results are vital in studying the nature and evolution of the QSOs, and in estimating their contribution to the cosmic diffuse X-ray background.

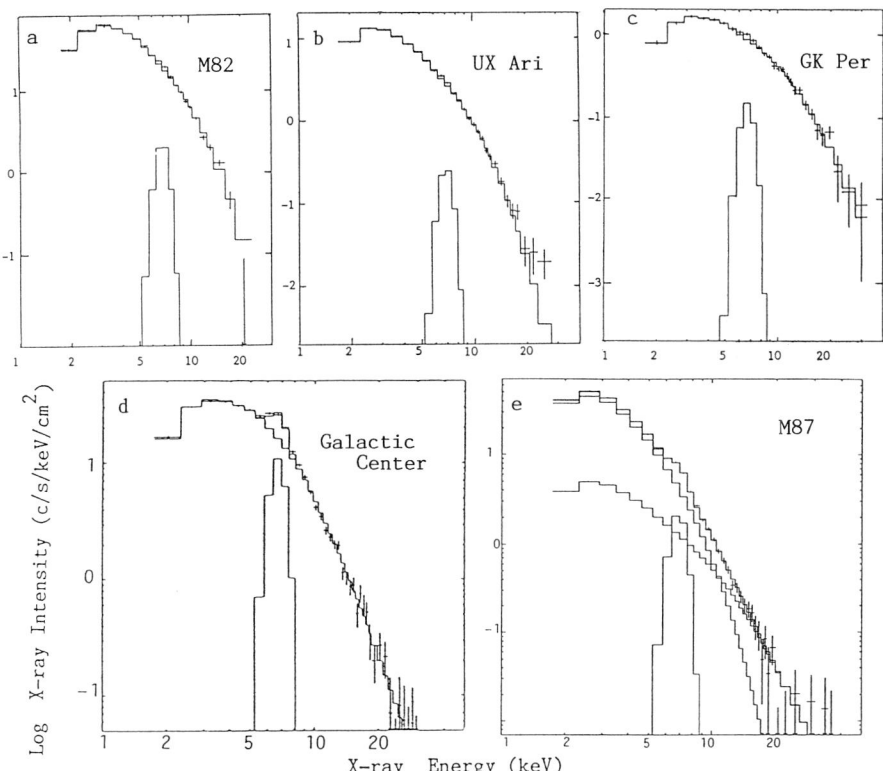

Fig.3 : X-ray pulse-height spectra for thin hot plasmas, fitted with thermal Bremsstrahlung continuum plus iron line. (a) Emission from hot plasma outflow in the starburst galaxy M82 (kT~6 keV; ref.11). (b) Flare X-rays from an RS-CVn binary UX Ari [kT~7 keV; PASJ]. (c) The mass-accreting magnetized white dwarf, GK Per (kT>20 keV). (d) A hot plasma emission with an intense iron line, discovered at the Galactic Center (kT~10 keV; ref.10). (e) The hot halo around the elliptical galaxy M87 (kT~2.3 keV), at the center of the Virgo cluster of galaxies. The harder X-ray emission from the active nucleus of M87 is seen as a hard tail in the spectrum.

2. THIN HOT PLASMAS IN ASTROPHYSICS

2.1 An overview

The 2-35 keV X-ray energy region, containing both Bremsstrahlung continuum and iron K-lines[7], is most suited to detection of optically-thin hot (10^{7-8} K) plasmas. GINGA discovered such hot plasmas in a wide range of astrophysical environments (Fig.3). Some of them are magnetically confined, while some are confined gravitationally or by external pressure. Others are expanding almost freely into space, with or without continuous energy input.

The GINGA LAC is a non-imaging instrument with a 1×2 degrees (FWHM) field of view. Therefore source confusion can often be a problem, but instead it has a superior sensitivity[8] to detect diffuse X-ray components distributed over large angular scales with low surface brightness. This merit is augmented by the fact that the spacecraft can easily be rotated around its Z-axis, so that the LAC view direction scans a region of sky at rather arbitrary slew rates[1]. Now evidence is accumulating which indicates that a considerable fraction of interstellar and intergalactic space is filled up with hot thin plasmas[8]. For example, Koyama et al., using the 6.7 keV K-line from Helium-like iron ions as a tracer, revealed a presence of a large amount of hot plasma along the Milky Way[9] and at the center of our Galaxy (Fig.3d)[10].

2.2 Magnetic heating and confinement

Figure 4 summarizes observed temperature and estimated density of various astrophysical plasmas (from GINGA and other missions). For example, magnetic poles of mass-accreting magnetized white dwarfs are a site of very hot plasma emission (10^{8-9} K; Fig.3c). There the magnetic field is strong (10^{6-7} G) enough for the plasma confinement. The gas from the companion star is accelerated by the gravitational field, funneled onto the magnetic poles, and then thermalized by standing shocks near the surface of the white dwarf.

Figure 3b shows typical example of flare X-ray emission from hot coronae (of density 10^{9-11} cm^{-3}) in interacting binary stars, in which magnetic loops may connect the two stars. The observed temperature, about 7×10^7 K, is among the highest coronal temperatures ever determined, and far exceeds the escape temperature of a star. In order for such coronal plasmas to be confined and heated, magnetic fields of up to 1 kG are required. The hot coronal X-rays thus serve as an indicator of very intense stellar magnetic fields. GINGA has been accumulating wealth of new information on this subject.

In interstellar spaces and galactic halos, magnetic fields have a typical intensity of several microgausses. Therefore plasmas in supernova remnants and those at the Galactic Center[10] (Fig.3d) cannot be affected by the magnetic field. However for thinner plasma components, e.g. galactic halos (Fig.3e), the magnetic field may play an important role. This subject should be paid more attention, because it leads to a very important issue of intensity, distribution and origin of intra- and inter-galactic magnetic fields, especially in relation to the cosmological (primordial) magnetic field.

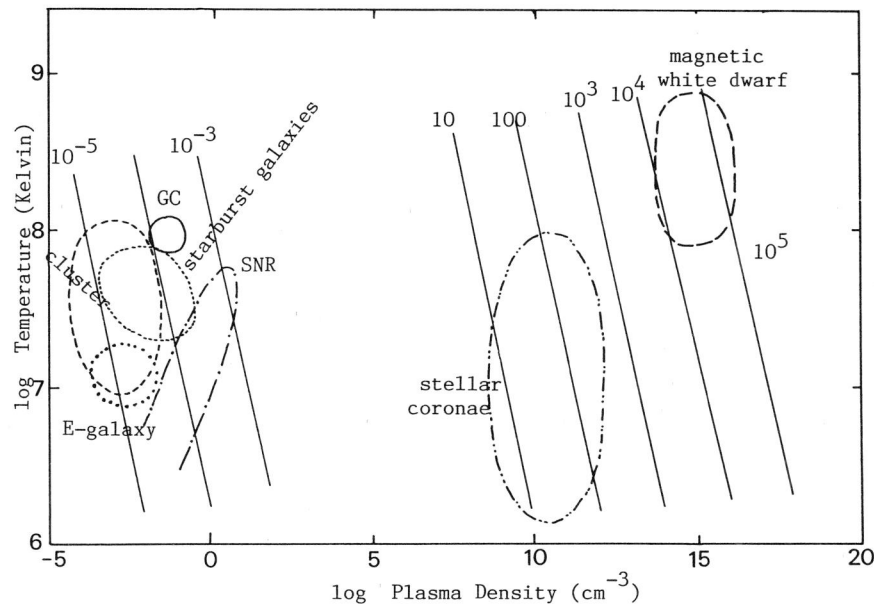

Fig.4 : Density and temperature of astrophysical plasma systems. Solid lines indicate the equipartition magnetic field strength in Gauss.

2.3 Gravitational confinement and dark matter

Another important quantity estimated from the cosmic plasma diagnostics is the "binding mass" necessary to confine the plasma by gravity. If a spherical plasma of radius r is assumed to be bound to the gravitational potential well produced by a gravitating mass M placed at the center, then $M \sim kTr/Gm$ is required. Here k is Boltzmann constant, T is plasma temperature, G is constant of gravity, and m is proton mass. Conversely, if the optically-thin plasma is produced by accretion onto compact gravitating objects, the temperature is expected to be $\lesssim GMm/kr$ (free-fall temperature).

Observed values of T and r, together with the required M, are summarized in Fig.5. The observed temperatures of white dwarfs (Fig.3c) are consistent with the free-fall temperature determined by their radius ($\sim 10^9$ cm) and mass (~ 1 $M_\odot$). On the contrary, the hottest stellar coronae need magnetic fields for confinement as already mentioned in 2.2. The hot gas in supernova remnants (with the total mass of at most a few solar masses) must be expanding into interstellar space, thus greatly affecting the environment. Similarly, the hot bubble at the Galactic center cannot be gravitationally bound, because the required binding mass (10^{10-11} solar masses) is comparable to the total mass of our Galaxy. This suggests that the Galactic-Center hot bubble is a result of some kind of energetic explosion. As well, the hot gas in starburst galaxies (Fig.3a)[11] may be streaming out of the system.

Hot plasmas are also associated with very large self-gravitating systems, e.g. elliptical galaxies (Fig.3e), clusters of galaxies, and possibly even super-clusters[8]. Recent GINGA observations[12] revealed that the Virgo cluster of galaxies is embedded in a hot ($T \sim 2 \times 10^7$ K) gas which is much larger in size than was previously thought. In these larger systems, there are a number of reasons to believe that the plasma is bound to the system by gravity. However the required binding mass often exceeds the "visible" mass (calculated from the optical brightness of the system) by 1-2 orders of magnitude. This fact provides an evidence for the presence of "massive dark matter" in these large systems. Since this missing mass problem is of profound importance in cosmology and elementary particle physics, X-ray observations of these plasmas are becoming of increasing importance.

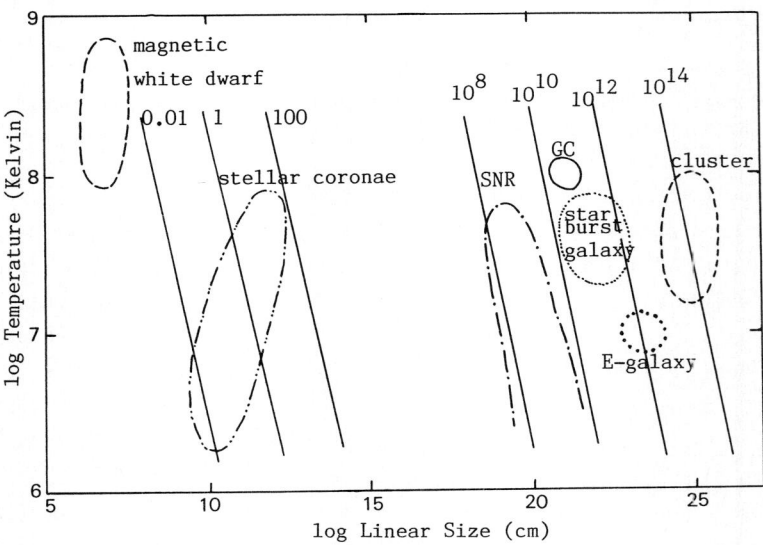

Fig.5 : Linear size and temperature of various astrophysical plasmas. Solid lines indicate the necessary binding mass in unit of solar masses.

2.4 Elemental abundance and iron K-line emission

Although the LAC detectors (Ar-Xe gas proportional counters) have a rather modest energy resolution (18% at 5.9 keV), they are capable of useful atomic-line spectroscopy of cosmic plasmas. The most important lines in the present energy range are the iron Kα lines at 6.4-6.9 keV[7], because iron is the most abundant heavy element serving as a tracer of nucleosynthesis activities. For a plasma of ionization equilibrium with temperature in the range 10^{7-8} K, the Helium-like iron is the most dominant ion species, from which K-line is emitted at 6.7 keV[7]. In Fig.3, iron lines are seen more or less in all the spectra. Supernova remnants usually exhibit very prominent iron K-lines, as a result of the explosive nucleosynthesis and the consequent iron overabundance. On the contrary, the spectra of UX Ari (Fig.3b) and M82 (Fig.3a) show much less intense iron line (1/3 - 1/2 in equivalent width) than theoretical predictions. This requires some mechanism of line suppression (other than iron underabundance) which is yet to be discovered. The emissions from the Galactic ridge[9] and the Galactic Center[10] exhibit the iron line to an equivalent width consistent with theoretical calculations assuming cosmical abundances and ionization equilibrium.

3. SELF-SIMILAR VARIABILITY IN COMPACT X-RAY SOURCES

3.1 Violent X-ray variations

In contrast to emission from diffuse plasma components as reviewed in section 2, X-ray emission induced by mass accretion onto "compact" celestial objects (white dwarfs, neutron stars and black holes) are highly variable on wide range of time scales. This is visualized in Fig.6. In fact, the analysis of periodic, quasi-periodic and aperiodic time variability in compact X-ray sources[13,14] forms an important research field of X-ray astronomy.

3.2 Various types of accreting compact objects

Here we briefly summarize classifications of "compact" X-ray sources.

(a) Accreting white dwarfs: An aged close binary system of a mass-donating non-degenerate primary and a mass-accreting white dwarf. The mass flow may form an accretion disk when the white dwarf is only weakly ($<10^6$ G) magnetized, or funneled onto magnetic polar caps of the white dwarf otherwise (Fig.3c). Both the accretion disk and the polar caps are optically thin.

(b) X-ray pulsars[15,16]: A close binary composed of a massive star and a strongly magnetized (10^{11-13} G) neutron star. The mass transfer from the primary onto the neutron star takes the form of either stellar-wind capture, Roche-lobe overflow, or accretion from circumstellar envelope. The matter flow onto the neutron star forms optically-thick hot accretion columns at the polar caps. The rotation of the neutron star in combination with anisotropic X-ray radiation due to cyclotron resonances (see section 4) produces periodic X-ray pulsations (Fig.6a).

(c) Low-Mass X-ray Binaries (LMXB) : A close binary composed of a low-mass primary and a neutron star without strong magnetic field ($B<10^8$ G). The mass transfer occurs via Roche-lobe overflow. X-ray spectrum is a superposition of optically-thick emission from accretion disk and a blackbody emission (2×10^7 K) from the neutron-star surface[17]. X-ray signal exhibits aperiodic (Fig.6b)[13] and often quasi-periodic[14] variations, but usually periodic pulsations are not detected. These sources often emit X-ray bursts.

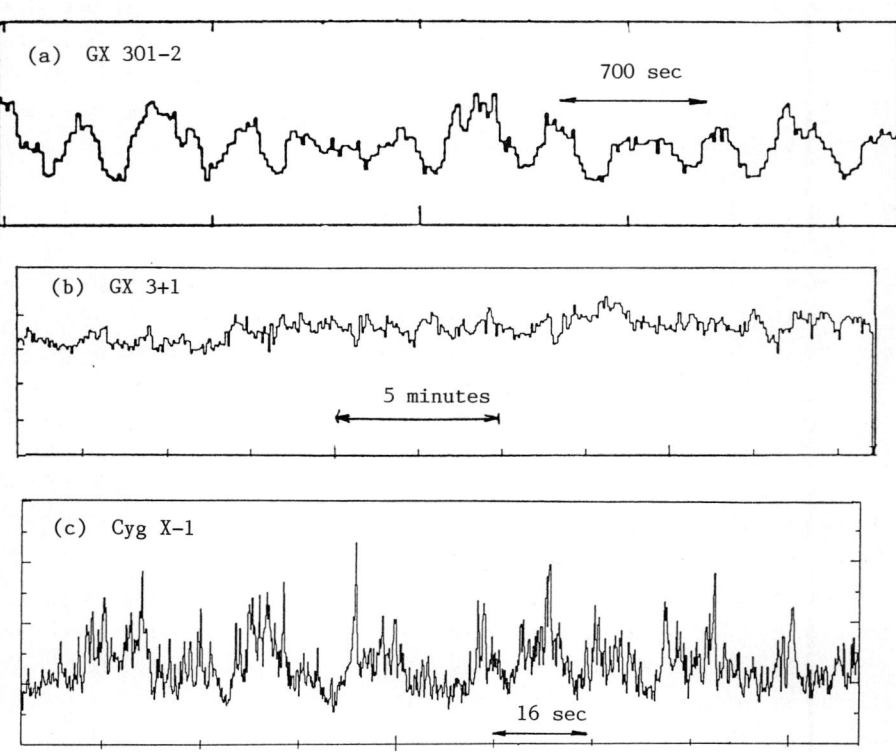

Fig.6: X-ray counting time series from three galactic compact X-ray sources. (a) An X-ray pulsar GX301-2 with the 700s periodic pulses. (b) A Low-Mass X-ray Binary GX3+1 showing random variations. (c) The black-hole candidate Cyg X-1, exhibiting violent random flickering.

(d) Black Hole Binaries : In a mass-exchanging binary containing a stellar-mass black hole secondary, an accretion disk will form around the black hole. The disk can either be optically-thick or optically-thin[18,19]. The optically-thick disk is expected to emit very soft spectrum[20], while the optically-thin disk may radiate in a hard spectrum via inverse Compton process[21]. There are so far a dozen of X-ray sources to be classified into this category, although their black-hole candidacy is in many cases yet to be confirmed. The leading candidate is Cyg X-1 (Fig.6c).

(e) Active galactic nuclei : So called active galaxies (Quasars, Seyfert galaxies, emission-line galaxies etc.) radiate enormous amount of energy in wide-band spectra from their nuclei. The X-ray intensity is variable on time scales from minutes to years. It is widely believed that there exists a massive (10^{6-8} $M_\odot$) black hole at the nucleus, which accretes matter from environment.

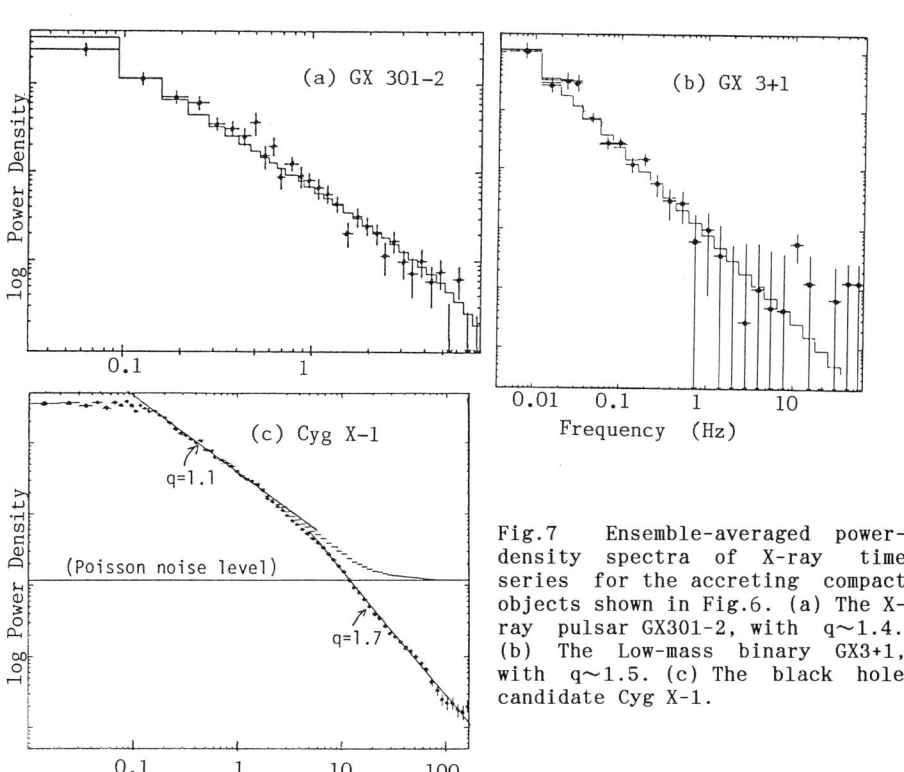

Fig.7 Ensemble-averaged power-density spectra of X-ray time series for the accreting compact objects shown in Fig.6. (a) The X-ray pulsar GX301-2, with $q\sim1.4$. (b) The Low-mass binary GX3+1, with $q\sim1.5$. (c) The black hole candidate Cyg X-1.

3.3 Aperiodic variability and self-similarity

Aperiodic variability is widely observed in nearly all types of accreting compact objects (Fig.6). Ensemble-averaged Fourier power-density spectra (PDS; Fig.7) provide a standard tool to analyze such variability. Although theory of the shape of the PDS continua is still quite primitive, we have in hand a useful concept called "self-similar variability"[13]. A random variation is called self-similar when the power-density of variation, P, is expressed as a power-law function of the frequency f, i.e. $P \propto f^{-q}$. Then the "characteristic time scale" of variation can no longer be defined, because the PDS is invariant under time-scale changes. As displayed in Fig.7, observations with GINGA[13] and EXOSAT[22] have discovered self-similar variability in X-ray data from various types of accreting compact objects. Moreover, the power-law index q was found mostly in the range 1.1-1.7 (typically around 1.4). Although we do not have adequate explanation of these results yet, we imagine that these self-similar time variability may be a result of plasma turbulence caused by strong thermalization of bulk kinetic energy of the accreting matter.

Miyamoto et al.[23], by analyzing the X-ray signal from Cyg X-1, discovered that variations in harder X-rays are delayed in Fourier phase as compared to those in softer X-rays. The lag was roughly constant in phase over the 0.1-100 Hz frequency range. Origin of this "phase lag" is yet to be discovered.

4. Cyclotron Resonance in X-ray Pulsars

4.1 Spectra of X-ray pulsars

X-ray pulsars have been known to exhibit very hard power-law like spectra (photon index $\alpha \sim 1$) with steep cutoff above 10-20 keV[15,16]. From observational viewpoint, little attempt has been made to clarify the origin of these spectral characteristics. Tanaka[24] argued that the cutoff may be caused by electron cyclotron resonance in the strong (10^{11-13} G) magnetic field of the pulsar, and showed that the observed spectral cutoff can be described by the classical formula for the electron cyclotron opacity. This idea was confirmed by Makishima et al.[25] through subsequent analysis of the TENMA data. These works suggested that the cyclotron resonance should take place widely among X-ray pulsars, although resonance features themselves had been observed only in two X-ray pulsars, Her X-1 and 4U0115+63.

4.2 New GINGA results

A series of new results on the cyclotron resonance features have been obtained from GINGA observations, using both the LAC and the GBD We briefly summarize them below.

(1) Murakami et al.[26] discovered absorption lines at 20 and 40 keV in the GBD spectra of two gamma-ray bursts, and interpreted them as the 1st and 2nd harmonics of the cyclotron resonance in magnetic field of strength 1.7×10^{12} G. This strongly supports the neutron star origin of gamma-ray bursts.

(2) A prominent absorption feature was discovered[27,16] at 20 keV in the spectra of the 530 second pulsar 4U1538-52. The overall spectral shape, including both the cyclotron feature and the cutoff at $\sim$15 keV, has been modeled successfully in terms of the classical cyclotron opacity.

(3) A hint of spectral feature was observed[28] at $\sim$6 keV from the 7 second pulsar X2259+586. The implied field strength, $\sim 5\times 10^{11}$ G, agrees with the value estimated from the spin-down trend of this pulsar.

(4) The LAC spectrum of Her X-1 resolved the well-known cyclotron feature of this pulsar with a high accuracy[29]. We determined the resonance energy to be 34 ± 1 keV, hence the field intensity to be $(2.9\pm 0.3)\times 10^{12}$ G. Similar to the case of 4U1538-52, we conclude that the spectral cutoff is a direct consequence of the cyclotron resonance.

(5) An X-ray outburst of the transient X-ray pulsar X0331+53 has been detected with the ASM in late September 1989. The LAC observations have revealed a prominent absorption structure at about 29 keV[30], in agreement with the prediction based on TENMA results.

4.3 Physical and astrophysical implications

There have been a fair number of theoretical calculations of X-ray pulsar spectra[31,32], taking into account the cyclotron resonance. The new GINGA results will give large impact upon these theoretical attempts. In addition, we have shown that the high-energy cutoff in pulsar spectra is produced as a direct consequence of cyclotron absorption. Therefore we will be able to used the cutoff energy as a reliable measure of the field strength for a number of X-ray pulsars. We expect these results to bring about significant progress on our understanding of the distribution and evolution of magnetic field of the neutron star.

ACKNOWLEDGEMENT

I would like to express my deepest thanks to the organizing committee of the conference, above all to Professor Setsuo Ichimaru, for this valuable opportunity. I acknowledge all the members of the GINGA team, as I owe much of the present talk to the collaborative effort of the team. I also thank Professor Toshiki Tajima for stimulating discussions.

References
1) F. Makino and the ASTRO-C Team, Astro. Lett. Comm., **25**, 223 (1987).
2) T. Dotani et al., Nature, **330**, 230 (1987).
3) R. Sunyaev et al., Nature, **330**, 227 (1987).
4) M. Itoh, S. Kumagai, T. Shigeyama, K. Nomoto and J. Nishimura, Nature, **330**, 233 (1987).
5) J. Kristian et al., Nature, **338**, 234(1989).
6) T. Ohashi, in "Physics of Neutron Stars and Black Holes", ed. Y. Tanaka (Universal Academy Press: Tokyo), p. 301 (1988).
7) K. Makishima, in "The Physics of Accretion onto Compact Objects", ed. K. Mason, M. Watson and N. White (Springer-Verlag: Berlin), p.249 (1986).
8) K. Koyama, ISAS Symposium on Cosmic Radiation, p.35, 1989 (in Japanese).
9) K. Koyama, K. Makishima, Y. Tanaka and H. Tsunemi, Publ. Astr. Soc. Japan **38**, 121 (1986).
10) K. Koyama, H. Awakai, H. Kunieda, S. Takano, Y. Tawara, S. Yamauchi, I. Hatsukade and F. Nagase, Nature, **339**, 603 (1989).
11) K. Makishima and T. Ohahsi, in "Physics of Neutron Stars and Black Holes", ed. Y. Tanaka (Universal Academy Press: Tokyo), p. 175 (1988).
12) S. Takano, H. Awaki, K. Koyama, Y. Tawara, S. Yamauchi, K. Makishima and T. Ohashi, Nature, **340**, 289 (1989).
13) K. Makishima, in "Physics of Neutron Stars and Black Holes", ed. Y. Tanaka (Universal Academy Press: Tokyo), p. 175 (1988).
14) W. Lewin, J. van Paradijs and M. van der Klis, Mon. No. R. Astr. Soc. **46**, 273 (1988).
15) N. E. White, J. H. Swank and S. S. Holt, Astrophys. J. **270**, 711 (1983).
16) F. Nagase, Publ. Astr. Soc. Japan **41**, 1 (1989).
17) K. Mitsuda et al., Publ. Astr. Soc. Japan **36**, 741 (1984).
18) S. Ichimaru, Astrophys. J. **214**, 840 (1977).
19) H. Inoue and R. Hoshi, Astrophys. J. **322**, 320 (1987).
20) K. Makishima, Y. Maejima, K. Mitsuda, H. V. Bradt, R. Remillard, I. R. Tuohy, R. Hoshi and M. Nakagawa, Astrophys. J. **308**, 635 (1986).
21) R. A. Sunyaev and L. G. Titarchuk, Astr. Astrophys. **86**, 121 (1980).
22) I. A. McHardy and B. Czerny, Nature **325**, 696 (1987).
23) S. Miyamoto, S. Kitamoto, K. Mitsuda and T. DOtani, Nature, **336**,450 (1988).
24) Y. Tanaka, in "Radiation Hydrodynamics in Stars and Compact Objects", ed. D. Mihalas and K. Winkler (Springer-Verlag: Berlin), p.198 (1985).
25) K. Makishima et al., Publ. Astr. Soc. Japan, submitted.
26) T. Murakami et al., Nature, **335**, 234 (1988).
27) G. Clark, J. Woo, F. Nagase, K. Makishima and T. Sakao, Astrophys. J. Letters, in press.
28) K. Koyama et al., Publ. Astr. Soc. Japan **41**, 461 (1989).
29) T. Mihara et al., Nature, submitted.
30) IAU Circular, No.4871 and No.4872, (1989).
31) P. Meszaros and W. Nagel, Astrophys. J. **298**, 147 (1985).
32) J. Wang, I. Wasserman, and E. Salpeter, Astrophys. J. **338** (1989).

PROTON ABUNDANCE IN HOT NEUTRON STAR MATTER

Tatsuyuki TAKATSUKA and Jun HIURA

College of Humanities and Social Sciences, Iwate University
Morioka 020 JAPAN

By solving the finite-temperature Hartree-Fock equations with effective interaction, the proton abundance in hot neutron star matter is investigated. It is found that the proton fraction increases remarkably as the temperature goes high, which is caused mainly by the entropy production due to the proton and electron contaminations.

1. INTRODUCTION

Neutron star matter is a kind of "plasma" in astropysical nuclear system, consisting of dominant neutrons and small admixture of protons and electrons. The properties have been extensively studied at zero temperature (T=0) in relation to usual (cold) neutron stars[1~3]. For the formation or the birth era of neutron stars, however, we are concerned with a "supernova matter", namely a hot neutron star matter with T = (10 ~ 50) MeV [4].

In this paper we investigate the properties of neutron star matter *at finite temperature* (T>0), for which very few works[5,6] have been done, especially at densities higher than nuclear density ρ_0 (0.17 nucleons/fm^3 ≃ 2.8×10^{14} g/cc). In particular, we discuss how the proton (and electron) population depends on the temperature of hot neutron star matter under β equilibrium. In different from the p-mixing at T=0, the effect of entropy also takes part in the problem at T>0.

2. APPROACH

We consider a hot neutron star matter composed of n, p and e$^-$ with the total nucleon density ρ and the temperature T, which is in thermodynamic and β equilibrium. Nucleons are taken to interact via two-nucleon interaction V, the Coulomb interaction being neglected because of the charge neutral system.

From the viewpoint of a variational theory, the ground state is determined by the minimization of the free energy under the subsidiary conditions, namely, conservation of the total nucleon number, charge neutrality and chemical equilibrium. This leads to a set of finite-temperature Hartree-Fock equations. In this treatment, we introduce the ρ-dependent effective interaction $\tilde{V}$ in place of V. The $\tilde{V}$ is constructed at T=0, as usual, based on the G-matrix calculation with V and thereby the short-range correlation (s. r. c.) has been incorporated in $\tilde{V}$. The validity to make an extended use of $\tilde{V}$ at T=0 into the T>0 case (namely, to neglect the T-dependence of the s. r. c.) has been shown by us [7].

The set of equations are given as

$$\varepsilon_n(\alpha) = t_n(\alpha) + \sum_\beta f_n(\beta)<\alpha\beta|\tilde{V}|\alpha\beta-\beta\alpha>_{nn} + \sum_\beta f_p(\beta)<\alpha\beta|\tilde{V}|\alpha\beta-\beta\alpha>_{np} \quad , \qquad (1)$$

$$\varepsilon_p(\alpha) = t_p(\alpha) + \sum_\beta f_p(\beta)<\alpha\beta|\tilde{V}|\alpha\beta-\beta\alpha>_{pp} + \sum_\beta f_n(\beta)<\alpha\beta|\tilde{V}|\alpha\beta-\beta\alpha>_{pn} + \delta m \quad , \qquad (2)$$

$$\rho = \rho_n + \rho_p = \sum_\alpha f_n(\alpha)/\Omega + \sum_\alpha f_p(\alpha)/\Omega \quad , \qquad (3)$$

$$\sum_\alpha f_p(\alpha)/\Omega = \rho_p = \rho_e = \sum_\alpha f_e(\alpha)/\Omega \quad , \qquad (4)$$

$$\mu_n = \mu_p + \mu_e \quad , \qquad (5)$$

where $\quad f_i(\alpha) = (1 + e^{(\varepsilon_i(\alpha)-\mu_i)/T})^{-1} \quad ; \quad i=n,\ p,\ e^- \quad . \qquad (6)$

In these expressions, α (or β) denotes the single-particle states; $\alpha\equiv$(momentum $\vec{q}_\alpha$, spin), ε_i (i=n, p. e$^-$) is the single-particle energy, f_i the occupation probability and μ_i the corresponding chemical potential. Also, $t_i(\alpha) = \vec{q}_\alpha^2/2m_i$ for i=n and p, $t_e(\alpha) = (\vec{q}_\alpha^2 + m_e^2)^{1/2}$ and $\delta m = m_p - m_n$, with m_i being the mass.*) The ρ_i denotes the density and Ω is the normalization volume.

Eq.(1) (Eq.(2)) (together with Eq.(6)) is the Hartree-Fock equation corresponding to n (p) sector which determines $\varepsilon_n(\alpha)$ ($\varepsilon_p(\alpha)$) for a given μ_n (μ_p). Note that $\varepsilon_e(\alpha)(=t_e(\alpha))$ is already known due to the absence of interaction. Eqs.(3), (4) and (5) (together with Eq.(6)) are implied by the subsidiary conditions. Since we have five equations (Eqs.(1)∼(5)) for five unknown quantities, the set of equations can be solved. From the solutions; ε_n, ε_p, f_n, f_p and f_e, we can derive directly various thermodynamic quantities, such as the internal energy E, entropy S and free energy F (=E-TS), expressed as per particle [7]. The proton mixing ratio y_p here discussed is given by $y_p = \rho_p/\rho$.

3. NUMERICAL RESULTS AND DISCUSSION

We adopt the approximation[1]; the use of $\tilde{V}=\tilde{V}(Z=0)$ for n-n part in Eq.(1), the neglect of p-p part in Eq.(2) because of the small contamination of protons and the use of $\tilde{V}=\tilde{V}(N=Z)$ for n-p parts in Eqs.(1) and (2). Here $\tilde{V}(Z=0)$ ($\tilde{V}(N=Z)$) represents the effective two-nucleon interaction with the state- and ρ-dependence for neutron matter (symmetric nuclear matter). As a suitable choice

---

*) Throughout the paper, the units $\hbar = c = \kappa_B = 1$ are used.

for $\tilde{V}(N=Z)$, we use the Go-force of Sprung and Banerjee [8]. As for $\tilde{V}(Z=0)$, we use the one in Ref. 9) which is a version of the Go-force applicable to neutron matter. The coupled equations (1)~(6) are solved numerically by the iterative method.

As a typical case, the y_p for $\rho \simeq 1.6\rho_0$ are shown in Fig.1 as functions of T. We note the following points.

(i) The attractive effect of n-p interaction, the main ingredient for the p-mixing at T=0, works for populating p even for the T>0 case, as is observed by comparing the solid line (case I with $\tilde{V}$) with the dashed one (case II without $\tilde{V}$). The effect, however, is relatively weakened with increasing T; (y_p(case I) $-y_p$(case II))/y_p(case II)$\simeq$3.6→1.8→0.8 as T=0→10→20 MeV. The reason is that as

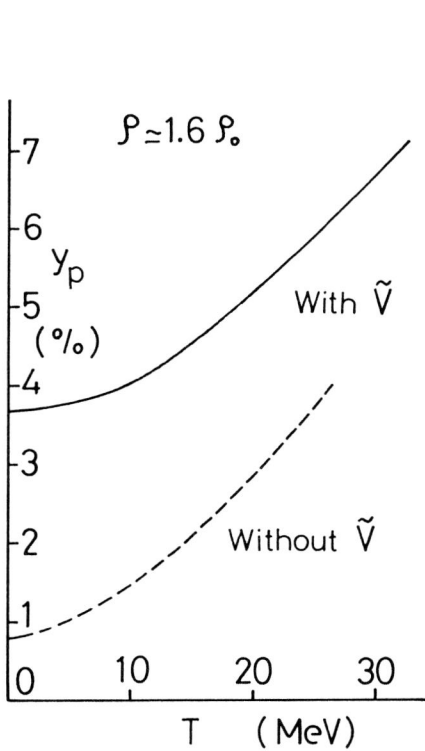
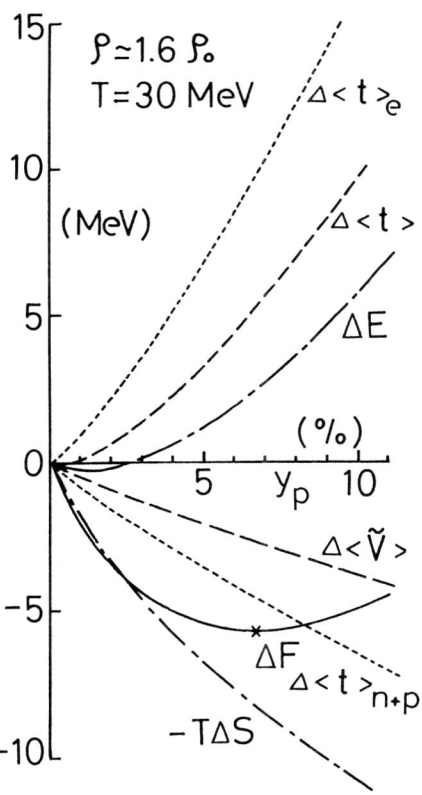

FIGURE 1
Proton fraction y_p in neutron star matter at the density $\rho \simeq 1.6\rho_0$ as functions of the temperature T.

FIGURE 2
Contribution to the free energy F (=E-TS) as functions of y_p. Notations are explained in the text.

T increases, high momentum components are enhanced due to the diffused f_n or f_p and so the contribution from n-p interaction gets less attractive.

(ii) The y_p is found to increase remarkably with increasing T; $y_p \simeq 3.7 \to 6.7\%$ as $T = 0 \to 30$ MeV. To clarify the reason, the components in F at T=30 MeV are shown in Fig.2 as functions of y_p, where $\Delta F = \Delta E - T\Delta S$, $\Delta E = \Delta \langle t \rangle + \Delta \langle \tilde{V} \rangle$ and the total kinetic energy $\Delta \langle t \rangle = \Delta \langle t \rangle_e + \Delta \langle t \rangle_{n+p}$, being measured from the $y_p = 0$ case. It is observed that $\Delta \langle t \rangle_{n+p}$ and $\langle \tilde{V} \rangle$ act for the increase of y_p whereas $\Delta \langle t \rangle_e$ acts against this. The net effect $\Delta E = \Delta \langle t \rangle_{n+p} + \Delta \langle \tilde{V} \rangle + \Delta \langle t \rangle_e$ is not effective to lower F. Most important is the effect of entropy part ($-T\Delta S$) which is always negative and grows according with y_p. Due to this effect, F can be minimum at a large y_p ($\simeq 6.7\%$).

4. CONCLUDING REMARKS

Protons and electrons are easy to appear in neutron star matter as the temperature increases. This originates from the fact that the effect of entropy, which is enhanced according with T and y_p, plays a significant role in lowering F at a given ρ.

Further investigations with getting y_p at various ρ, by improving the approximations used here and also under the situation of neutrino degeneracy, are in progress.

ACKNOWLEDGEMENT

We are thankful to Dr. S. Nishizaki for cooperative discussions. One of the author (T.T.) is indebted to the Grant-in-Aid for Scientific Research from the Ministry of Education, Science and Culture, No. 01652501.

REFERENCES
1) J.Nemeth and D.W.L.Sprung, Phys. Rev. 176 (1968) 1496.
2) S.Ikeuchi, S.Nagata, T.Mizutani and K.Nakazawa, Prog. Theor. Phys. 46 (1971) 95.
3) D.Ellis and D.W.L.Sprung, Can. J. Phys. 50 (1972) 2277.
4) T.Takatsuka, Prog. Theor. Phys. 82 (1989) No. 3.
5) H.A.Bethe, G.E.Brown, J.Applegate and J.M.Lattimer, Nucl. Phys. A324 (1979) 847.
6) J.M.Lattimer, C.J.Pethick, D.G.Ravenhall and D.Q.Lamb, Nucl. Phys. A432 (1985) 646.
7) T.Takatsuka and J.Hiura, Prog. Theor. Phys. 79 (1988) 268.
8) D.W.L.Sprung and P.K.Banerjee, Nucl. Phys. A168 (1971) 273.
9) T.Takatsuka, Prog. Theor. Phys. 72 (1984) 252.

THERMONUCLEAR REACTION RATES OF DENSE CARBON-OXYGEN MIXTURES IN WHITE DWARFS

Shuji OGATA, Hiroshi IYETOMI and Setsuo ICHIMARU

Department of Physics, University of Tokyo, Bunkyo-ku, Tokyo 113, Japan

Screening potentials between ions in dense C-O mixtures appropriate to white-dwarf interiors are analyzed by Monte Carlo simulation method, both in a fluid and in a bcc-crystalline state. Criteria for the validity of a classical approximation in the screening potential are derived; it is shown that the classical approximation applies only in a fluid state for a C matter. Enhancement rate of nuclear reactions in a dense C-O fluid is calculated.

White dwarf in a close binary systems is considered as a likely progenitor of Type-I supernova (SNI). Binary-ionic mixture (BIM) of carbon and oxygen is thought to make a typical internal constituent for such a white dwarf. Thermonuclear reaction rate in a dense BIM is therefore a crucial issue in the SNI mechanism.

The rate of fusion λ_{ab} between a pair of nuclei, a and b, is proportional to $g_{ab}(0)$, the value of the joint probability density $g_{ab}(r)$ at $r = 0$ for reacting pair averaged over the motion of their center-of-mass and the remainder of the particles. The probability density can be factored as

$$g_{ab}(r) = g_{ab}^0(r)\exp[\beta H_{ab}(r)] \tag{1}$$

where $g_{ab}^0(r)$ represents the joint probability in the partial system consisting only of the reacting pair; β is the inverse temperature in energy units. Equation (1) defines the screening potential, $H_{ab}(r)$. Since the contribution of the *direct* interaction between a and b has been factored in (1), $H_{ab}(r)$ results from many-body effects in scattering with the rest of the particles. It follows from (1) that the rate of fusion is enhanced by a factor, $\exp[\beta H_{ab}(0)]$, over the value λ_{ab}^0 arising from the direct binary-interaction between a and b.

The Monte Carlo (MC) method has provided the most accurate data on the joint probability functions in one-component plasmas[1] (OCP's) and in BIM's,[2] under the condition that the classical statistics is applicable. We begin, therefore, with a derivation of the conditions for the validity of a classical approximation in the analyses of the screening potentials. For simplicity, we consider the cases of an OCP (electric charge, Ze; mass, M; number density, n; temperature, T), though the results can be extended straightforwardly to multi-ionic cases.

The Coulomb-coupling constant for the OCP is $\Gamma = (Ze)^2/ak_BT$, where $a = (3/4\pi n)^{1/3}$ is the ion-sphere radius. We take the thermodynamic freezing condition[1,2] for the OCP at $\Gamma = 180$. In the fluid OCP with $\Gamma < 180$, the classical treatment applies when the interparticle separation is greater than the thermal de Broglie wavelength, so that

$$\Lambda \equiv \hbar(Mk_BT)^{-1/2}/a < 1\ . \tag{2}$$

In the crystalline OCP with $\Gamma > 180$, the classical approximation is valid when

$$k_B T > \hbar \omega_0 . \tag{3}$$

Here $\omega_0 = (4\pi \rho_e^2 / 3\rho_m)^{1/2}$ is the Einstein-oscillator frequency in the Wigner-Seitz sphere model ($\rho_e = Zen$, $\rho_m = Mn$).

Figure 1 depicts the freezing condition as well as the relations, (2) and (3), on the ρ_m–T plane for the carbon OCP. In the fluid state, the condition (2) for a classical treatment is clearly met. In the crystalline state, on the other hand, the condition (3) cannot be satisfied; the carbon OCP forms a quantum crystal.

Let us then investigate the nuclear reaction rate under the condition that the classical approximation, (2) or (3), is valid.

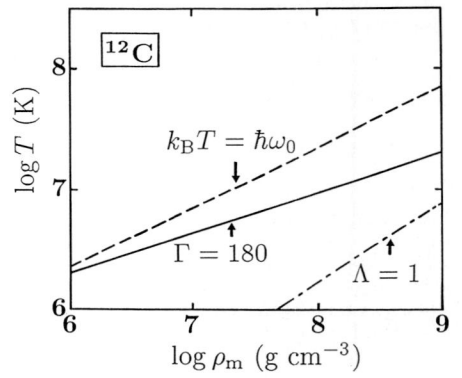

FIGURE 1
Comparison of parameters on ρ_m–T plane for carbon material.

In these circumstances, the spatial and velocity distributions are decoupled, so that the joint probability in the partial system of the reacting pair with relative velocity v is expressed as

$$g_c^0(r; v) = (\beta M / 4\pi)^{3/2} \exp\{-\beta [M v^2 / 4 + V(r)]\} , \tag{4}$$

and

$$g_c^0(r) = \int d^3 v\, g_c^0(r; v) = \exp[-\beta V(r)] . \tag{5}$$

Here the subscript c stands for "classical", and $V(r)$ refers to the potential of binary interaction. In an OCP, $V(r) = (Ze)^2/r$; the potential can further accommodate screening by quantum-mechanical objects such as electrons and muons as well. Substitution of (4) in (1) (deleting the subscripts, a and b) yields

$$H(r) = V(r) + \ln[g_c(r)]/\beta \tag{6}$$

The last term of (6) is minus the potential of mean force;[2] the screening potential (6) is represented diagrammatically in Fig. 2. It has been shown[4] that the value of $H(0)$ is equal to the increment in the interaction (or excess) part of the chemical potential for the reacting pair before and after the nuclear reaction. In Fig. 2 with $r_{12} \simeq 0$, the major contribution to $H(0)$ comes from the third and other particles (solid circles) located around the nearest neighbor (n.

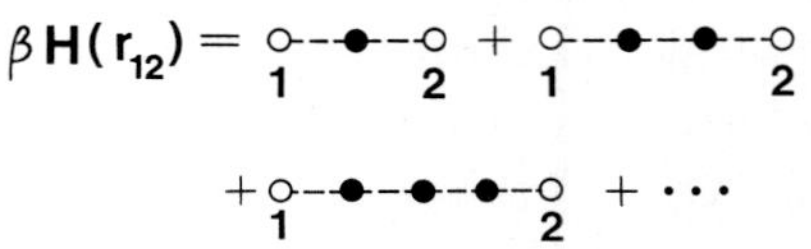

FIGURE 2
Diagrammatic representation of the screening potential. Dashed lines are the f-bonds, $f(r) = \exp[-\beta V(r)] - 1$, and the solid circles are the particle coordinates to be integrated.

n.) distances ($\sim 1.8a$) from one of the reacting particles. As long as (2) or (3) is met, one can therfore calculate the enhancement factor, $\exp[\beta H(0)]$, in accord with (6) or Fig. 2.

Since $g_c^0(r \to 0)$ vanishes identically due to divergence of $V(r \to 0)$, we must calculate $g^0(r \to 0)$ quantum-mechanically even when the classical condition, (2) or (3), is satisfied. The rate of fusion λ^0 per a pair with $V(r) = (Ze)^2/r$ is thus obtained approximately in a form proportional to the Gamow penetration factor,[2]

$$P(v) = \exp\{-(\tau/3)[(1/\zeta) + 2\zeta^{1/2}]\} ,\qquad(7)$$

where $\tau = [(27\pi^2/4)\beta M(Ze)^4 \hbar^{-2}]^{1/3}$ and $\zeta = (3\pi/\tau)^2 (Ze)^4/(\hbar v)^2$. The net reaction rate is $\lambda = \lambda^0 \exp[\beta H(0)]$.

We have performed a series of MC simulations[2] in dense C-O BIM's, both in fluid (4 cases) and in bcc-crystalline (5 cases) states, at various combinations of T and $x = N_O/(N_C + N_O)$. Figure 3 displays examples for $g_{ab}(r)$ (a, b=C, O), where T_C is the temperature at $\Gamma = 180$ in Fig. 1.

In Fig. 4 we show three different evaluations of $H(r)$ for C-C in a C-O fluid ($a_C = (Z_C/\langle Z \rangle)^{1/3} a$); dotted line, a straight extrapolation of the MC results (see Fig. 5 also) to $r = 0$; solid line, a quadratic extrapolation[4] of the MC results; and dashed line, the improved HNC evaluation.[2] The values of $H(0)$ in the latter two extrapolations agree within 1%. For C-C in a C-O fluid, we thus find

$$\beta H(0)/\Gamma = 1.10 ,\qquad(8)$$

where $\Gamma = (Z_C e)^2/a_C k_B T = 180 T_C/T$.

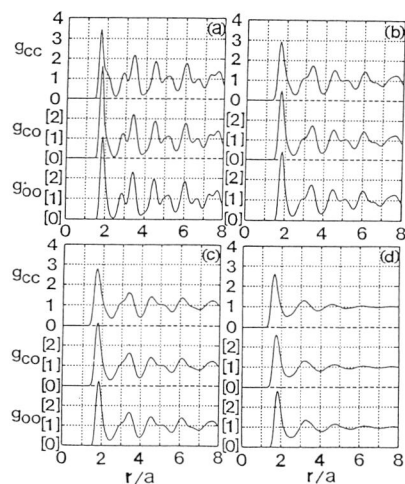

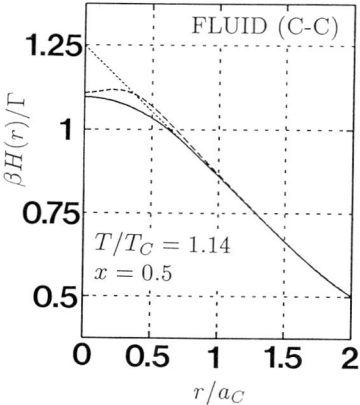

FIGURE 3
MC joint probability distributions: (a) crystal, $(T/T_C, x) = (0.5, 0.5)$; (b) crystal, $(0.9, 0.5)$; (c) crystal, $(1.1, 0.5)$; (d) fluid, $(1.1, 0.48)$.

FIGURE 4
Screening potentials between C-C in a C-O fluid.

In conjunction with the parametric survey in Fig. 1, we have concluded that a classical (e.g., MC simulation) evaluation of $H(0)$ cannot be used for the enhancement factor in a solid carbon. It is, nevertheless, of interest to investigate the values of $H(0)$ in a C-O solid predicted by an extrapolation of the MC results, analogously to the case of a C-O fluid. For this purpose, we decompose $g_{CC}(r)$ into separate contributions arising from those particles in the first n. n., in the second n. n., ..., as Fig. 6 illustrates. Separate contributions to $H_j(r)$ ($j = 1, 2, ...$) are then evaluated as in Fig. 5. We thus find

$$\beta H_1(0)/\Gamma = 1.04, \quad \beta H_2(0)/\Gamma = 0.98 \ . \tag{9}$$

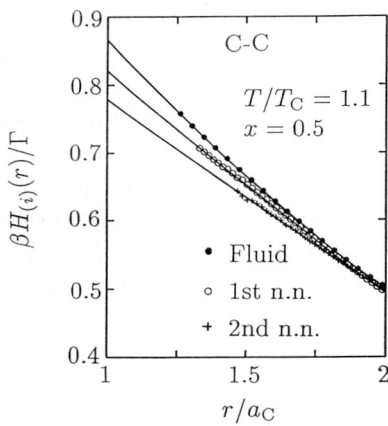

FIGURE 5
Screening potentials in MC simulations.

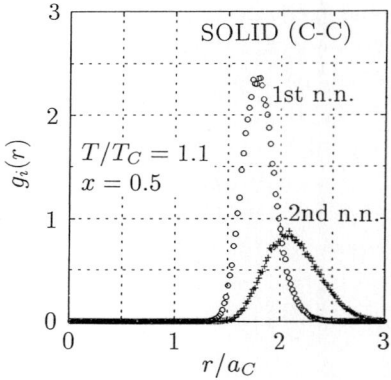

FIGURE 6
Decomposition of crystalline $g_{CC}(r)$ into separate nearest neighbor contributions.

In quantum crystal, wave-mechanical effects should act to increase the enhancement factor over that in a classical evaluation, as in λ_{ab}^0. The value of $\exp[\beta H_1(0)]$ in (9), therefore, represents a lower bound for the enhancement factor in the C-C reaction rate for C-O BIM solids. The solid/fluid ratio, $\exp[\beta H_1(0)]/\exp[\beta H(0)]$, takes on a value $\sim 10^{-4}$ at $T/T_C = 1.1$ and $x = 0.5$.

We thank Professor Hugh Van Horn for discussions on these and related subjects.

REFERENCES

1) For a review, S. Ichimaru, Rev. Mod. Phys. **54** (1982) 1017.

2) S. Ichimaru, H. Iyetomi, and S. Ogata, Astrophys. J. Lett. **334** (1988) L17; H. Iyetomi, S. Ogata, and S. Ichimaru. Phys. Rev. B **40** (1989) 309.

3) For a general reference, J. P. Hansen and I. R. McDonald, *Theory of Simple Liquids*, 2nd Ed. (Academic, London, 1986).

4) B. Widom, J. Chem. Phys. **39** (1963) 2803; B. Jancovici, J. Stat. Phys. **17** (1977) 357.

Chapter II:
Computer Simulations of Quantum and Classical Many-Body Systems

QUANTUM MONTE CARLO SIMULATION OF HYDROGEN PLASMAS

J. THEILHABER and B. J. ALDER

Lawrence Livermore National Laboratory, P.O. Box 808, Livermore, CA 94550, U.S.A

Feynman path-integral simulations of a system consisting of a few protons and distinguishable electrons under periodic boundary conditions at temperatures down to 0.6 eV and densities up to 1 gm/cm^3 are presented. The total energy, the potential energy, the pressure and the radial distribution functions are computed. For a two-proton, two-electron system, for which Fermi statistics can be ignored, the dissociation process of a single H$_2$ molecule with temperature and pressure into a completely ionized plasma has been observed without any evidence for a first-order phase transition. Results for an 8-electron, 8-proton system where the protons are free to move show that due to the Boltzmann statistics of the electrons a metastable cluster of protons forms at low temperature and high density. If the protons are fixed into an fcc molecular lattice, the properties of molecular hydrogen at high pressure can be compared to shock experiments and the neglect of exchange assessed.

1   INTRODUCTION

A central problem in plasma physics is to determine the properties of a plasma when the Coulomb particles are strongly interacting and the electrons are dealt with quantum-mechanically. One would like a method that automatically samples all regions of phase space, so as to account for the existence of all possible plasma components, including neutrals, partially ionized atoms and molecules, completely stripped ions and the continuum electrons. With this goal in mind, a numerical scheme based on the Feynman path-integral has been constructed to compute the equilibrium properties of a hydrogen plasma. Crucial to the success of the method is that it is able to demonstrate the formation, in a low density plasma, of atomic and molecular species as the temperature is lowered, and upon increasing the pressure, the disappearance of any such identifiable species.

In the present calculation, the electrons and protons are confined to a finite volume with periodic boundary conditions, and Fermi statistics are ignored, because the inclusion of exchange creates, so far, unsurmountable numerical difficulties. This limitation restricts calculations to a two-electron, two-proton system, for which Fermi statistics can be ignored. Nevertheless, this simple model qualitatively predicts the essential features of the hydrogen plasma which were outlined above. For larger numbers of particles our scheme can be used in its present form to investigate the question of the stability of matter, as well as the quantitative effect of statistics.

2 THE PATH-INTEGRAL ALGORITHM

A summary of the path-integral algorithm used in the computations is presented here, a more detailed description can be found in Ref.(1). Our model is three-dimensional and consists of electrons and protons interacting in a finite cubical volume with periodic boundary conditions. The equilibrium properties can be obtained from the partition function $Z = \text{Tr}(\rho)$, where the density operator $\rho = e^{-\beta H}$, where H is the Hamiltonian of the system, and $\beta = 1/T$ the inverse temperature. By resorting to the algebraic identity:

$$\rho = \left(e^{-\tau H}\right)^{m+1}, \qquad (1)$$

where $\tau = \beta/(m+1)$, the partition function can be written as:

$$Z = \int d\mathbf{R}_0 d\mathbf{R}_1 \cdots d\mathbf{R}_m \, \rho(\mathbf{R}_0, \mathbf{R}_1, \tau)\rho(\mathbf{R}_1, \mathbf{R}_2, \tau) \cdots \rho(\mathbf{R}_m, \mathbf{R}_0, \tau) \qquad (2)$$

where $\mathbf{R} = (\mathbf{x}_1, \mathbf{x}_2, \ldots, \mathbf{x}_n)$ comprises all the particle coordinates. The succession of points $\mathbf{R}_0, \mathbf{R}_1, \cdots, \mathbf{R}_m$ in Eq.(2) define for each particle a path in 3-dimensional space, hence the name path-integral. Because of its graphical appearance, the quantum path for each particle is often referred to as a ring "polymer".

If τ is made small enough, by taking m sufficiently large, an asymptotically exact expression for $\rho(\mathbf{R}_i, \mathbf{R}_{i+1}, \tau)$ can be obtained. Following Pollock and Ceperley[2], we use the high-temperature expansion:

$$\rho(\mathbf{R}, \mathbf{R}', \tau) = \rho_0(\mathbf{R}, \mathbf{R}', \tau)\tilde{\rho}(\mathbf{R}, \mathbf{R}', \tau) + O(\tau^3) \qquad (3)$$

where ρ_0 is the density matrix for the noninteracting particles, and

$$\tilde{\rho}(\mathbf{R}, \mathbf{R}', \tau) = \prod_{i<j} \frac{\rho_{ij}^{(2)}(\mathbf{x}_i, \mathbf{x}_j, \mathbf{x}_i', \mathbf{x}_j', \tau)}{\rho_{0i}(\mathbf{x}_i, \mathbf{x}_i', \tau)\rho_{0j}(\mathbf{x}_j, \mathbf{x}_j', \tau)} . \qquad (4)$$

In Eq.(4), ρ_{0i} is the free-particle density matrix for the i-th particle and $\rho_{ij}^{(2)}(\mathbf{x}_i, \mathbf{x}_j, \mathbf{x}'_i, \mathbf{x}'_j, \tau)$ is the pair density-matrix, which is determined from the numerical solution of the two-body Schrödinger equation[3]. The value of $\tau = 1/(81.4 \text{ eV})$ was used in most of the calculations, and thus at the lowest temperature of 0.6 eV, the quantum paths in Eq.(2) comprised $m = 135$ segments.

The multi-dimensional integral of Eq.(2) is evaluated by a classical Monte Carlo scheme using the Metropolis algorithm. Trial configurations are obtained by deforming the quantum path of a single particle at a time. For the electrons, at each trial move a section of path 5 segments long is deformed between fixed nodes, with the starting node chosen at random, the section length having been determined so that an acceptance ratio of roughly 0.5 is obtained. For the protons, whose large mass constrains the quantum path to a very small spatial volume, it was found necessary, in addition to deforming a section of path, to move the center-of-mass of the entire polymer so as to increase the rate of motion in phase space to an acceptable level. This was done by a molecular dynamics scheme in which a fictitious proton dynamics is introduced, with the proton mass equal to the electron mass and with the forces on the proton calculated from the charge distribution of all of the other particles. To generate a trial move, the proton center-of-mass is given an initial, random momentum, and then its motion is integrated through several time steps, bringing it to a new position.

3 TWO ELECTRONS AND TWO PROTONS

For a system composed of only two electrons and two protons, the replacement of Fermi statistics by Boltzmann statistics and the neglect of spin can be justified on the grounds that (i) the Boltzmann system will have the same ground state as the Fermi system, (ii) all excited energy levels will be the same and (iii) both systems will have the same high-temperature, classical limit. However, the Boltzmann statistics do not provide the proper statistical weights of some of the excited energy levels. This is not expected to be a serious approximation.

3.1 Radial Distribution Functions

For a low particle density, $\rho = 0.0225$ gm/cm^3, corresponding to a cubic box of side $L = 10\ a_0$ and an ion-sphere radius $r_s = 4.92 a_0$, the radial distribution functions for several temperatures are shown in Fig. 1, with the system becoming progressively colder from top to bottom. It should be noted that the radial distribution functions shown here include an *angular integration*, so that for instance $g_{ep}(r)dr$ is the fraction of electrons in a shell of volume $4\pi r^2 dr$ to be found, on average, at a distance r from a proton. In all cases, $\int_0^{\sqrt{3}L/2} g(r)\,dr = 1$, where L is a side of the box.

At the highest temperature displayed in Fig. 1, $T = 27.2$ eV, all the constituents of the system are completely dissociated, corresponding to an electron proton-plasma. Accordingly, the de Broglie wavelength for the electrons $\lambda_T = (\hbar^2/m_e T)^{1/2} = a_0$, much less than the ion-sphere radius $r_s = 4.92\,a_0$, and the coupling parameter $\Gamma = e^2/(r_s T) = 0.203$ is small, so that the plasma can be considered both nearly classical and weakly-coupled. The three different types of radial distribution functions are all similar to the noninteracting distribution function. This is because the kinetic energy dominates to such an extent that the particles can be considered noninteracting at all but small interparticle distances. The results for g_{pp} are considerably noisier than for the other two radial distribution functions, because in the Monte Carlo scheme, new proton configurations are generated less frequently than electron configurations, leading to poorer statistics for g_{pp}.

At $T = 3.88$ eV the strong electron-proton correlation is apparent, as indicated by the peak at $r = a_0$ in $g_{ep}(r)$, corresponding to the formation of atomic hydrogen with an electron in the 1s orbital. Note also that the shape of $g_{ee}(r)$ has changed very little, indicating that the electrons remain largely uncorrelated. $g_{pp}(r)$ gives weak indications of increased correlations between protons at small separation, however no clear peak corresponding to molecular formation can be observed.

At $T = 2.27$ eV the peak in $g_{ep}(r)$ at $r = a_0$ has greatly increased. This indicates a greater probability of H formation, and a decrease of the ionization probability. To make these observations more quantitative, one could define the relative populations of H and H$^+$ as being proportional to the areas under the first peak and the one beyond that, introducing a somewhat arbitrary separation distance $r \equiv r_i$. Here, r_i defines of how far an electron can separate itself from a proton before being considered completely ionized, which in this case could be taken as $r_i = 3a_0$. A second distinctive feature at $T = 2.27$ eV is the formation of a new peak in the proton-proton distribution function, at $r \approx 2a_0$, indicating the formation of H_2 or H_2^+. Nevertheless, the electron-electron distribution function has changed little, suggesting that there is still a significant fraction of ionized electrons.

Quantum Monte Carlo simulation of hydrogen plasmas

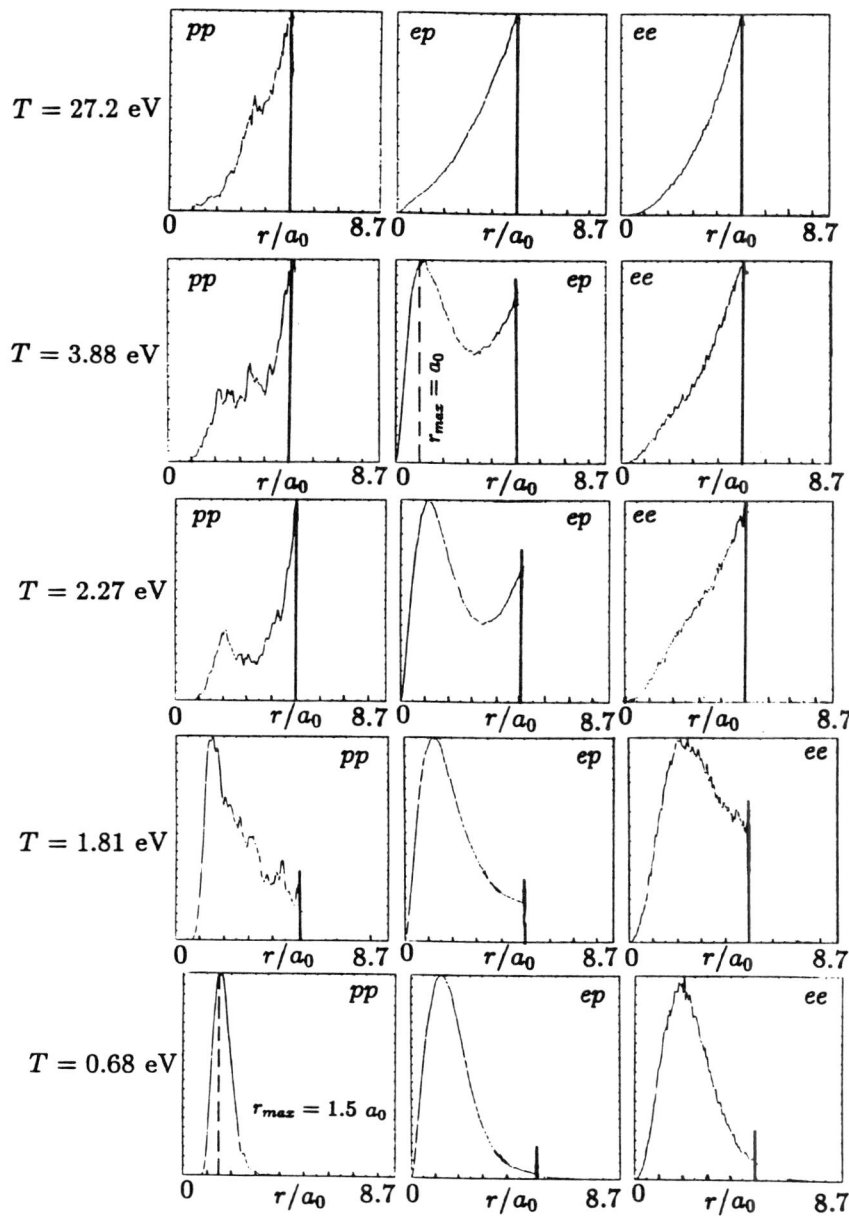

Figure 1: Radial distribution functions for the two-electron, two-proton system for a density $\rho = 0.0225$ gm/cm^3 ($r_s = 4.92 a_0$), at several temperatures indicated on the left; left column: $g_{pp}(r)$, center column: $g_{ep}(r)$, right column: $g_{ee}(r)$. The unphysical region $r > L/2 = 5\ a_0$ has been cut off.

At a lower temperature, $T = 1.81$ eV, the proton-proton distribution is strongly peaked about $r = 1.5a_0$, followed by steadily decreasing probability of proton separation with larger distances. This indicates that the hydrogen molecule can attain highly excited vibrational states, but with small probability of becoming unbound. A corresponding evolution in the electron-proton distribution function is observed. The electron-electron distribution under these conditions is peaked at a lower value of r, reflecting strong correlation between the two electrons, that are bound to the same molecule.

At the lowest temperature shown in Fig. 1, $T = 0.68$ eV, overwhelmingly the hydrogen molecule is formed with only moderate excitation of the vibrational states. Thus, the proton-proton distribution function is sharply peaked at $r = 1.5\ a_0$, a value close to the zero-temperature bond length $r = 1.4\ a_0$. Furthermore, the internal energy, obtained by subtracting from the total energy of the system ($E = -1.05e^2/a_0$) the kinetic energy of the H_2 molecule, ($E_{kin} = 3T/2 = 0.038e^2/a_0$), $E_{int} = -1.09e^2/a_0$ compares favorably to the ground-state energy $E_0 = -1.165e^2/a_0$ of the completely isolated H_2 molecule.

The above demonstrates that at low densities and at sufficiently low temperatures the plasma condenses into molecular hydrogen. At higher densities, however, the condensation into both the molecular and atomic phases is suppressed by the strong interaction with the nearest neighbors. In Figs. 2 the radial distribution functions for the density $\rho = 1$ gm/cm^3 ($r_s = 1.39a_0$) are displayed. At $T = 27.2$ eV the distribution functions are similar in shape to those of Fig. 1. The horizontal scale is, of course, changed, corresponding to the smaller box size at the higher density. However, at the lower temperatures no significant change in the radial distribution is observed, showing the lack of strong electron-proton and electron-electron correlations that were observed at lower densities. At $T = 0.91$ eV the proton-proton distribution function has only become slightly more localized, reflecting the strong plasma coupling parameter $\Gamma = 21.6$. Thus at high density the system is at all temperatures pressure-ionized, and ressembles a classical proton one-component plasma with an almost uniform electron background.

The same conclusions can be drawn from radial distribution functions obtained by keeping the temperature constant, $T = 12000°K = 1.03$ eV, and varying the density, as shown in Fig. 3. At the lowest density in Fig. 3, $\rho = 0.0225$ gm/cm^3, a bound though very hot hydrogen molecule is observed. As the density is inreased, first to $\rho = 0.044$ gm/cm^3, then to

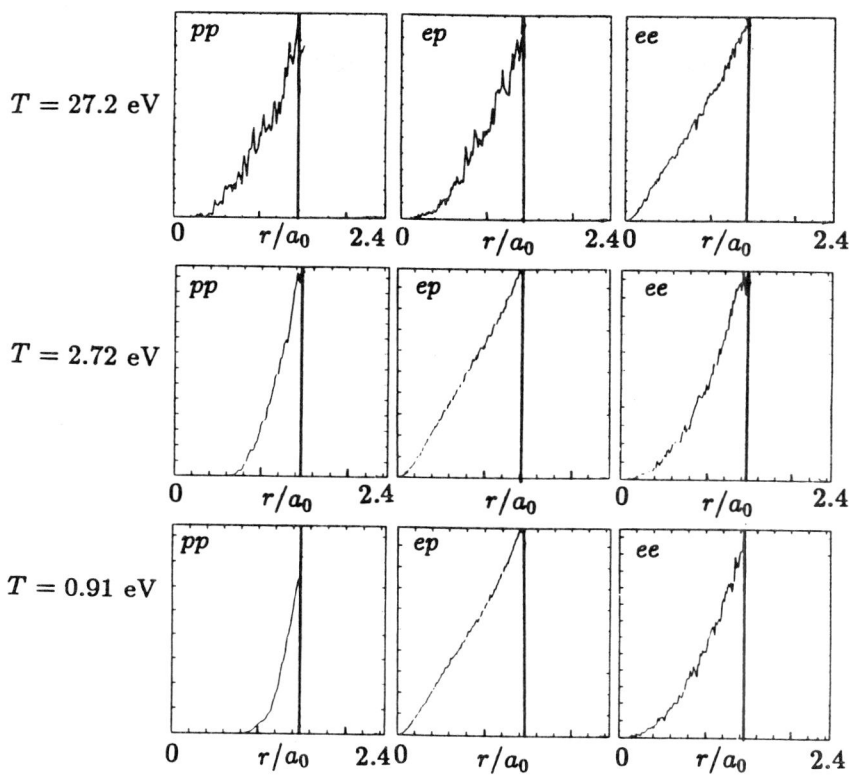

Figure 2: Radial distribution functions for the two-electron, two-proton system for a density $\rho = 1$ gm/cm^3 ($r_s = 1.39 a_0$), at several temperatures indicated on the left; left column: $g_{pp}(r)$, center column: $g_{ep}(r)$, right column: $g_{ee}(r)$. The unphysical region $r > L/2 = 1.41\ a_0$ has been cut off.

$\rho = 0.104$ gm/cm^3, pressure-dissociation of this H$_2$ molecule is observed, accompanied by pressure-ionization of the electrons. At the highest density, $\rho = 0.833$ gm/cm^3, the electrons have become almost completely delocalized, and the system now forms a classical proton one-component plasma. Note that at $\rho = 0.833$ gm/cm^3 the protons have regained some measure of localization, because the one-component plasma coupling parameter has become large, with $\Gamma = 17.8$.

3.2 Thermodynamic Properties

The internal energy curves displayed in Fig. 4 at the three densities $\rho = 0.0225$ gm/cm^3, 0.18 gm/cm^3 and 1 gm/cm^3 are similar, despite the different behavior of the radial distribution functions. At high temperature the energy curves correctly merge with the classical energy of the 4-particle electron-proton plasma, $E = 4(3T/2) = 6/\beta$, and at low temperature E tends to a value close to the ground-state energy of the isolated hydrogen molecule, $E_0 = -1.165\ e^2/a_0$.

The potential energy V has a greater dependence on density than the total internal energy. At the lowest density $\rho = 0.0225$ gm/cm^3 and at high temperatures the potential energy is small and negative as would be expected in a strongly shielded, classical electron-proton plasma, while at the lower temperatures the potential energy decreases to the value expected for the isolated H$_2$ molecule, $-2.33\ e^2/a_0$. Raising the density supresses shielding in the increasingly uniform electron background, and the potential energy due to the proton-proton interaction tends to zero more slowly as temperature increases. For all three densities it can be verified that in the weak coupling regime, that is $\Gamma \leq 1$, the potential energy is well approximated by the Debye-Hückel expression for an electron-proton plasma, $-2^{1/2}e^2 N_e/\lambda_d$ where $N_e = 2$ and where $\lambda_d = r_s^{3/2} T^{1/2}/(3e^2)^{1/2}$ is the Debye length.

The equation of state for the low density $\rho = 0.0225$ gm/cm^3 given in Fig. 5 shows that at high temperature the pressure correctly approaches the equation of state of a 4-particle electron-proton plasma ($PV = 4/\beta$), while at low temperature the equation of state approaches that of molecular hydrogen ($PV = 1/\beta$). At high density ($\rho = 1$ gm/cm^3), the system remains pressure-ionized and never reaches the equation of state of the molecular phase at low temperature, as shown in Fig. 6.

The isothermal pressure for $T = 12000^o$ K, shown in Fig. 7, was obtained in order to investigate the prediction of a possible "plasma phase

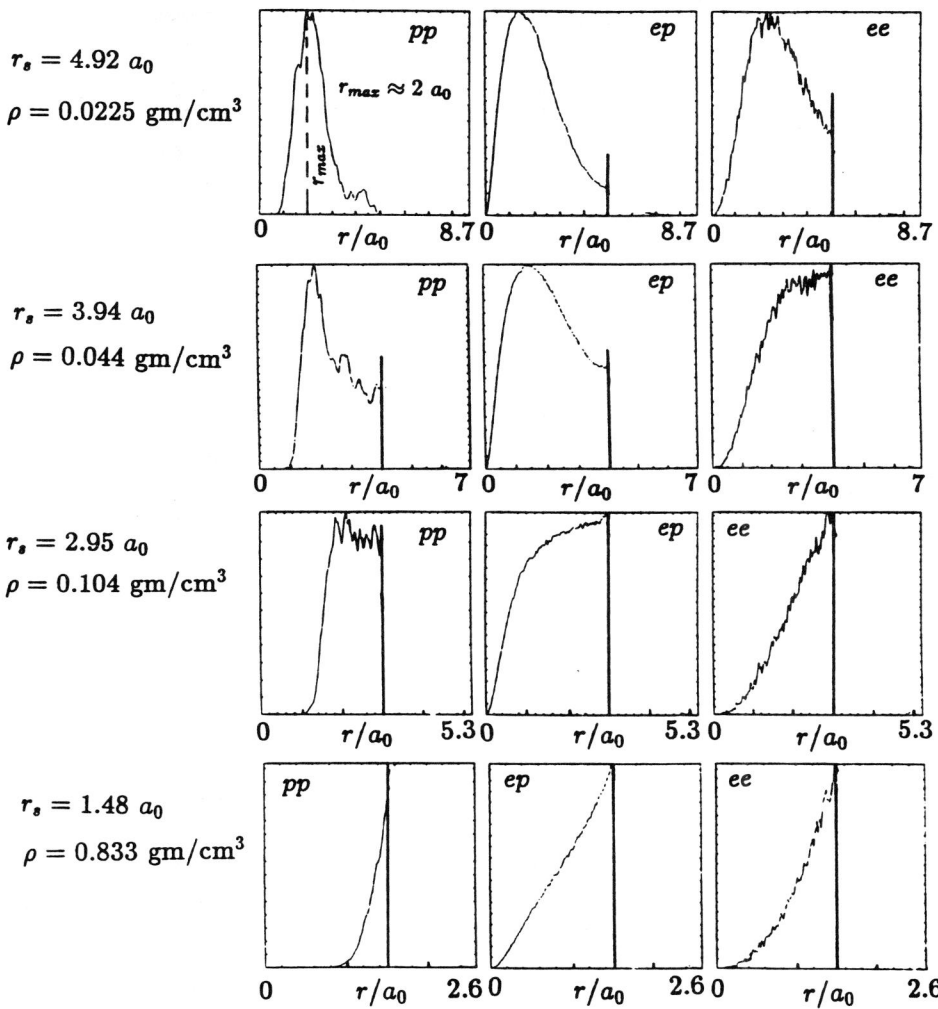

Figure 3: Radial distribution functions for the two-electron, two-proton system for a temperature $T = 12000°K$ ($T = 1.03$ eV) at various densities indicated on the left; left column: $g_{pp}(r)$, center column: $g_{ep}(r)$, right column: $g_{ee}(r)$. The unphysical regions $r > L/2$ have been cut off.

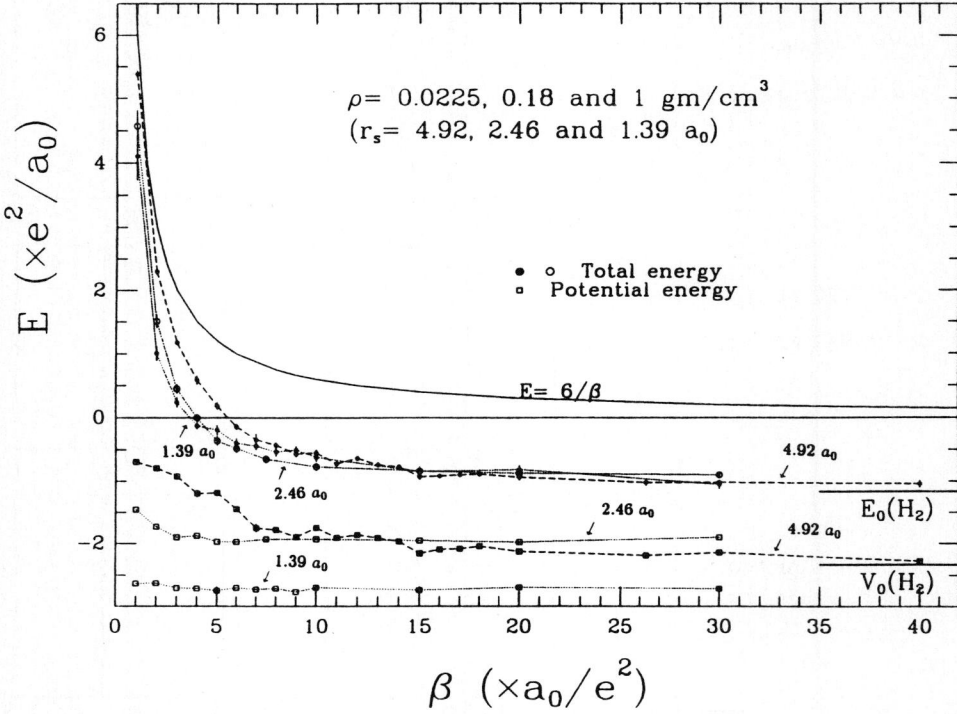

Figure 4: Internal and potential energies as a function of the reciprocal of the temperature for the various densities indicated in the figure for the two-electron, two-proton system. Error bars are indicated by vertical lines.

Quantum Monte Carlo simulation of hydrogen plasmas

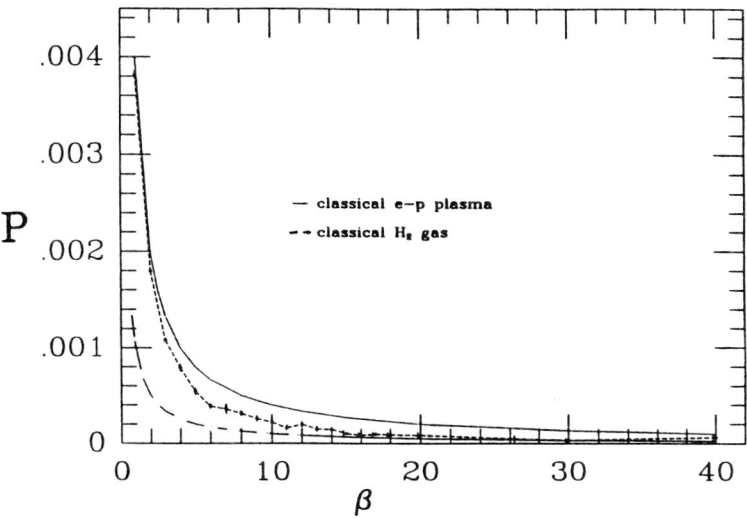

Figure 5: Pressure as a function of the reciprocal temperature at the density $\rho = 0.0225$ gm/cm^3 for the two-electron, two-proton system. Error bars are indicated by vertical lines.

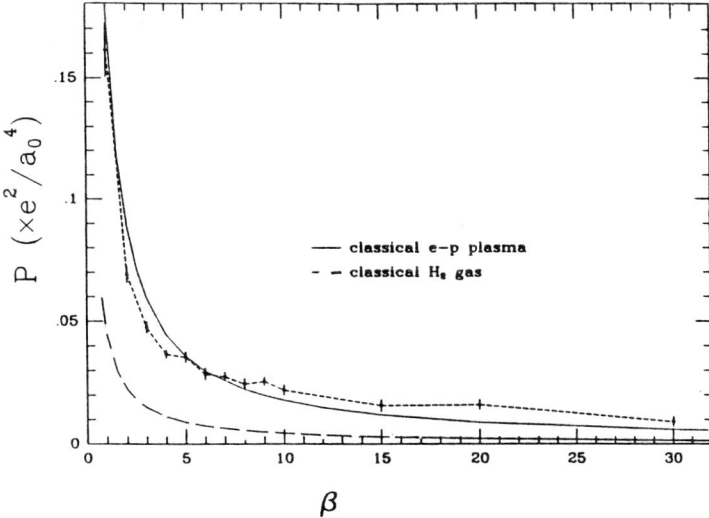

Figure 6: Pressure as a function of the reciprocal temperature at density $\rho = 1$ gm/cm^3 for the two-electron, two-proton system. Error bars are indicated by vertical lines.

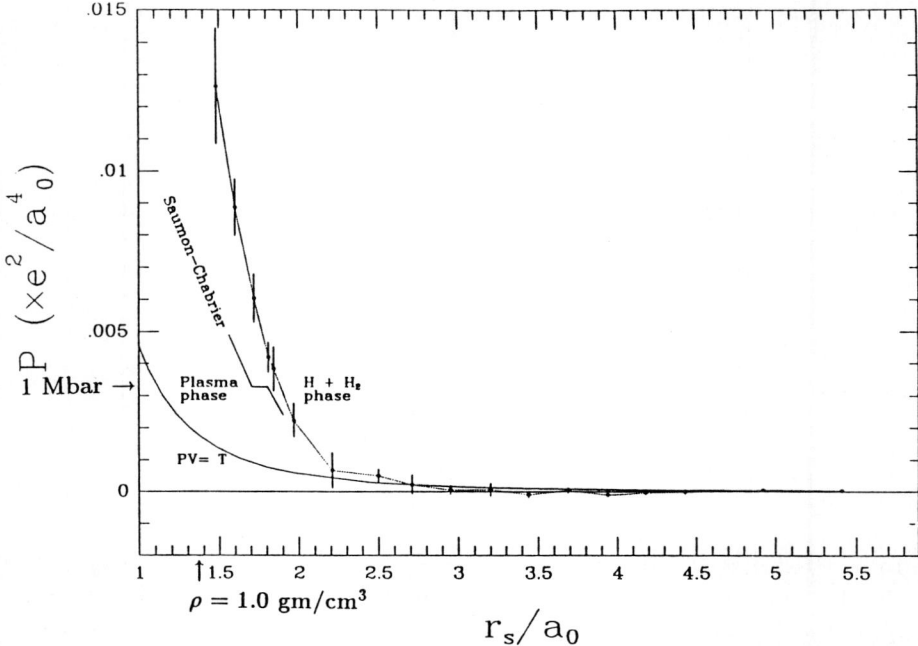

Figure 7: Pressure as a function of ion-sphere radius at the temperature of $T = 12000°K$ for the two-electron, two-proton system, compared to the phase-transition prediction of Saumon and Chabrier, Ref.(4). Error bars are indicated by vertical lines.

transition"[4], namely a first-order transition between neutral hydrogen, and an ionized plasma. No evidence for such a phase transition is obtained in these calculations. Nonetheless, both models predict comparable pressures for the isotherm at densities of the supposed coexistence region.

4 EIGHT ELECTRONS AND EIGHT PROTONS

For systems with more than two electrons, the path-integral algorithm leads to unphysical results if Fermi statistics are ignored, and the temperature sufficiently low for exchange effects to become significant. We plan however to exploit the Boltzmann algorithm in its present form to study the stability of matter. We show here that at sufficiently low temperatures the system collapses to form a cluster of protons bound by a cloud of electrons, instead of what one would physically expect under these conditions,

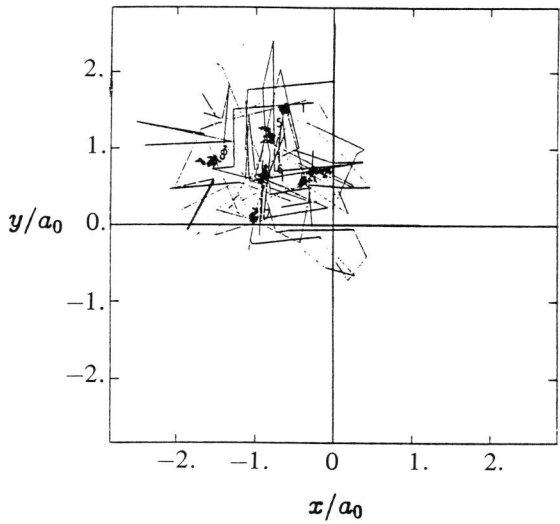

Figure 8: A two-dimensional projection of the coordinates of the Feynman paths representing the eight-electron, eight-proton Boltzmanon system at $T = 0.68$ eV, $\rho = 0.49$ gm/cm^3. The "polymers" of all eight protons are shown, but only a single electron "polymer" is displayed for the sake of clarity. Each path is composed of 135 discrete segments.

namely a homogeneous system of dense interacting H_2 molecules. A typical configuration of such a proton cluster at $T = 7000°K = 0.68$ eV and $\rho = 0.49$ gm/cm^3 is shown in Fig. 8. Also shown is a representation of a single electron quantum path, which in this calculation is composed of 135 discrete segments, illustrating that the electrons are distributed over the cluster. The corresponding proton-proton radial distribution function $g_{pp}(r)$ given in Fig. 9 shows that the protons have a high probability of being closer together than a hydrogen molecule bond length of $1.4a_0$.

Indications of the collapse of matter are also seen in that the total energy of the cluster is -123.6 eV below the total energy of four isolated hydrogen molecules, while the energy of the real molecular solid should be 14.4 eV *above* the energy of the four isolated hydrogen molecules[5], as shown in Table 1. The unphysical clustering observed in the eight-electron, eight-proton system is a consequence of gaining potential energy without having to pay the price of increased repulsion of the electrons due to the Pauli exclusion principle. By imposing Fermi statistics the electrons would be forced to spatially rearrange themselves in pairs that are far apart, making the system

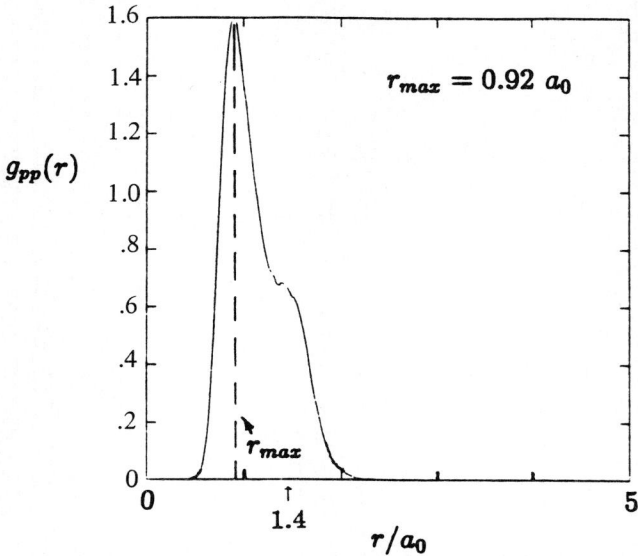

Figure 9: Proton-proton distribution function for the eight-proton, eight electron system; $\rho = 0.49$ gm/cm^3, $T = 7000°$ K.

prefer a state composed of four distinct hydrogen molecules.

In Table 1, the energies at $T = 0.68$ eV and $\rho = 0.49$ gm/cm^3 are compared for the various simulations with experiments. Also included is a simulation in which the protons are frozen into an fcc Pa3 lattice[6]. All energies are given in terms of pairs of protons, and relative to the energy of an isolated H$_2$ molecule in its ground state. In all cases shown in Table 1, the predicted energies and pressures are lower than the experimental ones. For the two-electron, two-proton system, the predicted energy and the pressure are of the right magnitude and sign, though considerably smaller than the experimental values. The table shows that even with the eight protons frozen so as to avoid collapse, the eight-electron eight-proton system still gives a metastable negative energy, though it is considerably smaller in magnitude than in the free-proton case.

Case	ΔE (eV)	P (Mbar)
EXPERIMENT (Ref.(5))	+3.26	0.75
QMC, 2e + 2p	+1.68 ± 0.5	0.2 ± 0.08
QMC, 8e + 8p (Frozen protons)	-3.4 ± 1.0	-0.87± 0.04
QMC, 8e + 8p (Mobile protons)	-30.9 ± 1.2	-0.89± 0.05

Table 1: Comparison of results for $T = 7000°K = 0.68$ eV and $\rho = 0.49$ gm/cm^3 ($r_s = 1.77\ a_0$). ΔE is the energy per proton pair relative to the ground-state energy of an isolated H_2 molecule.

5 CONCLUSIONS

Despite the drastic simplification to a two-electron, two-proton system, the path-integral scheme for the hydrogen plasma predicts essential features of the physical behavior of such a system, namely the appearance or disappearance of atomic and molecular species as the temperature and pressure conditions are changed. It is important to emphasize that the calculation is entirely based on first principles, with no *a priori* knowledge of molecular formation included in the algorithm. The basic limitation of the present calculation is that Fermi statistics are ignored, a limitation which we hope to remove in the near future by imposing the equivalent of the fixed node approximation, which has proven to be very accurate in ground state calculations.

ACKNOWLEDGMENTS

One of the authors (J. T.) wishes to thank Dr. E. L. Pollock for many detailed and valuable discussions regarding this work. Part of the computational work was done on a grant of the San Diego Supercomputer Center. This work was performed under the auspices of the U.S. Department of Energy by the Lawrence Livermore National Laboratory under contract No. W-7405-ENG-48.

References

[1] K. Theilhaber, E. L. Pollock, "'BBFT: a finite-temperature Monte Carlo code for Coulomb systems", UCID-21529, Lawrence Livermore National Laboratory, Oct. 1988.

[2] E.L. Pollock, D.M. Ceperley, Phys. Rev. B **30**, 2555 (1984).

[3] R. G. Storer, Phys. Rev. **176**, 326 (1968).

[4] D. Saumon, G. Chabrier, Phys. Rev. Let. **62**, 2397 (1989).

[5] W. J. Nellis et al., J. Chem. Phys. **79**, 1480 (1983).

[6] I. F. Silvera, Rev. Mod. Phys. **52**, 393 (1980).

QUANTUM SIMULATION OF SUPERCONDUCTIVITY

Masatoshi IMADA

Department of Physics, College of Liberal Arts, University of Saitama, Shimo-Okubo, Urawa 338, Japan

Recent numerical simulation studies on mechanisms of superconductivity are reviewed.

1. Introduction

Recent numerical analyses on the mechanisms of superconductivity have closely been connected with theoretical efforts to clarify the mechanism in high temperature superconductors discovered in 1986. In the high-Tc superconductors, strong electron-electron interaction is supposed to play an essential role in realizing the superconductivity. The difficulty is widely recognized in establishing the ground state phase of strongly correlated models solely from analytic calculation such as perturbational expansions with respect to the interaction and/or mean field theories, which sometimes results in the consequence that it is not clear whether the phase diagram obtained from various approximations represents the intrinsic phase diagram of the models or an artifact of approximations. Numerical simulation study has been applied to this problem to get insights without biased approximations. A disadvantage of the numerical simulation is that the tractable system size is finite. It is clear that this difficulty would be serious, if one wished to simulate BCS superconductivity with long coherence length. It has been, however, clarified that the high temperature superconductors have a character of extremely short coherence length. It may make the disadvantage less serious. These are the reasons why numerical simulation study is expected to play an important role in this field.

In this report, I will briefly sketch a historical overview of recent activities on numerical simulations of superconductivity, especially in relation to the high-Tc superconductors. The simulation study of superconductivity so far has been restricted to the lattice fermion models such as the Hubbard model derived as effective hamiltonians of real materials. In a broad sense of terminology, the simulation study includes exact diagonalization of a many-body hamiltonian of finite-sized cluster in addition to stochastic evaluation of the trace summation such as Monte Carlo simulation. I will survey recent studies done by the exact diagonalization as well as by the stochastic sampling method.

Numerical simulation of fermion systems in condensed matter phenomena is one of the most challenging issue with vast applicability. The methodology of the evaluation of thermodynamic quantities by a stochastic sampling has been greatly developed in the last decade. The range of applicability of fermion simulation method in condensed matter systems has been extended not only by the growth of computer ability but also in greater part by several methodological innovations. The largest difficulty presently unsolved in the fermion simulation is the notorious sign problem, which comes from the fact that the partition function in fermion models is, in genereal, expressed as the summation of samples with positive and negative signs. In some cases, this causes a serious difficulty in getting reliable estimate, because the statistical error overwhelms the net sum obtained from the subtraction of positive and negative samples. Unfortunately, generally speaking, this difficulty is serious in physically interesting region, which restricts the applicability of the stochastic sampling method. As I will discuss, however, several physically meaningful results have already been obtained from recent progress in the methodology.

Several numerical simulation methods have been developed for various purposes and choices of lattice fermion models in condensed matter theory. In one spatial dimension, a simple way of the Monte Carlo simulation is the checkerboard method[1], whose original form has been proposed in the study of quantum spin systems[2]. Although there are a few applications[1], the measurement of off-diagonal quantities such as the superconducting correlation in the checkerboard algorithm needs more complicated procedure with less efficiency than diagonal quantities such as the density correlation. In addition, the checkerboard algorithm has a serious difficulty of the sign problem in multi-dimensional systems.

By the Hubbard-Stratonovich transformation[3], the interaction terms in the partition function of Hubbard-like models may be replaced with the summation over the so-called Stratonovich variables, which transforms the partition function to that of a noninteracting hamiltonian. This formalism is now widely used in the numerical simulation of lattice fermion systems by replacing the sum over Stratonovich variales with the stochastic sampling[4-11]. Recently, the algorithm has been improved to allow the simulation at low temperatures and in the ground state as far as the sign problem is not serious[8-11]. In §2, simulation results of the Hubbard model are summarized. In §3, the application of the simulation method to extended-type Hubbard hamiltonians is summarized. Section 4 is devoted to results in strong coupling hamiltonians derived from the Hubbard-type models. In §5, numerical results to test the possibility of time reversal and parity symmetry breaking are discussed. In §6, a brief summary will be given with future problems.

2. Hubbard Model

When the conduction electrons are composed of single orbital, the Hubbard model defined from the hamiltonian

$$H = H_0 + H_1,$$
$$H_0 = -t \sum_{<ij>,\sigma} (c_{i\sigma}^\dagger c_{j\sigma} + h.c.)$$
$$H_1 = U \sum_i n_{i\uparrow} n_{i\downarrow}, \tag{1}$$

describes essential physics in a simplest form. Here $n_{i\sigma}$ is the number operator and $c_{i\sigma}^\dagger$ is the fermion creation operator with spin σ at the i-th site. The summation with respect to $<ij>$ is over the nearest neighbor pairs.

The simulation of the one-dimensional Hubbard model has been done first by the checkerboard algorithm[1]. The singlet-type superconducting correlation of the one-dimensional Hubbard model has been investigated in the ground state by using the Hubbard-Stratonovich transformation[9]. It has confirmed the absence of the growth of the superconducting correlation in the one-dimensional Hubbard model.

Magnetic properties of the two-dimensional Hubbard model have been studied at finite temperatures[5]. The results have shown that the antiferromagnetic order is likely to occur in the ground state of the half-filled band and this order disappears very rapidly when the system deviates from the half-filling. The presence of the antiferromagnetic order in the half-filled case has been supported from the system size dependence of later simulation results[10,11] in larger system size at lower temperatures. The short-ranged incommensurate structure has been observed in the ground state of the non-half-filled band[9]. These results in the two-dimensional Hubbard model have similarity to the experimental results in the high-Tc oxides[12,13].

The superconducting correlation of the two-dimensional Hubbard model has first been measured at finite temperatures[14]. The equal time correlation of the nearest neighbor pairing shows small enhancement as compared to the noninteracting system. This enhancement is also seen in the nearest neighbor s-wave pairing in the ground state[9]. It has, however, turned out that this enhancement has nothing to do with the superconductivity[9]. The enhancement is small even in the ground state with the absence of the system size dependence. The superconducting susceptibility obtained from the canonical correlation at finite temperatures always shows the suppression from the noninteracting results at any fillingness[15,16]. All the available data of equal time superconducting correlations remain small with the absence of the system size dependence in the ground state[9].

So far the on-site interaction is assumed to be repulsive to represent the realistic Coulomb repulsion at the same orbital. As a purely theoretical problem, however, we may think of the attractive on-site interaction by taking U negative. It is easily understood that the superconductivity is strongly favored due to the attractive force, which has, in fact, been confirmed in the simulation study[17,18]. Kosterlitz-Thouless transition of the superconductivity is expected in the two-dimensional attractive Hubbard model.

3. Extension of the Hubbard Model

The Hubbard model may be extended to include the nearest neighbor interaction. The hamiltonian is then

$$H = H_0 + H_1,$$
$$H_0 = -t \sum_{<ij>,\sigma} (c_{i\sigma}^\dagger c_{j\sigma} + h.c.)$$

$$H_1 = U \sum_i n_{i\uparrow} n_{i\downarrow} + V \sum_{<i,j>} n_i n_j, \qquad (2)$$

This hamiltonian has been studied in one-dimension to see the competition among charge-density wave, spin density wave and the superconductivity[1,19]. In one-dimension, excitonic mechanism has also been studied in the Monte Carlo calculation in the Hubbard-like models coupled to a polarizable localized electrons[20].

The superconducting correlation of the two-dimensional Hubbard model with the next nearest neighbor hopping term has been studied by the quantum simulation[21,22]. As compared to the Hubbard model, the superconducting susceptibility has slightly larger value in this case at low temperatures.

A realistic model of the high-Tc oxides represents the two-dimensional CuO_2 plane. This model is constituted from the $Cu - 3d$ orbital and the $O - 2p$ orbital and hence called the $d - p$ model, which is defined as

$$\begin{aligned}
H = & -t_1 \sum_{<ij>,\sigma} (d_{i\sigma}^\dagger p_{j\sigma} + d_{i\sigma}^\dagger q_{j\sigma} + h.c.) \\
& -t_2 \sum_{<ij>\sigma} (p_{i\sigma}^\dagger q_{j\sigma} + h.c.) \\
& + U_d \sum_i n_{di\uparrow} n_{di\downarrow} \\
& + U_p \sum_j (n_{pj\uparrow} n_{pj\downarrow} + n_{qj\uparrow} n_{qj\downarrow}) \\
& + V \sum_{<ij>} n_{di}(n_{pj} + n_{qj}) \\
& + \varepsilon_d \sum_i n_{di} + \varepsilon_p \sum_j (n_{pj} + n_{qj}),
\end{aligned} \qquad (3)$$

where $d_{i\sigma}^\dagger$ creates the fermion in the Cu d-orbital at the i-th site, while $p_{i\sigma}^\dagger$ and $q_{i\sigma}^\dagger$ create the fermion in the oxygen p_x and p_y orbital at the i-th site in the two-dimensional CuO_2 network. The level difference Δ is defined as $\Delta = \varepsilon_p - \varepsilon_d$. The quantum simulation of the $d - p$ model has been done for relatively small number of unit cells[15,23]. The model reduces to the Hubbard model in the so-called Mott-Hubbard region $\Delta \gg U, t$. However, the high-Tc oxides seem to belong to the charge transfer region $U > \Delta, t$. The superconducting susceptibility has shown small enhancement for spatially extended pairing as compared to the noninteracting case at finite temperatures in the charge transfer region[15]. Although this result shows qualitatively different tendency from the Hubbard model, the result is, so far, not conclusive as for the existence of the superconducting state, because of the limitation of the temperature range. The local magnetic moment at the Cu site and the antiferromagnetic correlation have been measured as a function of U, Δ and temperature[23]. It has been shown that the localized moment at the Cu site is developed in the region $\Delta > U/2$ for large U. In the situation of rather large V term in (3), possible excitonic origin of the pairing has been considered in the exact diagonalization of small cluster systems[24-26].

4. Strong Coupling Hamiltonians

To make clearer the possible mechanism of the superconductivity, it has been recognized that we need to extract the essence of the original $d-p$-type models, because the temperature range accessible from the present numerical study may not be low enough to establish the presence or the absence of the superconductivity in the relevant region. When the magnetic moment at the Cu site is preserved, the strong coupling hamiltonian may be reduced from the $d-p$ model. By the perturbational expansion with respect to t/U and t/Δ for small concentration of holes in the oxygen orbitals, the effective hamiltonian has been reduced in the form[27-31]:

$$H = -t \sum_{<i,j>\sigma} (c_{i\sigma}^\dagger c_{j\sigma} + c_{j\sigma}^\dagger c_{i\sigma})$$
$$- 2J_K^{(1)} \sum_i \vec{S}_i \cdot \vec{\sigma}_i^{(1)} - 2J_K^{(2)} \sum_i \vec{S}_i \cdot \vec{\sigma}_i^{(2)} - 2J_S \sum_{<l,m>} \vec{S}_l \cdot \vec{S}_m, \qquad (4)$$

where

$$\vec{\sigma}_i^{(1)} = \frac{1}{2} \sum_{\vec{\delta}} c_{\vec{i}+\vec{\delta},\sigma}^\dagger (\vec{\sigma})_{\sigma\sigma'} c_{\vec{i}+\vec{\delta},\sigma'}$$

$$\vec{\sigma}_i^{(2)} = \frac{1}{2} (\sum_{\vec{\delta}} c_{\vec{i}+\vec{\delta},\sigma}^\dagger)(\vec{\sigma})_{\sigma\sigma'} (\sum_{\vec{\delta}'} c_{\vec{i}+\vec{\delta}',\sigma'}) - \vec{\sigma}_i^{(1)},$$

$$n_{i\sigma} = c_{i\sigma}^\dagger c_{i\sigma}$$

It has been argued[31-34] that this effective hamiltonian may be reduced to the so-called $t-J$ model hamiltonian,

$$H = -t \sum_{<ij>} ((1-n_{i,-\sigma})c_{i\sigma}^\dagger c_{j\sigma}(1-n_{j,-\sigma}) + h.c.)$$
$$- 2J \sum_{<ij>} \vec{S}_i \cdot \vec{S}_j, \qquad (5)$$

which is originally derived from the essential terms in the strong coupling expansion of the Hubbard model. However, the mapping of the hamiltonian (4) to the $t-J$ model is, at present, a controversial issue[30]. The hole in the $t-J$ model may be identified with the $d-p$ singlet[35] in the model (4) in the mapping. However, the $d-p$ singlet in (4) necessarily has the spatially extended structure instead of the on-site singlet in the $t-J$ model. The possible importance of the next nearest neighbor and further hopping terms has also been pointed out to be added in the original $t-J$ model[31,36]. Although the intuitive correspondence between the $d-p$ singlet and the hole in the $t-J$ model seems to be clear, the relationship between Eq.(4) and the $t-J$ model has not been clarified in the low energy excitation. The hamiltonian (4) has also been simplified to the form of the coupled spin-fermion model by retaining the extended nature of the $d-p$ singlet as[37,38]

$$H = -t \sum_{<i,j>\sigma} (c_{i\sigma}^{\dagger} c_{j\sigma} + c_{j\sigma}^{\dagger} c_{i\sigma}) + U_h \sum_i n_{i\uparrow} n_{i\downarrow}$$
$$- 2J_K \sum_{<i,l>} \vec{S}_l \cdot \vec{\sigma}_i - 2J_S \sum_{<l,m>} \vec{S}_l \cdot \vec{S}_m, \qquad (6)$$

where

$$\vec{\sigma}_i = \frac{1}{2} c_{i\sigma}^{\dagger} (\vec{\sigma})_{\sigma\sigma'} c_{i\sigma'}$$
$$n_{i\sigma} = c_{i\sigma}^{\dagger} c_{i\sigma}$$

Exact diagonalization studies of small cluster systems of the hamiltonians (4)-(6) have shown the general tendency that the doped holes (or fermions) have mutually attractive interaction[29,37-43]. The binding energy of holes E_B is defined from the twice of the ground state energy of one-hole system subtracted from the sum of the ground state energies of zero-hole and two-hole systems:

$$E_B = 2E(1) - E(0) - E(2)$$

The origin of the attractive interaction may be interpreted from the formation of a cloud with reduced antiferromagnetic correlation around the $d-p$ singlet (or hole). However, the nature of the elementary excitation at the $d-p$ singlet has not been fully clarified yet. The $d-p$ singlet carries charge and the spin excitation is inherently coupled to this singlet through the formation of the spin distortion. The normal state properties of this system is hence not clear at this moment.

Because the $t-J$ model may be derived in the strong coupling expansion of the Hubbard model, the attractive interaction speculated in the $t-J$ model seems, at a first glance, to contradict with the absence of signs for superconductivity in the Hubbard model. The variational Monte Carlo calculation with a Gutzviller-type wave function has also predicted the superconducting state in the $t-J$ model near half-filling[44,45]. A possible origin of this seemingly controversial results is explained in the following way:

First, the attractive interaction in the $t-J$ model is a substantial overestimation of the corresponding Hubbard model even when the ratio $|J/t|$ is rather small($\sim 0.1 \sim 0.2$)[43]. This means that the convergence of the strong coupling expansion with respect to t/U is not fast in the Hubbard model. In the $t-J$ model, there seems to be a critical value $|J_c|$ only above which the interaction becomes attractive. It is conceivable that the Hubbard model is connected to the $t-J$ model only in the region below $|J_c|$.

The attractive interaction may not be necessarily be related to the superconducting state. However, the striking enhancement of the superconducting susceptibility has been observed even in the one-dimensional $t-J$ model for large $|J|$ [46]. Because the superconducting susceptibility directly measures the tendency to the off-diagonal long range order, it strongly suggests that the $t-J$ model may show the superconducting behavior for relatively large $|J/t|$.

Then the present data obtained from various approaches mentioned above seem to be reconciled each other. Large $|J/t|$ region may not be justified from the Hubbard model. Therefore, to discuss the superconductivity in a realistic model, we have to consider the relationship of the $t-J$ model at relatively large $|J/t|$ to the $d-p$ model or its effective hamiltonians. The superconducting correlation of the model (4) or (6) has yet to be clarified[5]. These remain for future studies.

The total momentum in the doped system has been calculated in systems (4) and (5) to clarify the mutual relationship of these models and also to get insight of normal state properties[47,48]. The interaction of doped fermions in eq.(6) with the ferromagnetic J_K has been investigated from the exact diagonalization study to examine the possibility of doped fermions in p_π or $d_{3z^2-r^2}$ orbitals in the high-Tc oxides[37,38,49,50]. The results show that the ferromagnetic J_K does not favor the attractive interaction. Single particle Green's function has been analyzed by the exact diagonalization study and compared with ample photoemission and inverse photoemission data to specify the relevant model and the location of doped fermions[34,41,51,52]. The information obtained from single particle Green's function is also expected to clarify the low energy excitation spectra, if the system size is taken sufficiently large.

5. On the Fractional Statistics

The possibility of the time reversal and the parity symmetry breaking in strongly correlated electron systems have been proposed[53-55]. The time reversal symmetry is broken in the external magnetic field. This is the case of two-dimensional electron gas showing fractional quantum Hall effect. The quasi particle excitation follows the anyon statistics. A similar situation may occur if the time reversal symmetry is spontaneously broken. Because this has raised a fundamental question on the nature of normal and superconducting properties in the high-Tc oxides, numerical studies done so far are summarized in this section. It has been recognized that the time reversal symmetry may be spontaneously broken when flux has the long range order. The flux order parameter is defined from

$$p_{123} = \chi_{12}\chi_{23}\chi_{31} - \chi_{13}\chi_{32}\chi_{21} \tag{7}$$

and

$$\chi_{ij} = \sum_\sigma c_{i\sigma}^\dagger c_{j\sigma} \tag{8}$$

for the elementary triangle with corners (1,2,3) in the lattice. The flux in the unit square is similarly defined by

$$p_{1234} = \chi_{12}\chi_{23}\chi_{34}\chi_{41} - \chi_{14}\chi_{43}\chi_{32}\chi_{21} \tag{9}$$

The flux state is defined from the long range order as $<p_{123}> \neq 0$ or $<p_{1234}> \neq 0$. It has been shown that the flux state is nothing but the chiral spin state in the half-filled limit[50], where the chirality order parameter is defined from the spin of fermions as

$$c_{123} = \mathbf{S}_1 \cdot (\mathbf{S}_2 \times \mathbf{S}_3) \tag{10}$$

for the elementary triangle in the lattice. The total uniform chirality C_+ and the staggered chirality C_- are defined as

$$C_\pm = \sum_{i_x,i_y}^{N} [\mathbf{S}_{l_1} \cdot (\mathbf{S}_{l_2} \times \mathbf{S}_{l_3}) \pm \mathbf{S}_{l_1} \cdot (\mathbf{S}_{l_3} \times \mathbf{S}_{l_4})], \tag{11}$$

where $l_1 = (i_x, i_y), l_2 = (i_x + 1, i_y), l_3 = (i_x, i_y + 1)$ and $l_4 = (i_x - 1, i_y + 1)$. The uniform flux P_+ and the staggered flux P_- defined from

$$P_\pm = \sum_{i_x,i_y} [p_{l_1 l_2 l_3} \pm p_{l_1 l_3 l_4}]. \tag{12}$$

have relationship to the chirality order $C_\pm$ in the half-filled band as

$$P_\pm = -4iC_\pm. \tag{13}$$

On the square lattice, the uniform and staggered flux of the plaquette is defined as

$$P_{S\pm} = \sum_{i_x,i_y} [p_{l_1 l_2 l_4 l_3} \pm p_{l_3 l_4 l_6 l_5}], \tag{14}$$

where $l_1 = (i_x, i_y), l_2 = (i_x + 1, i_y), l_3 = (i_x, i_y + 1), l_4 = (i_x + 1, i_y + 1), l_5 = (i_x, i_y + 2)$ and $l_6 = (i_x + 1, i_y + 2)$.

The variational argument has shown that the antiferromagnetic exchange coupling J in the $t - J$ model favors the appearance of the flux state. The renormalized mean field calculation indicates that the flux state is stabilized for unphysically large $|J/t|$ in the two-dimensional $t - J$ model[56]. The correlation of chirality has been calculated in various choices of lattices in the quantum spin systems defined by the Hamiltonian

$$H = -2J \sum_{<i,j>} \mathbf{S}_i \cdot \mathbf{S}_j, \tag{15}$$

with $\mathbf{S}_i$ being a localized spin-1/2 operator at the i-th site. The spin-1/2 antiferromagnetic system is the relevant model in the half-filled band. By the extrapolation to the thermodynamic limit from the exactly diagonalized results, the chirality correlation has turned out to be short-ranged in the square lattice and the triangular lattice systems[57]. The chirality correlation shows similar behavior even when the next nearest neighbor exchange coupling J' has the same strength with the nearest neighbor coupling J on the square lattice[58]. The triangular lattice with the next nearest neighbor interaction also shows similar behavior[58]. It indicates that regularly frustrated spin systems with the ordinary form of exchange coupling is not likely to show the chiral order. Similarly the correlation of the flux does not seem to show growth with the increase of system size in the two-dimensional Hubbard model at $U = 4$ away from the half-filling. The correlation of $\chi_{ij} - \chi_{ji}$ seems to remain short-ranged[59]. At this moment, we have no data supporting the appearance of the time reversal and the parity symmetry broken state in realistic models of high-Tc oxides. The enhancement of the superconducting correlation in the one-dimensional $t - J$ model also suggests that the two-dimensional system is not special. However, we do not have sufficient data to draw a conclusion on this problem at the moment. The flux correlation in strongly correlated systems such as the $t - J$ model or other effective hamiltonians in the non-half-filled band has to be calculated in detail.

6. Summary and Future Problems

At the present stage of the understanding, it is not possible to establish any of mechanisms recently proposed as the mechanism of high temperature superconductors. However, various interesting features have been clarified in the numerical studies. The attractive interaction of fermions in several strong coupling hamiltonians and the enhancement of the superconducting correlation are two of the most promising signatures of the superconductivity in models with purely electronic origin. Most of the results seem to indicate the essential importance of the antiferromagnetic superexchange interaction for the pairing mechanism. It is desired that numerical analyses contribute to specify a way to discriminate a mechanism from others in experiments.

The nature of normal state above the superconducting transition temperature has not been clarified yet. The important unsolved problem among others is whether the system is the fermi-liquid or not. Although we have no data supporting the anyon superconductivity, numerical data supporting the conventional fermi liquid behavior are also absent.

For a deeper understanding of the mechanism, the development of the numerical simulation algorithm has outstanding importance. As for the lattice fermion simulation, the sign problem is the only unsolved major difficulty.

References

1) J. E. Hirsch, R. L. Sugar, D. J. Scalapino and R. Blankenbecler : Phys. Rev. B 26 (1982) 5033.
2) M. Suzuki: Prog. Theor Phys. 56, (1976) 1454.
3) J. Hubbard: Phys. Rev. Lett. 3 (1959) 77.
   J. E. Hirsch: Phys. Rev. B 28 (1983) 4059.
4) R. Blankenbecler, D. J. Scalapino and R. L. Sugar: Phys. Rev. D 24 (1981) 2278.
5) J. E. Hirsch: Phys. Rev. B 31 (1985) 4403.
6) R. T. Scalettar, D. J. Scalapino and R. L. Sugar: Phys. Rev. B 36 (1987) 8632.
7) M. Imada: J. Phys. Soc. Jpn. 57, (1988) 2689

8) S. Sorella, E. Tosatti, S. Baroni, R. Car and M. Parrinello: Int. J. Mod. Phys. B 1 (1988) 993.
   S. Sorella, S. Baroni, R. Car and M. Parrinello: Europhys. Lett. 8 (1989) 663.
9) M. Imada and Y. Hatsugai: J. Phys. Soc. Jpn. 58 (1989) No.10 .
10) S. R. White, D. J. Scalapino, R. L. Sugar, E. Y. Loh, J. E. Gubernatis and R. T. Scalettar: Phys. Rev. B 40 (1989) 506.
11) J. E. Hirsch and S. Tang: Phys. Rev. Lett. 62 (1989) 591.
12) For example see Y. Kitaoka et al.: J. Phys. Soc. Jpn. 57 (1988) 734,737.
13) T. R. Thurston et al.: preprint.
    G. Shirane et al.: Phys. Rev. Lett. 63 (1989) 330.
14) J. E. Hirsch: Phys. Rev. Lett. 54 (1985) 1317.
15) M. Imada: J. Phys. Soc. Jpn. 57 (1988) 3128.
16) J. E. Hirsch H. Q. Lin: Phys. Rev. B 37 (1988) 5070.
    H. Q. Lin, J. E. Hirsch and D. J. Scalapino: Phys. Rev. B 37 (1988) 7359.
17) J. E. Hirsch: Proceedings of 2nd.NEC Symposium *"Mechanisms of High Temperature Superconductivity* " ed. by H. Kamimura and A. Oshiyama (Springer Verlag, 1989).
18) R.T. Scalettar et al.: Phys. Rev. B 62 (1989) 1407.
19) J. E. Hirsch and D. J. Scalapino: Phys. Rev. B 27 (1983) 7169.
20) J. E. Hirsch and D. J. Scalapino: Phys. Rev. Lett. 53 (1984) 706.
21) K. Saitoh and S. Takada: J. Phys. Soc. Jpn. 58 (1989) 783.
22) R. R. dos Santos: Phys. Rev. B 39 (1989) 7259.
23) G. Dopf, A. Muramatsu and W. Hanke: Proceedings of the Workshop *"Quantum Simulation of Condensed Matter Phenomena* " ed. by G. Gubernatis (World Scientific, Singapore) to be published
24) J. E. Hirsch, S. Tang, E. Loh and D. J. Scalapino: Phys. Rev. Lett 60 (1988) 1668.
25) W. H. Stephen, W.v.d. Linden and P. Horsch: Int. J. Mod. Phys. B 1 (1988) 1005.
26) C. A. Balseiro, A. G. Rojo, E. R. Gagliano and B. Alascio: Phys. Rev. B 38 (1988) 9315.
27) P. Prelovsek: Phys. Lett. A 126 (1988) 287.
28) K. Hida: J. Phys. Soc. Jpn. 58 (1988) 1387.
29) M. Ogata and H. Shiba: J. Phys. Soc. Jpn. 57 (1988) 3074.
30) M. Imada: Proceedings of 2nd.NEC Symposium *"Mechanisms of High Temperature Superconductivity* " ed. by H. Kamimura and A. Oshiyama (Springer Verlag, 1989)p.53.
31) H. Fukuyama, H. Matsukawa and Y. Hasegawa: J. Phys. Soc. Jpn. 58 (1989) 364.
    H. Matsukawa and H. Fukuyama: J. Phys. Soc. Jpn. 58 (1989) 2845.
32) F. C. Zhang and T. M. Rice: Phys. Rev. B 37, (1988) 3759.
33) A. Ramsek and P. Prelovsek: Phys. Rev. B 40, (1989) 2239..
34) H.-B. Schuttler and A. J. Fedro: preprint.
35) M. Imada: J. Phys. Soc. Jpn. 56, (1987) 3793

36) P. A. Lee: Phys. Rev. Lett. 63, (1989) 680.
See also numerical studies by H. Eskes, G. A. Sawatzky and L.F. Feiner:preprint, G. A. Sawatzky: to be published in Proc. IBM Int. School Mater. Sci and Techn. on "*Earlier and Recent Aspects of Superconductivity*" (Erice, Italy).
37) M. Imada, N. Nagaosa and Y. Hatsugai: J. Phys. Soc. Jpn. 57 (1988) 2901 .
M. Imada, Y. Hatsugai and N. Nagaosa: Int. J. Mod. Phys. B 1 (1988) 959.
38) Y. Hatsugai, M. Imada and N. Nagaosa: J. Phys. Soc. Jpn. 58 (1989) 1347.
39) J. Bonca, P. Prelovsek and I. Sega: Int. J. Mod. Phys. B 1 (1988) 943; Phys. Rev. B 39, (1989) 7074. .
40) E. Kaxiras and E. Manousakis: Phys. Rev. B 38 (1988) 866.
41) P. Horsch and W. H. Stephan:to be published in the proceedings of the NATO Advanced Research Workshop on "*Interacting Electrons in Reduced Dimensions*" ed. by D. Baeriswyl and D. Campbell (Plenum Press)
42) Y. Hasegawa and D. Poilblanc: preprint.
43) J. A. Riera and A. P. Young: Phys. Rev. B 39 (1989) 9697 .
44) H. Yokoyama and H. Shiba: J. Phys. Soc. Jpn. 57 (1988) 2482.
45) C. Gros: Phys. Rev. B 38, (1988) 931.
46) M. Imada: Proceedings of the Workshop "*Quantum Simulation of Condensed Matter Phenomena*" ed. by G. Gubernatis (World Scientific, Singapore) to be published
47) E. Dagotto, J. R. Schrieffer, A. Moreo and T. Barnes: preprint.
48) M. Ogata and H. Shiba: J. Phys. Soc. Jpn. 58 (1989) 2836.
49) M. Ogata and H. Shiba: Physica C 158 (1989) 355.
50) T. Nishino, M. Kikuchi and J. Kanamori: Solid State Comm. 68 (1988) 455.
51) Y. Kuramoto and H. J. Schmidt: to be published in Proc. IBM Japan International Symposium on "*Strong Correlation and Superconductivity* " ed. by H.Fukuyama et al. (Springer Verlag, Berlin)
52) P. Horsch, W. H. Stephan, K.v.Szczepanski, M. Ziegler and W. von der Linden: To be published in Physica C as the proceedings of International $M^2S - HTSC$ Conference, Stanford.
53) V. Kalmeyer and R. B. Laughlin: Phys. Rev. Lett. 59 (1987) 2095.
54) R. B. Laughlin: Science 242 (1988) 525.
55) W. G. Wen, F. Wilczek and A. Zee: Phys. Rev. B 39 (1989) 11413.
56) P. Lederer, D. Poilblanc and T. M. Rice: preprint.
57) M. Imada: J. Phys. Soc. Jpn. 58 (1989) 2650.
58) M. Imada: unpublished
59) M. Imada: unpublished

DYNAMIC SIMULATION OF MIXED QUANTUM-CLASSICAL SYSTEMS

Rajiv. K. KALIA and P. VASHISHTA

Materials Science Division, Argonne National Laboratory, Argonne, IL 60439, USA

S. W. DE LEEUW

Universiteit van Amsterdam, Amsterdam, The Netherlands

John HARRIS

Institut für Festkörperforschung der Kernforschungsanlage, D-5170 Jülich, West Germany

We discuss the simulation of mixed systems with quantum and classical particles based on the quantum molecular dynamics method. This technique is applied to the problem of an excess electron in a dense helium gas. The formation of an electron bubble and the results for the structure and dynamics of the bubble are discussed.

1. INTRODUCTION

There is a large variety of systems in which a small number of excess quantum carriers such as electrons (or positrons) are coupled strongly to a classical many-body system at finite temperatures. Examples of these mixed quantum-classical systems include excess electrons in amorphous semiconductors[1], in polar and non-polar solvents, and in dense gases[2].

Until recently there were two ways to simulate the behavior of mixed systems: 1) Path integral Monte Carlo[2] or molecular dynamics[3] and 2) dynamical simulated annealing[4]. Both of these methods provide structural characteristics of quantum particles in mixed systems. Attempts have also been made to study time correlation functions using the path integral Monte Carlo method[2,5]. The long-time behavior in path integral Monte Carlo simulations is not reliable.

In many systems the important experimental information is derived from mobility measurements. Therefore, in order to make a direct comparison with experiments the simulations must provide the dynamics of quantum particles, preferably in the presence of an applied field.

2. QUANTUM MOLECULAR DYNAMICS METHOD

In 1987 a new simulation method called quantum molecular dynamics (QMD) was proposed[6]. The QMD method simulates the dynamics of quantum and classical particles in mixed systems at finite temperatures through the simultaneous numerical solutions of the time-dependent Schrödinger equation for quantum particles and Newton's equations of motion for classical particles. To understand the QMD method, consider a quantum particle of mass m described by the wave function, $\psi(r,t)$, coupled to a classical system of N particles with masses M_i and positions $\{R_i(t)\}$. The time-dependent Schrödinger equation for the quantum particle reads ($\hbar = 1$)

$$i\frac{\partial \psi(\bar{r},t)}{\partial t} = \left(-\frac{\nabla^2}{2m} + V(\bar{r})\right)\psi(\bar{r},t); \quad V(\bar{r}) = \sum_{i=1}^{N} v(\bar{r} - \bar{R}_i) \quad (1)$$

where v is the interaction potential between the quantum and classical particles. As the classical particles evolve in time, they produce a time-dependent potential for the quantum particle. Provided the motion of the classical particles is much slower than that of the quantum particle so that the latter is always in the ground state of the instantaneous configuration of classical particles, the trajectories of the classical particles follow from the following equations of motion[6]:

$$M_i \ddot{\bar{R}}_i = -\nabla_i U(\{\bar{R}_i\}) - \nabla_i \int d\bar{r} \, |\psi(\bar{r},t)|^2 \, v(\bar{r} - \bar{R}_i) \quad (2)$$

where U is the interaction among the classical particles. For short increments of time Δt, the solution of Eq. (1) can be written as

$$\psi(\bar{r}, t+\Delta t) = e^{i\Delta t \nabla^2/4m} \, e^{-i\Delta t V} \, e^{i\Delta t \nabla^2/4m} \, \psi(\bar{r},t) + O[(\Delta t)^3] \quad (3)$$

Equation (3) is implemented with use of fast Fourier transform (FFT) techniques[7], since the potential and kinetic energies are diagonal in real and Fourier spaces, respectively.

There are three steps[7] involved in the time-stepping operation in Eq. (3). The first step is executed as follows:

$$e^{i\Delta t \nabla^2/4m}\psi(\vec{r},t) = \sum_{\vec{k}} \psi(\vec{k},t)\, e^{-i\Delta t k^2/4m} e^{i\vec{k}\cdot\vec{r}} \tag{4}$$

Equation (4) involves the Fourier transform of $\psi(r,t)$, i. e., $\psi(k,t)$ multiplied by $\exp(-i\Delta t\, k^2/4m)$, and followed by an inverse FFT. Next, the outcome of the first step is multiplied by $\exp(-i\Delta t\, V)$. The final step consists in taking the FFT after the second step, multiplying with $\exp(-i\Delta t\, k^2/4m)$ and then taking an inverse FFT. These three steps are repeated to obtain the time evolution of the wave function. The classical equations of motion can be integrated numerically with one of several available algorithms[8].

For finite systems or systems with surfaces, the use of periodic boundary conditions in the broken-symmetry direction is inappropriate. Therefore, in the broken-symmetry direction the first and third steps in the time-stepping operation cannot be executed with the FFT method. For these systems a new QMD algorithm[9] has been developed which is also appropriate for systems subject to an external electric field.

To understand this new QMD algorithm, consider a quantum particle interacting with a system of classical particles in a box of dimensions L_x, L_y, and L_z. In the y and z directions the system is infinite and hence periodic boundary conditions are imposed. We suppose that the system is finite in the x direction because of a uniform external electric field. In the execution of steps 1 and 3 in Eq. (3), FFT can still be used for propagation along the y and z directions. However, in the direction of the applied field, FFT cannot be used. Instead, we note

$$e^{i\frac{\Delta t}{4m}\frac{\partial^2}{\partial x^2}} \psi(x,y,z,t) \approx \left(1 - i\frac{\Delta t}{8m}\frac{\partial^2}{\partial x^2}\right)^{-1} \left(1 + i\frac{\Delta t}{8m}\frac{\partial^2}{\partial x^2}\right) \psi(x,y,z,t) \tag{5}$$

Corrections to the right-hand side of Eq. (5) are of the order of $(\Delta t)^3$ and also the product of operators on the right-hand side is unitary. Equation (5) can be rewritten as a set of tridiagonal equations which can be solved by standard numerical techniques[10].

3. AN EXCESS ELECTRON IN A DENSE HELIUM GAS

3.1 Background

The electron-helium interaction is strongly repulsive at short and intermediate range[11]. Therefore, an injected electron will move preferentially to a region of low density, push neighboring helium atoms aside and will lower the density further until the electron localizes in the form of a bubble. At sufficiently high density the bubble will be stable aginst temperature fluctuations whereas at low densities the pressure-volume term for helium will be too small to balance the energy of the localized electron.

The electron bubble formation in a dense helium gas has been investigated using the QMD scheme[12]. For the electron in helium problem, the electron-helium and helium-helium interactions are relatively well established. The helium particles interact with each other via a two-body Lennard-Jones potential with parameters ε = 10.22K and σ = 2.576 Å, and with the electron via a pseudopotential[11]. Because the helium atom is chemically inert, three-body forces are expected to be minute at the gas densities of interest.

3.2 Calculations

The QMD simulations[12] were performed at 77K for reduced helium densities $n = \rho\sigma^3$ = 0.1, 0.17, and 0.25 which correspond to ρ = 0.61, 1.0, and 1.46 x 10^{22} cm^{-3}. Systems with n = 0.1 and 0.17 contained 512 helium atoms while the simulations at the highest helium density were carried out with 64 and 140 particles. These gave similar results though the smaller system showed evidence of finite-size effects. The electron wave packet was propagated with 32^3 grid points and with time step, Δt, in the range of 0.2 - 0.5 a.u. The total energy was conserved to better than 0.1% over 10^6 time steps. Some simulations with 64^3 grid points gave identical results. Classical molecular dynamics for helium atoms was performed in the canonical ensemble[13].

The simulations at n = 0.17 and 0.25 were begun by first equilibrating the helium system at 77K, freezing the atomic configuration and propagating a gaussian wave packet in imaginary time until the electronic ground state is reached. Then, both the electron and helium atoms are propagated in real time for a few hundred thousand time steps to allow the system to equilibrate. The equilibrated system is run and averages accumulated over an additional 10^6 time steps at each temperature and density. From time to time, imaginary time propagations were carried out to ascertain that there was no significant departure from adiabacity.

3.3   Results

Figure 1 shows the participation ratio, $p(t) = (\Omega \int dr \mid \psi(r,t) \mid^4)^{-1}$, normalized to the volume of the largest MD cell (corrsponding to n = 0.10), as a function of time and for densities n = 0.1 and 0.17 (the inset). At n = 0.17 as well as at 0.25, the participation ratios remain small and almost constant over the entire simulation, indicating that the electron is localized. The behavior is very different at n = 0.10. This simulation was started by expanding the length scale of the final configuration of the n = 0.17 run to a value corresponding to n = 0.1 and projecting out the new electron ground state, which corresponds to an almost spherical bubble. Therefore the participation ratio starts out at a value determined primarily by the bubble size at n = 0.17 scaled by the increase in the cell size in changing the system density from n = 0.17 to 0.1. It is observed that the bubble quickly expands to a volume that is a sizeable fraction of the cell volume.

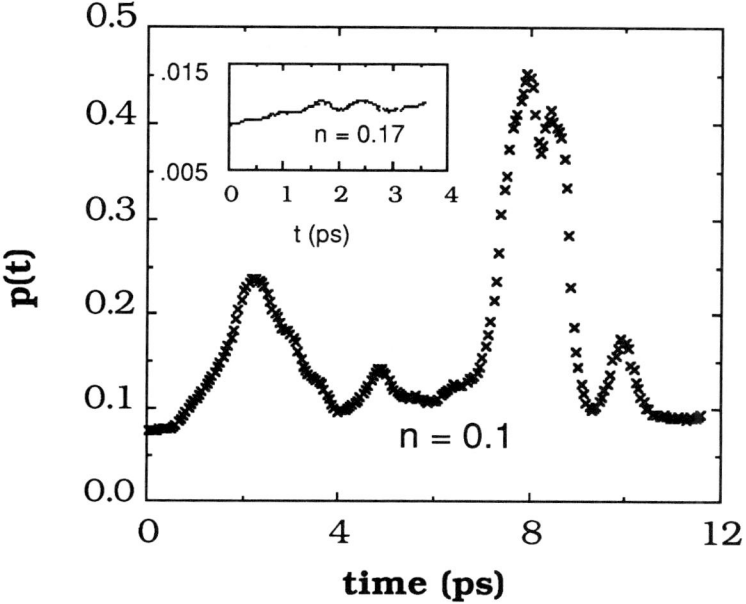

Figure 1
Participation ratio as a function of time in pico seconds. The inset shows the results for n = 0.17.

Furthermore, after 4 ps the electron wave packet undergoes dramatic expansions and contractions and finally after 8 ps the wave packet occupies almost the entire volume of the cell. This expansion indicates de-localization of the electron that is very likely limited by the size of the system. The wave packet attempts to find a region of the cell where the helium density is less. Thus, as the simulation proceeds, the wave packet localizes again temporarily because of thermal fluctuation in the system. This partial localization and de-localization is expected to continue indefinitely.

The electron-helium radial distribution function, G(r), for n = 0.17 is shown in Fig. 2. It is clear that there are no helium atoms up to r ~ 12 a.u. This excluded-volume effect is an indication of an electron bubble. The excluded volume is also present at n = 0.25. We estimate the size of the bubble to be ~ 12 a.u. For n = 0.1, G(r) showed a drastic change with time: at the beginning it displayed an excluded-volume behavior because of the way the simulation at this density was started. However, as the wave packet began to de-localize, the excluded-volume behavior disappeared and G(r) showed fluctuations about unity practically down to r = 0.

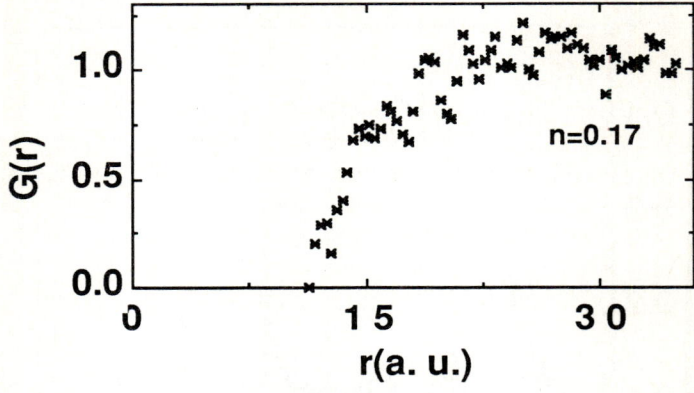

Figure 2
Electron-helium radial distribution function, G(r), measured relative to the center-of-mass of the electron wave packet.

Figure 3 shows the current-current correlation function of the electron, Im$\chi(\omega)$, obtained as a configuration average within the Franck-Condon[6] approximation by Fourier transforming the time correlation function $<\nabla\psi_t \cdot \nabla\psi_0>$. For the two higher

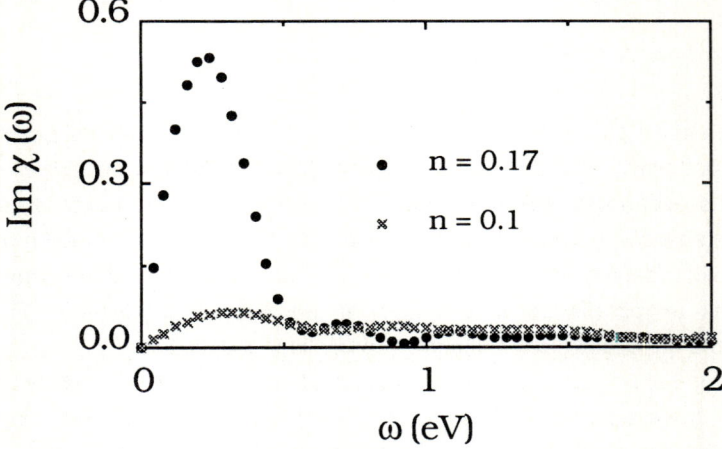

Figure 3
Imaginary part of the current-current correlation function for helium densities n = 0.17 and 0.1. The peak at n = 0.17 is due to intra-bubble transitions.

densities, Im$\chi(\omega)$, which is simply related to the optical absorption spectrum, displays a significant structure which shifts to a higher energy as the helium density increases. This structure is present throughout the simulation. At n = 0.1, when the wave packet is de-localized, the structure disappears and only a background is observed.

We have also calculated[9] the electron mobility at 77K and at a helium density of 1.25×10^{22} cm^{-3}. The applied electric field was 2.6×10^5 volt/cm in the x direction. We find that this electric field does not produce any noticeable changes in G(r). Also the voltage drop across the bubble is much smaller than the total bubble energy, indicating further that the applied field is indeed small. It is observed that the electron drifts along the x direction with a velocity of $(2.2 \pm 0.5) \times 10^4$ cm/s. This gives an electron mobility of (0.08 ± 0.02) cm^2/volt-s which is in good agreement with the extrapolated experimental value[14] (0.1 cm^2/volt-s).

In conclusion, we have discussed the QMD simulation technique for studying the dynamics of quantum particles in mixed systems at finite temperatures. This technique is used to simulate the behavior of an excess electron in helium gas at 77K. We find that the electron localizes in the form of a bubble above a critical density of 0.6×10^{22} cm^{-3}. The bubbles are nearly spherical for long periods of time and have radii ~ 7 Å. The bubble possesses quasi bound excited states and intra-bubble dipole transitions to these states give rise to pronounced structure in the optical absorption spectrum. Below the critical density the electron percolates through the helium gas and displays a featureless excitation spectrum.

ACKNOWLEDGEMENTS

We would like to thank L. H. Yang, M. H. Degani, and J. P. Rino for useful discussions. This work was supported by the U. S. DOE, BES-Materials Sciences Contract No. W-31-109-ENG-38. One of us (R.K.K) would also like to acknowledge grants of CPU time on the MFE Cray 2 at Livermore through a Grand Challenge proposal. R. K. K. and S. W. L. would like to acknowledge a travel grant from NATO.

REFERENCES

1) N. F. Mott and E. A. Davis, Electron Processes in Non-Crystalline Materials (Clarendon, Oxford, 1979).
2) B. J. Berne and D. Thirumalai in Annual Reveiew of Physical Chemistry, vol 37 eds. H. L. Strauss, G. T. Babcock, and C. Bradley Moore, (Annual Reviews Inc. Palo Alto, 1986) pp. 401-424.
3) M. Parrinello and A. Rahman, J. Chem. Phys. 80 (1984) 860.
4) R. Car and M. Parrinello, Phys. Rev. Lett. 55 (1985) 2471.
5) J. D. Doll and D. L. Freeman, J. Phys. Chem. 92 (1988) 3278.

6) A. Selloni, P. Carnevali, R. Car, and M. Parrinello, Phys. Rev. Lett. 59 (1987) 823.
7) M. D. Feit, J. A. Fleck, and A. Steiger, J. Comp. Phys. 47 (1982) 412.
8) A. Rahman, Correlation Functions and Quasiparticle Interactions in Condensed Matter, ed. J. Woods Halley, (Plenum, N.Y., 1978) pp. 417-433.
9) R. K. Kalia, P. Vashishta, and S. W. de Leeuw, J. Chem. Phys. 90 (1989) 6802.
10) W. H. Press, B. P. Flannery, S. T. Teukolsky, and W. T. Vetterling, Numerical Recipes (Cambridge University Press, Cambridge, 1986).
11) N. R. Kestner, J. Jortner, M. H. Cohen, and S. A. Rice, Phys. Rev. 140 (1965) A56.
12) R. K. Kalia and J. Harris, to be published.
13) S. Nosé, Mol. Phys. 52 (1984) 255.
14) A. Bartels, Appl. Phys. 8 (1975) 59.

MONTE CARLO SIMULATION STUDY OF DENSE PLASMAS: FREEZING, TRANSPORT AND NUCLEAR REACTION

Setsuo ICHIMARU and Shuji OGATA

Department of Physics, University of Tokyo, Bunkyo-ku, Tokyo 113, Japan

We report the results of the first microscopic study by Monte Carlo (MC) simulation ($N = 1458$) of how the particle correlations and bond-orientational orders develop in supercooled one-component plasmas (OCP's) at the freezing transitions. Starting with an equilibrated fluid state at $\Gamma = 160$, where Γ is the inverse temperature in units of the average Coulomb energy, we apply gradual quenches stepwise until $\Gamma = 200$, 300, or 400 is reached. The OCP at $\Gamma = 200$ remains in a supercooled fluidlike state. In the quenches to $\Gamma = 300$ or 400, we discovered an internal development of particle layers, which acted to expedite subsequent evolution of the OCP into a metastable solid state. The final state is found to be a bcc crystalline state with a small number of intralayer interstitials. The static structure factors calculated at $\Gamma = 160$, 200, 300, 400, and 800 are applied for explicit evaluation of the electric and thermal conductivities in dense Fe matter appropriate to the outer crustal material of neutron stars. The screening potentials of the OCP fluids and solids are expressed in parametrized forms, and are used for evaluation of the nuclear reaction rates in dense OCP matter. Dense plasmas more complex than the OCP has been studied by the MC method, two of such examples being dense C-O mixtures in white dwarfs and itinerant hydrogen atoms in metal hydrides.

1. INTRODUCTION

A physical process most remarkable in a dense plasma is the possibility of phase transition such as freezing. Rates of elementary and transport processes, such as nuclear fusion and conductivities, change drastically across such a transition, and act to influence plasma properties in laboratory[1,2] and astrophysical[3] settings.

A one-component plasma (OCP) model is applicable to a dense ionized material, such as the outer crustal matter of a neutron star.[4] One considers a crustal matter consisting mostly of Fe ($Z = 26$, $A = 56$) with a mass density ρ_m and a temperature T in the ranges of 10^4–10^8 g/cm^3 and 10^6–10^8 K, where Z and A refer to the charge and mass numbers. The r_s parameter of the electrons takes on a value,

$$r_\mathrm{s} = 1.8 \times 10^{-2} \left(\frac{Z}{26}\right)^{-1/3} \left(\frac{A}{56}\right)^{1/3} \left(\frac{\rho_\mathrm{m}}{10^6 \mathrm{g/cm}^3}\right)^{-1/3}, \tag{1}$$

ranging 10^{-2}–10^{-1}, so that the system of electrons may be regarded as a uniform background of negative charges neutralizing the average space charge of the positive ions.

The Coulomb coupling parameter of ions with a number density n is defined and given by

$$\Gamma = \frac{(Ze)^2}{ak_\mathrm{B}T} = 4 \times 10^2 \left(\frac{Z}{26}\right)^2 \left(\frac{A}{56}\right)^{-1/3} \left(\frac{T}{10^7 \mathrm{K}}\right)^{-1} \left(\frac{\rho_\mathrm{m}}{10^6 \mathrm{g/cm}^3}\right)^{1/3} \tag{2}$$

where $a = (3/4\pi n)^{1/3}$ is the ion-sphere radius. It has been predicted[5,6] that the OCP undergoes a first-order freezing transition (i.e., a Wigner transition) to a bcc-crystalline phase at $\Gamma = 178$–180.

When the OCP is in a fluid state, the ratio between the thermal de Broglie wavelength and a,

$$\Lambda = \frac{\hbar}{(Mk_BT)^{1/2}a} = 2 \times 10^{-2} \left(\frac{A}{56}\right)^{-5/6} \left(\frac{\rho_m}{10^6 \text{g/cm}^3}\right)^{1/3} \left(\frac{T}{10^7 \text{K}}\right)^{-1/2}, \quad (3)$$

measures the degree to which a quantum-mechanical description is necessitated in the behavior of the ions with mass M. For a crystalline OCP, the Einstein frequncy in the Wigner-Seitz sphere model,[7]

$$\omega_0 = [4\pi(Zen)^2/3\rho_m]^{1/2}, \quad (4)$$

goes into a description of the quantum states, so that the ratio

$$Y = \hbar\omega_0/k_BT \quad (5)$$

measures involvement of the quantum effects.

Figure 1 compares relative magnitude of those parameters on the ρ_m–T plane for Fe materials. Over the entire plane of Fig. 1, $\Lambda \ll 1$, so that quantum effects are negligible for Fe ions in a fluid state. We observe in Fig. 1 that Fe solids likewise behave classically over a significant domain on the ρ_m–T plane.

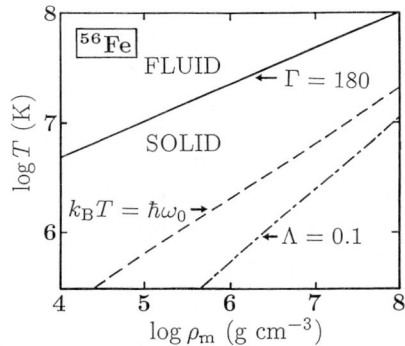

FIGURE 1
Comparison of parameters on ρ_m–T plane for iron material.

2. FREEZING TRANSITION

2.1. Quenches

We investigate the evolution of microscopic structures in a supercooled OCP by the Monte Carlo (MC) simulation method[5,6] at the number of MC particles, $N = 1458$. The top diagram of Fig. 2 shows the changes of temperature in the simulation runs: Starting with an equilibrated fluid state at $\Gamma = 160$, we applied gradual stepwise-quenches until $\Gamma = 300$ (dashed line) or $\Gamma = 400$ (solid line) was reached. Separately, we performed MC quenches to $\Gamma = 200$ and $\Gamma = 800$. The quantity c/N is the sequential number of MC configurations normalized by N. If the concept of "MC time" applies,[8] the total length of a simulation run, $c/N = 2.5 \times 10^5$, would correspond to a plasma time of $\omega_p t \simeq 3 \times 10^4$.

The middle and bottom diagrams of Fig. 2 exhibit evolution of the normalized excess internal-energy,

$$u = \frac{U}{Nk_BT} = \frac{n}{2k_BT} \int d^3r \frac{(Ze)^2}{r}[g(r) - 1], \qquad (6)$$

for the quenches to $\Gamma = 300$ and to $\Gamma = 400$, respectively. In Eq. (6), $g(r)$ refers to the radial distribution function (RDF). The double-dot-dashed lines indicate extrapolation to the respective values of the fluid internal-energy formulas (upper, ref. 5; lower, ref. 6). The dot-dashed lines analogously denote the bcc-crystalline internal energies[6] at those Γ values. The levels of u in the final states of the simulations still exceed slightly the crystalline values by the amounts $\Delta u \simeq 0.08$ ($\Gamma = 300$) and $\Delta u \simeq 0.21$ ($\Gamma = 400$).

It is instructive to compare those increments with the differences between the Coulombic Madelung energies:

$$\frac{E_M}{Nk_BT} = \begin{cases} -0.895929\Gamma, & \text{(bcc)} \\ -0.895874\Gamma, & \text{(fcc)} \\ -0.895838\Gamma, & \text{(hcp)}. \end{cases} \qquad (7)$$

At $\Gamma = 400$, we thus find $(E_{\text{fcc}} - E_{\text{bcc}})/Nk_BT = 0.022$ and $(E_{\text{hcp}} - E_{\text{bcc}})/Nk_BT = 0.036$; these are smaller by an order of magnitude than the values of Δu cited above.

Figure 2 also defines the stages of evolution, (a)–(h) for the quench to $\Gamma = 300$, and (α)–(δ) for the quench to $\Gamma = 400$. The stages (a) and (α) correspond approximately to the points where the values of u start to deviate significantly from the fluid extrapolation; (c), (e), and (γ) mark stages of a steep transition; and (f)–(h) and (δ) may be regarded as in metastable steady states.

Figure 3 shows sequential evolutions of the RDF at various stages of Fig. 2. We observe that $g(r)$ in (a), (b), (α), and (β) are quite fluidlike (cf. Fig. 1(a) in ref. 9), while those of (f)–(h) and (δ) closely resemble $g(r)$ obtained in the bcc-crystalline simulation (cf. Fig. 1(b) in ref. 9). Degrees of imperfections in the metastable states will be quantified in the subsequent sections.

Figures 4 and 5 show two-dimensional (x, y) mapping of the real-space trajectories for the MC particles located within thickness $\Delta z = 2a$ in the final states with $\Gamma = 400$ and 200. The MC particles "move" from the open circles to the solid circles over intervals of c/N designated in the figures. In Fig. 4, we find that the particles positions are virtually locked near equilibrium positions (excepting the center-of-mass motion inherent in the MC simulation[9]) while the translational symmetry is not maintained in the directions of observation. On the other hand, Figure 5 exhibits random diffusive motion of the MC particles; the OCP quenched to $\Gamma = 200$ appears to remain in a supercooled *fluid* state. These findings are essentially the same as those obtained earlier[9] in a smaller system ($N = 432$).

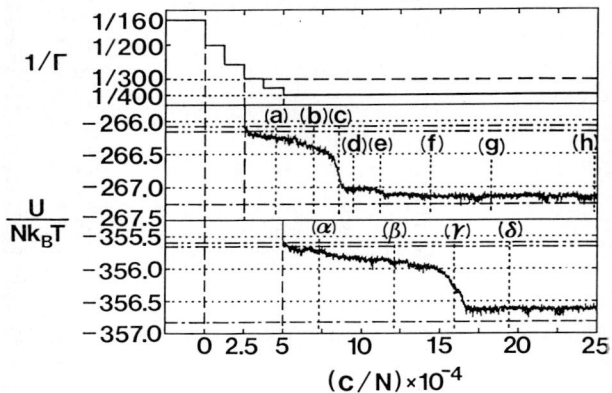

FIGURE 2
Variation of $1/\Gamma$ in the MC simulation runs (top). Evolution of the normalized excess internal energy for the quenches to $\Gamma = 300$ (middle) and to $\Gamma = 400$ (bottom). Stages (a)–(δ) are specified with a width $\Delta(c/N) = 6.9 \times 10^2$.

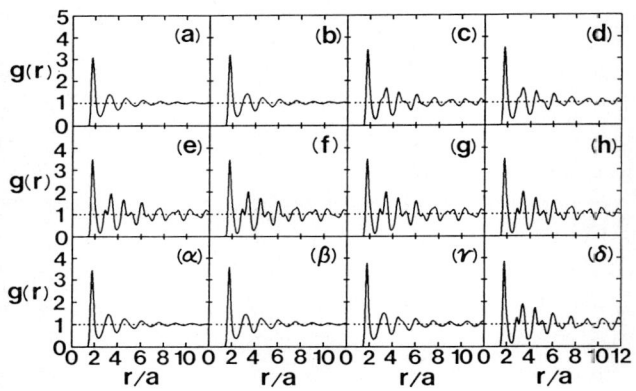

FIGURE 3
Evolution of radial distribution functions.

2. 2. Shape spectroscopy

We analyze sequential evolution of the local and extended bond-orientational symmetries by the method of *shape spectroscopy*.[9,10] It is a method to quantify similarity or closeness of a given *cluster* of particles to any of crystalline (e.g., fcc, hcp, bcc, icosahedral) clusters. A cluster consists of a particle plus its adjacent particles within the radius of $2.3a$; number of the adjacent particles defines the coordination number, N_c. For the fcc, hcp, icosahedral clusters, $N_c = 12$, while for the bcc clusters $N_c = 14$. The third-order rotationally invariant quantities, W_4 and W_6, and the extended bond-orientational correlation functions, $G_4(r)$ and $G_6(r)$, play key parts in the cluster shape spectroscopy in liquids and glasses. (See ref. 9 for mathematical definition and implication of these quantities.)

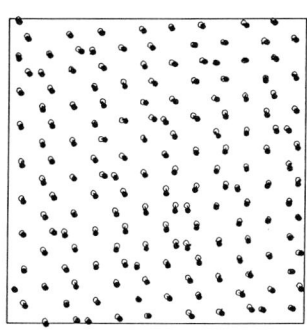
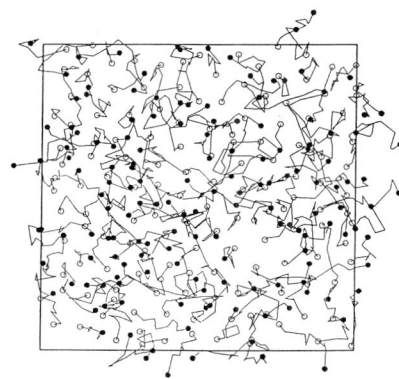

FIGURE 4
The x–y mapping of particle trajectories over an interval of $c/N = (18.0\text{–}19.7) \times 10^4$ at $\Gamma = 400$.

FIGURE 5
The x–y mapping of particle trajectories over an interval of $c/N = (5.7\text{–}7.3) \times 10^4$ at $\Gamma = 200$.

Figures 6 and 7 display sequential distributions on the (W_4, W_6) planes with $N_c = 12$ (upper) and $N_c = 14$ (lower) for the quenches to $\Gamma = 300$ and 400. In the stages (a) and (b), we observe fluidlike configurations, closely resembling Fig. 27 in ref. 9. In the stage (c), the number of $N_c = 12$ clusters in the fcc quadrant (and to a lesser extent that in the hcp quadrant) decreases suddenly, and the number of bcc-like clusters increases. Such a trend develops further in the stages (d) and (e). The stages (f)–(h) are seen to be dominated by the bcc-like clusters, while a small number of clusters remain in the fcc and hcp quadrants.

Evolution of the cluster distributions for the quench to $\Gamma = 400$ exhibits features quite analogous to the above. In particular, as we shall remark in conjunction with an emergence of layered structures in the subsequent section, we interpret the stage (γ) as in the midst of a formation of layered structures inside the plasma, which expedites a subsequent establishment of the bcc-crystalline structures.

Sequential evolution of $G_6(r)$ is shown in Fig. 8 for the quench to $\Gamma = 400$. In the fluidlike stages (α) and (β), the spatial correlation of bond-orientational symmetries remains short-ranged. In the stage (γ) $G_6(r)$ begins to fill the entire MC cell, and in the stage (δ) it closely resembles the result (cf. Fig. 4(b) in ref. 9) of a bcc-crystalline simulation. It appears that a long-ranged bond-orientational order is established in the final stages.

2. 3. Layered structures

To investigate microscopic configurations of the particles contained in the sphere inscribed to the cube of MC cell, we set the Cartesian axes (x, y, z) along the cube and rotate the sphere as a whole by an angle ζ around the y axis and by an angle η around the z axis.[10,11] Particle positions, averaged over $\Delta(c/N) = 6.9 \times 10^2$ in a stage, are then projected onto a y–z plane. Figure 9, (a)–(h), shows a sequential evolution of ordered structures for the quench to $\Gamma = 300$ viewed at $\zeta = \pi/4$ and $\eta = \pi/5$; Fig. 9, (α)–(δ), to $\Gamma = 400$ at $\zeta = 11\pi/40$ and $\eta = 2\pi/5$. The stages (a), (b), (α), and (β), which are apparently fluidlike in Figs. 3, 6, 7, and 8, are now seen to have a nucleation of layered structures in the plasma. We observe that the stage (γ) is in the midst of a formation of particle layers extending over the entire cell, as indicated in

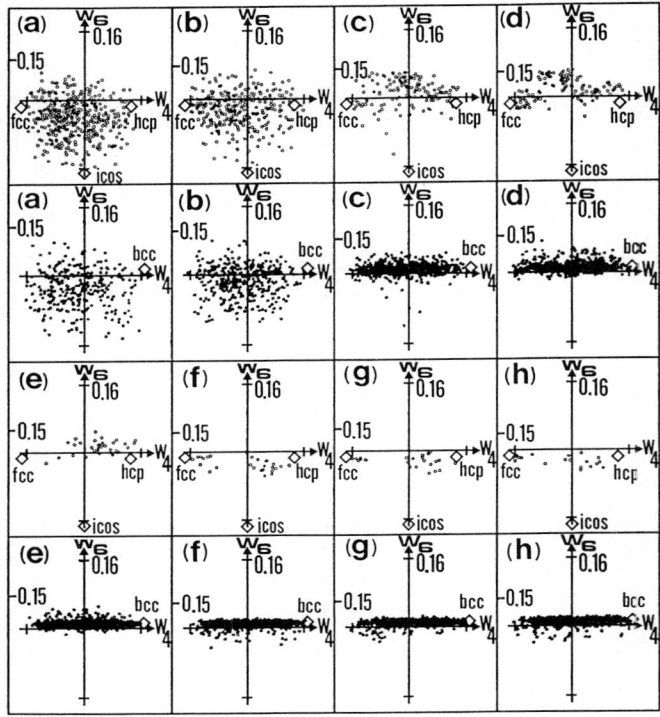

FIGURE 6
Evolution of cluster distributions with $N_c = 12$ (upper) and $N_c = 14$ (lower) for the quench to $\Gamma = 300$.

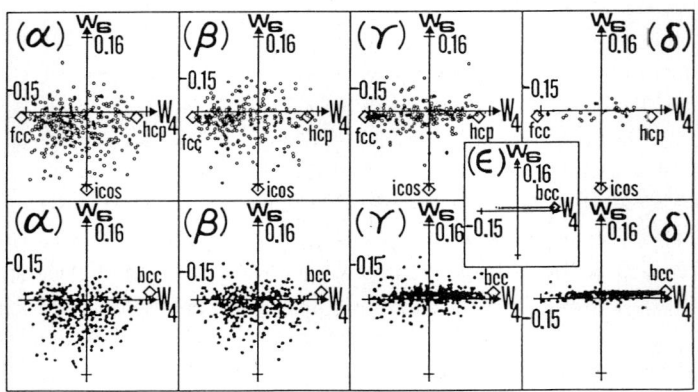

FIGURE 7
Evolution of cluster distributions with $N_c = 12$ (upper) and $N_c = 14$ (lower) for the quench to $\Gamma = 400$. (ϵ) is the result[9] of a bcc-crystalline simulation with $N = 432$.

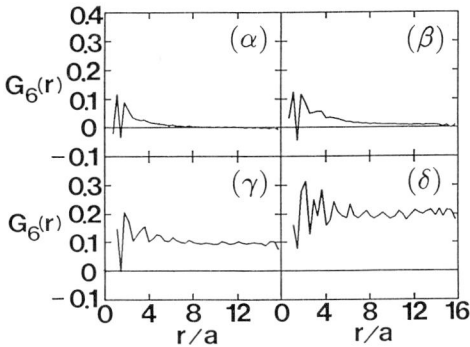

FIGURE 8
Evolution of a bond-orientational-order correlation function.

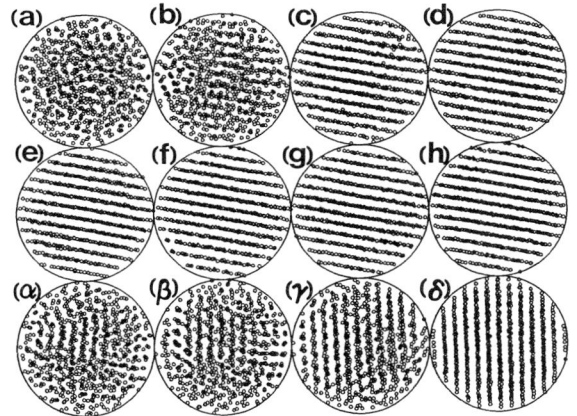

FIGURE 9
Evolution of the layered structures for the quench to $\Gamma = 300$, (a)–(h), and for the quench to $\Gamma = 400$, (α)–(δ).

Fig. 8. Such a formation of layered structures appears to be completed in the stages (c)–(h) and (δ).

Figure 10 describes correlations between particles within a layer in terms of the bond-angle distributions[10,11] $P(\theta)$, and the two-dimensional RDF[10,11] $g_2(r)$.

The results of all these analyses lead us to conclude that the final states in the present simulations with $N = 1458$ for $\Gamma = 300$ and 400 are the bcc-monocrystalline states with admixture of a few imperfections. These conclusions differ significantly from the polycrystalline structures observed in the earlier simulations[9] at $N = 432$; in the latter cases, the residuals Δu took on larger values (0.24 for $\Gamma = 300$ and 0.35 for $\Gamma = 400$). We attribute these differences as stemming from the MC-cell boundary effects which hinder (rather than assist) a nucleation of ordered structures in a smaller system.

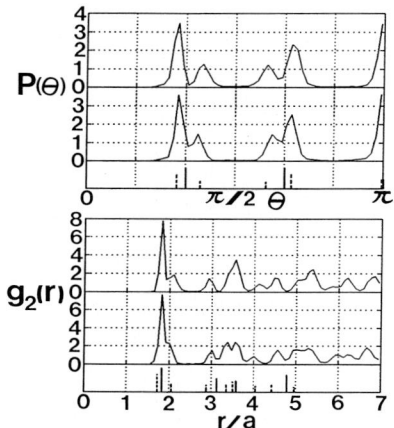

FIGURE 10
Bond-angle distribution function and two-dimensional RDF in the stage (h) for $\Gamma = 300$ (top), and in the stage (δ) for $\Gamma = 400$ (middle). The bottom figures show the corresponding quantities for the bcc lattice (dashed lines) and for the fcc (hexagonal) lattice (solid lines).

To investigate the character of the imperfections further in detail, we have singled out one of the layers as exemplified in Fig. 9, and studied the motion of the particles during each of the following three periods of c/N in units of 10^4: (1) 12.6–16.7, (2) 16.7–20.8, (3) 20.8–24.9. Specification of the layer was made at the solid circles.

Samples of such diagrams are shown in Fig. 11. The imperfections found in these analyses can be classified into three types: (A) a particle outside the layer (belonging mostly to one of the adjacent layers) in the open-circle configurations, (B) an interstitial (or an extra particle) in the layer, and (C) a vacancy in the open-circle configurations. During the period (1) (the top diagrams of Fig. 11), we find real transitions in the forms of merges between (A) and (C) and between (B) and (C), and of a transformation (or a settling) from (A) to (B), as well as virtual transitions within (B); the virtual transitions usually ceases at a return to the original interstitial configurations. We attribute these real transitions as the transient behaviors found in Fig. 2 near the stage (f). During the periods (2) and (3) (the bottom diagrams of Fig. 11), however, no imperfections of the types (A) or (C) appear to remain in the system; transient behaviors arise only through the virtual transitions within (B). The jitters observed in the steady states of the quenched OCP are attributed to those incidents of the virtual transitions.

The final states thus contain a few imperfections in the form of intralayer interstitials. An annihilation of such an isolated interstitial by the MC sampling processes would call for a slight but homogeneous compression of a configuration of particles that surround the interstitial in the layer, concurrent with appropriate compression of the particle configurations in the neighboring layers, in such a way as to preserve the overall bcc-crystalline structures. We speculate that the probability of such a sampling would be extremely small. Further quenches of a $\Gamma = 400$ OCP stepwise to $\Gamma = 800$ with a subsequent maintenance of the system over a period of $\Delta(c/N) = 9 \times 10^4$, in fact, have not produced any additional changes in the microscopic structures.

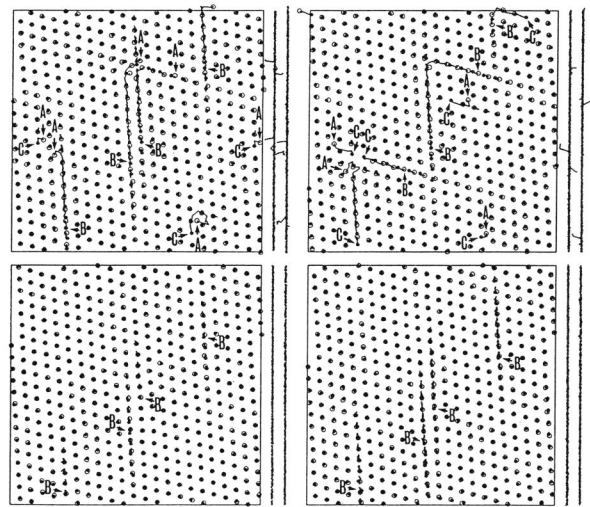

FIGURE 11
Motion of MC particles in and near a layer projected onto a x–z plane during the periods specified in the text. Between the open circles and the solid circles, 19 particle positions are followed at an equal interval of 3×10^6 configurations. Shown in the diagrams are examples of the types (A), (B), and (C) imperfections. The pair of vertical noisy lines (left for $x < 0$, right for $x > 0$) depict the projections of the particle positions onto a y–z plane, and display settling of the type (A) imperfections.

3. CONDUCTIVITIES

Knowledge on electric and thermal conductivities of the crustal matter is essential in theoretical estimates for the decay rate of the magnetic field and the cooling rate in the neutron star. The conductivities in a crystalline OCP were approached by Flowers and Itoh[12] in the single-phonon scattering approximation; those in a glassy OCP were considered by Ichimaru et al.[13,14]

Here we newly evaluate the directionally averaged static structure factor $S(k)$ in the final states of the quenched OCP at $\Gamma = 160$, 200, 300, 400, and 800 in addition to the bcc-crystalline states at $\Gamma = 300$ and 800 (Fig. 12). The conductivities are calculated by the same method as in refs. 13 and 14, where we take into account the screening effect of relativistically degenerate electrons[15] with exchange local-field corrections.[16] In Figs. 13 and 14, we compare the computed results with other theoretical calculations. The present results predict the conductivities lower by a factor of three than the single-phonon approximation calculation at $\Gamma = 400$.

4. RATES OF NUCLEAR REACTIONS

4.1. Screening potentials

Rates of nuclear reactions in dense steller material depend sensitively on the screening potential[7,17] defined by

$$H(r) = (Ze)^2/r + k_B T \ln[g(r)]. \tag{8}$$

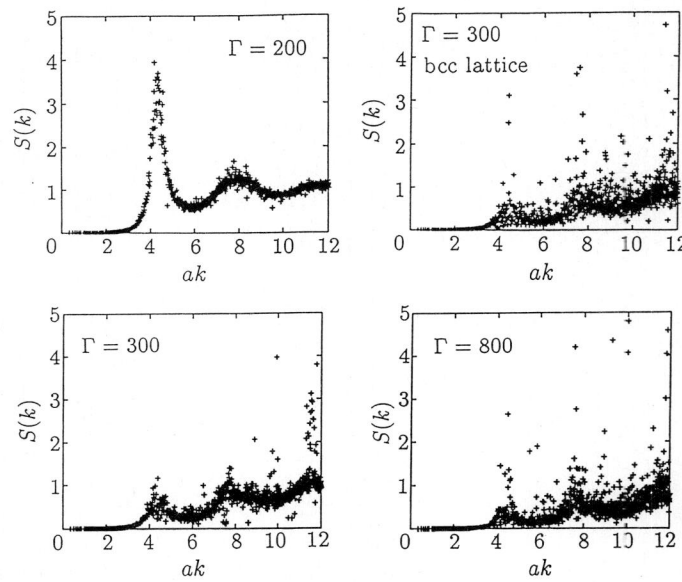

FIGURE 12
Structure factors for a fluid (top left), a bcc solid (top right), and quenched solids (bottom).

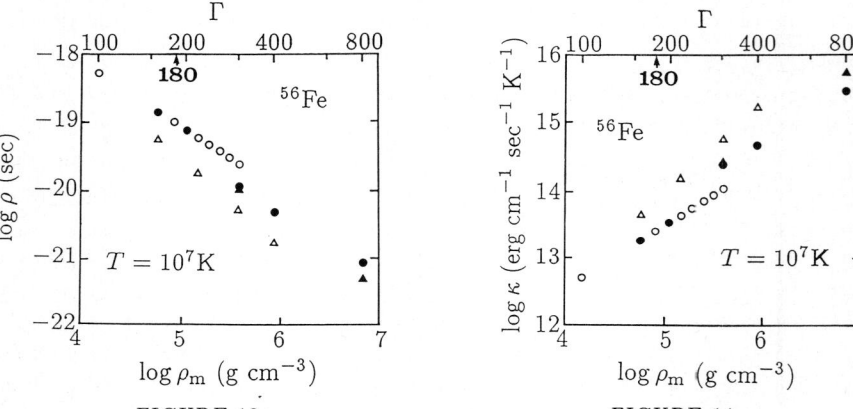

FIGURE 13
Electric resitivity of Fe material: present (solid circles), crystalline simulations (solid triangles), ref. 13 (open circles), ref. 12 (open triangles).

FIGURE 14
Thermal conductivity of Fe material: present (solid circles), crystalline simulations (solid triangles), ref. 13 (open circles), ref. 12 (open triangles).

This quantity for the OCP has been analyzed by the MC method at $\Gamma = 10$, 40, 80, 160, and 200. Figure 15 shows the MC data for $H(r)$ at $\Gamma = 200$ in the fluid and bcc-lattice simulations. In the lattice simulation, the contributions to $g(r)$ from the 1st, 2nd, ... nearest-neighbor particles are distinguished. We find that the OCP screening potentials for $0.8a < r < 2a$ are accurately parametrized by the formulas:

$$\frac{H(r)}{(Ze)^2/a} = 1.262 - 0.0025 \ln \Gamma - 0.390 \left(\frac{r}{a}\right) + 1.5 \times 10^{-6} \exp\left[5.0 \left(\frac{r}{a}\right)\right] \quad (9)$$

for the fluid OCP, and

$$\frac{H_1(r)}{(Ze)^2/a} = 1.183 - 0.350 \left(\frac{r}{a}\right) + 2.0 \times 10^{-5} \exp\left[3.3 \left(\frac{r}{a}\right)\right] \quad (10)$$

for the 1st nearest-neighbor particles in a crystalline OCP.

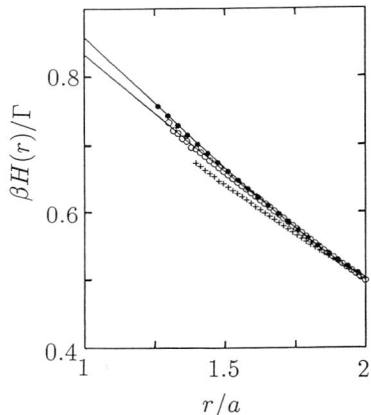

FIGURE 15

OCP screening potentials at $\Gamma = 200$. Solid circles are fluid simulation data; open circles, 1st nearest neighbors (solid); and crosses, 2nd nearest neighbors (solid). The lines represent Eqs. (9) and (10).

4. 2. Enhancement of nuclear reaction rate

The rate of reaction between a pair of nuclei is proportional to $g_N(0)$, the value of the joint probability density $g_N(r)$ at $r = 0$ for the nuclei averaged over the motion of their center-of-mass and the remainder of the particles. In a classical fluid or solid (see Sec. 1 for the criteria), the enhancement factor of the reaction rate due to strong Coulomb correlations is given by[7,17] $\exp[H(0)/k_B T]$. Taking account of the short-range quadratic corrections,[7] we find from Eqs. (9) and (10)

$$\frac{H(0)}{k_B T} = 1.110 \Gamma - 0.0025 \Gamma \ln \Gamma \quad (11)$$

for the fluid OCP, and $H_1(0)/k_B T = 1.060 \Gamma$ for the classical solid OCP.

For the pycnonuclear reaction in the quantum solid, one considers the reaction processes between the nearest neighbor particles, so that $g_N(0)$ is evaluated directly from the solution to the Schrödinger equation with the effective potential given by $(Ze)^2/r - H_1(r)$. We thus find approximately

$$g_N(0) = \frac{1}{3 x_0^2 \Delta \sqrt{\pi}} \exp\left(-\frac{x_0^2}{\Delta^2}\right) \quad (12)$$

where $\Delta^2 = [\hbar^2/(Ze)^2 aM_{\rm r} v'']^{1/2}$, $x_0 = 1.743$, $v'' = 0.3094$, and $M_{\rm r}$ is the reduced mass.

5. MULTI-COMPONENT PLASMAS

Dense plasmas more complex than the OCP have been studied by the MC method, two of such examples being binary ionic mixtures of carbon and oxygen in white dwarfs,[17,18] and itinerant hydrogen atoms in metal hydrides.[19] Equation of state, the phase diagram, and nuclear reaction rates are elucidated in the former case. The screening potentials and the resulting rates of "cold fusion" are investigated in the latter.

REFERENCES

1) J. J. Bollinger, S. L. Gilbert, D. J. Heinzen, W. M. Itano, and D. J. Wineland, Observation of correlations in finite strongly coupled ion plasmas, this volume.

2) J. P. Schiffer, Static and dynamic properties of confined cold ion plasmas: MD simulations, this volume.

3) H. W. Van Horn, Phase transitions in dense astrophysical plasmas, this volume.

4) For a review, S. Ichimaru, H. Iyetomi, and S. Tanaka, Phys. Rep. **149** (1987) 91.

5) W. L. Slattery, G. D. Doolen, and H. E. DeWitt, Phys. Rev. A **26** (1982) 2255.

6) S. Ogata and S. Ichimaru, Phys. Rev. A **36** (1987) 5451.

7) For a review, S. Ichimaru, Rev. Mod. Phys. **54** (1982) 1017.

8) S. Ogata and S. Ichimaru, Phys. Rev. A **38** (1988) 1457.

9) S. Ogata and S. Ichimaru, Phys. Rev. A **39** (1989) 1333.

10) S. Ogata and S. Ichimaru, Phys. Rev. Lett. **62** (1989) 2293.

11) S. Ogata and S. Ichimaru, J. Phys. Soc. Jpn. **58** (1989) 356.

12) E. Flowers and N. Itoh, Astrophys. J. **206** (1976) 218.

13) S. Ichimaru, H. Iyetomi, S. Mitake, and N. Itoh, Astrophys. J. Lett. **265** (1983) L83.

14) H. Iyetomi and S. Ichimaru, Phys. Rev. A **27** (1983) 1734.

15) B. Jancovici, Nuovo Cimento **25** (1962) 428.

16) K. Sato and S. Ichimaru, J. Phys. Soc. Jpn. **58** (1989) 787.

17) S. Ogata, H. Iyetomi, and S. Ichimaru, Thermonuclear reaction rates of dense carbon-oxygen mixtures in white dwarfs, this volume.

18) S. Ichimaru, H. Iyetomi, and S. Ogata, Astrophys. J. Lett. **334** (1988) L309.

19) S. Ichimaru, S. Ogata, A. Nakano, H. Iyetomi, and T. Tajima, Statistical-mechanical effects on cold nuclear fusion in metal hydrides, this volume.

STATIC AND DYNAMIC PROPERTIES OF CONFINED, COLD ION PLASMAS: MD SIMULATIONS

John P. SCHIFFER

Argonne National Laboratory, Argonne, Illinois 60439-4843 and University of Chicago, Chicago, Illinois 60637

Some four years ago it was suggested that in the new generation of heavy ion accelerator storage rings for multiply charged ions, being planned in Europe, one may well attain internal temperatures that would correspond to very cold plasmas[1]. Since that time, the techniques of electron or laser cooling of such beams has evolved and it may well be possible to reach temperatures corresponding to a plasma coupling parameter $\Gamma \gg 100$. I was fortunate to have had an opportunity to collaborate during 1986-87 with my former colleague Aneesur Rahman, of Molecular Dynamics fame, and we adapted the MD method to the calculation of the properties of cold confined plasmas[2]. After Rahman's premature death two years ago I have continued the exploration of these systems and would like to summarize the results here.

Ions in a storage ring are confined to a mean orbit by focusing elements. To a first approximation, these may be described by a constant harmonic restoring force: $F = -Kr$. If the particles in the frame moving along with the beam have small random thermal energies, then they will occupy a cylindrical volume around the mean orbit and the focusing force will be balanced by that from the mutual repulsion of the particles. Inside the cylinder only residual two-particle interactions will play a significant role and some form of ordering might be expected to take place.

The results of some of the first MD calculations[2] showed a surprising result: not only were the particles arranged in the form of a tube, but they formed well-defined layers: concentric shells, with the particles in each shell arranged in a hexagonal lattice that is characteristic of two-dimensional Coulomb systems. These patterns are shown in figure 1.

The radial size of these cold systems grows gradually with particle number. Defining λ as the linear density: the number of particles per unit length 'a' (the Wigner-Seitz radius), at low linear densities the particles do not repel each other sufficiently and sit on the axis. For $\lambda \gtrsim 0.7$ they push each other away from the axis to a finite radius, forming a series of different patterns. Some of these patterns are shown in figure 2, plotted as if on the mantle of a cylinder with appropriately increasing radius. The approximate hexagonal arrangement shows up rather quickly, and various helical

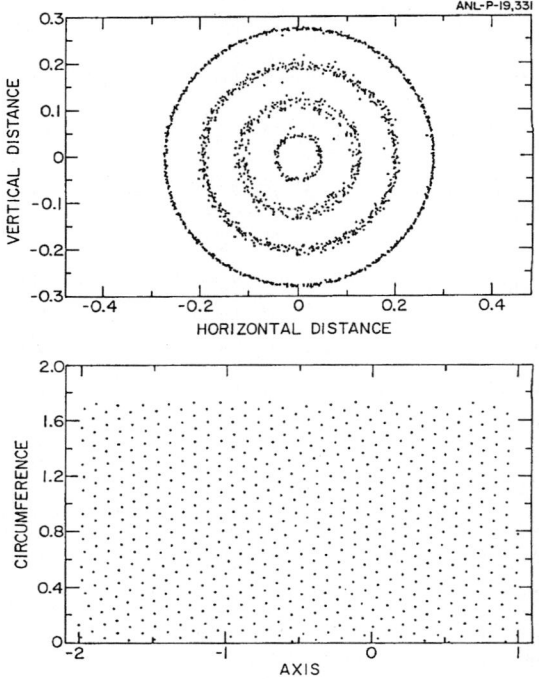

FIGURE 1
Arrangement of particles from a MD simulation of a cold beam with constant focusing, for λ = 27. The top shows the projection of the beam onto a perpendicular plane, the bottom shows the particles in one (the outer) shell along the mantle of the cylinder.

patterns appear as the number of hexagons per turn increases. These helical patterns have been studied further analytically[3]. At $\lambda \approx 3.1$, the radius of the cylinder has reached a sufficiently large value such that particles may start appearing on the axis; as λ increases further these start forming a second inner cylindrical layer. The number of cylindrical shells has been found empirically to be $n_s \approx 0.8\ \lambda^{1/2}$. These shells continue; up to eight distinct shells have been observed in the MD calculations. The shell structure does not appear at a well defined temperature, but seems to emerge gradually, as may be seen from the beam profiles at different temperatures in figure 3.

This property of confined ionic systems to form an outer layer parallel to an equipotential surface, with further parallel layers on the inside, seems to be quite general[2,4]. For a three-dimensional confining force that is

Static and dynamic properties of confined, cold ion plasmas

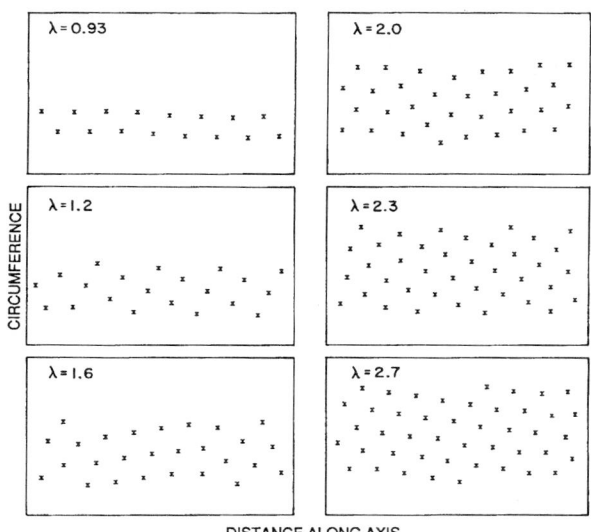

FIGURE 2
Arrangement of particles from MD simulations for several small values of λ plotted on the mantle of the cylinders that represent the particle shells.

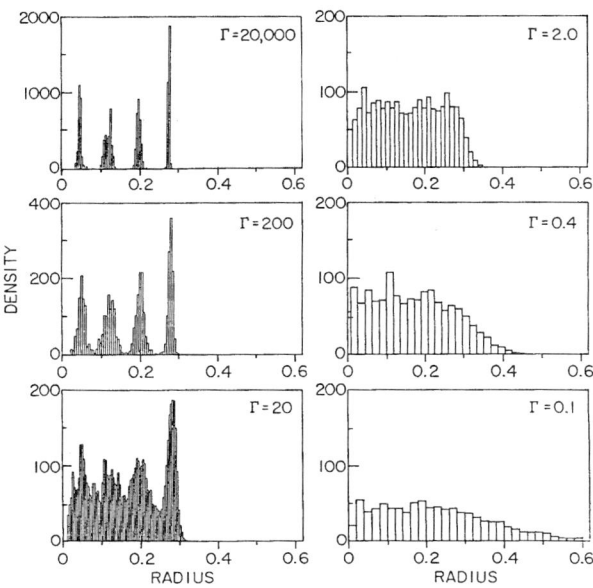

FIGURE 3
The radial density profile of particles at different temperatures from MD simulations for a beam with $\lambda = 27$.

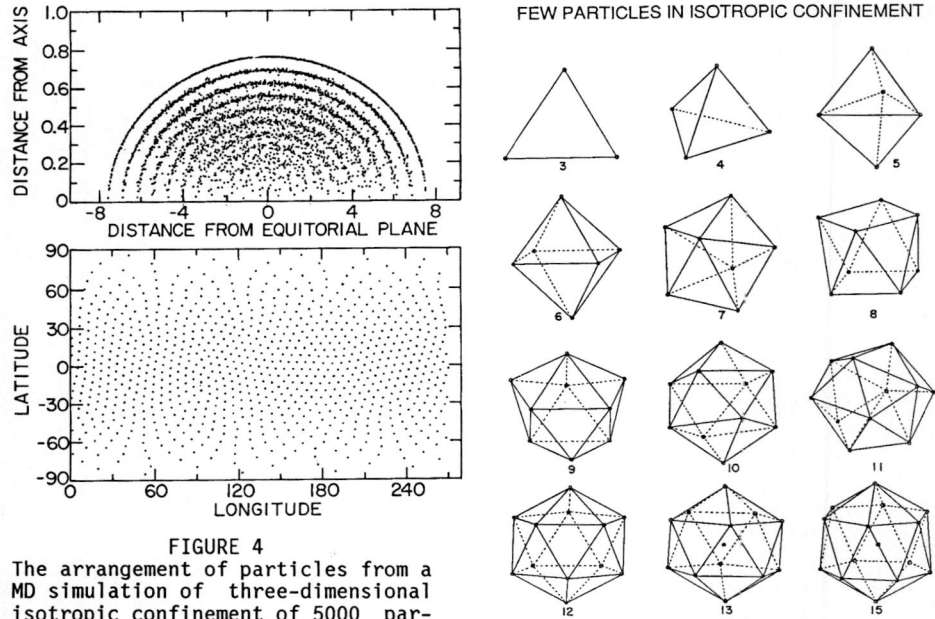

FIGURE 4
The arrangement of particles from a MD simulation of three-dimensional isotropic confinement of 5000 particles at a low temperature. The top shows the projection onto a half-plane intercepting the origin, the bottom the projection of particles in one shell.

FIGURE 5
Arrangement of few particle systems in isotropic confinement.

isotropic, one also obtains layers: spherical shells as shown in figure 4, again with hexagonal order within each shell. With few particles, isotropic confinement produces the geometrical shapes shown in figure 5, simple patterns at a fixed distance from the origin (note that these shapes are not necessarily trivial, for instance the minimum configuration for eight particles is not a cube, as was predicted by some). The Hamiltonian for this system is the same as for J. J. Thompson's "plum pudding" model of the atomic electrons, that preceded quantum mechanics. A new particle first appears in the center for a system with 13 particles, a second one for 20, and these gradually grow into a new central object. The number of spherical shells is given by the empirical relationship $n_s = (N/4)^{1/3}$, where N is the total number of particles. These spherical shells were also obtained in simulations by Dubin and O'Neil[5], using a slightly different Hamiltonian to simulate the confinement in the magnetic fields corresponding to Penning traps - in the limit of low temperatures the two are equivalent. Recent experiments in Penning traps[6] provide beautiful confirmation of the predictions of layered structures.

All the cold confined systems have in common the layered structures and the hexagonal pattern within the layers. Yet, when one tries to determine the overall structure of how particles in the different layers arrange themselves with respect to each other, the correlation function suggests an approximate bcc symmetry, that is seen in the crystallization of infinite cold one-component plasmas[7]. The hexagonal pattern does not map into a bcc lattice -- so the two must be approximate and these confined structures have mixed, approximate symmetries.

If the confining forces are spacially anisotropic, similar structures are obtained, elliptically or spheroidally deformed.

DYNAMIC EFFECTS

For a beam of particles, the system is quite elastic for radial distortions and oscillations can be induced[8]. Such a mode is shown in figure 6. Its frequency is different from the betatron oscillation (the motion of an isolated particle in the focusing field) because of the effect of space charge. Empirically the frequency seems constant at $\sqrt{2}$ times the betatron frequency and independent of the linear particle density. Although the damping of this mode has not been studied carefully, it is clearly small on the time scale of many betatron periods.

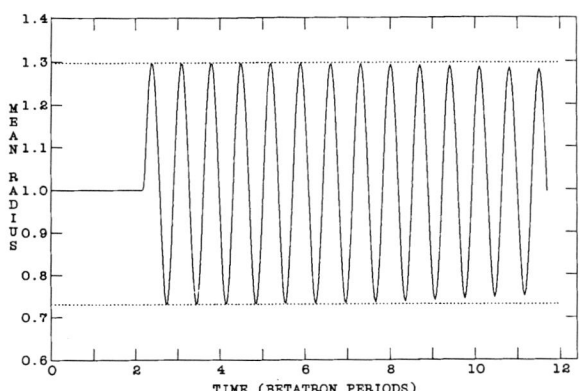

FIGURE 6
MD simulation of the behavior of a cylindrically confined beam after a single radial perturbation in the focusing.

Space-charge effects couple the horizontal and vertical size of the beam, giving rise to a radial quadrupole mode that also persists for long times with a period that is very close to the betatron period, and independent of particle number.

These radial modes may be relatively easy to detect with the diagnostic tools in a storage ring. They seem to be independent of temperature up to $\Gamma \approx 1$, where the overall size is space-charge limited - they appear to be macroscopic effects, not clearly related to ordering.

There are similar modes in the three-dimensional system with isotropic confinement. The monopole breathing mode has a frequency that is $\sqrt{3}$ times (to 0.2%) the one-particle frequency in the confining field, and two uncoupled quadrupole modes (two axes with respect to each other, and the pole with respect to the equator) both at 1.095 times the one particle frequency, or .6315 times the monopole frequency. When a system with an anisotropic confining potential is used, with the retaining force a factor of two smaller in the z direction than in x and y, a spheroidal system appears, with the ratio of major to minor axes in the outer shell of 1.821. The monopole mode, when rescaled for the same density, remains the same (within 0.5 %) as for the spherical cloud. The quadrupole modes have frequencies of about .554 times that of the monopole mode for the oscillation of the poles against the equator, and .671 times for the oscillation of the equator.

The problems with attaining the ordered patterns seen in the MD calculations are associated with the real features of storage rings[9]. Three major problems are:

a) that the cooling in a storage ring is not constant, but is applied only once per revolution, and that cooling forces act only in the direction of the beam and not three-dimensionally as was done in the MD calculations discussed so far,

b) the fact that the focusing lattice of an accelerator is not constant in time but consists of a series of alternating focusing elements,

c) the shear associated with the motion of particles in a closed loop in the storage ring.

The fact that the particles in a beam are moving within a storage ring and have to follow a closed path, means that for a pair of particles travelling side by side at the same velocity (this is what the "beam cooling" does), the outer one is at a slightly larger radius and will lag behind its inner neighbor. The motion along a closed path therefore introduces a shear in the beam and the strength of the elastic constant of a condensed beam against such distortion determines whether the ordered system of particles can withstand the shear or not. Since the driving force in a real system is fixed by the cyclotron frequency, and the parameter scaling the calculated system is the focusing field that represents the betatron frequency, the question becomes whether the elastic properties are sufficient for a given cyclotron frequency.

As shown in figure 7, this is a function of the particle number, and for a typical ring the system will not survive the shear without slipping. Only for a very large ring, or very strong focusing, such as RHIC with a betatron tune of 30, would larger systems survive without slipping.

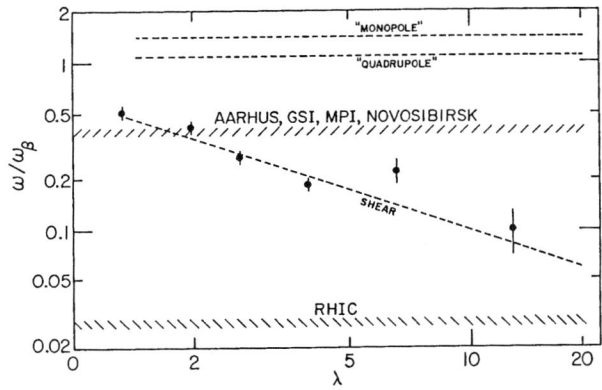

FIGURE 7
Calculated response of particle beams from MD simulations to shear induced by storage rings as a function of linear density. The imposed shear from various storage rings is also indicated.

A MD calculation with beams of finite width was allowed to proceed with appropriate shear imposed. This was done with cooling only along the direction of motion. The system settled only very slowly, and gradually a pattern of threads of particles shown in figure 8 emerged. The particles are still roughly along the cylindrical layers seen in the static case, but they are segregated into individual threads to minimize their interaction as they slip past each other. It is interesting to note that the arrangement between the threads is again trying to be a hexagonal one.

Other possibilities remain. If one could introduce a slight gradient in the cooling, to have particles on the outside travelling somewhat faster than those on the inside, then this shear would be absent. Alternatively, if the beam could be induced to rotate about its own axis, path differences for different particles would disappear for a rotational frequency that is the same as (or an integral multiple of) the cyclotron frequency. It is not clear whether such rotation could be induced, it would imply centrifugal forces on the beam particles that are comparable to space-charge forces.

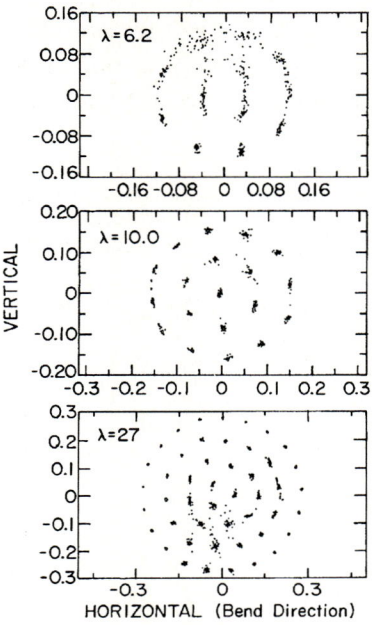

FIGURE 8
Projection of particles from MD simulations for beams subjected to shear, with various linear densities. The projection is onto a plane perpendicular to the beam axis, the applied shear, to simulate the bending of the beam in a storage ring, is across the horizontal direction.

The fact that the beam is not cooled as ideally as was assumed in the MD program needs to be considered. The 'cooling force' from either laser or electron cooling is applied only along the direction of the beam and over only a short stretch in the particle's trajectory. MD calculations indicate that the rate of cooling a system will be slower, but the rate of equilibration of kinetic energy is such that the system can still settle in a few tens of turns.

Another complication of storage rings is the time dependent focusing field of an accelerator lattice. A full simulation has not been attempted and is not very practical, but an approximation to the time dependence has been explored. When a periodic focusing field is introduced the particles oscillate but eventually settle into an ordered pattern as before. Apparently the system settles into a mode in which the radial layers oscillate coherently and smoothly and the ordering is almost as good as for static focusing. The same is true if cooling is allowed to act only once per cyclotron orbit. Beam profiles corresponding to a simulation of a beam allowed to settle in a constant focusing field, then subjected to periodic focusing with once-a-turn axial cooling is shown in figure 9.

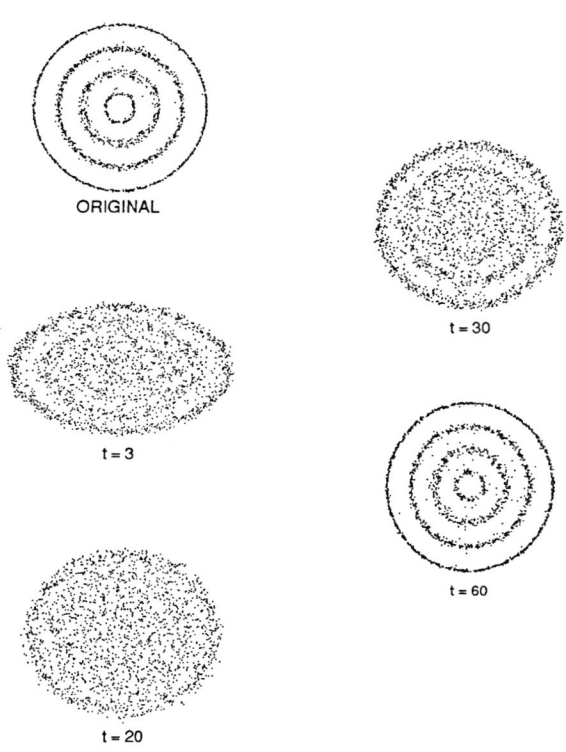

FIGURE 9
MD simulation of a beam that had been previously cooled in a constant confining field, which then was changed at T=0 to fluctuate in time while the beam was cooled only once per turn (taken as 2.5 betatron periods). Beam profiles at various times (in units of the betatron period) are shown.

In the case of the beautiful stratified ion clouds that were reported by the Boulder group[6] and that will be discussed in more detail here, MD calculations successfully predicted the layered nature of the ion cloud, including the number of shells that are to be expected for a given number of total ions. In addition, when the Boulder results first appeared, it seemed a puzzle that they observed roughly cylindrical shells, while the MD calculations kept showing spheroidal ones. After some trial and error, we followed up on one aspect of the experiment: the presence of heavy impurity ions that would be carried along with the rotating Be+ cloud, but not visible with the lasers. A MD simulation shown in figure 10, indicates that the heavier impurities ions form an equatorial belt around the cloud, and the tightening effect of this belt produces cylindrical shells over a substantial

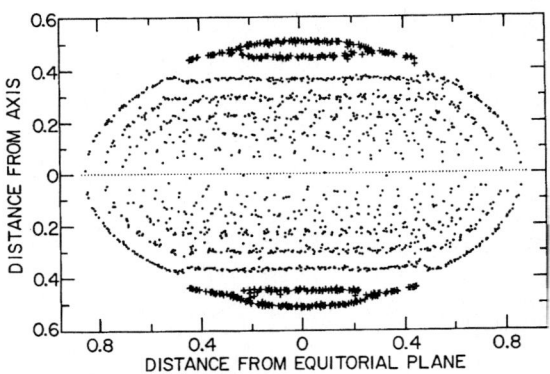

FIGURE 10
MD simulation of 2000 particles confined in a three dimensional potential with the confining force in one direction half that of the other two, simulating the Penning trap confinement of ref [6]. 500 of the particles were assumed to have a heavier mass and are shown with '+' symbols.

length of the cloud. It is not yet clear whether this is the correct explanation of the observed results.

Since the MD calculations, both with a harmonic force, and the 'guiding center approximation' used by Dubin and O'Neil[6] have been remarkable successful in predicting the data in ion traps, it seems reasonable to assume that they will continue to be successful in predicting properties of particle beams.

The brute force technique of including the fine details of an accelerator lattice and electron cooling in the simulation, and computing the properties of a beam, is not likely to be sensible with the present generation of super-computers; the number of time steps is likely to have to increase by several orders of magnitude. However, many features of the process could be modelled better than has been done so far, and I suspect that considerable quantitative information remains to be extracted from modelling using the MD techniques for these very cold beams. Examples are the spacing of focusing elements to minimize the excitation of radial modes, the question of the placement of cooling sections with respect to the dipole and focusing elements, better understanding of the quantitative aspects of electron or laser cooling, etc.

If it were possible to detect the radial modes of a beam discussed above, this could be a rather useful technique for determining the temperature regime of the beam. Beyond that, the shell structure may be observable directly, as it was in the Boulder experiments, with fluorescent light. The finer details of structure would have to be observed with diffraction measurements with visible light, which does not seem to be unreasonable considering the size of spacings one has to deal with. But diagnostic techniques that involve the scattering of light would require that the particles not be fully ionized, and this may present some problems at particular facilities. Alternatively one might be able to "see" the ions in the cooling section when they are immersed in an electron beam travelling with the same velocity.

One can speculate of more far fetched techniques; of extracting the beam and detecting its structure directly, even of superchannelling between two colliding condensed beams -- but these ideas seem perhaps a little premature at present.

Condensed layered structures have now been observed in ion traps, and many of the features that one is anticipating for storage rings should have their counterpart in these systems. The competition and interplay between the two rather different techniques should have a beneficial effect on the pursuit of our understanding of this new form of condensed matter.

This research was supported by the U. S. Department of Energy, Nuclear Physics Division, under Contract W-31-109-Eng-38.

REFERENCES

1) J. P. Schiffer and P. Kienle, Zeitschr. f. Phys. A 321, 181 (1985).

2) A. Rahman and J. P. Schiffer, Phys. Rev. Letts. 57, 1133 (1986).

3) D. Habs, MPI-H-1987-V10; R. Hasse, private communication and Proceedings of the Workshop on Crystalline Ion Beams, GSI-89-10 April 1989, ISSN 0171-4546.

4) J. P. Schiffer, Phys. Rev. Letts. 61, 1843 (1988).

5) D. H. E. Dubin and T. M. O'Neil, Phys. Rev. Letts. 60, 511 (1988).

6) E. L. Pollock and J. P. Hansen, Phys. Rev. A8, 3110 (1973); W. L. Slattery et al., Phys. Rev. A21, 2087 (1980); S. Ichimaru, H. Iyetomi and S. Tanaka, Physics Reports 149 (2 & 3), 91 (1987).

7) A. Rahman and J. P. Schiffer, Physica Scripta T22, 133 (1988); and Zeitschr. f. Phys. A 331, 71 (1988).

8) J. P. Schiffer Proceedings of the Workshop on Crystalline Ion Beams, GSI-89-10, April 1989, ISSN 0171-4546.

9) S. L. Gilbert, J. J. Bollinger, and D. J. Wineland, Phys. Rev. Letts. 60, 2022 (1988).

MOLECULAR DYNAMICS STUDY OF RAPIDLY QUENCHED OCP

Minoru TANAKA

Department of Engineering Science, Faculty of Engineering, Tohoku University, Sendai 980 Japan

Topological characteristics in the short-range oder of the classical one-component plasma (OCP) is investigated in a metastable glassy state around $\Gamma=8800$ quenched rapidly from equilibrium fluid at $\Gamma=110$ by the molecular dynamics method. The OCP system underwent gradual structural change near $\Gamma=300$ from supercooled fluid to a glassy state, and in the final quenched state the second peak of both PDF and TDF splits into two sub-peaks. Topological statics of the Voronoi polyhedron also shows that the short-range order of the rapidly quenched OCP is similar to that of amorphous monatomic systems with an ordinary pair potential.

1. INTRODUCTION

The static and the dynamic properties of OCP are investigated in detail and known accurately in thermal equilibrium over a wide range of the coupling constant Γ ($= e^2/akT$). For non-equilibrium or metastable OCP, characteristics of supercooled fluid around $\Gamma \approx 300 \sim 400$ and process of solidification are investigated recently by the molecular dynamic (MD)[1] and the Monte Calro (MC)[2] methods.

In this paper we report MD simulation of rapid-quenching process of OCP from equilibrium fluid at $\Gamma = 110$ to an almost frozen state around $\Gamma = 8800$. We concentrate on analysis of topological characteristics in the short-range order of OCP particles in the final state, and show that pair (PDF) and the triplet (TDF) distribution functions and distribution of the Voronoi polyhedron exhibit very similar features to those found in a monatomic amorphous system with an ordinary pair potential, e.g. the Lennard-Jones type in spite of the softest purely repulsive nature of the Coulomb interaction between OCP particles.

2. MD SIMULATION OF RAPID-QUENCHING PROCESS

The OCP system consists of 864 point-charge particles with the mass of argon atom in the uniform background of opposite charge. The number density of particles is kept at $8.080 \times 10^{21} cm^{-3}$ ($a = 3.029 Å$, $r_s = 5.482$) during the whole stages of simulation. The force acting on each particle is computed by the Ewald method, and the coupled equations of motion are integrated by using Verlet's algorithm with $\Delta t = 0.094 \omega_p^{-1}$ ($\omega_p = 1.879 \times 10^{13} s^{-1}$). The initial equilibrium fluid at $\Gamma = 110.28$ (490.1K) is simulated by the microcanonical MD over the period of $282 \omega_p^{-1}$, and found consistent with other MD and MC results at near values of Γ.

MD simulation of rapid-quenching process is done by subtracting a portion of the kinetic energy of the whole particles intermittently during 28000 MD steps ($\simeq 2600\omega_p^{-1}$). After each subtraction of energy, the system is allowed 3000 steps for internal relaxation from disturbance and stablizes quickly into a stationary intermediate state. The whole process of quenching is simulated by the microcanonical MD algorithm[3], and monitored by inspecting the value of *instant temperature*, i.e. (2/3) of the mean kinetic energy per particle at every step. The OCP system is thus quenched stepwise from the initial fluid state to the final state around 6.1K (Γ = 8800), via intermediate states at 389K (Γ = 139), 302K (179), 208K (260), 103K (527) and 51K (1063). The rate of quenching $\Delta\Gamma/\Delta t \simeq 4.6 \times \omega_p$. The final state is left free for internal relaxation over 15000 steps, and after first 1000 steps with increase of 0.1K in temperature the system becomes stationary without further structural relaxation. The MD records of the last 10000 steps are used to investigate the short-range order in the quenched state. PDF and TDF as well as the self-diffusion constant D of OCP particles are calculated in each state including five intermediate states. Changes in PDF and the equilateral TDF are shown in Figures 1 and 2, respectively. It should be noted that the second peak splits into two sub-peaks in states at higher values of Γ.

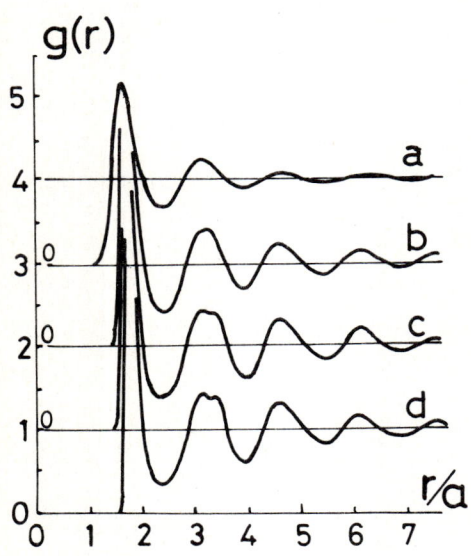
FIGURE 1
Change in PDF. a:Γ=110, b:Γ=527, c:Γ=1063, d:Γ= 8800. PDF changes monotonously between a and b.

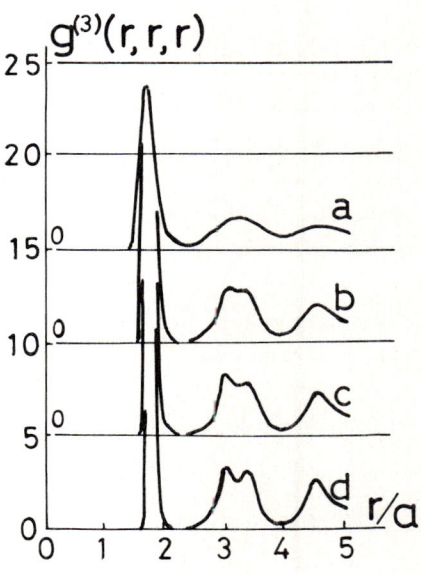
FIGURE 2
Change in equilateral TDF. Identification is the same as in Figure 1.

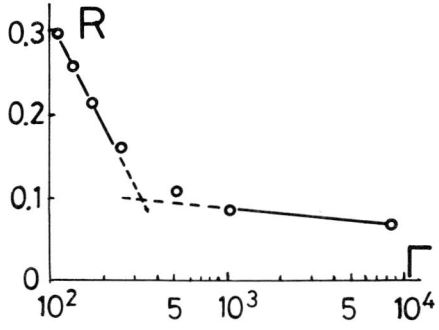

FIGURE 3
Change in $R=g_{min}/g_{max}$, the ratio of the first minimum to the first maximum of PDF. Packing in the first-shell around each OCP particle starts to saturate around $\Gamma \approx 300$. A similar saturation in R vs. T plot is found in rapid-quenching of other monatomic liquids.

Changes in the Wendt-Abraham parameter R and the normalized self-diffusion constant D^* are shown in Figures 3 and 4. These are very similar to those found in MD simulation of rapid-quenching process of classical monatomic liquid with an ordinary pair potential such as argon.

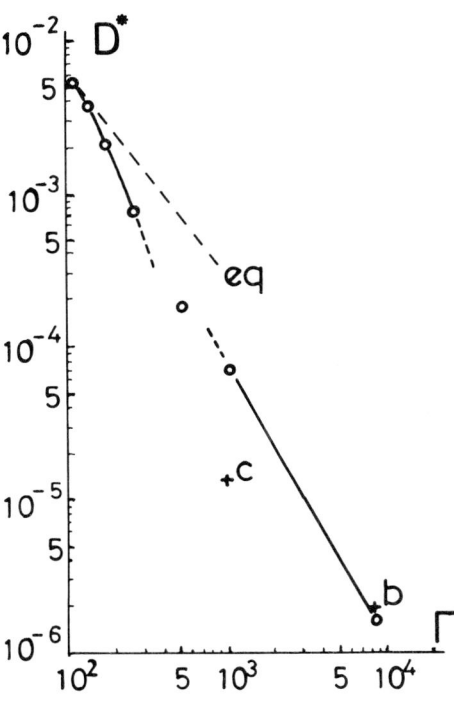

FIGURE 4
Change in $D^*=D/\omega_p a^2$. Open circles: along rapid-quenching process, crosses b and c: annealed states(see Figure 5), dashed line eq: $D^*=2.95\Gamma^{-1.34}$.

3. TOPOLOGICAL CHARACTERISTICS OF THE SHORT-RANGE ORDER IN THE QUENCHED STATE

Characteristics of PDF and TDF at Γ = 8800 such as the very high peak (4.78 for PDF, 82.41 for the equilateral TDF) and splitting of the second peak are very similar to those calculated in rapidly-quenched monatomic amorphous systems. The short-range order of particles in a rapid-quenched state is determined by a certain common mechanism irrespectively to details of the pair potential for a monatomic system. Three-dimensional aspects of local structure in the quenched state can be analized through the distribution of the Voronoi polyhedron.[4] In Table 1, the quintuplet of face-index of the Voronoi polyhedron are shown for major polyhedra of F = 12 to 15 with the percentage of population among total 864. The ditribution of major polyhedra is also similar to those obtained in a rapidly-quenched amorphous state of argon.[4] The short-range order in the quenched OCP at Γ = 8800 have many common features with other monatomic amorphous

TABLE 1. Major Voronoi polyhedron

F	n_3	n_4	n_5	n_6	n_7	Γ=8800	Γ=110
12	0	2	8	2	0	1.04%	0.68%
	0	0	12	0	0	0.64	0.14
	0	3	6	3	0	0.30	0.69
13	0	3	6	4	0	7.27	2.42
	0	1	10	2	0	4.34	0.76
	0	4	4	5	0	2.28	0.78
	0	2	8	3	0	1.94	0.57
	0	5	2	6	0	1.75	0.24
14	0	4	4	6	0	11.64	1.50
	0	3	6	5	0	9.63	2.03
	0	2	8	4	0	7.73	1.90
	1	3	4	5	1	2.50	1.91
	0	6	0	8	0	2.40	0.03
	0	1	10	3	0	1.94	0.60
	1	3	5	3	2	1.59	1.54
	1	1	8	3	2	1.46	0.61
15	0	4	4	7	0	4.35	0.60
	0	3	6	6	0	3.15	1.06
	0	2	8	5	0	2.58	0.80
	1	3	5	4	2	1.82	1.35
	1	4	3	5	2	1.68	0.95
	0	5	2	8	0	1.52	0.09

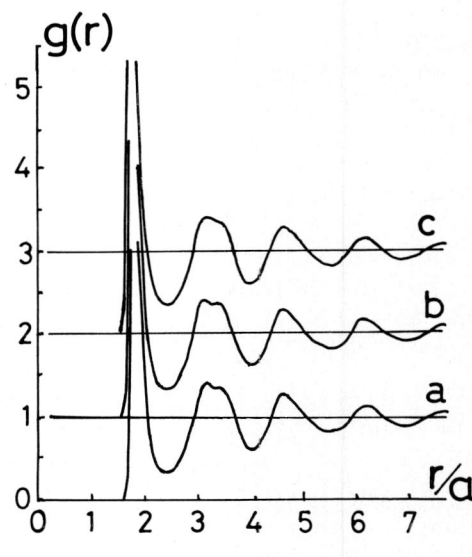

FIGURE 5
Change in PDF in the annealing process.
a: quenched state at Γ=8800(d in Fig.1),
b: annealed 9000 steps at T=6.1K(Γ=8800),
c: heated up to T=54.05K(Γ=1000) and annealed 9000 steps(to be compared to c in Fig.1).

systems. The metastable nature of the quenched state at Γ = 8800 is examined briefly by simulating the annealing process in contact with the heat-bath using canonical MD algorithm. At first, the system is made contact with the heat-bath of T = 6.1K over 9000 steps, and then is heated up to 54.05K (Γ = 1000) and kept temperature constant over 9000 steps. PDF in these annealed states are shown in Figure 5 and D* shown with crosses in Figure 4. No structural change toward solidification is seen during these annealing processes.

REFERENCES

1) H. Totsuji, Static and dynamic properties of strongly-coupled one-component plasmas, in: Strongly Coupled Plasma Physics, eds. F.J. Rogers and H.E. Dewitt (Plenum, New York, 1987) pp. 19-33.

2) S. Ichimaru, Monte Carlo simulation study of dense plasma, this volume.

3) M. Tanaka, J. Phys. Soc. Jpn. 51 (1982) 3802 ; ibd. 52 (1983) 1270.

4) M. Tanaka, J. Phys. Soc. Jpn. 55 (1986) 3108 & 3428.

MONTE-CARLO SIMULATIONS FOR THE SURFACE PROPERTIES OF THE STRONGLY COUPLED ONE-COMPONENT PLASMA

Akira ISHIDA[†], Toshihiko SHIRAKAWA[†], Masayuki HASEGAWA[*] and Mitsuo WATABE[†]

[†]Faculty of Integrated Arts and Sciences, Hiroshima University, Hiroshima 730, Japan
[*]Faculty of Engineering, Iwate University, Morioka 020, Japan

The surface properties of the classical one-component plasma (OCP) are investigated by Monte-Carlo simulation for a wide range of the plasma coupling parameter Γ including the bulk freezing point, $\Gamma_m \approx 180$. It is shown that the OCP surface has many interesting features characteristic of a strongly coupled Coulomb system: oscillatory or layered structures, getting more and more pronounced with increasing Γ, of the density profile near the surface, a novel phase appearing at $\Gamma \sim 130$ in which only the outermost layer crystallises into a hexagonal lattice while inner layers still remain liquid-like, etc.

1. INTRODUCTION

We have been studying the surface properties of the one-component classical plasma (OCP) both theoretically and by computer simulations (for a review see reference 1). Our original aim of this study was to obtain precise enough information about the OCP surface to be used as a reference in thermodynamic perturbation theory for investigating the surface properties of real liquid metals. It has been shown that the OCP may serve fairly well as a reference system for liquid alkali metals, but, on the other hand, our simulation results indicate that the surface of the very strongly coupled OCP has very peculiar features and behaves quite in contrast with real liquids. Actually the study of a finite OCP system is of interest itself and is also relevant to recent studies of cold, condensed plasmas confined in external fields (storage ring or ion traps);[2,3] a charged background of the OCP plays a role of a confining external field for ions.

In this report we present some results of our Monte-Carlo (MC) study of the surface properties of the strongly coupled OCP, especially the characteristic change of the ion density distribution near the surface with increasing coupling strength.

2. SYSTEM AND METHOD

For simulating a planar surface of the OCP we use an OCP slab, which consists of identical ions with charge Ze moving in the charged background, the density of which is $-n_b e$ for $|z|<D$ and zero for $|z| \geq D$. The system is indefinitely extended in the x and y directions parallel to the surface by the periodic boundary condition with period L_x and L_y respectively. Let N be the number of ions in a unit cell; then the ionic number density defined by $\rho_b = N/(2DL_xL_y)$ satisfies the condition $Z\rho_b = n_b$, the

requirement of the total charge neutrality. We define the bulk plasma parameter by $\Gamma=(Ze)^2/ak_BT$, where a is the ionic-sphere radius defined by $4\pi a^3/3=\rho_b^{-1}$. In the following all the lengths are given in units of a.

In an earlier paper,[4] we have reported the results of our MC simulations for this system in the relatively weak coupling range of $\Gamma\leq 30$. Here we wish to present the results of our more recent MC simulations in the stronger coupling range $30\leq\Gamma\leq 400$. The system size we used is N=416, D=9.33, L_x=11.49 and L_y=8.12.

3. RESULTS

3.1 Γ=30~130

It has been shown in our previous work[4] that the density profile $\rho(z)$ normal to the surface shows stratified structure near the surface for $\Gamma\geq 5$ which is more pronounced with increasing Γ. The present results of $\rho(z)$ for Γ=30 ~130 show that the ion configurations in this range are qualitatively similar to those for Γ<30, although the stratified structure becomes much more pronounced. Figure 1(a) shows

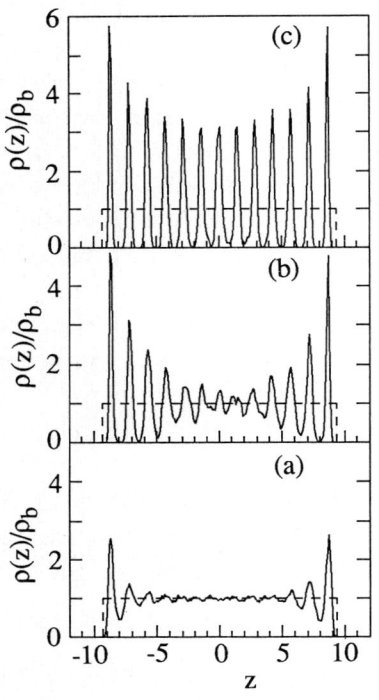
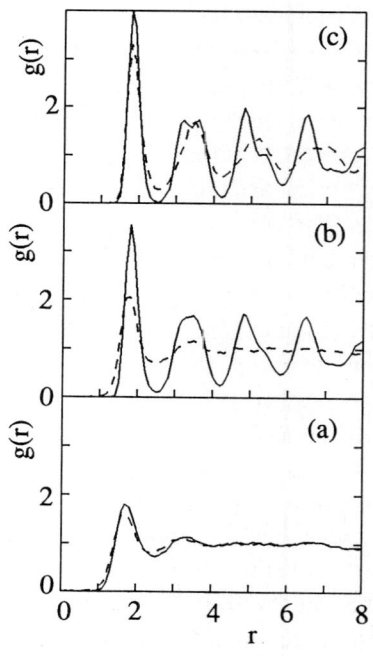

FIGURE 1
Ion density profile: (a) Γ=50; (b) Γ=160; (c) Γ=200.

FIGURE 2
Two-dimensional radial distribution function g(r) within the outermost layer (——) and a layer near the centre of the slab (----): (a) Γ=50; (b) Γ=160; (c) Γ=200.

a typical result of $\rho(z)$ for $\Gamma=50$. For $\Gamma\geq100$ there is a region with very little ions present between the outermost and the next layers. However, the density profile is still more or less uniform near the centre of the slab.

Figure 2 shows the two-dimensional (parallel to the surface) radial distribution function $g(r)$ in the outermost layer and the one in a near-centre layer. For the case of $\Gamma=50$ shown in Figure 2(a) both of them show liquid-like structures, but interionic correlation is seen to be stronger near the surface than in the inner part. The first layer behaves rather like a two-dimensional liquid which has different features from the bulk liquid.

3.2 $\Gamma=140\sim170$

For $\Gamma=140\sim170$, the stratified structure appears not only near the surface but extends throughout the slab as shown in Figure 1(b) for $\Gamma=160$. It is seen from the figure that the outermost layer becomes independent of the inner part with no exchange of ions between the two regions. Actually the ions in the outermost layer are arranged into a two-dimensional lattice of hexagonal symmetry, as is seen from the two-dimensional $g(r)$ of the outermost layer, which has the features of the two-dimensional hexagonal lattice. This is more directly seen in Figure 3 which displays the actual configuration of ions in the outermost layer. At $\Gamma=170$, the second layer also crystallises similarly to the first layer, but the arrangement of ions further inside still remain more or less liquid-like. From these facts it is seen that the two-dimensional liquid-like layer formed on the surface crystallises in the hexagonal lattice independently of the interior at $\Gamma\sim130$ which is far below the bulk freezing point $\Gamma_m\approx180$.[5] This phenomenon, which may be called surface crystallisation, is quite in contrast to the surface melting observed in actual systems.

3.3 $\Gamma=180\sim400$

A few hexagonal layers are formed progressively from the surface in the early stage of structural relaxation and then the whole system starts to solidify. In the final phase ions in several layers near the centre of the slab arrange on the bcc lattice with each layer being a bcc (110) plane (see Figure 2(c)). This is consistent with the fact that the bulk OCP crystallises in the bcc structure. From the obvious

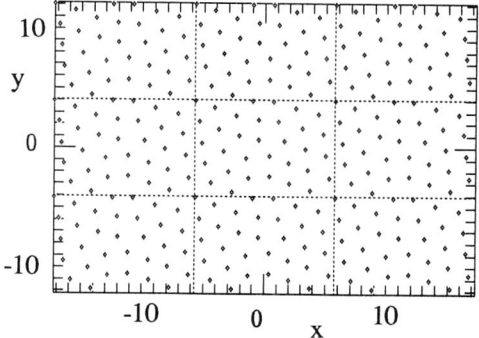

FIGURE 3
Configuration of the ions in the outermost layer for $\Gamma=160$. Configurations in the neighbouring cells extended periodically are also included to show more clearly the hexagonal arrangement of ions.

reason of incompatible stacking there are a few layers, whose structures are undefinable, between the surface region of the hexagonal layers and the central region of the bcc layers.

4. DISCUSSION

The results described in the previous section are those for the simulations performed starting with a configuration of ions distributed randomly throughout the background. As the random distribution may be taken as an actual configuration of the OCP at a sufficiently high temperature, our MC calculations with this initial configuration is considered as simulating a sudden cooling process to the final temperature specified by the respective value of Γ studied. Although we have not tried to do simulations of a gradual cooling process, we consider that the results for such a process will be the same as our results for a sudden cooling process; in face, we have observed that the system, suddenly cooled to a certain value of Γ, follows the final equilibrium states of the systems with smaller coupling constants than Γ successively on the way of relaxation process.

We have also performed MC calculations, taking a bcc array of ions with (110) planes parallel to the surface as an initial configuration. This corresponds to a 'sudden heating' process. The results obtained for this initial configuration show: (a) the whole system retains its initial bcc structure and does not melt within our MC run down to Γ as low as 140, although the amplitude of the thermal motion of each ion becomes larger with decreasing Γ; (b) no dependence of the behaviour of the system after a sufficiently long run on the initial configurations is observed for $\Gamma \leq 130$. The above result for $\Gamma \geq 140$ is quite different from that obtained starting with the random configuration, suggesting that the MC simulation in this coupling range depends significantly on the initial configuration. As the result is obviously dependent on the size of the system (i.e. the thickness of the slab), further MC studies with varying the system size is necessary to clarify the real meaning of the above point, and work along this line is under way.

REFERENCES

1) M. Hasegawa and M. Watabe, Surface properties of Coulomb liquid: from the classical one-component plasma to liquid metals, this volume.

2) A. Rahman and J. P. Schiffer, Phys. Rev. Lett. **57** (1986) 1133.
   J. P. Schiffer, Phys. Rev. Lett. **61** (1988) 1843.

3) S. L. Gilbert, J.J. Bollinger and D.J. Wineland, Phys. Rev. Lett. **60** (1988) 2022.

4) A. Ishida, M. Hasegawa and M. Watabe, J. Phys. C: Solid State Phys. **20** (1987) L599.

5) W.L. Slattery, G.D. Doolen and H.E. DeWitt, Phys. Rev. A **26** (1982) 2255.
   S. Ogata and S. Ichimaru, Phys. Rev. A **36** (1987) 5451.

Chapter III:
Glass and Freezing Transitions

FREEZING OF COULOMB LIQUIDS

Mario P. TOSI

Department of Theoretical Physics of the University of Trieste and International Centre for Theoretical Physics, Trieste, Italy

Density functional theory has found useful application in the treatment of liquid - solid coexistence as approached from the liquid phase in a variety of systems, starting with the work of Ramakrishnan and Yussouff in 1977. We briefly review progress in the theory of Wigner crystallization of the electron gas in both the classical and the fully degenerate limit, the latter having been evaluated in recent work by Pastore and Senatore. We then discuss liquid structure and freezing in a pseudoclassical version of the bond charge model for elemental semiconductors, as an example where the DFT approach, though difficult to implement quantitatively, provides useful insights.

1. INTRODUCTION

The idea that structural correlation functions in a simple liquid near freezing reflect properties of its solid near melting has a long history. An early example is the Kirkwood - Monroe theory of the liquid - solid transition[1], which was framed within the BBGKY hierarchy in terms of the pair potential and the pair distribution function in the liquid. An alternative line of approach to inhomogeneous classical systems is provided by the cluster expansion[2], which is more closely related to the modern density functional formalism[3]. The general aim for an implementation of the theory is an approximate evaluation of the Helmholtz free energy functional $F[n(r)]$ for the inhomogeneous system from knowledge of thermodynamic and correlation - response properties of a corresponding homogeneous system. This route to the theory of freezing was opened by the work of Ramakrishnan and Yussouff[4,5].

Within the above context, crystallization of the one-component classical plasma (OCP) into a bcc lattice has become a test case for the quantitative validity of the theory[6-9]. The quantal extension has recently been given by Pastore and Senatore[10], who have used it to evaluate the phase transition for the fully degenerate electron system. These recent developments on Wigner crystallization in classical and degenerate jellium are briefly reviewed in section 2 below for an illustration of the essential ideas and the quantitative limitations of the theory[11].

The value of the theory, however, is not limited to the possibility of quantitative predictions on liquid - solid coexistence. At the lowest level of approximation for a classical system, it only invokes knowledge of the liquid pair structure and hence allows inferences on hot solid properties by the simple expedient of matching the main features of the liquid structure with the location of reciprocal lattice points in the crystal. As an example, we briefly review in section 3 below the structural evolution induced with increasing coupling strength in a liquid - state version of the bond charge model for semiconductors and discuss the insights that can be gained from such data on the freezing of these systems[12, 13]. The main point at issue in this case is how tetrahedrally constrained localization of bond particles may lead, on the one hand, to freezing accompanied by volume expansion, and on the other to supercooled disordered states having characteristic structural features of amorphous materials.

2. FREEZING OF JELLIUM

The single-particle density profile $n(r)$ changes across the phase transition from a constant (n_l, say) in the liquid phase to a periodic function of position in the crystalline phase, where it is represented by the Fourier series

$$n(\mathbf{r}) = n_s + \sum_{\mathbf{K} \neq 0} n_\mathbf{K} \exp(i\mathbf{K}\cdot\mathbf{r}) \tag{1}$$

the $\mathbf{K}$'s being the reciprocal lattice vectors (RLV). A stable density profile must minimize the grand potential $\Omega = F - \mu N$, namely

$$\frac{\delta F[n(\mathbf{r})]}{\delta n(\mathbf{r})} = \mu \quad . \tag{2}$$

It is customary in the case of jellium to look for coexistence of fluid and crystal at the same average density ($n_s = n_l = n$, say), coexistence being determined by the equality of the Helmholtz free energy in the two phases. Equilibrium between the two phases in contact then requires charge separation at their interface to balance an interfacial drop in chemical potential μ.

The free energy functional for the jellium model has the form

$$F[n(\mathbf{r})] = F_o[n(\mathbf{r})] + \frac{e^2}{2} \iint d\mathbf{r} d\mathbf{r}' \frac{n_Q(\mathbf{r})n_Q(\mathbf{r}')}{|\mathbf{r} - \mathbf{r}'|} + F_c[n(\mathbf{r})] \tag{3}$$

where F_o is the ideal term, $n_Q(\mathbf{r}) = n(\mathbf{r}) - n$ is the charge density determining the Hartree contribution, and F_c contains all the effects due to correlations and - in the quantal case - to exchange. The main idea from the work of Ramakrishnan and Yussouff is to consider directly the free energy difference $\Delta F_c = F_c[n(\mathbf{r})] - F_c(n)$ between the two phases, rather than the separate free energies, and to base its estimation on knowledge on the homogeneous fluid phase. Indeed, if a functional expansion of $F_c[n(\mathbf{r})]$ around the homogeneous phase exists, then the phase transition can be evaluated from sole knowledge of fluid phase properties at freezing. In practice, the simplest approximation which is already worth exploring truncates the expansion at second order terms, when the only fluid phase property that is needed is the static, linear density response function.

2.1 Classical jellium

The appropriate expression for the free energy difference ΔF between the two phases of the OCP is

$$\frac{\Delta F}{Nk_B T} = \int d\mathbf{r}\, n(\mathbf{r}) \ln\left[\frac{n(\mathbf{r})}{n}\right] - \frac{1}{2}\iint d\mathbf{r} d\mathbf{r}' n_Q(\mathbf{r}) n_Q(\mathbf{r}')\, c(|\mathbf{r} - \mathbf{r}'|) + \ldots \tag{4}$$

The first contribution on the rhs of eqn (4) is the ideal free energy difference, while the second contains the Hartree term and the leading term in the expansion of the correlation free energy difference, expressed through the Ornstein - Zernike correlation function c(r) of the fluid. The latter is related in Fourier transform to the liquid structure factor S(k) by

$$n\, c(k) = 1 - \frac{1}{S(k)} \quad (5)$$

The liquid structure factor of the OCP is accurately known as a function of the coupling strength from both simulation and liquid structure theory.

Because of the periodicity of $n_Q(r)$, the function c(k) or equivalently S(k) is needed at wave numbers corresponding to the various RLV stars appropriate to the Bravais lattice of the crystal structure into which freezing is being considered. Explicitly, the equilibrium condition for the density profile can be written from eqns (2) and (6) as

$$n(\mathbf{r})/n = \exp\left[\beta\Delta\mu^* + \int d\mathbf{r}'\, c(|\mathbf{r} - \mathbf{r}'|)\, n_Q(\mathbf{r}') + ...\right] \quad (6)$$

and the coexistence condition reads

$$\beta\, \Delta F/N = \beta\, \Delta\mu^* + \frac{1}{2} V \sum_{\mathbf{K} \neq 0} c(K)\, |n_\mathbf{K}|^2 + ... = 0\ , \quad (7)$$

$\Delta\mu^*$ being the interfacial drop in chemical potential between the two phases at coexistence.

The main results that have been obtained on the crystallization of the OCP into a bcc lattice are[7-9] (i) a lower limit $\Gamma \approx 150$ for the coupling strength at the phase transition, against the currently accepted value $\Gamma = 180 \pm 1$ from simulation[14], (ii) good agreement with simulation data for the entropy change on freezing and for the interfacial potential drop at coexistence, and (iii) evidence for an important role of higher order correlations in the fluid phase. Equation (4) is easily supplemented by higher order terms arising from the dependence of the Ornstein - Zernike function on the coupling strength. However, the fluid structure, while appropriately soft to modulation in the (110) star of RLV for the bcc lattice, is quite rigid against modulation in the (200) star. Microscopic couplings between the Fourier components of the crystalline density at different RLV stars are needed to stabilize the bcc lattice, as explicitly demonstrated by Barrat et al[8] through an evaluation of three-body correlations in the OCP.

2.2 Degenerate jellium

The equivalent of eqn (4) for the jellium of electrons at zero temperature is[10]

$$\Delta F = T_0[n(\mathbf{r})] - \frac{3}{5}N\varepsilon_F + \frac{1}{2}\iint d\mathbf{r}d\mathbf{r}' n_Q(\mathbf{r}) n_Q(\mathbf{r}') \left[\frac{e^2}{|\mathbf{r}-\mathbf{r}'|} - K_{xc}(|\mathbf{r}-\mathbf{r}'|)\right] + ... \quad (8)$$

where $T_0[n(\mathbf{r})]$ is the kinetic energy of noninteracting electrons at density $n(\mathbf{r})$, ε_F is the Fermi energy of the fluid phase and $-K_{xc}(r)$ denotes the second functional derivative of the exchange and correlation energy with respect to $n(\mathbf{r})$, evaluated on the fluid phase. This function also enters the static dielectric screening function of the homogeneous electron gas.

The well known argument of Kohn and Sham allows an exact treatment of the ideal kinetic energy term in eqn (8) by conversion of the equilibrium condition (2) into a single-particle effective Schrödinger equation. The effective selfconsistent potential in this equation can be written in Fourier transform as

$$v_{eff}(\mathbf{K}) = n_{\mathbf{K}} \frac{4\pi e^2}{K^2} [1 - G(K)] \qquad (9)$$

where G(K) is the local field factor due to exchange and short-range correlations in the dielectric function of the electron gas[15], evaluated at the RLV. Explicitly, the Kohn-Sham equation is

$$\left[-\frac{\hbar^2}{2m} \nabla^2 + v_{eff}(\mathbf{r}) \right] \psi_i(\mathbf{r}) = \varepsilon_i \psi_i(\mathbf{r}) \qquad (10)$$

and is solved selfconsistently for Bloch orbitals $\psi_i(\mathbf{r})$ determining the equilibrium density profile $n(\mathbf{r}) = \Sigma_i |\psi_i(\mathbf{r})|^2$ and for Bloch energies ε_i entering the coexistence condition

$$\Delta E_{ground} = \frac{1}{N}\sum_i \varepsilon_i - \frac{3}{5}\varepsilon_F - \frac{1}{2} V \sum_{K \neq 0} \frac{4\pi e^2}{K^2} [1 - G(K)] |n_K|^2 + ... \quad , \qquad (11)$$

the sums being over the occupied electron states.

It can be seen from eqns (8) and (9) that Wigner crystallization requires in the present theoretical framework that, with increasing coupling strength r_s, G(K) should become appreciably greater than unity at relevant RLV stars, and in particular at the first star in approximate correspondence with the diameter of the Fermi sphere. Under this condition, the gain in potential energy from exchange and correlation can balance the loss in kinetic energy on crystallization. This behaviour is found by Pastore and Senatore on evaluating G(K) by the STLS approximation, which relates this function to the structure factor of the electron gas. They consider in their calculations the case of a fully spin-polarized system, for which the phase transition has been reported at $r_s = 100 \pm 20$ by quantal computer simulation[16]. Full solution of the Kohn - Sham equation yields crystallization of the spin-polarized fluid into a bcc lattice at $r_s \approx 130$. The calculations also show that the fcc and the bcc lattice are competitive in the immediate vicinity of the phase transition, with the bcc lattice becoming clearly the stable one at somewhat larger values of the coupling strength, and that the electrons are well localized at their lattice sites in the Wigner crystal, the calculated Lindemann ratio being in substantial agreement with the value of about 0.30 obtained from quantal simulation[16].

3. BOND PARTICLE MODEL FOR SEMICONDUCTORS

The bond charge model, as proposed a long time ago by Phillips[17] for crystalline Si and Ge, represents the electronic charge distribution in each covalent bond as a point-like charge located at mid distance between each pair of neighbouring atoms and of amount $z \cong 2/\sqrt{\varepsilon_0}$, ε_0 being the static dielectric constant of the material. The bond particles participate in the lattice dynamics and indeed the model has found useful applications in the evaluation of both bulk and surface phonon dispersion curves[18]. The broad relevance of the model to melting and freezing of these materials is immediately brought out by noticing that its Coulomb coupling strength can be measured by the quantity $z^2e^2/(ak_BT)$, with $a = (4\pi n)^{-1/3}$ in terms of the atomic number density n, and that similarity of melting behaviour for group-IV and III-V semiconductors suggests the melting criterion

$$4e^2 / (\varepsilon_o a k_B T_m) \cong \text{constant} \qquad (12)$$

with T_m the melting temperature at atmospheric pressure. This criterion is indeed empirically satisfied[12], the constant in eqn (12) being equal to 20. An earlier melting criterion for semiconductors[19] relates $k_B T_m$ to the valence-conduction band gap E_g.

Melting at standard pressure brings these materials from tetrahedrally coordinated open structures into metallic liquids having higher density than the solid and first neighbour coordination number close to 7. Their liquid structure is nevertheless quite distinct from that of other liquid metals[20]. Specifically, the coordination number in the liquid is still relatively low and a second shell of neighbours is found to lie in the region of interatomic distance where the pair distribution function g(r) in other liquid metals shows its main minimum. Similarly, the liquid structure factor S(k) shows a distinctive shoulder on the large - k side of its main peak. Furthermore, Si and Ge can be prepared in an amorphous state, which shares with other glassy and amorphous materials the characteristic of having a first sharp diffraction peak (FSDP) indicating medium - range order associated with the connectivity of basic tetrahedral units. The FSDP in amorphous Si[21] and Ge[22] lies at a wave number in approximate correspondence with the (111) star of RLV of the crystal, whereas the subsequent main peak in the diffraction pattern is in approximate correspondence with the location of the shoulder in the liquid structure factor.

Simple classical bond-particle models for liquid structure, which appear to bear qualitative relevance to the observed behaviour of real Ge, have been recently studied by Ferrante and Tosi[12]. Technically, the models are of the two-component non-additive hard sphere type. The liquid is regarded as a mixture of hard-sphere atoms (A) and point-like bond particles (B), in concentration ratio 1:2 and with mutual attractive interactions which can induce localization of bond particles between pairs of atoms under steric constraints limiting the coordination of an atom by bond particles to a maximum of 4 in tetrahedral configuration. The constraints are most simply imposed by fixing the B-B distance of closest approach from tetrahedral coordination geometry. The localizing interactions are alternatively chosen as Coulombic or represented by a narrow attractive well attached to the surface of each atom. In either case there is a simple coupling strength parameter, which is continuously varied through the liquid phase in equilibrium conditions and in supercooled states. It is given, respectively, by $\Gamma = z^2e^2/(ak_BT)$ or by $V^* = V/k_BT$, where V is the localized well depth.

The above models involve only A-A, A-B and B-B pair potentials and have been solved for liquid structure by the standard hypernetted-chain integral equations, tested by Monte Carlo simulation. A similar structural evolution with increasing coupling strength is found under the exclusive effect of Coulomb interactions or as a result of localized attractions, although all structural changes are appreciably smoother in the former model. In brief, increasing localization of bond particles in the immediate vicinity of atoms is found to occur with increasing coupling strength at constant density, as gauged from the steady decrease of the value of the $g_{AB}(r)$ distribution function at its main minimum towards zero and from the approach of the A-B coordination number to the value 4. At values of the coupling strength approaching those that are relevant for liquid Ge near freezing (i. e. for $\Gamma \cong 20$ or for $V^* \cong E_g/k_BT_m \cong 7$), the localization of the bond particles is sufficiently developed to split the first A-A coordination shell into two neighbouring shells. The shape of the $g_{AA}(r)$ distribution function thus starts to resemble the observed Ge-Ge g(r) and correspondingly the $S_{AA}(k)$ structure factor acquires the characteristic shape of a main peak with a shoulder on its large - k side. The A-A coordination number is approximately 7 at this point and slowly decreases on further increase of the coupling strength, but a volume expansion of the order of the observed $\Delta V/V$ on freezing in Ge is necessary for it to move towards the value 4. Clearly, the unusual sign of the volume change on freezing in Ge is related to the need to accomodate the sharp decrease in the Ge-Ge coordination number from 7 to 4, which is frustrated by the relatively high density maintained in the metallic liquid by the conduction electrons.

3.1. Freezing of the bond particle model

We show in Figure 1 the partial structure factors of the model at a coupling strength appropriate to liquid Ge near freezing. $S_{AA}(k)$ has the characteristic structure of a peak with a shoulder, as already noted. The strong peak in $S_{BB}(k)$ and the deep valley in $S_{AB}(k)$ reflect strong short-range order in the subsystem of bond particles and in their alternation with atoms in space. We stress that the location of these main structural features matches the position of the shoulder in $S_{AA}(k)$.

The above observations suggest the following interpretation for the structural results in Figure 1. A part of the calculated partial diffraction patterns, consisting primarily of the shoulder in $S_{AA}(k)$, the valley in $S_{AB}(k)$ and the peak in $S_{BB}(k)$, could in essence be associated with relatively long-lived units consisting of an atom surrounded by bond particles and by first-neighbour atoms. The remaining part, consisting primarily of the peak in $S_{AA}(k)$ after separation from the shoulder, would then represent the interference between the centres of such units. We shall for convenience refer below to this crude partition of the partial structure factors as $S_{\alpha\beta}^{(1)}$ and $S_{AA}^{(2)}$, respectively.

The above qualitative interpretation is confirmed by the behaviour of the model on further increase of the coupling strength at constant density. The shoulder in $S_{AA}(k)$ grows in intensity without shift in position, and indeed the whole of $S_{\alpha\beta}^{(1)}(k)$ just shows the usual sharpening and increase in peak intensity that are normally associated with a decrease in temperature. Instead, $S_{AA}^{(2)}(k)$ decreases in intensity and moves towards lower wave numbers, thus acquiring the nature of a FSDP. The position of such FSDP is very sensitive to the detailed shape of the second-neighbour A-A coordination shell, showing that indeed the FSDP directly reflects the

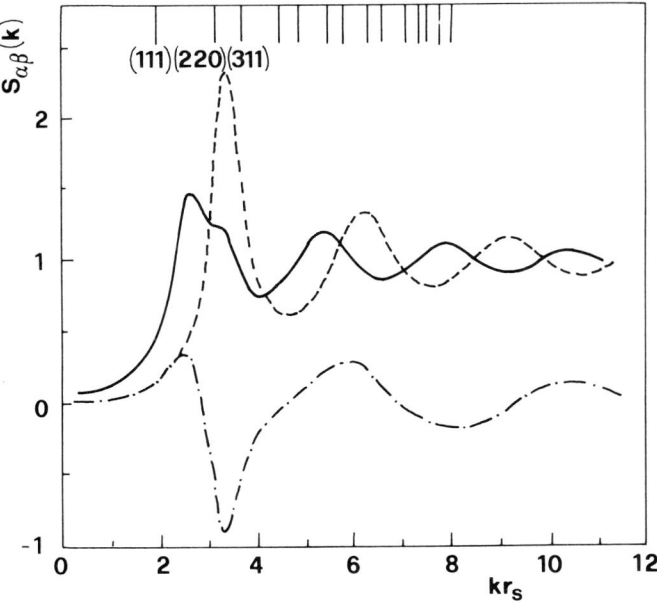

FIGURE 1
Partial structure factors for a bond-particle model of liquid Germanium near freezing. The various curves show the atom - atom (full line), atom - bond (dot - dashed) and bond - bond (dashed) structure factors. The vertical bars at the top mark the location of the allowed Bragg reflections from the diamond structure.

nature of the medium-range order in the disordered material. This behaviour of the model mimicks observations on amorphous Ge[22], as recalled earlier in this section. In this viewpoint, the low density of amorphous Ge allows a degree of medium-range order which is strongly reduced in the more dense liquid.

Let us now use the above insight to discuss equilibrium freezing of the bond particle model, and by implication of Ge, into the diamond structure[13]. The microscopic order parameters of the phase transition are the Fourier components $n_K^{(A)}$ and $n_K^{(B)}$ of the periodic single-particle densities for atoms and bond particles in the crystal, the **K**'s being the RLV of the fcc lattice. The location of the allowed Bragg reflections from the diamond structure are superposed on the top of Figure 1 on the partial liquid structure factors. There clearly is good correspondence between the (220) and (311) stars of RLV and the main features in $S_{\alpha\beta}^{(1)}(k)$. Thus, the Fourier components of the periodic crystalline densities at the above RLV stars describe "freezing of bonds" driven by tetrahedrally constrained attractions between ionic cores and valence electrons and leading to regular tetrahedra as components of the crystal. The connectivity that these tetrahedral units have in the diamond structure is to be described by the microscopic order parameters associated with the (111) star. This is inferred from the above interpretation of

$S_{AA}^{(2)}(k)$ and from the fact that the FSDP at a wave number in correspondence with the (111) star is well developed in amorphous Ge. However, formation of such connectivity in the liquid is frustrated by its relatively high density. Hence, crystallization of the liquid into the diamond structure requires volume expansion. On the contrary, crystallization of compacted amorphous Ge is accompanied by volume contraction.

The main order parameters of the equilibrium freezing transition, in addition to the percentual density difference $\eta = (n_s - n_l)/n_l$ between crystal and liquid, are thus expected to be (a) charge density waves (opposite signs for $n_K^{(A)}$ and $n_K^{(B)}$) at the (220) and (311) stars, and (b) number density waves (equal signs for $n_K^{(A)}$ and $n_K^{(B)}$) at the (111) star. This is immediately checked from the linearized form of the equilibrium conditions for a two-component system. These read[23]

$$\frac{n_K^{(A)}}{n_K^{(B)}} \approx \frac{\sqrt{2}\, c_{AB}(K)}{(n_1/n_s) - c_{AA}(K)} \approx \frac{(n_1/n_s) - c_{BB}(K)}{c_{AB}(K)/\sqrt{2}} \tag{13}$$

The values for the partial Ornstein-Zernike functions $c_{\alpha\beta}(K)$ are obtained from the partial structure factors in Figure 1 by means of the Pearson-Rushbrooke equations. In particular we note, as is evident from Figure 1, that at the (111) star c_{AA} and c_{BB} are strongly negative and c_{AB} is strongly positive, yielding a positive sign for the ratio in eqn (13). The shape of $S_{\alpha\beta}^{(1)}$ at the (220) and (311) stars, on the other hand, is qualitatively similar to that of ideal 2:1 molten salts, which primarily freeze by the onset of a charge density wave.

A numerical determination of the order parameters and of the coexistence point is currently in progress and is proving to be quite delicate[24]. Preliminary results have been obtained by representing the microscopic density profiles of atoms and bond particles in the diamond-type crystal as superpositions of Gaussians centred on the various lattice sites. Within the uncertainties of the values for the $c_{\alpha\beta}(K)$'s, and with an estimated isothermal compressibility $K_T \approx 2.5 \times 10^{-12}$ cm^2/dyn for molten Ge, the liquid structure shown in Figure 1 appears to be consistent with freezing of the atoms into a diamond-type crystal with (i) a volume expansion of the order of several percent (real Ge has $\eta = -5\%$), (ii) a Lindemann ratio of order 0.2 for the A particles, as expected from the Lindemann criterion for melting, and (iii) a Lindemann ratio of order $0.3 \div 0.4$ for the B particles.

4. CONCLUDING REMARKS

We have briefly reviewed recent progress in the statistical mechanical theory of freezing, focussing on models with Coulomb interactions which illustrate its two main aspects of usefulness. On the one hand, the theory yields for well defined simple models predictions which already have semiquantitative value at the simplest level of approximation and are well worthy of further effort and improvement. On the other hand, however, for complex systems the relationship between liquid pair structure and freezing can be usefully exploited even at a level of qualitative understanding. Needless to say, a number of very important questions remain open.

To mention just one of them, a reliable scheme for predicting the crystal structure that a given liquid takes on freezing should appeal to more information than just the liquid pair structure.

ACKNOWLEDGEMENT

I am indebted to Prof. G. Senatore for many useful discussions concerning in particular the freezing of jellium. I also wish to acknowledge continued sponsorship by the Ministero della Pubblica Istruzione and the Consiglio Nazionale delle Ricerche of Italy.

REFERENCES

1) J.G. Kirkwood and E. Monroe, J. Chem. Phys. 8 (1940) 845 and 9 (1941) 514.

2) C. De Dominicis, J. Math. Phys. 3 (1962) 983; F.H. Stillinger and F.P. Buff, J. Chem. Phys. 37 (1962) 1; J.L. Lebowitz and J.K. Percus, J. Math. Phys. 4 (1963) 116.

3) See e. g. R. Evans, Adv. Phys. 28 (1979) 143; S. Lundqvist and N.H. March, Theory of the Inhomogeneous Electron Gas (Plenum Press, New York, 1983).

4) T.V. Ramakrishnan and M. Yussouff, Solid State Commun. 21 (1977) 389 and Phys. Rev. B19 (1979) 2775.

5) A.D.J. Haymet and D.W. Oxtoby, J. Chem. Phys. 74 (1981) 2559.

6) A.D.J. Haymet, Phys. Rev. Lett. 52 (1984) 1013.

7) M. Rovere and M.P. Tosi, J. Phys. C18 (1985) 3445.

8) J.L. Barrat, Europhys. Lett. 3 (1987) 523; J.L. Barrat, J.P. Hansen and G. Pastore, Mol. Phys. 63 (1988) 747.

9) H. Iyetomi and S. Ichimaru, Phys. Rev. B38 (1988) 6761.

10) G. Pastore and G. Senatore, in the course of publication.

11) For a brief review with references to studies on other systems, see M.P. Tosi, M. Rovere and B. D'Aguanno, Z. Physik. Chem. 156 (1988) 411.

12) A. Ferrante and M.P. Tosi, J. Phys.: Cond. Matter 1 (1989) 1679.

13) Z. Badirkhan, A. Ferrante, M. Rovere and M.P. Tosi, N. Cimento, in press.

14) S.G. Brush, H.L. Sahlin and E. Teller, J. Chem. Phys. 45 (1966) 2101; E.L. Pollock and J.P. Hansen, Phys. Rev. A8 (1973) 3110; W.L. Slattery, G.D. Doolen and H.E. DeWitt, Phys. Rev. A21 (1980) 2087 and A26 (1982) 2255; S. Ogata and S. Ichimaru, Phys. Rev. A36 (1987) 5451.

15) See e. g. K.S. Singwi and M.P. Tosi, Solid State Phys. 36 (1981) 177.

16) D.M. Ceperley, Phys. Rev. B18 (1978) 3126; D.M. Ceperley and B.J. Alder, Phys. Rev. Lett. 45 (1980) 567.

17) J.C. Phillips, Covalent Bonding in Crystals and Molecules (University of Chicago Press, Chicago, 1970).

18) R.M. Martin, Phys. Rev. 186 (1969) 871; W. Weber, Phys. Rev. B15 (1977) 4789); U. Harten, J.P. Toennies, C. Wöll, L. Miglio, P. Ruggerone, L. Colombo and G. Benedek, Phys. Rev. B38 (1988) 3305, and references given therein.

19) L.R. Godefroy and P. Aigrain, Proc. Int. Conf. on Physics of Semiconductors (Institute of Physics, Exeter, 1962) p. 234.

20) Y. Waseda, Structure of Non-Crystalline Materials (McGraw-Hill, New York, 1980).

21) T.A. Postol, C.M. Falco, R.T. Kampwirth, I.K. Schuller and Y.B. Yelon, Phys. Rev. Lett. 45 (1980) 648.

22) G. Etherington, A.C. Wright, J.T. Wenzel, J.C. Dore, J.H. Clarke and R.N. Sinclair, J. Non-cryst. Solids 48 (1982) 265.

23) N.H. March and M.P. Tosi, Phys. Chem. Liquids 11 (1981) 79 and 89.

24) Z. Badirkhan, M. Rovere and M.P. Tosi, to be published.

DENSITY FUNCTIONAL THEORY OF QUANTUM WIGNER CRYSTALLIZATION

Giorgio PASTORE[†] and Gaetano SENATORE[‡]

[†]International School for Advanced Studies, Trieste, Italy.
[‡]Dipartimento di Fisica "A. Volta", Università di Pavia, Pavia, Italy.

By working within the framework of density functional, the so-called density wave theory (DWT) of freezing is extended to treat the quantum Wigner crystallization. A basic ingredient of the theory is the static local field factor $G(q)$ of the electron liquid, which is extracted from the Quantum Monte Carlo (QMC) static structure by an approximate decoupling scheme. Predictions for the freezing of the electron liquid into a regular lattice are examined.

1. INTRODUCTION

The modern theory of freezing for classical systems, which is due to Ramakrishnan and Yussouff[1,2], constructs the appropriate free energy difference needed to study phase coexistence by taking advantage of the knowledge of properties of the liquid phase. The excess Helmholtz free energy in the solid, in fact, is expanded around that in the liquid in powers of the density difference between the two phases, thus calling into play the static structure of the liquid. Applications of this theory to a variety of systems have been presented to date, ranging all the way from hard spheres[3] to a model description of water[4]. In most cases they have enjoyed a satisfactory success, one notable exception[5] however being the one-component classical plasma (OCP).

The basic idea on which the approach of Ramakrishnan and Yussouff is based is clearly applicable to quantum systems as well. Of course there are differences between the classical and the quantum case. These arise[6] in: (i) the manner in which the relevant functional can be constructed starting from the known properties of the liquid phase; (ii) the need of the static response function, which for quantum liquids does not reduce to the static structure factor. Here we briefly consider the application of a quantum version of DWT to the freezing of jellium at $T = 0$.

2. DWT FOR JELLIUM

Jellium is a model system of electrons on a uniform neutralizing background whose density n is usually specified by the Wigner-Seitz to Bohr radius ratio r_s, $r_s = (3/4\pi n)^{1/3}/a_B$. With decreasing density the system of electrons undergoes a transition from the liquid to a crystalline phase, as it was pointed out first by Wigner[7]. This is due to the increasing importance of Coulomb interactions with increasing r_s. Many different estimates of the critical r_s for freezing appeared in the years[8], spanning about three orders of magnitude.

Recently QMC simulations of Ceperley and Alder[9] located the crystallization into a BCC lattice at $r_s = 100 \pm 20$, from a fully spin polarized liquid.

At zero temperature the relevant free energy characterizing a system of many particles is just the ground state energy. This, by virtue of the Hohenberg-Kohn theorem[10], is a functional of the density $n(\mathbf{r})$ and it is minimum in correspondence to the equilibrium density distribution. For jellium this functional can be written as[11]

$$E[n(\mathbf{r})] = T_0[n(\mathbf{r})] + \frac{e^2}{2} \iint d\mathbf{r} d\mathbf{r}' \frac{n_Q(\mathbf{r}) n_Q(\mathbf{r}')}{|\mathbf{r} - \mathbf{r}'|} + E_{xc}[n(\mathbf{r})], \qquad (1)$$

with $n_Q(\mathbf{r}) = n(\mathbf{r}) - n$ the charge density, T_0 the independent particle kinetic energy, and E_{xc} the so-called exchange and correlation energy. The exact form of this last functional is not know at present, therefore practical applications of DFT require approximations. Specializing to freezing, one can try to approximate directly the energy difference $\Delta E_{xc} = E_{xc}[n(\mathbf{r})] - E_{xc}(n)$ between solid and liquid phase, rather than the separate energies. If a functional expansion around the homogenous phase is meaningful, one can study the phase transition from the sole knowledge of fluid properties.

To investigate the stability of a solid phase with respect to the liquid one has to calculate the difference $\Delta E = E_s - E_l$ of ground state energies between the two phases. Truncating the expansion of ΔE_{xc} to second order terms yields

$$\Delta E = T_0[n(\mathbf{r})] - \frac{3}{5} N \epsilon_F + \frac{1}{2} \iint d\mathbf{r} d\mathbf{r}' \, n_Q(\mathbf{r}) n_Q(\mathbf{r}') \left[\frac{e^2}{|\mathbf{r} - \mathbf{r}'|} - K_{xc}(\mathbf{r} - \mathbf{r}') \right]. \qquad (2)$$

Here, ϵ_F is the Fermi energy of the liquid phase and $-K_{xc}$ denotes the second functional derivative of E_{xc} with respect to $n(\mathbf{r})$ evaluated in the liquid. In the crystalline phase the density is represented by the Fourier series

$$n(\mathbf{r}) = n + \sum_{\mathbf{K} \neq 0} n_{\mathbf{K}} \, e^{i\mathbf{K} \cdot \mathbf{r}}, \qquad (3)$$

where the **K**'s are reciprocal lattice vectors (RLV). Thus, for a given crystalline structure, the energy difference of Eq. (2) can be written as

$$\Delta E = T_0[n(\mathbf{r})] - \frac{3}{5} N \epsilon_F + \frac{V}{2} \sum_{\mathbf{K} \neq 0} v(K) [1 - G(K)] |n_{\mathbf{K}}|^2, \qquad (4)$$

with V the total volume and $G(K)$ the static local field factor in the dielectric function of the homogeneous electron gas[12]. Above $v(K)$ denotes the Fourier transform of the Coulomb coupling, i.e., $v(K) = 4\pi e^2/K^2$.

The solid-liquid coexistence point is characterized by the vanishing of ΔE, which locates the critical r_s for crystallization. At larger r_s one should find $\Delta E < 0$, since the crystal should be stable, while $\Delta E > 0$ at smaller r_s, the liquid being stable. Since the kinetic energy difference in Eq. (4) is always positive, it is evident that in the present approach Wigner crystallization is possible only if the local field factor exceeds 1 at some RLV and the contribution from such vectors is dominant in the potential energy sum appearing in ΔE. In other words, the Kohn-Sham effective potential[11]

$$v_{eff}(\mathbf{K}) = n_{\mathbf{K}} v(K)[1 - G(K)], \tag{5}$$

felt by the electrons must balance the kinetic energy cost of modulation by density waves if the electrons are to form a solid. Crystallization is thus arising from the softening of the electron fluid against modulation by density waves with wave numbers equal to RLV. Clearly, the local field factor $G(q)$ enters the quantum DWT of freezing in a crucial way.

3. LOCAL FIELD FACTOR AND FREEZING OF JELLIUM

Practical calculations require the knowledge of the local field factor $G(q)$, which however is not exactly known. A possible way to proceed is to use the STLS decoupling scheme[12],

$$G(q) = -\frac{1}{n} \int \frac{d\mathbf{k}}{(2\pi)^3} \frac{\mathbf{k} \cdot \mathbf{q}}{q^2} \frac{v(k)}{v(q)} [S(\mathbf{q} - \mathbf{k}) - 1], \tag{6}$$

expressing $G(q)$ in terms of the static structure factor $S(q)$, which is available from QMC. The local field factor that one obtains in this manner for the spin polarized electron fluid has the following features[6]: (i) with increasing r_s it develops a peak for wavevectors slightly less than $2q_F$; (ii) it exceeds 1 in this region of wavevectors—where the first star of RLV of both BCC and FCC structures also occurs. Clearly, such features are consistent with the possibility of predicting freezing within the present approach.

The calculation of the energy difference ΔE requires the knowledge of the equilibrium density $n(\mathbf{r})$, which can be determined by taking advantage of the minimum principle obeyed by the ground state energy functional[10]. This leads[11] to a problem of independent particles in a self-consistent effective potential, whose Fourier components are given in Eq. (5). Due to the assumed periodicity of the density (see, Eq. (3)), the effective potential is also periodic. Thus, one has to perform a self-consistent band structure calculation. In particular, restricting to the freezing of the fully spin polarized liquid into a solid with one electron per unitary cell, one has a single fully occupied energy band.

The Bloch orbitals that naturally arise in this calculation have been treated in two alternative ways[6], being either (a) systematically expanded in plane waves or (b) parametrized by constructing Bloch sums from a single gaussian orbital per site. Since the electronic density is expected to be well localized at cristallization, in case (a) many plane are needed, whereas in case (b) there is only one variational parameter, i.e., the width of the gaussian orbital . On the other hand, the plave-wave expansion should ensure more flexibility compared with the simpler *gaussian ansatz*.

With both treatments of Bloch orbitals one obtains reasonable estimates of the freezing. One finds[6] a critical r_s for the Wigner crystallization of 128 and 107, in case (a) and (b) respectively. These numbers compare favourably with the value $r_s = 100 \pm 20$ yielded by QMC[9]. Also, good electronic localization is found[6] at freezing, in accord with QMC[9]. An investigation on the relative stability of the BCC and FCC structure reveals their substantial equivalence around freezing, with the BCC structure becoming more stable at larger r_s.

4. CONCLUSIONS

The satisfactory results for the freezing of jellium that we have summarized above suggest some confidence in the quantum DWT. Evidently the present results rely on (i) the second order expansion of E_{xc} and (ii) the use of STLS decoupling to construct the static response function of the electron liquid from $S(q)$. The importance of these two approximations deserves some farther investigation. One should also contrast the present results with those for the classical analog of jellium, the OCP, for which the second order density wave theory is not able to predict freezing[5]. At present the precise reasons for such a difference between the quantum and the classical one-component plasma remain unclear.

ACKNOWLEDGMENTS

This work has been supported in part by the Ministero della Pubblica Istruzione and by the Consiglio Nazionale delle Ricerche of Italy.

REFERENCES

1) T.V. Ramakrishnan and M. Yussouf, Phys. Rev. 19B, 2775 (1979).

2) A.D.J. Haymet and D.W. Oxtoby, J. Chem. Phys. 74, 2559 (1981).

3) A.D.J.Haymet, J. Chem. Phys. **78**, 4641 (1983).

4) K. Ding, D. Chandler, S.J. Smithline and A.D.J. Haymet, Phys. Rev. Lett. 59, 1698 (1987).

5) A.D.J. Haymet, Phys. Rev. Lett. 52, 1013 (1984); M. Rovere and M.P. Tosi, J. Phys. 18C, 3445 (1985); J.L. Barrat, Europhys. Lett. 3, 523, (1987); J.L. Barrat, G. Pastore and J.P. Hansen, Mol. Phys. 63, 747; H. Iyetomi and S. Ichimaru, Phys. Rev 38B, 6761 (1988).

6) G. Senatore and G. Pastore, in the course of publication.

7) E.P. Wigner, Phys. Rev. 46, 1002 (1934) and Trans. Faraday Soc. 34, 678 (1938).

8) C. Care and N.H. March, Adv. Phys. 24, 101 (1975).

9) D.M. Ceperley and B.J. Alder, Phys. Rev. Lett. 45, 567 (1980); see, also, D. Ceperley, Phys. 18B, 3126 (1978).

10) P. Hohenberg and W. Kohn, Phys. Rev.136B, 864, (1964).

11) W. Kohn and L.J. Sham, Phys. Rev. 140, 1133 (1965).

12) See, for instance, K.S. Singwi and M.P. Tosi, *Solid State Physics*, edited by H. Ehrenreich, F. Seitz and D. Turnbull (Academic, New York, 1981),

MOLECULAR DYNAMICS STUDIES OF GLASSY STATES: SUPERCOOLED LIQUIDS AND AMORPHIZED SOLIDS

Sidney YIP

Department of Nuclear Engineering
Massachusetts Institute of Technology, Cambridge, MA 02139 (USA)

1. INTRODUCTION

In recent attempts to understand the formation of glassy states at a conceptual level, quite different questions arise depending upon whether the initial state of the system is a liquid or a crystal. In the case of rapid cooling or compression of a liquid (vitrification), it is natural to focus attention on the changes in dynamical behavior since the liquid and the glass have qualitatively similar structure as measured by the radial distribution function. In the case of crystalline transition to a glass (amorphization) by irradiation or other means, the relevant question is the mechanism by which a lattice becomes structurally disordered. Each problem is of current interest, but because different measurements and communities of investigators are involved in these studies, little is known about the underlying connections between the two processes and how to exploit their commonalities.[1]

Molecular dynamics is a technique ideally suited for studying both the structural and dynamic aspects of a strongly coupled system of interacting particles. There have been a number of simulation studies of supercooled liquids with the goal of elucidating the liquid to glass transition.[2] Simulations of amorphization, on the other hand, have begun relatively recently and only a few results have been reported thus far.[3-6]

In this brief review we discuss selected results on both problems to call attention to some of the insights that have emerged from several recent simulation studies. We find that molecular dynamics data provide evidence for a dynamical transition in metastable fluids which is distinct from the glass transition observed in the laboratory. A second result is that in the case of amorphization induced by the introduction of point defects, simulation is able to show that the transition threshold is dependent on both the rate of introduction and the defect concentration. It is also our intention here to suggest that vitrification and amorphization may have connections which are relevant to the fundamental understanding of transitions between a liquid and a crystal, namely, melting and crystallization.

2. A DYNAMICAL TRANSITION IN METASTABLE FLUIDS

Although it may seem more natural to picture the liquid to glass transition as a phenomenon induced by quenching,[7] it is generally acknowledged that the transition also can be induced by pressure.[8] From the standpoint of interpretation it is particularly useful to have results in which the effects of density change are separated from the effects of thermal motion. We consider here a simulation study in which an atomic fluid is systematically compressed isothermally beyond the normal freezing point.[9,10] For this study standard molecular dynamics technique is employed using the (N,V,E) ensemble. The system is a periodic cubic cell containing 500 particles interacting through the repulsive part of the Lennard-Jones potential. Compression is achieved by changing the volume of the cell, and temperature is controlled by velocity rescaling. The system is considered to be in thermal equilibrium if the temperature does not drift when velocity rescaling is turned off.

Typically, a run extends over 10^4 time steps for equilibration, followed by 2×10^4 steps for generating the trajectories for property calculations. All the simulation results will be quoted as dimensionless quantities, energy in units of ϵ, length in σ, and time in $\tau = (m\sigma^2/\epsilon)^{1/2}$, where ϵ and σ are the usual parameters of the Lennard-Jones potential. The time step size is $\Delta t = 0.005$. We also define dimensionless density $n* = n\sigma^3$, temperature $T* = k_B T/\epsilon$, and pressure $p* = p\sigma^3/\epsilon$.

Fig. 1 shows the variation of pressure with density at $T* = 0.6$.[9,10] It can be seen that p* is essentially piece-wise linear in n*, and there is an increase in the slope around n* = 1.02 which means a somewhat abrupt

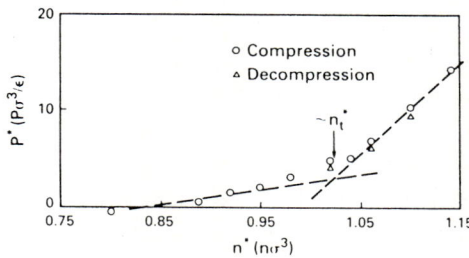

Fig. 1 Pressure-density variation along an isotherm $T* = 0.6$. Circles and triangles denote compression and decompression data respectively.

decrease in the compressibility. In the quenching simulation one would see a corresponding behavior in the volume variation with temperature indicating a decrease in the thermal expansion. This is conventionally taken to signify the glass transition.[7] We will refrain from such an interpretation and refer to the characteristic density where the compressibility changes as n^*_x. As we will see below the subscript is chosen to denote a crossover transition of dynamical origin.

If the equation-of-state data indeed reveal a transition of some kind, one expects there should be corresponding manifestations in the other basic properties of the fluid. We will look for structural changes in the radial distribution function g(r), and for dynamical responses in the mean square displacement $\langle \Delta r^2 \rangle$ and the density correlation function F(k,t).

An examination of the g(r) results at densities ranging from n* = 0.884 to 1.24 shows only a gradual change in the region of the second peak. Around n* = 1.02 a flattening at the top of this peak is discernible, and at higher densities a splitting occurs which is the characteristic signature of random close packing. Aside from this rather subtle feature there is little variation in g(r) over a considerable range of fluid densities.

The detailed behavior of $\langle \Delta r^2 \rangle$ over the time interval of simulation, up to 48τ, also is not unusual.[10] One finds that the portion of the data following an essentially linear variation with time shows a decreasing slope as the density is increased, thus the effective diffusion coefficient is reduced with increasing density as one would expect. At about n* = 1.02 there appears to be some delay in the onset of diffusive behavior, also there is more scatter in the data at this density and beyond. This suggests that a change may be taking place in the physical mechanism by which atoms can move over an appreciable fraction of the interatomic separation. Also it means that the determination of the diffusion coefficient from such data is less certain than at lower densities. At n* = 1.24 the particles are well localized as can be seen from the magnitude of $\langle \Delta r^2 \rangle$. Although an increase in $\langle \Delta r^2 \rangle$ with time is still observable in the data, the diffusion coefficient extracted at this density is at best a qualitative estimate.

The effective diffusion coefficients[10] are shown in Fig. 2. One sees clearly a change in the density variation, from a linear behavior up to about n* = 0.98 to a slower decrease at higher densities. This is quite characteristic of how diffusivity and viscosity behave in supercooled liquids. For example, it has been shown for a number of glass-forming liquids that the experimental value of shear viscosity η follows an inverse square-law temperature variation down to a temperature where η has increased by about two orders of magnitude over its value at normal liquid temperature, and at lower temperatures η increases much more rapidly.[11]

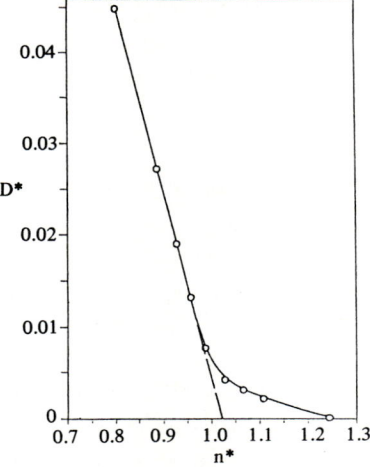

Fig. 2 Variation of effective diffusion coefficient $D* = D\tau/\sigma^2$ with density at $T* = 0.6$.

Moreover, if one extrapolates the low-η data to infinite viscosity, one obtains a temperature T_o which is considerably higher than the glass transition temperature T_g, defined to be that temperature at which η has the value of 10^{13} poise (some 14 to 15 orders of magnitude greater than its liquid value). If we now extrapolate linearly the data in Fig. 2 in the low-compression region to zero diffusivity, we find a value of $n* = 1.01$, which is about the same density where the compressibility change is observed in Fig. 1.[9] This would suggest that the density $n*_x$ should be identified with the characteristic temperature T_o. Notice that the actual value of D at $n* = 1.02$ is still quite appreciable, just as the viscosity at T_o is only two to three orders of magnitude greater than the liquid value.

The density correlation function $F(k,t)$ provides dynamical information about collective behavior in the system. The results shown in Fig. 3 are obtained by averaging over different $\underline{k}$ vectors all having a magnitude within about 10% of the value 2 Å^{-1}, the position of the diffraction maximum in the static structure factor $S(k)$. As the fluid is more compressed, the decay of the density fluctuations is seen to extend to longer

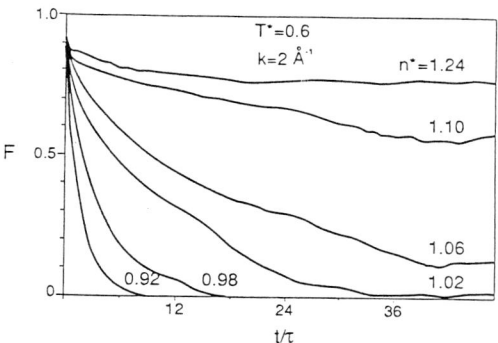

Fig. 3 Temporal decay of density fluctuations at wave number $k = 2$ Å^{-1}.

times. Around $n* = 1.02$ there is the appearance of a slowly decaying component, possibly a behavior signifying the onset of structural arrest.[9] At high compressions such that $n* > 1.10$, it is evident that $F(k,t)$ can be decomposed into two components, a rapid decay followed by a much slower relaxation. Although these data have not been analyzed to see if they are consistent with a stretched exponential decay at long times, similar results from a simulation of quenched binary fluids[12] are found to show this behavior. Recently nonexponential decay has been derived from mode-coupling theory[13] and observed in neutron scattering measurements.[14]

Since the density correlation function is the central quantity in the formulation of self-consistent mode-coupling theory, a direct confrontation of calculation and simulation results on $F(k,t)$ is clearly desirable. At present only the numerical calculations based on the original formulation which considers only coupling to the density fluctuations are available for comparison.[15,10] One finds a number of similar features in the decay of $F(k,t)$, specifically the separation of the fast and slowly decaying component is seen to occur at about the same time and the same value of the correlation function. The main discrepancy revealed in this comparison lies in the high-compression regime. The theory describes the onset of structural arrest at a value of $n*$ that is too low by comparison to the simulation. Thus the behavior seen in Fig. 3 at $n* = 1.24$ is obtained theoretically at $n* = 1.10$. We interpret this discrepancy to arise at least partly from the neglect in the theory of the coupling to current modes.[16] It is known that when this coupling is taken into account the ergodic to nonergodic transition is cutoff and correlation functions such as $F(k,t)$ take on a more rapid decay[17,18]; however, it still remains to be determined whether mode-coupling theory can be brought into quantitative agreement with the simulation results.

Given the foregoing considerations we conclude that the property changes we have observed around the density $n*_x$ reflect a transition of dynamical origin.[10] Our interpretation is that these results signal a change in the fundamental character of the atomic motions, from continuous collisions among particles which still retain a certain degree of mobility, to hopping motions by particles which find themselves increasingly trapped in positions of local potential minima. It is also clear from the simulation results that this transition is distinct from the glass transition where the fluid is structurally arrested over times very much longer than the simulation and in which state the diffusivity is many orders of magnitude lower than those observed here.

While we have examined the simulation results only on fluids under compression, it is expected that for supercooled liquids a corresponding transition would exist at a characteristic temperature T_x which is distinctly above the glass transition temperature T_g.[19] It is perhaps not surprising that among the many studies of viscous liquids and the glass transition one finds arguments and discussions pointing to the plausibility of such a transition. For example, from the point of view that at low temperatures viscous flow is dominated by potential barriers up to the point where significant fluidity sets in, a crossover transition would seem to be a natural consequence.[20] Besides the viscosity data mentioned above several recent findings appear to support this general picture. From a molecular dynamics study of local stress fluctuations in quenched liquids, it is found that below a characteristic temperature T_1 which is also well above T_g, the shear stresses become spatially correlated.[21] In another simulation of quenched liquids, the diffusion coefficient was found to extrapolate to zero at a temperature which is interpreted to signify the onset of effective spinodal nucleation.[22] From an analysis of spin-glass models, it is shown that in such systems one can identify two transition temperatures T_A and T_K, with $T_A > T_K$.[23] At the former one has a dynamical transition which involves the onset of barriers in the local energy surface, and at the latter there is an equilibrium transition associated with the vanishing of the configurational entropy.

3. DEFECT-INDUCED AMORPHIZATION

In recent years it has been found that amorphous solids can be produced by a variety of processes other than the rapid solidification of a melt. In particular, atomic disordering of crystalline solids by irradiation has become a prominent process having considerable fundamental as well as technological interests.[24] As an attempt to model radiation-induced amorphization we consider the problem of inserting point defects into an atomic lattice and studying the structural and elastic responses at varying insertion rate and defect concentration.[2,3] The objective is to understand better the nature of the disordering process and the factors that determine the threshold condition for amorphization.

The crystal under study is an fcc lattice of Lennard-Jones particles into which self-interstitials are randomly and sequentially inserted at constant temperature and pressure. Prior to any defect insertion the system contains 576 particles with the z axis oriented along the [111] direction. Each self-interstititial is randomly introduced at the center of a cell formed by three nearest neighbors of a particle. A time-step size of $\Delta t = 0.01$ is used in the simulation; however, following every insertion the time step is decreased by a factor of 20 to allow for dissipation of local heating, and then over the next 20 steps Δt is increased back to normal.

It is useful to regard each simulation as consisting of two stages, an irradiation stage where a total of N_i interstitials are inserted at a constant rate r, followed by an annealing or relaxation stage where the system evolves without further perturbation. Throughout the simulation the system is subjected to zero external stress, and its temperature is maintained within certain limits about a mean value of $T* = 0.2$ by velocity rescaling. This temperature is about 30% of the bulk melting point.

We will consider the data on three systems, which will be called A, B, and C, each corresponding to a different combination of (N_i, r). The values are respectively (160, 5.95), (80, 14.71), and (160, 14.71). Thus, the defect insertion rate for system A is slower than that for B and C by a factor of about 2, while the defect concentration in B is lower by a factor of 2 relative to that for A and C. As we will see below, these differences are sufficient to demonstrate the existence of a threshold condition for amorphization.

Fig. 4 shows the temporal behavior of the system density during the irradiation and relaxation stages. One sees an initial decrease which indicates that the lattice overexpands when disorder is introduced; however, the effect saturates after about fifty insertions. Once the insertion stops, the system shows a tendency to densify which can be interpreted

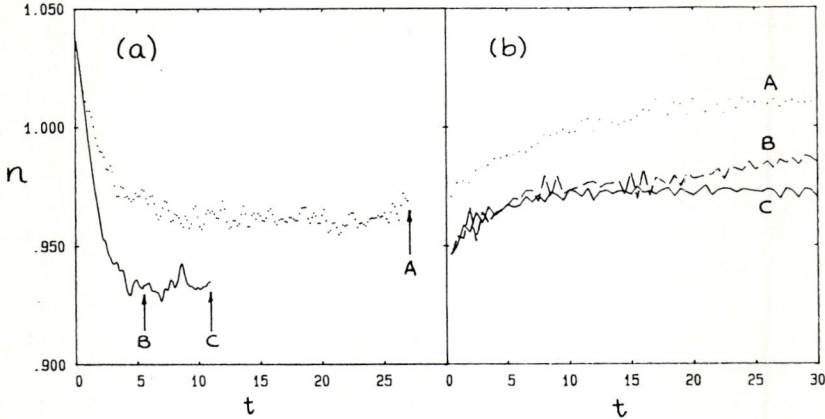

Fig. 4 System density during interstitial insertion (a) and relaxation (b). Arrows indicate times at which insertions are stopped.

as a tendency to regain crystalline order. It is seen that the extent of the density recovery depends on the extent of the perturbation, both in rate and magnitude. The potential energy responses (not shown) are consistent with the pattern of the density behavior.

To follow the structural evolution we show in Fig. 5 the radial distribution functions at the end of the two stages. It is quite clear that at the end of the respective insertions, systems A and B still retain the structure of an fcc lattice, although there is considerable broadening of the near-neighbor peaks, and during relaxation the distributions evolve toward a more crystalline structure. Particularly significant during the relaxation process is the growth of the second-neighbor peak at r = 1.56. By contrast, system C relaxes to a random close-packed structure (identifiable by the absence of the second-neighbor peak of the fcc lattice and by the split in the second-neighbor peak in a liquid at the same density), although some remnants of the fcc structure are still discernible at the end of the insertion stage.

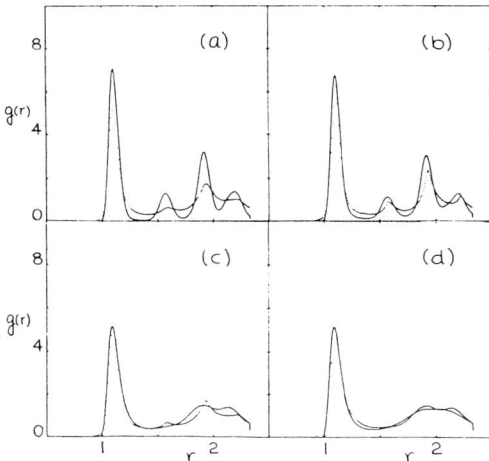

Fig. 5 Radial distribution function after insertion (dotted curve) and after relaxation (solid curve) for system A (a), B (b), and C (c). (d) Comparison of system C with a liquid at $T* = 0.75$ (dotted curve).

To see if the structure of system C is robust we have heated the system up to $T* = 0.75$ ($T* = 0.68$ is the melting point of a Lennard-Jones solid) at constant volume to obtain a well equilibrated liquid for which the $g(r)$ is also shown in Fig. 5. The liquid is then quenched to $T* = 0.2$ at constant volume giving a resulting $g(r)$ which is virtually indistinguishable from that of system C.

On the basis of these results one can conclude that the threshold for defect-induced amorphization is both insertion rate and defect concentration dependent. Given that systems A and B are defective but still crystalline structures while C is amorphous, we can compare their elastic responses by calculating the elastic constants for these structures.[4,25] The results, given in Table 1, show a pronounced softening in the shear modulus C_{44} and a reduction of about 20% in the bulk modulus B_t in going from the single crystal to the amorphous structure. This behavior can be traced to the contribution due to fluctuations in the stress tensor as opposed to static or structural effects described by the so-called Born term. Notice also in Table 1 that $C_{11} - C_{12}$ is approximately equal to C_{44} in the case of system C, an indication of system isotropy which is not seen in systems A and B.

Table 1. Elastic constants averaged over the three orthogonal directions of the simulation cell and bulk modulus. All values are at $T^* = 0.2$.

System	C_{11}	C_{12}	C_{44}	B_t
perfect lattice	533	146	136	275
A	469	175	111	273
B	424	180	105	261
C	254	199	39	217
quenched	264	219	39	243

We have analyzed the simulation results to better understand the failure of systems A and B to amorphize. It was found that a strong tendency exists for the interstitial atoms to aggregate and form pieces of (111) planes.[3,26] To overcome the kinetics of this athermal migration requires a sufficiently fast insertion rate, and on the basis of the results for system B a sufficient concentration of defects is also needed.

4. DISCUSSION

In this brief review we have examined molecular dynamics results on the diffusion and density fluctuation properties of a fluid compressed beyond its freezing point, and on the structural and elastic responses of a crystal disordered by the introduction of interstitial particles. While the two studies are of interest for different reasons, they both illustrate a significant advantage of the simulation approach, namely, access to a variety of properties systematically determined under well-controlled conditions.

Our interpretation that the transition around n^*_x signifies a crossover in the character of the atomic dynamics, or equivalently the onset of potential barriers in which the particles can be trapped at significantly longer periods, should be further verified.[27] The density or temperature region of interest is one where a test of the mode-coupling theory, with inclusion of current fluctuations, is meaningful.[28] Also it is a region where both simulation and neutron scattering measurements are feasible.

The amorphization problem is at present mostly concerned with the nature of the driving force which destabilizes the lattice. To make contact with the considerable body of experimental data on intermetallic compounds, it is necessary to extend the study to binary lattices and take into consideration the role of chemical disorder. Two studies have been reported recently. In one simulation which employs an n-body potential fitted to the properties of a specific binary alloy, it is shown that amorphization is induced by the strains resulting from the chemical disorder.[5] In another simulation using Lennard-Jones potentials with only size difference, a new driving force for structural disordering is found in the form of residual local stresses.[6] It is fair to say that there are many fundamental issues outstanding in this area[24] which can be usefully addressed by atomistic simulations.

Further simulation studies exploring different ways of producing glass states would be of interest for still another reason. Recently it has been observed that thermodynamic parallels between melting and solid-state amorphization can be established which further suggest a fundamental connection between these two phenomena.[29] It would seem that a systematic, unified investigation of the structural, kinetic, and elastic properties of the various states involved in melting, crystallization, and vitrification/amorphization, which is possible through simulation, could yield physical insights of considerable value.

ACKNOWLEDGMENT

I would like to acknowledge the collaboration of J. J. Ullo in the study of compressed fluids and that of H. Hsieh, A. Rahman (deceased), and Y. Limoge in the study of defect-induced amorphization. Recent discussions with D. Wolf and P. Okamoto on melting and amorphization have been most stimulating. This work was supported by the U. S. National Science Foundation through grants CHE-8415078 and 8806767.

REFERENCES

1) See the overview lectures in Annals N.Y. Acad. Sci., C. A. Angell and M. Goldstein, eds., vol. 484 (1986).

2) See C. A. Angell, J. H. R. Clark, and L. V. Woodcock, Adv. Chem. Phys. $\underline{48}$, 397 (1981), and J. R. Fox and H. C. Andersen, J. Phys. Chem. $\underline{88}$, 4019 (1984), and references given therein; for more recent discussions see J. N. Roux, J. L. Barrat, and J. P. Hansen, Phys. Rev. A, to be published, and Ref.10, and the citations therein.

3) Y. Limoge, A. Rahman, H. Hsieh, and S. Yip, J. Non-Cryst. Solids 99, 75 (1988).

4) H. Hsieh and S. Yip, Phys. Rev. Lett. 59, 2760 (1987).

5) C. Massobrio, V. Pontikis, and G. Martin, Phys. Rev. Lett. 62, 1142 (1989).

6) H. Hsieh and S. Yip, Phys. Rev. B 39, 7476 (1989).

7) A recent discussion of quenching is given by K. Shinjo, J. Chem. Phys. 90, 6627 (1989)

8) See F. F. Abraham, J. Chem. Phys. 72, 359 (1980) for a discussion of simulation of quenching and compression.

9) J. J. Ullo and S. Yip, Phys. Rev. Lett. 54, 1509 (1985).

10) J. Ullo and S. Yip, Phys. Rev. A 39, 5877 (1989).

11) R. Taborek, R. N. Kleiman, and D. J. Bishop, Phys. Rev. B 34, 1835 (1986).

12) G. Pastore, B. Bernu, J. P. Hansen, and Y. Hiwatari, Phys. Rev. A 38, 454 (1988).

13) W. Götze and L. Sjogren, J. Phys. C 20, 879 (1987).

14) F. Mezei, W. Knaak, and B. Farago, Phys. Rev. Lett. 58, 571 (1987).

15) U. Bengtzelius, Phys. Rev. A 34, 5059 (1986).

16) S. Yip, J. Stat. Phys. in press.

17) S. P. Das and G. M. Mazenko, Phys. Rev. A 34, 2265 (1986); S. P. Das, Phys. Rev. A 36, 211 (1987).

18) W. Götze and L. Sjogren, Z. Phys. B 65, 415 (1987).

19) C. A. Angell, J. Phys. Chem. Solids 49, 863 (1988).

20) M. Goldstein, J. Chem. Phys. 51, 3728 (1969).

21) S.-P. Chen, T. Egami and V. Vitek, Phys. Rev. B 37, 2440 (1988).

22) J.-x. Yang, H. Gould, and W. Klein, Phys. Rev. Lett. 60, 2665 (1988).

23) T. R. Kirkpatrick and D. Thirumalai, Phys. Rev. 37, 5342 (1988); T. R. Kirkpatrick, D. Thirumalai, and P. J. Wolynes, Phys. Rev. 40, 1045 (1989).

24) R. B. Schwarz and W. L. Johnson, J. Less-Common Met. 140, 1 (1988), and other contributions in this issue; D. E. Luzzi and M. Meshii, Res. Mechanica 21, 207 (1987); D. M. Parkin and R. O. Elliott, J. Mater. Res. 3, 453 (1988).

25) See also J. L. Barrat, J. N. Roux, J.-P. Hansen, and M. L. Klein, Europhys. Lett. 7, 707 (1988).

26) S. Yip and H. Hsieh, in Science of Advanced Materials, M. Meshii and H. Wiedersich, eds. (ASM, Metals Park, OH), in press.

27) Preliminary results of binary fluids of Lennard-Jones particles show the same behavior as those presented here, J. J. Ullo and S. Yip, work in progress.

28) S. P. Das, work in progress.

29) D. Wolf, P. R. Okamoto, S. Yip, J. F. Lutsko and M. Kluge, J. Mater. Res., submitted.

STOCHASTIC DYNAMICS OF ATOMS NEAR A GLASS TRANSITION POINT

Yasuaki HIWATARI

Department of Physics, Kanazawa University, Kanazawa 920, Japan

Takashi ODAGAKI

Department of Physics, Kyoto Institute of Technology, Kyoto 606, Japan

Dynamical behaviors of atoms in a glass-forming system is studied by making use of a stochastic trapping model on a lattice. The frequency-dependent diffusion constant is determined within the coherent medium approximation, from which the dynamical behavior of various physical quantities in an intermediate time window is obtained.

Dynamical behaviors of atoms in glass-forming systems near the glass transition point have extensively been studied in recent years.[1] In particular, Miyagawa et al. reported recently that atoms perform occasionally stochastic jump motions and change their average positions.[2] According to their study, the characteristic frequency of the vibration is of the order of 10^{11} sec^{-1} and the jump rate is of the order of 10^{10} sec^{-1}. The jump rate from a given location is determined by the dynamical state of surrounding atoms and thus will be widely distributed. The long time behavior of dynamical quantities is determined by the stochastic motion of atoms.

We have developed a trapping model on a lattice for the glass transition which focuses on the stochastic motion of atoms.[3] In this model, the random structure is represented by a power law distribution of jump rate $w_\mathbf{s}$ from a given site s to its nearest neighbors

$$P(w_\mathbf{s}) = \begin{cases} (\rho+1)w_\mathbf{s}^\rho/w_0^{\rho+1} & 0 \leq w_\mathbf{s} \leq w_0 \\ 0 & \text{otherwise} \end{cases}. \qquad (1)$$

When $\rho > 0$ ($\rho < 0$), larger jump rates appear more (less) frequently than smaller ones, and hence ρ can be considered as a phenomenological parameter representing thermodynamic state of the system. We showed that when $\rho < 0$ the static diffusion constant vanishes and the mean square displacement increases sublinearly in time and when $\rho > 0$ the static diffusion exists and the mean square displacement increases linearly in time in the asymptotic region.[3] These results are somewhat different from observations by molecular dynamics studies in that (i) the static diffusion constant shows a residual diffusion, that is $D(0)$ does not vanish completely below the glass transition point. (ii) The sublinear dependence of the mean square displacement on time is seen before the glass transition takes place. In this report, we address how these discrepancies can be explained from the stochastic model.

First, we note that the observations in molecular dynamic studies have been made in an intermediate time window and not in the infinite time limit. For example, Miyagawa et al observed a few tens of jumps during their simulation, which corresponds to $w_0 t \sim$ a few tens, a much shorter time. Therefore, in order to compare directly the stochastic model with the results found by the molecular dynamic study, we have to evaluate physical quantities at the corresponding time scale.

The stochastic dynamics of an atom (tracer) is described by the conditional probability, $P(\mathbf{s},t|\mathbf{s}_0,0)$, that the atom is at site s at time t when it was at site $\mathbf{s}_0$ at time $t=0$. We assume that the time dependence of $P(\mathbf{s},t|\mathbf{s}_0,0)$ is determined by the trapping model master equation

$$\frac{\partial P(\mathbf{s},t|\mathbf{s}_0,0)}{\partial t} = \sum_{\mathbf{s}'}[w_{\mathbf{s}'}P(\mathbf{s}',t|\mathbf{s}_0,0) - w_{\mathbf{s}}P(\mathbf{s},t|\mathbf{s}_0,0)], \qquad (2)$$

where the summation is taken over the nearest neighbors of site s and the jump rate $w_\mathbf{s}$ is distributed according to Eq. (1). We employ the coherent medium approximation[4] to evaluate the ensemble average of $\tilde{P}(\mathbf{s},u|\mathbf{s}_0)$, the Laplace transform of $P(\mathbf{s},t|\mathbf{s}_0,0)$: $\tilde{P}(\mathbf{s},u|\mathbf{s}_0) = \int_0^\infty P(\mathbf{s},t|\mathbf{s}_0,0)e^{-ut}dt$. We further assume that $\bar{P}_{00} \equiv <\tilde{P}(\mathbf{s}_0,u|\mathbf{s}_0)>$ is given by

$$\bar{P}_{00} = 2[u + 6w_c + \sqrt{u(u+12w_c)}]^{-1}, \qquad (3)$$

where w_c (a function of the Laplace parameter u) is the coherent jump rate to be determined by the self-consistency condition

$$[\Omega + w_c]^{-1} = <[\Omega + w_\mathbf{s}]^{-1}> \qquad (4)$$

($\Omega = -w_c[1 + (u\bar{P}_{00} - 1)^{-1}]$). The assumption (3) to $\bar{P}_{00}$ is equivalent to assume a semi-elliptic density of state for the underlying lattice. In the coherent medium approximation, the frequency-dependent diffusion constant $D(u)$ ($u = i\omega$) is given by $D(u) = a^2 w_c$, a being the lattice constant.

Figure 1 shows the time dependence of the mean square displacement $R_2(t)$ for various ρ, which is obtained from the inverse Lapalce transformation of $6D(u)/u^2$

$$R_2(t) = (2\pi i)^{-1}\int_{0^+ - i\infty}^{0^+ + i\infty}[6D(u)/u^2]e^{ut}du. \qquad (5)$$

We define the apparent diffusion constant D_{app} by

$$D_{\text{app}} = \frac{1}{6}\frac{dR_2(t)}{dt}. \qquad (6)$$

The ρ dependence of the apparent diffusion constant determined at $w_0 t = 20$ is shown in Fig. 2. The apparent diffusion constant decreases rapidly near $\rho = 0$, but it does not vanish even for $\rho < 0$, showing a residual diffusion. This behavior of the diffusion constant is exactly what the molecular dynamics studies have shown.[2,5] As the observation time is increased, D_{app} approaches $w_0 a^2 \rho/(\rho + 1)$, the true static diffusion constant.

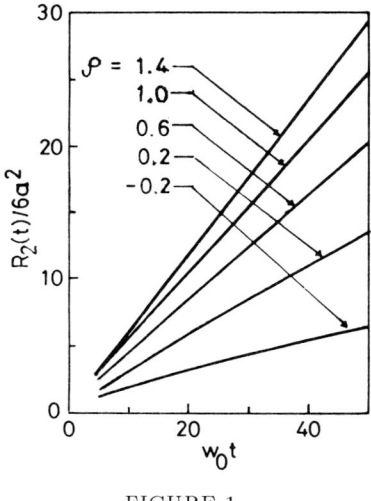

FIGURE 1

The mean square displacement in the intermediate time window.

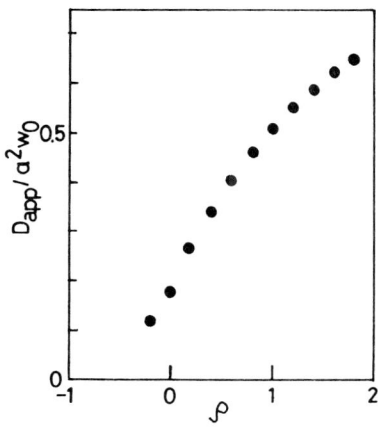

FIGURE 2

The apparent diffusion constant D_{app} determined at $w_0 t = 20$ is plotted against the parameter ρ.

We also define the apparent exponent θ_{app} which describes the long time behavior of the mean square displacement

$$\theta_{\text{app}} = \frac{d \log R_2(t)}{d \log t}. \qquad (7)$$

Figure 3 shows the ρ dependence of the apparent exponent θ_{app} determined at $w_0 t = 20$. The apparent exponent becomes less than unity before the glass transition takes place at $\rho = 0$. That is, the mean square displacement may show time dependence weaker than the linear dependence if it is determined at intermediate time windows. The incoherent scattering function $F_S(\mathbf{k}, t)$ for small wave vectors is determined by the mean square displacement. Therefore, $F_S(\mathbf{k}, t)$ is expected to begin to show a stretched exponential decay in time before the transition takes place. These results agree with the observations made by molecular dynamic studies.[2,5]

Figure 4 shows the time dependence of the non-Gaussian parameter

$$A(t) = \frac{3 R_4(t)}{5 R_2(t)^2} - 1, \qquad (8)$$

where $R_4(t)$ is the mean quartic displacement. As one can see from Fig. 4, $A(t)$ remains nonzero even at $t = \infty$ when $\rho < 0$. The long time behavior of $A(t)$ also agrees qualitatively with computer experiments.[5]

In summary, we have calculated various physical quantities at an intermediate time window using the stochastic model for the glass transition. The physical quantities determined

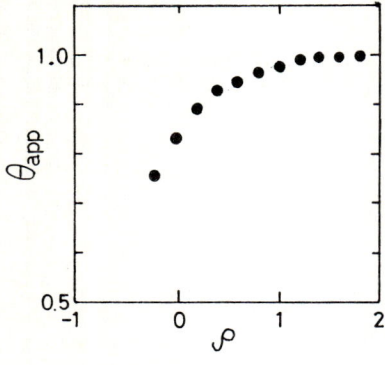

FIGURE 3

The apparent exponent θ_{app} determined at $w_0 t = 20$ is shown as a function of ρ.

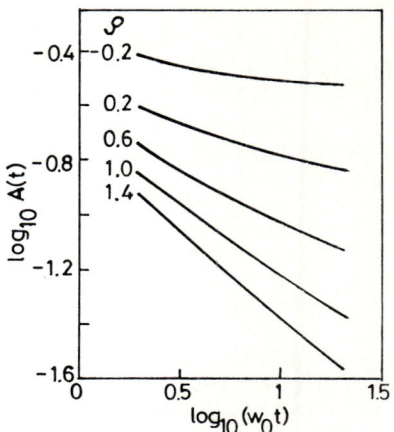

FIGURE 4

The non-Gaussian parameter in the intermediate time window.

at the intermediate time scale show behaviors different from the ones observed in the asympotic region. The present results are in good agreement with the computer experiments for soft sphere fluids, and therefore the stochastic model is considered to be very useful in understanding the dynamical behaviors of atoms in glass-forming systems.

REFERENCES

1) W. Götze, Z. Phys. Chem. **156** (1988) S3.

2) H. Miyagawa, Y. Hiwatari, B. Bernu, and J. P. Hansen, J. Chem. Phys. **88** (1988) 3879.

3) T. Odagaki and Y. Hiwatari, Proceedings of the 7th International Conference on Liquid and Amorphous Metals – LAM-7, ed. H. Endo (Elsevier, Amsterdam) (in press); T. Odagaki and Y. Hiwatari, Phys. Rev. A. **41** (1990) (in press).

4) T. Odagaki, J. Phys. A**20** (1987) 6455; T. Odagaki, Phys. Rev. B**38** (1988) 9044.

5) Y. Hiwatari, B. Bernu and J. P. Hansen, in *Condensed Matter Physics*, eds. P. Vashista, B. K. Kalia and R. F. Bishop (Plenum, New York, 1987), p. 19; S. Kambayashi and Y. Hiwatari, J. Phys. Soc. Jpn. **56** (1987) 2788.

MOLECULAR-DYNAMICS STUDY OF BINARY ALLOYS: DYNAMICAL CORRELATIONS OF THE
SUPERCOOLED LIQUIDS NEAR THE GLASS TRANSITION OF BINARY SOFT-SPHERE MIXTURES

Hiroh MIYAGAWA and Yasuaki HIWATARI

Department of Physics, Faculty of Science, Kanazawa University, Kanazawa, Ishikawa 920, Japan

With the aid of molecular-dynamics simulations of soft-sphere mixtures we have computed the longitudinal and transverse collective modes (sound waves). The shear stress autocorrelation function is also computed. It is shown that the stress autocorrelation function exhibits characteristic slow relaxation phenomena (long-time tails) near and beyond the glass transition, compatible with the previous result obtained from the density autocorrelation functions. However, the decaying form of both functions at long times appears to be significantly different.

1. INTRODUCTION

We examine the dynamical properties of supercooled liquids in the vicinity of the glass transition, using MD simulations on a binary soft-shpere model. We compute the Fourier transforms of both longitudinal and transverse current correlation functions and the shear stress autocorrelation function. We argue on the behaviors of the collective excitation modes (sound waves) and stress correlations, which can be strongly influenced by slow relaxation mechanisms near the glass transition. The transverse sound waves at small wavenumbers and the long-time tail of the stress autocorrelation function are particularly of interest in discussing the dynamical origin of the glass transition.

2. MODEL AND MD SIMULATIONS

We consider a binary soft-sphere alloy in which atoms interact through purely repulsive pair potentials:

$$v_{\alpha\beta}(r) = \epsilon(\sigma_{\alpha\beta}/r)^{12} \quad , \tag{1}$$

where $1 \leq \alpha, \beta \leq 2$ are species indices, and we assume that $\sigma_{\alpha\beta} = (\sigma_\alpha + \sigma_\beta)/2$

With the scaling property of the inverse power potentials and the effective one-component approximation, it is easily shown that all reduced equilibrium properties of such mixtures, in excess of their ideal gas counterparts, depend only on the following coupling constant:[1,2]

$$\Gamma_{eff} = \rho^*(T^*)^{-1/4}(\sigma_{eff}/\sigma_1)^3 \quad , \tag{2}$$

$$\sigma_{eff}^3 = \sum_\alpha \sum_\beta x_\alpha x_\beta \sigma_{\alpha\beta}^3 \qquad (3)$$

where $\rho^* = N\sigma_1^3/V$ denotes the reduced number density with the total number of atoms N($=N_1+N_2$, where N_1 or N_2 is the number of atoms of the respective species), and the total volume V of the system. The reduced temperature T^* equals to $k_B T/\epsilon$, $x_1 = N_1/N$, and $x_2 = 1 - x_1$. All present MD simulations (constant T) were performed with the core-size ratio of $\sigma_2/\sigma_1 = 1.2$, the mass ratio of $m_2/m_1 = 2$, $x_1 = 0.5$ and $\rho^* = 0.8$, using the 7-th order Gear algorithm as well as periodic boundary conditions. Most of simulations were carried out with the system size of N=500, but some with N=4000.

The microscopic time scale is measured in units of $\tau = \sqrt{m_1 \sigma_1^2/\epsilon}$ which turns out to be of the order of the inverse of the Einstein frequency associated with the two species.

Based on earlier MD simulations,[1,2] we expect the glass transition to take place around $\Gamma_{eff} = 1.5$ and the freezing point at $\Gamma_{eff} = 1.15$. Supercooled samples were prepared by quenching an equilibrium liquid of $\Gamma_{eff} = 0.8$ at some different rates.

3. RESULTS

3.1. Sound waves

The sound velocity and attenuation coefficient of longitudinal sound waves were determined from the peak of the dynamical mass structure factor $S(k,\omega)$ for several different couplings Γ_{eff}. The result obtained shows that the sound velocity C is nearly constant over $1.40 \leq \Gamma_{eff} \leq 1.55$, and the sound attenuation coefficient γ at the respective k's is also nearly constant for such highly supercooled liquids.[3] The latter apparently disagrees with the result of the linearized Navier-Stokes equations. The standard expression for the sound attenuation coefficient would predict a dramatic increase of sound damping with decreasing temperature, since the viscosity increases sharply. Our result is compatible with the result of the recent MD simulations for the same model, in which only the smallest wavenumbers has been taken into consideration.[4]

We have also computed the Fourier transform of the transverse current autocorrelation function.[3] No sound peak was observed for $\Gamma_{eff} = 0.8$, even at the smallest wavenumber k_1 ($=2\pi/L$) compatible with the size of the simulation cell L, whereas a clear sound peak is observed for $\Gamma_{eff} \geq 1.4$. The sound velocity of the transverse mode turns out to be less than half that of the longitudinal mode. This behavior can not be explained by the simplest version of the visco-elastic theory. The visco-elastic theory also predicts that shear waves will disappear for wavenumbers k smaller than a critical wavenumber $k_c \sim 1/\eta$

Table 1. Lists of the values of the respective components of $<\{\dot{G}(0)\}^2>/N$, in units of $m_1^2\sigma_1^4/\tau^4$, for various Γ_{eff} : The total (T) and kinetic (K), cross (C) and potential (P) parts with N=500 except for the second and seventh rows with N=4000. The sample of the last-row was prepared by a ten-time slower quenching rate than others.

Γ_{eff}	T	K	C	P
0.80	23.6	11.2	-0.06	12.4
	24.1	11.3	0.05	12.7
1.40	0.738	0.122	-0.002	0.618
1.45	0.684	0.093	0.004	0.586
1.50	0.547	0.076	-0.001	0.471
1.55	0.629	0.055	0.001	0.573
	0.509	0.054	0.001	0.454
	0.404	0.054	0.001	0.349

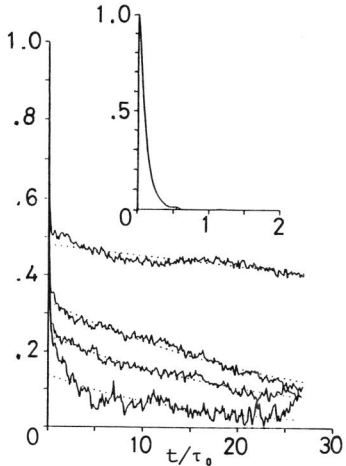

FIGURE 1
The normalized stress autocorrelation function for Γ_{eff}=1.40, 1.45, 1.50 and 1.55 from bottom to top. The inset is for Γ_{eff}=0.8 . $\tau_0=\tau(\rho^*)^{-1/3}(T^*)^{-1/2}$.

(inverse shear viscosity). However, near the glass transition k_c becomes much smaller than the minimum wavenumber k_1 available to the size of our simulations, which renders our discussions difficult with such a small system size as N=500 or N=4000.

3.2. Stress autocorrelation

The normalized stress autocorrelation function is defined by:

$$Z(t) = <\dot{G}(t)\dot{G}(0)>/<(\dot{G})^2> \qquad (4)$$

$$G = \sum_{i=1}^{N} m_i \dot{x}_i y_i \qquad (5)$$

We have computed three separate terms of the stress autocorrelation function, i.e., the kinetic, potential and cross terms. As shown in Table 1, the cross term is found to be much smaller than the other two terms except for the equilibrium liquid. It is also obtained that the kinetic part damps fastly around the time of the order of τ, which are much shorter than that of the potential part. Moreover, normalized kinetic correlation turns out to be little dependent on the temperature. Therefore, the main feature of the stress autocorrelation is determined by the potential part. The stress autocorrelation function obtained for $\Gamma_{eff} \geq 1.45$ exhibits long time tails, whereas it decays almost completely for $\Gamma_{eff} \leq 1.4$, in the time ranges shown in Fig. 1.

We have attempted to fit the stress autocorrelation function at large t's in terms of the simple exponential function, i.e., $Z(t) = A_\eta \exp[-(t/t_\eta)]$. The result of the best fits are summarized in Table 2. It turns out that the characteristic relaxation time t_η increases for increasing Γ_{eff} and such fittings work well. This result exhibits a remarkable contrast to the result obtained for the density autocorrelation function. This indicates that these two correlations are affected by different slowing-down mechanisms in the vicinity of the glass transition.

Table 2. Results of best fittings for $Z(t)$ in terms of the simple exponential function, with N=500 except for the second and seventh rows with N=4000. The sample of the last-row was prepared by a ten-time slower quenching rate than others.

Γ_{eff}	A_η	t_η
0.80	1.125	0.101
	1.126	0.106
1.40	0.138	14.40
1.45	0.240	24.00
1.50	0.330	27.00
1.55	0.480	160.0
	0.373	850.0
	0.218	32.20

Overall the results shown here indicate that the relaxation of the stress autocorrelation extremely slows down when Γ_{eff} approaches around the range of 1.45-1.50, which may be identified the liquid-glass transition of the present model. This conclusion is compatible with the MD results on $F_s(k,t)$,[2] shear modulus[5] and van Hove functions[4].

ACKNOWLEDGMENTS

This work is partly supported by the Grand-in-Aid from the Ministry of Education, Science and Culture.

REFERENCES

1. S. Kambayashi and Y. Hiwatari, J. Phys. Soc. Jpn. 56 (1987) 2788.

2. B. Bernu, J. P. Hansen, Y. Hiwatari and G. Pastore, Phys. Rev. A 36 (1987) 4891. G. Pastore, B. Bernu, J. P. Hansen and Y. Hiwatari, Phys. Rev. A 38 (1988) 454. H. Miyagawa, Y. Hiwatari, B. Bernu and J. P. Hansen, J. Chem. Phys. 88 (1988) 3879.

3. Y. Hiwatari and H. Miyagawa, Proceedings of the 7-th International Conference of Liquid and Amorphous Metals, September 4-8, Kyoto, to be published.

4. J. N. Roux, J. L. Barrat and J. P. Hansen, private communication

5. J. L. Barrat, J. N. Roux, J. P. Hansen and M. L. Klein, Europhys. Lett. 7 (1988) 707.

EFFECT OF THE QUANTUM ELECTRONS TO FORMATION OF A CRYSTALLINE ORDER IN ALKALI METALS

Seido NAGANO and Shuhei OHNISHI

Fundamental Research Laboratories, NEC Corporation, 34 Miyukigaoka, Tsukuba, Ibaraki 305, Japan

We propose a new Lagrangean to unify the molecular dynamics and the density-functional theory without introducing the electron wave function. It gives us a possibility of handling the many electron atom systems. Furthermore, we present a preliminary numerical study of the effect of the quantum electron gas to formation of a crystalline order in alkali metals, especially a hydrogen plasma, by this method.

1. INTRODUCTION

With the recent rapid advance of the supercomputer, various computer simulation studies have been conducted very actively. The computer experiment at the molecular level can be done by the Monte Carlo method or the molecular dynamics method. The former is the stochastic method. On the other hand, the latter is the deterministic method in the sense that it computes phase space trajectories of a collection of particles which obey the Newton's equation of motion. Historically, the molecular dynamics method was used by Fermi, Pasta, and Ulam[1] for a study of randomization of vibrational energy in a one-dimensional chain of atoms. However, the first application to the condensed matter was done by Alder and Wainwright[2]. They have studied the hard core system by this method. Furthermore, Rahman[3] applied it to the Lenard-Jones system. Since the development of the simple algorithm by Verlet[4], wide variety of problems in the condensed matter physics have been investigated by the molecular dynamic method. Especially, the charged system was studied by Hansen[5] very extensively.

Since the molecular dynamics is essentially for the classical systems, the special treatment is needed for a study of the quantum systems. For example, Stillinger and Weber[6], and Tersoff[7] have devised the effective potential for silicon such that they can reproduce the various results of the energy band calculation. Then, these potentials have been adopted for the molecular dynamics study of silicon. First serious treatment of the quantum electrons was undertaken by Car and Parrinello[8] in a study of the amorphous silicon. They have proposed a Lagrangean to unify the molecular dynamics and the density-functional theory[9,10] such that they can reproduce the Kohn-Sham equation at the minimization of the energy functional. In order to investigate large molecules, constructing a Lagrangean without introducing the wave function is much more favorable. In this paper, we propose such a Lagrangean. In order to construct a new Lagrangean, it is also important to find the proper form of the kinetic energy functional $T(\rho)$ and the exchange energy functional $E_{xc}(\rho)$ of the electron number density $\rho(r)$. Various kinds of study in this direction have been undertaken by many researchers[11]. We utilize those results for our work. Since Ichimaru and co-researchers[12] have extensively studied the hydrogen plasma theoretically, we will apply our Lagrangean to the same system for the confirmation of our theory.

2. THEORY

Our Lagrangean for the system of quantum electrons and classical ions is:

$$L = \frac{1}{2}\lambda \int \dot{\rho}^2 d^3r + \frac{1}{2} M \sum_I \dot{R}_I^2 + \mu \left[\int \rho d^3r - N \right]$$

$$- T(\rho) - U(\rho, \{R_I\}), \tag{1}$$

where $\{R_I\}$, the atomic coordinate vectors, M, atomic mass, μ, the chemical potential, ρ, the number density function of electrons, and, N, the total number of electrons. $T(\rho)$ denotes the kinetic-energy functional of ρ. We have introduced the pseudo-mass, λ, for the electron number density. The potential energy term is:

$$U = \frac{1}{2}\sum_{I \neq J} \frac{(Ze)^2}{|R_I - R_J|} + \frac{1}{2}\int \frac{\rho(r)\rho(r')}{|r - r'|} d^3r d^3r'$$

$$- \sum_I \int \frac{Ze^2 \rho(r)}{|R_I - r|} d^3r + E_{xc}(\rho), \tag{2}$$

where E_{xc} is the exchange energy functional of $\rho(r)$. Therefore, Lagrange's equation of motions become:

$$\frac{d}{dt}(M\dot{R}_I) = -\frac{\partial}{\partial R_I}\left[\sum_{J \neq I} \frac{(Ze)^2}{|R_I - R_J|} - \int \frac{Ze^2 \rho(r)}{|R_I - r|} d^3r\right] \tag{3}$$

$$\frac{d}{dt}(\lambda \dot{\rho}) = \mu - \frac{\partial T}{\partial \rho} - \frac{\partial U}{\partial \rho} \tag{4}$$

As we can see from Eq.(4) easily, we are able to obtain the physical quantities when ρ takes the extreme value, and the pseudo-mass controls the converging speed to the thermal equilibrium state. If we adopt the uniform electron distribution, Eq.(3) becomes that of the classical systems.

In order to implement the numerical calculation, let us introduce the periodic boundary condition. Then, Eq.(1) can be transformed to that of ρ_G, and ρ_G as far as the electron distribution is concerned. Where G is the reciprocal lattice vector, and ρ_G is the corresponding Fourier component.

3. NUMERICAL RESULTS

We have applied our theory to the study of a hydrogen plasma. Generally, the ion system is characterized with $\Gamma = e^2 (4\pi\rho/3)^{1/3}/k_BT$, and the electron system with $r_s = me^2/\hbar^2(3/4\pi\rho)^{1/3}$, where k_B is the Boltzmann's constant, $\hbar$, Planck's constant, and T is the temperature. We have used 108 ions for a numerical study. Furthermore, 1331 reciprocal lattice vectors are utilized for the electron system. In Fig.1, we show the electron-ion interaction energy, $-E_{12}/Nk_BT$, as a function of the number of the time steps of the molecular dynamics calculation for $\Gamma=0.01$. Fig.2 shows $-E_{12}/Nk_BT$ as a function of Γ. Both figures are for $r_s=0.0184$. Since r_s is small here, we have adopted the Thomas-Fermi-Slater approximation for $T(\rho)$ and $E_{xc}(\rho)$.

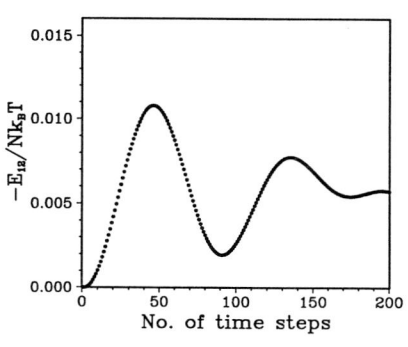

FIGURE 1
The electron-ion interaction energy, $-E_{12}/Nk_BT$, as a function of the number of the time steps for $\Gamma=0.01$ and $r_s=0.0184$. Unit of the time step is 4.84×10^{-19} second.

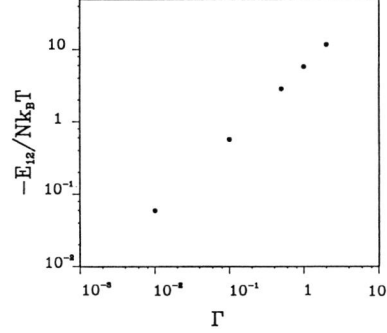

FIGURE 2
$-E_{12}/Nk_BT$ as a function of Γ, where $r_s=0.0184$.

In the case of the uniform electron distribution, E_{12} always takes zero value. Since we started the molecular dynamics calculation from the face-centered cubic configuration for ions and the uniform electron distribution, $-E_{12}$ increases gradually to the equilibrium value. Finite values of E_{12} reflect the effect of the non-uniformity of the electron distribution. Judging from the comparison with the theory[12], we may need more computing time to reach the equilibrium state. We plan to report a more detailed study very soon.

REFERENCES
1) E. Fermi, J. R. Pasta, and S. M. Ulam, in *Collected Work of Enrico Fermi*, Vol. 2 (University of Chicago Press, Chicago, 1965).
2) B. J. Alder and T.E. Wainwright, J. Chem. Phys. *27*, 1208(1957); J. Chem. Phys. *31*, 459 (1959).
3) A. Rahman, Phys. Rev. *136*, A405(1964); J. Chem. Phys. *45*, 258 (1966).
4) L. Verlet, Phys. Rev. *159*, 98(1967).
5) J. P. Hansen, I. R. McDonald and E. L. Pollock, Phys. Rev. A*11*, 1025 (1975).
6) F. H. Stillinger and T. A. Weber, Phys. Rev. B*31*, 5262 (1985).
7) J. Tersoff, Phys. Rev. Lett., 56, 632(1986); B. W. Dowson, Phys. Rev. B*35*, 2795 (1987).
8) R. Car and M. Parrinello, Phys. Rev. Lett. *55*, 2471(1985); Phys. Rev. Lett. *60*, 204 (1988).
9) W. Kohn and L. J. Sham, Phys. Rev. *140*, 1133 (1965).
10) N. H. March and B. M. Deb, *The Single-Particle Density in Physics and Chemistry* (Academic Press, London, 1987).
11) A. Cedillo, J. Robles, and J.L. Gazquez, Phys. Rev. A38, 1697 (1988).
12) S. Ichimaru, S. Mitake, S. Tanaka, and X.-Z. Yan, Phys. Rev. A*32*, 1768 (1985); S. Mitake, S. Tanaka, X.-Z. Yan, and S. Ichimaru, Phys. Rev. A*32*, 1775 (1985);S. Tanaka, S. Mitake, X. -Z. Yan, and S. Ichimaru, Phys. Rev. A*32*, 1779 (1985).

Chapter IV:
Strong-Coupling Theories and Experiments in Specific Geometries

OBSERVATION OF CORRELATIONS IN FINITE, STRONGLY COUPLED ION PLASMAS*

J.J. BOLLINGER, S.L. GILBERT, D.J. HEINZEN, W.M. ITANO, and D.J. WINELAND

National Institute of Standards and Technology, 325 Broadway, Boulder, CO 80303

1. INTRODUCTION

We have observed[1] spatial correlations with up to 15 000 Be^+ ions in a Penning trap with a coupling (defined below) of $\Gamma > 100$. These correlations are strongly affected by the boundary conditions and take the form of concentric shells as predicted by computer simulations.[2-4] In this paper we briefly describe the experimental confinement geometry and the method of producing low temperature ions. The relatively large spacings between the ions (~ 20 μm) permit the shells to be directly viewed by imaging the Be^+ laser-induced fluorescence onto a photon-counting camera. Diagnostic techniques capable of measuring the ion diffusion are then discussed. Qualitative observations of the ion diffusion are compared with theoretical predictions.

2. CONFINEMENT GEOMETRY

The Penning trap uses a static, uniform magnetic field and a static, axially symmetric electric field for the confinement of charged particles. The magnetic field, which is directed along the z axis of the trap, provides confinement in the radial direction. The ions are prevented from leaving the trap along the z axis by the electric field. In the work described here, the electric field was provided by three cylindrical electrodes as shown in Fig. 1. The dimensions of the trap electrodes were chosen so that the first anharmonic term (i.e. fourth order term) in the expansion of the trapping potential was zero. Over the region near the trap center, the potential can be expressed (in cylindrical coordinates) as $\Phi \simeq AV_o(2z^2-r^2)$ where $A = 0.146$ cm^{-2}. A background pressure of 10^{-8} Pa ($\approx 10^{-10}$ Torr) was maintained by a triode sputter-ion pump. The confinement geometry is similar to that used by the group of the University of California at San Diego (UCSD)[5] with the exception that our trap is smaller than the UCSD traps.

*Contribution of the U.S. Government, not subject to copyright.

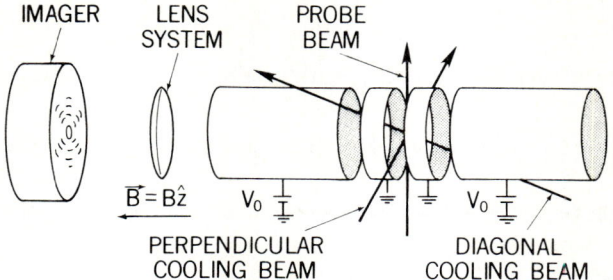

FIGURE 1
Schematic drawing of the trap electrodes, laser beams, and imaging system (not to scale). The overall length of the trap is 10.2 cm. The trap consists of two end cylinders and two electrically connected central cylinders with 2.5 cm inner diameters. Ion clouds are typically less than 1 mm in both diameter and axial length. The diagonal cooling beam crosses the cloud at an angle of 51° with respect to the z axis In the experiments, B = 1.92 T or 0.82 T and V_o ranged between 20 V and 200 V.

The stored ions can be characterized by a thermal distribution where the "parallel" (to the z axis) temperature $T_\parallel$ is approximately equal to the "perpendicular" temperature $T_\perp$. This thermal distribution is superimposed on a uniform rotation of the cloud[6-9] at frequency ω which, at the low temperatures of this experiment, is due to the $\vec{E} \times \vec{B}$ drift, where $\vec{E}$ is the electric field due to the trap voltage and the space charge of the ions. In a frame of reference rotating with the ions, the static thermodynamic properties of an ion cloud confined in a Penning trap are identical to those of a one-component plasma (OCP).[7] An OCP consists of a single species of charge embedded in a uniform-density background of opposite charge. For the system of ions in a Penning trap, the trapping fields play the role of the neutralizing background charge. An OCP is characterized by the Coulomb coupling constant,[7,10]

$$\Gamma \equiv q^2/(a_s k_B T),$$

which is a measure of the nearest-neighbor Coulomb energy divided by the thermal energy of a particle. The quantities q and T are the ion charge and temperature. The Wigner-Seitz radius a_s is defined by $4\pi a_s^3 n_0/3 = 1$, where $-q n_0$ is the charge density of the neutralizing background. In the Penning trap the background density n_0 depends on the rotation frequency ω and the cyclotron frequency Ω and is given by[6-9]

$$n_0 = m\omega(\Omega-\omega)/(2\pi q^2). \tag{1}$$

3. LASER COOLING AND COMPRESSION

The ion density that can be achieved in a Penning trap is limited by the magnetic field strength that is available in the laboratory. Consequently to obtain large values of Γ and therefore strong couplings, a technique to obtain low ion temperatures is necessary. In our work, radiation pressure from lasers is used to reduce the temperature of the stored ions to less than 10 mK. This technique, known as laser cooling,[11-13] uses the resonant scattering of laser light by atomic particles. The laser is tuned to the red, or low-frequency side of the atomic "cooling transition" (typically an electric dipole transition like the D lines in sodium). Ions with a velocity component opposite to the laser beam propagation ($\vec{k}\cdot\vec{v} < 0$) will be Doppler shifted into resonance and absorb photons at a relatively high rate. Here, $\vec{k}$ is the photon wave vector ($|\vec{k}|=2\pi/\lambda$, where λ is the wavelength of the cooling radiation). For the opposite case ($\vec{k}\cdot\vec{v} > 0$), the ions will be Doppler shifted away from the resonance and the absorption rate will decrease. When an ion absorbs a photon, its velocity is changed by an amount $\Delta\vec{v} = \hbar\vec{k}/m$ due to momentum conservation. Here $\Delta\vec{v}$ is the change in the ion's velocity, m is the mass of the ion, and $2\pi\hbar$ is Planck's constant. The ion spontaneously reemits the photon symmetrically. In particular, when averaged over many scattering events, the reemission does not change the momentum of the ion. The net effect is that for each photon scattering event, the ion's average velocity is reduced by $\hbar\vec{k}/m$. To cool an atom from 300 K to millikelvin temperatures takes typically 10^4 scattering events but, since scatter rates can be 10^8/s, the cooling can be rapid.

In our work with Be$^+$, the 2s $^2S_{1/2} \rightarrow$ 2p $^2P_{3/2}$ "D_2" transition was used as the cooling transition as indicated in Fig. 2. Cooling laser beams were directed both perpendicularly and at an angle with respect to the magnetic field as indicated in Fig. 1. This enabled us to control the cloud size and obtain the lowest possible temperatures.[14,15] The 313 nm radiation required to drive this transition was obtained by frequency doubling the output of a continuous wave, narrow band (3 MHz) dye laser. The 313 nm power was typically 50 μW. The theoretical cooling limit, due to photon recoil effects,[11-13] is given by a temperature equal to $\hbar\gamma/(2k_B)$ where γ is the radiative linewidth of the atomic transition in angular frequency units. For the Be$^+$ cooling transition ($\gamma = 2\pi \times 19.4$ MHz), the theoretical minimum temperature is 0.5 mK.

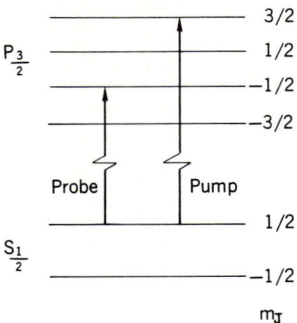

FIGURE 2

Energy level structure of the $^9\text{Be}^+$ $^2S_{1/2}$ ground state and the first excited $^2P_{3/2}$ state. The magnetic field splits each state into its m_J sublevels. The laser cooling (pump) and depopulation (probe) transitions are shown.

Laser scattering can also be used to change the angular momentum and compress the stored ion plasma.[8,9] The z component of the canonical angular momentum for an individual ion in the plasma is

$$\ell_z = mv_\theta r + \frac{qBr^2}{2c}. \qquad (2)$$

The two terms in Eq. (2) are the ion's mechanical angular momentum and the field angular momentum. The total z component of the angular momentum of the plasma is

$$L_z = m(\Omega/2-\omega)N\langle r^2 \rangle. \qquad (3)$$

Here N is the total number of ions and $\langle r^2 \rangle$ is the mean-squared radius of the plasma. For most of the work described in this paper $\omega \ll \Omega$ and

$$L_z \approx \frac{m\Omega N}{2}\langle r^2 \rangle > 0. \qquad (4)$$

Therefore the total angular momentum is dominated by the field angular momentum. Suppose the cooling laser beam is directed normal to the z axis but at the side of the plasma which is receding from the laser beam due to the plasma rotation. Because the rotation of the positive ions is in the $-\hat{\theta}$ direction, the torque of the laser on the ions will also be negative. Consequently, angular momentum is removed from the plasma and according to Eq. (4) the radius of the plasma must decrease. In general, the plasma is compressed until the torque due to the cooling laser is balanced by another external torque. As the radius decreases, the density of the plasma increases.

Even in the absence of external torques, there is a limit to how far the plasma can be compressed. From Eq. (1), the maximum density, known as the Brillouin density, occurs when the rotation frequency $\omega = \Omega/2$. The Brillouin density is given by

$$n_{max} = \frac{m\Omega^2}{8\pi q^2}.$$

We have recently been able to achieve densities at or near the Brillouin limit. In fact we have also been able to achieve rotation frequencies $\omega > \Omega/2$ where according to Eq. (1) the ion density decreases. In these experiments, the temperature was not determined. At the magnetic field of 1.92 T used in some of the work discussed here, the Brillouin density is 1.1×10^9 cm^{-3}. This density with the theoretical minimum 0.5 mK temperature, results in a coupling $\Gamma \sim 5500$. For the work reported here, we have been able to obtain ion temperatures in the 1-10 mK range with densities 5-10 times less than the Brillouin density. This results in couplings Γ of a few hundred.

We measured[8,14,15] the ion density and temperature by using a second laser, called the probe laser, to drive the "depopulation" transition as indicated in Fig. 2. The cooling laser optically pumps the ions into the $^2S_{1/2}$ $m_J=+1/2$ state.[8] The resonance fluorescence (i.e. laser light scattered by the ions) from this transition is used as a measure of the ion population in the $m_J=+1/2$ state. The probe laser drives some of the ion population from the $^2S_{1/2}$ $m_J=+1/2$ state to the $^2P_{3/2}$ $m_J=-1/2$ state where the ions decay with 2/3 probability to the $^2S_{1/2}$ $m_J=-1/2$ state. This causes a decrease in the observed ion fluorescence because the $^2S_{1/2}$ $m_J=-1/2$ state is a "dark" state a (state which does not fluoresce in the cooling laser). The ion temperature is obtained from the Doppler broadening of the resonance lineshape when the probe laser is scanned through the depopulation transition. The ion rotation frequency is measured from the shift in the depopulation transition frequency as the probe laser is moved from the side of the plasma rotating into the laser beam to the side of the plasma rotating with the laser beam. From the measured rotation frequency, the density is calculated from Eq. (1). The measured density and temperature is used to calculate the coupling Γ.

4. OBSERVED CORRELATIONS

With measured couplings $\Gamma > 100$, we anticipate that the ions will exhibit correlated behavior. If the number of stored ions is large enough for infinite volume behavior, the ions may be forming a bcc lattice.[10] Until now we have cooled and looked for spatial correlations with up to 15 000 Be$^+$ ions stored in the Penning trap of Fig. 1. A currently unanswered question is how many stored ions are required for infinite volume behavior, i.e. the appearance of a bcc lattice for $\Gamma > 178$. For a finite plasma consisting of a hundred to a few thousand ions, the boundary conditions are predicted to have a significant effect on the plasma state. Simulations involving these numbers of ions in a spherical trap potential predict that the ion cloud will separate into concentric spherical shells.[2-4] Instead of a sharp phase transition, the system is expected to evolve gradually from a liquid state characterized by short-range order and diffusion in all directions, to a state where there is diffusion within a shell but no diffusion between the shells (liquid within a shell, solid-like in the radial direction), and ultimately to an overall solid-like state.[4] These conclusions should apply to a nonspherical trap potential as well if the spherical shells are replaced with shells approximating spheroids. Independent theoretical investigations[16,17] of the nonspherical case support this conjecture.

We have observed shell structures with ^{9}Be$^+$ ions stored in a Penning trap by imaging the laser induced fluorescence from the cooling transition. This technique is sensitive enough to observe the structures formed with only a few ions in a trap.[18-21] About 0.04% of the 313-nm fluorescence from the decay of the $^2P_{3/2}$ state was focused by f/10 optics onto the photocathode of a resistive-anode photon-counting imaging tube (see Fig. 1). The imager was located along the z axis, about 1 m from the ions. The imaging optics was composed of a three-stage lens system with overall magnification of 27 and a resolution (FWHM) of about 5 μm (specifically, the image of a point source when referred to the position of the ions was approximately 5 μm in diameter). Counting rates ranged from 2 to 15 kHz. Positions of the photons arriving at the imager were displayed in real time on an oscilloscope while being integrated by a computer. The probe laser could be tuned to the same transition as the cooling laser and was directed through the cloud perpendicularly to the magnetic field. With the probe laser turned on continuously, the cooling laser could be chopped at 2 kHz (50% duty cycle) and the image signal integrated only when the cooling laser was off. Different portions of the cloud could then be imaged by the translation of the probe beam, in a calibrated fashion, either parallel or perpendicular to the z axis. Images were also obtained from the ion fluorescence of all three laser beams.

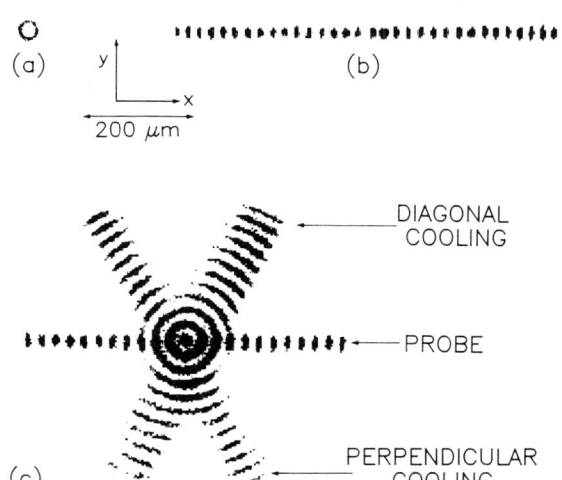

FIGURE 3
Images of shell structures obtained with B = 1.92 T. (a) A single shell in a cloud containing approximately 20 ions. Trap voltage V_0 = 14 V and cloud aspect ratio a_r (axial length/diameter) ≈ 6.5. This image was obtained from the ion fluorescence of the perpendicular and diagonal cooling beams. (b) Sixteen shells (probe-beam ion fluorescence only) in a cloud containing about 15 000 ions with V_0 = 100 V and a_r ≈ 0.8. (c) Eleven shells plus a center column in the same cloud as (b), with V_0 = 28 V and a_r ≈ 2.4. This image shows the ion fluorescence from all three laser beams. Integration times were about 100 s for all images.

We have observed shell structure in clouds containing as few as 20 ions (one shell) and as many as 15 000 ions (sixteen shells). Images covering this range are shown in Fig. 3. Even with 15 000 ions in the trap there is no evidence for infinite volume behavior. We measured the coupling constant Γ for several clouds containing about 1000 ions. Drift in the system parameters was checked by verifying that the same images were obtained before and after the cloud rotation frequency and ion temperatures were measured. Figure 4 shows examples of shell structures at two different values of Γ. The first image is an example of high coupling (Γ ≈ 180) and shows very good shell definition in an intensity plot across the cloud. The second image is an example of lower coupling (Γ ≈ 50) and was obtained with cooling only perpendicular to the magnetic field. Variations in peak intensities equidistant from the z axis are due to signal-to-noise limitations and imperfect alignment between the imager x axis and the probe beam.

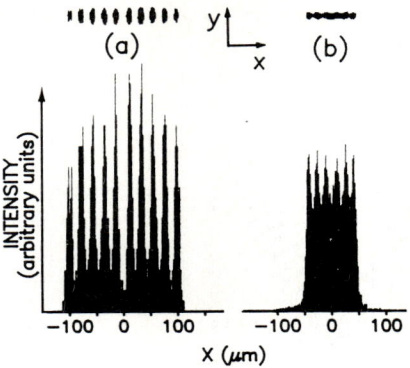

FIGURE 4
Intensity plots along the imager x axis (parallel to the probe beam) through the center of the ion cloud with corresponding images (above). (a) $\Gamma = 180^{+90}_{-70}$ (T = 6^{+4}_{-2} mK, $n_0 \simeq 7 \times 10^7$ ions cm^{-3}). Cloud aspect ratio $a_r \simeq 3.5$. (b) $\Gamma = 50^{+30}_{-20}$ (T = 33^{+17}_{-13} mK, $n_0 = 2 \times 10^8$ ions cm^{-3}), $a_r \simeq 5$. The clouds contained about 1000 ions and B = 1.92 T in both cases.

We obtained three-dimensional information on the shell structure by taking probe images at different z positions; two types of shell structure were present under different circumstances. The first type showed shell curvature near the ends of the cloud, indicating that the shells may have been closed spheroids. Shell closure was difficult to verify because of a lack of sharp images near the ends of the cloud where the curvature was greatest. This may have been due to the averaging of the shells over the axial width of the probe beam. In the other type of shell structure, it was clear that the shells were concentric cylinders with progressively longer cylinders near the center. An example of these data is shown in Fig. 5. Other evidence for cylindrical shells was obtained from the observation that shells in the diagonal-beam images occurred at the same cylindrical radii as those from the perpendicular beams. This can be seen in the three-beam images such as that shown in Fig. 3(c). Systematic causes of these two different shell configurations have not yet been identified.

One comparison which can be made between the theoretical calculations and our experimental results is the relationship between the number of shells and the number of ions, N, in a cloud. For a spherical cloud, two independent approaches[2,22] estimate $(N/4)^{1/3}$ and $(3N/4\pi)^{1/3}$ shells. For the nearly spherical cloud of Fig. 3(b) (N $\simeq$ 15 000), these formulae predict 15.5 and

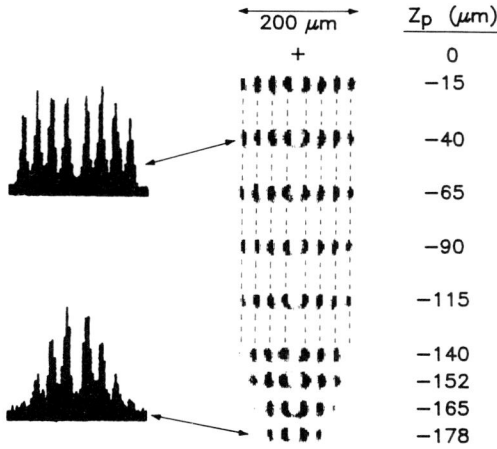

FIGURE 5
Data showing evidence for concentric cylindrical shells. On the right is a series of images obtained with the probe beam for different z positions z_p of the probe beam (lower half of the cloud only). Intensity plots for z_p = -40 μm and z_p = -178 μm are shown on the left. The cloud aspect ratio a_r was about 1.9 and B ≈ 1.92 T.

15.3 shells and we measure 16. At present, it is difficult to make further quantitative comparisons between our data and the theoretical calculations. For example, there is substantial uncertainty in our measurement of Γ due to uncertainty in the temperature measurement. Our data do agree qualitatively with the simulations with the exception, in some cases, of the presence of an open-cylinder shell structure as opposed to the predicted closed spheroids. Shear (that is, different rotation frequencies) between the shells could possibly account for this discrepancy. In our experiment, shear could be caused by differential laser torque or the presence of impurity ions.[8] For the data here, we have determined that the rotation frequency does not vary by more than 30% across the cloud.

5. ION DIFFUSION

The probe laser can be used to optically tag ions and observe the ion diffusion.[14] With the probe laser tuned to the depopulation transition (see Fig. 2), ions in the path of the probe laser beam are put into the "dark" m_J = -1/2 ground state. These ions will not fluoresce when they pass through the cooling laser until they are optically pumped back into the m_J=+1/2 ground state. This repumping time is typically on the order of 1.0 s. By pulsing the probe laser on and measuring the length of time it takes the dark ions to diffuse from the probe beam to the cooling laser beam, it should be possible to measure the ion diffusion. By directing the probe laser beam to the radial edge of the plasma so that only the outer shell is intersected by the probe beam, it should be possible to observe the diffusion of ions in the radial direction or between shells. By directing the probe beam to the axial edge of the cloud it should be possible to observe the diffusion of the ions in the axial direction or within a shell. According to the simulations of Dubin and O'Neil,[4] for intermediate values of the coupling ($\Gamma \sim 100$) we expect to observe that the diffusion between shells is much slower than the ion diffusion within a shell (solid-like behavior between shells, liquid-like within a shell). As the temperature is lowered and the coupling Γ increases, the diffusion within a shell should smoothly slow down. At high enough couplings (i.e. $\Gamma > 400$) the diffusion should be very slow both between and within shells, indicative of solid-like behavior.

We have qualitatively observed the ion diffusion at intermediate values of Γ ($\Gamma \sim 100\text{-}200$). We observed that the diffusion of ions between shells is slow compared to the optical repumping time (~ 1 s) but that the diffusion within a shell (i.e. from the axial end of a shell to the $z = 0$ plane) is fast compared to this repumping time. We have also observed states with higher couplings (the couplings Γ were not measured) where the diffusion of ions both between and within a shell was slow compared to the optical repumping time. In the future we plan to make quantitative measurements of the ion diffusion.

ACKNOWLEDGEMENT

We gratefully acknowledge the support of the U.S. Office of Naval Research and the Air Force Office of Scientific Research. We thank M. Raizen and F. Moore for carefully reading the manuscript.

REFERENCES
1) S.L. Gilbert, J.J. Bollinger, and D.J. Wineland, Phys. Rev. Lett. 60 (1988) 2022.

2) A. Rahman and J.P. Schiffer, Phys. Rev. Lett. 57 (1986) 1133; J.P. Schiffer, Phys. Rev. Lett. 61 (1988) 1843.

3) H. Totsuji, in Strongly Coupled Plasma Physics, eds. F.J. Rogers and H.E. DeWitt (Plenum, New York, 1987) pp. 19-33.

4) D. Dubin and T. O'Neil, Phys. Rev. Lett. 60 (1988) 511.

5) C.F. Driscoll, J.H. Malmberg, and K.S. Fine, Phys. Rev. Lett. 60 (1988) 1290; J.H. Malmberg and J.S. deGrassie, Phys. Rev. Lett. 35, 577 (1975).

6) T.M. O'Neil, in Non-Neutral Plasma Physics, eds. C.W. Roberson and C.F. Driscoll (American Institute of Physics, New York, 1988) pp. 1-25.

7) J.H. Malmberg and T.M. O'Neil, Phys. Rev. Lett. 39 (1977) 1333.

8) L.R. Brewer, J.D. Prestage, J.J. Bollinger, W.M. Itano, D.J. Larson, and D.J. Wineland, Phys. Rev. A38 (1988) 859.

9) D.J. Wineland, J.J. Bollinger, W.M. Itano, and J.D. Prestage J. Opt. Soc. Am. B2 (1985) 1721.

10) S. Ichimaru, H. Iyetomi, and S. Tanaka, Phys. Rep. 149 (1987) 91 and references therein.

11) D.J. Wineland and W.M. Itano, Phys. Rev. A20 (1979) 1521.

12) W.M. Itano and D.J. Wineland, Phys. Rev. A25 (1982) 35.

13) D.J. Wineland and W.M. Itano, Phys. Today 40(6) (1987) 34; S. Stenholm, Rev. Mod. Phys. 58 (1986) 699.

14) L.R. Brewer, J.D. Prestage, J.J. Bollinger, and D.J. Wineland, in Ref. 3), pp. 53-64.

15) J.J. Bollinger and D.J. Wineland, Phys. Rev. Lett. 53 (1984) 348.

16) J.P. Schiffer, Argonne Natl. Lab., Argonne, IL, private communication.

17) D.H.E. Dubin, Dept. of Physics, UCSD, La Jolla, CA, private communication.

18) F. Diedrich, E. Peik, J.M. Chen, W. Quint, and H. Walther, Phys. Rev. Lett. 59 (1987) 2931.

19) D.J. Wineland, J.C. Bergquist, W.M. Itano, J.J. Bollinger, and C.H. Manney, Phys. Rev. Lett. 59 (1987) 2935; D.J. Wineland, W.M. Itano, J.C. Bergquist, S.L. Gilbert, J.J. Bollinger, and F. Ascarrunz, in Ref. 6), pp. 93-108.

20) J. Hoffnagle, R.G. DeVoe, L. Reyna, R.G. Brewer, Phys. Rev. Lett. 61 (1988) 255.

21) Th. Sauter, H. Gilhaus, I. Siemers, R. Blatt, W. Neuhauser, P.E. Toschek, Z. Phys. D10 (1988) 153.

22) D.H.E. Dubin, Phys. Rev. A40 (1988) 1140.

THEORY OF STRONGLY-CORRELATED PURE ION PLASMA IN PENNING TRAPS*

D.H.E. DUBIN and T.M. O'NEIL

Physics Department, University of California at San Diego, La Jolla, CA 92093, USA

1. INTRODUCTION

In a recent series of experiments[1] at the National Institute of Standards and Technology in Boulder, Colorado a cloud of ions is trapped and confined for long periods of time. The number of ions trapped, N, is sufficiently large so that the cloud may be regarded as a plasma, i.e., a nonneutral or pure ion plasma. The ions are subsequently cooled to extremely low temperature T at sufficiently high density n_0 so that the correlation parameter, $\Gamma = q^2/a_{ws} kT$ is much larger than unity (here q is the ion charge and $a_{ws} \equiv (3/(4\pi n_0))^{1/3}$ is the Wigner-Seitz radius). The ions therefore become strongly correlated and it is possible to study in detail the effects of strong correlation, including formation of liquid and even crystalline states. Furthermore, as we will see, there is a direct correspondence between the thermal equilibrium properties of these trapped ions and those of the so-called one-component plasma (OCP). All this is quite exciting since there is a large body of theoretical work which has been generated over several decades on the one-component plasma, and so it may be possible to test several of the theoretical predictions. For instance, it has been predicted that a first-order phase transition should occur from a liquid to a body-centered cubic (bcc) crystal at $\Gamma \cong 180$ in the infinite homogeneous OCP.[2]

Does this result apply to the experiments? We will see that in fact the result does not apply, because in present experiments the number of ions trapped is relatively small ($N \lesssim 10^4$) so that surface effects are important. However, because the number of ions is small, computer simulation of the system in realistic geometry becomes possible. We will discuss the results of such simulations,[3] which predict correlation behavior in the ion clouds which is quite different from that of the homogeneous one-component plasma. Rather than undergoing a simple first-order transition from a liquid to a bcc crystal, the system passes through an intermediate regime rather like the smectic phase of a liquid crystal, in which the ion cloud forms concentric spheroidal shells. Ions are confined to the shells but move randomly within each shell. At larger Γ values, this diffusion is suppressed and a distorted 2-D hexagonal lattice forms in each shell.

---

*Supported by NSF grant PHY87-06358, ONR Contract N00014-82-K-0621 and a grant of cpu time from the San Diego Supercomputer Center.

Some of these properties can be understood by means of a simple slab model of the ion cloud. This model will provide us with a simple first estimate for the size of system required before the results of the infinite homogeneous computer calculations should apply to the pure ion crystal.[4]

In Section 2 we discuss several important properties of nonneutral plasmas and consider the confinement characteristics of such plasmas when trapped in the so-called Penning trap used in the experiments. In Section 3 the results of computer simulations are discussed, and in Section 4 an analytic model is presented which is based on a finite-temperature slab geometry model of a bounded ion crystal.

2. THERMAL EQUILIBRIUM OF STRONGLY-CORRELATED NONNEUTRAL PLASMAS

Nonneutral plasmas, that is, plasmas consisting of an unneutralized collection of charged particles, have many properties in common with neutral plasmas. For instance, they exhibit the phenomenon of Debye shielding and also exhibit collective effects such as plasma oscillations. However, there are several differences between neutral and nonneutral plasmas. For instance, when a nonneutral plasma is cooled to low temperature it suffers no recombination since there is no oppositely charged species with which to recombine. (We will focus here on single-species nonneutral plasmas—the experiments at NIST usually involve plasmas consisting of Be^+ ions.) At sufficiently low temperature and high density, the kinetic energy per particle is less than the interaction energy and the plasma becomes strongly correlated.

Another difference between the neutral and nonneutral plasma is that nonneutral plasmas can be confined for very long periods of time using only static electric and magnetic fields. In the experiments long-time confinement is provided by means of the cylindrical Penning trap geometry shown schematically in Figure 1. This trap in its simplest form consists of three electrically isolated conducting cylinders whose axes of symmetry are oriented parallel to a uniform magnetic field. Confinement in the axial direction is provided by a potential well induced by a voltage difference between the central cylinder and the end cylinders. Confinement in the radial direction is provided by the uniform magnetic field **B**. The radial confinement can be understood by analyzing the balance of forces in the radial direction. There is a large radial electric field E_ρ due to the unneutralized collection of ions in the trap. This electric field is balanced by the magnetic component of the Lorentz force,

$$E_\rho + \frac{v_\theta B}{c} \cong 0 , \qquad (1)$$

where v_θ is the velocity of the particles in the azimuthal direction. Solving for the velocity v_θ, we find that the entire plasma rotates about the axis of symmetry. This rotation is just the familiar $\vec{E} \times B$ drift.

Another way to understand the radial confinement of the nonneutral plasma is to consider the constants of the motion. Cylindrical symmetry implies that the total angular momentum in the axial direction is a constant of the motion:

$$L = \sum_i (m\mathbf{v}_{\theta i} + \frac{q}{c} A_\theta(\rho_i))\rho_i \ , \tag{2}$$

where $A_\theta(\rho) = B\rho/2$ is the vector potential associated with the magnetic field. If the magnetic field is sufficiently strong the vector potential contribution to the angular momentum dominates over the kinetic contribution and the angular momentum can be written in the following form:

$$L \cong \frac{qB}{2c} \sum_i \rho_i^2 \ . \tag{3}$$

This equation implies that the mean square radius of the plasma is a constant of the motion and thus the plasma cannot expand. This simple argument has been made more rigorous by O'Neil.[5] In the actual experiments small imperfections in the trap cause slight cylindrical asymmetries which allow the plasma to slowly expand. However, by careful construction, these asymmetries can be reduced and confinement times on the order of several days have been achieved in experiments on pure electron plasmas at the University of California in San Diego.[6]

These confinement times are much longer than any internal time scales in the dynamics and so the particles can come to a state of confined thermal equilibrium. The thermal equilibrium is characterized by the angular momentum L and the energy H, where

$$H = \sum_{i=1}^{N} \frac{1}{2} m\mathbf{v}_i^2 + e\Phi(\mathbf{x}_1, \cdots \mathbf{x}_N) \tag{4}$$

and Φ is the potential energy of the system of charges, including electrostatic interactions and external confining potentials. If one then assumes that the system is thermally isolated, a microcanonical (constant H and L) ensemble describes the statistics of the system. However, it is often useful to use a canonical ensemble based on constant temperature T and rotation frequency ω. For large N the two ensembles predict averages which differ only $O(1/N)$. Such differences are unimportant for the N values considered here even though we are interested in effects stemming from the boundedness of the cloud. The probability density associated with a given state is then given by the Gibb's distribution f, where

$$f(\mathbf{x}_1, \mathbf{v}_1, \cdots \mathbf{x}_N, \mathbf{v}_N) = Z^{-1} e^{-\beta(H + \omega L)} \ . \tag{5}$$

For given external confining fields N, f is characterized by the parameters T and ω and these parameters in turn determine the average values of the energy and angular momentum of the system. Substituting Eq. (2) and (4) into Eq. (5) yields, after some simple algebra, the following form for f:

$$f = Z^{-1} e^{-\beta m \sum_i (\mathbf{v}_i - \omega \rho_i \hat{\theta}_i)^2/2} \times e^{-\beta(\Phi + \sum_i \frac{m\omega}{2}(\Omega_c - \omega)\rho_i^2)} \tag{6}$$

One can see that the velocity dependence is Maxwellian in a frame rotating with frequency ω. As one expects, the thermal equilibrium distribution corresponds to a shear free flow. We call such a flow a rigid rotor.

In Eq. (6), the spatial distribution of ions is determined by three potentials: the total electrostatic potential Φ, the centrifugal potential $-m\omega^2\rho^2/2$ and the potential $m\omega\Omega_c\rho^2/2$. This latter potential is associated with the electric field induced by rotation through a magnetic field. It is this field that provides the radial confinement. It dominates over the deconfining effect of the centrifugal potential, since the cyclotron frequency is larger than the rotation frequency in the experiments. These two potentials form an effective potential which is proportional to ρ^2 and thus they can be interpreted as the potential energy of ions in a cylinder of uniform negative charge. The density of this effective neutralizing background charge n_0 may be found by substituting the effective potential into Poisson's equation:

$$n_0 = \frac{\nabla^2}{4\pi q^2} \frac{m\omega}{2} (\Omega_c - \omega)\rho^2 = \frac{m\omega(\Omega_c - \omega)}{2\pi q^2} \quad . \tag{7}$$

Thus, the static thermal equilibrium properties of the magnetically confined plasma are the same as those of a one-component plasma, that is, a system of charges embedded in a uniform neutralizing background charge. This OCP resides in the potential well that is produced by the cylinder of uniform neutralizing charge and the end electrodes. For given potentials on the end electrodes and a given value of $\omega(\Omega_c - \omega)$ one can determine the shape of this confining potential well. Of course, the shape of this well is important in determining the shape of the plasma. For the simple case of a small plasma at the bottom of the well the effective confining potential is approximately quadratic and can be written as

$$\Phi_{conf} = \frac{m\omega_z^2}{2} (z^2 + \alpha\rho^2) \tag{8a}$$

where ω_z is the single-particle axial bounce frequency in the external trap potential, and

$$\alpha = \frac{\omega(\Omega_c - \omega)}{\omega_z^2} - \frac{1}{2} \tag{8b}$$

is the "trap parameter" which determines the shape of the effective confining potential. For this case Bollinger[7] and Turner[8] have shown that the plasma takes the shape of an ellipsoid as $T \to 0$, neglecting correlation effects. Correlations between charged particles in the plasma then are set up within this overall shape.

The correlation properties of strongly correlated one-component plasmas have been the subject of several theoretical investigations. Computer simulations of unbounded homogeneous one-component plasmas predict that for $\Gamma \geq 2$ the system begins to exhibit the local order characteristic of a liquid and for $\Gamma \sim 180$ there is a phase transition to a bcc crystal.[2] Current experiments with cryogenic pure ion plasmas in a Penning trap have achieved Γ values in the range of several hundred, but the experiments involve a relatively small number of particles so the theoretical studies of an infinite homogeneous one-component plasma cannot be trusted. On the other hand, these small plasmas are ideally suited for numerical simulations with a realistic number of particles. In order to study the correlation properties of these bounded plasmas we have carried out molecular dynamic (MD) simulations and Monte Carlo (MC) calculations with boundary conditions motivated by the experiments.

In the MD simulation, the equations of motion for N interacting charges in a Penning trap are integrated forward in time until the charges come into thermal equilibrium with each other. Average properties of the system such as the local density $n(\mathbf{x})$ are then determined as long-time averages: for instance,

$$n(\mathbf{x}) = \sum_{i=1}^{N} \int_0^{\tau} \frac{dt}{\tau} \delta(\mathbf{x} - \mathbf{x}_i(t)) \ .$$

In contrast, the MC calculation is a statistical game of chance based on the distribution for a canonical ensemble. We find that the MD simulation and the MC calculation yield the same answer for average quantities such as the local density, provided that $N \geq 100$ within the statistical error of the simulations. This is in general agreement with the previously stated result that the canonical and microcanonical ensembles yield the same average properties for large N.

There is one subtlety which should be noted in our molecular dynamics simulations. In the experiments the cyclotron radius typically is much smaller than the distance between particles, or equivalently, the cyclotron period is much shorter than an interaction time. Under these circumstances it is useful to average out the high frequency cyclotron dynamics before turning to the computer. This is accomplished by using the guiding center equations of motion rather than the exact equations of motion. Although the guiding center equations are only approximate, the thermal equilibrium structure obtained with them is not. By substituting the guiding center Hamiltonian and guiding center angular momentum into the Gibbs distribution (see Ref. 3), one finds the guiding center system has the same thermal equilibrium structure as the exact system for a slightly shifted magnetic field strength.

What follows are numerical results for the case of a small plasma at the bottom of the effective potential well given by Eq. (8).

3. NUMERICAL RESULTS

For convenience in displaying results, we typically choose conditions so that the well and plasma have spherical symmetry. The fact that the plasma radius is small compared to the radius of the conducting cylinder also implies that the force due to an image charge is small, so we take the interaction potential to be simply $e^2/|\mathbf{x}_i - \mathbf{x}_j|$. For the molecular dynamics simulation the guiding center equations of motion are solved using the fourth order Runga-Kutta algorithm and a fourth-order corrector algorithm, both with a variable time step. The codes have been tested against one another and have been vectorized to run efficiently on a CRAY X-MP computer. In the code, times are normalized to ω_z^{-1} and distances to $a_0 \equiv (3q^2/m\omega_z^2)^{1/3}$. [From Eqs. (7) and (8b), $a_0 = (2\alpha + 1)^{1/3} a_{ws}$.] We typically integrate for times of order $10^4 \omega_z^{-1}$, and in all cases energy is conserved to better than 1% of the total kinetic energy, and angular momentum to one part in 10^5. The MC calculation follows a standard Metropolis-Rosenbluth algorithm. Figure 2 shows the results of a MC calculation for Γ values ranging from 1 to 10. The average density $n(r)$ is plotted as a function of the spherical radius r for r values near the plasma edge. The effective potential well and plasma have spherical symmetry. For $\Gamma = 1$ the

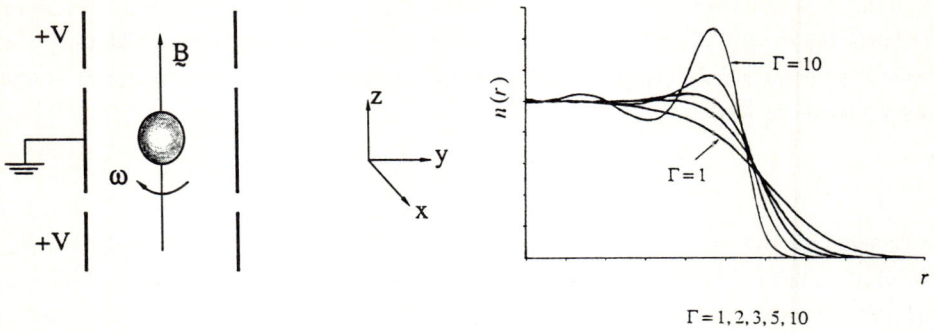

FIGURE 1
Cylindrical Penning Trap

FIGURE 2
MC results for density at the edge of a trapped cloud for various values of Γ.

density falls smoothly to zero, as it does for weak correlation, but for higher values of Γ oscillations are present near the plasma edge. These oscillations are evidence of local order. The damping length for the oscillations is a measure of the correlation length. One may think of the density maxima as embryonic lattice planes or more precisely spherical lattice shells. Such oscillations have also been observed in previous Monte Carlo studies of the so-called one-component plasma with an edge,[9] and were also observed by Totsuji for trapped ions.[10] Schiffer has also studied the correlation properties of cold ion clouds trapped in heavy ion storage rings.[3]

As Γ is increased, the oscillations increase in magnitude until the density between peaks goes to zero. For a spherical cloud with 100 particles, this occurs at $\Gamma \sim 140$ (Fig. 3). Thus the ion cloud separates into concentric spheres. For $N = 100$ there are three spheres with four ions in the innermost sphere, 26 in the middle sphere, and 70 in the outermost sphere. The areas under each peak are about equal implying that the number of ions per unit area in each sphere is the same, being set by the background density n_0. Thus the number of ions per sphere roughly scales as the surface area of the sphere.

If one tags an individual particle on one of these shells, one finds that the particle is localized to the shell, but is not localized on the shell. For this value of Γ, the particles still diffuse over the surface of the shell, that is, the system behaves like a crystal in the radial direction but like a liquid along the surface of the shell as in a smectic liquid crystal. For the $\Gamma = 140$ and $N = 100$ case of Figure 3 we study the particle diffusion further by considering the mean square displacement of the ions in time. For instance, we determine the average of $\delta z^2(t)$ where

$$<\delta z^2>(t) = \frac{1}{mN} \sum_{i=1}^{N} \sum_{j=1}^{m} [z_i(t_j + t) - z_i(t_j)]^2 ,$$

and $t_j - t_{j-1}$ is a constant time increment and $t \leq t_j - t_{j-1}$. This function increases linearly in time for $<\delta z^2>^{1/2}$ small compared to the cloud radius, so that we may obtain the average diffusion coefficient in the z direction, D_z, through the definition $<\delta z^2(t)>= 2D_z t$. For $\Gamma = 140$ this diffusion is shown in Figure 4 and may be contrasted to a similar plot of $<\delta r^2>$ in the same figure, where r is the spherical radius from the center of the cloud. While $<\delta z^2>$ does indeed increase linearly with time, $<\delta r^2>$ is almost constant, showing that there is little diffusion of ions from sphere to sphere.

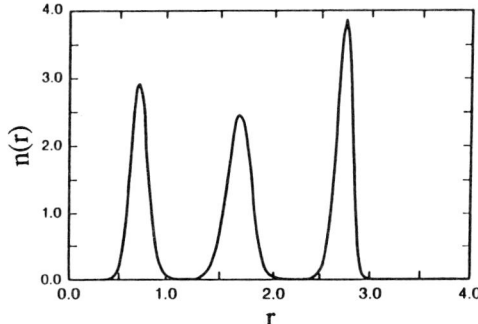 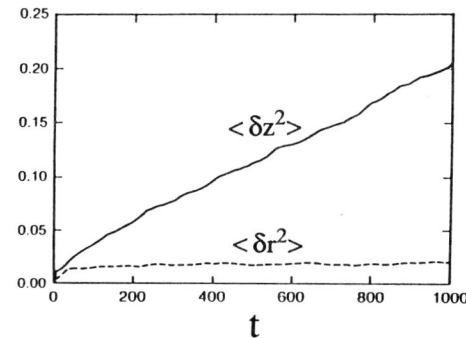

FIGURE 3
MD result for density of a cloud with $\alpha = 1$, $N = 100$, $\Gamma = 140$.

FIGURE 4
Diffusion in r and z for ions in cloud of Fig. 3; $\omega_z/\Omega_c = 0.1$.

For substantially higher values of Γ the particle diffusion along the surface of a shell also goes to zero and an imperfect 2-D hexagonal crystal is formed on the shell. Figure 5 shows a projection onto a plane of one-half of the outer shell for a $N = 256$ spherical plasma. For $\Gamma \geq 380$, the ions were confined to the lattice sites shown; the sites form a local equilibrium which may be thought of loosely as going to the $T = 0$ or $\Gamma = \infty$ limit. However, one should note that there are many such local equilibria in the N particle potential energy. Nevertheless there is a tendency toward a hexagonal crystal which one can identify in Figure 5. One may confirm this intuition by calculating the spatial correlation $c(s)$ function of all ions within a particular sphere. This function is defined as the probability density that an ion is at a distance s from another ion, counting only ions on a given shell. For $\Gamma = 140$ this correlation displays decaying oscillations characteristic of a fluid. However, for larger values of Γ the correlations become more highly peaked and the peaks correspond in position to those of a 2-D hexagonal crystal, and furthermore the number of ions in each peak correspond to that expected for 2-D hexagonal symmetry (see Fig. 6).

Another useful correlation function which further characterizes the crystalline order is the bond angle correlations $c_\theta(\theta)$ in the shell. This correlation function is defined as a probability density that a bond angle is at angle θ, averaged over all ions in a shell. The set of bond

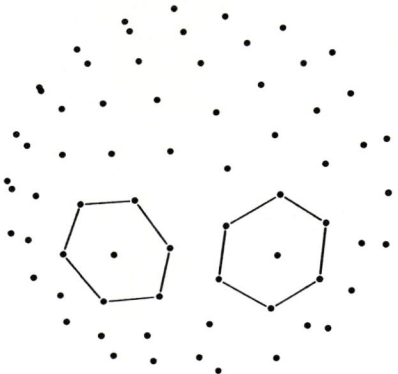

FIGURE 5
$T = 0$ equilibrium state for 1/2 of outer shell of an $N = 256$ spherical cloud.

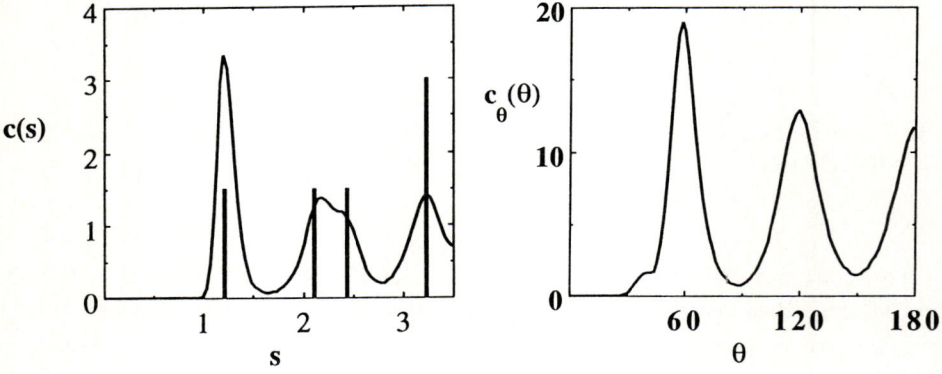

FIGURE 6
Correlations within outer shell of $N = 256$ cloud at $\Gamma = 380$. a) Spatial correlations. Vertical lines give positions and weights for $T = 0$ 2-D hexagonal lattice; b) Bond angle correlations, using 6 nearest neighbors in shell.

angles for a given ion are those angles subtended by any two of the ion's M nearest neighbors, using the given ion as a vertex; there are $M(M-1)/2$ such angles for each ion. As plotted in Figure 6 for the case of the outer shell of the $N = 256$, $\Gamma = 380$ crystallized cloud, one sees that the bond angle correlation function corresponds well to that expected for a 2-D hexagonal lattice.

This lattice structure is quite different from the bcc lattice predicted for an infinite homogeneous one-component crystal. Indeed, if one determines the three-dimensional spatial correlation function within a large Γ cloud one finds that even for large N the correlations bear little

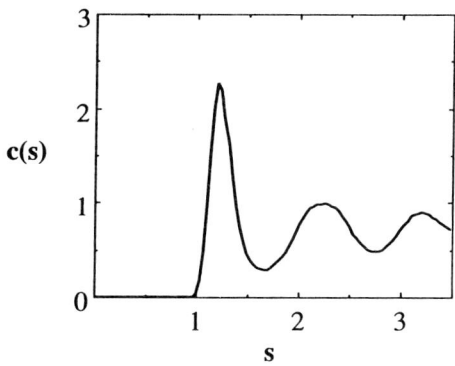
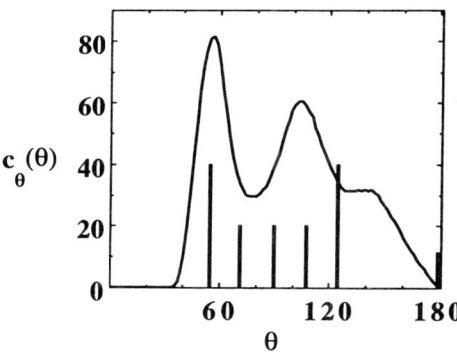

FIGURE 7a
3-D spatial correlations in $N = 2048$, $\alpha = 1$, $\Gamma = 290$ cloud, counting ions only within $r < 4.8$ (cloud extends to $r = 8$). First peak contains 14 ions.

FIGURE 7b
3-D bond angle correlations for $N = 256$ $\Gamma = 380$ cloud, using 14 nearest neighbors in cloud for each ion within inner shells. Vertical lines are $T = 0$ bcc lattice bond angles (14 n.n.).

resemblance to those of a body-centered cubic crystal. The three-dimensional bond angle correlations are also quite different (see Fig. 7).

We have performed computer simulations for up to 2,048 ions and we observe no convincing evidence of a body-centered cubic structure in the bulk of the ion clouds. Presumably, however, as $N \to \infty$, the system becomes infinite and homogeneous and the body-centered cubic crystal structure should appear. So the question arises, how large must an ion cloud be before this bulk behavior is observed?

4. SLAB MODEL OF THE BOUNDED COULOMB SYSTEM

In this section we consider in more detail how the boundedness of an ion system can affect its lattice structure. In the process, an estimate is obtained for the size of system required before infinite volume behavior is achieved, and several other relations are derived, including the approximate spacing between shells in the ion clouds, a result which can be compared to numerical and experimental results.

In order to make theoretical progress we consider a slab model of a bounded Coulomb crystal, which neglects the effect of curvature but still incorporates the effects of boundedness. The model consists of a collection of ions trapped in planar geometry in a 1-D quadratic well of the form $m\omega_0^2 z^2/2$. Note that this potential is that due to a uniform density background (here $n_0 = m\omega_0^2/(4\pi q^2)$), so this system is an OCP. The system is infinite and homogeneous in the x–y plane but bounded in the z-direction. This model allows us to make predictions concerning the lattice structure of ion clouds which are large enough so that shell curvature is small compared to the inter-ion spacing.

We find that the $T = 0$ equilibria for this system consist of a series of 2-D lattice planes oriented parallel to the $x-y$ plane. Sufficiently far into the bulk from the surfaces, these planes become evenly spaced in z, setting up a 3-D bulk lattice (see Fig. 8). The free energy F of this system then depends on the density n_0, the total number of ions per unit $x-y$ area $\bar{\sigma}$, the temperature T, the type of 3-D bulk lattice (e.g. bcc, fcc, hcp, etc.) and the orientation of the bulk lattice with respect to the surfaces. These latter two thermodynamic parameters are written in the language of solid state physicists by stating which bulk lattice plane lies parallel to the surface; for instance an fcc(111) lattice is the bounded lattice consisting of an fcc lattice in the bulk with the (111) plane oriented parallel to the surface.

In general the thermodynamically stable state is that which has minimum free energy. The free energy per ion F can be written as a sum of bulk and surface terms:

$$F = F_b + 2F_s/P \tag{9}$$

where F_b is the free energy per ion of the bulk lattice (including the "Vlasov" energy per ion of a uniform slab of charge in the external quadratic well), P is the number of lattice planes in the system (a function of $\bar{\sigma}$ and the lattice type), and F_s is the (positive) surface free energy. For a given T, n_0, and lattice type, as $\bar{\sigma}$ approaches infinity the number of planes P also approaches infinity, and by Eq. (9) $F \to F_{bulk}$. In this limit the system becomes infinite and homogeneous and, as is well known, the lattice with minimum free energy is body-centered cubic (bcc). However, for finite P surface effects are important; ion-ion correlations in the z direction are disrupted by the finite system size and bcc symmetry in the bulk is no longer necessarily the minimum free energy state.

We have determined F for this system as a function of $\bar{\sigma} n_0^{2/3}$ and Γ for various lattice types and orientations, in the "harmonic approximation," in which the temperature is assumed to be sufficiently small so that ions move only slightly from their lattice sites and the interion force may then be linearized (i.e., the system is assumed to be an ideal gas of phonons). Anharmonic effects, which are important near the liquid-solid phase transition, have not yet been included. In this approximation the free energy per ion may be written as

$$F = \phi_v + E_{corr} + \frac{kT}{N} \sum_{r=1}^{3N-3} \ln\left[\frac{\omega_r}{\omega_0}\right] + 3kT \ln\left[\frac{\hbar\omega_0}{kT}\right] \tag{10}$$

where $E_{corr} = E_b + 2E_s/P$ is the $T=0$ electrostatic correlation energy per ion. Here, E_b is the correlation energy per ion of the infinite homogeneous lattice and E_s is the contribution to the energy due to the two surfaces. The "Vlasov energy" ϕ_v gives the energy per ion of a uniform slab of charge in the quadratic well; E_{corr} is the additional energy due to the fact that the slab is actually made up of lattice planes.[4] The frequencies ω_r are the normal mode frequencies of the bounded lattice; the sum over normal modes may be interpreted for $N \to \infty$ as an average over the reciprocal cell of the lattice, and is written as $\frac{1}{N}\sum_r \ln(\omega_r/\omega_0) \equiv \langle \ln(\omega_r/\omega_0) \rangle$. This average approaches a constant "bulk" value as $P \to \infty$; the remainder for finite P is a surface term, which may be written as $\langle \ln(\omega_r/\omega_0) \rangle = \langle \ln(\omega_r/\omega_0) \rangle_b + 2\langle \ln(\omega_r/\omega_0) \rangle_s/P$. Comparison of Eqs. (9) and (10) then lead to the following identification for F_b and F_s:

$$F_b = \Phi_v + E_b + 3kT \ln\left(\frac{\hbar\omega_0}{kT}\right) + kT <\ln(\omega_r/\omega_0)>_b ,$$

$$F_s = E_s + kT <\ln(\omega_r/\omega_z)>_s$$

We have determined E_b, E_s, $<\ln(\omega_r/\omega_0)>_b$ and $<\ln(\omega_r/\omega_0)>_s$ for various lattices. The values of E_b for bcc, hcp and fcc lattices are well-known, and $<\ln(\omega_r/\omega_0)>_b$ has been determined for the bcc lattice.[2] (A previously-published value for the fcc lattice [11] is incorrect due to numerical error.) Values for E_b and E_s are found in Ref. 4 for these lattices as well as for other lattices, and we include in Table 1 some of our results for $<\ln(\omega_r/\omega_z)>_b$ and $<\ln(\omega_r/\omega_z)>_s$. Details of this calculation will appear elsewhere.

These results allow us to compare the free energy of the various lattice types as a function of $\bar{\sigma}n_0^{2/3}$ and Γ. We find that for $\bar{\sigma}n_0^{2/3} \geq 53$ (corresponding to about 60 bcc(110) planes), the minimum free energy lattice always has bcc(110) symmetry, regardless of Γ. However for $\bar{\sigma}n_0^{2/3}$ below this value, a competition occurs between the fcc(111) and bcc(110) lattices; the winner is usually fcc(111). (The rather complex phase diagram is shown in Fig. (9). Precise details of the diagram in the small Γ region should not be taken too seriously since our free energies neglect anharmonic effects. However, the general structure—bands of alternating fcc

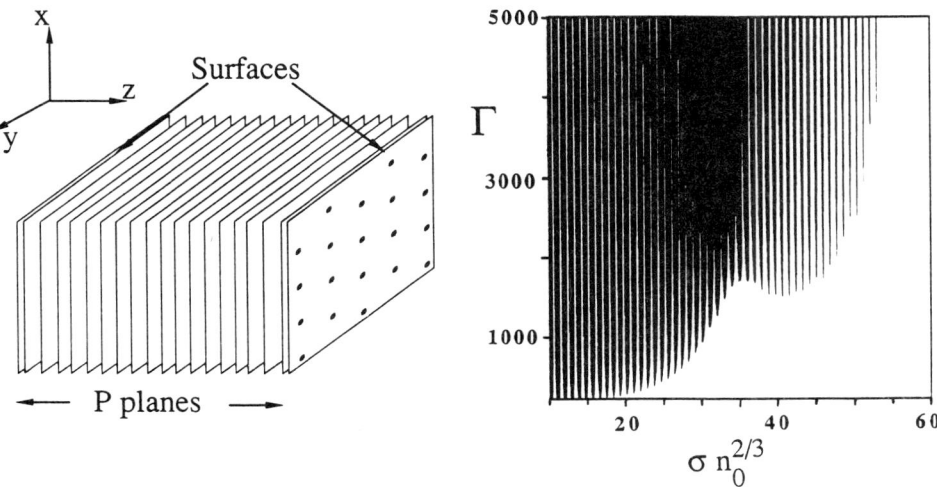

FIGURE 8
Schematic of $T=0$ ion slab equilibrium for 20 planes.

FIGURE 9
Phase diagram for bounded ion slab in regime $200<\Gamma<5000$, $10<\bar{\sigma}n_0^{2/3}<60$. fcc (111) lattice has lower free energy than bcc (110) in dark region.

Table 1

lattice type	$\langle ln(\omega_r/\omega_0)\rangle_b$	$\langle ln(\omega_r/\omega_0)\rangle_s$
fcc (001)	-2.45373(1)	0.105(1)
fcc (111)	-2.45373(1)	0.240(2)
bcc (001)	-2.49384(1)	-0.20(1)
bcc (110)	-2.49384(1)	0.233(1)

and bcc symmetry—is probably correct. It should also be noted that for large Γ metastable equilibria may exist for long times before thermodynamic equilibrium is achieved.) The dominant fcc structure may be understood through the fact that fcc(111) lattice planes are 2-D hexagonal lattices, which is the most efficient 2-D packing method, so planes are spaced far apart and correlations between planes are minimized. Surface effects, which raise F, depend on interplane correlations and are therefore also minimized. These surface effects are important even for large $\bar{\sigma}$ because F_b is almost the same for fcc and bcc lattices, so only small differences in F_s are needed to affect the lattice structure.

Note that in the simulations a distorted 2-D hexagonal lattice also appears on each shell. The distance D between (111) planes in the fcc lattice is, for $\bar{\sigma}$ large and $T=0$, given by $Dn_0^{1/3} = 2^{2/3}/\sqrt{3} = 0.9165$, which corresponds closely to the numerical results, which give slightly larger values, of approximately .92 - .93 (see Fig. 3). However, the spheroidal shells are not as closely correlated as are fcc(111) planes (see Fig. 7). This is because shell curvature causes a loss of correlation from shell to shell since 2-D lattices on different shells get "out of phase" as one moves from point to point on the shell surfaces. If one entirely neglects correlations between shells one finds[4] that for large clouds the intershell distance is now $Dn_0^{1/3} = 0.956$; the intrashell lattice remains 2-D hexagonal. The simulation results lie between no intershell correlation and perfect fcc(111) correlation.

REFERENCES

1) S. Gilbert, J. Bollinger and D. Wineland, Phys. Rev. Lett. **60** (1988) 2022.
2) E. Pollock and J. Hansen, Phys. Rev. A **8** (1973) 3110; W. Slattery, G. Doolen and H. DeWitt, ibid. **26** (1982) 2255; S. Ogata and S. Ichimaru, ibid. **36** (1987) 5451.
3) D. Dubin and T. O'Neil, Phys. Rev. Lett. **60** (1988) 511; J. Schiffer, ibid. **61** (1988) 1843.
4) D. Dubin, Phys. Rev. A **40** (1989) 1140.
5) T.M. O'Neil, Phys. Fluids **23** (1980) 2216.
6) J. Malmberg et al., *Proc. of 1984 Sendai Symposium on Plasma Nonlinear Phenomena* (1984) 31.
7) J. Bollinger and D. Wineland, Phys. Rev. Lett. **53** (1984) 348.
8) L. Turner, Phys. Fluids **30** (1987) 3196.
9) See, for example, S. Ichimaru, H. Iyetomi and S. Tanaka, Phys. Rep. **149** (1987) 91.
10) H. Totsuji, Static and dynamic properties of strongly-coupled classical one-component plasmas: Numerical experiments on supercooled liquid state and simulation of ion plasma in the Penning trap, in: *Strongly Coupled Plasma Physics,* eds. F. Rogers and H. Dewitt (Plenum, New York, 1987) p. 19.
11) H. Helfer, R. McCrory and H. Van Horn, J. Stat. Phys. **37** (1984) 577.

SURFACE PROPERTIES OF THE COULOMB LIQUIDS: FROM THE CLASSICAL ONE-COMPONENT
PLASMA TO LIQUID METALS

Masayuki HASEGAWA* and Mitsuo WATABE**

*Faculty of Engineering, Iwate University, Morioka 020, Japan
**Faculty of Integrated Arts and Sciences, Hiroshima University, Hiroshima
730, Japan

The results of the recent theoretical and computer simulation studies on
surface properties of the classical one-component plasma (OCP) provide a
useful starting point for understanding a liquid metal surface. Most difficult problems in the studies of a liquid metal surface are how to achieve
self-consistency between the electronic and ionic distributions and how to
treat screening effect in the inhomogeneous liquid-vapor transition zone.
Recently, an important progress has been made in these problems. We summarize these results of theoretical and computer simulation studies for the
OCP and liquid metal surfaces.

1. INTRODUCTION

In this paper we are primarily concerned with the surface properties of liquid simple metals such as the alkali metals, which may be considered as two-component systems consisting of the conduction electrons and ions. One of the simplest analyses of such a system assumes that the ions play no essential role other than providing a neutralizing charge background. This model is called an electron gas (or a jellium model of electrons) and its surface properties have been most extensively studied on the basis of the density functional theory (DFT)[1]. This model for a metal becomes more realistic if we introduce the discrete-ion effect in terms of the pseudopotential. Lang and Kohn[1] actually took this approach and calculated the surface energies of solid simple metals to first order in the pseudopotential. Their results for the densely packed crystallographic surfaces are in reasonable agreement with experiment.

The classical one-component plasma (OCP) may be another simplest model for a metal: in this case the conduction electrons are treated as the frozen charge background in which the ions move. Recently, theoretical and Monte Carlo (MC) simulation studies have been made systematically for the fluid OCP surface[2-7] and these results have provided a useful starting point for understanding a liquid metal surface. A theory of the inhomogeneous OCP also plays an important role in developing a microscopic theory of a liquid metal surface. Evans and Hasegawa developed such a theory[8], in which the electron-ion interaction was taken into account to first order in the pseudopotential and the treatment of the electronic and ionic distributions was completely self-consistent. However, surface tensions of the liquid alkali metals predicted by this first-order theory are about twice as large as experiment, which is in contrast to the reasonable success of the Lang and Kohn theory for a solid metal surface[1]. These results indicate that higher order terms in the pseudopotential (i.e. screening effect) play an important role in determining the surface properties of a liquid metal.

If we wish to take into account screening effect in the theoretical and computer simulation studies of a liquid metal surface, we have to face the difficult problems of how to achieve self-consistency between the ionic and electronic distributions and of how to treat screening effect in the liquid-vapor transition zone. Recently, some important progress has been made in these problems. In the following sections we summarize the recent developements in the studies of a liquid metal surface, which have been made along the route starting from the OCP surface.

2. OCP SURFACE

The OCP is a system of point-like charged particles (which we call ions) in a rigid compensating charge background. The bulk properties of a uniform OCP are characterized by the plasma parameter defined by $\Gamma = (Ze)^2/R_s k_B T$, where Ze is the charge of the ion and R_s is the ion-sphere radius defined by $R_s = (3/4\pi\rho)^{1/3}$, ρ being the ion number density. The bulk properties of the OCP are now rather well understood in the wide range of Γ by the extensive theoretical and computer simulation studies[9,10]. Hereafter, we refer Γ_b and ρ_b to the bulk values.

In the studies of the OCP surface the step-function has most often been assumed for the charge background. To simulate a planar surface of such a semi-infinite OCP, Badiali et al[3] used a finite system consisting of several hundred ions in a spherically symmetric charge background. However, the surface energies obtained by the MC simulations for such a system are in serious contradiction to the theoretical predictions in the weak coupling regime ($\Gamma_b < 1$)[4], where the approximate theory based on the DFT is expected to work well. We have demonstrated in later theoretical investigations[11] that these discrepancies between theory and "experiment" can be explained by the effect of finite curvature and the influence of a hard wall, the introduction of which is essential in the simulations of a finite system for preventing the particles from escaping to infinity[3]. To avoid such difficulties a system of slab geometry, which is extended indefinitely in two dimensions parallel to the surface by the periodic boundary condition, was used in later MC simulations[6,7].

The density profiles $\rho(z)$ and the surface energies U_s obtained by the MC simulations for a slab and spherical systems are shown in figs. 1 and 2. We find that $\rho(z)$ exhibits an oscillatory or stratified structure for $\Gamma_b > 5$ and this structure becomes prominent as Γ_b increases. For $\Gamma_b < 10$, the MC results of $\rho(z)$

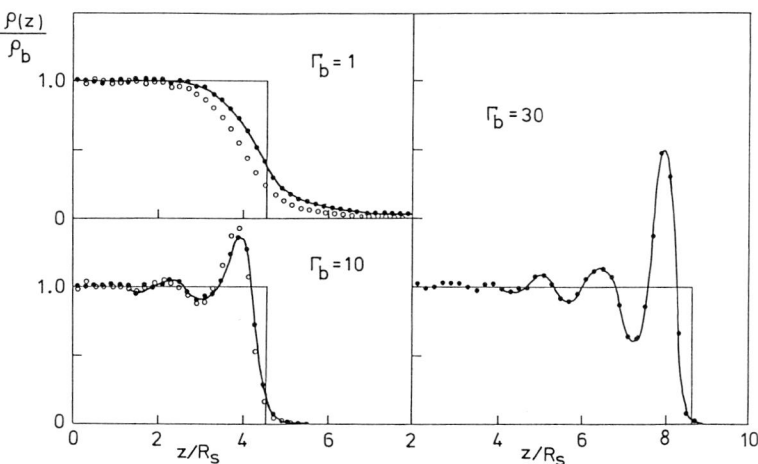

FIGURE 1
MC results for the ion density profile of the OCP. Full curves (smooth interpolations of the full circles), results for a slab (ref. 7)); open circles, results for a spherical system (ref. 3)).

for a spherical system differ from those for a slab and, as we have discussed in the above, a spherical system is in adequate for simulating a planar surface.

Recently, we have extended MC simulation studies to higher Γ_b and found that the outermost layer crystallizes into a two-dimensional hexagonal lattice at $\Gamma_b \simeq 130$, which is somewhat smaller than the bulk critical value $(\Gamma_c \simeq 178)$[4]. The crystallized phase grows towards the inside as Γ_b increases. This phenomenon, which we may call a "surface crystallization", is quite in contrast to the "surface melting" which is believed to occur in the actual systems[12]. The details of our MC simulation studies for higher Γ_b are given elsewhere[13].

Theoretical investigations of the OCP surface have been based on either the DFT or the integral equation method. We find that the DFT is more useful for studying the structural and thermodynamic properties on the same footing. The free energy functional of an inhomogeneous OCP is conveniently written as

$$F_{OCP}[\rho] = G_{OCP}[\rho] + \frac{e^2}{2} \int d\mathbf{r} \int d\mathbf{r}' \frac{[Z\rho(\mathbf{r}) - n(\mathbf{r})][Z\rho(\mathbf{r}') - n(\mathbf{r}')]}{|\mathbf{r} - \mathbf{r}'|} \quad (1)$$

where G_{OCP} is the non-coulombic contribution defined by this equation and $-en(\mathbf{r})$ is the charge background density. The basic problem of the DFT for any system is how to devise a functional form of the free energy and one of the practical methods is to use the truncated density-gradient expansion:

$$G_{OCP}[\rho] = \int d\mathbf{r} [g_0(\rho(\mathbf{r})) + g_2(\rho(\mathbf{r})) |\nabla \rho(\mathbf{r})|^2] \quad (2)$$

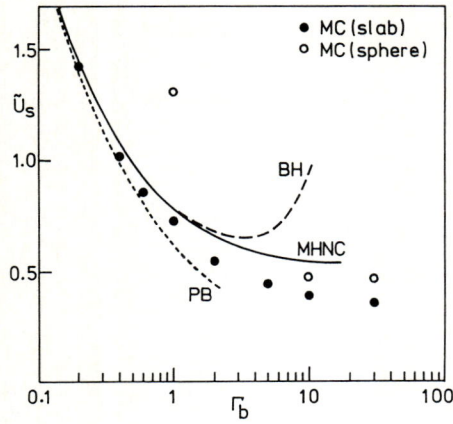

FIGURE 2
Surface energy of the OCP. Circles, MC results; full and dashed curves, variational calculations with the use of $Y(\Gamma)$ calculated in the modified hyper-netted chain (MHNC) and Baus-Hansen (BH) approximations; dotted curve, Poisson-Boltzmann (PB) approximation (eq. (5)). ($\tilde{U}_s = U_s/R_s\rho_b k_B T$).

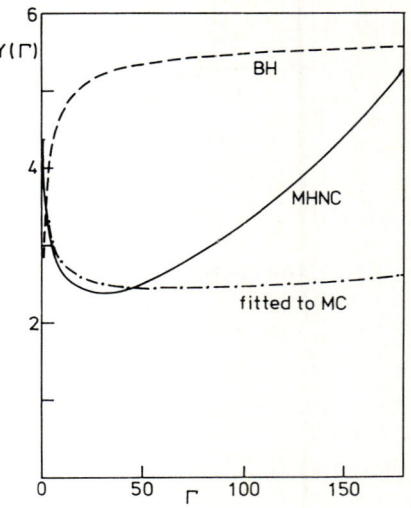

FIGURE 3
The results for $Y(\Gamma)$ calculated in the MHNC and BH approximations and fitted to the MC simulations for the first-order model.

where $g_0(\rho(\mathbf{r}))$ is the free energy density of a uniform OCP of the density equal to $\rho(\mathbf{r})$. The coefficient g_2 of the square-gradient term can be determined by requiring that eq. (2) should produce the result of linear response theory in an appropriate limit. The result is conveniently written as[5]

$$g_2(\rho) = \frac{(Ze)^2}{528}(3/4\pi)^{1/3}\rho^{-4/3}Y(\Gamma) \tag{3}$$

where

$$Y(\Gamma) = \frac{132}{\Gamma}\int_0^\infty x^4[c(x,\Gamma) + \frac{\Gamma}{x}]dx \quad (x = r/R_s) \tag{4}$$

$c(x,\Gamma)$ being the direct correlation function. The numerical factor in eq. (3) is nothing but the historical consequence and has no particular physical meaning.

Slattery et al[10] obtained an analytic expression for $g_0(\rho)$ which provides a very accurate fit to their MC simulations. On the other hand, the coefficient $g_2(\rho)$ or $Y(\Gamma)$ has been of some uncertainty. Figure 3 shows the results for $Y(\Gamma)$ obtained by using $c(x,\Gamma)$ in the modified hyper-netted chain (MHNC) and Baus-Hansen (BH) approximations[5]. The validity of these $Y(\Gamma)$ has been tested against the MC simulations: the results of variational calculations for the surface energy U_s are are compared with the MC results in fig. 2. We find that the MHNC result for $Y(\Gamma)$ produces the surface energy in reasonable agreement with the MC simulations for a slab.

The poisson-Boltzmann (PB) approximation, in which G_{OCP} is replaced by that of an ideal gas, provides a useful check on the validity of the MC simulations and variational calculations. In this approximation U_s can be calculated exactly and is given by[4,11]

$$U_s/R_s\rho_b k_B T = 1.08312/(3\Gamma_b)^{1/2} \tag{5}$$

This result should be exact in the weak coupling limit ($\Gamma_b \to 0$). The results of MC simulations for a slab and variational calculations in fact approach the PB result quite naturally as Γ_b decreases (see fig. 2). We note that the MHNC result for $Y(\Gamma)$ behaves as $\sim \Gamma^{-1/2}$ in the limit of small Γ and as $\sim \Gamma^2$ for large Γ. These asymptotic behaviors of $Y(\Gamma)$ have also been confirmed by Rosenfeld[14].

We may conclude from these results that the square-gradient approximation provides a reasonable description of the OCP surface in the relatively weak coupling regime $\Gamma_b < 10$. For $\Gamma_b > 10$, it seems that the stratified structure of $\rho(z)$ is too prominent for the square-gradient approximation to be applicable. In fact, variational calculations for $\Gamma_b > 10$ with the use of the MHNC result for $Y(\Gamma)$ produce too high U_s. One possible explanation for this is that higher order gradient terms become important in the strong coupling regime. Alastuey made an exhaustive attempt to incorporate the forth order gradient term in the surface calculations of the Lenard-Jones fluids[15]. We do not find it useful to make such an attempt for the OCP partly because of the convergence problem, for which almost nothing is known except our experiences in the relatively weak coupling regime. In stead of challenging such an exhausting attempt of calculating higher order gradient terms, we took another approach to this problem and determined an effective $Y(\Gamma)$ which reproduces the MC results for the surface energy of a model liquid metal. This result for $Y(\Gamma)$ is compared with the other results in fig. 3 (fitted to MC) and the method of this analysis is given in §3.

3. LIQUID METAL SURFACES

In the usual studies of a liquid metal we first rewrite the total Hamiltonian by adding and subtracting a charge background of density $Ze\rho(\mathbf{r})$, where Z is the valence of the ion and $\rho(\mathbf{r})$ is an assumed ion number density. Next, we make a Born-Oppenheimer separation of the total Hamiltonian, which enables us to reduce the problem to that of an one-component system. The effective Hamiltonian for the ion system is then given by[16]

$$H_{eff} = H_{OCP} + F_{el}(\{\mathbf{R}_\ell\}) \tag{6}$$

where H_{OCP} is the Hamiltonian of an OCP with charge background density $-Ze\rho(\mathbf{r})$ and F_{el} is the free energy of an electron system in the presence of the ions in a fixed configuration $\{R_\ell\}$.

In the usual perturbation calculations of F_{el}, the electron gas is adopted as the unperturbed system and the external field, which is treated as the perturbation, is given by

$$\delta v(r) = \sum_\ell v_{ps}(|r - R_\ell|) + \int dr' \frac{Ze^2\rho(r')}{|r - r'|} \tag{7}$$

where we have assumed that the electron-ion interaction can be described by a local pseudopotential $v_{ps}(r)$. For an inhomogeneous liquid metal, however, it is much more advantageous to take into account an average effect of $\delta v(r)$ in the unperturbed system. The unperturbed system we have chosen is given by[16]

$$H_0 = H_{eg} + \sum_i <\delta v(r_i)> \tag{8}$$

where H_{eg} is the Hamiltonian of an electron gas with charge background density $Ze\rho(r)$ and $<...>$ represents the average over an assumed ion distribution, i.e.

$$<\delta v(r)> = \int dr' v_{ps}(|r - r'|)\rho(r') + \int dr' \frac{Ze^2\rho(r')}{|r - r'|} = \int dr' w(|r - r'|)\rho(r') \tag{9}$$

Here $w(r)$ is the so-called repulsive part of $v_{ps}(r)$ defined by $v_{ps}(r) = -Ze/r + w(r)$. The external field which is treated as the perturbation is then given by $\Delta v(r) = \delta v(r) - <v(r)>$ and the perturbation series in $\Delta v(r)$ can be used in the calculations of F_{el}.

We note some important implications of the above perturbation expansion. Firstly, an assumed ion density $\rho(r)$ is arbitrary and may be taken to be the true one, which has yet to be determined. Then, self-consistency between $n_0(r)$ (the electron density of the unperturbed system) and $\rho(r)$ is achieved within the first-order coupling between the electrons and ions. The electron density $n_0(z)$ at the surface determined in this way is expected to relax more than the usual ones because $\phi_{ps}(z) = <\delta v(z)>$, which is large and positive in the bulk and vanishes outside the surface, acts as softening the steep dipole barrier created at the surface. We also find that the present scheme of the perturbation expansion is very efficient[16].

3.1. First-order model

If we neglect second and higher order terms in $\Delta v(r)$ in the perturbation calculations of F_{el}, the free energy of a liquid metal is given by

$$F = G_{OCP}[\rho] + G_{eg}[n] + E_{es}[n, \rho] + E_{ps}[n, \rho] \tag{10}$$

where G_{eg} is the non-coulombic part of the free energy of an electron gas, E_{es} the electrostatic energy (the second term on the r.h.s. of eq. (1)) and E_{ps} the pseudopotential term (i.e. first order term in the pseudopotential) given by

$$E_{ps} = \int dr \int dr' w(|r - r'|)n(r)\rho(r') \tag{11}$$

In the above we wrote $n_0(\mathbf{r})$ as $n(\mathbf{r})$ because it is the total electron density in this first-order model. Evans and Hasegawa[8] developed a self-consistent theory of a liquid metal surface based on this model and derived an exact expression for the surface tension. However, as we have discussed in §1, this model is still not realistic enough to describe an actual liquid metal surface, although it is useful as a starting point for understanding the screening effect.

Combined theoretical and MC simulation studies for the first-order model also provide a useful knowledge of an inhomogeneous OCP. We first note that the surface free energy (i.e. the surface tension) of this model is written as

$$\sigma = \sigma_{OCP}[\rho] + \sigma_{eg}[n] + \sigma_{es}[n, \rho] + \sigma_{ps}[n, \rho] \tag{12}$$

where each term on the r.h.s. corresponds to that of eq. (10). We performed MC simulations for this model and used both density profiles to calculate each term of eq. (12) except σ_{OCP}, which enabled us to separate u_{OCP}, the surface energy part of σ_{OCP}, from the MC result for the surface energy. We made these analyses using the parameters of liquid Na in the temperature range T = 773 - 1273, for which the bulk plasma paramer of the corresponding OCP is in the range Γ_b = 56 - 97. Then, we determined an effective $Y(\Gamma)$ which yields the results of u_{OCP} deduced from the MC simulations. The result can be fitted to analytic form

$$Y_{eff}(\Gamma) = 2.3125\Gamma^{-1/2} + 2.037 + 0.2153 \times 10^{-2}\Gamma \tag{13}$$

where we have required that it fits to the MHNC result for $\Gamma < 5$. The result of eq. (13) is shown in fig. 3 and we see that it is almost constant in the range $50 < \Gamma < 100$. This result will be useful as an effective $Y(\Gamma)$ which includes the effect of higher order gradient terms.

3.2. Second-order model: screening effect

As we have discussed in the above, the inclusion of higher order terms (i.e. screening effect) is essential in the studies of the actual liquid metal surface. We first consider MC simulation studies which have been made in the linear response approximation. In this approximation the effective potential for the ion system is written as

$$U_{eff} = U_0 + \sum_{\ell} \phi_1(\mathbf{R}_\ell) + \sum_{\ell<\ell'} \phi_{eff}(\mathbf{R}_\ell, \mathbf{R}_{\ell'}) \tag{14}$$

where U_0, ϕ_1 and ϕ_{eff} are respectively the constant term, one-body potential and effective (screened) pair potential between the ions and all these are functionals of the electron density $n_0(r)$ and (assumed) ion density $\rho(r)$. The expressions for U_0 and ϕ_1 depend on the scheme of perturbation expansion[16] and our results are different from the previous ones used by Rice and his co-workers[17]. The present method of achieving self-consistency between the electron and ion density profiles is also different, in accordance with the above difference, from that of the previous approach. Hence, it is interesting to compare the results of MC simulations based on the different approaches.

Figure 4 shows the comparison of the two results for the ion density profile of liquid Na. We see that two results are somewhat different from each other: whereas our result for $\rho(z)$ shows an oscillation which damps rather fast, the previous result has a tendency to oscillate even in the deep inside of the surface. The behaviors of the outermost layer are also different. These differences may be attributed primarily to the difference of the one-body potential ϕ_1. The detailed analyses of our MC simulations will be given elsewhere[18].

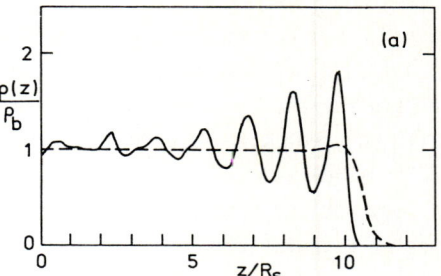

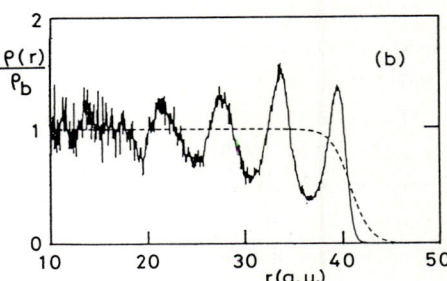

FIGURE 4
MC simulation results for the ion density profile of liquid Na at the melting point obtained by (a) Ishida et al for a slab (ref. 18)) and (b) Gryko and Rice for a spherical system (ref. 17)).

Next, we proceed to the theoretical investigations of a liquid metal surface in the second-order model. In the early stage of these studies, theoretical predictions of the surface tension have been of primary interest and various type of theories have been developed. These theories are based on either the usual scheme of the perturbation expansion[19] or more efficient one discussed in the above[20,21] and generally successful in predicting the surface tension of the liquid simple metals. However, all these theories are too much complicated or too crude to be used for studying the density profile at the surface. Recently, we have developed a simple theory which fully utilizes the theory of an inhomogeneous OCP and is fairly successful in predicting the surface density profile as well as the surface tension[22].

Our theory starts from the formal expression for the excess free energy F^{ex} of a liquid metal in the linear response approximation:

$$F^{ex} = E_M - TS^{ex} + G_{eg} + E_{es} + E_{ps} + E^{sc}_{pair} + E^{sc}_{self} \qquad (15)$$

where E_M is the Madelung energy, S^{ex} is the excess entropy and E^{sc}_{pair} and E^{sc}_{self} are the screening contributions to the pair interaction energy and self-energy of the ions respectively. Variational calculations based on the Gibbs-Bogoliubov inequality provide a practical method of calculating F^{ex} of a bulk liquid metal (in this case, $E_{es} = 0$)[22]. For an inhomogeneous liquid metal, such a method is unpractical because it is difficult to specify a reference system appropriate to an inhomogeneous system and, even if that is done, calculations of each term of eq. (15), especially of the pair interaction energy $E_M + E_{pair}^{sc}$ and entropy S^{ex}, are generally outside our capability. We have bypassed these difficulties and devised a useful method of evaluating these quantities of an inhomogeneous liquid metal. The results are given by

$$E_M + E_{pair}^{sc} = \int d\mathbf{r}\, \tilde{U}_{OCP}^{ex}(\mathbf{r}; [\rho], Z_{eff}^2(n_0(\mathbf{r}))) \tag{16}$$

and

$$S^{ex} = \int d\mathbf{r}\, \tilde{S}_{OCP}^{ex}(\mathbf{r}; [\rho], Z_0^2(n_0(\mathbf{r}))) \tag{17}$$

where $\tilde{U}_{OCP}^{ex}$ and $\tilde{S}_{OCP}^{ex}$ are the excess internal energy density and excess entropy density of an inhomogeneous OCP respectively, each with the density dependent effective coupling constant Z_{eff}^2 and Z_0^2. The square-gradient approximation may be used to obtain the expressions for $\tilde{U}_{OCP}^{ex}$ and $\tilde{S}_{OCP}^{ex}$. These effective coupling constants can be determined by performing variational calculations of the free energy for a uniform bulk liquid metal of arbitrary density (at fixed temperature) and then by equating the resulting $E_M + E_{pair}^{sc}$ and S^{ex}, each per unit volume, to $\tilde{U}_{OCP}^{ex}(Z_{eff}^2)$ and $\tilde{S}_{OCP}^{ex}(Z_0^2)$ respectively. We note that a local treatment is made for the effective coupling constants in eqs. (16) and (17). We further note that both Z_{eff}^2 and Z_0^2 should be viewed as functions of the electron density because the deviations of the values of these coupling constants from the bare value Z^2 are due to screening effect.

Figure 5 shows the density dependences of the effective coupling constants of three liquid metals determined in this way. The above method may be applicable to a liquid whose effective plasma parameters $\Gamma_{eff} = (Z_{eff}e)^2/R_s k_B T$ and $\Gamma_0 = (Z_0 e)^2/R_s k_B T$ at the bulk density n_b don't too much exceed the bulk critical value $\Gamma_c \approx 178$ (see table 1).

If we use eqs. (16) and (17), the surface free energy is given by

$$\sigma = u_{OCP}[Z_{eff}^2] - T s_{OCP}[Z_0^2] + \sigma_{eg}$$
$$+ \sigma_{es} + \sigma_{ps} + \sigma_{self} \tag{18}$$

where u_{OCP} and s_{OCP} are the surface energy and surface entropy of an OCP with the density dependent coupling constants $Z_{eff}^2(n_0(z))$ and $Z_0^2(n_0(z))$ respectively, and σ_{self} is the contribution from E_{self}^{sc}.

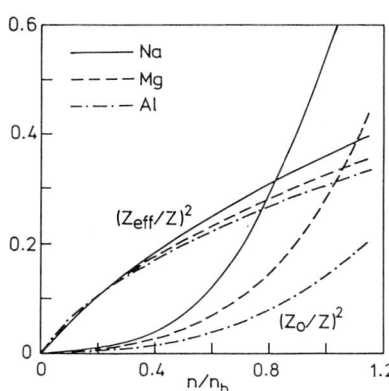

FIGURE 5
Density dependences of the effective coupling constants.

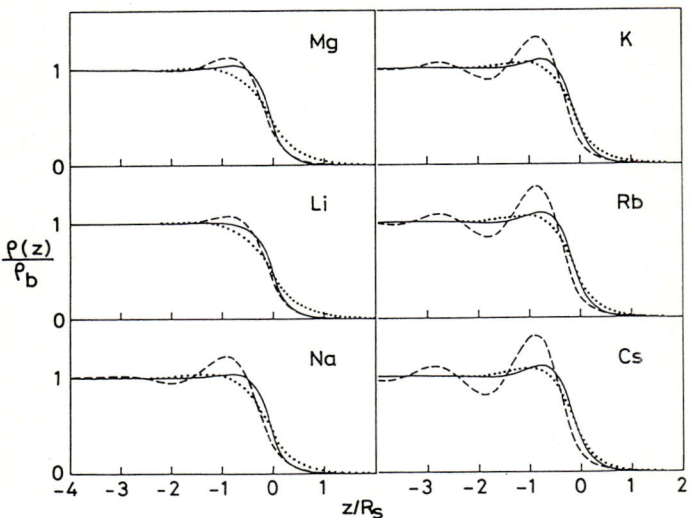

FIGURE 6
Ion density profiles $\rho(z)$ obtained by using the MHNC result for $Y(\Gamma)$: full curves, a monotone profile is assumed for $n_0(z)$; dashed curves, an oscillation is allowed for in $n_0(z)$ (dotted curves are the resulting $n_0(z)$).

Using the above formulation we performed variational calculations, in which we used the empty-core model potential for $v_{ps}(r)$ and Ichimaru-Utsumi screening and allowed for possible oscillations in the parametrized $n_0(z)$ and $\rho(z)$. In these calculations $n_0(z)$ is determined for a given $\rho(z)$, following the present scheme of perturbation expansion, by the minimum condition of the surface energy of the unperturbed electron system, $\sigma_0 = \sigma_{eg} + \sigma_{es} + \sigma_{ps}$. We then minimize σ with respect to $\rho(z)$ to have the opimum value, i.e. the surface tension. The results of these calculations are summarized in figs. 6 and 7 and in table 1.

Figure 6 shows the density profiles of the liquid alkali metals and Mg at their melting points obtained by using the MHNC result for $Y(\Gamma)$ to calculate the effective OCP contributions u_{OCP} and s_{OCP} in eq. (18)[22]. If we allow for an oscillation in $n_0(z)$, its effect on $\rho(z)$ is significant, especially for the heavy alkali metals, and the resulting $\rho(z)$ show more pronounced oscillations (dashed curves) than otherwise (full curves). These results for the alkali metals are in qualitative agreement with the MC simulations for Na and Cs[17,18]. The predicted surface tensions are also in good agreement with experiment (see table 1). It is easy to see from fig. 3 and table 1 that the use of the effective $Y(\Gamma)$ given by eq. (13) doesn't change so much the above results.

The ion density profiles of the divalent metals show only small oscillation even if the effective $Y(\Gamma)$ is used (see fig. 7). These results are somewhat

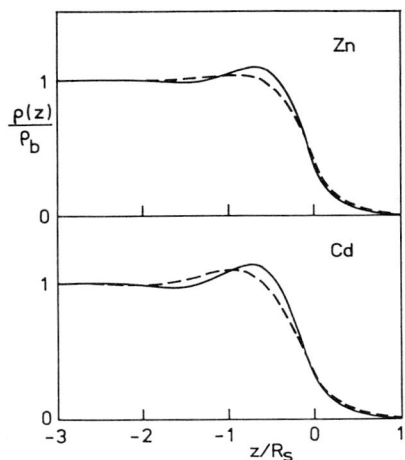

FIGURE 7
Ion density profiles of Zn and Cd: dashed curves, the MHNC $Y(\Gamma)$ is used; full curves, $Y_{eff}(\Gamma)$ is used.

different from that for Hg: for this metal an oscillatory $\rho(z)$ similar to that of Na (fig. 4) was obtained by the analysis of the X-ray reflectance experiment[17]. On the other hand, the recent analyses of the RHEED experiments suggest that monotone density profiles are plausible for liquid In and Sn[23]. We also note that approximate treatments made in the present theory for the inhomogeneous OCP and for the screening effect are more crucial for the polyvalent metals than for the alkali metals. As Foiles and Ashcroft pointed out, a nonlocal treatment of the screening in the calculations of σ_{self} is also important for the polyvalent metals[24].

4. CONCLUSION

The surface properties of the OCP are now rather well understood by theoretical and computer simulation studies and these results provide a useful starting point in developing a theory of liquid metal surfaces. The present theory, one of such theories, actually provides a reasonable description of the surface properties of the liquid alkali metals. On the other hand, our theory is less satisfactory for the polyvalent metals and we need further theoretical and experimental investigations for these metals.

REFERENCES

1) N.D. Lang and W. Kohn, Phys. Rev. B 1 (1970) 4555.

2) P. Ballone, G. Senatore and M.P. Tosi, Physica 119A (1983) 356.

3) J.P. Badiali, M.L. Rosinberg, D. Levesque and J.J. Weis, J. Phys. C 16 (1983) 2183.

4) M.L. Rosinberg, J.P. Badiali and J. Goodisman, J. Phys. C 16 (1983) 4487.

5) M. Hasegawa and M. Watabe, J. Phys. C 18 (1985) 2081.

6) H. Totsuji, J. Phys. C 19 (1986) L573.

7) A. Ishida, M. Hasegawa and M. Watabe, J. Phys. C 20 (1987) L599.

8) R. Evans and M. Hasegawa, J. Phys. C 14 (1981) 5225.

TABLE 1. Calculated surface tensions of liquid simple metals.

Metal	T(K)	r_c	Γ_b	Γ_{eff}	Γ_0	u_{OCP}	$-Ts_{OCP}$	σ_{eg}	σ_{es}	σ_{ps}	σ_{self}	σ	σ_{exp}
Li	454	1.37	210	79	120	120	-57	84	154	-627	63	339	398
Na	371	1.70	210	77	117	257	-26	128	85	-244	15	215	191
K	337	2.20	187	69	116	92	-15	90	51	-105	1	114	115
Rb	312	2.38	188	70	122	56	-13	70	47	-74	-2	91	85
Cs	302	2.59	181	67	122	26	-12	64	42	-47	-4	69	70
Mg	922	1.30	398	129	112	1689	-90	-278	608	-1506	282	705	559
Zn	693	1.04	613	188	134	2812	-109	-1033	975	-2261	502	886	782
Cd	594	1.25	635	205	168	1931	-80	-383	661	-1668	319	780	570
Al	933	1.12	977	298	138	—	—	—	—	—	—	—	914

r_c is the empty-core radius (given in atomic units) of the model potential and σ and its component parts (see eq. (18)) are given in units of dyn/cm. We have used the MHNC $Y(\Gamma)$ for the alkali metals and $Y_{eff}(\Gamma)$ for the divalent metals.

9) M. Baus and J.P. Hansen, Phys. Rep. 59 (1980) 1.

10) W.L. Slattery, G.D. Doolen and H.E. DeWitt, Phys. Rev. A 21 (1980) 2087.

11) M. Hasegawa and M. Watabe, J. Stat. Phys. 41 (1985) 281.

12) e.g. P. Stoltze, J.K. Norskov and U. Landman, Phys. Rev. Lett. 61 (1988) 440.

13) A. Ishida, T. Shirakawa, M. Hasegawa and M. Watabe, Monte Carlo simulations for the surface properties of the strongly coupled one-component plasma, this volume.

14) Y. Rosenfeld, private communication.

15) A. Alastuey, Molec. Phys. 52 (1984) 637.

16) M. Hasegawa and M. Watabe, Can. J. Phys. 65 (1987) 348.

17) S.A. Rice, Z. Phys. Chem. 157 (1988) S.445, and references therein.

18) A. Ishida, M. Hasegawa and M. Watabe, Proc. 7th Intern. Conf. Liquid and Amorphous Metals, in print.

19) M. Hasegawa and M. Watabe, J. Phys. C 15 (1982) 353.

20) D.M. Wood and D. Stroud, Phys. Rev. B 28 (1983) 4374.

21) E. Chacon, F. Flores and J. Navascues, J. Phys. F 14 (1984) 1587.

22) M. Hasegawa, J. Phys. F 18 (1988) 1449.

23) M. Hasegawa and T. Ichikawa, Proc. 7th Intern. Conf. Liquid and Amorphous Metals, in print.

24) S.M. Foiles and N.W. Ashcroft, Phys. Rev. A 30 (1984) 3136.

CLASSICAL CHARGED PARTICLE SYSTEMS WITH INTERFACES

Hiroo TOTSUJI

Department of Electrical and Electronic Engineering,
Faculty of Engineering, Okayama University,
Tsushimanaka 3-1-1, Okayama 700, Japan

Layered structures of finite systems of classical charged particles at low temperatures, confined by external forces of various dimensionalities, are analyzed on the basis of a model which takes the main part of the effect of discreteness of charges into account. The results of previous numerical experiments are reproduced to a good accuracy. Structural transitions at critical values of the characteristic parameter are also predicted and confirmed by our numerical experiments. The roles of intra-layer and inter-layer Coulomb interactions are discussed by comparing our model with known three-dimensional lattices of one-component plasma.

1. INTRODUCTION

Recently, structures in finite nonneutral systems of charged particles have been observed by experiments on ions in the Penning trap.[1] The possibilities to confine nonneutral plasmas by magnetic field[2] or in Penning and Paul traps have long been known but these structures have appeared after strong coupling of charges is attained by cooling down trapped ions.[3,4] The main feature of this kind of structure is the existence of a well-defined boundary and layers which reflect the symmetry of the confining field.

These systems of charged particles are equivalent to a finite nonneutral system of one component, embedded in the uniform background of opposite charges with appropriate symmetry. In the case of one-component plasma (OCP) with the semi-infinite background, similar structures have also been observed at low temperatures.[5] We may thus regard these layered structures as a result of the existence of the boundaries or interfaces in finite or semi-infinite systems.

These structures at low temperatures may also be considered as a kind of lattice of charges under the constraint of the symmetry inherent in the system. In the case of confinement around an axis, charges make up a solid whose structure is consistent with the axial symmetry. In the case of one-dimensional confinement, such a solid will have translational invariances in the plane perpendicular to the confining force. By analogy with the usual lattice which is composed of lattice planes, the lattice of axially confined charges may be composed of lattice planes which are cylinders concentric with the symmetry axis. In the case of semi-infinite background, the orientation of the lattice will be determined by the existence of the interface and may lead to appearance of lattice planes parallel to the interface.

For theoretical analyses of these structure, numerical experiments[6-8] with well defined parameters have been particularly useful. In simulations, we are also able to change the dimensionality of the problem and the results with reduced dimensions are also helpful to clarify the nature of these systems.

In this paper, we first describe a model[9] which reproduces these structures in the case of axially symmetric confinement to a good accuracy and compare its most interesting prediction, discontinuous transitions between different structures, with the results of our numerical experiments.[10] We then generalize the model to different dimensionalities and discuss the roles of Coulomb interaction within and between the layers.

2. SHELL STRUCTURE OF CHARGES UNDER AXIALLY SYMMETRIC CONFINEMENT

2.1. Theory

Shell structures appeared in numerical experiments on axially confined charges[6] have provided, though for a restricted value of characteristic parameters, a very useful information (the position of the shells and their populations) for theoretical analyses of a model proposed by the present author and J. -L. Barrat.[9]

The potential energy of this system is written as

$$V = \sum_{i>j} q^2/|\mathbf{r}_i - \mathbf{r}_j| + (k/2) \sum_i R_i^2, \tag{2.1}$$

where q is the charge and $\mathbf{r}_i = (x_i, \mathbf{R}_i)$ is the coordinates of the particle i. Denoting the number density per unit length along the x-axis by n, we rewrite (2.1) as

$$V = q^2 n \left[\sum_{i>j} 1/n|\mathbf{r}_i - \mathbf{r}_j| + (k/2q^2 n^3) \sum_i (nR_i)^2 \right]. \tag{2.2}$$

Thus our system at low temperatures is characterized by a single parameter

$$k' = k/q^2 n^3. \tag{2.3}$$

We note that the effect of the confining field is equivalent to the existence of the uniform distribution of the charge of opposite sign with the charge density $-k/2\pi q$. When the discreteness of particles is completely neglected, the ground state is therefore a uniformly charged cylinder in which the electric field is canceled. Thus the appearance of structures in this system, radial shells and ordering within each shell, is a result of the discreteness of particles.

In order to take the effect of discreteness in the radial direction into account, we assume that our charges form N shells which are concentric with the axis and denote the radius and the linear number density by R_i and n_i, respectively. The potential energy of our system is then evaluated as

$$u = -(q^2/n)[2 \sum_{i>j} n_i n_j \ln(R_i/L) + \sum_i n_i^2 \ln(R_i/L)]. \qquad (2.4)$$

With f_i, i=0, 1, 2, ..., N, defined by

$$f_i = \sum_{j \leq i} n_j/n \qquad (f_0 = 0, \ f_N = 1), \qquad (2.5)$$

the total energy is written as

$$v_0 = (q^2 n/2) \sum_i [(k'/2)(f_i - f_{i-1}) R_i'^2 - (f_i^2 - f_{i-1}^2) \ln(R_i')], \qquad (2.6)$$

where $R_i' = nR_i$.

When the number of shells N is fixed, the position and population of the ground state of the system described by (2.6) approximately reproduce the results of numerical experiments.[6] When optimized with respect to N, however, we have again a uniform distribution: The number of shells becomes infinite. We thus expect that the discreteness within each shell plays a major role to determine N in the ground state.

The effect of discreteness in shells appears as the cohesive (or correlation) energy of two-dimensional charge distribution. In estimating this energy, we note that the correlation energy of OCP at low temperatures is approximately given by the Madelung energy of appropriate lattices both in two[11] and three dimensions.[12]

In order to evaluate this energy, we have calculated the Madelung energy of some lattices (one-dimensional along the axis) on the shell.[13] The Madelung energy of such a lattice with radius R_i and linear number density n_i is expressed as $q^2 n_i e(n_i R_i)$ where $e(n_i R_i)$ is a dimensionless function. In the case where the radius of the shell is much larger than the mean distance on the shell, the correlation energy reduces to the known values of two-dimensional system of charges. In the opposite case of very small radius, the lattice almost reduces to a linear chain. The results of calculations indicate that the correlation energy of lattices on a cylinder is approximately expressed by either of these two limiting expressions as

$$e_p(n_i R_i) = -0.78213(n_i R_i)^{-1/2} \tag{2.7}$$

when the radius is large, or

$$e_\ell(n_i R_i) = (\gamma - \ln 2) + \ln(n_i R_i) = -0.23186/2 + \ln(n_i R_i) \tag{2.8}$$

when the radius is small. The former is the Madelung energy of a triangular lattice and the latter is that of a linear chain of charges.

Using these two limiting values as correlation energy in shells to be added to the total energy (2.6), we write the energy as

$$v = v_0 + \sum_{i \geq 1} (f_i - f_{i-1})^2 e_p[(f_i - f_{i-1})R_i'] + \ln(Ln) \tag{2.9}$$

or

$$v = v_0 + \sum_{i \geq 2} (f_i - f_{i-1})^2 e_p[(f_i - f_{i-1})R_i'] + f_1^2 e_\ell(f_1) + \ln(Ln). \tag{2.10}$$

The above expressions are optimized with respect to the number, the radii, and the populations of shells. The results are summarized in FIGURES 1-3.

The number of shells in the ground state is shown in FIGURE 1. Here we see that the number of shells increases with the decrease of the parameter k'. The critical values are shown in TABLE I.

The radii of shells are summarized in FIGURE 2. With the decrease of k', the new shell appears at the center with very small radius and then expands outwards. Outer shells expand discontinuously when a new shell appears.

In FIGURE 3, populations on the shells are shown. We see that the population of the new born shell is small but finite: For k' a little larger than the critical value, the energy with smaller number of shells is lower than that with a new shell at the center with small population. When we sweep the parameter k', the structure thus experiences successive discontinuous transitions.

For large values of k', our system reduces to a helical lattice around the axis. The number of particles in the unit cell along the axis also changes in the order of 1 (equally spaced linear chain), 2 (helix with 2-fold symmetry), 3 (helix with 3-fold symmetry), ... when k' is decreased.

2.2. Comparison with Numerical Experiments

The results of experiments by Rahman and Schiffer,[6] Schiffer,[7] and ourselves[10] are also shown in FIGURES 1-3. We observe that our model reproduces them with sufficient accuracy.

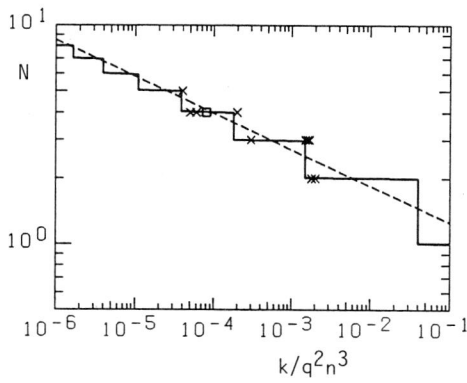

FIGURE 1
Number of shells N vs. characteristic parameter k/q^2n^3. Experimental results are shown by square[6] and crosses.[10] Broken line is the average behavior obtained by experiments.[7]

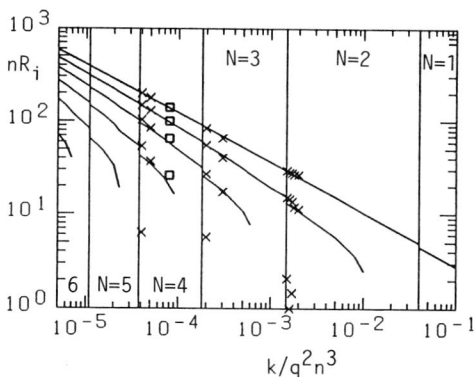

FIGURE 2
Positions of shells nR_i vs. k/q^2n^3. Experimental results are shown by squares[6] and crosses.[10]

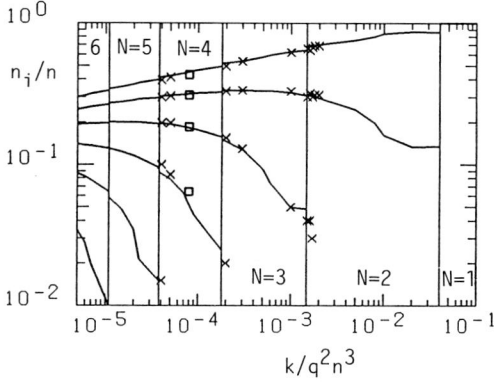

FIGURE 3
Populations of shells n_i/n vs. k/q^2n^3. Experimental values are shown by squares[6] and crosses.[10]

TABLE I
Critical values of k/q^2n^3 for transitions $N \to N+1$.[9]

N	k/q^2n^3
1	$4.00 \cdot 10^{-2}$
2	$1.48 \cdot 10^{-3}$
3	$1.80 \cdot 10^{-4}$
4	$3.80 \cdot 10^{-5}$
5	$1.09 \cdot 10^{-5}$
6	$3.95 \cdot 10^{-6}$
7	$1.65 \cdot 10^{-6}$
...	...

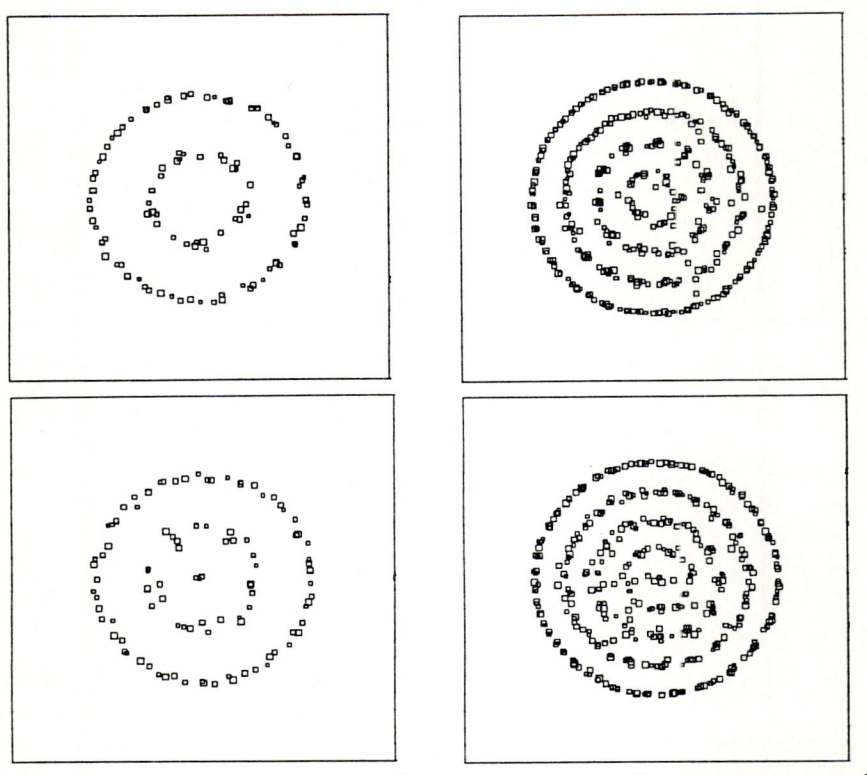

a) N=2(k'=1.8•10⁻³) → N=3(k'=1.7•10⁻³) b) N=4(k'=5.0•10⁻⁵) → N=5(k'=4.0•10⁻⁵)

FIGURE 4
Examples of transition:[10] a) $N=2(k'=1.8 \cdot 10^{-3}) \to N=3(k'=1.7 \cdot 10^{-3})$, and b) $N=4(k'=5.0 \cdot 10^{-5}) \to N=5(k'=4.0 \cdot 10^{-5})$.

Examples of the change of number of layers around critical values of the characteristic parameter are shown in FIGURE 4. Drastic changes of the structure with small change of the parameter may be observed. As for the critical values, our numerical experiments give

$$1.7 \cdot 10^{-3} < k'(2 \to 3) < 1.8 \cdot 10^{-3},$$
$$2.0 \cdot 10^{-4} < k'(3 \to 4) < 3.0 \cdot 10^{-4},$$
$$4.0 \cdot 10^{-5} < k'(4 \to 5) < 5.0 \cdot 10^{-5},$$

for transitions which are predicted at $1.48 \cdot 10^{-3}$, $1.80 \cdot 10^{-4}$, and $3.80 \cdot 10^{-5}$, respectively. While predicted values are a little smaller than experimental one, we may conclude our model works in these predictions.

3. ONE-DIMENSIONAL CONFINEMENT

3.1. Theory

Let us now consider the confinement of a system of nonneutral charges in a one-dimensional external field parallel to the x-axis derived from the potential $(k/2)x^2$. In this case, the potential energy of our system is given by

$$V = \sum_{i>j} q^2/|r_i-r_j| + (k/2) \sum_i x_i^2. \tag{3.1}$$

The effect of the confining field is equivalent to the existence of a slab of the uniformly distributed opposite charges, symmetrical with respect to the plane x=0, with sufficient thickness along the x-direction. The charge density in the slab is related to the parameter k by $-(k/4\pi q)$.

We confine n charges per unit area perpendicular to the x-axis. Rewriting (3.1) as

$$V = q^2 n^{1/2} [\sum_{i>j} 1/n^{1/2}|r_i-r_j| + (k/2q^2 n^{3/2}) \sum_i (n^{1/2} x_i)^2], \tag{3.2}$$

we see that this system at low temperatures is characterized by a single parameter proportional to $k/q^2 n^{3/2}$.

When we neglect the discreteness of charges and regard them as a continuum, the lowest energy configuration is clearly the uniform distribution of thickness $nq/(k/4\pi q)=4\pi nq^2/k$ around the plane x=0: The uniform opposite charges are exactly canceled, giving no electric field for $|x|<2\pi nq^2/k$.

The discreteness of charges leads to structures parallel and perpendicular to the x-axis. We first take a part of this effect into account by assuming that our system is organized into N sheets which are perpendicular to the x-axis, and denote the position and the population (areal number density) of the sheet i by x_i and n_i, respectively. Due to symmetry of the external confining field, our system is symmetrical with respect to x=0 and we have, with natural indices to sheets,

$$x_{-i} = -x_i, \quad n_{-i} = n_i, \quad \text{for } i = 1, 2, \ldots M, \quad \text{and} \quad x_0 = 0. \tag{3.3}$$

The population of the central sheet n_0 is either 0, when N is even (N=2M), or positive when N is odd (N=2M+1).

The first term on the right hand side of (3.1), the electrostatic energy, is then evaluated as (per particle)

$$u = 4\pi(q^2/n)[\sum_{i>0} n_i^2(L-x_i) + 2\sum_{i>j>0} n_i n_j(L-x_j)]$$

$$+ 4\pi(q^2/n)[(1/4)n_0^2 L + \sum_{i>0} n_0 n_i(L-x_i)]. \qquad (3.4)$$

Here we have taken the zero of the potential at $x = \pm L$.

Introducing n_i' and f_i defined respectively by

$$n_i' = n_i/n, \qquad (3.5)$$

$$f_i = \sum_{0<j\leq i} n_j' + n_0'/2, \qquad (3.6)$$

we rewrite (3.4) into

$$u = 4\pi n^{1/2} q^2 [L'/4 - \sum_{i\geq 0} (f_i^2 - f_{i-1}^2) x_i']. \qquad (3.7)$$

Here $L' = n^{1/2} L$, $x_i' = n^{1/2} x_i$, $f_0 = n_0'/2$, $f_1 = n_1' + n_0'/2$, ..., and $f_M = 1/2$.

Total energy per particle is thus given by

$$v = 4\pi q^2 n^{1/2} \{ \sum_{i>0} [k' x_i'^2 - (f_i^2 - f_{i-1}^2) x_i'] + L'/4 \}, \qquad (3.8)$$

with the characteristic parameter k' is defined by

$$k' = k/4\pi q^2 n^{3/2}. \qquad (3.9)$$

Minimization with respect to x_i' ($i = 1, 2, \ldots$ M) and f_i' ($i = 0, 1, 2, \ldots$ M) leads to the results

$$f_i = i(1/N), \; n_i = n/N, \; x_i' = (i-1/2)(1/k'N) \quad \text{for } N = 2M, \qquad (3.10)$$

$$f_i = (2i+1)(1/2N), \; n_i = n/N, \; x_i' = i(1/k'N) \quad \text{for } N = 2M+1. \qquad (3.11)$$

Thus the ground state is simply realized by equally charged sheets with equal spacing $x_i - x_{i-1} = 1/k'N$. We here note that the total thickness of our system is given by $1/k'$ which is equal to the thickness in the case of continuum.

As in the case of cylindrical symmetry, the number of sheets is not determined by the above argument: The state with infinite number of sheets or the state of uniformly charged continuum has the minimum energy, a situation similar to the one without discreteness effects. This indicates that we have to take other aspects of the latter effect to obtain a finite number of sheets.

Next step may be to include those effects within each sheet which manifests themselves as the negative correlation energy in the two-dimensional system of charges with ordinary three-dimensional Coulomb interaction. In order to take this correlation energy in each sheets into account, we note the fact that it is approximately expressed by the Madelung energy of the triangular lattice for a wide domain of the coupling parameter of this two-dimensional system. We may thus approximate the correlation energy (per particle) in each sheet with the number density n_i by $e_p^0 q^2 n_i^{1/2}$ where e_p^0 is a constant given as

$$e_p^0 = -2.106711(3^{1/2}/2)^{1/2} = -1.960515. \quad (3.13)$$

The total energy with the correlation energy is written as

$$v = 4\pi q^2 n^{1/2} \{ \sum_{i \geq 0} [k'x_i'^2 - (f_i^2 - f_{i-1}^2)x_i' + (e_p^0/4\pi)(f_i - f_{i-1} + \delta_{i,0} f_0)^{3/2}] + L'/4 \}. \quad (3.14)$$

Now we minimize the above expression with respect to the position and population of each sheet. The results for the positions and populations are the same as in the case of no correlations in sheets when the number of sheets is fixed. Resultant minimum energy, however, does include the effect of intra-sheet correlation and the optimum number of sheets is determined by the characteristic parameter of our system k'.

The spacing and the optimum number of sheets are shown in FIGURES 5 and 6. The critical values for transitions are given in TABLE II. When the characteristic parameter k' exceeds these critical values, charges are reorganized into new structure with increased number of sheets. The spacing between sheets is decreased discontinuously, while the total thickness kept unchanged.

3.2. Experiments

It has been reported[7] that the spacing, population, and the number of sheets obtained by numerical experiments are approximately expressed as

$$x_i - x_{i-1} = 1.38 \, (3q^2/k)^{1/3} = 1.38 \, (3/4\pi)^{1/3} \, k'^{-1/3} = 0.86 \, k'^{-1/3},$$

$$n_i = [1/\pi(1.74/2)^2] \, (3q^2/k)^{-2/3} = [1/\pi(1.74/2)^2] \, (3/4\pi)^{2/3} \, k'^{2/3} = 1.09 \, k'^{2/3},$$

$$N = (1/0.43) \, n \, (3q^2/k)^{2/3} = (1/0.43)(3/4\pi)^{2/3} \, k'^{-2/3} = 0.89 \, k'^{-2/3}.$$

These results of experiments are also plotted in FIGURES 5 and 6 and we see that they are roughly reproduced by our model. We here note that the distance between adjacent sheets is not a monotonous function of the parameter: In addition to the systematic change with the parameter k', the distance experiences jumps when the number of sheets changes. Therefore it is not possible to simply extrapolate the above experimental results obtained for the cases $N \sim 6$ to the cases $N \gg 1$. This may explain the apparent discrepancy between the total thickness obtained by using the above expressions formally, $0.86 \cdot 0.89/k' = 0.77/k'$, and the expected value $1/k'$.

TABLE II
Critical values of $k/4\pi q^2 n^{3/2}$ for transitions $N \to N+1$ sheets.
$1 \to 2$: $6.84 \cdot 10^{-1}$, $2 \to 3$: $2.86 \cdot 10^{-1}$, $3 \to 4$: $1.68 \cdot 10^{-1}$, $4 \to 5$: $1.14 \cdot 10^{-1}$
$5 \to 6$: $8.38 \cdot 10^{-2}$, $6 \to 7$: $6.50 \cdot 10^{-2}$, $7 \to 8$: $5.23 \cdot 10^{-2}$, ...

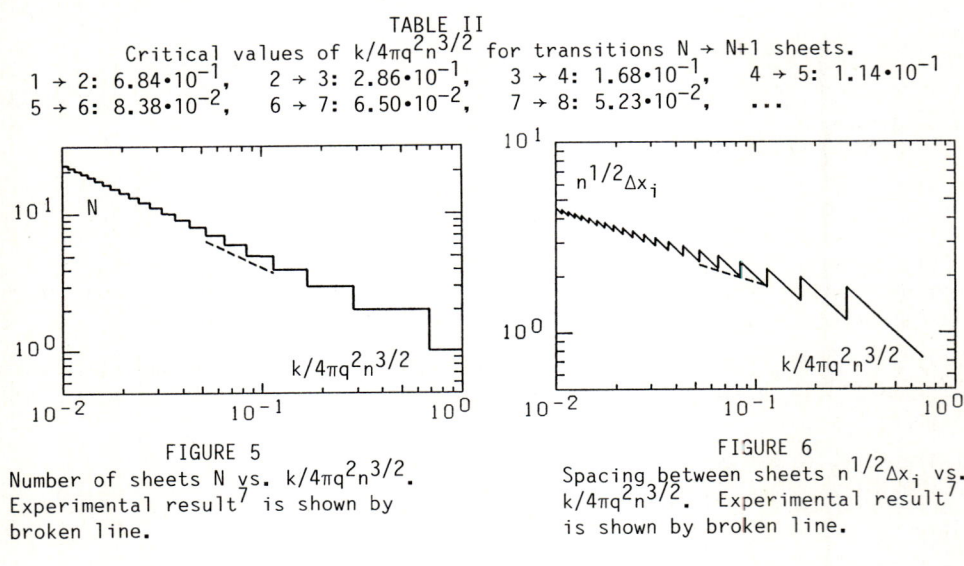

FIGURE 5
Number of sheets N vs. $k/4\pi q^2 n^{3/2}$. Experimental result[7] is shown by broken line.

FIGURE 6
Spacing between sheets $n^{1/2}\Delta x_i$ vs. $k/4\pi q^2 n^{3/2}$. Experimental result[7] is shown by broken line.

4. THREE-DIMENSIONAL OCP LATTICE AS COLLECTION OF TWO-DIMENSIONAL OCP LATTICES

Here we discuss the roles of intra-layer and inter-layer interactions in layered structures. For this purpose it may be useful to describe known three-dimensional OCP lattices in the same approximation as of our model.

Any lattice is expressed as a collection of lattice planes. Let us now construct a lattice by first (a) placing uniformly charged sheets at the positions of lattice planes, and then (b) forming some ordered configuration of point charges in each sheet.

The process (a) requires a positive electrostatic energy (per particle)

$$(\pi/6)q^2 n_s d \tag{4.1}$$

where n_s is the surface number density and d is the distance between sheets.

The process (b) requires the correlation energy in the plane, or that of two-dimensional OCP with three dimensional Coulomb interaction, and the correlation energy due to inter-layer correlations. It is known that the former is negative and approximately given by (per particle)

$$-1.9605 q^2 n_s^{1/2}. \tag{4.2}$$

We here note that, since we have already taken into account the effect of average charge distribution in planes in the process (a), the residual part of the inter-layer interaction energy related to the process (b) comes from the two-dimensional Fourier component of the charge density (in the plane) ρ_K with nonzero wave numbers **K**. The electric potential due to these Fourier components at a distance x from the plane is given by

$$\sum_{\mathbf{K}} \exp(i\mathbf{K}\cdot\mathbf{R} - K|x|)(2\pi/K)\rho_{\mathbf{K}}. \tag{4.3}$$

Here the smallest value of K is of the order of $2\pi/L_{yz}$ where L_{yz} is the period of order in the plane. We thus expect that, as long as $|x|$ is not much smaller than L_{yz}, these potential may be smaller than the contribution of uniform, smeared-out distribution by a factor $\exp(-K|x|)$. It may be also clear that their contribution to the total energy of the lattice is usually positive. In our model, this part of inter-layer interaction is neglected and inter-layer interactions are evaluated only for averaged uniform distributions.

The correlation (Madelung) energies of the simple, body-centered, and face-centered cubic lattices with number density n evaluated in this approximation are shown below in comparison with exact values.

[lattice]	[model]	[exact]
simple cubic	$-1.437 q^2 n^{1/3}$	$-1.41865 q^2 n^{1/3}$
body centered cubic	$-1.435 q^2 n^{1/3}$	$-1.44423 q^2 n^{1/3}$
face centered cubic	$-1.437 q^2 n^{1/3}$	$-1.44414 q^2 n^{1/3}$

We see that our model works as a good approximation. (The Madelung energy of the triangular lattice is used in these calculations. The correlation energy of the simple cubic lattice is overestimated since the triangular lattice is stable compared with the square lattice. With the Madelung energy of the latter gives $-1.427 q^2 n^{1/3}$.)

We may thus, though not completely, justify our neglect of a part of the inter-layer interaction in our model.

5. CONCLUSION

We have shown that the structures of finite charged particle system at low temperatures are reproduced by a simple model which is simple but takes the essential part of the discreteness effect of charges into account. One of the most interesting features of this kind of system, the discontinuous change of the structure, is predicted and confirmed by numerical experiments. It is also shown that the model is applicable for the case of different dimensionalities.

The roles of intra-layer and inter-layer Coulomb interactions in these systems are discussed and the approximate treatment of the inter-layer interaction based on the neglect of Fourier components with nonzero wave number is shown to be useful.

ACKNOWLEDGMENTS

Part of this article is based on the joint work with J. -L. Barrat. The author would like to thank him and J. -P. Hansen for stimulating discussions which resulted in that work. Thanks are also due to J. P. Schiffer and J. J. Bollinger for enlightening conversations which he enjoyed during the meeting.

REFERENCES

1) S. L. Gilbert, J. J. Bollinger, and D. J. Wineland, Phys. Rev. Lett. 60 (1988) 2022.
2) For example, J. M. Malmberg and T. M. O'Neil, Phys. Rev. Lett. 39 (1977) 1333.
3) J. J. Bollinger and D. J. Wineland, Phys. Rev. Lett. 53 (1984) 348.
4) H. Totsuji, in: Strongly Coupled Plasma Physics, eds. F. J. Rogers and H. E. DeWitt (Plenum, New York, 1987), p.19.
5) For example, H. Totsuji, J. Phys. C: Solid State Phys. 19 (1986) L573.
6) A. Rahman and J. P. Schiffer, Phys. Rev. Lett. 57 (1986) 1133.
7) J. P. Schiffer, Phys. Rev. Lett. 61 (1988) 1843.
8) D. H. E. Dubin and T. M. O'Neil, Phys. Rev. Lett. 60 (1988) 511.
9) H. Totsuji and J. -L. Barrat, Phys. Rev. Lett. 60 (1988) 2484.
10) H. Totsuji, unpublished.
11) H. Totsuji, Phys. Rev. A17 (1977) 399; R. C. Gann, S. Chakravarty, and G. V. Chester, Phys. Rev. B20 (1979) 326.
12) W. L. Slattery, G. D. Doolen, and H. E. DeWitt, Phys. Rev. A26 (1982) 2255.
13) H. Totsuji, Phys. Rev. A38 (1988) 5444.
14) Analysis on one-dimensional confinement has also been done by Dubin (Phys. Rev. A40 (1989) 1140) which was brought to the author's notice after completion of this work.

SURFACE CORRELATIONS IN CLASSICAL FINITE COULOMB SYSTEMS

Ph. CHOQUARD

Institute de Physique Théorique, Ecole Polytechnique Fédérale de Lausanne, PHB-Ecublens, CH-1015 Lausanne, Switzerland

We review analytical and numerical studies of surface correlations in finite, homogeneously polarizable, classical Coulomb systems placed in an insulating or conducting environment.

The original motivation of this work was the lack of understanding, from the point of view of statistical mechanics, of the shape-dependent laws of phenomenological electrostatics which govern the relations connecting the dielectric constant (a shape-independent quantity) of a macroscopic piece of an homogeneously polarizable system, its susceptibility tensor (a quantity proportional to the equilibrium fluctuations of the system's instantaneous polarization per unit volume), the depolarization tensor associated with its shape, and the dielectric constant of the surrounding medium.

This gap has been filled in several cases and the overall picture is (i) that the shape-dependent part of the susceptibilities results from the action of unbounded observables (the products of the components of the system's instantaneous polarization) on long-range surface correlations, both angular and radial, and (ii) that the relations of electrostatics are verified by means of shape-dependent thermodynamic limits, a crucial and novel prescription in the statistical mechanics of Coulomb systems.

The case of the Clausius-Mossotti relation will serve as an enlightening illustration of these features.

1. Introduction.

In a series of three papers (refs. 1, 2, 3, henceforth referred to as I, II and III) which deal with the theory of the size and shape dependence of the dielectric susceptibility of finite, homogeneously polarizable, classical Coulomb systems surrounded by a medium of arbitrary dielectric constant, it has been shown that the contact with the laws of electrostatics can be achieved through a systematic study of surface correlations in model systems.

The first exact results on the size and shape dependence of the dielectric susceptibility of a classical conductor appeared in I. Direct calculation of the susceptibility of a two-dimensional one-component plasma (O.C.P.) in a neutral disk containing N particles and for the particular value of the coupling constant $q^2/k_BT = 2$, for which the model is exactly solvable, produced the value π^{-1} $(1 - (\pi N/2)^{-1/2} + 2 N^{-1} + O(N^{-3/2}))$, i.e. π^{-1} in the thermodynamic limit, as required by *the Clausius-Mossotti relation* (see below) but which is twice as large as the value given by *the second moment Stillinger-Lovett sum rule* (idem and also ref. 4, section II.D.1). Why is $\pi^{-1} = (2\pi)^{-1} + (2\pi)^{-1}$ is a question to which the second paper is addressed. For the same solvable model it has been shown there that the correlation function of two particles close to the boundary of the disk decays as the inverse square of their distance (a typical dipole–dipole interaction due to the asymmetric screening of the charges when they are close to the wall of the system) whereas it has a gaussian decay when the particles are in the interior of the disk. Furthermore, in introducing the concept of partial second moment, an object which permits us to examine the contributions to the total susceptibility of pair of particles of increasing separation, it has been shown that the susceptibility is made up of a bulk contribution which saturates over the scale of the interparticle distance toward the Stillinger–Lovett value $1/2\pi$ and of a surface contribution which varies on the scale of the disk diameter and is described by the law π^{-2} arcsin (r/2R), originally suggested by computer simulations and which saturates to $1/2\pi$ at the disk diameter. Computer simulations of a two-component Coulomb system in a disk and at high temperature confirmed the generality of the "arccsine law" followed by the partial susceptibility of the system. The susceptibility of the 2D O.C.P. in a strip geometry and on the surface of a sphere have also been investigated in I and II. We found : saturation of the perpendicular susceptibility to $1/2\pi$ and divergence of the parallel susceptibility for the strip as required by electrostatics and saturation of the isotropic susceptibility toward the Stillinger–Lovett value of $1/2\pi$ for the boundary free spherical surface.

So far we had been dealing with 2D systems surrounded by the vacuum i.e. $\varepsilon' = 1$.

At this stage, a series of questions emerged naturally, namely : What about the Clausius-Mossotti relation which says that the susceptibility of a disk-shaped conductor surrounded by an insulator of dielectric constant ε' is, in thermodynamic limit, given by $2/\pi + 2\varepsilon'/\pi$? About the size dependence of this law ? About the origin of its divergence if the surrounding medium is a conductor, i.e. $\varepsilon' = \infty$? What about other shapes such as elliptic ones ? About three-dimensional systems ? These are the questions addressed in the third paper.

The problem of providing answers to these questions is that there are no solvable model of finite Coulomb systems of given size and shape and subjected to image forces resulting from the presence of a surrounding medium with arbitrary dielectric constant. For that reason, we have 1°) had recourse to *the linear Debye-Hückel approximation* which had already, and successfully, been applied to the investigation of surface correlations of semi-infinite conductors (ref. 5, p. 55) for $\varepsilon' = 1$, and 2°) conducted a series of computer experiments with the O.C.P. on a disk in different environments to guide us and to enable confrontation with the results of the Debye-Hückel theory.

We found that, for a conducting disk and sphere of radius R in an ε' environment, the D.H. approximation was not only capable of reproducing in the thermodynamic limit the predictions of electrostatics but it did so for both situations of a permeable and impermeable wall. In the first case a "grand canonical" version of the D.H. theory has been used whereas for the second case a novel "canonical" version has been developed. In both cases it has been shown (e.g. Eqs.(4.17) & (4.30) of III) that, for finite ε', the dominant term of the surface correlation function is proportional to ε' and consists of an angular-dependent part which decays with the inverse square (cube) of the chord of the disk (sphere) plus an angular-independent contribution which is proportional to the inverse square (cube) of the radius of the disk (sphere). It turns out that, in the "canonical" limit of the theory, the amplitude of this contribution vanishes for the disk whereas it persists for the sphere. The three-dimensional analogue of the "arcsine law" reflects this feature: the partial surface susceptibility contains indeed two terms and reads (e.g. Eq.(5.16) of III) $\varepsilon' (4\pi)^{-1} \left(\frac{r}{2R} + \left(\frac{r}{2R}\right)^4 \right)$. In adding the bulk contribution of $(4\pi)^{-1}$ we find the correct value of the susceptibility for the conducting sphere, namely $(1 + 2\varepsilon')/4\pi$.

So far, ε' was finite. We considered next the case of a metallic environment ($\varepsilon' = \infty$: Dirichlet boundary conditions) : it is the situation of a conductor in a conductor with permeable or impermeable walls. In both cases the susceptibility is shown to diverge linearly with the radius of the disk or of the sphere (e.g. Eq.(3.19a) of III). In the first case this divergence is due to charge fluctuations. In the second case, we found that the canonical surface correlation function (e.g. Eq. (6.5) of III) is given by the grand canonical one with Dirichlet boundary condition (a function of short range) plus an angular-independent contribution proportional to R^{-1} for the disk and to R^{-2} for the sphere. This peculiar property can be appreciated if we recall that in an ordinary fluid the difference between a canonical and grand canonical correlation function would be proportional to R^{-2} for the disk and R^{-3} for the sphere. In the appendix B of III, it is shown that the radial pair distribution function possesses a dip of amplitude R^{-1}, respectively R^{-2}, close to the boundary of the disk and of the sphere and that the second moment of these radial distributions diverges linearly with R. A remarkable fit is observed between the radial pair distribution function obtained by computer simulations and by the canonical D.H. theory, as shown in Fig.6 and Fig.8 of III. It would be interesting to establish the $R^{-(\nu - 1)}$ behavior of the canonical pair distribution function on the basis of the BBGKY hierarchy.

Anisotropic shapes are studied in Sections 8 and 9 of III. In Section 8, the particular case of an elliptic geometry is considered. The solution of the D.H. equation exhibits again a slow decay of the correlation functions along the wall. The new result is that the strength of the correlations in regions of high curvature is increased. This feature is reminiscent of the "needle effect" known in electrostatics and provides an intuitive explanation to the problem of the shape-dependent susceptibilities in terms of correlation functions.

A fresh look at the problem of the surface correlations of a conductor in a vacuum is proposed in Section 9 of III. The idea is to approach the problem of the surface correlation of a conductor in the vacuum from outside, i.e. from a point source located at a point close to the boundary of the system but exterior to it. An intuitive description of the situation is presented, and proved in the Appendix D of III for the disk, that, for sufficiently smooth shapes (i.e. of curvatures larger than the Debye length), the surface correlation of the conductor is given by the bi-gradient of the exterior Coulomb kernel satisfying Dirichlet boundary conditions. This proposal is applied to the disk, to the ellipse, to the sphere, to the cylinder and reproduces exactly the results known for these smooth shapes.

At this stage of the investigations we feel that more work needs to be done if the system is in a dielectric state, if it can undergo phase transitions such as dielectric-plasma or para-ferroelectric ones and, of course, if it is not classical. More work needs also to be done on the theory of image forces.

In section 2 of this paper we formulate the problem of the shape dependance of the susceptibility in more precise terms and in section 3 we sketch the derivation of the D.H. approximation for the disk and for the sphere and summarize the results obtained in sections 3 to 6 of III.

2. Statement of the problem.

The problem can be stated as follows. We consider an assembly of n species of charged particles confined in a domain Λ, of volume $|\Lambda|$, in the v-dimensional euclidean space ($\nu = 2,3$). The system is assumed to be in thermal equilibrium at the inverse temperature $\beta = 1/K_B T$, where K_B is Boltzmann's constant, and to be surrounded by vacuum.

We know that much information regarding the equilibrium properties of such systems can be derived from the truncated charge-charge correlation function

$$S_\Lambda(x,y) = <Q(x)Q(y)> - <Q(x)><Q(y)> \qquad (2.1)$$

where x and y are v-dimensional vectors in Λ, Q(x) is the instantaneous charge density

$$Q(x) = \sum_{\alpha;j} q_{\alpha;j} \delta(x - x_{\alpha;j}) \qquad ; \alpha = 1, 2, ..., n \; ; j = 1, 2, ..., N_\alpha \qquad (2.2)$$

and where the bracket means ensemble average.

In terms of the one- and two-particle correlation functions

$$\rho_{\alpha;\Lambda} = \sum_j <\delta(x - x_{\alpha;j})> \quad \text{and} \quad \rho_{\alpha\beta;\Lambda} = \sum_{j;k} <\delta(x - x_{\alpha;j})\delta(y - x_{\beta;k})>$$

with $j \neq k$ if $\alpha = \beta$ and of the truncated two-particle correlation functions

$$\rho^T_{\alpha\beta;\Lambda}(x,y) = \rho_{\alpha\beta;\Lambda}(x,y) - \rho_{\alpha;\Lambda}(x)\rho_{\beta;\Lambda}(y)$$

$S_\Lambda(x,y)$ becomes

$$S_\Lambda(x,y) = \sum_\alpha q_\alpha^2 \rho_{\alpha;\Lambda}(x)\delta(x-y) + \sum_{\alpha;\beta} q_\alpha q_\beta \rho^T_{\alpha\beta;\Lambda}(x,y) \qquad (2.3)$$

We consider next the response of the system to an homogeneous external field $E_{1;ext}$ applied in the x direction. According to the linear response theory, the dielectric susceptibility component $\chi_{11;\Lambda}$ takes the form

$$\chi_{11;\Lambda} = \frac{\beta}{|\Lambda|} \left(<P_1^2> - <P_1>^2 \right) \tag{2.4}$$

where P_1 is the x_1 component of the instantaneous polarization of the system

$$P = \sum_{\alpha;j} q_{\alpha;j} x_{\alpha;j} = \int_\Lambda dx \, x \, Q(x) \tag{2.5}$$

Using Eqs. (2.5) and (2.1), Eq. (2.4) becomes

$$\chi_{11;\Lambda} = \frac{\beta}{|\Lambda|} \int_\Lambda dx \, dy \, x_1 \, y_1 \, S_\Lambda(x,y)$$

which can also be written, using Eq. (2.3)

$$\chi_{11;\Lambda} = \frac{\beta}{|\Lambda|} \int_\Lambda dx \, dy \, x_1 \, y_1 \left\{ \sum_\alpha q_\alpha^2 \rho_{\alpha;\Lambda}(x) \delta(x-y) + \sum_{\alpha;\beta} q_\alpha q_\beta \rho^T_{\alpha\beta;\Lambda}(x,y) \right\} \tag{2.6}$$

This is the polarization fluctuation (P.F.) formula for the susceptibility. In using the identity

$$x_1 y_1 = -\frac{(y_1-x_1)^2}{2} + \frac{x_1^2 + y_1^2}{2}$$

we can decompose the P.F. susceptibility in two parts: one is called *the second moment (S.M.) susceptibility*

$$\chi_{11;\Lambda}(S.M) = -\frac{\beta}{2|\Lambda|} \int_\Lambda d^v x \, d^v y \, (y_1 - x_1)^2 S_\Lambda(x,y) \tag{2.7}$$

and the other, *the excess susceptibility*

$$\Delta\chi_{11;\Lambda} = \frac{\beta}{2|\Lambda|} \int_\Lambda d^v x \, d^v y \, (x_1^2 + y_1^2) S_\Lambda(x,y) \tag{2.8}$$

We notice here that whenever

$$\int_\Lambda dy \, S_\Lambda(x,y) = 0 \tag{2.9}$$

the excess susceptibility is zero. The relation (2.9) is called *perfect screening* or *monopole sum rule*. This sum rule is true whenever there are no charge fluctuations in the system : the P.F. susceptibility is then equal to the S.M. susceptibility and, for isotropic systems $(y_1 - x_1)^2$ can be replaced by $(y-x)^2/\nu$. In the cases where this sum rule is violated we shall call total charge fluctuation the volume integral of eq.(2.9) and, pictorialy, "charge fluctuation" density the l.h.s. of eq.(2.9).

The interest of the dielectric susceptibility resides in the fact that it gives us information on the state of the system considered. This occurs via the relation between the susceptibility and the dielectric constant ε of the system. We recall that a dielectric state is characterized by $0 < \varepsilon^{-1} \leq 1$ and a plasma state by $\varepsilon^{-1} = 0$. According to the phenomenological laws of electrostatics the relation between the dielectric constant and the susceptibility is shape-dependent.

If Λ is a ν-dimensional sphere immersed in a vaccum then the susceptibility is isotropic and its relation to the dielectric constant should be governed by *the Clausius-Mossotti equation*, namely

$$\varepsilon^{-1} = \frac{1 - \left(\frac{\nu-1}{\nu}\right) 2\pi \chi_\oplus}{1 + \left(\frac{(\nu-1)^2}{\nu}\right) 2\pi \chi_\oplus} \quad ; \nu = 2, 3 \quad (2.10)$$

Here the index $\oplus$ means disk or sphere.

It follows from Eq. (2.10) that the value of the susceptibility in the plasma state, $\chi_\oplus^P$, is given by

$$\chi_\oplus^P = \frac{\nu}{(\nu-1) 2\pi} \quad ; \nu = 2, 3 \quad (2.11)$$

On the other hand, the statistical mechanics of infinite Coulomb systems (e.g. ref. 4, section II.C & II.D) tells us that, under the assumptions that (1) the state of the finite system converges to a state of the infinite system defined by correlation functions $\rho_\alpha(x)$, $\rho_{\alpha\beta}(x,y)$, which are stationnary solutions of the BBGKY hierarchy with Maxwellian velocity distribution and that (2) $\rho_{\alpha\beta}^T(x, y)$ decays faster than $|x-y|^{-(\nu+2)}$ then, it is proved that, for a conductor, the susceptibility tensor obtained from the inverse dielectric function in the limit of vanishing wave vector (e.g. ref.4, eqs.(1.29) & (1.30))is isotropic and is given by *the second moment Stillinger-Lovett sum rule*

$$\chi_{SL}^P = -\frac{\beta}{2\nu} \sum_{\alpha;\beta} q_\alpha q_\beta \int_{R^\nu} dr\, r^2\, \rho_{\alpha\beta}^T(r) = \frac{1}{(\nu-1)\, 2\pi} \quad (2.12)$$

We notice at this point that $\chi_\oplus^P$ of Eq. (2.11) is ν times χ_{SL}^P of Eq. (2.12) !

It is thus clear that assumption (2) is incompatible with electrostatics : long range correlations have to come into play !

To resolve this contradiction it appeared to us imperative to calculate $\chi_{11;\Lambda}$ explicitly for a given model and to proceed to the thermodynamic limit in a way preserving the shape of the system. This is precisely what we did in I and II, as reported above. The extension of these investigations to other shapes and to other environments ($\varepsilon' \neq 1$) implied the derivation of a generalization of Eq. (2.10) . This was the purpose of the second section of III.

The presence of a dielectricum surrounding the systems confined in their domain Λ modifies their hamiltonians in two ways : i) through the kernel of the interaction potential which conveys the effect of the image charges and ii) through the coupling of their instantaneous polarization $\mathcal{P}$ with the external field E_0 which occurs now via the field E_a acting in Λ. This field is linearly related to the applied field with coefficients which depend upon ε' and upon the size-invariant but shape- dependent depolarization tensor T_Λ defined by eq.(2.7) of II. Considering for simplicity but without loss of generality a coordinate system in which T_Λ is diagonal and consequently the susceptibility tensor as well, we find for each component of the corresponding tensors, the relation :

$$\varepsilon = \frac{\varepsilon'+(1-\varepsilon')T_\Lambda + \varepsilon's_v(1-T_\Lambda)\chi_\Lambda^{\varepsilon'}}{\varepsilon'+(1-\varepsilon')T_\Lambda - s_vT_\Lambda \chi_\Lambda^{\varepsilon'}} \qquad (2.13)$$

where $s_v = 2^{v-1}\pi$. The synthetic form of eq.(2.13) is new. Clearly, eq.(2.10) is recovered for $\varepsilon'=1$. Strictly speaking, macroscopic electrostatics demands that on the r.h.s. of eq.(2.13), the thermodynamic limit of $\chi_\Lambda^{\varepsilon'}$ be taken. We have nevertheless used the above equation as a definition of $\varepsilon(|\Lambda|)$ in order to examine its size dependence, notably for conductors.

We proceeded with a detailed analysis of interesting special cases of eq.(2.13) in taking care of undetermined limiting ratios on its r.h.s.. To do so, we distinguished i) $\varepsilon'= \infty$ and 0 from $0<\varepsilon'<\infty$ and ii) regular from singular shapes of Λ. We defined the latters as follows: A regular (singular) shape is such that all (one at least of) the components of $T_\Lambda \in \,]\,0,1[$ (=0 or =1), their sum being of course equal to 1. In this sense ellipses and ellipsoids have regular shapes since $0<T_{ii}<1$ (i=1,...,v) while strips, slabs and cylinders have singular shapes since their depolarization tensors are given by (1,0), (1,0,0) and (1/2,1/2,0) respectively. Let us consider here only the cases where the system is a conductor, i.e. $\varepsilon = \infty$ and refer to III for the discussion of the other cases. Of particular interest are the critical values reached by the components of χ_Λ in this state. Setting the denominator of eq.(2.13) equal to zero results in

$$\chi_\Lambda^{\mathcal{P};\varepsilon'} = \frac{1}{s_v} + \frac{\varepsilon'(1-T_\Lambda)}{s_vT_\Lambda} \qquad (2.14)$$

This relation is valid without further precautions for regular shapes and $0\leq\varepsilon'\leq\infty$ and for singular shapes and $0<\varepsilon'<\infty$. Eq.(2.14) is particularly interesting for $\varepsilon' = 0$. Indeed, its r.h.s. is immediately identified as the Stillinger-Lovett value of the bulk susceptibility of an infinite conductor and its l.h.s. tells us that it is with and only with Neumann boundary conditions that the susceptibility of a finite conductor converges to the S.L. value and this for all shapes compatible with the requirement that the system be homogeneously polarizable. In fact, these boundary conditions minimize the surface effects and, although unphysical, may be of practical interest notably in computer experiments. In summary, the problem is to understand eq.(2.14) with the help of specific models.

3. Debye-Hückel approximation.

In the absence of exactly solvable models of multicomponents classical Coulomb systems with image charges we have recourse to the linearized D. H. approximation which has already been successfully applied to the study of the long range transverse pair correlation near the surface of semi-infinite conductors filling a half space (ref. 5) or a cylinder (ref.6). However, whereas infinite or semi-infinite systems are necessarily neutral, finite systems can bear finite charges which spread over their walls. This feature introduces a new parameter into the theory and it is an interesting property of the D.H. approximation that it can mimic this situation.

This approximation is defined as follows: Let ρ_α and q_α be the number density and the particles charge of the species α. A unique mean Debye wave number κ occurs, defined through $\kappa^2 = \beta s_v \sum_\alpha q_\alpha^2 \rho_\alpha$. We define also $\rho = \sum_\alpha \rho_\alpha$ and $\Gamma = \beta \sum_\alpha q_\alpha^2 \rho_\alpha \rho^{-1}$ in such a way that $\kappa^2 = s_v \Gamma \rho$. In the D.H. approximation the one particle densities $\rho_{\alpha;\Lambda}^{\varepsilon'}(x)$ are to be taken constant and equal to ρ_α for $x \in \Lambda$ and zero otherwise. The truncated two-particle correlation functions are approximated by their high temperature and asymptotic form:

$$\rho_{\alpha\beta;\Lambda;\varepsilon'}^T(x,y) = -\beta\, q_\alpha q_\beta \rho_\alpha \rho_\beta \, G_\Lambda^{\varepsilon'}(x,y) \tag{3.1}$$

where $G_\Lambda^{\varepsilon'}(x,y)$ is the D.H. kernel discussed below and $\beta = (K_B T)^{-1}$.

In this approximation the charge-charge correlation function $S_\Lambda^{\varepsilon'}(x,y)$ becomes:

$$S_\Lambda^{\varepsilon'}(x,y) = \frac{\Gamma\rho}{\beta}\left(\delta(x-y) - \Gamma\rho G_\Lambda^{\varepsilon'}(x,y)\right) \tag{3.2}$$

and the D.H. kernel satisfies the self-consistent equation:

$$-\Delta_x G_\Lambda^{\varepsilon'}(x,y) = \Xi_\Lambda(x)\, \frac{s_v \beta}{\Gamma\rho} S_\Lambda^{\varepsilon'}(x,y) \tag{3.3}$$

where Ξ_Λ is the characteristic function of the domain Λ and y is a point source in Λ. For $x \in \Lambda$ we shall speak of $G_{\Lambda;in}^{\varepsilon'}(x,y)$ and for $x \notin \Lambda$, where the r.h.s. of eq.(3.3) vanishes, we shall speak of $G_{\Lambda;out}^{\varepsilon'}(x,y)$. We note here that whereas the effective interactions as well as the correlation functions are usually defined for x and $y \in \Lambda$, the D.H. kernel like the Coulomb kernel can be defined for both x and $y \notin \Lambda$, a situation considered in section 9 of III. The boundary conditions satisfied by the inner and outer kernels are discussed below. It is convenient to define here the dimensionless "charge fluctuation" density:

$$p_\Lambda^{\varepsilon'}(y) = \frac{\beta}{\Gamma\rho} \int_\Lambda d^\nu x\, S_\Lambda^{\varepsilon'}(x,y) \tag{3.4}$$

In terms of the above functions, the P.F., S.M., and excess susceptibilities read as follows:

$$\chi_{11;\Lambda}^{\varepsilon'}(\text{P.F.}) = \frac{\Gamma\rho}{|\Lambda|}\int_\Lambda d^\nu x\, d^\nu y\, x_1 y_1 \left(\delta(x-y) - \Gamma\rho G_\Lambda^{\varepsilon'}(x,y)\right) \tag{3.5}$$

$$\chi_{11;\Lambda}^{\varepsilon'}(S.M.) = \frac{\Gamma^2 \rho^2}{2|\Lambda|} \int_\Lambda d^\nu x \, d^\nu y \, (x_1 - y_1)^2 \, G_\Lambda^{\varepsilon'}(x,y) \tag{3.6}$$

and

$$\Delta\chi_{11;\Lambda}^{\varepsilon'} = \frac{\Gamma\rho}{|\Lambda|} \int_\Lambda d^\nu y \, y_1^2 \, p_\Lambda^{\varepsilon'}(y) \tag{3.7}$$

In section 4 of III the D.H. kernel is constructed explicitly for circular and spherical domains of radius R by means of expansions in terms of Bessel functions of the second kind for $G_{\Lambda;in}^{\varepsilon'}(x,y)$ and harmonic functions for $G_{\Lambda;out}^{\varepsilon'}(x,y)$ with coefficients satisfying the following boundary conditions : continuity of the inner and outer part of $G_{\Lambda;in}^{\varepsilon'}(x,y)$ and $G_{\Lambda;out}^{\varepsilon'}(x,y)$ and continuity of the normal derivative of $G_{\Lambda;in}^{\varepsilon'}(x,y)$ and ε' times the normal derivative of $G_{\Lambda;out}^{\varepsilon'}(x,y)$, for x or y on $\partial\Lambda$. The coefficients of the angular-independent part of these expansions depend upon a parameter L/R through the amplitude of the surface Coulomb potential $\ln\frac{L}{R} \equiv \xi_D$ for the disk and through $(L-R)/R \equiv \xi_S$ for the sphere. As to the boundary conditions they can be understood if we write $G_{\Lambda;in}^{\varepsilon'}(x,y)$ in integral form. In the D.H. approximation, the effective interaction between two particles $q_\alpha q_\beta G_{\Lambda;in}^{\varepsilon'}(x,y)$ is determined by *the Ornstein–Zernicke equation* in which the direct correlation function is approximated by $-\beta$ times the bare Coulomb interaction $q_\alpha q_\beta C_\Lambda^{\varepsilon'}(x,y)$, where $C_\Lambda^{\varepsilon'}(x,y)$ is the Coulomb kernel which satisfies the standard boundary conditions on $\partial\Lambda$ owing to the dielectric constant ε' of the surrounding medium being generally $\neq 1$ (cf. eq.(7.1) of III for example). It is then easy to show that the boundary conditions satisfied by the D.H. kernel derive from the following integral equation

$$G_\Lambda^{\varepsilon'}(x,y) = C_\Lambda^{\varepsilon'}(x,y) + \frac{\kappa^2}{s_\nu} \int_\Lambda d^\nu z \, C_\Lambda^{\varepsilon'}(x,z) \, G_\Lambda^{\varepsilon'}(z,y) \tag{3.8}$$

Considering again circular or spherical systems it can be shown that, for the solution of eq.(3.8) to be everywhere regular, $L/R \geq 1$. It transpires from the integral equation (3.8) that the excess density $p_\Lambda^{\varepsilon'}(y)$ will depend upon ξ_D and ξ_S and consequently the monopole sum rule, the excess and S.M. susceptibilities as well, whereas χ(P.F.) does not. Introducing the parameter $\lambda_\Lambda = \ln(L/R) \equiv \xi_D$ for $\nu = 2$ and $\lambda_\Lambda = L/R \equiv \xi_S + 1$ for $\nu = 3$ and defining $Z = \kappa R$, we find from eq.(5.30) of III that, for $\Lambda = \oplus$: a disk or a sphere,

$$p_\Lambda^{\varepsilon'}(|y|) \approx \frac{\varepsilon'}{\varepsilon' + \lambda_\Lambda Z} \left(\frac{Z}{\kappa|y|}\right)^{\frac{\nu-2}{2}} \exp(-(Z - \kappa|y|)) \tag{3.9}$$

and that

$$\int_\Lambda d^\nu y \, p_\Lambda^{\varepsilon'}(|y|) \approx \frac{\varepsilon'}{\varepsilon' + \lambda_\Lambda Z} \, 2^{\nu-2} \pi$$

Eq.(3.9) shows that the "charge fluctuation" density is concentrated in a surface layer of width κ^{-1} and that its amplitude, which is proportional to the total charge fluctuation, is zero whenever $\varepsilon'/\lambda_\Lambda \to 0$, for finite Z. It follows that if we set *a priori*, $\varepsilon' = 0$ or $\lambda_\Lambda = \infty$, the D.H. approximation satisfies the monopole sum rule. Thus in these two cases, it mimics the properties of a *canonical ensemble*. In *all* other cases, the D.H. approximation mimics the properties of a *grand canonical ensemble*.

In order to give an idea of the consequences of the D.H. approximation applied to finite systems, we give a summary, limited to the susceptibilities, of the results established for the disk ($\nu = 2$) and for the sphere ($\nu = 3$) in the sections 3 to 6 of III.

A) $\varepsilon' = 0$, all λ_Λ. In this case we have :

$$\chi^0_\oplus (\text{P.F.}) = \frac{1}{2\pi(\nu-1)} + O(Z^{-1}) \tag{3.10.a}$$

$$\Delta \chi^0_\oplus = 0 \tag{3.10.b}$$

$$\chi^0_\oplus (\text{S.M.}) = \chi^0_\oplus (\text{P.F.}) \tag{3.10.c}$$

As expected from eq.(2.14) the P.F. susceptibility produces the S.L. value. The saturation is reached with the inverse radius, in contrast with the case of the O.C.P. on a sphere at $\Gamma = 2$ for which we found in eq.(7.8) of II a saturation inversely proportional to the surface of the sphere.

B) $\varepsilon' = \infty$, all λ_Λ. Here we find :

$$\chi^\infty_\oplus (\text{P.F.}) = \frac{Z}{2\pi(\nu-1)} - \frac{\nu+1}{4\pi(\nu-1)} + O(Z^{-1}) \tag{3.11.a}$$

$$\Delta \chi^\infty_\oplus = \frac{Z}{2\pi(\nu-1)} - \frac{\nu+3}{4\pi(\nu-1)} + O(Z^{-1}) \tag{3.11.b}$$

$$\chi^\infty_\oplus (\text{S.M.}) = \frac{1}{2\pi(\nu-1)} + O(Z^{-1}) \tag{3.11.c}$$

The divergence of the P.F. susceptibility and consequently that of ε according to eq.(2.13) and to eq.(2.5) of III goes linearly with the radius of the ν-dimensional sphere. It is not due to the S.M. susceptibility which produces the S.L. value, but to the second moment of the excess charge density. The limit $\lambda_\Lambda \to \infty$ with the "grand canonical" prescription $\varepsilon'/\lambda_\Lambda \to \infty$ is included in the cases described above.

C) $\varepsilon' = \infty$, $\lambda_\Lambda = \infty$, with the "canonical" prescription $\varepsilon'/\lambda_\Lambda \to 0$. In this case we have :

$$\chi^\infty_\oplus (\text{P.F.}) = \chi^\infty_\oplus (\text{S.M.}) = \frac{Z}{2\pi(\nu-1)} + \frac{1}{2\pi(\nu-1)} + O(Z^{-1}) \tag{3.12.a}$$

D) $0 < \varepsilon' < \infty$, $\lambda_\Lambda > 0$. In these cases, the results are :

$$\chi^{\varepsilon'}_\oplus (\text{P.F.}) = \frac{1}{2\pi(\nu-1)} + \frac{\varepsilon'}{2\pi} + O(Z^{-1}) \tag{3.13.a}$$

$$\Delta\chi^{\varepsilon'}_{\oplus} = \frac{\varepsilon'}{2\pi}\left(\frac{1}{\lambda_\Lambda(\nu-1)} + O(\frac{1}{\lambda_\Lambda Z})\right) \qquad (3.13.b)$$

$$\chi^{\varepsilon'}_{\oplus}(S.M.) = \frac{1}{2\pi(\nu-1)} + \frac{\varepsilon'}{2\pi}\left(1 - \frac{1}{\lambda_\Lambda(\nu-1)} + O(\frac{1}{\lambda_\Lambda Z})\right) + O(Z^{-1}) \qquad (3.13.c)$$

The ε' - contribution to the P.F. susceptibility results from long range surface correlations as reported in the introduction. The canonical limit $\lambda_\Lambda \to \infty$ can be taken in the above relations. Notice that eq.(3.13.a) reproduces exactly the results expected from eq.(2.14).

E) $0 < \varepsilon' < \infty$, $\lambda_\Lambda = 0$. Here we find :

$$\chi^{\varepsilon'}_{\oplus}(P.F.) = \frac{1}{2\pi(\nu-1)} + \frac{\varepsilon'}{2\pi} + O(Z^{-1}) \qquad (3.14.a)$$

$$\Delta\chi^{\varepsilon'}_{\oplus} = \Delta\chi^{\infty}_{\oplus} = \frac{Z}{2\pi(\nu-1)} - \frac{\nu+3}{4\pi(\nu-1)} + O(Z^{-1}) \qquad (3.14.b)$$

$$\chi^{\varepsilon'}_{\oplus}(S.M.) = -\frac{Z}{2\pi(\nu-1)} + \frac{\nu+3}{4\pi(\nu-1)} + \frac{1}{2\pi(\nu-1)} + \frac{\varepsilon'}{2\pi} + O(Z^{-1}) \qquad (3.14.c)$$

We notice that the P.F. susceptibility given by eq.(3.14.a) equals that given by eq.(3.13.a), i.e is independent of λ_Λ, as expected. We notice furthermore that the excess susceptibility given by eq.(3.14.b) is equal to that given by eq.(3.11.b). This property is explained in subsection 5.2 of III. We remark lastly that while in the case B) the divergence of $\Delta\chi^{\varepsilon'}_{\oplus}$ entails that of $\chi^{\varepsilon'}_{\oplus}$(P.F.), this is no longer true in E) since the divergence of $\Delta\chi^{\varepsilon'}_{\oplus}$ is compensated by that of $\chi^{\varepsilon'}_{\oplus}$(S.M.).

The main conclusion to be drawn from these results is that, for a spherical geometry at least, the D.H. approximation applied to finite systems is capable of reproducing in thermodynamic limit the predictions of classical electrostatics.

REFERENCES.

[1]   Ph. Choquard, B. Piller and R. Rentsch, J. Stat. Phys. , **43**, Nos 1/2, p. 197-205, (1985)

[2]   Ph. Choquard, B. Piller and R. Rentsch, J. Stat. Phys. , **46**, Nos 3/4 , p. 599-633, (1986)

[3]   Ph. Choquard, B. Piller, R. Rentsch and P. Vieillefosse, J. Stat. Phys., **55**, Nos 5/6, p. 1185-1262, (1989)

[4]   Ph. A. Martin , Rev. Mod. Phys. 60(4) (1988)

[5]   B. Jancovici, J. Stat. Phys., **28**, No 1 (1982)

[6]   B. Jancovici and X. Artru, Molecular Physics, 49, 2, 487-497 (1983)

PATTERN FORMATION PROCESSES IN BINARY MIXTURES WITH SURFACTANTS

Toshihiro KAWAKATSU and Kyozi KAWASAKI

Department of Physics, Kyushu University 33, Fukuoka 812, Japan

In order to investigate the pattern formation processes in a binary mixtures with surfactants (e.g. oil-water-surfactant system), we propose a dynamical model which is suitable for computer experiments. In the model, we represent the binary mixture by a continuous field, while the surfactants are treated as discrete molecules. The cell dynamics method for continuum and the molecular dynamics method for surfactant molecules are combined to integrate the equations of motion. Bicontinuous and micellar pattern formation processes are demonstrated.

1. INTRODUCTION

A surfactant is a molecule which has two chemically distinct bases and, therefore, has the amphiphilic nature. In an immiscible two component mixture, the surfactants align on the interfaces and lower the surface free energy. An oil-water mixture with detergent is a well-known example.

When the surface tension of the interfaces is almost reduced to zero due to the presence of surfactants, the interfaces become highly ramified on the semi-microscopic level. In an oil-water-surfactant mixture, such a structure has the characteristic length of the order of 100 Å and is called the "microemulsion".[1]

In this paper we propose a model and a computational scheme for such systems, which is a hybrid of the cell dynamics method and the molecular dynamics method, and show some results of our computer experiments.

2. MODEL SYSTEM

The dynamics of phase separation in a binary mixture is often studied by the so-called "time-dependent Ginzburg-Landau (TDGL) model", in which the binary mixture is represented by a continuous field.[2] If we define the field as $X(\mathbf{r}) \equiv \rho_A(\mathbf{r}) - \rho_B(\mathbf{r})$, where ρ_A and ρ_B are the densities of each of the components of the binary mixture, the time evolution of this field $X(\mathbf{r})$ is expressed as

$$\frac{\partial}{\partial t} X(\mathbf{r}) = L_X \nabla^2 \frac{\delta F}{\delta X(\mathbf{r})} \tag{2.1}$$

where L_X is the Onsager kinetic coefficient and F is the total free energy of the system whose explicit form will be given later. In (2.1) we temporarily neglected the hydrodynamic interaction, which can be easily introduced.[2] However, we may suppose that eq. (2.1) is valid for the cases where the fluid motion is suppressed, such as a thin fluid layer between two parallel plates.

Now we introduce surfactants into the system. In modeling the surfactant molecules, we must take into account the fact that they possess two (in some cases three or more) chemical bases. We model such a surfactant molecule by a "dumbbell" or a rigid rod of length ℓ which has two interaction centers located at the two ends, which is analogous to an electric dipole. One of the two interaction centers prefers the component A and the other prefers the component B. For simplicity, we consider only the symmetric case where the system is invariant under the simultaneous exchanges of A and B components and of the two interaction centers of surfactant molecules (Extension to the asymmetric case is obvious but will not be given in this article). We further assume that the A-philic (B-philic) interaction center has the same chemical nature as that of the A (B) component of the mixture. Thus we have to introduce only two interaction potential functions $\Phi(r)$ and $\Psi(r)$, where the central forces are assumed. The potential $\Phi(r)$ expresses the interaction between chemically similar components (A-A and B-B) and $\Psi(r)$ expresses the interaction between chemically dissimilar components (A-B). Introducing the strength of the interaction center of the surfactant denoted as q, which corresponds to the electric charge in the electric dipole case, we can calculate the interaction potential energy between the surfactant molecules and the field $X(\mathbf{r})$ and that between the surfactants. Assuming that ℓ is small, we expand the above interaction energy in power series in ℓ (multipole expansion) and retain up to the quadratic terms. Then we obtain the following expressions for the total free energy F;

$$F \equiv F_{XX} + F_{XS} + F_{SS} \tag{2.2}$$

$$F_{XX} = \int d\mathbf{r} [\tfrac{1}{2} D_X (\nabla X)^2 + f_0(X)] \quad \text{where} \quad f_0(X) = -\tfrac{c}{2} X^2 + \tfrac{u}{4} X^4 \tag{2.3}$$

$$F_{XS} = \mu_S N_S + \tfrac{q\ell}{2} \sum_i \int d\mathbf{r}\, V_-(\mathbf{r}-\mathbf{r}_i)\, \hat{\mathbf{s}}_i \cdot \nabla X(\mathbf{r}) \tag{2.4}$$

$$F_{SS} = q^2 \sum_{i<j} [2V_+(r_{ij}) + \tfrac{\ell^2}{4}(\mathbf{s}^-_{ij}\mathbf{s}^-_{ij}):\nabla\nabla\Phi(r_{ij}) + \tfrac{\ell^2}{4}(\mathbf{s}^+_{ij}\mathbf{s}^+_{ij}):\nabla\nabla\Psi(r_{ij})]. \tag{2.5}$$

where $\mathbf{r}_i$ and $\hat{\mathbf{s}}_i$ are the center of mass and the unit vector along the direction from B-base to the A-base (usually called the director) of the i-th surfactant molecule, respectively. F_{XX} is the bare free energy of the field $X(\mathbf{r})$ of the binary mixture in the absence of surfactants, F_{XS} is the interaction between the field and the surfactants and F_{SS} is the interaction between the surfactants. D_X, c, u and μ_S are constants and $V_\pm(r) \equiv \Phi(r) \pm \Psi(r)$, $\mathbf{s}^\pm_{ij} \equiv \hat{\mathbf{s}}_i \pm \hat{\mathbf{s}}_j$ and $r_{ij} \equiv |\mathbf{r}_i - \mathbf{r}_j|$. In this model the surfactants are regarded as particles with spins which are represented by only the two variables, their center of mass positions

$\{r_i\}$ and directors $\{\hat{s}_i\}$. We assume the equations of motion for r_i and $\hat{s}_i$ as follows;

$$\frac{d}{dt} r_i = - L_R \frac{\partial F}{\partial r_i} \quad \text{and} \quad \frac{d}{dt} \hat{s}_i = - L_S [\frac{\partial F}{\partial \hat{s}_i} - (\frac{\partial F}{\partial \hat{s}_i} \cdot \hat{s}_i) \hat{s}_i], \qquad (2.6)$$

where L_R and L_S are Onsager's kinetic coefficients and the second term in the second equation of (2.6) arises from the constraint $|\hat{s}_i| = 1$. In eq. (2.6) we again neglected the hydrodynamic effects.[3] Equations (2.1)-(2.6) constitute a complete set of equations which describe the time evolution of the system.

3. COMPUTER EXPERIMENT

The TDGL equation (2.1) can be integrated numerically by the well-known cell dynamics method.[4] On the other hand, the molecular equations of motion (2.6) are integrated by the usual molecular dynamics method.[5] We combine these different methods to integrate the whole equations of motion.

Our system has 30×30 meshes for the field $X(r)$ and contains 500 surfactant molecules which overlap with the field $X(r)$ and the interaction potentials are chosen as $\Phi(r) = a \exp(-2r) - b \exp(-r)$ and $\Psi(r) = a \exp(-2r)$, where a and b are properly chosen constants and we initially started from the completely random distributions of the field $X(r)$ and the surfactants. In Fig.1 some results of our computations are shown. The A-rich cells and B-rich cells are represented by asterisks and blanks, respectively, and a surfactant molecule is represented by a small square with a tail which is directed along $-\hat{s}_i$. Fig 1a) shows the results for the case with 50% composition of A and Fig.1b) shows those for 25% composition of A, respectively. The bicontinuous domain structure (mutually percolated domains of A and B) can be seen in Fig.1a), while we observe micelles (globular domains surrounded by the surfactants) in Fig.1b). The characteristic length at time t, say $R(t)$, is evaluated from the first moment of the structure function and we found the growth behavior $R(t) \sim t^{0.35}$ after the transient regime of $R(t) \sim t^{0.5}$ for the bicontinuous case, while for the micellar case the characteristic length $R(t)$ was found to grow very slowly. The exponent 0.35 of the former case is the well-known exponent of the spinodal decomposition in alloys. In the micellar case, however, the slow growth of the micelles is due to the surfactants on the interface which prevent the micelles to grow.

4. CONCLUSION

As was shown in the previous section, the combination of the cell dynamics method and the molecular dynamics method is a useful tool in investigating the systems where amphiphiles are immersed in a normally immiscible fluid, such as the oil-water-surfactant system.

ACKNOWLEDGEMENT

The authors would like to thank Dr. Ken Sekimoto for valuable discussions.

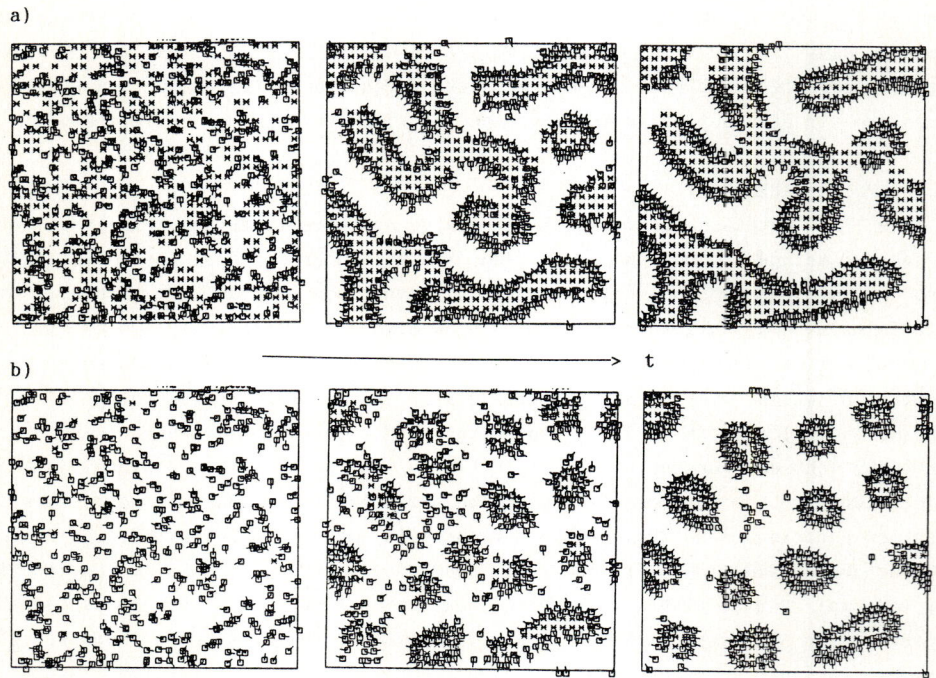

FIGURE 1

Snapshots of the computed ordering processes are shown in a) for A:B = 1:1 and in b) for A:B = 1:3, respectively.

REFERENCES

1) P.G. de Gennes and C. Taupin, J. Chem. Phys. 86 (1982) 2294.

2) K. Kawasaki, Physica 119A (1983) 17.

3) P.G. de Gennes, The Physics of Liquid Crystals. (Oxford,1974).

4) Y. Oono and S. Puri, Phys. Rev. A38 (1988) 434.

5) M.P. Allen and D.J. Tildesley, Computer Simulation of Liquids (Oxford, 1987).

Chapter V:
Charged Particles in Lower Dimensions and/or in Magnetic Fields

EXCITATIONS IN CONDUCTING POLYMERS*

Yasushi WADA

Department of Physics, University of Tokyo, Bunkyo-ku, Tokyo 113, Japan

Quasi-one-dimensional systems used to be an ideal object of theoretical studies. After synthesis of the simplest conjugated polymer polyacetylene, $(CH)_x$, they have been extensively studied experimentally as well as theoretically. It is a system made of electrons and ions. The ions make a lattice. It is not uniform but dimerized due to the one-dimensional nature. The dimerization gives two degenerate ground states. We get a nonlinear excitation named soliton as a boundary between the two states. In addition, there is another one named polaron which is not topological. They have characteristic excitations of electrons and phonons. The latter is closely associated with the photo-induced infrared absorption experiments. We discuss the role of electron correlations in the phonon excitations and how to identify the induced carriers with the soliton or polaron. The absorption experiments turn out to be ineffective. Photo-induced Raman experiments should be effective. Recent experimental findings are to be discussed.

1. INTRODUCTION

Conducting polymers are not necessarily systems of the strongly coupled plasmas. It has, however, been known that the simplest conducting polymer, trans-polyacetylene, has soliton excitation. The soliton is a stable local excitation which is mobile. When two solitons collide each other, the same type of solitons are generated after the collision. The soliton excitation was first found and discussed in hydrodynamics a century ago. It was represented by a solution of Korteweg-de Vries (KdV) equation.

Interest in solitons was revived in the plasma studies. In 1958, Sagdeev[1] postulated that plasma can have propagating excitations similar to the hydrodynamic systems. Gardner and Morikawa[2] noticed a similarity between certain waves in a plasma in a magnetic field and waves on the surface of water of finite depth. In 1966, Washimi and Taniuti[3] showed that ion waves with a small but finite amplitude have the same similarity. They are all represented by the solutions of the KdV equation. In 1970, Ikezi, Taylor, and Baker[4] observed the formation and propagation of ion-acoustic solitons in a plasma with density $10^9 cm^{-3}$, electron temperature 1.5 - 3ev, and ion temperature 0.2ev.

Importance of the soliton excitation was then realized at various fields of physics. Solid state physics was rather late for joining it. A pioneering work was carried out by Krumhansl and Schrieffer[5]. They showed that the solitons in quasi-one-dimensional systems are elementary excitations in the terminology of many-body problem and statistical mechanics. At present, the trans-polyacetylene is the material which is most likely to involve the soliton excitations. It is denoted by $(CH)_x$. Each CH group contributes one π-electron. The electronic levels are half-filled. As it is predominantly one-dimensional, a lattice distortion takes place to have bond alternation along the polymer chain. It has

---
*This is based on the works supported by a Grant-in-Aid for Scientific Research from the Ministry of Education, Science and Culture.

a single bond and a double bond alternately. A simple but effective theoretical model was proposed by Su, Schrieffer, and Heeger[6] (SSH). It is a one-dimensional model for electrons and lattice. Interactions between them are induced by modulations in electron transfers due to changes in the bond lengths. Electron-electron interactions are not taken into account. It was soon realized that the soliton extends over many CH groups. This fact enabled Takayama, Lin-Liu, and Maki[7] to derive a continuum (TLM) model from the SSH model. The lattice constant was regarded as infinitesimal in comparison with the extent of the soliton. The continuum model made analytical approaches feasible. Campbell and Bishop[8] used this model to find the polaron excitation which is another type of stable local excitations originated with the help of nonlinear interactions in the system.

On the other hand, infrared absorption experiments showed that there are many infrared active modes of lattice vibrations. The SSH model was too simple, since it intended to reproduce the electronic level structures in the dimerized states. Horovitz[9] pointed out that three types of molecular vibrations are relevant and the infrared absorption data can be understood consistently with the Raman scattering data when parameters are suitably chosen for the three types of the vibrations. His theory was named amplitude mode formalism. Mele and Hicks[10] later showed that the amplitude mode formalism can be incorporated into the TLM model by a simple generalization.

The purposes of the present paper are to review how the lattice vibration modes, which are called phonons, are determined when there is a soliton or a polaron and to apply the results to the photoinduced absorption. It is an infra-red absorption by the soliton or polaron which is induced by pumping light. We will find that the soliton is hardly distinguishable from the polaron without the help of very high sensitivity of detection. We then propose the possibility that the photoinduced Raman scattering would be an effective experiment of distinguishing them. Recent field-induced Raman experiment by Lawrence, Burroughes, and Friend[11] clearly indicates the possibility.

We would like to give some impressions on how the soliton study has developed in the solid state physics, looking forward to having revived interests in solitons in the field of strongly coupled plasmas which have very nonlinear interactions.

2. SOLITON, POLARON, AND MODELS

The SSH Hamiltonian is

$$H_{\text{SSH}} = -\sum_{n,s}[t_0 - \alpha(u_{n+1} - u_n)](c^\dagger_{n+1,s}c_{n,s} + c^\dagger_{n,s}c_{n+1,s}) + \frac{K}{2}\sum_n(u_{n+1} - u_n)^2 + \frac{M}{2}\sum_n \dot{u}_n^2, \quad (1)$$

where u_n is the displacement of the n-th CH unit from its equilibrium point of undimerized state. Its mass is M, $c_{n,s}$ ($c^\dagger_{n,s}$) annihilates (creates) an electron with spin s at the n-th site, and $\alpha = 4.1 \text{ev/A}$ represents the coupling between electrons and lattice motions, which changes the electron hopping constant $t_0 = 2.5\text{ev}$. $K = 21\text{evA}^{-2}$ is the spring constant. The idealized (infinite chain) structures at the ground states are shown in Fig.1. The chain consists of alternating single and double bonds. It is said to be dimerized. The two structures in Fig.1 give the same energy. The ground states are degenerate. A soliton excitation is expected in such a system. The bond alternation pattern on one side of the soliton is, say, that of the A phase. On the other side, it is the B phase. The transition takes place gradually at the region where the soliton is. It is schematically shown in Fig.2.

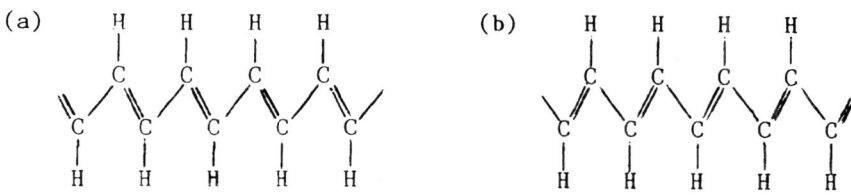

FIGURE 1
The two degenerate ground states of trans-polyacetylene. The single and double lines indicate the single and double bonds, respectively.

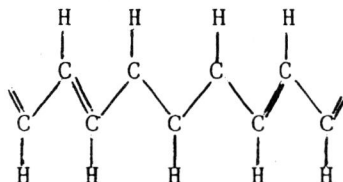

FIGURE 2
A schematic view of the bond distribution for a soliton. It is a boundary between the two ground state distributions.

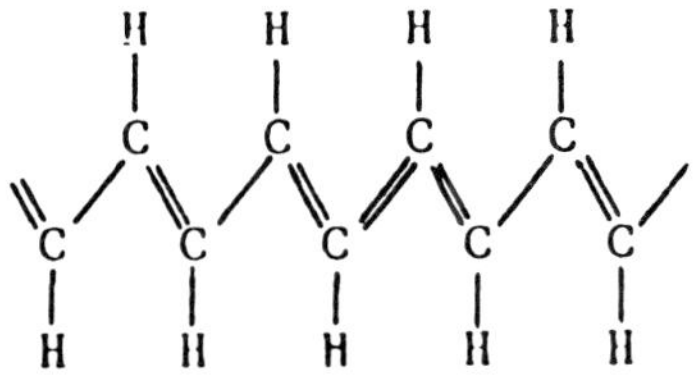

FIGURE 3
A schematic view of the bond distribution for a polaron. Both sides of the polaron have the same ground state distributions.

Since the transition is gradual enough, the TLM continuum model is effective. It has the Hamiltonian

$$H_{\text{TLM}} = \frac{1}{2\pi v_F \lambda} \int dx (\Delta^2(x) + \dot{\Delta}^2(x)/\omega_Q^2) + \sum_s \int dx \psi_s^\dagger(x)(-iv_F \sigma_3 \partial/\partial x + \sigma_1 \Delta(x)) \psi_s(x), \quad (2)$$

where the electron creation and annihilation operators are replaced by a two component field $\psi_s(x)$. The two components reflect the two distinct parts ($\pm k_F$) of the Fermi surface, k_F being given by $\pi/2a$ in terms of the lattice constant a. The order parameter $\Delta(x)$ is defined by

$$\Delta(x) = 2\alpha \tilde{y}(n), \quad \tilde{y}(n) = (-1)^n (u_n - u_{n+1}), \quad \text{at } x = na. \quad (3)$$

The Fermi velocity v_F is given by $2t_0 a$, $\lambda = 2\alpha^2/\pi K t_0 \simeq 0.19$, $\omega_Q^2 = 4K/M$, and σ_1 and σ_3 are the Pauli matrices. A polaron excitation was found, using this model[8]. Its configuration is schematically shown in Fig.3.

Mele and Hicks[10] generalized the TLM Hamiltonian (2) into

$$H_{\text{HMH}} = \sum_{\beta=1}^{3} \frac{1}{2\pi v_F \lambda_\beta} \int dx (\Delta_\beta^2(x) + \dot{\Delta}_\beta^2(x)/\omega_\beta^2) + \sum_s \int dx \psi_s^\dagger(x)(-iv_F \sigma_3 \partial/\partial x + \sigma_1 \Delta(x)) \psi_s(x), \quad (4)$$

with $\Delta(x) = \sum_{\beta=1}^{3} \Delta_\beta(x)$ and $\lambda = \sum_{\beta=1}^{3} \lambda_\beta$. Three components Δ_β of the order parameter correspond to three relevant molecular vibrations. The six parameters λ_β and ω_β were determined by Horovitz[9], using Raman scattering data.

It is straightforward to get the SSH solutions for the ground states. We have

$$\tilde{y}(n) = \pm 0.079 \text{A}, \tag{5}$$

which gives

$$\Delta(x) = \pm \Delta_0, \quad \Delta_0 = 0.65 \text{ev}. \tag{6}$$

The electronic band gap is given by $2\Delta_0$. It is remarkable to find that a simple numerical iteration calculation works very well in studying the soliton and polaron solutions. These static solutions are determined as minimum energy configurations for each set of N and N_e where N is the number of the lattice sites and N_e the number of electrons. Using the periodic boundary condition, we get the perfectly dimerized ground states when N_e is even and $N_e = N$. A charged spinless soliton state is obtained as a ground state with an even N_e and $N_e - N = \pm 1$. When N_e is odd and $N_e = N$, we obtain a neutral soliton with spin 1/2. The polaron states are found for an odd N_e and $N_e - N = \pm 1$.

The soliton solution is shown in Fig.4 for $N = 201$ and $N_e = 200$. The abscissa is the site number and the ordinate is the optical component of the bond variable $\psi(n)$ which is defined by

$$\psi(n) = (\tilde{y}(n) + \tilde{y}(n-1))/4, \tag{7}$$

the unit being angstrom. The polaron solution is shown in Fig.5 for $N = 200$ and $N_e = 199$. In addition to these lattice configurations, we have complete knowledge of electronic wave function $\phi_i(n)$ and its energy ε_i for the soliton as well as the polaron.

3. PHONONS AROUND THE SOLITON OR THE POLARON

Phonons are the small vibrations around the static soliton or polaron solution. The displacement of the n-th CH unit is written

$$u_n(t) = u_n + g(n,t), \tag{8}$$

where u_n is the static solution. The electronic wave function is also written

$$\phi_i(n,t) = (\phi_i(n) + \delta\phi_i(n)) \exp(-i\varepsilon_i t), \tag{9}$$

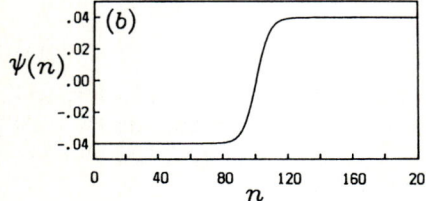
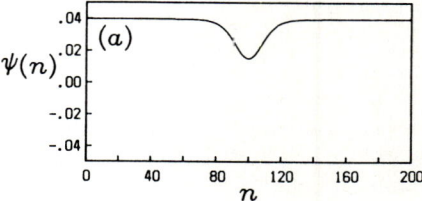

FIGURE 4
Optical compotent of the bond variable $\psi(n)$ of the soliton, defined by (7)[12]. The abscissa is the site number.

FIGURE 5
Optical component of the bond variable $\psi(n)$, of the polaron[12].

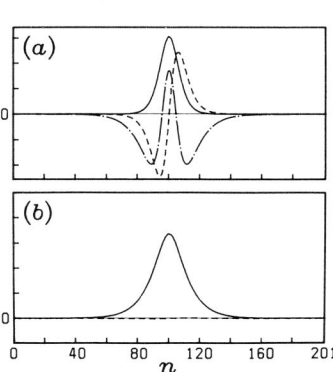

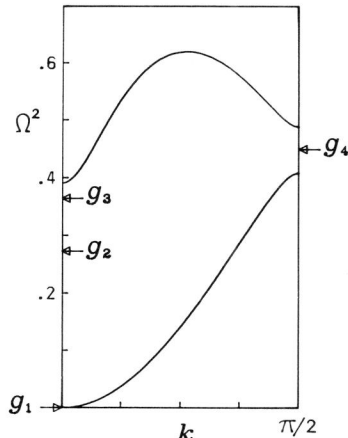

FIGURE 6
Localized modes around the soliton[12]. Abscissa is the site number. (a) The solid curve is g_1, the broken g_2, and the dash-dotted g_3. (b) The solid curve shows $(-1)^n g_4(2n+1)$ and the broken $g_4(2n)$.

FIGURE 7
The phonon frequency Ω with the soliton[12]. The upper and lower curves are for the optical and acoustic branches, respectively. The frequencies of the localized modes are indicated by the arrows. The ordinate is scaled by ω_Q^2 and the abscissa by a.

where $\phi_i(n)$ is the i-th electronic wave function with energy ε_i of the static lattice solution and the adiabatic approximation is used, since the energy gap $2\Delta_0$ is much larger than the phonon frequencies. Nonlinear equations of motion are linearized with respect to g and $\delta\phi_i$. Elimination of $\delta\phi_i$ leads to an eigenvalue equation for the Fourier transform of $g(n,t)$

$$\Omega^2 g_\Omega(n) = \sum_m A_{nm} g_\Omega(m), \qquad (10)$$

where the kernel is determined by the static solution. Equation (10) is numerically solved and gives three solutions localized around the soliton and a staggered mode[12]. They are illustrated in Fig.6 for $N = 201$ and $N_e = 200$. The solid line in Fig.6a is the Goldstone mode g_1 associated with the soliton translation. The third mode g_3 was first found by a numerical solution of the TLM equations[13]. It was, however, controversial and its existence was confirmed by the SSH solution. In Fig.6b, the solid curve indicates $(-1)^n g_4(2n+1)$ while $g_4(2n) = 0$. The eigenvalues are shown in Fig.7 as a function of the wave number k. The ordinate is scaled by ω_Q^2 and the abscissa by a. The symmetry of the modes indicates that g_1 and g_3 are infrared active while g_2 is Raman active.

Table 1 shows how the independent calculations have given the result consistent each other. Chao-Wang[14] and Terai-Ono[12] have used the parameters given in Sec.2. Sun et al [15] used $t_0 = 2.5$ev, $\alpha = 7.3$evA^{-1}, and $K = 52$evA^{-2}. Hicks and Gammel[16] have used a modified TLM equation to confirm the existence of g_3.

TABLE 1
Frequencies of the localized modes Ω_i^2/ω_0^2 for the soliton.

g_i	Sun-Wu-Shen[15]	Chao-Wang[14]	Terai-Ono[12]
g_1	0	0	0
g_2	0.68	0.71	0.70
g_3	0.90	0.94	0.93
	0.99*		
g_4	0.93	1.15	1.15
N	101	101	201

*An additional mode has been obtained, since a very strong electron-phonon coupling strength has been used.

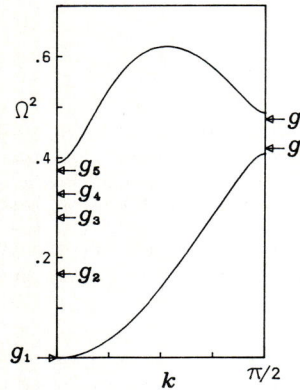

FIGURE 8

The phonon frequency Ω with the polaron[12].

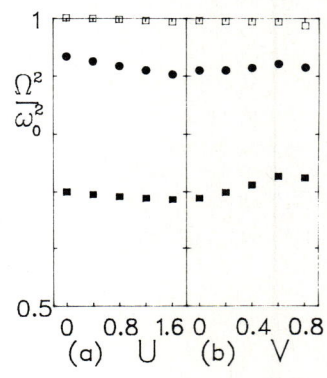

FIGURE 9

The frequencies of the g_2-, g_3-, and g_{ex} modes for the charged soliton by the closed squares, closed circles, and open squares[18]. The ordinate is Ω^2/ω_0^2 with $\omega_0^2 = 2\lambda\omega_Q^2$. The abscissa is U with $V = 0$ in (a). It is V with $U = 1.2t_0$ in (b). They are scaled by t_0.

Eq.(10) is solved also for a polaron to give five localized modes and two staggered modes in the system of $N = 200$ and $N_e = 199$[12]. The dispersion relation is shown in Fig.8. The two modes, g_1 and g_4, are infrared active and the other three, g_2, g_3, and g_5, are Raman active.

4. ELECTRON INTERACTIONS

Electron interactions are not taken into account in the SSH model. It has been known that the interactions are playing a role in some properties of polyacetylene. It is important to know that the obtained phonon modes can survive the interactions. Particularly, the controversial third mode g_3 of the soliton and the fifth mode g_5 of the polaron have frequencies so close to the extended phonon frequency ω_0 that the electron interactions might shift them into the continuum region. This problem has been studied, using the SSH-Hubbard model with the help of Hartree-Fock calculations. The Hamiltonian is written

$$H_{\text{SSHH}} = H_{\text{SSH}} + U \sum_i n_{i\uparrow} n_{i\downarrow} + V \sum_i n_i n_{i+1}, \qquad (11)$$

where $n_{i,s} = c_{i,s}^\dagger c_{i,s}$ and $n_i = n_{i\uparrow} + n_{i\downarrow}$. The second term is the on-site interactions and the third term is the nearest neighbor interactions. Wu and Sun[17] studied a positive soliton with $N = 101$ and $N_e = 100$. The parameters were $t_0 = 2.2\text{ev}$, $\alpha = 6.61\text{evA}^{-1}$, $K = 52.7\text{evA}^{-2}$, and $0 \le U/2t_0 \le 1$. The nearest neighbor interactions were neglected, $V = 0$. They found that the frequencies of g_2 and g_3 modes decrease more strongly than ω_0 as U increases. These modes survive the electron interactions. Yonemitsu, Ono, and Wada[18] studied the problem more extensively. They took $t_0 = 2.5\text{ev}$, $\alpha = 4.1\text{evA}^{-1}$, $K = 21\text{evA}^{-2}$, $\lambda = 0.204$, $0 \le U/2t_0 < 0.9$ for $V = 0$ and $U/2t_0 = 0.6$ for $0 \le V < 1.05t_0$. Fig.9 shows the frequencies of the two localized modes g_2, g_3 by the closed squares and circles, respectively, and of the lowest extended mode g_{ex} by the open squares for the system of a charged soliton with $N = 101$ and $N_e = N \pm 1$. The abscissa is U/t_0 for (a) and V/t_0 for (b). The frequency square Ω^2 is normalized by ω_0^2. Fig.9(a) roughly reproduces Wu and Sun's result. Fig.10 shows these frequencies for a neutral soliton with $N = N_e = 101$. They change with U and V differently from the charged soliton, but, survive the electron interactions as well. At $U = 1.6t_0$, they abruptly decrease since the bond alternation phase is nearly unstable and SDW fluctuations are effective.

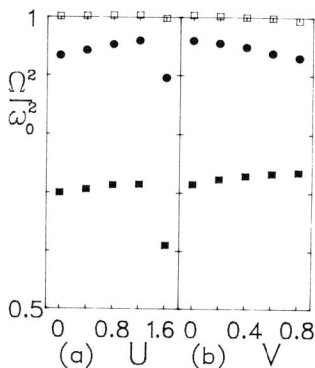

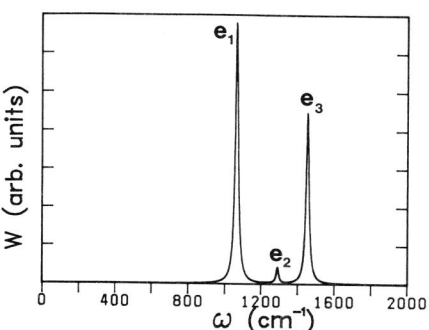

FIGURE 10
The frequencies of the g_2-, g_3-, and g_{ex} modes for the neutral soliton[18]. The notations are the same as Fig.9.

FIGURE 11
Calculated Raman spectrum of $(\text{CH})_x$ without the pumping light[25]. The incident photon frequency is $2\Delta_0$. The abscissa is the Stokes shift ω. The peaks are due to the E modes, originated from g_{ex}.

Yonemitsu, Ono, and Wada[19] have also studied how reliable the Hartree-Fock approximations are. Fluctuations around the H-F solution are taken into account for the system with the neutral soliton. Generally speaking, the fluctuations reduce the overestimations of various quantities by the H-F approximation.

5. PHOTO-INDUCED ABSORPTION

To discuss infrared activities of transpolyacetylene, theoretical model has to be realistic also for the lattice motions. As discussed in the Introduction, the simplest model was proposed by Horovitz[9] and represented by the Hamiltonian (4). It leads to the conclusion that there are three physical phonon modes which correspond to each mode in the TLM model. Each one of the three modes, g_1, g_2, and g_3, with the soliton produces three modes. Those by g_1 are named T modes and denoted by T_1, T_2, and T_3. The second mode g_2 gives A_1, A_2, and A_3. The third mode g_3 makes B_1, B_2, and B_3. Finally, the lowest extended mode g_{ex} produces E_1, E_2, and E_3. The frequencies Ω of the three modes are determined by

$$-\left(1 - \frac{2\lambda\Omega_i^2}{\omega_0^2}\right)^{-1} = \sum_{\beta=1}^{3} \frac{\lambda_\beta \omega_\beta^2}{\lambda} (\Omega^2 - \omega_\beta^2)^{-1}, \qquad (12)$$

where Ω_i is the frequency of the g_i mode in the TLM model.

The photoinduced absorption experiment measures the transmittances of probe infrared light by a polyacetylene film with and without pumping light whose frequency ω_p is larger than $2\Delta_0$. Since the pumping light makes solitons and polarons, the difference

TABLE 2
Frequencies of the localized modes in cm^{-1} for $(CH)_x$ by theoretical estimations and photoinduced absorption.

Model,	Sample Mode	Theories		Experiments	
		Terai et al [22] soliton	Hicks et al [23] polaron	Vardeny et al [20] 30% trans	Schaffer et al [21] pure trans
g_1	T_1	488	463		536
	T_2	1278	1281		1288
	T_3	1364	1360		1365
g_2	A_1	937		928	
	A_2	1288		1286	
	A_3	1404		1405	
g_3*	B_1	1049	1041		1034
g_4*	B_2	1292	1291		—
	B_3	1445	1441		1438
g_{ex}	E_1	1074			
	E_2	1293			
	E_3	1460			

*The mode g_3 is for the soliton and g_4 for the polaron.

TABLE 3
Relative intensities of photo-induced absorption peaks for $(CH)_x$.

Model, Sample Mode	Theories		Experiment
	Terai et al [22] soliton	Hicks et al [23] polaron	Schaffer et al [21] pure trans
T_1	11	11	6.9
T_2	0.24	0.24	0.24
T_3	1	1	1
B_1	0.062	0.095	0.048
B_2	0.002	0.004	—
B_3	0.045	0.067	0.038

ΔT in the transmittance between with and without the pumping light is due to the absorption by these nonlinear excitations.

Experiments have been independently performed by two groups. One is Technion-Nagoya[20] and the other UCSB-Cavendish. Technion-Nagoya have used samples with cis-polyacetylene components and observed the A modes of the soliton. They are inactive in a trans sample. However, the cis-component restricts the region where the soliton can move, breaking the translational symmetry. It changes the symmetry property of the A modes, making them infrared active. The B modes have not been observed by them.

UCSB-Cavendish have used the cis-rich samples, too, and the pure trans sample as well and found the B modes. Observed frequencies are listed on Table 2 for the pure sample.

There have been also two theoretical groups. One has been Terai, Ono, and Wada[22] who have calculated the infrared absorption by a soliton. The other has been Hicks and Mele[23] who have calculated the absorption by a polaron. Calculated frequencies are also listed on Table 2. There are practically no difference in the frequencies between the soliton and the polaron except the T_1 mode. This is due to the particular form of Eq.(12). The solutions of Eq.(12) are rather insensitive to Ω_i^2. For T_1 we assume a nonvanishing value for Ω_1 to reproduce the observed T_1 peak, taking into account the possible pinning of the soliton or polaron by impurities and defects. Observed peak positions in the photoinduced absorption cannot distinguish the soliton from the polaron.

Table 3 shows the relative intensities of the T and B peaks. The second column is the theoretical result for the soliton[22] and the third column for the polaron[23]. The fourth column is the observed value[21]. The intensity of the T_3 peak is taken as the standard. The three columns agree each other fairly well. We thus get the conclusion that the photo-induced absorption cannot determine whether the induced carriers are solitons or polarons.

6. PHOTO-INDUCED RAMAN SCATTERING

We would like to suggest that photo-induced Raman scattering experiment should be able to determine the carrier. It is to measure the Raman spectra of the probe light

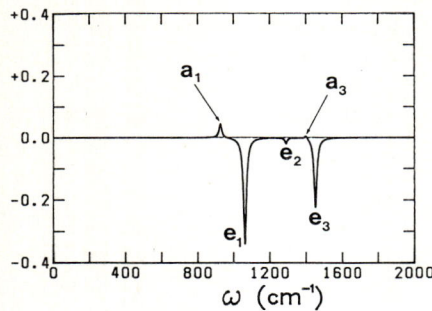 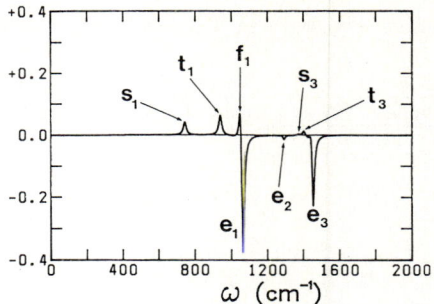

FIGURE 12
Calculated DRS by two solitons in the 200 sites[25]. The ordinate is scaled by the e_1-peak height in Fig.11.

FIGURE 13
Calculated DRS by a polaron in the 200 sites[25].

with the frequency Ω with and without the pumping light. The scattered light has the frequency $\Omega - \omega$, where ω is the Stokes shift which is the frequency of a phonon. Difference Raman spectrum (DRS) is the difference in the Raman spectra between with and without the pumping light. The Raman spectrum without the pumping light was first calculated by Horovitz[9] in his amplitude mode formalism. The spectra with the soliton and polaron were calculated by Ono and Ito[24] in the TLM model. The discussion has been generalized by Terai, Ono, and Wada[25], using the Hamiltonian (4). Fig.11 shows the calculated Raman spectrum of $(CH)_x$ without the pumping light as a function of the shift ω when the incident energy is $\Omega = 2\Delta_0$. The ordinate is in an arbitrary scale. The peaks are due to the E modes which are given by g_{ex}. The system size is $N = N_e = 200$. The DRS by two solitons is shown in Fig.12. The system size is $N = 200$ and $N_e = 198$. There are two solitons in the two hundred sites. The ordinate is scaled by the e_1-peak height in Fig.11. The e-peaks are reduced by bleaching effect of the solitons. Integrated intensity of the peaks reduces by the amount of thirty per cent. We have a small peak at the low frequency side of each e_i dip. It is due to the A modes by g_2. The a_2 peak is too weak to be seen.

Fig.13 shows the calculated DRS by a polaron in the system with $N = 200$ and $N_e = 199$. The e-peaks are again bleached. The s-, t-, and f-peaks are due to the g_2, g_3, and g_5 modes in the TLM model, respectively. There are three small peaks at the low frequency side of each e_i dip. If it is possible to observe these small peaks to count the number of them, we should be able to determine the type of the photoinduced carriers.

7. A RAMAN EXPERIMENT USING MISFET

Recently, Lawrence, Burroughes, and Friend[11] have performed a Raman scattering experiment, using MISFET which stands for metal-insulator-semiconductor field effect transistor. They have used the polyacetylene made by the Durham route which prepares the sample while the polymer is in soluble precursor form. The device is constructed on a Si substrate, with an n-silicon gate layer, an insulator layer of silicon dioxide. Precursor form of polymer is then coated on the substrate and transformed to transpolyacetylene by a heating process. Applying a gate voltage across the insulating larger, they have been able to introduce charges into the polyacetylene. The charges have turned solitons or polarons. Raman spectra have been measured with and without the gate voltage. The difference between the two is the DRS. They have observed two main features in the DRS. One of them is the bleaching of the two modes which are dominant without the gate voltage. They are at 1080cm^{-1} and 1475cm^{-1} with the excitation wave length 584nm. Bleaching intensities are in agreement with the theoretical result shown in Fig.12. The second feature is that there is a small peak at the low frequency side of each of the bleaching features. They occur at 1065cm^{-1} and 1455cm^{-1}. These findings clearly show that the gate voltage is creating solitons and the soliton A mode is actually observable. There remains a problem. The observed phonon frequencies are higher than the corresponding calculated values in Table 2. Durham polyacetylene as an unoriented film is poorly crystalline. It has been pointed out[11] that this could readily give the higher frequencies.

8. CONCLUSIONS

It is quite remarkable to find that the simple models as SSH are working so effectively in the physics of phonons around solitons and polarons. In other words, the success does show how reliable the models are. Electron-electron interactions would be important in order to understand associated properties of soliton. They include, for instance, broadening and energy increase for the charged soliton along with narrowing and energy reduction for the neutral soliton. Spin density alternation of electrons in the soliton is another example. However, for other properties as the phonon excitations, the electron-electron interaction only renormalizes the parameters in the electron-phonon models.

It is encouraging to know that the solitons are experimentally distinguishable from the polarons with the help of the Raman spectrum measurements. This method should be reliable, since the energy transfer to the polyacetylene is small. Electronic optical absorption is another method. It involves interband transitions and mid-gap transitions. They create holes in the valence band and particles in the conduction band, which would lead to new solitons or polarons. Final states, after the absorption, would be rather messy and the analyses should not be so straightforward. On the other hand, the DRS experiments only count the number of small peaks at the low frequency side of the bleaching features.

There still remain some inconsistencies among the observed data of photoinduced absorption experiments and DRS. It is highly desirable that a series of experiments are to be carried out, using the same good samples.

We look forward to a resurgence of the soliton physics in the field of strongly coupled plasmas in the near future so that the soliton physicists can actively participate in the study of the field with their ideas and methods.

ACKNOWLEDGEMENT

The another would like to thank Y. Ono, A. Terai, H. Ito, and K. Yonemitsu for their contributions to the series of the works.

REFERENCES

1) R.Z. Sagdeev, Plasma Physics and Problem of Controlled Fusion Reactions, AN SSSR Publ. Moscow, 4: 384 (1958).

2) C.S. Gardner and G.K. Morikawa, Courant Inst. of Math. Sc. Rep., NYO-9082 (1960).

3) H. Washimi and T. Taniuti, Phys. Rev. Lett. 17 (1966) 966.

4) H. Ikezi, R.J. Taylor, and D.R. Baker, Phys. Rev. Lett. 25 (1970) 11.

5) J.A. Krumhansl and J.R. Schrieffer, Phys. Rev. B11 (1975) 3535.

6) W.P. Su, J.R. Schrieffer, and A.J. Heeger, Phys. Rev. Lett. 42 (1979) 1698; Phys. Rev. B22 (1980) 2099.

7) H. Takayama, Y.R. Lin-Liu, and K. Maki, Phys. Rev. B21 (1980) 2388.

8) D.K. Campbell and A.R. Bishop, Nucl. Phys. B200 (1982) 297.

9) B. Horovitz, Solid State Commun. 41 (1982) 729.

10) E.J. Mele and J.C. Hicks, Phys. Rev. B32 (1985) 2703.

11) R.A. Lawrence, J.H. Burroughes, and R.H. Friend, Proc. Winterschool on Cond. Polymers, to be published in Springer Series on Solid State Sciences.

12) A. Terai and Y. Ono, J. Phys. Soc. Japan 55 (1986) 213.

13) H. Ito, A. Terai, Y. Ono, and Y. Wada, J. Phys. Soc. Japan 53 (1984) 3520.

14) K.A. Chao and Y. Wang, Physica Scripta 34 (1986) 177.

15) X. Sun, C. Wu, and X. Shen, Solid State Commun. 56 (1985) 1039.

16) J.C. Hicks and J.T. Gammel, Phys. Rev. Lett. 57 (1986) 1320.

17) C. Wu and X. Sun, Phys. Rev. B33 (1986) 8722.

18) K. Yonemitsu, Y. Ono, and Y. Wada, J. Phys. Soc. Japan 56 (1987) 4400.

19) K. Yonemitsu, Y. Ono, and Y. Wada, J. Phys. Soc. Japan 57 (1988) 3875.

20) Z. Vardeny, E. Ehrenfreund, O. Brafman, B. Horovitz, H. Fujimoto, J. Tanaka, and M. Tanaka, Phys. Rev. Lett. 57 (1986) 2995.

21) H.E. Schaffer, R.H. Friend, and A.J. Heeger, Phys. Rev. B36 (1987) 7537.

22) A. Terai, Y.Ono, and Y. Wada, J. Phys. Soc. Japan 55 (1986) 2889.

23) J.C. Hicks and E.J. Mele, Phys. Rev. B34 (1986) 1091.

24) H. Ito and Y. Ono, J. Phys. Soc. Japan 54 (1985) 1194.
Y. Ono and H. Ito, J. Phys. Soc. Japan 54 (1985) 4828.

25) Terai, Y. Ono, and Y. Wada, to be published in J. Phys. Soc. Japan.

DOPING DISORDER AND BAND STRUCTURES IN CONJUGATED POLYMERS*

Kikuo HARIGAYA, Yasushi WADA, and Klaus FESSER[†]

Department of Physics, University of Tokyo, Bunkyo-ku, Tokyo 113, Japan
[†]Physikalisches Institut, Universität Bayreuth, D-8580 Bayreuth, Federal Republic of Germany

Doping of polyacetylene, $(CH)_x$, gives qualitative changes in electrical and optical properties at dopant concentrations above a few percent. It now makes materials as conducting as copper. The mechanism of this change has not been so clear. Disorder effects by the doping are analyzed, using the coherent potential approximation. Two types of impurities are taken into account. One is bond-type. They give rise to backward scatterings of electrons. The other is site-type with forward scatterings. The order parameter, electronic band structure, and density of states are obtained as functions of impurity concentration and strengths of impurity potentials. Due to the one-dimensional nature, the formation of impurity band in the gap is suppressed, if strength of the bond impurity is larger than that of the site impurity. When the site impurity is strong enough, a gapless structure takes place at high impurity concentrations. Phase diagrams are extensively given for various impurity strengths.

1. INTRODUCTION

Trans-polyacetylene, $(CH)_x$, is a typical quasi one-dimensional conjugated polymer. It contains a dimerized carbon backbone consisting of alternating single and double bonds. The motion of the lattice backbone of CH units is coupled with π-electrons which can move along the backbone chain. Configuration of the backbone is determined in a consistent way with the electron distribution. Su, Schrieffer, and Heeger[1] (SSH) proposed a model with the discrete lattice. Takayama, Lin-Liu, and Maki[2] (TLM) derived a continuum model as the limit of a small lattice constant. These have provided the basis on which nonlinear excitations, solitons and polarons, have been extensively studied. The single-particle excitations have also been investigated for the electrons and phonons with and without the nonlinear excitations. The electrons, however, have strong Coulomb interactions between them. It has been known that the Coulomb interactions do not give a qualitative change for the ground state configuration and nonlinear excitations. Optical properties are remarkably different.

The *trans*-polyacetylene films show a dramatic increase in electrical conductivity when doped with various dopants. Mechanisms of the high conductivity are not yet clear. It is, thus, important to study the electronic level structure in the doped system. We have used the coherent potential approximation (CPA) to study the effects of two types of impurities. One is bond-type and the other site-type. We assume the electronic states are half-filled and a uniform dimerization is represented by a constant order parameter. We have reported the CPA result that the formation of impurity bands in the energy gap is suppressed, if strength of bond impurity is larger than that of the site

---

*This work was supported by a Grant-in-Aid for Scientific Research from the Ministry of Education, Science, and Culture, by Research Association for Basic Polymer Technology under the sponsorship of NEDO, Japan, and by Deutsche Forschungsgemeinschaft through SFB213 (TOPOMAK, Bayreuth).

impurity (region I). A phase diagram has been given on the space of impurity concentration and the site-impurity strength with the vanishing bond strength.[3] Three phases have been shown. One is the phase where the impurity band is connected either to the valence or to the conduction band (region II). In region III, the impurity band is separated and isolated. In region IV, the order parameter and the energy gap vanish. The present paper will show how the phase diagram changes with the bond impurity strength. We find rich changes with a new phase III'.

2. COHERENT POTENTIAL APPROXIMATION

We start from the TLM model with the impurity term

$$H_{\text{TLM}} = \sum_s \int dx \Psi_s^\dagger(x)[-iv_F\sigma_3(\partial/\partial x) + \Delta\sigma_1]\Psi_s(x) + (2\pi v_F \lambda)^{-1}\int dx \Delta^2, \tag{1}$$

$$H_{\text{imp}} = \sum_{is} \int dx \Psi_s^\dagger(x) 2aV\delta(x - x_i)\Psi_s(x), \quad V = I\sigma_1 + J, \tag{2}$$

where $\Psi_s(x)$ is a two-component field operator of electrons with spin s, v_F the Fermi velocity, σ_i the Pauli matrices, Δ the order parameter, λ the electron-phonon coupling constant, a the undimerized lattice constant in the SSH model, I and J the bond- and site-impurity strengths, respectively, and x_i the location of the i-th random impurity.

The single-site version of CPA[4] determines the self-energy part Σ of the effective medium Green function by

$$c(V - \Sigma)[1 - \bar{g}(V - \Sigma)]^{-1} - (1 - c)\Sigma(1 + \bar{g}\Sigma)^{-1} = 0, \tag{3}$$

where c is the impurity concentration and

$$\bar{g} = (1/N)\sum_k [iE_n + \mu - v_F k\sigma_3 - \sigma_1\Delta - \Sigma]^{-1}. \tag{4}$$

Here, $N = L/2a$, L being the system length, $E_n = (2n+1)\pi T$ the Matsubara frequency, μ the chemical potential, and Σ is the function of the frequency. The self-consistency equation for Δ is

$$\Delta = -(\pi\lambda v_F T/a)\sum_n \text{tr}(\sigma_1 \bar{g}(iE_n)). \tag{5}$$

3. ORDER PARAMETER, IMPURITY BANDS, AND PHASE DIAGRAMS

Eqs. (3-5) are numerically solved with $v_F/a = 2t_0 = 5.0\text{eV}$, $\lambda = 0.183$ at $T \to 0$. It has been known that the minimum of the total energy occurs at the side $I\Delta > 0$ and we can take positive Δ, I, and J without loss of generality.[3] Figures 1(a-c) show the concentration dependence of Δ. It increases with I and c, indicating that the dimerization is stabilized with increasing number of stronger bond impurities. The gapless region (region IV) is found in Fig. 1(a) around $c \sim 0.5$ when $J/t_0 \sim 1.0$. Fig. 1(b) shows a larger J is necessary to get the region when $I/t_0 = 0.1$. Electronic density of states is calculated by

$$\rho(\omega) = -(1/\pi)\text{Im}(\text{tr}\bar{g}), \tag{6}$$

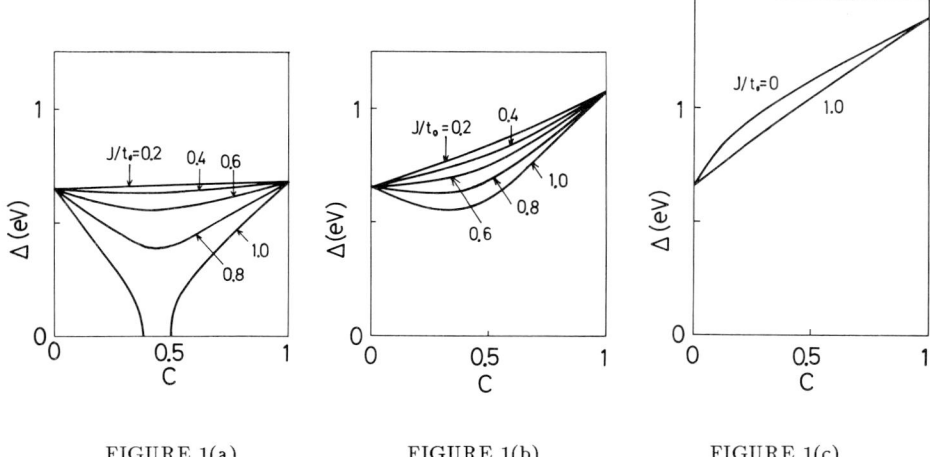

FIGURE 1(a)   FIGURE 1(b)   FIGURE 1(c)

The concentration dependence of Δ; (a) with $I/t_0 = 0.01$, (b) with $I/t_0 = 0.1$, and (c) with $I/t_0 = 0.5$

where $iE_n + \mu$ is replaced by $\omega + i\delta$. Figures 2(a-b) show the top of the valence band by the solid lines and the bottom of the conduction band by the dashed lines. There are no electronic levels in the regions with a cusp. One of them is close to the valence top at $c \sim 0$ in Fig. 2(a). It indicates there is an isolated impurity band. Another isolated band is close to the conduction bottom at $c \sim 1$ (region III). For a smaller J in Fig. 2(b), two isolated impurity bands emerge at $c \sim 1$ (region III'). One is closer to the valence band and the other to the conduction band.

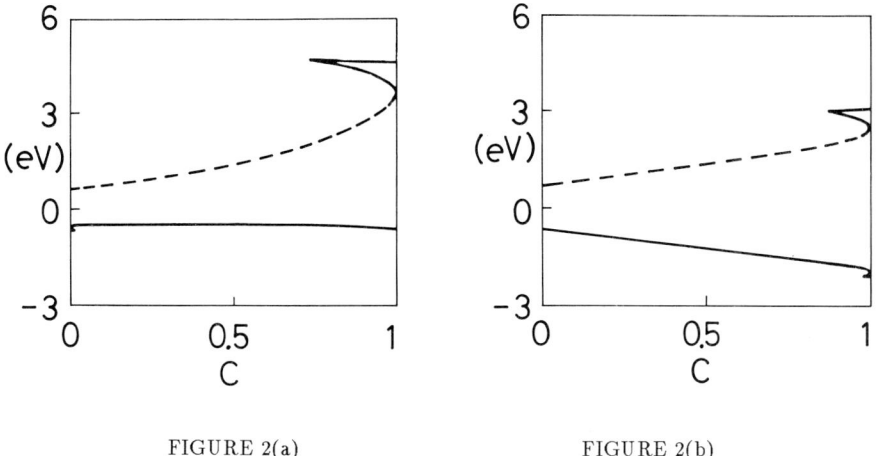

FIGURE 2(a)   FIGURE 2(b)

The solid line is the top of the valence band and the dashed line the bottom of the conduction band; (a) with $I/t_0 = 0.5$, $J/t_0 = 0.8$, and (b) with $I/t_0 = 0.5$, $J/t_0 = 0.2$.

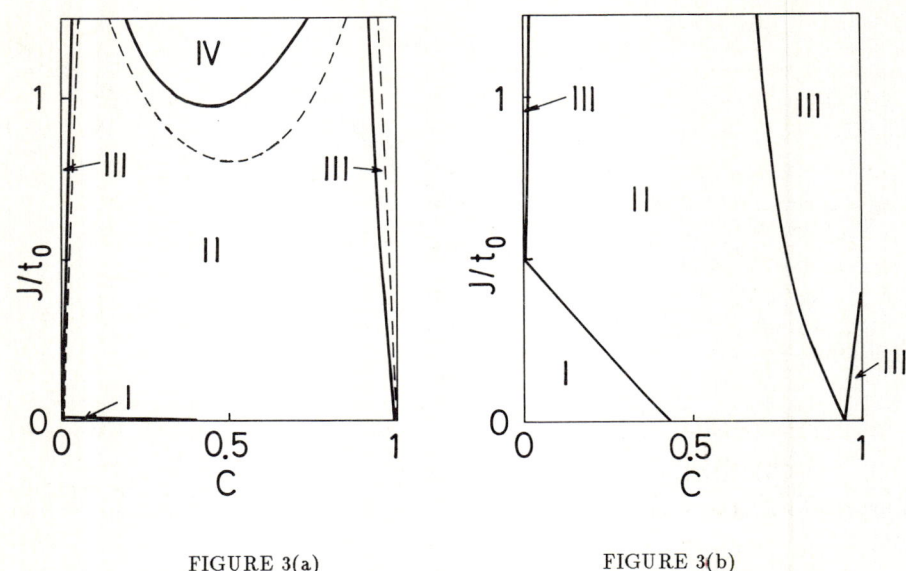

FIGURE 3(a) FIGURE 3(b)

The phase diagram; (a) with $I/t_0 = 0$ (dashed lines) and 0.01 (solid lines) and (b) with $I/t_0 = 0.5$; I–no impurity bands, II–impurity band connected with the valence or conduction band, III–an isolated impurity band, III'– two isolated impurity bands, IV–no dimerization.

These results are summarized on the phase diagrams Figs. 3(a-b). The dashed lines in Fig. 3(a) are the phase boundaries without the bond impurities $I = 0$. The phase diagram is sensitive to the bond strength. As it increases, the region I becomes appreciable while the region IV goes out of the relevant domain. We find a new phase III' at high concentrations.

4. CONCLUSIONS

The CPA is applied for wide parameter ranges of c, I, and J to show how sensitive the physical properties are. The doping effects on polyacetylene could be understood definitely only after the characters of the dopants are well specified. The shortcoming of the present theory is the assumption of a uniform order parameter Δ. A self-consistent treatment of a space dependent $\Delta(x)$ is desirable.

REFERENCES

1) W. P. Su, J. R. Schrieffer, and A. J. Heeger, Phys. Rev. Lett. 42 (1979) 1698; Phys. Rev. B 22 (1980) 2099.
2) H. Takayama, Y. R. Lin-Liu, and K. Maki, Phys. Rev. B 21 (1980) 2338.
3) K. Harigaya, Y. Wada, and K. Fesser, submitted for publication.
4) See, for example, F. Yonezawa and K. Morigaki, Suppl. Prog. Theor. Phys. 53 (1973) 1; J. M. Ziman, Models of Disorder (Cambridge University Press,1979).

STRONGLY COUPLED ONE-DIMENSIONAL SYSTEM AND THE POLYMER

Xin SUN *, Chang-qin WU *, Rouli Fu §, Duo Liang LIN †, and T.F. GEORGE †

* Lab for Solid State Microstructure, Nanjing University, Nanjing 210008; and Department of Physics, Fudan university, Shanghai 200433, China (a)
§ Lab for Infrared Physics, Academia Sinica, Shanghai 200081, China
† Department of Physics and Astronomy, State University of New York at Buffalo, Buffalo, New York 14260

What is the effect of the electron interaction on the lattice instability of one-dimensional systems with strong electron correlation is a controversial problem. One side believes the electron interaction enhances the instability; but the other side declares the instability is reduced by the electron interaction. The origin of this controversy is analysed in this paper and we find that the range of the electron interaction plays a decisive role. When the range is long, the electron interaction enhances the lattice instability; whereas, if the range is short, the electron interaction reduces the instability. Based on this conclusion, we can clarify the dispute about the effect of the electron interaction on the dimerization of the conducting polymers.

1. INTRODUCTION

For one-dimensional system, due to the electron-lattice interaction, a proper modulation of the lattice can always make the boundary of the Brillouin zone coincide with the Fermi surface, then, a gap is opened on the Fermi surface. This gap will lower the energy and a lattice distortion results from such modulation. It is so-called Peierls instability, which is the peculiarity of one-dimensional systems. If the band is half filled, this instability will cause the dimerization, which produces the bond alternation of the polymers. In many one-dimensional systems, such as the conducting polymers, the electron correlation is strong, then, an important question is asked: is the lattice instability enhanced or reduced by the electron-electron interaction? It has attracted considerable attention.[1-6] So far the extended Hubbard model

$$H_H = U \sum_{i,s} n_{i,s} n_{i,-s} + V \sum_{i,s,s'} n_{i,s} n_{i+1,s'} \qquad (1)$$

is used to describe the electron interaction. And many different techniques, such as renormalization group,[1] Gutzwiller variational calculation,[2] Monte Carlo simulation,[3] Valence bond method,[4] etc, have been used to investigate the effect of electron interaction on the dimerization. These studies reach the same result that the dimerization is greatly enhanced by electron interaction. Furthermore they conclude that the main origin of the dimerization is the electron-electron interaction rather than the electron-lattice interaction.

2. DISPUTE

However, this conclusion is challenged recently by Kivelson, Su, Schrieffer and Heeger (KSSH),[7] who argue that the repulsive interaction between electrons should be unfavorable to deviate from the uniform structure. Then they declare an opposite opinion that the electron interaction has to reduce the dimerization. KSSH point out that the extended Hubbard model only contains site-charge repulsion, which is the diagonal part of the Coulomb interaction, missed the bond-charge repulsion

$$W C^+_{i,s} C_{i+1,s} C^+_{i,s'} C_{i+1,s'} \tag{2}$$

which is one of the off-diagonal terms of the Coulomb interaction. By adding such W term to the extended model, they show the dimerization can be suppress by the electron interaction. But the other side presents opposite Comments on KSSH's theory and insists the original opinion.[8]

Obviously, the divergence of these two sides roots in the different descriptions of the electron interaction. One side only takes the diagonal terms U, V of Coulomb interaction, the other side adds one more term W. Since the site-charge repulsion U, V and the bond-charge repulsion W are some terms of Coulomb interaction, the full interaction between electrons contains many other terms also. Therefore, in order to clarify this dispute, we should start the study from a general electron interaction — the screened Coulomb potential

$$v(r) = \frac{U_0}{\sqrt{1 + (r/a)^2}} \exp(-\beta r/a) \tag{3}$$

where U_0 and β are the strength and the screening factor of the interaction. This general interaction includes all diagonal and off-diagonal terms. Actually, in the second quantized representation, the interaction (3) can be written as

$$H' = \sum_{i,j,l,m,s,s'} V(i,j,l,m) C^+_{i,s} C^+_{j,s'} C_{l,s'} C_{m,s} , \tag{4}$$

$$V(i,j,l,m) = \int dx \int dx' \phi^*_i(x) \phi^*_j(x') v(x - x') \phi_l(x') \phi_m(x) . \tag{5}$$

It is easy to see that $V(i,i,i,i) = U$, $V(i,j,j,i) = V$, and $V(i,j,i,j) = W$. So, the extended Hubbard model and the KSSH's model are different approximations of the interaction (3), and the general interaction (3) can give a reliable answer to the dispute.

3. CORRELATED BASIS FUNCTIONS (CBF) THEORY

The key quantity of many-electron system is the electron correlation function $g(1,2)$, from which the energy and other properties, including the dimerization u, of the system can be obtained. A powerful method to get the correlation function is the CBF theory.[10]

Let us denote the non-interacting electron orbit as $\phi_k(x)$, then, the ground state of the interacting system can be written as

$$\Psi(1, 2, \ldots, N) = D[\phi_k] \cdot \exp[u(1, 2, \ldots, N)] \tag{6}$$

where $D[\phi_k]$ is the Slater determinant and $u(1,2,\ldots,N)$ is the correlation factor, which can be determined by the variational principle. In the case of half-filled band, each atom has only one electron, the electron density is not high. Meanwhile, there is no electron condensation due to the repulsion between electrons. Then, it is rare for three or more electrons to be gathered closely, and the two-body correlation factor u_{ij} is dominant. In such case, the correlation function $g(1,2)$ can be calculated by solving the following equations[9]

$$g(1,2/\xi) = g(1,2/0)\exp\int_0^\xi d\xi' K(1,2/\xi') , \qquad (7)$$

$$K(1,2/\xi) = [u_{12} + \alpha(1,2) + \alpha(2,1) + \int d3P(3)h_{12}\alpha(2,3)]$$
$$\cdot[\delta(1,2) + \delta(2,1) + \varsigma(1,2) + \gamma(1,2)]/g(1,2) , \qquad (8)$$

$$h_{12} = g(1,2) - 1 , \qquad \alpha(1,2) = \int d3P(3)h_{13}u_{23} , \qquad (9)$$

$$\delta(1,2) = \int d3P(3)h_{13}h_{23}[u_{13} + \alpha(1,3) + \alpha(3,1) + \int d4P(4)h_{14}\alpha(3,4)] , \qquad (10)$$

$$\varsigma(1,2) = \int d3P(3)h_{13}\delta(3,2) , \qquad (11)$$

$$\gamma(1,2) = \int d3P(3)h_{13}h_{23}\int d4P(4)[u_{34}g(3,4)$$
$$+\alpha(3,4)(1 + h_{34}/2)h_{34}\int d5P(5)h_{45}(u_{45} + \alpha(4,5)/2)]. \qquad (12)$$

4. RESULTS AND CONCLUSIONS

By using the variational principle and the obtained $g(1,2)$, we can get the dependence of the dimerization $u(U_0)$ on the interaction strength U_0 for different screening factor β. Our results show there exists a critical value β_c,
1. if $\beta < \beta_c$, the dimerization is initially enhanced by the electron-electron interaction.
2. if $\beta > \beta_c$, the dimerization is reduced by the electron-electron interaction.

In other words, the effect of the electron-electron interaction on the lattice instability depends on the interaction range $\Lambda = a/\beta$. The long-ranged interaction will enhance the instability, but the short-ranged interaction will suppress it. For the conducting polymers, $\beta_c \sim 2$.

It is not difficult to understand why the interaction range is the decisive factor to determine the behavior of the lattice instability. First, let us look at the dependence of the ratio W/V on the screening factor β, which is shown in the following table. The ratio W/V monotonically

β	1.0	3.0	5.0	7.0
W/V	0.02	0.10	0.26	0.43

increases with increasing β. Therefore, when the interaction range Λ is long (β is small), the off-diagonal terms are very small and can be neglected, then the extended Hubbard model is valid, so the dimerizasion is enhanced by the electron-electron interaction. In the opposite case, Λ is short (β is big), the off-diagonal terms become important, the extended Hubbard model fails, then the bond-charge repulsion and other off-diagonal terms make the dimerization reduced. For the polyacetylene, it can be estimated from the optical gap that its screening factor $\beta_{PA} = 1.7$. Since $\beta_{PA} < \beta_c$, the dimerization of the polyacetylene is enhanced about 25% by the electron interaction.

ACKNOWLEDGEMENTS

This work was supported by the National Science Foundation of China, Grant 863-715-22, the office of Naval Research, NSF Grant CHE-8620274 and the Air Force Offices of Scientific Research (AFSC), United States Air Force, under Contract F 49620-86-C-0009.

REFERENCES

(a) The permanent address.

1. G.W. Hayden and E.J. Mele, Phys. Rev. B34 (1986) 5484.

2. D. Baeriswyl and K. Maki, Phys. Rev. B31 (1985) 6633.

3. J.E. Hirsch, Phys. Rev. Lett. 51 (1983) 296; D.K. Campbell, T. DeGrand and S. Mazumdar, Phys. Rev. Lett. 52 (1984) 1717.

4. Z. Soos and S. Ramasesha, Phys. Rev. Lett. 51 (1983) 2374; S. Mazumdar and S. N. Dixit, Phys. Rev. Lett. 51(1983) 292; P. Tavan and K. Schulten, Phys. Rev. B36 (1987) 4337.

5. S. Kivelson and D.Heim, Phys. Rev. B26 (1982) 4278; H. Fukutome and M. Sasai, Prog. Theor. Phys. 61 (1983) 1.

6. W. Wu and S. Kivelson, Phys. Rev. B33 (1986) 8546.

7. S. Kivelson, W.P. Su, J.R. Schrieffer and A.J. Heeger, Phys. Rev. Lett. 58 (1987) 1899.

8. D. Baeriswyl, P. Horsch and K. Maki, Phys. Rev. Lett. 60 (1988) 70; J. Gammel and D.K. Campbell, Phys. Rev. Lett. 60 (1988)71.

9. C. Wu, X. Sun and K. Nasu, Phys. Rev. Lett. 59 (1987) 831.

10. E. Feenberg, *Theory of Quantum fluids* (Academic, New York, 1969); S. Chakravarty and C.W. Woo, Phys. Rev. B13 (1976) 4815.

MANY-BODY EFFECTS IN QUANTUM WELLS

T. ANDO

Institute for Solid State Physics, University of Tokyo
7-22-1 Roppongi, Minato-ku, Tokyo 106, Japan

A review is given of roles of many-body electron-electron interactions in two-dimensional systems at semiconductor heterostructures. One of the most typical examples is the band-gap renormalization in modulation-doped quantum wells and photo-excited electron-hole plasmas and another is excitonic final-state interactions giving rise to Fermi-edge singularities of absorption and emission spectra. A special emphasis will be given on oscillatory change of many-body effects due to complete quantization of orbital motion in high magnetic fields.

1. INTRODUCTION

Two-dimensional (2D) systems made at semiconductor heterostructures are one of the ideal systems for the study of electron-electron interactions. These systems are considered as an ideal electron liquid. Further, results of theoretical calculations can be directly compared with experiments including powerful spectroscopic means such as optical absorptions, emissions, and scatterings. The purpose of this paper is to give a brief review of various effects of electron-electron interactions which have been predicted theoretically and manifest themselves in observed phenomena.

Figure 1 shows schematically the structure of a modulation-doped AlGaAs/GaAs/AlGaAs quantum well. The motion of electrons in the direction perpendicular to the interface is quantized by the strong barrier potential and electrostatic potential of ionized donors and electron themselves, while the motion parallel to the interface remains free. We have a series of subbands each having a 2D degree of freedom. The typical electron concentration is $n \sim 10^{12}$ cm^{-2} per each well. Only the lowest subband is occupied by electrons usually, but the first excited subband becomes occupied for large well-width and high electron concentrations. Because ionized donors are spatially separated, they have only a negligible influence as scatterers and the electron system with extremely high quality is achieved[1,2].

Electron-electron interactions manifest themselves in optical properties of quantum wells through the band-gap renormalization and excitonic enhancement of the absorption and emission edge. These topics will be reviewed in Section 2. Effects of high magnetic fields, especially oscillations due to discrete density of states, are reviewed in Section 3.

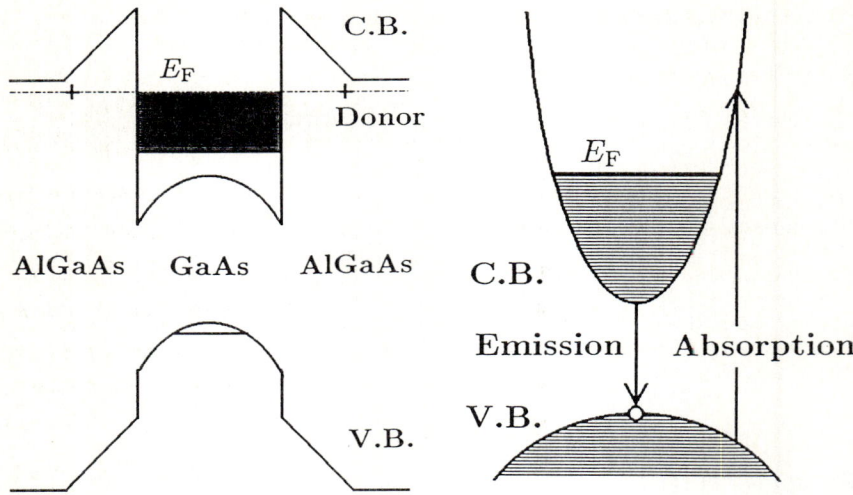

FIGURE 1 A schematic illustration of a modulation doped AlGaAs/GaAs/AlGaAs quantum well.

FIGURE 2 A schematic illustration of optical processes in a doped quantum well. The transition between the lowest conduction-band state and the highest valence-band state is dominant in emission processes, giving the effective band gap. The absorption is allowed only for wave vectors larger than the Fermi wave vector k_F of the conduction band and is shifted to higher energy side in doped systems (the Moss-Burstein shift).

2. OPTICAL PROPERTIES

2.1 Photoluminescence – Band-Gap Renormalization

In photoluminescence experiments, after a light illumination photo-excited holes are quickly relaxed into their quasi-equilibrium states (at the top of the valence band at absolute zero temperature) and then recombine with electrons in the conduction band by emitting a photon. Therefore, the photoluminescence gives a direct information on the band gap. In such optical processes, the effect of interactions among electrons and holes can usually be divided into self-energy effects and vertex corrections. The latter describes excitonic correlations between electrons and photo-excited holes. The former is essentially modification of the one-particle electron and hole energies. In highly modulation-doped quantum wells the vertex correction does not modify the band edge and the photoluminescence energy reflects purely self-energy shifts of electrons and holes[3].

The so-called local density approximation is often the method which can be used to estimate effects of such exchange and correlation in the inhomogeneous electron gas. It can be extended to calculate self-energy shift of a photo-excited hole in dense electron liquid[3]. Other methods all come down to an RPA type of calculation, taking into account the lowest subbands via a form factor in the electron-electron interaction[3,4]. These two different methods give results quite similar to each other[3].

Figure 3 gives a comparison of calculated[3] and observed[5,6] effective band-gaps.

The figure demonstrates the excellent agreement between calculations and experiments. The wave function of electrons and holes in a quantum well is primarily determined by the barrier potential and consequently the effect of the electrostatic potential on the photoluminescence energy is canceled to the lowest order between electrons and holes. As a matter of fact, the result of the Hartree approximation is nearly independent of the electron concentration except in the case of high concentrations. It can be concluded that the renormalization of photoluminescence energy reflects purely the strength of the electron-electron and electron-hole interactions.

Much more experimental and theoretical works have been performed more recently, which have also given a further support on the band-gap renormalization[7-11]. Works have been extended to electron-hole systems created from undoped quantum wells by a strong light illumination. Various experimental[12-15] and theoretical[16-19,9] works have been reported so far. Figure 4 gives a comparison of observed[13] and calculated[17] effective band-gaps in electron-hole systems. Experiments, in general, seem to show a slightly larger shift in comparison with results of density-functional calculations.

A comment is worthwhile on the effective dimensionality of the present systems. In GaAs, the effective Bohr radius is about 100Å and the typical three-dimensional (3D) electron concentration is 10^{18}cm^{-3} which gives $r_0 \sim 60$Å where r_0 is the effective electron radius defined by $n^{-1} = (4\pi/3)r_0^3$ with n the 3D electron density. Considering the fact that the typical thickness of quantum wells is 200Å, we can conclude that the system is more three-dimensional rather than two-dimensional and in the high-density regime ($r_s \sim 0.6$). Therefore, calculations made in RPA type approximations should give results in good agreement with experiments.

2.2 Absorption – Excitonic Effects

The lineshape of the optical spectra can be strongly modified by excitonic effects[20]. Density-functional theory can also be employed if it is extended to time-dependent cases[17] as well as diagrammatic perturbations[21]. Calculations have demonstrated pronounced enhancements of oscillator strengths of absorptions at the Fermi level in comparison with those in the absence of interactions. An enhancement of the absorption at the Fermi level has been observed in AlGaAs/GaAs/AlGaAs systems[22], which is qualitatively explained by calculations. Figure 5 gives an example.

When holes are localized by potential inhomogeneities, a similar enhancement can be seen in photoluminescence spectra. The situation is quite similar to the case of singularities observed in soft X ray emission spectra in metals. An enhancement has been observed in InP/InGaAs/InP quantum wells in which existing alloy disorder in the well InGaAs layer tends to trap a photo-excited hole into localized bound states[23]. Similar results have also been observed in AlGaAs/GaAs/AlGaAs quantum wells[24,22]. Exact calculations have been reported within a simplified model of short-range interactions between a hole and electrons[25]. Figure 5 contains an example of emission spectra observed in AlGaAs/GaAs/AlGaAs quantum wells also[22]. In this example, photo-excited holes are trapped in bound localized states and a strong

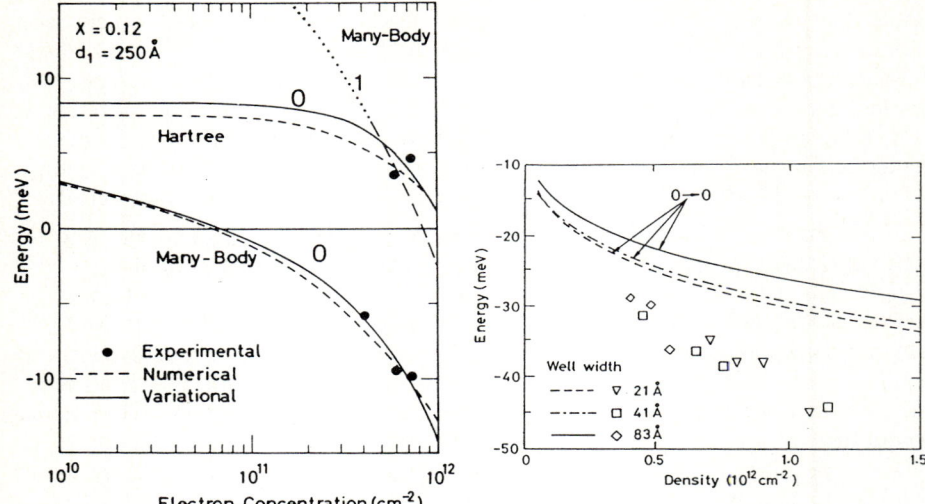

FIGURE 3 An example of calculated band-gap renormalization and observed photoluminescence energies in modulation-doped AlGaAs/GaAs/AlGaAs quantum wells. All energies are measured relative to the band gap of bulk GaAs. The curves labeled "Hartree" display the results calculated in the Hartree approximation [solid curve—variational (Gaussian wave function; dashed—numerical wave function]. The Hartree results are modified by exchange-correlation as shown by the curves marked "many-body" [solid curve—self-energy added; dashed—self-consistent results using the local exchange-correlation potential]. The second electron subband is not occupied for the densities corresponding to the dotted line. [After Ref. 3]

FIGURE 4 Band-gap renormalization of electron-hole plasmas in quantum wells as a function of pair density for different well widths calculated in the local density approximation[17]. The markers denote the experimental results of Tränkle et al.[13]. [After Ref. 17]

enhancement of luminescence at the Fermi edge can be observed. Experiments have recently been extended to the case in high magnetic fields and given various unexpected and interesting results such as reappearance of exciton lines[26–29]. A review on optical properties of quantum wells appeared quite recently[30].

3. MAGNETIC OSCILLATION

3.1 Landau-Level Broadening

When a strong magnetic field H is applied perpendicular to the 2D system, the orbital motion is quantized into discrete Landau levels with energy $E_N = (N + 1/2)\hbar\omega_c$, where $N = 0, 1, \cdots$ and $\omega_c = eH/mc$ is the cyclotron frequency. Each Landau level is degenerate with respect to positions of the center of cyclotron motion. The degeneracy is $1/2\pi l^2$ per unit area, where l is the magnetic length corresponding to

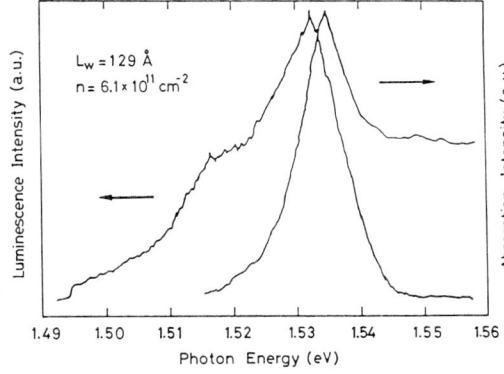

FIGURE 5 An example of absorption and emission spectra observed in modulation-doped AlGaAs/GaAs/AlGaAs quantum wells. The electron concentration is 6.1×10^{11} cm^{-2} and the well-width is 129 Å. [After Ref. 22]

the radius of the lowest ($N = 0$) Landau level, defined by $l = \sqrt{c\hbar/eH}$. In actual systems, however, the δ-function-like density of states is broadened into a Gaussian-like density of states due to the presence of scatterers like charged donors, interface roughness, etc.

The discrete density of states in high magnetic fields gives rise to oscillations of effects of electron-electron interactions. One of the most typical examples is the broadening of Landau levels or the density of states itself. The broadening of the Landau level is determined by strength of potentials of scatterers. The impurity potential is strongly screened by electrons in the quantum well. On the other hand, the screening depends on the broadening through the density of states at the Fermi level. Therefore, the broadening of the Landau level and the screening should be determined self-consistently. Since the screening also depends on the Fermi level position, this self-consistency gives rise to an oscillation of the broadening as a function of the Fermi level.

Such a self-consistent calculation was first performed in inversion layers on Si surfaces[31]. The calculation has demonstrated that the level broadening actually exhibits a strong oscillatory behavior: When the Fermi level lies at the center of Landau levels, the broadening is small. When the Fermi level lies in the tail region, however, the broadening becomes even larger than the separation of adjacent Landau levels, i.e., the density of states of Landau levels overlap with each other appreciably. Figure 6 schematically shows this oscillation. Similar phenomena has been predicted also in semiconductor heterostructures[32–34].

Experimental evidence of the oscillation of the broadening was provided by Kukushkin and Timofeev quite recently in Si inversion layers[35–37]. They have observed photoluminescence spectra corresponding to recombination of electrons in the inversion layer and holes trapped in acceptors in the vicinity of the Si/SiO$_2$ interface. The experiments have shown a beautiful Landau level density of states, whose width exhibits an oscillation consistent with the theoretical prediction, as is shown in Fig. 7.

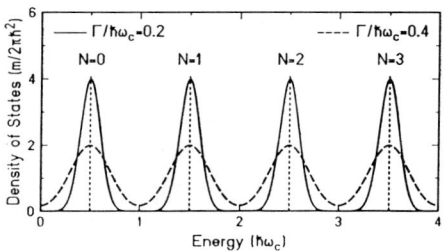

FIGURE 6 A schematic illustration of the density of states in a 2D system in strong magnetic fields. When the Fermi level lies in the central part of the broadened Landau level, the broadening is small. When it lies in the tail region of the density of states, the broadening becomes large, causing overlapping of the density of states of adjacent Landau levels

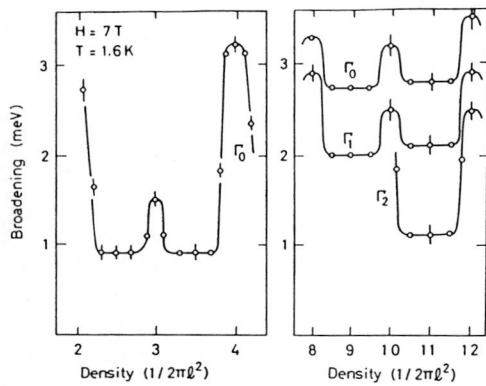

FIGURE 7 An example of observed broadenings of Landau levels in Si inversion layers. At electron concentrations where each Landau level is completely filled (integers in units of $1/2\pi l^2$), the broadening increases considerably. [After Ref. 35]

Englert et al. observed the cyclotron-resonance broadening depending strongly on the Fermi-level position in GaAs/AlGaAs heterostructures[38]. More convincing cyclotron-resonance experiments have been provided by Heitmann et al. in InAs quantum wells[39]. Figure 8 gives an example of observed lineshapes. The amplitude of the oscillation can be as large as 2–6 times of the typical broadening.

Quantitative explanation of experiments involves complicated problems[40,41]. For example, nonlinearity of screening is presumably very important because of the discrete density of states. Further, states are all localized except just at the center within each Landau level[42] (the Anderson localization) and the localization can strongly modify the screening properties.

3.2 Exchange Enhancement of Spin Splitting

Another example is the exchange enhancement of the spin Zeeman splitting. In strong magnetic fields the difference in the number of electrons with ↑ and ↓ spins depends strongly on the position of the Fermi level, as is schematically shown in Fig. 9. When the Fermi level lies at the middle of the gap between ↑ and ↓ levels, the

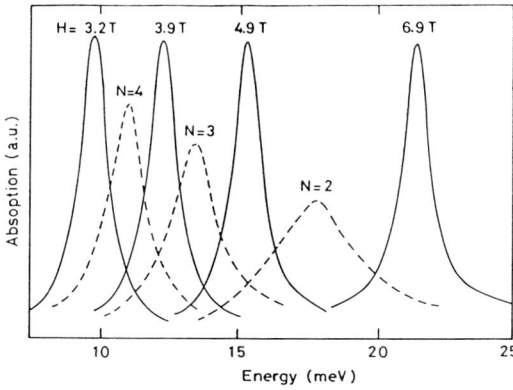

FIGURE 8 Examples of cyclotron resonance absorptions observed in an InAs quantum well. The dashed curves represent absorptions when the Fermi level lies just at the middle of adjacent Landau levels (the index of the top most filled Landau-level is given). The solid lines are the absorption when the Fermi level lies in the vicinity of the center of Landau levels. [After Ref. 39]

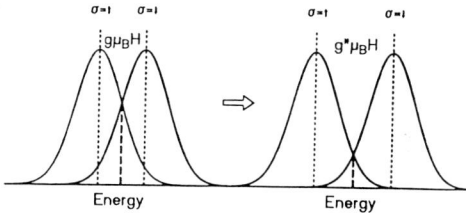

FIGURE 9 A schematic illustration of density of states of spin-split Landau levels and enhancement of the spin splitting due to exchange effect

difference becomes a maximum and the g factor has a maximum value. When it lies midway between Landau levels with different level index N, the difference vanishes and the g factor has a minimum value unaffected by the exchange effect.

First experimental results which suggested the enhancement were provided by Fang and Stiles in 1968[43]. Later, Englert and von Klitzing have provided a convincing experimental proof that the spin splitting does show oscillatory behavior[44]. This oscillatory g factor was first calculated in 2D systems in Si inversion layers, which has shown that the g factor when enhanced can be as large as $3 \sim 4$ in comparison with undressed $g = 2$ depending on magnetic fields and broadening of Landau levels[45]. It has been later applied to the valley splitting[46,47]. Quite recently, this oscillatory enhancement of the spin splitting was suggested also by photoluminescence spectra[35].

3.3 Band-Gap Renormalization

When a strong magnetic field is applied, the many-body renormalization of the photoluminescence energy can be modified through the change in the dielectric response of the system. Such effects were studied quite recently by two groups independently[48,49]. In the following, a brief review will be given following Katayama and Ando[48] on origins of such oscillations.

In the random-phase approximation (RPA), the quasi-particle self-energy for the electron Green's function Σ^e is given by a sum of the screened exchange (Σ^e_{xs}) and the Coulomb hole term (Σ^e_{ch}) as

$$\Sigma^e(N,\omega) = \Sigma^e_{xs}(N,\omega) + \Sigma^e_{ch}(N,\omega), \tag{3.1}$$

where the first and second term in the right hand side describe, respectively, the exchange interaction screened dynamically and the correlation effects, *i.e.* the creation of the Coulomb hole or electron-density deficiency around an electron through virtual excitations of density fluctuations. The two terms play complemental roles in reducing the electron energy. Explicitly, we have

$$\Sigma^e_{xs} = \sum_{N'}\sum_q J_{NN'}(q)^2 F^e(q) \int \frac{d\omega'}{\pi} f_F(\omega') \frac{V(q)}{\varepsilon(q,\omega-\omega')} \mathrm{Im} G^e_{N'}(\omega') \tag{3.2}$$

and

$$\Sigma^e_{ch} = -\sum_{N'}\sum_q J_{NN'}(q)^2 F^e(q) \int \frac{d\omega'}{\pi} n_B(\omega') \mathrm{Im}\left[\frac{V(q)}{\varepsilon(q,\omega')}\right] G^e_{N'}(\omega-\omega'), \tag{3.3}$$

where $V(q) = 2\pi e^2/q\kappa$ with κ the background dielectric constant, f_F and n_B stand for the Fermi and Bose distribution, respectively, and F^e denotes the form factor for electron-electron interaction, which is evaluated by using the wave function corresponding to the ground subband in quantum wells. The electron Green's function G^e_N is evaluated by using the self-consistent Born approximation to take account of impurity scatterings[50].

A similar formula can be obtained for a photo-excited hole except that Σ^h_{xs} does not exist because of the negligibly small number of holes and that an appropriate form factor should be used for electron-hole interaction. The Coulomb-hole term Σ^h_{ch} describes the reduction of the hole energy by attracting a 2D electron cloud around the hole.

The dynamical dielectric function $\varepsilon(q,\omega)$ of the 2D electron gas describes the screening effect on the exchange in Eq. (3.2) and virtual excitations of density fluctuations in Eq. (3.3). We employ the plasmon-pole approximation, which is known to work quite well in the absence of a magnetic field in spite of its simplicity. Figure 10 shows examples of calculated quasi-particle self-energies for an AlGaAs/GaAs/AlGaAs quantum well with $n = 4 \times 10^{11}$ cm^{-2} at zero temperature as a function of magnetic field. Each of the self-energies exhibits clear oscillations as a function of the magnetic field. One of the most prominent features is that the phase of the oscillation of Σ^e_{xs} and Σ^e_{ch} is exactly opposite. When the Fermi level E_F sits on the Landau-level center, intra-Landau-level excitations give rise to a large polarization in the dielectric function $\varepsilon(q,\omega)$. Large $\varepsilon(q,\omega)$ means importance of the Coulomb hole effect, *i.e.*, enhancement of $-\Sigma^e_{ch}$, and at the same time a reduction of the exchange interaction, $-\Sigma^e_{xs}$, due to large screening. As E_F approaches the edge of the density of states, on the other hand, such intra-Landau-level polarization is reduced and only small inter-Landau-level polarization remains, giving rise to enhancement of the exchange term and reduction of the Coulomb hole term. Therefore, there is a

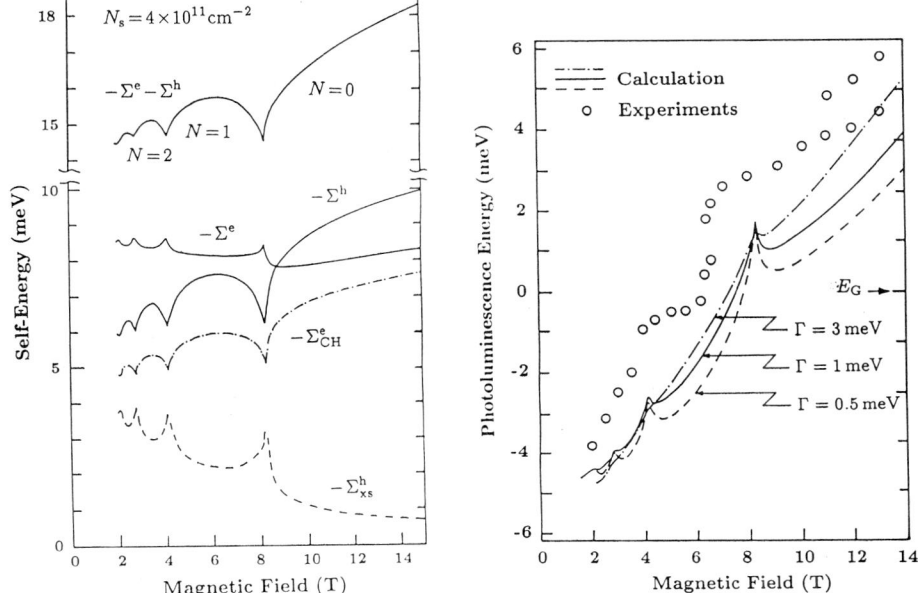

FIGURE 10 Calculated quasi-particle self-energies as a function of the magnetic field in a modulation-doped AlGaAs/GaAs/AlGaAs quantum well. [After Ref. 48]

FIGURE 11 Photoluminescence energy measured from the band gap of bulk GaAs as a function of the magnetic field in a modulation-doped AlGaAs/GaAs/AlGaAs quantum well. The dashed, solid, and dash-dotted lines represent calculations for three different values of the broadening Γ, and open circles experimental results of Perry et al.[51]. [After Ref. 48]

strong tendency for the cancellation of the oscillation between Σ_{xs}^e and Σ_{ch}^e. There exists another cancellation between Σ_{xs}^e and Σ_{ch}^e: Apart from the oscillation, $-\Sigma_{xs}^e$ decreases as H increases, while $-\Sigma_{ch}^e$ increases. Consequently the total self-energy shift $-\Sigma^e$ is almost independent of the magnetic field.

Such cancellation does not occur for holes, because only the Coulomb hole term is present. The self-energy shift of holes has a magnetic-field dependence similar to the Coulomb-hole term for electrons, *i.e.*, it exhibits a strong oscillation and increases almost linearly on the average as a function of magnetic field. Note that it is comparable to the shift of electrons and can be even larger in sufficiently strong magnetic fields. Therefore, the net magnetic-field dependence of the band-gap renormalization $-\Sigma^e-\Sigma^h$, especially its oscillation, is dominated by that of the self-energy of holes. In the limit of vanishing field, the gap renormalization approaches the value (about 14 meV) estimated in the absence of a field.

Perry *et al.* have performed photoluminescence experiments in doped AlGaAs/GaAs/AlGaAs quantum wells in large perpendicular magnetic fields and found unexpected oscillatory behavior of the luminescence energy[51]. Figure 11 gives the photoluminescence energy measured from the band gap of bulk GaAs at the electron density corresponding to the experiments of Perry et al. The open circles are experimental results at 2 K. The calculation explains the overall feature of experiments very well. In particular, the calculation gives both period and phase of the oscillation in excellent accord with experiments. There appears a splitting of photoluminescence peaks observed experimentally above $H=11$T, which is due to many-body enhancement of the g factor discussed in Section 2.

There is a difference in shape of magnetic oscillations between calculations and experiments. A possible explanation is the oscillation of the broadening Γ discussed in the previous section. If we consider this effect, there is a crossover of the photoluminescence energy between small Γ and large Γ as a function of magnetic field. The sharp peaks near $H=3.6$, 4.1, and 8.2 T tend to be broadened and the calculated energy becomes much closer to the observed step-like dependence.

4. SUMMARY AND CONCLUSION

A review has been given of roles of electron-electron interactions mainly in optical properties of modulation-doped quantum wells. It has been demonstrated that many-body exchange and correlation manifest themselves in various optical properties and that the shrinkage in the band-gap is quantitatively explained by calculations in RPA type approximations. This should be expected since the system is considered as more three-dimensional and in the high-density regime although spatially confined strongly. The discrete density of states in high magnetic fields has been shown to give rise to various oscillations of many-body effects. The oscillations predicted theoretically are in qualitative agreement with experiments. However, the quantitative comparison cannot be made until difficult problems related to broadenings of Landau levels and localization due to the presence of random potential fluctuations are well understood.

We have left out the fractional quantum Hall effect, which is one of the most typical manifestations of many-body effects in high magnetic fields. Since its first observation[52,53] and the proposal of ground state wavefunctions by Laughlin[54] and hierarchy structure of different fractions[55], there have been various developments especially on the experimental side. Typical examples are observation of optical gaps in photoluminescence spectra[35], "even" denominator for excited Landau levels[56,57]. Unfortunately, theoretical understanding of these experiments is still in unsatisfactory stage. Several reviews have already been published already[58,59].

REFERENCES

1) R. Dingle, H.L. Störmer, A.C. Gossard and W. Wiegmann, Appl. Phys. Lett. **37**, 665 (1978);
2) H.L. Störmer, R. Dingle, A.C. Gossard, W. Wiegmann and R.A. Logan, in: Proc. 14th Int. Conf. Phys. Semiconductors, Edinburgh, 1978, edited by B.L.H. Wilson (The Institute of Physics, Bristol, 1978), p. 557.
3) G.E.W. Bauer and T. Ando, Phys. Rev. B **31**, 8321 (1985); J. Phys. C **19**, 1537 (1986).

4) D.A. Kleinman and R.C. Miller, Phys. Rev. B **32**, 2266 (1985).
5) J.M. Worlock, A.C. Maciel, A. Petrou, C.H. Perry, R.L. Aggarwal, M.C. Smith, A.C. Gossard and W. Wiegmann, Surf. Sci. **142**, 486 (1984).
6) A. Pinczuk, J. Shah, R.C. Miller, A.C. Gossard and W. Wiegmann, Solid State Commun. **50**, 735 (1984); Surf. Sci. **142**, 492 (1984).
7) M.H. Meynbadier, J. Orgonasi, C. Delalande, J.A. Brum, G. Bastard, M. Voos, G. Weimann and W. Schlapp, Phys. Rev. B **34**, 2482 (1985).
8) D.A. Kleinman, Phys. Rev. B **33**, 2540 (1986).
9) G.E.W. Bauer and T. Ando, Phys. Rev. B **34**, 1300 (1986).
10) C. Delalande, G. Bastard, J. Orgonasi, J.A. Brum, H.W. Liu, M. Voos, G. Weimann and W. Schlapp, Phys. Rev. Lett. **59**, 2690 (1987).
11) H. Yoshimura, G.E.W. Bauer and H. Sakaki, Phys. Rev. B **38**, 10791 (1988).
12) S. Tarucha, Y. Horikoshi and H. Okamoto, Jpn. J. Appl. Phys. **23**, 874 (1984).
13) G. Tränkle, H. Leier, A. Forchel, H. Haug, C. Ell and G. Weimann, Phys. Rev. Lett. **58**, 419 (1987).
14) G. Bongiovanni and J.L. Staehli, Phys. Rev. B **39**, 8359 (1989).
15) Ch. Weber, C. Klingshirn, D.S. Chemla, D.A.B. Miller, J.E. Cunningham and C. Ell, Phys. Rev. B **38**, 12748 (1988).
16) A. Tomita and A. Suzuki, IEEE J. Quantum Electron. **23**, 1155 (1987).
17) G.E.W. Bauer, in *Proc. 19th Int. Conf. Phys. Semiconductors*, edited by W. Zawadzki (Institute of Physics, Polish Academy of Science, Warsaw, 1988), p. 143.
18) S. Das Sarma, R. Jalabert and S.-R. Eric Yang, Phys. Rev. B **39**, 5516 (1989).
19) P. Hawrylak, Phys. Rev. B **39**, 6264 (1989).
20) G.D. Mahan, Phys. Rev. **153**, 882 (1967).
21) S. Schmitt-Rink, C. Ell and H. Haug, Phys. Rev. B **33**, 1183 (1986).
22) J.S. Lee, N. Miura, A. Iwasa, Semicond. Sci. Technol. **2**, 675 (1987); Surf. Sci. **196**, 534 (1988).
23) M.S. Skolnick, J.M. Rorison, K.J. Nash, D.J. Mowbray, P.R. Tapster, S.J. Bass and A.D. Pitt, Phys. Rev. Lett. **58**, 2130 (1987).
24) G. Livescu, D.A.B. Miller and D.S. Chemla, Superlatt. Microstruc. (1988).
25) K. Ohtaka and Y. Tanabe, in *Proc. 19th Int. Conf. Phys. Semiconductors*, edited by W. Zawadzki (Institute of Physics, Polish Academy of Science, Warsaw, 1988), p. 315.
26) N. Miura, J.S. Lee and T. Ando, in *Proc. 19th Int. Conf. Phys. Semiconductors*, edited by W. Zawadzki (Institute of Physics, Polish Academy of Science, Warsaw, 1988), p. 111.
27) J.S. Lee, N. Miura and T. Ando, J. Phys. Soc. Jpn. (submitted for publication).
28) H. Yoshimura and H. Sakaki, Phys. Rev. B **39**, 13024 (1989).
29) M. Potemski, J.C. Maan, K. Ploog and G. Weimann, in *Proc. 19th Int. Conf. Phys. Semiconductors*, edited by W. Zawadzki (Institute of Physics, Polish Academy of Science, Warsaw, 1988), p. 119.
30) S. Schmitt-Rink, D.S. Chemla and D.A.B. Miller, Adv. Phys. **38**, 89 (1989).
31) T. Ando, J. Phys. Soc. Jpn. **43**, 1616 (1977).

32) R. Lassnig and E. Gornik, Solid State Commun. **47**, 959 (1983).
33) T. Ando and Y. Murayama, in *Proc. 17th Int. Conf. Phys. Semiconductors, San Francisco, 1984*, edited by J.D. Chadi and W.A. Harrison (Springer, New York, 1986), p. 317; J. Phys. Soc. Jpn. **54**, 1519 (1985).
34) Y. Murayama and T. Ando, Surf. Sci. **170**, 311 (1986); Phys. Rev. B **35**, 2252 (1987).
35) I.V. Kukushkin and V.B. Timofeev, JETP Lett. **40**, 1231 (1984); JETP Lett. **43**, 499 (1986); Sov. Phys. JETP **65**, 146 (1987); Sov. Phys. JETP **66**, 613 (1987); Surf. Sci. **196**, 196 (1988).
36) I.V. Kukushkin, K. von Klitzing and V.B. Timofeev, JETP Lett. **47**, 598 (1988); Phys. Rev. B **37**, 8509 (1988).
37) V.B. Timofeev, in *Proc. 19th Int. Conf. Phys. Semiconductors*, edited by W. Zawadzki (Institute of Physics, Polish Academy of Science, Warsaw, 1988), p. 11.
38) Th. Englert, J.C. Maan, Ch. Uihlein, D.C. Tsui and A.C. Gossard, Solid State Commun. **46**, 545 (1983).
39) D. Heitmann, M. Ziesmann and L.L. Chang, Phys. Rev. B **34**, 7463 (1986).
40) B.I. Shklovskii and A.L. Efros, JETP Lett. **44**, 669 (1986).
41) R.R. Gerhardt and V. Gudmundsson, Phys. Rev. B **34**, 2999 (1986).
42) See, for example, T. Ando, Prog. Theor. Phys. Suppl. **84**, 69 (1985).
43) F.F. Fang and P.J. Stiles, Phys. Rev. **174**, 823 (1968).
44) Th. Englert and K. von Klitzing, Surf. Sci. **73**, 70 (1978).
45) T. Ando and Y. Uemura, J. Phys. Soc. Jpn. **37**, 1044 (1974).
46) F.J. Ohkawa and Y. Uemura, J. Phys. Soc. Jpn. **43**, 925 (1977).
47) H. Rauh and Kümmel, Surf. Sci. **98**, 370 (1980); Solid State Commun. **35**, 731 (1980).
48) S. Katayama and T. Ando, Solid State Commun. **70**, 97 (1987).
49) T. Uenoyama and L.J. Sham, Phys. Rev. B **39**, 11044 (1989).
50) T. Ando and Y. Uemura, J. Phys. Soc. Jpn. **36**, 959 (1974); T. Ando, J. Phys. Soc. Jpn. **36**, 1167 (1974); **37**, 622, 1233 (1975); **38**, 989 (1975).
51) C.H. Perry, J.M. Worlock, M.C. Smith and A. Petrou, *Application of High Magnetic Fields in Semiconductor Physics*, ed. by G. Landwehr (Springer, Heidelberg, 1986), p. 202.
52) D.C. Tsui, H.L. Stormer and A.C. Gossard, Phys. Rev. Lett. **48**, 1559 (1982).
53) H.L. Stormer, D.C. Tsui, A.C. Gossard and J.C.M. Hwang, Phys. Rev. Lett. **50**, 1953 (1983).
54) R.B. Laughlin, Phys. Rev. Lett. **50**, 349 (1983).
55) F.D.M. Haldane, Phys. Rev. Lett. **51**, 605 (1983)
56) R.G. Clark, R.J. Nicholas and A. Usher, Surf. Sci. **170**, 141 (1986).
57) R. Willett, J.P. Eisenstein, H.L. Stormer, D.C. Tsui, A.C. Gossard and J.H. English, Phys. Rev. Lett. **59**, 1776 (1987).
58) *The Quantum Hall Effect*, edited by R.E. Prange and S.M. Girvin (Springer, New York, 1987).
59) D. Yoshioka, Prog. Theor. Phys. Suppl. **84**, 97 (1985).

Strongly Coupled Plasma Physics
S. Ichimaru (Editor)

STRONGLY CORRELATED TWO DIMENSIONAL ELECTRONS FORMED ON DIELECTRIC MATERIALS

Koji KAJITA

Department of Physics, Toho University, Miyama 2-2-1, Funabashi 274, Japan

The properties of electron systems formed on dielectric materials such as liquid helium, solid or liquid neon and hydrogen are briefly reviewed stressing the effect of the electron correlation. They constitutes the classical two dimensional electron systems with a nearly ideal uniform positive background. The correlation effect comes up to the experiments when the correlation parameter $\Gamma = (\pi N_e)^{1/2}e^2/kT$ exceeds 10. The electron crystal is formed at $\Gamma = 135$. The change in the coduction phenomena with increasing correlation is discussed.

1. INTRODUCTION

Electrons trapped on some dielectric materials are known to form nearly ideal two dimensional systems[1]. For such electron systems to be formed ,the substrate is required to have a dielectric constant close to unity and to have a surface potential barrier for the electrons . Only a few materials such as helium, neon and hydrogen meet these conditions and can be used for the substrates. Figure 1 gives a schematic picture of the potential for the surface electrons and the resultant electronic bound state.

A practical experimental set up is shown in Fig.2. Electrodes placed above and below the substrates and the battery to charge them are inevitable to ascertain the charge neutrality of the system. Since the positive background which is located on the electrodes is at a distance

	$-E_1$ (meV)	b (Å)
^{3}He (liq)	0. 39	103
^{4}He (liq)	0. 68	78
Ne (sol)	14. 67	19
H_2 (sol)	15. 76	17
D_2 (sol)	20. 67	15

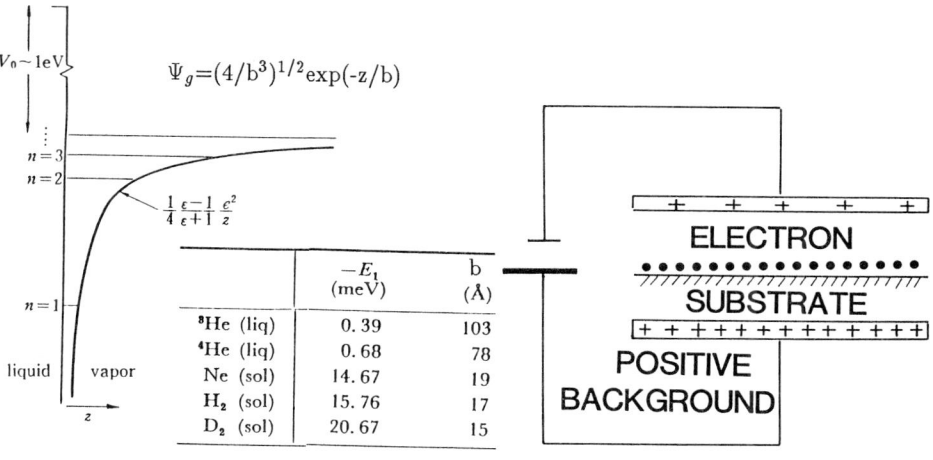

FIGURE 1
Schematic figure of the surface electron potential and the bound states.

FIGURE 2
Experimental system of the surface electrons.

much longer than the electron spacing in the system, the system can be considered to be the one component (electron) plasma in the sea of the uniform background.

The substrate on which electron systems are formed is chosen from the liquid helium, solid or liquid hydrogen and liquid or solid neon. Sometimes, another interesting substrate, liquid helium film, is used. When a solid materials is inserted in the helium gas atmosphere, the surface is covered by a very thin helium film. The thickness of the film can be varied by changing the gas atom pressure. When the gas pressure is very near the saturate vapor pressure, the film thickness becomes several tens angstroms. It is known that electrons can be trapped on such a helium film.

Those systems are interesting in that they provide us a good testing ground for strongly correlated electron systems with two dimensional nature. Usually, these systems cover the electron density from 10^7/cm^2- 3×10^{10}/cm^2. Since the experiments are done in the temperature region above 0.1K, electrons with those density are non-degenerate or in other word, they are in the classical state.

The strength of the electron correlation of the classical electron systems is represented by a dimensionless parameter $\Gamma =(\pi N_e)^{1/2}e^2/kT$, where Ne is the electron density, e, the electronic charge and T is the temperature. In the present systems, the value of Γ ranges from very low value to high value going up to several hundreds. As will be discussed later, the electron correlation effect are predominant when the value of Γ exceed 10. The next and the most important value of Γ is 135. At this point the electron systems undergo the transition from the gaseous or liquid state to the crystalline state[2].

In this paper, we deal with the above electronic systems mainly referring to the one formed on solid neon.

2. BASIC ELECTRONIC PROPERTIES

The energy of the electron is written as

$$E=-((\epsilon+1)/4(\epsilon-1))^2 \times (me^4/2h^2)\times 1/n^2, (n=1,2,3....),$$

where ϵ is the dielectric constant of the substrate and m is the electron mass. The first term represents the binding energy due to the surface potential, and the second term is the kinetic energy of the electrons.

As shown in the inset of Fig.1, the binding energy of the surface electrons ranges from about 0.39meV (electrons on liquid He^3) to 20.67meV (electrons on solid deuteron). The excitation energy of the electrons on liquid He^4 measured by the light absorption experiment[3] is in good agreement with the calculation. When the experiments are done at liquid helium temperatures, the binding energy is comparable to or a little larger than the temperature, so that the most surface electrons stay in the lowest bound state. However, the energy gap to the excited state is so low that the electrons are easily excited. Hot electron effect has been observed for the electrons on liquid helium[4].

Since the electrons are in the outer surface of the materials, the interaction of electrons and the substrate is weak so that the two dimensional mass m is almost equal to the free electron mass m_0.

Electrons move nearly freely on the surface of materials. The scattering centers are 1) gas atoms in the space (Since the temperature is low, the atoms remaining in the gaseous state are only helium.) and 2) the imperfections of the surface (For the liquid helium, the surface imperfection is the ripplon on the surface. If the substrate is solid, two kinds of imperfections are possible: surface roughness and ions trapped on the surface.). In the next section, we briefly discuss the electron scattering.

3. ELECTRON MOBILITY

Figure 3 gives the electron mobility against the helium gas atom density for electrons on solid neon and on liquid helium[5,6]. In both systems, the electron mobility decrease, as we go to the right of the figure where the density of scattering center (helium gas) increases . At the same gas atom density, the mobility of electrons on liquid helium is always higher than that on solid neon. These are understood in terms of the following expression of the scattering life time given by Saitoh[7] which state that the life time of an electron is proportional to the width of the electronic wave function and inversely proportional to the gas atom density. $\tau_G = 1.47 \times 10^{-13} b/N_G$ sec. , where b is the width of the electron wave function normal to the

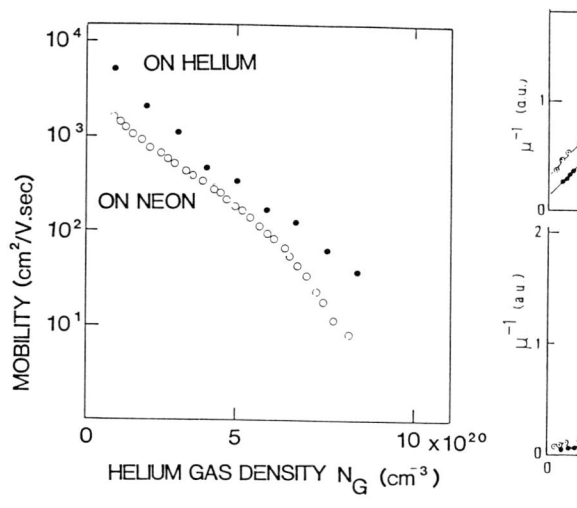
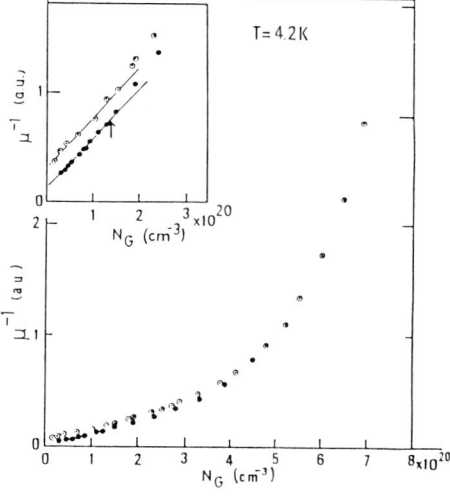

FIGURE 3
Mobility of electrons trapped on liquid helium and solid neon. The horizontal axis is the helium gas density. Gas atoms penetrate into the electron channel and act as the scattering centers.

FIGURE 4
The inverse of the electron mobility, μ^{-1}, against the helium gas density N_G. The inset shows the enlarged figure of low N_G part. Two curves correspond to two electron systems formed on different neon surfaces. (electrons on solid neon)

surface measured in angstrom and N_G is the gas atom density in a unit of $10^{20}/cm^3$.

When the gas atom density becomes low, other scattering mechanisms comes up. For electrons on liquid helium, it is ripplons on the surface. Because the ripplon scattering is weak, the electron mobility in this system goes up to an extremely high value[5], $10^7 cm^2/V \cdot sec$. On the other hand for electrons on solid materials, the surface imperfections are important. The surface of solid hydrogen can be made rather smooth[8], and the highest value of the mobility reaches $6.5 \times 10^4 cm^2/V \cdot sec$. Solid neon surface, on the other hand, is not so good[6] and the mobility is limited to about $10^4 cm^2/V \cdot sec$.

As we see in Fig.3, one of the interesting character of these systems is that we can handle the density of scattering centers in the run of the experiment, and the scattering center density can be used as a parameter in describing the properties of these systems. For the case of electrons on liquid helium, we have to change the temperature to vary the gas atom density. However, for electrons on solid neon or solid hydrogen, the scattering centers can be varied at the fixed temperature only by feeding helium gas into the experimental cell or by evacuate it.

Figure 4 gives the inverse of the electron mobility as a function of the helium gas atom density for electrons on solid neon. In the low gas atom density region below about $2 \times 10^{20}/cm^3$, the mobility which is defined as $\mu \equiv (\sigma/N_e)$ is expressed as

$$\mu^{-1} = \mu_S^{-1} + \alpha \cdot N_G.$$

Here, μ_S represents the mobility for zero gas atom density, which is determined by the surface imperfection scattering, and the second term gives the increment of μ^{-1} with N_G. The surface scattering depends on the surface perfectness and thus, the value of μ_S changes with runs of experiments as shown in the inset of Fig.4. The parameter α, on the other hand, is independent of the experimental runs and represents the strength of the electron scattering with the gas atoms. The value of α agrees with the calculation of Saitoh[7] within a factor of 2. In the higher gas density region, μ^{-1} is seen to rise nonlinearly. This implies that in high gas density region, the localization of electron becomes important.

4. EFFECT OF ELECTRON CORRELATION IN THE INTERMEDIATE Γ REGION

When the value of Γ exceeds 10, the effect of the correlation comes up on many electronic properties. In this section, we give a few example of the correlation effect.

4.1 Electron Escape from the Surface States to the Three Dimensional States

In the finite temperatures, two dimensional electrons on the surface will leave there evaporating into the three dimensional space.

In the usual conditions, the *escape rate*, τ^{-1}, defined as $\tau^{-1} \equiv (dN_e/dt)/N_e$ is independent of the electron density N_e and is related to the binding energy of the surface state as $\tau^{-1} \propto \exp(E_b/kT)$. This relation has been confirmed by the experiment[9] which gave a reasonable value of E_b.

When the escape rate measurement is extended to high electron density region, this was found to be no more valid. As shown in Fig.5[10], the escape rate is strongly reduced when the electron density becomes high.

This is the effect of the electron correlation. Nagano[11] has shown that the escape rate is written as $\tau^{-1} \propto \exp(\mu_c - E_b)/kT$, where E_b is the binding energy of the electron and μ_c is the change of the chemical potential of electrons due to the correlation effect. He has found that $\mu_c \simeq kT(-1.67\Gamma + 3.19\Gamma^{1/4} - 0.38\ln\Gamma - 2.5)$. From the experiment in Fig.5, μ_c has been determined as $\mu_c = kT(-1.4\Gamma + \beta)$, $\beta \geq 6$, which agree qualitatively with the theory. In Fig.5, the appearance of the correlation effect is recognized as the drop of the escape rate from the value at the low Γ limit, which begins at about $\Gamma = 10$.

4.2 Electron Mobility

The electron correlation is an inner force of the electron system, so that the effect will not be observed in the conductivity measurements. However, we have found that the correlation effect appears with the aid of the interaction of electrons with the scattering centers[12]. As the correlation gets stronger, the electron system becomes like a greasy liquid and the interaction with the scattering centers gets stronger resulting in the reduction of the conductivity. Figure 6 demonstrates the effect of the correlation on the electron conductivity. In this experiment, the conductivity has been measured at fixed temperatures changing the electron densities. In the

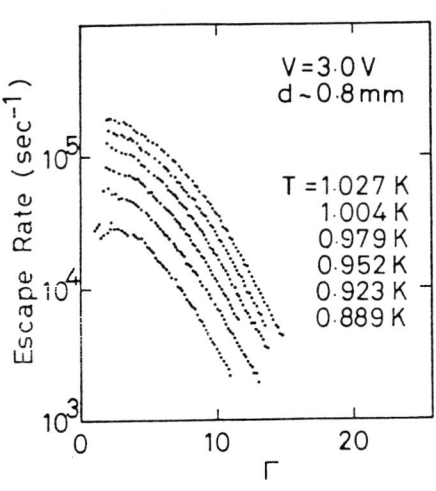

FIGURE 5
The escape rate of electrons against the correlation Γ. (electrons on liquid helium)

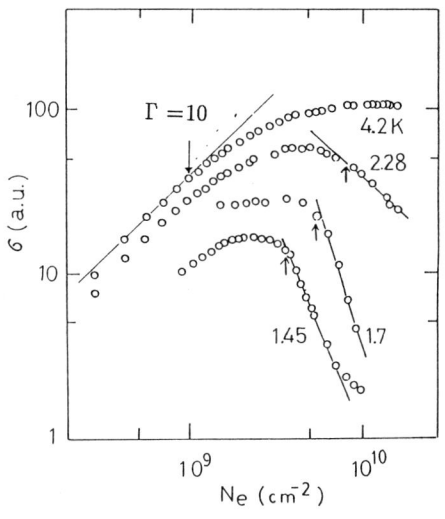

FIGURE 6
The electron conductivity against the electron density for several temperatures. The position $\Gamma=10$ is indicated for the curve at T=4.2K. The arrows in the curves for T=2.28, 1.7 and 1.45K indicate the electron crystallization. (electrons on solid neon)

left end of the figure, the electron conductivity is a linearly increasing function of N_e indicating that the correlation is not important in this region. The discrepancy of the curve from the linear line begins around the point where Γ exceeds 10 (See the curve for T=4.2K.). In this experiment, too, the effect of the correlation is prominent in the region $\Gamma \geq 10$ as it was in the escape rate experiment.

The following analysis shows how the correlation effect affects the electron scattering.

In the highly correlated electron systems, the single electron picture is not valid. However, if we define a *single electron mobility*, μ, as $\mu \equiv \sigma/N_e$, we have found that the dependence of μ^{-1} on the gas atom density N_G is of the same form with that in the low correlation region[13]. It is demonstrated in Fig.7. Here, however, the parameters μ_S and α are no more constants but depend on the electron correlation. The Γ dependence of the parameters shown in Fig.8 represents the effect of the correlation on the electron conduction. The figure tells that gas atom scattering and the surface scattering are enhanced by the electron correlation. According to this figure, both α and μ_S^{-1} are nearly proportional to the parameter Γ in the intermediate region.

4.3 Electrons on the Surface of Thin Liquid Helium Film

In the above sections, we have treated electron systems formed on bulk liquid or bulk solid materials. In this subsection, we mention the electron systems formed on very thin helium film. This system is of interest, because the interaction of electrons with the helium film may become important when electron correlation is strong.

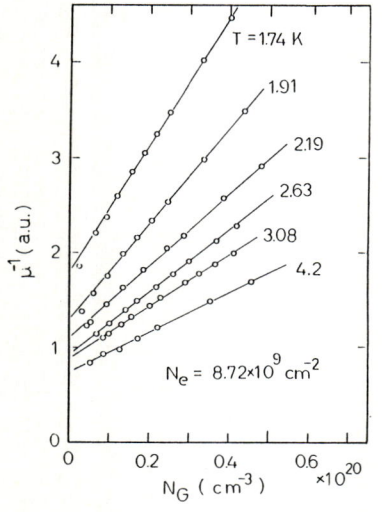

FIGURE 7
The inverse of the electron mobility as functions of helium gas atom density in high correlation region. (electrons on solid neon)

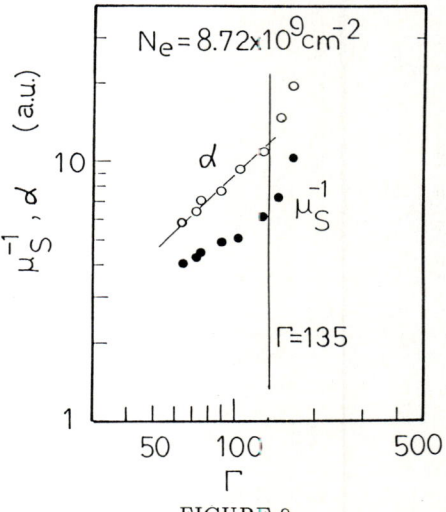

FIGURE 8
Parameters α and μ_S^{-1} derived from the results in Fig.7 as functions of electron correlation Γ. (electrons on solid neon)

Figure 9 gives the electron conductivity against the helium film thickness for two electron densities. For the low electron density system, it is noted that the conductivity increases with increasing film thickness. This is understood in terms of the change in the electronic wave function with the change of the film thickness[14]. When the helium film is thin, the binding potential on the surface electrons are not only from the helium film but the neon substrate also exerts the attracting force. Since the dielectric constant of solid neon is larger than that of helium, the binding energy of the electron is the larger, the thinner the helium film thickness is. And the range of the wave function becomes smaller as the film gets thinner. As has been pointed before, the electron mobility is proportional to the wave function width so that the conductivity becomes small as the film becomes thin.

On the other hand, when the electron correlation is high, we found that the electron mobility does not increase with increasing helium film thickness but it rather decreases.

It implies that when the electron density is high, the interaction between electrons and the film becomes strong and moreover, that the interaction is stronger for thicker films.

5 ELECTRON CRYSTAL

5.1 Phase Diagram

In this section, the crystallization of the electrons and the property of the crystal are discussed.

In Fig.8, a drastic change in the slope of α-Γ and μ_S^{-1}-Γ curves takes place at the point where $\Gamma=135$, which indicates that some thing has happened to the electron system at this

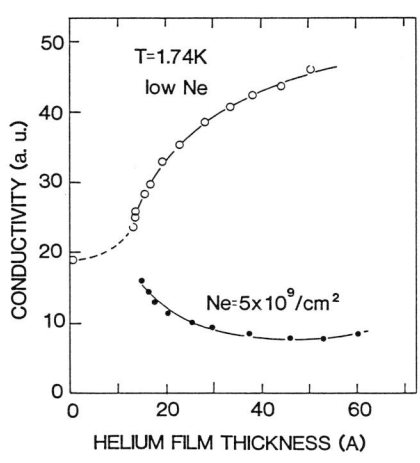

FIGURE 9
The mobility of electrons trapped on liquid helium film for two different electron densities. (electrons on helium film)

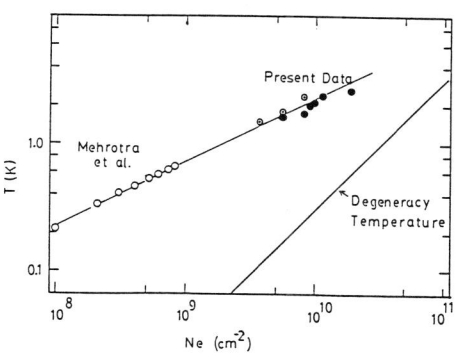

FIGURE 10
The phase diagram of the surface electron crystal. The data for electrons on liquid helium and those on solid neon are plotted.

point. Anomalous change of the electron property at $\Gamma=135$ is also noticed in Fig.6. Above this point, the slope of the curves becomes drastically steep. These are considered to be due to the crystallization of electrons.

As has been mentioned before, the correlated electron liquid is greasy and tends to stick to the random potential in the surface resulting in the reduction of the conductivity. Such tendency becomes stronger with increasing correlation and lastly when the system undergoes the transition to a hard solid, it will be pinned by the random potential and the electrons stop to move. The electron crystal formed on solid neon is ,thus , considered to be a pinned crystal.

The fact that the crystallization takes place at $\Gamma=135$ and that electrons form a triangular lattice was first definitely confirmed on the electron system formed on the surface of liquid helium[2]. After that, electron crystal has been recognized on the electron systems formed on the solid neon[15] and electrons on thin liquid helium film[16]. Figure 10 shows the phase boundary of the electron crystal for two systems, electrons on liquid helium and those on solid neon. In both systems, the solid phase is characterized by a relation $\Gamma=135$.

The melting of the electron crystal has been investigated by Williams's group[17] from the point of the transverse phonon mode and has been ascribed to the Kosterlitz-Thouless transition.

5.2 Low Field Conductivity in the Crystal State

In the preceding subsection, pinning of electron crystal has been ascribed to as the origin of the drop of the electron conductivity. If this is the case, the electron conduction is not to be observed. However, in practice, the conductivity is not zero but finite in this region. Two mechanisms are possibly responsible: 1) as the conductivity is measured by rf electric field, the electron crystal can vibrate at the pinned position , or 2)

some electrons are excited to the higher energy state and contribute to the conduction. If we assume that the second is important , we can estimate the activation energy as follows.

In the low gas density region, we have found that the relation between the conductivity and the gas atom density has the same form with that in low correlation region.
$$\mu^{-1} \equiv (\sigma/N_e)^{-1} = \mu_S^{-1} + \alpha \cdot N_G.$$

Figure 11 presents the parameters α and μ_S^{-1} not as a function of Γ but as a function of the inverse of temperature. From the slope of the curve in the crystal region , we can determine the activation energy . It is about 14.4K for the system with $N_e=1.9\times 10^{10}/cm^2$ and about 5.5k for the one with $N_e=8.7\times 10^9/cm^2$.

5.3 Nonlinear Transport Phenomena

If electron crystal is pinned , nonlinear transport phenomena due to the depinning of the crystal are expected , which in fact is discovered in the experiment shown in Fig.12 [18]. The pinning effect is stronger w hen the electrons are on a helium film than on a bare solid neon. So, the electron system trapped on a substrate -liquid helium film on solid neon- is used in the experiment. At this temperature, the critical density is $4\times 10^9/\text{cm}^2$. Thus, the data of the upper most curve is for the liquid electron system and all the others are in crystal states. This figure gives the evidence of the existing of the nonlinear transport in the solid phase of the system and the absence of it in the liquid phase. In the nonlinear region, the I-V curves are not smooth but have many structures and sometimes we have observed hysteresys in the curve as the electric field is first increased and then decreased. These features are characteristic to pinning and depinning effect.

6. CONCLUSION

We have studied the effect of the electron correlation on the properties of surface electrons on dielectric materials. Electron correlation becomes recognizable when the parameter Γ exceeds 10. The effect appears in many aspect such as a decrease of the escape rate of electrons from the surface to the three dimensional space or the reduction of the electron mobility. At the point $\Gamma=135$, the electron system undergoes the transition to the crystalline state. Passing through the phase boundary, the effect of the correlation changes qualitatively. For the electrons on solid neon, the electron crystal is pinned by surface imperfections and the nonlinear transport phenomenon due to the depinning is observed.

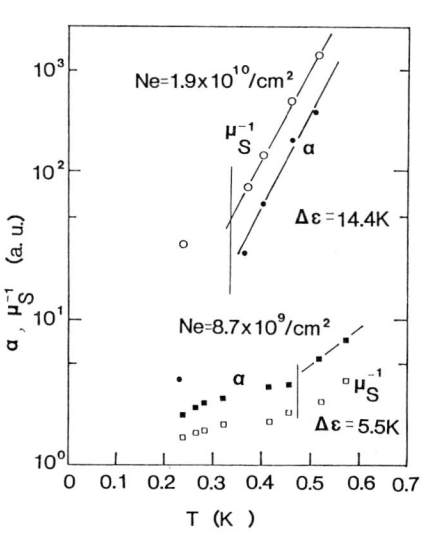

FIGURE 11
Parameters α and μ_S^{-1} against the inverse of the temperature. The parameters are determined by fitting the experimental results of the conductivity to a relation $\mu^{-1}=\mu_S^{-1}+\alpha\cdot N_G$. (electrons on solid neon)

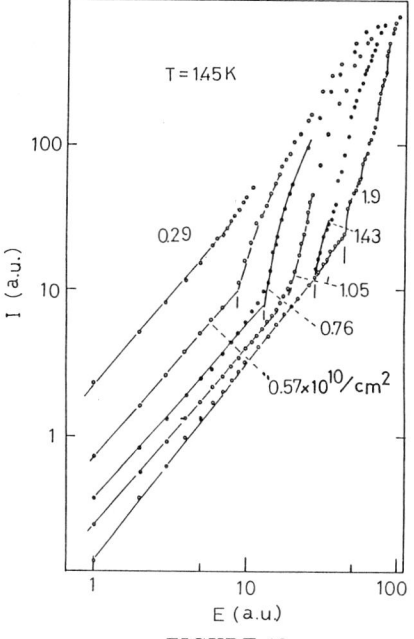

FIGURE 12
Nonlinear transport phenomena of surface electron crystal formed on liquid helium film adsorbed on solid neon. (electrons on liquid helium film)

REFERENCES

1) M. W. Cole, Phys. Rev. B3 (1971) 4418.
2) C. C. Grimes and G. Adams, Phys. Rev. Letters, 42 (1979) 785.
3) C. C. Grimes and G. Adams, Surf. Science 73 (1978) 379.
4) V. S. Edel'man, Pis'ma Zh. Eksp. Teor. Fiz. 25 (1977) 422.
5) Y. Iye, J. Low Temp. Phys. 40 (1980) 441.
6) K. Kajita and W. Sasaki, Surf. Science 113 (1982) 419.
7) M. Saitoh, J. Phys. Soc. Jpn. 42 (1977) 201.
8) M. A. Paalanen and Y. Iye, Surf. Science 170 (1986) 80.
9) K. Kono, Y. Iye, K. Kajita, S. Kobayashi and W. Sasaki, Surf. Science 98 (1980) 17.
10) K. Kono, K. Kajita, S. Kobayashi and W. Sasaki, Surf. Science 113 (1982) 438.
11) S. Nagano, S. Ichimaru, H. Totuji and H. Itoh, Phys. Rev. B19 (1979) 2449.
12) K. Kajita, Surf. Science, 142 (1984) 86.
13) K. Kajita, Y. Nishio and W. Sasaki, Surf. Science, 170 (1986) 88.
14) K. Kajita, J. Phys. Soc. Jpn. 52 (1983) 372.
15) K. Kajita, J. Phys. Soc. Jpn. 54 (1985) 4092.
16) H. W. Jiang and A. J. Dahm, Jpn. J. Appl. Phys. 26 (1987) supplement 26.
17) G. Deville, a. Valdes, E. Y. Andrei and F. I. B. Williams, Phys. Rev. Letters 53 (1984) 588.
18) K. Kajita, Surf. Science 196 (1988) 29.

TWO-DIMENSIONAL COULOMB SYSTEMS : SOLVABLE MODELS AT Γ = 2

Bernard JANCOVICI

Laboratoire de Physique Théorique et Hautes Energies, Bât. 211,
Université de Paris-Sud, 91405 Orsay, France*

1. INTRODUCTION

At the previous conference, in Santa Cruz, A. Alastuey gave a review talk[1] with a similar title. This review was centered on the equilibrium statistical mechanics of the classical two-dimensional one-component plasma; at the special value Γ = 2 of the coupling constant, many exact explicit results had been obtained for both bulk and surface properties.

Two especially tantalizing problems were unsolved at that time. First, although the one-component plasma had been solved for a large class of inhomogeneous cases, one had not succeeded in dealing with a background having the double periodicity of a two-dimensional crystal; this would have been a model for (classical) electrons in a crystalline field or for a superionic conductor. Second, although a lattice version of the two-component plasma[2] had just been solved, the bulk and surface properties of the two-component plasma in the continuum had not been studied.

I am glad to report that these two problems (which actually are related) have now been solved, and more generally, that we have a better overall understanding of the two-dimensional Coulomb systems at Γ= 2, largely thanks to the fresh insight of a graduate student, Françoise Cornu[3]. The present review will include the recent progress in today's perspective.

The two-dimensional Coulomb systems which we consider are made of particles of charge $\pm e$. The interaction potential between two identical particles is $-e^2 \ln(r/L)$, where r is the distance between the particles and L a length scale which only fixes the zero of energy. Using this logarithmic interaction, which is the two-dimensional version of the Coulomb potential, insures that our two-dimensional systems will be qualitatively very similar to the three-dimensional Coulomb systems with an e^2/r interaction. In the two-dimensional case, the relevant dimensionless coupling constant is $\Gamma = e^2/k_B T$ (where T is the temperature and k_B is Boltzmann's constant); for point-particles, the equation of state[4] has the trivial form

$$p = \rho\, k_B T (1 - \frac{\Gamma}{4}) ,$$

---
*Laboratoire Associé au Centre National de la Recherche Scientifique.

where p is the pressure and ρ the number density. For the other thermodynamical quantities and the correlations, both the one-component plasma (particles of one sign in a continuous background of opposite charge) and the two-component plasma (positive and negative particles) become solvable models for the special temperature such that $\Gamma = 2$. We shall study these models at $\Gamma = 2$, starting with the two-component plasma.

2. TWO-COMPONENT PLASMA

For a long time, it had been known to field theorists that the two-dimensional two-component plasma (Coulomb gas) should be simple at $\Gamma = 2$, because it is then equivalent to a free Fermi field. Indeed, the Coulomb gas is equivalent[5] to a Sine-Gordon field, which in turn is equivalent[6,7] to the massive Thirring model; the latter one reduces to a free Fermi field when $\Gamma = 2$. In the following, the equivalence to a free Fermi field will be shown directly. The two-component plasma is described in the grand-canonical ensemble.

For $\Gamma \geq 2$, a two-component plasma made of point particles is unstable against collapse. For overcoming this difficulty, Gaudin[2] introduced a lattice model and solved it at $\Gamma = 2$. This model is a convenient starting point.

2.1. Lattice model

We represent the position $\vec{r} = (x,y)$ of a particle by the complex number $z = x + iy$. Two interwoven lattices U and V are introduced. The positive (negative) particles sit on the sublattice U(V); each lattice site is occupied by zero or one particle. Let $u_i (u_i \in U)$ be the complex coordinate of the ith positive particle and $v_j (v_j \in V)$ be the complex coordinate of the jth negative particle. For one positive and one negative particle, the canonical partition function is

$$\sum_{\substack{u_1 \in U \\ v_1 \in V}} \exp[-2\ell n \frac{|u_1 - v_1|}{L}] = \sum_{\substack{u_1 \in U \\ v_1 \in V}} \frac{L^2}{|u_1 - v_1|^2} = \sum_{\substack{u_1 \in U \\ v_1 \in V}} \det \begin{vmatrix} 0 & \frac{L}{u_1 - v_1} \\ \frac{L}{\overline{v_1 - u_1}} & 0 \end{vmatrix}.$$

A similar expression holds for n positive and n negative particles. It is convenient to introduce a more compact notation : each lattice site is characterized by its complex position z and by an isospinor, $\binom{1}{0}$ for a positive site, $\binom{0}{1}$ for a negative one. In terms of Pauli matrices σ operating in the isospinor space, the grand partition function (with only neutral configurations) is

$$Z = 1 + \lambda^2 \sum_{\substack{u_1 \in U \\ v_1 \in V}} \frac{L^2}{|u_1 - v_1|^2} + \dots ,$$

where λ is the fugacity per site; it can be recognized as the expansion of the determinant

$$Z = \det [1 + \frac{\sigma_x + i\sigma_y}{2} \frac{\lambda L}{z-z'} + \frac{\sigma_x - i\sigma_y}{2} \frac{\lambda L}{\overline{z-z'}}] ;$$

the lines and the columns of this determinant are labeled by the positions and charge signs of all the lattice sites.

Gaudin went on with this lattice model, and he solved it in terms of elliptic functions. A somewhat simpler, although less symmetrical, version can be obtained by constraining the particles to sit on an array of parallel equidistant lines rather than on a lattice; this parallel line model[8] can be solved in terms of trigonometric functions. Still simpler however is the continuum limit.

2.2. Small charged hard discs[8]

In the continuum limit, the factor $1/(z-z')$ which occurs in the grand partition function Z can be viewed as the (z,z') element of an infinite matrix. The key point for a simple proof of the equivalence between our Coulomb gas and a free Fermi field is that the inverse of the matrix with elements $1/(z-z')$ is the differential operator $(2\pi)^{-1}(\partial_x + i\partial_y)$. Indeed

$$\frac{1}{2\pi}(\partial_x + i\partial_y) \frac{1}{z-z'} = \frac{1}{2\pi}(\partial_x + i\partial_y)(\partial_x - i\partial_y) \ln|\vec{r}-\vec{r}'| = \delta(\vec{r}-\vec{r}') .$$

Thus, in terms of a rescaled fugacity $m = 2\pi L\lambda/S$, where S is the area per site, in the continuum limit the grand partition function can be rewritten as

$$Z = \det [(\sigma_x \partial_x + \sigma_y \partial_y + m)(\sigma_x \partial_x + \sigma_y \partial_y)^{-1}]$$

and

$$\ln Z = \text{Tr}[\ln(\sigma_x \partial_x + \sigma_y \partial_y + m) - \ln(\sigma_x \partial_x + \sigma_y \partial_y)] .$$

The fugacity m becomes the mass in the two-dimensional Dirac operator $\sigma_x \partial_x + \sigma_y \partial_y + m$.

Before proceeding to the calculation of the pressure p from $\ln Z$, it is convenient to discuss the one-body and n-body densities. Replacing m for a while by a charge and position dependent fugacity $[(1+\sigma_z)/2] m_+(\vec{r}) + [(1-\sigma_z)/2]m_-(\vec{r})$, and taking functional derivatives of $\ln Z$ with respect to $m_\pm(\vec{r})$, one obtains the densities in terms of the propagator

$$G_{s_1 s_2}(\vec{r}_1, \vec{r}_2) = < \vec{r}_1 \, s_1 | \frac{1}{\sigma_x \partial_x + \sigma_y \partial_y + m} | \vec{r}_2 \, s_2 > ,$$

where $s_i = \pm 1$ is the sign of the particle at $\vec{r}_i$. More explicitly,

$$G_{++}(\vec{r}_1, \vec{r}_2) = G_{--}(\vec{r}_1, \vec{r}_2) = \frac{1}{2\pi} K_0(m \, r_{12}) ,$$

$$G_{-+}(\vec{r}_1, \vec{r}_2) = -G_{+-}(\vec{r}_2, \vec{r}_1) = -\frac{m}{2\pi} e^{i\Theta} K_1(m r_{12}) ,$$

where $\vec{r}_{12} = \vec{r}_2 - \vec{r}_1$, Θ is the polar angle of $\vec{r}_{12}$, and K_0 and K_1 are modified Bessel functions. The 2-body truncated densities are

$$\rho_{++}^{(2)T}(\vec{r}_1, \vec{r}_2) = \rho_{--}^{(2)T}(\vec{r}_1, \vec{r}_2) = -[mG_{++}(\vec{r}_1, \vec{r}_2)]^2 = -(\frac{m^2}{2\pi})^2 [K_0(mr_{12})]^2 ,$$

$$\rho_{+-}^{(2)T}(\vec{r}_1, \vec{r}_2) = \rho_{-+}^{(2)T}(\vec{r}_1, \vec{r}_2) = |mG_{-+}(\vec{r}_1, \vec{r}_2)|^2 = (\frac{m^2}{2\pi})^2 [K_1(mr_{12})]^2 ,$$

and higher-order truncated densities can be expressed as sum of products of factors G_{++} and G_{-+}. The Bessel functions behave like $\exp(-mr_{12})$ at large distance, and therefore the correlations have an exponential decay, with an inverse correlation length of the order of m, the rescaled fugacity.

A remarkable feature of the truncated n-body densities ($n \geq 2$) is that they are finite, for a given value of m and for non-zero separations, even for a point-particle system. The collapse at $\Gamma = 2$ is however apparent on the one-body densities : for a point-particle system, the densities of positive and negative particles would be $\rho_+ = \rho_- = mG_{++}(\vec{r}, \vec{r}) = (m^2/2\pi) K_0(0) = \infty$. This divergence originates in the small distance divergence of the one-pair canonical integral $\int d^2\vec{r}(L/r)^2$. In order to suppress it, one may replace the point particles by charged hard discs of small diameter R. The one-body densities can then be obtained by using the perfect-screening sum rule

$$\frac{1}{2}\rho = \rho_{\pm} = \int_{r>R} [\rho_{+-}^{(2)T}(r) - \rho_{++}^{(2)T}(r)] d^2\vec{r}$$

with the result

$$\rho_{\pm} \sim \frac{m^2}{2\pi} K_0(mR) \sim \frac{m^2}{2\pi} [\ln \frac{2}{mR} - \gamma] ,$$

where $\gamma = 0.5772...$ is Euler's constant. By integration with respect to m, one obtains the pressure p :

$$\frac{p}{k_B T} \sim \frac{m^2}{2\pi} [\ln \frac{2}{mR} - \gamma + \frac{1}{2}] .$$

This behavior of the pressure can also be obtained directly by evaluating $\ln Z$ in the momentum representation, with some high-momentum cutoff of the order of R^{-1}; the precise value to be taken for the cutoff is determined by the perfect-screening rule.

The above expressions of the density and the pressure are the leading-order terms as $R \to 0$: the theory is valid for hard cores small in the sense $mR \ll 1$ or $\rho R^2 \ll 1$. For point particles, the equation of state becomes $p = (1/2) \rho k_B T$, describing a gas of collapsed neutral pairs. The knowledge of the correlation functions allows the calculation of other thermodynamical quantities, such as the excess internal energy per particle

$$u = \frac{1}{4} e^2 [\ln \frac{2R}{mL^2} - \gamma]$$

or the excess specific heat at constant volume, per particle,

$$c_v = \frac{1}{6}(\ln \frac{2}{mR} - \gamma)^2 - \frac{1}{4}(\ln \frac{2}{mR} - \gamma) - \frac{1}{8} .$$

The thermodynamical properties are close to what would be obtained for a system of independent neutral pairs. However, as $R \neq 0$, there are a few "free" particles which give screening effects, i.e. correlations characteristic of a conducting system : the correlations decay exponentially, and the Stillinger-Lovett second-moment sum rule

$$2 \int d^2\vec{r} \; r^2 [\; \rho_{++}^{(2)T}(r) - \rho_{+-}^{(2)T}(r)] = - \frac{2k_BT}{\pi e^2} = - \frac{1}{\pi}$$

is indeed satisfied. This is in agreement with the known fact that the Kosterlitz-Thouless transition to a dielectric phase occurs only at a lower temperature such that $\Gamma \sim 4$.

2.3. Electrical double layers [9]

The electrical double layer which forms at the interface between two conducting media can be mimicked by inhomogeneous two-dimensional plasmas. Here we consider two-component plasmas. External potentials such as walls can be taken into account by using a charge and position dependent fugacity. The propagator has now the more general form

$$G_{s_1 s_2}(\vec{r}_1, \vec{r}_2) = <\vec{r}_1 \; s_1| \frac{1}{\sigma_x \partial_{x} + \sigma_y \partial_{y} + m_+(\vec{r})\frac{1+\sigma_z}{2} + m_-(\vec{r})\frac{1-\sigma_z}{2}} |\vec{r}_2 \; s_2> .$$

The one-body and n-body densities can still be expressed in terms of G. In particular

$$\rho_+(\vec{r}) = m_+(\vec{r}) G_{++}(\vec{r},\vec{r}) \quad , \quad \rho_-(\vec{r}) = m_-(\vec{r}) G_{--}(\vec{r},\vec{r}) \; ;$$

here too we deal with small hard discs, and a cutoff has to be used for the densities to be finite.

In general, the propagator G is the solution of a system of coupled partial differential equations, which are, in a 2x2 matrix notation

$$[\sigma_x \partial_{x_1} + \sigma_y \partial_{y_1} + m_+(\vec{r}_1) \frac{1+\sigma_z}{2} + m_-(\vec{r}_1) \frac{1-\sigma_z}{2}] G(\vec{r}_1,\vec{r}_2) = \delta(\vec{r}_1-\vec{r}_2) .$$

Here, we only deal with plane interfaces, and the fugacities depend only on x, the coordinate in the direction normal to the interface : $m_\pm(\vec{r}) = m_\pm(x)$. The standard technique then is to introduce the Fourier transform $\hat{G}$ with respect to y_{12}, the coordinate difference along the interface :

$$\hat{G}(x_1,x_2,\ell) = \int_{-\infty}^{\infty} dy_{12}\, G(x_1,x_2,y_{12}) e^{i\ell y_{12}}$$

This transformation leads to systems of two coupled <u>ordinary</u> differential equations, such as

$$m_+(x_1)\hat{G}_{++}(x_1,x_2,\ell) + (\frac{d}{dx_1} + \ell)\hat{G}_{-+}(x_1,x_2,\ell) = \delta(x_1-x_2),$$
$$(\frac{d}{dx_1} - \ell)\hat{G}_{++}(x_1,x_2,\ell) + m_-(x_1)\hat{G}_{-+}(x_1,x_2,\ell) = 0.$$

In the cases which have been considered, $m_+(x)$ and $m_-(x)$ are constant or exponential functions with discontinuities; the differential equations for $\hat{G}$ are easily solved by splicing exponentials together, and an inverse Fourier transformation provides a simple integral representation for G and the densities. The following interfaces have been studied :

A. Charged hard wall (primitive electrode)

The hard wall occupies the half-space x<0 and carries a "surface" charge density (actually charge per unit length) -eσ. In the half-space x>0, the wall creates a constant external electrical field represented by fugacities $m_\pm(x) = m \exp(\mp 4\pi\sigma x)$. The density profiles are drawn on Fig.1. Note the drop in density near the wall : the pressure is close to half the ideal gas value, and therefore the density at the wall (which generates the pressure) has to be small, at least for a moderate surface charge.

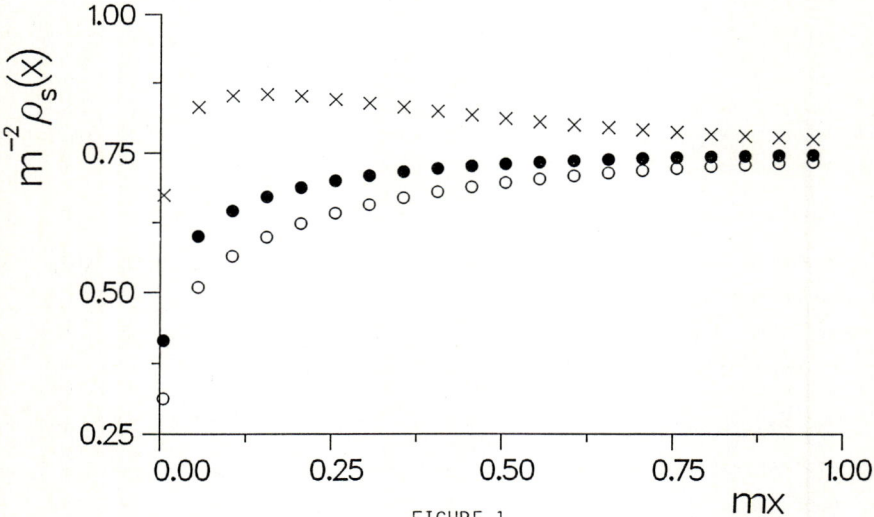

FIGURE 1

The density profiles near a hard wall. For an uncharged wall (σ=0), $\rho_+(x)=\rho_-(x)$ (black circles). For a charged wall (2πσ=m), $\rho_+(x)$ (crosses) and $\rho_-(x)$ (white circles). The cutoff is mR = 0.01.

B. Polarizable interface

A polarizable interface (impermeable membrane separating two media) is described by giving different values m_+^a, m_-^a, m_+^b, m_-^b to the fugacities in regions $a(x>0)$ and $b(x<0)$. Actually, the physics depends only on three parameters (the bulk density on each side and the potential drop across the interface), and only three combinations of the fugacities are relevant quantities: $m_a = (m_+^a m_-^a)^{1/2}$, $m_b = (m_+^b m_-^b)^{1/2}$, and $m_-^a m_+^b / m_+^a m_-^b$. The density profiles are drawn on Fig.2. Note the discontinuities in the densities at the interface.

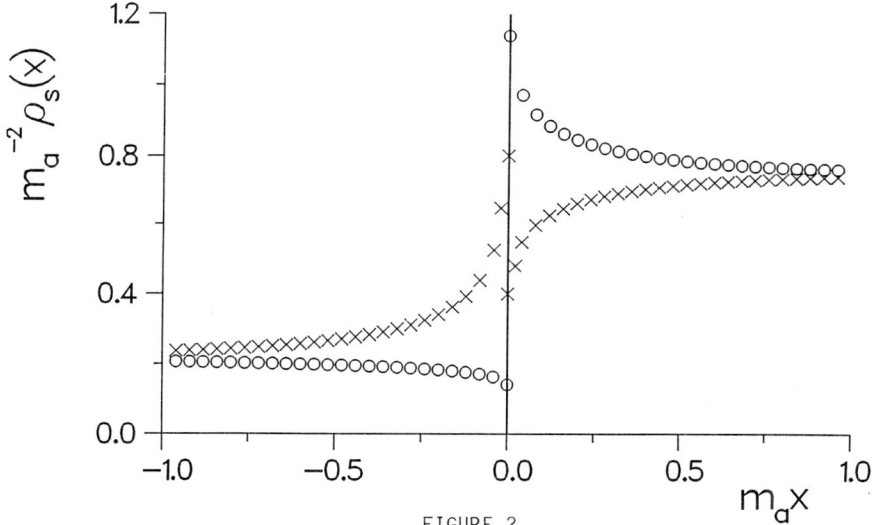

FIGURE 2
The density profiles near a polarizable interface, for $m_+^b m_-^b = 0.25\ m_+^a m_-^a$ and $m_-^a m_+^b / m_+^a m_-^b = 16$: $\rho_+(x)$ (crosses) and $\rho_-(x)$ (white circles). The cutoff is $mR = 0.01$.

C. Semipermeable membrane

It is also possible to describe a membrane permeable to, say, the positive particles and impermeable to the negative ones. Now there are two relevant parameters which can be chosen independently, the bulk density on each side, or alternatively $m_a = (m_+^a m_-^a)^{1/2}$ and $m_b = (m_+^b m_-^b)^{1/2}$. The density profiles are drawn on Fig. 3. Note the continuity of the density of positive particles which can freely cross the membrane.

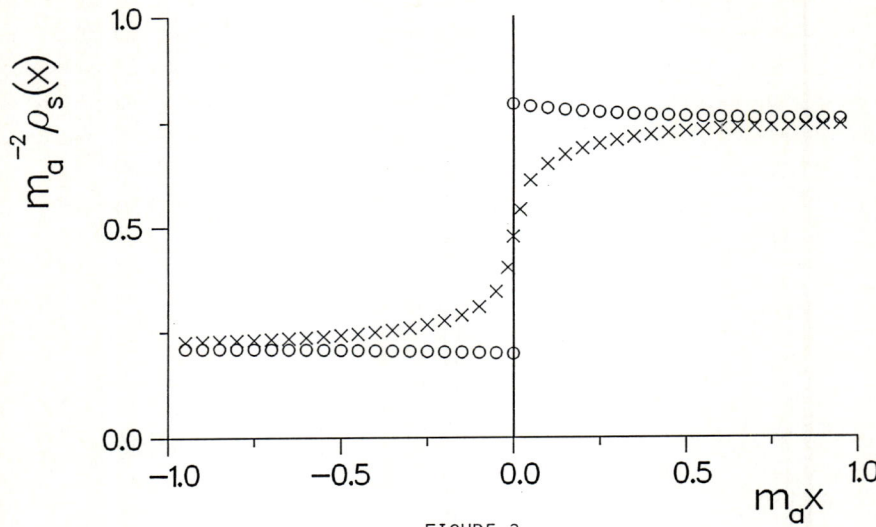

FIGURE 3
The density profiles near a semipermeable membrane, for $m_+^b m_-^b = 0.25\ m_+^a m_-^a$: $\rho_+(x)$ (crosses) and $\rho_-(x)$ (white circles). The cutoff is $mR = 0.01$.

For both the charged hard wall and the polarisable interface, it is possible to compute exactly quantities of interest to the electrochemists, such as the differential capacity and the surface tension. The "surface" charge density $e\sigma$ on the right-hand side of the interface, the potential drop $\Delta\phi$ (electrical potential at $x = +\infty$ minus electrical potential at $x = -\infty$) and the surface tension γ are related by the exact Lippmann equation

$$\frac{\partial \gamma}{\partial (\Delta\phi)} = - e\sigma\ .$$

2.4. Two-component plasma with a background

The general propagator G of section 2.3. can be used for studying a system made of positive and negative particles immersed in a continuous charged background. This could be a model for a metal-salt liquid mixture[10] of the type alkali metal plus halide of this metal. This model is obtained by including in the fugacities $m_+(\vec{r})$ and $m_-(\vec{r})$ the electrostatic potential $eV(\vec{r})$ created by the background. Explicit expressions can be obtained for bulk and surface properties of this system[3,11].

For future reference, let us note that, if there are only electrostatic interactions, the fugacities are of the form $m_\pm(\vec{r}) = m\ \exp\ [\mp\ 2V(\vec{r})]$ and the G_{++} part of the propagator can be shown to be alternatively given by

$$me^{-V(\vec{r}_1)} G_{++}(\vec{r}_1,\vec{r}_2) e^{-V(\vec{r}_2)} = <\vec{r}_1| \frac{m^2}{m^2+A^+A} |\vec{r}_2>,$$

where $A = \partial_x + i\partial_y + \partial_x V(\vec{r}) + i\partial_y V(\vec{r})$.

3. ONE-COMPONENT PLASMA REVISITED

The (in general inhomogeneous) one-component plasma can be recovered[11] by starting with a two-component plasma with a, say, negative background, and letting the fugacity m in the previous formula go to zero. Then, the negative particles disappear, and only those positive particles which are necessary for neutralizing the background remain : one obtains a canonical description of the one-component plasma.

In this limit $m \to 0$, the operator $m^2/(m^2 + A^+A)$ becomes the projector P on the eigenstates of A^+A with eigenvalue zero, i.e. the projector on the space E of the functions of the form $\psi = \exp[-V(\vec{r})]$ times an entire function of $z = x+iy$. The densities are

$$\rho(\vec{r}) = <\vec{r}|P|\vec{r}>, \quad \rho^{(2)T}(\vec{r}_1,\vec{r}_2) = -|<\vec{r}_1|P|\vec{r}_2>|^2, \text{ etc...}$$

and the one-component plasma problem amounts to compute the projector P, i.e. to find an orthogonal basis for the space E.

Alternatively, the same can be shown by starting directly with the canonical Boltzmann factor of the one-component plasma[12]. The Hamiltonian of N charges e is

$$H = e^2 \sum_i V(\vec{r}_i) - e^2 \sum_{i<j} \ell n(|z_i-z_j|/L) + cst$$

and the Boltzmann factor is, at $\Gamma = 2$,

$$\exp(-\beta H) = C \left|\det\{\exp[-V(\vec{r}_i)] z_i^{j-1}\}_{i,j=1,\ldots,N}\right|^2,$$

where C is some constant. The functions $\exp[-V(r)]z^{j-1}$ are not necessarily orthogonal to one another. However, in the thermodynamical limit, they span the space E; we can choose in this sapce an orthogonal basis ψ_j, and rewrite the Boltzmann factor with a new determinant

$$\exp(-\beta H) = C|\det\{\psi_j(\vec{r}_i)\}|^2.$$

It is then easy to see that the densities are related to the projector

$$<\vec{r}_1|P|\vec{r}_2> = \sum_j \frac{\psi_j(\vec{r}_1)\overline{\psi_j(\vec{r}_2)}}{\int d^2r |\psi_j(\vec{r})|^2}$$

as stated above. The whole formalism much resembles the quantum-mechanical description of a system of independent fermions in an external field.

This new approach to the one-component plasma allows to make full use of the symmetries of the problem when choosing the basis ψ_j. A related important remark is that, for a given background charge density, the background potential $V(\vec{r})$ is not fully determined by Poisson's equation, and this freedom can be used for choosing $V(\vec{r})$ in the most convenient way, depending on the problem to be solved.

For instance, for a constant background charge density $-e\,\rho_B$, it is convenient to work with circular symmetry, choosing $\beta e^2 V(\vec{r}) = \pi\rho_B r^2$; the functions $\exp[-V(r)]\,z^{j-1}$ are orthogonal, and one recovers at once the known Gaussian correlation function $\rho^{(2)T}(r) = -\rho_B^2 \exp(-\pi\rho_B r^2)$. It may however be more appropriate to choose $V(\vec{r}) = V(x)$, as illustrated by the following two examples.

3.1. Electrode with adsorption sites

The model under consideration mimicks the electrical double layer at an electrode-electrolyte interface, the crystalline structure of the metallic electrode being taken into account. The present model has been previously studied[13] by very difficult graph resummations. The new method[14] is easier and gives more general results.

Let us assume that the interface is along the y axis. When there are no adsorption sites, the potential created by the background and the walls can be chosen of the form $e^2 V_0(x)$, and an orthogonal basis in the space E can be defined by the function $\psi_k(\vec{r}) = \exp[-V_0(x)+k(x+iy)]$, with $k \in \mathbb{R}$; these functions are indeed orthogonal because of the factor $\exp(iky)$. Incidentally, one can recover in this way all the known results[15] about inhomogeneous backgrounds with a charge density of the form $-e\rho_B(x)$.

Let us now add a line of equidistant adsorption sites along the y axis, creating an adsorption potential of the Baxter type, i.e.

$$\exp(-\beta V_{ad}) = 1 + \lambda\delta(x)\sum_m \delta(y - mb) ,$$

where b is the distance between adjacent sites. It is convenient to write $k = 2\pi(\xi+n)/b$, with $\xi \in [0,1]$ and n integer. In presence of the adsorption potential, the functions $\exp(-\beta V_{ad}/2)\,\psi_k$ are not orthogonal. However, after these functions have been properly normalized, the matrix of their scalar products takes the simple form

$$M(\xi,n;\xi',n') = \delta(\xi-\xi')[\delta_{nn'} + \mu N(n)\,N(n')] .$$

It is a simple matter to complete the diagonalization of this matrix which has a separable form, and to obtain explicit expressions for the projector P, the densities, and the correlations, both for an externally charged hard wall and for an impermeable polarizable membrane.

Here too, the basic electrochemical quantities can be computed. A generalized contact theorem[16] has been verified : the kinetic pressure on the interface is increased by a contribution from the gradient of the mobile charge density at an adsorption site.

3.2. One-component plasma in a doubly periodic background[11,12]

This model can be understood as made of mobile (classical) "electrons" interacting between themselves and with a lattice of extended fixed "ions" (the background). It can also be regarded as a two-component plasma in which the particles of one species have been frozen into a lattice.

The background density is the sum of its average value ρ_0 plus a periodic modulation; if the lattice unit cell is a rectangle of sides a (along x) and b (along y), $\rho_0 ab = 1$, i.e. there is one unit ionic charge per unit cell. The background potential can be chosen of the form

$$e^2 V(\vec{r}) = e^2 [\pi\rho_0 x^2 + \phi(\vec{r})],$$

where $\phi(\vec{r})$ has the double periodicity of the background. In the functional space E, one can start with a basis defined by functions ψ_k which are the same as in section 3.1., except for a different normalization :

$$\psi_k(\vec{r}) = \exp[-\phi(\vec{r})] \exp[-\pi\rho_0(x - \frac{k}{2\pi\rho_0})^2 + iky].$$

Again, $k = 2\pi(\xi+n)/b$, with $\xi \in [0,1]$ and n integer. These functions ψ_k are not orthogonal. However, since they depend on x only through x-na, this suggests introducing Bloch's functions

$$\tilde{\psi}_{\xi,n}(r) = \sum_n \exp(-2\pi i \eta n) \psi_{\xi,n}(\vec{r}),$$

and these functions $\tilde{\psi}$ do indeed form an orthogonal basis. From the functions $\tilde{\psi}$ one can build the projector P, the densities, and the correlations. The method can be generalized to more complicated lattices.

The simplest example is for a square lattice (a=b) with a background potential modulation such that

$$\exp[-2\phi(\vec{r})] = 1 + \lambda(\cos 2\pi X + \cos 2\pi Y),$$

where $X = x/a$, $Y = y/a$. Then, the one-particle density has the double integral representation

$$\rho(\vec{r}) = \langle\vec{r}|P|\vec{r}\rangle = \rho_0 \sqrt{2} \exp[-2\phi(\vec{r})] \int_{-\infty}^{\infty} d\zeta \int_{-\infty}^{\infty} d\eta$$

$$\times \frac{\exp[-\pi(X-\zeta)^2 - \pi(Y-\eta)^2] \cos[2\pi(X-\zeta)(Y-\eta)]}{1+\lambda[\exp(-\pi/2)](\cos 2\pi\zeta + \cos 2\pi\eta)}.$$

Whatever the oscillations of the background may be, the correlations are found to decay exponentially and the Stillinger-Lovett sum rule is satisfied : The present model, considered as a two-component plasma with one species fixed on a lattice, is in its conducting phase at $\Gamma = 2$. This is in agreement with the arguments[17] indicating that the Kosterlitz-Thouless transition occurs at $\Gamma \gtrsim 4$ for this fixed-ion model also.

4. CONCLUSION

We have now reached a rather good understanding of both the one-component and the two-component plasmas, in two dimensions, at $\Gamma = 2$. After the difficulty about the divergence in the two-component plasma has been circumvented, the two-component plasma turns out to be the simplest one. A variety of inhomogeneous situations can be dealt with. The key of the solvability is the transformation of a problem about classical interacting particles into a problem about quantum-mechanical non-interacting particles, possibly in an external field.

REFERENCES

1) A. Alastuey, Solvable models of Coulomb systems in two dimensions, in : Strongly Coupled Plasma Physics, eds. F.J. Rogers and H.E. DeWitt (Plenum, New York, 1987) pp 331-347.

2) M. Gaudin, J. Phys. (France) 46 (1985) 1027.

3) F. Cornu, Thèse n°887 (Orsay, 1989).

4) E.H. Hauge and P.C. Hemmer, Phys. Norv. 5 (1971) 209.

5) J. Fröhlich, Commun. Math. Phys. 47 (1976) 233.

6) S. Coleman, Phys. Rev. D11 (1975) 2088.

7) S. Mandelstam, Phys. Rev. D11 (1975) 3026.

8) F. Cornu and B. Jancovici, J. Stat. Phys. 49 (1987) 33.

9) F. Cornu and B. Jancovici, J. Chem. Phys. 90 (1989) 2444.

10) G. Chabrier and J.P. Hansen, Molec. Phys. 59 (1986) 1345.

11) F. Cornu and B. Jancovici, Europhys. Lett. 5 (1988) 125. There are sign mistakes at the bottom of p.127 of this reference.

12) F. Cornu, B. Jancovici, and L. Blum, J. Stat. Phys. 50 (1988) 1221.

13) M.L. Rosinberg, J.L. Lebowitz, and L. Blum, J. Stat. Phys. 44 (1986) 153.

14) F. Cornu, J. Stat. Phys. 54 (1989) 681.

15) A. Alastuey and J.L. Lebowitz, J. Phys. (France) 45 (1984) 1859.

16) L. Blum, M.L. Rosinberg, and J.P. Badiali, J. Chem. Phys. 90 (1989) 1285.

17) A. Alastuey, F. Cornu, and B. Jancovici, Phys. Rev. A38 (1988) 4916.

APPROXIMATE THERMODYNAMIC FUNCTIONS FOR THE TWO-DIMENSIONAL TWO-COMPONENT COULOMB GAS

Michel LAVAUD & Stéphane BROCHOT

G.R.E.M.I., C.N.R.S., Université d'Orléans, 45067 Orléans Cedex 2, FRANCE

We compute numerically the configuration integral Q_N of the two-dimensional two-component Coulomb gas for small values of N. We find a simple approximation in terms of Gamma functions, that fits these results and satisfies exact constraints for any N at high and low temperatures. This gives approximations for thermodynamic functions that fit reasonably well the results of numerical simulations of Hansen and Viot. We compute additional exact constraints at high and low temperatures that could help in finding improved approximations.

1. Introduction.

We consider a two-dimensional system of N point ions and N point electrons with unit charge, included in a disk of radius R. The potential energy of the system is:

$$U_{2N} = -e^2 \sum_{i,j=1}^{2N} q_i q_j \ln |\vec{r}_i - \vec{r}_j|/L, \quad q_i = \pm 1 \qquad (1)$$

L is an arbitrary length that fixes the zero of energy. We take $L = a$, where a is the mean interionic spacing. The configuration integral is defined by:

$$Q_N(\gamma, R) = \int_{r_i \leq R} e^{-\beta U_{2N}} \, d\vec{r}_1 \ldots d\vec{r}_{2N} \qquad (2)$$

where $\gamma = e^2/kT$ is the plasma parameter. $\gamma = 2$ corresponds to the critical temperature T_c where the plasma collapses in pairs of particles of opposite charges. The reduced configuration integral is defined by $Q_N(\gamma) = Q_N(\gamma, 1)$, and is related to $Q_N(\gamma, R)$ by the scaling relation:

$$Q_N(\gamma, R) = L^{N\gamma} R^{2(2N - N\gamma/2)} Q_N(\gamma) \qquad (3)$$

Systems of particles with logarithmic interactions are very interesting on a theoretical point of view, because they are complicated enough to provide realistic models of certain physical systems, and at the same time simple enough so that some exact results can be obtained. For the two-dimensional two-component system, one has for example [1,2,3]:

$$Q_1(\gamma) = \frac{2\pi^2}{2-\gamma} \frac{\Gamma(3-\gamma)}{\Gamma(3-\frac{\gamma}{2})\Gamma(2-\frac{\gamma}{2})}, \qquad (4)$$

$$Q_N(\gamma) \sim N!(Q_1)^N, \qquad \gamma \to 2, \qquad (5)$$

$$u(\gamma) = -\frac{1}{2}(\frac{1}{2}\ln\gamma + C), \qquad \gamma \to 0 \qquad (6)$$

where $C = 0.577\ldots$ is Euler's constant, and $u(\gamma)$ is the internal energy. For the one-dimensional one-component plasma made of point electrons on a unit circle, one has [4]:

$$Q_N(\gamma) = (2\pi)^N \frac{\Gamma(1 + \frac{N\gamma}{2})}{[\Gamma(1 + \frac{\gamma}{2})]^N} \qquad (7)$$

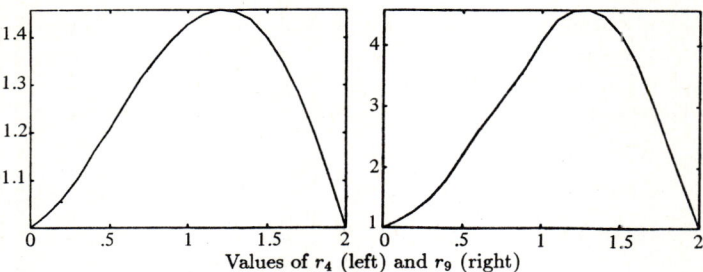

Values of r_4 (left) and r_9 (right)

and, for the two-dimensional one-component plasma at $\gamma = 2$ [5]:

$$Q_N \sim \pi^{2N} N^{-N(N+1)/2} \prod_{n=1}^{N} n!, \quad N \to \infty \tag{8}$$

Exact results have also been obtained for distribution and correlation functions [3,6,7]:

$$g_{+,+}(r_{12}) \sim e^{\beta\mu^{ex}} \frac{4[\Gamma(2-\frac{\gamma}{2})]^2}{(2-\gamma)(\gamma-1)\Gamma(\gamma)\Gamma(3-\gamma)} \left(\frac{r_{12}}{a}\right)^{2-\gamma}, \quad r_{12} \to 0, \quad 1 < \gamma < 2, \tag{9}$$

$$\rho_{+,+}(r_{12}) = -\left(\frac{m^2}{2\pi}\right)^2 [K_0(mr_{12})]^2, \qquad \rho_{+,-}(r_{12}) = \left(\frac{m^2}{2\pi}\right)^2 [K_1(mr_{12})]^2, \quad \gamma = 2 \tag{10}$$

We note that all the results on the configuration integrals are expressed in terms of Gamma functions only. Further, the Debye-Hückel approximation (6) expresses the logarithmic derivative of Q_N in terms of the logarithmic derivative $\psi(x)$ of the Gamma function since $C = -\psi(1)$. This leads us to conjecture that the configuration integral of the two-dimensional two-component Coulomb gas can also be expressed exactly in terms of Gamma functions only. We propose to try to guess this expression by computing numerically Q_N for small values of N, and finding constraints on Q_N valid for any N. Indeed, knowing the exact result (if it exists!) would be a considerable help to devise an analytical proof.

2. Numerical computation of the configuration integral.

We have computed the canonical configuration integral $Q_N(\gamma)$ and its first nine derivatives for systems with up to 32 particles. We use a Monte Carlo method, adapted to deal with the condensation of pairs of particles with opposite charges. This gives a piecewise polynomial approximation via its Taylor development in the whole domain. On the basis of these numerical results and known exact results (4)-(8), we consider the trial function:

$$Q_N(\gamma) \sim \left(\frac{2\pi^2}{2-\gamma}\right)^N \Gamma(1+\frac{N\gamma}{2})[\Gamma(a+b\gamma)]^{-Nc(N)} \tag{11}$$

Exact values for $\gamma = 0$ and $\gamma = 2$ imply:

$$a + b\gamma = 1 + \frac{\gamma}{2} \quad \text{or} \quad a + b\gamma = 2 - \frac{\gamma}{2}$$

The first solution can be eliminated because of the shape of the functions $r_N = Q_N \times [2\pi^2/(2-\gamma)]^{-N}[\Gamma(1+\frac{N\gamma}{2})]^{-1}$. Indeed, they all have the shape indicated in the following figure. In particular, the abscissa of the maximum of r_N is always larger than 1, while it is equal to 0.92 for $1/\Gamma(1+\frac{\gamma}{2})$, and to 1.08 for $1/\Gamma(2-\frac{\gamma}{2})$.

The numerical analysis of the data has been done with the computer program Calcas [8]. This program is an aid in theoretical computations, devised to manipulate, without any programming, formulas involving special functions and functions defined by sequences of points (arising from experiences or numerical simulations).

We have also computed the free energy in the immediate vicinity of $\gamma = 2$, for a system with 200 particles. This is possible because our algorithm is more and more accurate as one approaches the value $\gamma = 2$. This domain had not been investigated before, because of large fluctuations of energy. We obtain a numerical fit that is valid in the whole range $1 \leq N \leq 100$:

$$Q_N(\gamma) \sim (\frac{2\pi^2}{2-\gamma})^N N! \left[1 + N(1 - \frac{\gamma}{2})\right] \tag{12}$$

3. Approximate thermodynamic functions.

We find numerically that $c(N)$ is a steadily increasing function of N. It is expected to go to a finite value $c = c(\infty)$ since thermodynamic functions do exist[9]. This limit is obtained by identifying the resulting internal energy to its Debye-Hückel value (6) for $\gamma \sim 0$. This gives:

$$c = \frac{2C + \ln 2 - 1}{1 - C} = 2.004\ldots \tag{13}$$

and:

$$Q_N(\gamma) \sim \left(\frac{2\pi^2}{2-\gamma}\right)^N \frac{\Gamma(1 + \frac{N\gamma}{2})}{[\Gamma(2 - \frac{\gamma}{2})]^{cN}}, \quad (N \text{ large}) \tag{14}$$

The corresponding excess internal energy and excess chemical potential read:

$$u = -\frac{1}{4}(\ln\frac{\gamma}{2} + \frac{2}{2-\gamma}) - \frac{c}{2}\psi(2 - \frac{\gamma}{2}) \tag{15}$$

$$\frac{\mu}{e^2} = -\frac{1}{2}\ln\frac{\gamma}{2} + \frac{1}{\gamma}[\ln(1 - \frac{\gamma}{2}) + c\ln\Gamma(2 - \frac{\gamma}{2})] \tag{16}$$

We test the validity of our approximation by comparing the values of u and μ/e^2 to results of Hansen and Viot [3] obtained by Monte Carlo simulation (indexed by MC) and with a model of Bjerrum-type (indexed by BJ):

γ	u	u_{MC}	u_{BJ}	μ/e^2	μ_{MC}/e^2	μ_{BJ}/e^2
1	−0.35	−0.40	−0.47	−0.59	−0.48	−0.78
1.5	−0.81	−0.80	−0.80	−0.91	−0.9	−1.08
1.8	−2.26	< −1.4	−1.98	−1.28	−1.3	−1.38
1.9	−4.74			−4.29	−1.58	−1.64

4. Derivatives of the configuration integral for $\gamma = 0$ and $\gamma = 2$.

To find better approximations of thermodynamic functions, we look for additional exact constraints on the configuration integral $Q_N(\gamma)$. We have computed the value of its first four derivatives for $\gamma = 0$:

$$Q'_N(0) = \frac{N}{4}\pi^{2N} \tag{17}$$

$$Q''_N(0) = \pi^{2N}\frac{N}{48}[(8\pi^2 - 43)N - (4\pi^2 - 25)] \tag{18}$$

$$Q'''_N(0) = \pi^{2N}N[-(2N^2 - 3N - \frac{1}{2})\zeta(3) + \frac{\pi^2}{16}(10N^2 - 13N + 4)$$
$$- \frac{1}{192}(763N^2 - 997N + 576)] \tag{19}$$

$$Q''''_N(0) = \pi^{2N}N[(-\frac{\pi^4}{120} - 6\zeta(3) - \frac{43\pi^2}{16} + \frac{48781}{960}) + N(\frac{53\pi^4}{240} - \frac{107}{2}\zeta(3) + \frac{267\pi^2}{32} - \frac{1105799}{11520})$$
$$+ N^2(-\frac{17\pi^4}{60} + 99\zeta(3) - \frac{803\pi^2}{96} + \frac{6663}{128}) + N^3(\frac{3\pi^4}{20} - 38\zeta(3) + \frac{121\pi^2}{48} - \frac{151643}{11520})] \tag{20}$$

$\zeta(x)$ is Riemann's zeta function. We have generalized the preceding results to a non-neutral system, with N_+ ions and N_- electrons. The first two derivatives read:

$$Q'_{N_+,N_-}(0) = \pi^M(M - m^2) \tag{21}$$

$$Q''_{N_+,N_-}(0) = \frac{\pi^M}{192}[M(M-1)(8\pi^2 - 42) - (M - m^2)(M + 3m^2 - 8)] \tag{22}$$

with $M = N_- + N_+$ and $m = N_- - N_+$. The two others are too long to be given here.

We have also computed the first two derivatives of the function $q_N(\gamma)$, defined by $Q_N(\gamma) = (2\pi^2/(2-\gamma))^N q_N(\gamma)$, for $\gamma = 2$. This function measures the difference between real plasma and a system of N independent atoms:

$$q_N(2) = N! \tag{23}$$

$$q'_N(2) = \frac{N}{2}N! \tag{24}$$

$$q''_N(2) = \frac{N}{4}(N + \frac{\pi^2}{3})N! \tag{25}$$

5. Perspectives.

We are currently looking for a better fit on $Q_N(\gamma)$, that would satisfy also the new constraints (17)-(25).

REFERENCES.
1) G. KNORR, Phys. Lett. 28A (1968) 166.
2) C. DEUTSCH and M. LAVAUD, Phys. Rev. A9 (1974) 2598.
3) J.P. HANSEN and P. VIOT, J. Stat. Phys. 38 (1985) 823.
4) F. DYSON, J. Math. Phys. 3 (1962) 140; K.G. WILSON, J. Math. Phys. 3 (1962) 1040; J. GUNSON, J. Math. Phys. 3 (1962) 752
5) A. ALASTUEY and B. JANCOVICI, J. Physique 42 (1981) 1
6) M. LAVAUD J. Stat. Phys. 49 (1987) 1191.
7) F. CORNU and B. JANCOVICI, J. Stat. Phys. 49 (1987) 33; J. Chem. Phys. 90 (1989) 2444
8) M. LAVAUD, Calcas Version 1.2 for IBM/PC (1989)
9) J. FRÖHLICH Commun. Math. Phys. 47 (1976) 233

STRONGLY COUPLED 2D OCP IN A MAGNETIC FIELD

Akira ISIHARA

Department of Physics, State University of New York at Buffalo,
Buffalo, New York, 14260, USA

Several important Coulomb effects in 2D electron systems are
discussed. First, a theory of electron correlations which is
applicable to a wide range of density is presented. This
theory yields the exact exchange specific heat and high and low
density series for the ring energy. It explains well the
strong density dependence of the effective g-factor and also
the valley degeneracy transition in Si [100] inversion layers.
Its generalization to the case with a strong magnetic field
gives an important prediction that the phase of the dHvA oscillation will change at a point at which the ground state energy
becomes negative. Moreover, the amplitude in the gaseous
phase is expected to decrease with increasing r_s. A self-
consistent approach to the dynamical counductivity which goes
beyond the RPA explains density variations of the cyclotron
resonance parameters. A consideration of many-body coupling
results in a new ground state which improves Laughlin's state
for the FQHE.

1. ELECTROSTATIC ANALOGUE

It is known that electron correlations are relatively strong in two dimensions. Since the density of 2D electrons such as in Si inversion layers can be varied in a wide range, correlation effects can be studied well in these 2D systems. In fact, a number of interesting properties which are attributable to electron correlations have been observed. In what follows, we shall discuss some theoretical aspects of electron correlations in these systems, particularly in a perpendicular magnetic field. For this purpose, it is convenient to start with an electrostatic analogue of the RPA dielectric function.

We consider a 2D electron gas near absolute zero. Its dielectric function can be expressed as
$$[\varepsilon(q, \omega)]^{-1} = 1 - u(q)\lambda(q, \omega). \qquad (1.1)$$
Here, $u(q) = 2\pi e^2/q$ is the Coulomb potential, and $\lambda(q, \omega)$ is the RPA polarization function. This function can be expressed in terms of new variables defined by
$$\rho = 1/q, \quad z = -i\omega/(q^2 p_F), \quad a = 1/2p_F \qquad (1.2)$$

as follows[1]
$$\lambda(\rho, z) = Q\int_0^\infty \frac{2\pi r dr}{\exp[\eta(r^2/\rho^2)]+1}\int_a^\infty \frac{dk}{a}\sin ka J_0(kr)e^{-k|z|}, \quad (1.3)$$
where
$$Q = 1/(2\pi)^2, \quad \eta = \beta p_F^2, \quad p = p_F r/\rho. \quad (1.4)$$
p is an auxiliary momentum variable which enters the Fermi distribution function in the original expression of the polarization function. Equation (1.3) shows that for absolute zero $\lambda(\rho, z)$ represents the electric flux at height z coming out of a uniformly charged disc at z = 0. Q is the charge of the disc and ρ is a radial distance measured from the center.

This electrostatic analogue suggests that there is a coordinate system which represents the flux flow in a natural way. This is an oblated spheroidal coordinate system given by
$$\rho = \cosh\xi\sinh\phi, \quad z = \sinh\xi\sin\phi, \quad (1.5)$$
where $\pi/2 \geq \phi \geq 0$ and $\infty > \xi \geq 0$. In this new coordinate system, the polarization function $\lambda(\omega, z)$ becomes
$$\lambda(\phi) = (1 - \cos\phi)/2\pi. \quad (1.6)$$
Note that this is independent of ξ.

The above simple form is very advantageous for evaluating Coulomb effects. For instance, the exchange specific heat can be exactly evaluated[1]. The ring diagram contribution to the correlation energy can also be expressed by a single ϕ integral in a finite domain[2]. From this integral, we learn that
$$r_s = 2^{1/2} \quad (1.7)$$
is the dividing point for high and low density regions. We obtain
$$\varepsilon_r = \begin{cases} -0.6137 - 0.1726 r_s \ln r_s + 0.8653 r_s + O(r_s^2 \ln r_s), & r_s \leq \sqrt{2}; \\ -\frac{1.2935}{r_s^{2/3}} + \frac{1.2004}{r_s} - \frac{0.14018}{r_s^{4/3}} + O((\frac{\ln r_s}{r_s})^2). & r_s \geq \sqrt{2} \end{cases} \quad (1.8)$$
Note that the second term in the low density series cancels exactly the exchange energy. This means that the Hartree-Fock approximation is invalid for $r_s \geq \sqrt{2}$.

The above energy has been obtained for a simple electron gas model. On the other hand, in [100] Si inversion layers, there are two equivalent valleys in which electrons can nest. These valleys represent two-fold degeneracy of the ground state. However, this symmetry can be broken due to electron correlations at a low but finite density. This can be shown by evaluating the energy difference between the two- and one-valley ground states. We find[3]

$$\Delta \varepsilon_g = \frac{1}{2r_s^2} - \frac{0.07}{r_s^{4/3}} + \cdots . \qquad (1.9)$$

Here, the first and second terms are due respectively to the kinetic and correlation energies. The first order exchange energy does not enter this difference. As is clear, the ground state energy difference between the two states is positive for small r_s but becomes negative for large r_s. A precise calculation shows that this transition takes place at $r_s = 8.011$. This translates into $n = 1.2 \times 10^{11}$ cm^{-2} in electron density for Si inversion layers. A slightly larger value of 3×10^{11} cm^{-2} is obtained by taking into consideration the higher order exchange energies in an approximate way. These values are not very far from the experimental value of 5×10^{11} cm^{-2}.

We expect that a similar mechanism will lift spin degeneracy. The corresponding transition point is found to be $r_s = 13$ or $n = 4.5 \times 10^{10}$ cm^{-2} for Si inversion layers. Below this critical density, the system becomes magnetic. However, in the absence of a magnetic field, a domain structure would be formed. Besides, the system would crystallize. Therefore, it can be difficult to observe such a magnetic phase.

2. COULOMB EFFECTS IN MAGNETIC FIELD

The presence of strong Coulomb interaction in 2D electron systems was first indicated by the sharp increase of the effective g-factor toward low densities in Si inversion layers. This increase can be explained by extending the above theory of the correlation energy to the case with a magnetic field. For this purpose, we evaluate the paramagnetic susceptibility because it is related to the g-factor. We find[4]

$$\frac{\chi_0}{\chi} = 1 - \frac{(2g_v)^{1/2}}{\pi} r_s + g_v r_s^2 A(r_s), \qquad (2.1)$$

where $\chi_0 = g_v m \mu_B^2 / (\pi \hbar^2)$ is the ideal susceptibility, g_v being the valley degeneracy factor of Si inversion layers. $A(r_s)$ consists of the contributions from the kinetic, exchange and correlation energies. The former two can be obtained easily. The correlation contribution is more difficult particularly when it should be evaluated for a wide density range. Fortunately, we have succeeded in expressing this contribution in terms of a simple angle integral based on the electrostatic analogue discussed in Section 1. We have found

$$A(r_s) = \frac{1}{g_v r_s^2} + \frac{\sqrt{2g_v}}{\pi r_s g_v^{3/2}} \int_0^\infty \int_0^\infty \frac{dzd\rho}{\rho} \frac{\partial}{\partial \rho}(\frac{\partial \lambda}{\rho \partial \rho}) \frac{2\pi}{1+\sqrt{2}\pi r_s g_v^{3/2} \rho \lambda} . \quad (2.2)$$

Here, the first term comes from the kinetic energy, while the second term is a part of the correlation contribution. The exchange contribution is cancelled out by a part of the correlation contribution. As a result, the effective g-factor which is given by the ratio in Eq.(2.1) increases strongly toward low densities. We have found numerically that our theoretical result is in very good agreement with experimental data.

Even more interesting Coulomb effects can be expected as the magnetic field becomes stronger. In a strong magnetic field, the chemical potential jumps from one Landau level to another. Hence, it becomes necessary to work in intervals in which the chemical potential varies continuously if we want analytical results. An appropriate density parameter is the Landau level filling factor $\nu = n/(eH/ch) = 1/\gamma_o$. We assume that γ_o defined here is not very large. This ensures that the Fermi distribution is sharp. For the interval $2i+1 > 1/\gamma_o > 2i -1$, where i is an integer, we have found[5]

$$\varepsilon_g(H)/4\varepsilon^o = [1 + r_s^2 \varepsilon(r_s)]i\gamma_o - [1 + \frac{(2i-1)(2i+1)}{(2i)^2} r_s^2 \varepsilon(r_s)]i^2\gamma_o^2, \quad (2.3)$$

where ε^o is the kinetic energy in the absence of the magnetic field, and

$$\varepsilon(r_s) = -\frac{8\sqrt{2}}{3} \frac{1}{r_s} + \frac{8}{\pi}\phi, \quad (2.4)$$

where ϕ is an angle integral associated with $\lambda(\phi)$ in Eq.(1.6). That is, it is the sum of the first-order exchange and ring energies in the absence of the magnetic field.

As a function of $1/\gamma_o$, the energy varies parabolically, indicating a linear variation of the susceptibility in the above interval. It is very interesting to observe that the parabola is convex for small r_s and this convexity decreases as r_s increases. That is, the dHvA oscillations are suppressed in the gaseous phase by Coulomb interaction. However, at a point where the ground state energy in the absence of magnetic field becomes negative, the curvature flips so that the energy curve becomes concave. As a result, the dHvA oscillation changes its phase.

The ground state energy of an electron gas has been evaluated by several authors. Although their calculations differ from each other in detail, the resulting energy curves are qualitatively similar in that the energy changes from positive values to negative values. However, it seems that no direct experimental verification of this general result has been achieved. Our present theoretical prediction that the phase of dHvA oscillations alters at the point where the energy changes its sign is important in this respect.

It is encouraging that at least some of the major experimental difficulties have been overcome and dHvA oscillations have been observed in 2D systems. Störmer et al[6] stacked many layers of GaAs heterostructures to increase the surface area because the signal is proportional to this area and observed dHvA oscillations. Very interestingly, their simultaneous measurements of the magnetoresistance showed the presence of localized states, while the magnetization measurements did not. Even though both the dHvA and quantized Hall effects are due to electron's orbital motion, there is apparently a basic difference between the two. Moreover, they found that the density of states evaluated for short range interaction did not quite agree with their dHvA data.

3. DYNAMICAL CONDUCTIVITY

It has been shown elsewhere[7] that the dynamical conductivity can be given exactly in terms of a force-force correlation function. To first order in the strength of impurity scattering, the conductivity is given by

$$\sigma(\omega) = \frac{ine^2/m}{\omega - \omega_c + M(\omega)}, \quad (3.1)$$

where $M(\omega)$, called memory function, is given in terms of the density-density correlation function as follows:

$$M(\omega) = \frac{n_i}{4\pi nm} \int_0^\infty d\vec{q}\, q^3 |v(q)|^2 C(q,\omega). \quad (3.2)$$

Here, n_i is the impurity concentration, $v(q)$ is the impurity potential, and $C(q,\omega)$ is the correlation function which has the general form:

$$C_{ij}(\omega) = i\int_0^\infty dt\, e^{i\omega t} (A_i(t), A_j). \quad (3.3)$$

The dynamical variable A_i represents the Fourier transform of the density ($i = 0$) or current ($i = 1$). Such a correlation function satisfies a matrix equation:

$$[\omega \cdot 1 - \Omega + M] \cdot C(\omega) = -1, \quad (3.4)$$

where

$$\Omega_{ij} = i(\dot{A}_i, A_j). \quad (3.5)$$

In the first approximation, M in Eq. (3.4) can be ignored, resulting in the 0th order solution $C^{(o)}$. By introducing this solution back to the term with M on the left side of Eq. (3.4), we obtain the next approximation:

$$C_{oo}^{(1)} = \frac{C_{oo}^{(o)}(\omega + M(\omega))}{1 + M(\omega) C_{oo}^{(o)}(\omega + M(\omega))}. \quad (3.6)$$

When introduced into Eq. (3.3), this result yields a nonlinear integral equation. We have solved this equation numerically and computed the real and imaginary parts, M_1 and M_2 respectively, of the memory function $M(\omega)$.[8] When the filling factor ν is larger than 1, both $-M_1(\omega)$ and $M_2(\omega)$ decrease with ν, in good agreement with experimental data.[9] Their frequency dependences are also explained well. In order to obtain these results, we used

$$C_{oo}^{(o)}(q,\omega) = \frac{1}{\omega}[\chi_o(q,\omega) - \chi_o(q)], \quad (3.7)$$

where

$$\chi_o(q,\omega) = \frac{\lambda(q,\omega)}{1 + u(q)\lambda(q,\omega)} \quad (3.8)$$

with the RPA $\lambda(q,\omega)$ which is given in terms of the contributions from the Landau levels. Therefore, our results are valid for $\nu > 1$. Indeed, our numerical results deviate from the data for small ν.

4. FRACTIONAL QUANTIZED HALL EFFECT

For the fractional quantized Hall effect, Laughlin[10] introduced a Jastrow-type trial wave function which has an important relation with the potential energy of a classical OCP and found that the ground state energy of this wave function is -0.4156 in units of $e^2/\kappa \ell$ for $\nu = 1/3$, where κ is the background dielectric constant and ℓ is the magnetic length. This energy is lower than the corresponding energy of the CDW state. Since the OCP analogue suggests that the density is uniform, Laughlin concluded that his state represents a new liquidlike state.

Since then, the ground-state energy of the FQHE was evaluated by several investigators. For example, Yoshioka et al[11] tried to diagonalize the Hamiltonian of small systems and reported - 0.4152, - 0.4127, and - 0.4128 for N = 4, 5 and 6, respectively. These values are higher than the limiting energy - 0.4100 reported by Levesque et al[12] and Morf and Halperin.[13] Moreover, the variation of the energy values with the total number N of electrons is not monotonic. There are some other energy values reported. Although the ground state energy depends on models and boundary conditions, it is worth to make an exact approach to the Laughlin state.

The Laughlin state for $\nu = 1/m$ is expressed by

$$\psi_m = \prod_{i<j} (z_i - z_j)^m \exp(-|z_j|^2/4), \qquad (4.1)$$

where $z_i = x_i + iy_i$ is the coordinate of the $\underline{i\text{th}}$ electron in units of the magnetic length. We can expand ψ_m in angular momentum space in terms of states $|\ell_1, \ldots, \ell_N\rangle$ such that[14]

$$\psi_m = \sum_{\ell_1 < \ldots < \ell_N} A(\ell) |\ell_1, \ldots, \ell_N\rangle, \qquad (4.2)$$

where

$$m(N-1) \geq \ell_N > \ell_{N-1} > \ldots > \ell_1 \geq 0,$$

$$\sum_j \ell_j = mN(N-1)/2. \qquad (4.3)$$

The probability amplitude $A(\ell)$ is given by

$$A(\ell) = (\ell_1! \ldots \ell_N!)^{1/2} C(\ell), \qquad (4.4)$$

$$C(\ell) = \sum_{n_{jk}=0}^{N-1} \prod_{j=1}^{N} \delta(\sum_{k=1}^{m} n_{jk}, \ell_j) \prod_{k=1}^{m} \varepsilon(n_{1k}, \ldots, n_{Nk}),$$

where j and k in the first sum run, respectively, from 1 to N and from 1 to m. $\delta(\ldots)$ is Kronecker's delta and $\varepsilon(\ldots)$ represents the antisymmetric unit tensor of elements $n_{1k}, \ldots, n_{NK}$.

The ground state energy ε_g of the Laughlin state consists of four terms:

$$\varepsilon_g = \varepsilon_{ee}^d + \varepsilon_{ee}^x + \varepsilon_{eb} + \varepsilon_{bb}, \qquad (4.5)$$

where the right side terms represent resepctively the direct- and exchange-type electron-electron interaction energies, the electron-background interaction energy, and the background-background interaction energy. Takano and Isihara[14] have given the exact expressions for these energies. For N = 6, these energies are given by 0.556933, - 0.079534, - 1.717742 and 0.848826 respectively. Hence, the exchange-type energy is small and the electron-background coupling is large. This indicates that the ground state energy depends on models sensitively. The total electron-electron energy is 0.4777399 which is approximately one-half of the magnitude of the electron-background and background-background interaction energies. The total ground state energy for N = 6 is - 0.391517, and all these energies correspond to the case m = 3. This ground state energy is lower than - 0.389 of the CDW state but is higher than - 0.4128 of Yoshioka et al and - 0.4156 of Laughlin. Since the result of Yoshioka et al is based on a diagonalization of the Hamiltonian, a direct comparison with ours is difficult. Nevertheless, since their values are not monotonic when N changes, and because the above value is lower than the well established asymptotic value of - 0.4100, there can be some errors in their numerical analyses. On the other hand, Laughlin's value depends on the HNC distribution function. A limiting value of - 0.4100 which has now been considered as the correct aymptotic value for N → ∞ has been given by Levesque et al.[12] Since the ground state energy is expected to be lower for a larger N, our above result is reasonable.

The Laughlin state adequately describes the fractional quantized Hall effect. It can be used to generate the hierarchy of the fractional states. It is close but not the exact ground state. Moreover, it has been pointed out that the true ground state may deviate the m-fold zero character of the Laughlin state. Also, the true ground state may have m-fold degeneracy. Therefore, apart from the ground state energy calculation, it is very important to try to approach the true ground state.

We recognize that the Laughlin state represents a Jastrow-type trial function which consists of pair functions. It is a many-body wave function, but can be improved if we include many-electron correlations. We have tried to include such correlations such that our new state does not deviate significantly from the Laughlin state. The construction of this new state is based on the conservation of the total momentum L = mN(N-1)/2. Since this momentum is determined by the power of the z_j, the conservation law is rather restrictive. In addition, the usual restriction due to Fermi statistics must be satisfied.

The FQHE occurs usually for odd integers m. Although an even denominator fractional state has recently been found, such a state can be considered as exceptional. Hence, let $m = 2p + 1$, where p is an arbitrary integer. We recognize that all odd integers share unity so that we associate 1 with Fermi statistics and p with a given fractional state. Under these considerations, we have arrive at a new variational wave function given by[15]

$$\Psi = \prod_{i<j} (z_i - z_j)(f_{ij})^P (f_{ji})^P \exp(-\sum_k |z_k|^2/4), \qquad (4.6)$$

where

$$f_{ij} = z_i - z_j + c \sum_{k \neq i,j} z_k. \qquad (4.7)$$

The linear combination of z_j ensures the momentum conservation. The constant c is a variational parameter which may be complex. Note that in the limit $c \to 0$, the above wave function is reduced to the Laughlin state.

The c term in f_{ij} includes the coordinates of all the electrons but the chosen pair i and j. It is symmetric with respect to these coordinates, and may be interpreted as follows. Suppose that the electron at z_i approaches the one at z_j in the presence of all other electrons. If the center of mass of these other electrons coincides with the origin about which the entire electron system is symmetric, the disk geometry of the back ground positive charges will cancel the effect of these charges. In this case, the sum in f_{ij} automatically vanishes. If the center of these charges is shifted from the origin, the charge cloud of these electrons would act on the ith electron as if the charge cloud of the jth electron has become asymmetric about z_j. Such a shift is due to long range Coulomb repulsion so that c may include e^2.

The role played by the c term in the new variational wave function can be interpreted in terms of the OCP analogue. We can assume that c is small, because otherwise deviations from Laughlin's state becomes large. We can also assume that none of $z_i - z_j$ vanishes. To first order in c, the OCP potential energy associated with $|\Psi|^2$ is given by

$$\Phi = -m^2 \sum_{i<j} \ln|z_i - z_j| + \frac{m}{4} \sum_i |z_i|^2 + \frac{m-1}{2m} \sum_{i<j} \frac{|c|^2 |\sum_k' z_k|^2}{|z_i - z_j|^2} \qquad (4.8)$$

Hence, the c term can be interpreted to play the role of adding a repulsive potential to the system. This potential depends on the location of the center of gravity of the electrons other than the chosen pair. It decreases rather fast with electron distances and is proportional to the density $(1-\nu)$ of the background positive charges.

We have introduced a new variational wave function in which c is a variational parameter. Although this introduction is in conformity with the long range character of the Coulomb potential, the new wave function may be useful to short-range potentials as well. In order to see this, we consider a family of short-range potentials which can be expanded in powers of a range parameter b such that[16]

$$V(r) = \sum_j a_j b^{2j} \nabla^{2j} \delta(\vec{r}). \qquad (4.9)$$

We note that due to the repulsiveness, $<V(r)>$ is nonnegative. Since our variational wave function is antisymmetric, $<a_o \delta(\vec{r})> = 0$. We can choose c such that it is proportional to b, and investigate the limit $b \to 0$. Since Ψ is a finite polynomial with terms of the form $b^n r^{m-n}$, where n is an integer or zero, $<V(r)>$ will be of order b^{2m}. This is the lowest possible order. Hence, Ψ will be asymptotically exact in the limit $b \to 0$.

We have evaluated the ground state energy of the 1/3 fractional state as a function of c. For the background charges, we have assumed the standard disc geometry. In this case, we can use the exact electron-background and background-background energies. We have adopted the standard Metropolis algorithm and performed 20 Monte Carlo simulations, each consisting of 250,000 Monte Carlo steps for a given c. For improved reliability of several points, we have repeated the same 20 Monte Carlo simulations a few extra times. For $N = 30$, each Monte Carlo simulation took approximately 7,000 seconds with the IBM 3081. On the average, it took about 39 hours for one average point out of 20 simulations.

We have performed Monte Carlo calculations for $N = 10, 20$ and 30 and found that the energy becomes lower as N increases. More interestingly, the energy is lowered if c is small but finite. Even more interesting is that the case $N = 20$ indicates that there are three nearly equal energy minima at $c = 0.0325, 0.0425$ and 0.05. This is even more clearly demonstrated in the case $N = 30$, where three cusp-type energy minima of values $-0.400795, -0.400798$ and -0.400865 appear at $c = 0.02125, 0.03$ and 0.035 respectively. These values are lower than -0.400549 corresponding to $c = 0$. Since they are very close to each other, they might represent three-fold degeneracy in the limit $N \to \infty$.

We have carefully examined the approach to this limit, particularly because we have noticed that the minimum point c tends to decrease as N increases, even though the energy minima become clearer. Within our limited numerical analyses, it appears as if this point c decreases in approximate proportion to $N^{-1/2}$. However, the sum in the c term includes the coordinates of $N - 2$ electrons. Hence, we expect that the average $\langle |\Sigma' z_k|^2 \rangle \sim N^{1/2}$. Thus, we think that the c term yields a net effect even in the limit $N \to \infty$. In fact, we have obtained an empirical formula for the difference in the ground state energies between Laughlin's state and ours as follows:

$$\Delta\varepsilon_g = 0.000350 + 0.0445 N^{-2.6} \qquad (4.10)$$

The Laughling state is definite and simple. However, there is no variational parameter and no degeneracy. It has been shown that the Laughlin wave function is exact for short range potentials. In comparison, our new wave function is also exact for such potentials and yields a lower ground state energy. Moreover, this new state for the 1/3 fractional state appears to be three-fold degenerate. Even though the need of having such degeneracy has been pointed out in the literature, no corresponding explicit wave function has been introduced before. Our wave function has very interesting features to approach the true ground state.

5. CONCLUDING REMARKS

We have discussed that an electrostatic analogue of the dielectric function enables to identify the point which divides the high and low density regions for the ground state energy of a 2D electron system. Moreover, it enables to derive the exact high and low density series for the correlation energy associated with the ring diagrams. An immediate application of this theory provides a satisfactory explanation of the valley degeneracy transition in Si [100] inversion layers. In the presence of a magnetic field, an extension of the theory yields strong density dependences of the effective g-factor in agreement with experiment. Moreover, a phase reversal of the dHvA oscillations is predicted to take place at a point at which the ground state energy in the absence of a field changes its sign.

We have developed an exact theory of the dynamical conductivity which displays explicitly cyclotron resonance. To first order in impurity scattering, the memory function entering this conductivity is reduced to the well established expression. Starting with the RPA dielectric function, we have used Götze's[17] feedback method to derive a self-consistent integral equation for the memory function. A numerical solution of this equation explains the density variations of the real and imaginary parts of the memory function, in good agreement with the data due to Wilson et al.[9]

For the FQHE, we have succeeded in obtaining the exact ground state energy of the Laughlin state. Moreover, we have introduced a new variational wave function which not only yields lower energy but also shows three-fold degeneracy for the 1/3 fractional state.

REFERENCES

1) A. Isihara and L. C. Ioriatti, Jr., Physica 103A (1980) 621.
   L. C. Ioriatti, Jr. and A. Isihara, Phys. Stat. Sol.(b) 97 (1980) K65.
2) L. Ioriatti, Jr. and A. Isihara, Z. Phys. B44 (1981) 1.
3) A. Isihara and L. C. Ioriatti, Jr., Phys. Rev. B29 (1982) 5534.
4) A. Isihara and L. C. Ioriatti, Jr., Physica 113B (1982) 42.
5) Y. Shiwa and A. Isihara, Phys. Rev. B27 (1983) 4743.
6) H. L. Störmer et al, J. Vac. Sci. Tech. B1 (1983) 423. J. P. Eisenstein et al, Surf. Sci. 170 (1986) 271.
7) Y. Shiwa and A. Isihara, J. Phys. C16 (1983) 4853.
8) Y. Shiwa and A. Isihara, Solid State Commun. 53 (1985) 519.
9) B. A. Wilson et al, Phys. Rev. Lett. 44 (1980) 479. Phys. Rev. B24 (1981) 5887. S. J. Allen et al, Phys. Rev. B26 (1982) 5590.
10) R. B. Laughlin, Phys. Rev. Lett. 50 (1983) 1395.
11) D. Yoshioka et al, Phys. Rev. Lett. 50 (1983) 1219.
12) D. Levesque et al, Phys. Rev. B30 (1984) 1056.
13) R. Morf and B. I. Halperin, Phys. Rev. B33 (1986) 2221.
14) K. Takano and A. Isihara, Phys. Rev. B34 (1986) 1399.
15) K. Takano and A. Isihara, J. Phys. C20 (1987) L281. Degenerate ground state for fractional quantum Hall effect, in: Anderson Localization, eds T. Ando and H. Fukuyama (Springer Verlag, Berlin, 1988) pp268-272.
16) S. A. Trugman and S. Kivelson, Phys. Rev. B31 (1985) 5280.
17) W. Götze, Phil. Mag. B43 (1981) 219. A. Gold and W. Götze, Phys. Rev. B33 (1986) 2495.

COLLISIONAL RELAXATION OF A STRONGLY MAGNETIZED PURE ELECTRON PLASMA (THEORY AND EXPERIMENT)*

T.M. O'NEIL, P.G. HJORTH[†], B. BECK, J. FAJANS[‡] and J.H. MALMBERG

Physics Department, University of California at San Diego, La Jolla, CA 92093, USA

1. INTRODUCTION

We say that an electron plasma is strongly magnetized when the cyclotron period is short compared to the duration of a close collision. The gyroangles for the electrons may then be thought of as a collection of high frequency oscillators and the remaining variables (the guiding center variables) as slowly varying parameters that modulate the high frequency oscillators. Loosely speaking, one expects the high frequency oscillators to resonantly exchange quanta (or action) with each other, but not with the slowly varying variables. More precisely, one expects the total action associated with the cyclotron motion (i.e., $\sum_j m v_{\perp j}^2 / 2\Omega$) to be an adiabatic invariant.[1,2] Here, $\Omega = eB/mc$ is the cyclotron frequency and $v_{\perp j}$ is the component of the jth electron velocity that is perpendicular to the magnetic field. For the simple case of a uniform magnetic field, one may equivalently say that the total perpendicular kinetic energy is an adiabatic invariant, and this paper discusses the influence of the invariant on the collisional relaxation of the electron velocity distribution.

On a short time scale, the adiabatic invariant is well conserved, and there is negligible exchange of energy between the parallel and the perpendicular degrees of freedom. The distribution of parallel velocities and the distribution of perpendicular velocities relax separately to Maxwellian, with the parallel temperature $(T_\parallel)$ not necessarily equal to the perpendicular temperature $(T_\perp)$. However, the evolution does not stop at this stage, since an adiabatic invariant is not strictly conserved; it suffers exponentially small changes. In the present case, each collision produces an exponentially small exchange of energy between the parallel and the perpendicular degrees of freedom, and these act cumulatively in such a way that $T_\parallel$ and $T_\perp$ relax to a common value. The time for this relaxation (the second time scale) is exponentially long, or equivalently, the rate is exponentially small.

This paper presents a calculation of this exponentially small equipartition rate.[2] It also presents the results of molecular dynamics simulations that corroborate the existence of the adi-

---

*Supported by NSF grant PHY87-06358 and a grant of cpu time from the San Diego Supercomputer Center.
[†]Mathematical Institute, D.T.H., DK-2800 Lyngby, Denmark.
[‡]Dept. of Physics, Univ. of California, Berkeley, CA 94720.

abatic invariant and the value of the calculated rate,[3] and it presents the results of recent experiments (with cryogenic pure electron plasmas) that are in agreement with the theory.[4]

In Section 2, we consider the isolated collision of two electrons in a strong magnetic field and calculate the exponentially small exchange of energy between the parallel and the perpendicular degrees of freedom. In Section 3, this energy exchange is used to calculate the equipartition rate for a plasma with $T_\| \neq T_\perp$. The analysis in this section employs a Boltzmann-like collision operator, which treats collisions as well separated binary interactions. The justification for this is that the most important collisions (most effective in producing energy exchange) are close collisions, and these tend to be well separated binary interactions—at least for a weakly correlated plasma. We consider only the case of weak correlation, and the experiments are carried out for this case.

The molecular dynamics simulations are presented in Section 4; the dynamics of 50 point charges that interact electrostatically in the presence of a uniform magnetic field is followed numerically for various values of the plasma parameters (density, temperature, and field strength). In Section 5, the results of the experiments are presented; a magnetically confined pure electron plasma is cooled to the cryogenic temperature range, and the equipartition rate is measured as a function of magnetic field strength and plasma temperature. For both the simulations and the experiments, the equipartition rate drops dramatically in accord with theory as the plasma enters the parameter regime of strong magnetization.

2. BINARY INTERACTION

In this section, we analyze an isolated collision between two electrons that interact electrostatically in the presence of a strong and uniform magnetic field, $\mathbf{B} = \hat{z}B$. The equations of motion for the electrons are

$$\frac{d\mathbf{v}_1}{dt} + \Omega \mathbf{v}_1 \times \hat{z} = \frac{e^2}{m} \frac{(\mathbf{r}_1 - \mathbf{r}_2)}{|\mathbf{r}_1 - \mathbf{r}_2|^3} , \qquad (1)$$

$$\frac{d\mathbf{v}_2}{dt} + \Omega \mathbf{v}_2 \times \hat{z} = \frac{e^2}{m} \frac{(\mathbf{r}_2 - \mathbf{r}_1)}{|\mathbf{r}_1 - \mathbf{r}_2|^3} , \qquad (2)$$

where $\mathbf{r}_j$ and $\mathbf{v}_j$ are the position and velocity of electron j. By adding and subtracting these equations, we obtain the two equations

$$\frac{d\mathbf{V}}{dt} + \Omega \mathbf{V} \times \hat{z} = 0 , \qquad (3)$$

$$\frac{d\mathbf{v}}{dt} + \Omega \mathbf{v} \times \hat{z} = \frac{e^2}{\mu} \frac{\mathbf{r}}{|\mathbf{r}|^3} , \qquad (4)$$

where $\mathbf{V} = d/dt\,(\mathbf{r}_1 + \mathbf{r}_2)/2$ is the velocity of the center of mass, $\mathbf{r} = \mathbf{r}_2 - \mathbf{r}_1$ is the position of electron 2 relative to that of electron 1, $\mathbf{v} = d/dt\,(\mathbf{r})$ is the relative velocity, and $\mu = m/2$ is the reduced mass. The center of mass motion is equivalent to that of an electron in a uniform magnetic field, and the relative motion is equivalent to that of an electron in a uniform magnetic field and the field of a fixed charge. The solution for the center of mass motion is trivial, and the solution for the relative motion is simplified by the existence of an adiabatic invariant.

The condition for strong magnetization insures that the cyclotron frequency is much larger than any other frequency that characterizes the collisional dynamics [i.e., $\Omega \gg v_\parallel/b$, $v_\perp/b$, where $|\mathbf{r}| \geq b$]; so the cyclotron action $\mu v_\perp^2(t)/2\Omega$ is an adiabatic invariant. Since Ω is a constant, one can equivalently say that $\mu v_\perp^2(t)/2$ is an adiabatic invariant. Incidently, the fact that $\mu v_\perp^2(t)/2$ is an invariant implies that the distance of closest approach is given by $b = e^2/(\mu v_\parallel^2/2)$.

Note that $\mu v_\perp^2(t)/2\Omega$ is a new adiabatic invariant associated jointly with the two electrons; neither $m v_{1_\perp}^2(t)/2\Omega$ nor $m v_{2_\perp}^2(t)/2\Omega$ are valid invariants. For example, in Eq. (1), $\mathbf{r}_2(t)$ is a time-dependent function that varies at the cyclotron frequency, and this breaks the adiabatic invariant of electron 1 [i.e., $m v_{1_\perp}^2(t)/2\Omega \neq$ const]. Likewise, the temporal variation of $\mathbf{r}_1(t)$ breaks the adiabatic invariant of electron 2. By introducing the relative position and velocity (i.e., $\mathbf{r}$ and $\mathbf{v}$), we have removed the explicit time dependence from the interaction and uncovered a new adiabatic invariant, $\mu v_\perp^2(t)/2\Omega$.

From Eq. (3), one can see that $V_\perp^2(t)$ is an exact constant of the motion; consequently, the relation $m v_{1_\perp}^2(t)/2 + m v_{2_\perp}^2(t)/2 = \mu v_\perp^2(t)/2 + (2m) V_\perp^2(t)/2$ implies that the sum of the perpendicular kinetic energies for the two electrons is an alternative expression for the adiabatic invariant. This expression can be generalized to the case where many electrons interact simultaneously, that is, the quantity $\sum_j m v_{j_\perp}^2(t)/2$ is also an adiabatic invariant.[1,2]

An adiabatic invariant is not strictly conserved but suffers exponentially small changes. For the case of a weakly correlated plasma, we will argue that the overall invariant [i.e., $\sum_j m v_{j_\perp}^2(t)/2$] suffers changes primarily through close two-particle collisions; therefore we calculate the change that occurs in $\mu v_\perp^2(t)/2$ during a collision. From Eq. (4), it follows that

$$\frac{d}{dt}\frac{\mu v_\perp^2(t)}{2} = \frac{e^2 \mathbf{v}_\perp(t) \cdot \mathbf{r}_\perp(t)}{|\mathbf{r}(t)|^3} \ ; \tag{5}$$

thus, the quantity $\Delta(\mu v_\perp^2/2) = \mu v_\perp^2(\infty)/2 - \mu v_\perp^2(-\infty)/2$ is given by the time integral

$$\Delta\left[\frac{\mu v_\perp^2}{2}\right] = \int_{-\infty}^{+\infty} dt \frac{e^2 \mathbf{v}_\perp(t) \cdot \mathbf{r}_\perp(t)}{|\mathbf{r}(t)|^3} \ . \tag{6}$$

Following the usual practice in the theory of adiabatic invariants, we use the lowest-order orbits in evaluating the time integral, that is, we rewrite Eq. (6) as

$$\Delta\left[\frac{\mu v_\perp^2}{2}\right] \cong e^2 v_\perp \rho \int_{-\infty}^{+\infty} \frac{dt \cos(\Omega t + \delta)}{[\rho^2 + z^2(t)]^{3/2}} \ , \tag{7}$$

where (ρ, z) is the guiding center approximation for $(\mathbf{r}_\perp, z)$, δ is a constant, and $z(t)$ is determined by

$$\dot{z}^2 + \frac{2e^2/\mu}{[\rho^2 + z^2(t)]^{1/2}} = \dot{z}^2(t = -\infty) = v_\parallel^2 \ . \tag{8}$$

The origin of time may be shifted simply by changing the value of the constant δ; so, without loss of generality, we choose the origin of time so that $z^2(t)$ is an even function of t, that is, so that the electron either passes $z=0$ or reflects at the time $t=0$. Equation (8) then reduces to the form

$$\Delta\left[\frac{\mu v_\perp^2}{2}\right] = e^2 v_\perp \rho \cos(\delta) \int_{-\infty}^{+\infty} \frac{dt \cos(\Omega t)}{[\rho^2 + z^2(t)]^{3/2}} . \qquad (9)$$

In terms of the scaled variables

$$\zeta = z/b , \quad \xi = v_\parallel t/b , \quad \eta = \rho/b , \quad \varepsilon = v_\parallel/\Omega b , \qquad (10)$$

the time integral in Eq. (9) can be written as

$$\int_{-\infty}^{+\infty} \frac{dt \cos(\Omega t)}{[\rho^2 + z^2(t)]^{3/2}} = \frac{1}{v_\parallel b^2} \int_{-\infty}^{+\infty} \frac{d\xi \cos(\xi/\varepsilon)}{[\eta^2 + \zeta^2(\xi)]^{3/2}} , \qquad (11)$$

and Eq. (8) can be rewritten as

$$\left[\frac{d\zeta}{d\xi}\right]^2 = 1 - \frac{1}{[\eta^2 + \zeta^2(\xi)]^{1/2}} . \qquad (12)$$

Strong magnetization implies that $\varepsilon \ll 1$, so the ξ integral in Eq. (11) involves the product of a rapidly oscillating function and a slowly varying function and turns out to be exponentially small. The integral can be evaluated by deforming the contour of integration into the complex ξ plane and is of the form[2]

$$\int_{-\infty}^{+\infty} \frac{d\xi \cos((\xi/\varepsilon)}{[\eta^2 + \zeta^2(\xi)]^{3/2}} \equiv h(\varepsilon,\eta) e^{-g(\eta)/\varepsilon} , \qquad (13)$$

where $h(\varepsilon,\eta)$ is neither exponentially small nor exponentially large in the range of ε and η of interest, and $g(\eta)$ is given by

$$g(\eta) = \left| \int_1^\eta \frac{x^{3/2} dx}{\sqrt{(x-1)(\eta^2 - x^2)}} \right| . \qquad (14)$$

From the plot of $g(\eta)$ in Fig. 1, one can see that $g(\eta) \cong \pi/2$ for $\eta = \rho/b \ll 1$ and that $g(\eta) \cong \eta$ for $\eta = \rho/b \gg 1$. Thus, the ξ integral is of order $\exp[-\pi\Omega b/2v_\parallel]$ for $\rho \ll b$ and of order $\exp[-\Omega\rho/v_\parallel]$ for $\rho \gg b$. These exponentials are each of the form $\exp(-\Omega\tau)$, where τ characterizes the duration of the collision. Also, we note that the ξ integral is largest for collisions characterized by small impact parameter and large relative velocity.

3. CALCULATION OF THE EQUIPARTITION RATE

Next we turn to the question of how such collisions act cumulatively to produce the relaxation of $T_\parallel$ and $T_\perp$ in a strongly magnetized plasma. For simplicity, we consider the case of a

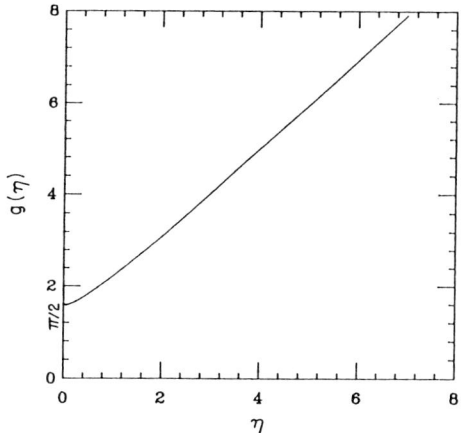

FIGURE 1
A plot of the function $g(\eta)$, defined in Eq. (14).

weakly correlated plasma, that is, a plasma in which $e^2 n^{1/3} \ll T_\|$, where n is the electron density. This inequality can be rewritten as the condition that $\bar{b} = e^2/(\mu \overline{v_\|^2}/2) = 2e^2/T_\|$ is small compared to the mean interparticle spacing (i.e., $\bar{b} \ll n^{-1/3}$). One can easily verify that correlations are determined by $T_\|$ in the strongly magnetized parameter regime.

We have just seen that for the class of two-particle collisions it is the close collisions (i.e., $|\mathbf{r}_1 - \mathbf{r}_2| \cong \bar{b}$) that are most effective in producing an exchange of parallel and perpendicular energy. If we add a third particle into the dynamics, we obtain a small perturbation on the close two-particle collision unless all three particles are close simultaneously. We assume that the energy exchange caused by a close three-particle collision is of the same order as that for a close two-particle collision; this is reasonable considering time scale arguments concerning the durations of the collisions. Of course, the overall adiabatic invariant, $\sum_j m v_{j\perp}^2(t)/2$, exists for many-electron collisions. Since the inequality $\bar{b} \ll n^{-1/3}$ implies that close two-particle collisions are more frequent than close three-particle collisions, we neglect the close three-particle collisions and, similarly, all higher-order collisions. Also, we note that the close two-particle collisions are well-separated events (i.e., $|\mathbf{r}_1 - \mathbf{r}_2| \cong \bar{b} \ll n^{-1/3}$).

Such well-separated binary collisions can be treated with a Boltzmann-like collision operator.[1,2,5] In particular, we evaluate the integral

$$\frac{dT_\perp}{dt} = \int d\mathbf{v}_1 \frac{m v_\perp^2}{2} \frac{\partial f}{\partial t}(\mathbf{v}_1, t) \qquad (15)$$

by replacing the time derivative of the distribution function with the Boltzmann-like operator

$$\frac{\partial f}{\partial t}(\mathbf{v}_1,t) = n \int 2\pi\rho\, d\rho \int d\mathbf{v}_2 |\hat{z}\cdot(\mathbf{v}_2-\mathbf{v}_1)| \times [f(\mathbf{v}'_1,t)f(\mathbf{v}'_2,t) - f(\mathbf{v}_1,t)f(\mathbf{v}_2,t)] \, . \tag{16}$$

In relation to the usual form of the Boltzmann collision operator, the integral over $2\pi\rho\, d\rho$ is equivalent to the integral over the impact parameter (or, scattering cross section), and the quantity $|\hat{z}\cdot(\mathbf{v}_2-\mathbf{v}_1)|$ replaces $|\mathbf{v}_2-\mathbf{v}_1|$, since the two electrons stream toward one another along field lines. In the usual manner $(\mathbf{v}'_1,\mathbf{v}'_2)$ are velocities that evolve into $(\mathbf{v}_1,\mathbf{v}_2)$ during a scattering.

By substituting the collision operator into Eq. (15) and by using detailed balance, we obtain

$$\frac{dT_\perp}{dt} = \frac{n}{4}\int 2\pi\rho\, d\rho \int d\mathbf{v}_1 \int d\mathbf{v}_2 |\hat{z}\cdot(\mathbf{v}_2-\mathbf{v}_1)|$$
$$\times [f(\mathbf{v}'_1,t)f(\mathbf{v}'_2,t) - f(\mathbf{v}_1,t)f(\mathbf{v}_2,t)]$$
$$\times \left[\frac{m v_{1\perp}^2}{2} + \frac{m v_{2\perp}^2}{2} - \frac{m v'^2_{1\perp}}{2} - \frac{m v'^2_{2\perp}}{2}\right] . \tag{17}$$

The distribution functions are assumed to be of the form

$$f(\mathbf{v}_j,t) = \left[\frac{m}{2\pi T_\parallel}\right]^{1/2} \left[\frac{m}{2\pi T_\perp}\right] \exp\left[-\frac{m v_{j\parallel}^2}{2T_\parallel} - \frac{m v_{j\perp}^2}{2T_\perp}\right] . \tag{18}$$

To evaluate the multiple integral, it is convenient to introduce the center of mass velocity and the relative velocity (i.e., $\mathbf{V}$, and $\mathbf{v}$). With the aid of the relations

$$\frac{m v_{1\perp}^2}{2} + \frac{m v_{2\perp}^2}{2} = \frac{\mu v_\perp^2}{2} + \frac{2m V_\perp^2}{2} ,$$
$$\frac{m v_{1\parallel}^2}{2} + \frac{m v_{2\parallel}^2}{2} = \frac{\mu v_\parallel^2}{2} + \frac{2m V_\parallel^2}{2} , \tag{19}$$
$$V'^2_\perp = V_\perp^2 , \quad d\mathbf{v}_1\, d\mathbf{v}_2 = d\mathbf{v}\, d\mathbf{V} .$$

Eq. (17) can be rewritten as

$$\frac{dT_\perp}{dt} = \frac{n}{4}\int 2\pi\rho\, d\rho \int d\mathbf{v}\, |v_\parallel|\, [f_r(\mathbf{v}'_\parallel,\mathbf{v}'_\perp) - f_r(\mathbf{v}_\parallel,\mathbf{v}_\perp)]\times \Delta(\mu v_\perp^2/2) , \tag{20}$$

where the integral over $d\mathbf{V}$ has been evaluated and

$$f_r(\mathbf{v}_\parallel,\mathbf{v}_\perp) = \left[\frac{\mu}{2\pi T_\parallel}\right]^{1/2} \left[\frac{\mu}{2\pi T_\perp}\right] \exp\left[-\frac{\mu v_\parallel^2}{2T_\parallel} - \frac{\mu v_\perp^2}{2T_\perp}\right] \tag{21}$$

is the distribution of relative velocities.

By using $\Delta(\mu v_\perp^2/2) = \mu v_\perp^2/2 - \mu v'^2_\perp/2 = \mu v'^2_\parallel/2 - \mu v_\parallel^2$, Eq. (20) can be rewritten as

$$\frac{dT_\perp}{dt} = \frac{n}{4}\int 2\pi\rho\, d\rho \int d\mathbf{v}\, |v_\parallel| f_r(\mathbf{v}_\parallel,\mathbf{v}_\perp)\Delta\left[\frac{\mu v_\perp^2}{2}\right] \times \left\{\exp\left[\left[\frac{1}{T_\perp}-\frac{1}{T_\parallel}\right]\Delta\left[\frac{\mu v_\perp^2}{2}\right]\right]-1\right\} . \tag{22}$$

Taylor-expanding the exponential, substituting for $\Delta(\mu v_\perp^2/2)$ from Eq. (9), and integrating over $d\mathbf{v}_\perp$ yields the expression

$$\frac{dT_\perp}{dt} = \left[T_\perp\left[\frac{1}{T_\perp} - \frac{1}{T_\parallel}\right]\right]\frac{n}{4}\int 2\pi\rho\, d\rho \int dv_\parallel |v_\parallel|$$

$$\times \frac{\exp(-\mu v_\parallel^2/2T_\parallel)}{(2\pi T_\parallel/\mu)^{1/2}} \frac{e^4\rho^2}{\mu}\left[\int_{-\infty}^{+\infty} \frac{dt\,\cos(\Omega t)}{[\rho^2 + z^2(t)]^{3/2}}\right]^2. \qquad (23)$$

By using Eqs. (11) and (13) to reexpress the time integral, Eq. (23) reduces to the form

$$\frac{dT_\perp}{dt} = (T_\parallel - T_\perp)n\bar{b}^2\bar{v}_\parallel I(\bar{\varepsilon}),$$

$$I(\bar{\varepsilon}) = \frac{\sqrt{2\pi}}{8}\int_0^\infty \frac{d\sigma}{\sigma}\exp\left[-\frac{\sigma^2}{2}\right] \times \int_0^\infty \eta^3 d\eta\, h^2\left[\frac{1}{\bar{\varepsilon}\sigma^3}, \eta\right]\exp\left[-\frac{2g(\eta)}{\sigma^3\bar{\varepsilon}}\right], \qquad (24)$$

where we have introduced the scaled variables

$$\bar{v}_\parallel = \sqrt{T_\parallel/\mu}, \quad \sigma = v_\parallel/\bar{v}_\parallel, \quad \bar{b} = 2e^2/\mu\bar{v}_\parallel^2, \quad \bar{\varepsilon} = \bar{v}_\parallel/\Omega\bar{b}. \qquad (25)$$

Since $\bar{\varepsilon}$ is assumed to be small and $g(\eta)$ is an increasing function of η, the main contribution to the η integral comes from small η. By making the approximations[2]

$$g(\eta) \cong g(0) + g''(0)\eta^2/2 = 1.57 + (0.675)\eta^2,$$

$$h(1/\bar{\varepsilon}\sigma^3, \eta) \cong h(1/\bar{\varepsilon}\sigma^3, 0) = 2.79(1/\bar{\varepsilon}\sigma^3), \qquad (26)$$

we obtain

$$I(\bar{\varepsilon}) \cong (0.67)\int_0^\infty d\sigma\, \frac{e^{-\sigma^2/2}}{\sigma} e^{-(3.14)/(\bar{\varepsilon}\sigma^3)}. \qquad (27)$$

For small $\bar{\varepsilon}$, the σ integral may be carried out by the saddle point method, and the result is

$$I(\bar{\varepsilon}) = (0.47)\bar{\varepsilon}^{1/5} e^{-(2.04)/\bar{\varepsilon}^{2/5}}. \qquad (28)$$

In Fig. 2, this small $\bar{\varepsilon}$ asymptotic expression for $I(\bar{\varepsilon})$ is compared to the results of a numerical evaluation of $I(\bar{\varepsilon})$. The curve is a plot of the asymptotic expression, and the points are the result of numerical integration for various values of $\bar{\varepsilon}$. One can see that the agreement is good for small enough values of $\bar{\varepsilon}$.

The main point to note here is that the equilibration rate is larger than one might have guessed. Since the exchange of parallel and perpendicular energy for an isolated collision between two electrons is exponentially small in $1/\varepsilon$, one might have guessed that the equilibration rate would be exponentially small in $1/\bar{\varepsilon}$. However, the equilibration rate turns out to be exponentially small in $1/\bar{\varepsilon}^{2/5}$, and this distinction is important since $\bar{\varepsilon}^{2/5} \gg \bar{\varepsilon}$ for $\bar{\varepsilon} \ll 1$.

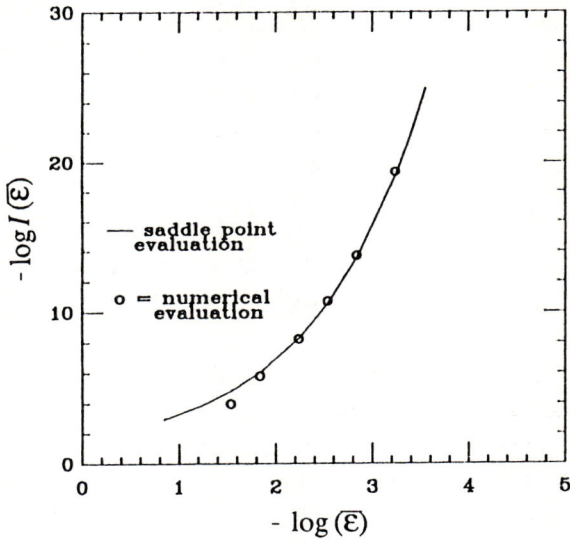

FIGURE 2

The saddle point evaluation of $I(\bar{\mathcal{E}})$ compared to the result of a numerical evaluation of $I(\bar{\mathcal{E}})$ for various values of $\bar{\mathcal{E}}$.

The $1/\bar{\mathcal{E}}^{2/5}$ dependence is determined by a competition between the velocity dependence of $\exp(-2\pi/\mathcal{E}) = \exp(-4\pi e^2 \Omega/m v_\parallel^3)$ and the velocity dependence of the distribution of relative velocities, $\exp(-\mu v_\parallel^2/2T_\parallel)$. Collisions characterized by large relative velocity are particularly effective at producing an exchange of parallel and perpendicular energy, but there are relatively few such collisions.

4. MOLECULAR DYNAMICS SIMULATION

This section presents the results of molecular dynamics simulations.[3] The dynamics for $N = 50$ point changes that interact electrostatically and move in a uniform magnetic field is followed numerically, and the equipartition rate is determined for various values of the plasma parameters (density, temperature, and field strength).

For computational convenience, it is useful to scale velocities by $\bar{v}_\parallel = \sqrt{T_\parallel/\mu}$, distances by $\bar{b} = 2e^2/\mu \bar{v}_\parallel^2$, and times by $\bar{b}/\bar{v}_\parallel$. With these units, the equations of motion take the form

$$\dot{\mathbf{x}}'_i = \mathbf{v}'_i, \quad \dot{\mathbf{v}}'_i = \frac{1}{\bar{\mathcal{E}}} \mathbf{v}'_i \times \hat{z} + \frac{1}{4} \sum_{i \neq j}^{N} \frac{\mathbf{x}'_i - \mathbf{x}'_j}{|\mathbf{x}'_i - \mathbf{x}'_j|^3}, \quad i = 1, \cdots, N, \quad (29)$$

where primes signify scaled variables. Because we need to evaluate the full Coulomb force term, the number of floating point operations associated with each reference to this set of equations scales roughly quadratically with the number N of particles in the system; this is the main obstacle to a simulation involving a realistically large number of particles. Nevertheless, a glimpse of the phenomena has been obtained with a simulation involving 50 interacting particles, performed on the SDSC CRAY X-MP.

As initial conditions, we take the 50 particles to be uniformly distributed spatially inside a box $-L/2 < (x,y,z) < +L/2$, of volume L^3, and with initial velocities picked from a bi-Maxwellian velocity distribution with $(T_\perp/T_\parallel)=0.2$. As the system evolves, the particles are confined in the z direction by specular reflections at the walls at $z = \pm L/2$. Confinement in the radial direction is ensured, since the total canonical angular momentum P_θ is a constant of the motion and provides a constraint on the allowed radial positions of the electrons.[6] Another constant of the motion is the total energy; we employ a high precision Bulirsch-Stoer ODE solver[7] and find during a typical run that the total energy is conserved to order $\Delta E/E \sim 10^{-7}$ and that the total canonical angular momentum is conserved to order $\Delta P_\theta/P_\theta \sim 10^{-5}$.

From a statistical (or macroscopic) perspective, the scaled system is characterized by two parameters: $\bar{\varepsilon}$ and L'. Here, we have in mind fixed $N=50$, fixed initial $T_\perp/T_\parallel=0.2$, and fixed initial $<v'^2_\parallel> = (T_\parallel/m)/(T_\parallel/\mu)=1/2$. The parameter $\bar{\varepsilon}$ is a measure of the magnetic field strength, and the parameter L' is a measure of density $n'=N/L'^3$. By decreasing L', we increase the collision frequency (v_{ee}), or equivalently, we increase the correlation strength ($\Gamma = e^2/aT_\parallel$, where $4\pi na^3/3 = 1$). Not all of the ($\bar{\varepsilon},\Gamma$) parameter plane is physically accessible. For a nonneutral electron plasma, the Brillouin limit[8] (i.e., $\omega_p < \Omega/\sqrt{2}$) specifies the maximum density that can be confined for a given magnetic field strength, and in terms of $\bar{\varepsilon}$ and Γ this limit takes the form $\bar{\varepsilon}\Gamma^{-3/2} < 1/\sqrt{12}$.

Some sample results are shown in Figs. 3(a) and 3(b), where $W_\perp/W$ and $W_\parallel/W$ are plotted as functions of time. Here, $W_\perp$ and $W_\parallel$ refer to the total perpendicular and parallel kinetic energies, respectively, and $W=W_\perp+W_\parallel$. The dashed lines mark the values predicted by the equipartition theorem: $W_\perp/W=2/3$ and $W_\parallel/W=1/3$. Figure 3(a) shows the results of a run with low magnetic field strength ($\bar{\varepsilon}=14$) and Fig. 3(b) shows the results of a run with large field strength ($\bar{\varepsilon}=0.14$) but the same correlation strength (i.e., $\Gamma=0.03$). One can see that the increased field suppresses the relaxation. To investigate the rate equation numerically [i.e., Eq. (28)], we examined the initial rate of temperature equilibrium for a bi-Maxwellian velocity distribution with $T_\perp=0.2T_\parallel$. In order to suppress statistical fluctuations without going to a large number of particles, we averaged the change in perpendicular and parallel kinetic energy over 20 different sets of initial conditions. These were advanced forward a time Δt short compared with the equipartition time, but long enough for a least-squares fit to the slope of the evolving $W_\perp(t)=NT_\perp(t)$. In Fig. 4, the analytic prediction $(dT_\perp/dt)[(T_\parallel-T_\perp)n\bar{b}^2\bar{v}_\parallel]^{-1}=I(\bar{\varepsilon})$ is compared to numerical values of $(\Delta T_\perp/\Delta t)[(T_\parallel-T_\perp)n\bar{b}^2\bar{v}_\parallel]^{-1}$ for various values of $\bar{\varepsilon}$ and Γ. One can see that the numerical values are insensitive to the value of Γ and follow the $I(\bar{\varepsilon})$ curve quite well, although the agreement is best in the small $\bar{\varepsilon}$ limit. This is to be expected, since $I(\bar{\varepsilon})$ was obtained in the small $\bar{\varepsilon}$ asymptotic limit.

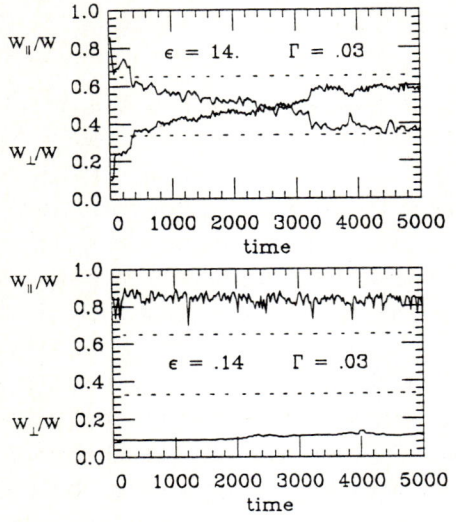
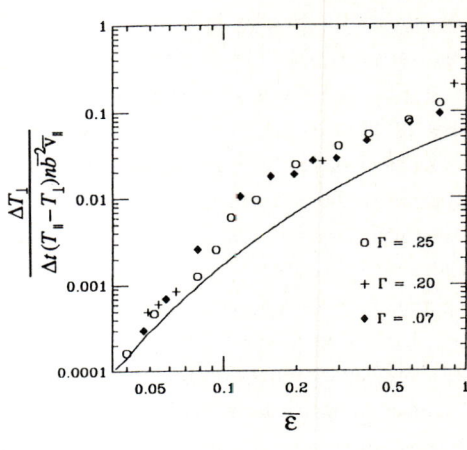

FIGURE 3
Evolution of $W_\perp/W$ and $W_\parallel/W$ (a) for weak magnetic field (i.e., $\bar{\varepsilon}=14$) and (b) for strong magnetic field (i.e., $\bar{\varepsilon}=0.14$).

FIGURE 4
Comparison of the function $I(\bar{\varepsilon})$ (solid curve) to values of $(\Delta T_\perp/\Delta t)[(T_\parallel - T_\perp)n\bar{b}^2\bar{v}_\parallel]^{-1}$ obtained from the simulations.

V. EXPERIMENT

The condition for strong magnetization can be written as $\Omega \gg \bar{v}/\bar{b}$, or equivalently, as $T^{3/2}(eV) \ll 10^{-7} B(G)$, where for simplicity we have set $T_\parallel = T_\perp = T$. One can see that strong magnetization requires low temperature as well as large magnetic field; even for B as large as 100 kG, the required temperature $[T \ll (.05)eV]$ is such that a neutral plasma would recombine. However, a pure electron plasma cannot recombine, since there are negligibly few ions in the confinement region. Recent experiments have succeeded in cooling a magnetically confined pure electron plasma to the cryogenic temperature range (where the plasma is strongly magnetized) and in measuring the equipartition rate as a function of plasma temperature and magnetic field strength.[4]

A schematic diagram of the confinement apparatus used in these experiments is shown in Fig. 5. A conducting cylinder is divided into several electrically isolated sections (only three of which are shown), and the whole apparatus is immersed in a magnetic field that is nearly axial and uniform in the region of the plasma. The plasma resides in the central grounded cylinder, with radial confinement provided by the axial magnetic field and axial confinement provided by electrostatic fields. The two end cylinders are biased sufficiently negative to provide this axial confinement. The plasma density for these experiments is of order 10^9 cm^{-3} and the initial plasma temperature is of order 10 eV. The magnetic field is provided by a superconducting coil, and field strengths up to 60 kG are used.

Immediately after the plasma is trapped, it begins to cool by cyclotron radiation, and the radiated energy is absorbed by the walls, which are maintained at 4°K. Each electron radiates according to the Larmor formula[9] (the plasma is optically thin), and this leads to an exponential decrease in plasma temperature with a time constant $\tau = 4 \times 10^8/B^2$ sec, where B is in Gauss. For example, for $B = 60$ kG, the time constant is .1 sec. This time is long compared to the equipartition time over the whole parameter range explored in the experiments, so $T_\parallel$ remains nearly equal to $T_\perp$ even though cyclotron radiation extracts only perpendicular kinetic energy.

The temperature $T_\parallel$ is measured by letting the plasma gradually escape to the collector [see Fig. 5]. As the potential on the cylindrical section in front of the collector is gradually increased (toward zero), electrons in the tail of the Maxwellian velocity distribution begin to make it over the potential barrier and reach the collector. The temperature is determined by measuring the current collected as a function of the potential and fitting to a Maxwellian. Of course, dumping the plasma in this manner is a destructive procedure, and the experiments rely on shot to shot reproducibility. By measuring the temperature as a function of time after injection, one can follow the radiative cooling of the plasma, and the measured temperature tracks the theoretical expectation down to about 50°K, which is very likely the present limit of the temperature diagnostic.

To determine the equipartition rate, a small oscillating component is added to the potential on one of the end cyclinders. This acts like an electrostatic piston that alternately compresses and expands the plasma in the axial direction. If the oscillation frequency is low compared to the equipartition rate, the compression and expansion cycle is a reversible process (a 3-D compression and expansion characterized by $c_p/c_v = (f+2)/f = 5/3$). Likewise, if the oscillation frequency is large compared to the equipartition rate, the cycle is a reversible process (a 1-D adiabatic compression and expansion characterized by $c_p/c_v = (f+2)/f = 3$). However, when the oscillation frequency is comparable to the equipartition rate, the process is not reversible. More work is done during the compression stroke than is given back during the expansion stroke, and the excess appears as plasma heat. One can easily show that the heat per cycle is a maximum for $\omega = 3\nu$, where ω is the angular frequency of the oscillation and ν is the equipartition rate [i.e., $dT_\perp/dt = \nu(T_\parallel - T_\perp)$]. To understand the factor of 3, note that in the absence of heating the definition of ν implies that $d(T_\perp - T_\parallel)/dt = 3\nu(T_\parallel - T_\perp)$.

In the experiments, the oscillation frequency is adjusted so that the heating per cycle is maximum. Also, the oscillation amplitude is adjusted so that the heating just balances the cooling due to radiation, and the plasma temperature remains constant. This procedure is then repeated for various values of the temperature and magnetic field strength, and the equipartition rate is measured as a function of these parameters. In Fig. 6, the points are measured values and the curves are theoretical predictions. The dashed curve applies to a weakly magnetized

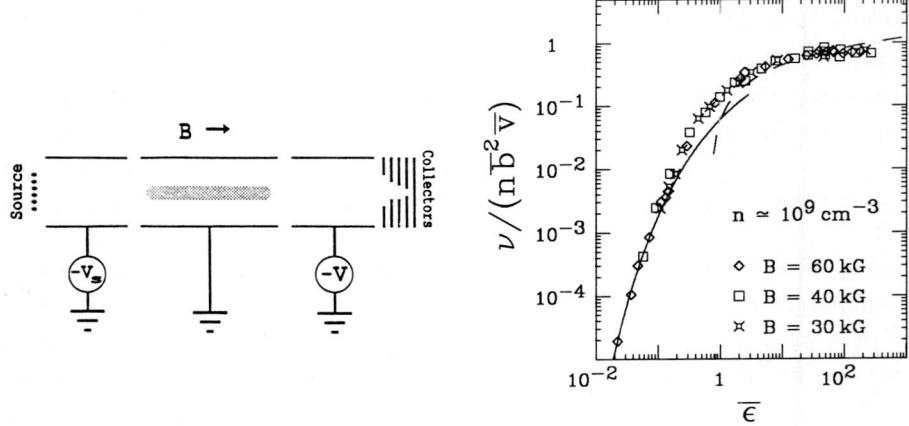

FIGURE 5
Schematic diagram of the confinement apparatus.

FIGURE 6
Comparison of measured equipartition rate to theory.

plasma ($\bar{\varepsilon} > 1$); it is the prediction of Ichimaru and Rosenbluth[10] as modified by Montgomery, Joyce and Turner's[11] prescription for the Coulomb logarithm [i.e., $\ln(\lambda_D/b) \to \ln(\bar{\varepsilon})$]. For strong magnetization ($\bar{\varepsilon} < 1$), the solid curve is the function $I(\bar{\varepsilon})$ given in Eq. (28). One can see that the measured rate drops dramatically for $\bar{\varepsilon} \ll 1$ in good agreement with theory.

REFERENCES
1) T.M. O'Neil, Phys. Fluids **26** (1983) 2128.
2) T.M. O'Neil and P.G. Hjorth, Phys. Fluids **28** (1985) 3241.
3) P.G. Hjorth and T.M. O'Neil, Phys. Fluids **30** (1987) 2613.
4) B. Beck, J. Fajans, and J.H. Malmberg, Bull. Am. Phys. Soc. **33** (1988) 2004.
5) G. Uhlenbeck and G. Ford, *Lectures in Statistical Mechanics* (AMS, Providence, Rhode Island, 1963) 118.
6) T.M. O'Neil, Phys. Fluids **23** (1980) 2216.
7) W.H. Press, B.P. Flannery, S.A. Teukolsky, and W.T. Vetterling, *Numerical Recipes* (Cambridge University Press, London, England, 1966).
8) R.C. Davidson, *Theory of Nonneutral Plasmas* (Benjamin, Reading, Massachusetts, 1974) 4.
9) J.D. Jackson, *Classical Electrodynamics* (Wiley, New York, 1975) 659.
10) S. Ichimaru and M.N. Rosenbluth, Phys. Fluids **13** (1970) 2778.
11) D. Montgomery, G. Joyce, and L. Turner, Phys. Fluids **17** (1974) 2201.

LONG-TIME TAILS OF TIME CORRELATION FUNCTIONS FOR AN IONIC MIXTURE IN A MAGNETIC FIELD AND THE VALIDITY OF MAGNETO-HYDRODYNAMICS

L.G. SUTTORP

Institute for Theoretical Physics, Valckenierstraat 65, 1018 XE Amsterdam, The Netherlands

Mode-coupling theory is used to determine the long-time behaviour of the Green-Kubo integrands for the heat conductivity and the diffusion coefficients of an ionic mixture in a magnetic field. It is shown that the presence of several species of particles with a different ratio of charges and masses is a prerequisite for the validity of dissipative magnetohydrodynamics.

1. INTRODUCTION

In recent years the collective modes for a classical one-component plasma in a magnetic field have been studied by using a projection operator formalism[1,2]. With the help of mode-coupling theory the long-time behaviour of the velocity autocorrelation function[3] and the heat-conductivity time correlation function (or Green-Kubo integrand for the heat conductivity)[4] have been determined. The long-time tail of the heat-conductivity time correlation function has been found to decay as $t^{-1/2}$, owing to a coupling of two 'gyro-plasmon' modes. The same slowly-decaying long-time tail has been obtained by using a method based on kinetic theory[5,6]. This slow decay implies that the static heat conductivity coefficient for a one-component plasma in a magnetic field is divergent.

To investigate whether this divergency is a peculiarity of the one-component plasma we have recently studied[7] the long-time tails of the heat-conductivity time correlation function for an ionic mixture in a uniform magnetic field. The mixture consists of particles of several species, all with charges of the same sign, that move in an inert uniform background of opposite charge.

2. MODE SPECTRUM OF AN IONIC MIXTURE

The collective modes of a multicomponent ionic mixture are linear combinations of the partial particle densities $n_\sigma(\mathbf{k})$, with σ labelling the s components, the total momentum density $\mathbf{g}(\mathbf{k})$ and the total energy density $\epsilon(\mathbf{k})$ in Fourier space. The time development of these quantities is governed by the Liouville operator L acting in phase space. Employing a projection operator $P = 1 - Q$ that projects an arbitrary function in phase space on the set of linear combinations of n_σ, $\mathbf{g}$ and ϵ, one finds the collective modes as the eigenvectors of the $(s+4) \times (s+4)$-dimensional matrix

$$\Omega_{ij}(\mathbf{k}, z) = -\frac{1}{V} < \bar{a}_i^*(\mathbf{k}) L a_j(\mathbf{k}) > + \frac{1}{V} < \bar{a}_i^*(\mathbf{k}) L Q \frac{1}{z + QLQ} Q L a_j(\mathbf{k}) >, \qquad (1)$$

for small values of the wave vector $\mathbf{k}$. The angular brackets denote an equilibrium ensemble average, with V the volume. Furthermore, $a_j(\mathbf{k})$are chosen from $\{n_\sigma, \mathbf{g}, \epsilon\}$, while $\bar{a}_i(\mathbf{k})$ are the adjoints satisfying the relation $V^{-1} < \bar{a}_i^*(\mathbf{k}) a_j(\mathbf{k}) > = \delta_{ij}$. The mode frequencies that govern the time evolution of the modes follow as the eigenvalues z of the frequency matrix. For an unmagnetized mixture the dispersion relation that determines the mode frequencies reads:

$$z^{s+2}\left[z^2 - zc(z) - \omega_\mathrm{p}^2\right] = 0, \qquad (2)$$

in lowest order of the wavenumber. Here $\omega_\mathrm{p} = q_v/m_v^{1/2}$ is the collective plasma frequency, with q_v the total charge density and m_v the total mass density. Furthermore, $c(z)$ is given by:

$$c(z) = \beta \lim_{\mathbf{k}\to 0} \frac{1}{V} < \frac{q_v^*(\mathbf{k})}{k} LQ \frac{1}{z+QLQ} QL \frac{q_v(\mathbf{k})}{k} >, \qquad (3)$$

with β the inverse temperature and $q_v(\mathbf{k}) = \sum_\sigma e_\sigma n_\sigma(\mathbf{k})$ the fluctuation of the charge density. As (2) shows the system supports $s+2$ modes with vanishing frequency in the long-wavelength limit. Two of these are viscous modes; the remaining s are mixed heat-diffusion modes. Furthermore, (2) possesses two complex solutions z_ρ, with $\rho = \pm 1$, which are generalized plasmon modes. Both modes are damped as a result of friction between the various components that oscillate out of phase. If all components have equal ratios of charge and mass this damping mechanism is absent. Indeed, it follows from (3) that $c(z) = 0$ in this case, since the electric current density is then proportional to the total momentum density. As a consequence the frequencies of the plasmon modes are then simply $\rho\omega_\mathrm{p}$, as in the one-component plasma. We shall call a mixture consisting of particles with equal charge-mass ratios a 'well-poised' mixture.

If the ionic mixture is magnetized the dispersion relation that determines the mode frequencies in order k^0 gets a more complicated form:

$$z^s\left\{(z^2 - zc - \omega_\mathrm{p}^2\hat{k}_\parallel^2)[z^2 - (\omega_B + ia)^2] - z^2[(\omega_\mathrm{p} + b)^2 - (b')^2]\hat{k}_\perp^2 - 2iz(\omega_B + ia)(\omega_\mathrm{p} + b)b'\hat{k}_\perp^2\right\} = 0, \qquad (4)$$

with $\omega_B = q_v B/(m_v c)$ the collective Larmor frequency. The symbols $\hat{k}_\parallel$ and $\hat{k}_\perp$ denote the components of the unit vector $\hat{\mathbf{k}} = \mathbf{k}/k$ parallel and perpendicular to the magnetic field, respectively. The coefficients a, b, b' and c depend on z. The last of these has been defined already in (3), while the other ones have a similar form, with one or both of the factors $q_v(\mathbf{k})/k$ replaced by $\mathbf{g}(\mathbf{k})$. In the magnetized case only s modes, viz. the mixed heat-diffusion modes, have a vanishing frequency in the long-wavelength limit. The viscous modes merge with the generalized plasmon modes so as to yield four generalized 'gyro-plasmon' modes. Their frequencies $z_{\lambda\rho}$, with $\lambda = \pm 1$ and $\rho = \pm 1$, are complex in general, so that these modes are damped. As in the unmagnetized case the dispersion relation becomes simpler if the mixture is 'well-poised'. In that case the coefficients a, b, b' and c all vanish, so that the gyro-plasmon mode frequencies are the solutions of the relation:

$$z^4 - (\omega_\mathrm{p}^2 + \omega_B^2)z^2 + \omega_\mathrm{p}^2\omega_B^2\hat{k}_\parallel^2 = 0. \qquad (5)$$

This relation has the same form as that valid for the one-component plasma in a magnetic field. Again the modes are no longer damped in the long-wavelength limit.

The modes are linear combinations of the particle densities, the total momentum density and the energy density. For the unmagnetized mixture the plasmon modes are:

$$a_\rho(\mathbf{k}) = \beta^{1/2}\frac{q_v(\mathbf{k})}{k} + \left(\frac{\beta}{m_v}\right)^{1/2}\frac{\omega_\mathrm{p}}{z_\rho}\hat{\mathbf{k}}\cdot\mathbf{g}(\mathbf{k}). \qquad (6)$$

It should be noted that $q_v(\mathbf{k})$ is divided by a factor k, so that its contribution seems to grow with decreasing values of the wavenumber. However, large-scale charge fluctuations are strongly suppressed owing to the long-range Coulomb forces.

The gyro-plasmon modes in a magnetized mixture also contain a term proportional to $q_v(\mathbf{k})/k$:

$$a_{\lambda\rho}(\mathbf{k}) = \beta^{1/2}\frac{q_v(\mathbf{k})}{k} + \left(\frac{\beta}{m_v}\right)^{1/2} \mathbf{v}_{\lambda\rho}(\hat{\mathbf{k}}) \cdot \mathbf{g}(\mathbf{k}). \qquad (7)$$

The vector $\mathbf{v}_{\lambda\rho}$ depends on the coefficients a, b and b', for $z = z_{\lambda\rho}$.

To determine the mode frequencies of the mixed heat-diffusion modes in order k^2 one may use perturbation theory with respect to the wavenumber. In this way one proves that these frequencies are proportional to the eigenvalues of the $(s \times s)$–dimensional matrix:

$$M_{ij}(\hat{\mathbf{k}}, z) = \sum_{n=1}^{s} N_{in} \lim_{k \to 0} \frac{1}{k^2 V} < \left[a_n^{(0)}(\mathbf{k})\right]^* LQ \frac{1}{z+L} QL a_j^{(0)}(\mathbf{k}) > . \qquad (8)$$

for $z \to i0$, provided this limit exists. The elements of the matrix N are trivial thermodynamic derivatives which need not be specified here. The basis set $a_i^{(0)}(\mathbf{k})$ is given by:

$$a_1^{(0)}(\mathbf{k}) = \epsilon(\mathbf{k}) - \frac{h_v}{q_v} q_v(\mathbf{k}) , \quad a_\sigma^{(0)}(\mathbf{k}) = n_\sigma(\mathbf{k}) - \frac{n_\sigma}{q_v} q_v(\mathbf{k}) , \qquad (9)$$

with $\sigma = 2, \cdots, s$ and h_v the enthalpy per unit of volume. The derivation of (8) for a magnetized mixture is rather complicated[7]. In fact, perturbation theory first yields an expression containing the resolvent of QLQ and a few supplementary terms depending on the coefficients $a(z)$, $b(z)$ and $b'(z)$. Upon introducing the resolvent of L one finds that these supplementary terms drop out.

3. LONG-TIME TAILS OF TIME CORRELATION FUNCTIONS

To assess whether the limit of (8) for $z \to i0$ exists one studies the asymptotic behaviour of the related time correlation functions:

$$F_{\alpha\beta}(\hat{\mathbf{k}}, t) = \lim_{k \to 0} \frac{1}{V} < \left[Q\hat{\mathbf{k}} \cdot \mathbf{j}_\alpha(\mathbf{k})\right]^* e^{iLt} Q\hat{\mathbf{k}} \cdot \mathbf{j}_\beta(\mathbf{k}) > . \qquad (10)$$

Here α en β take the values $1, \cdots, s$, with $\mathbf{j}_1 \equiv \mathbf{j}_\epsilon$ the energy-current density and $\mathbf{j}_\sigma \equiv \mathbf{g}_\sigma/m_\sigma$ the particle-current density of component $\sigma = 2, \cdots s$. The long-time behaviour of these correlation functions can be obtained by using mode-coupling theory. In fact, we may write for large t:

$$F_{\alpha\beta}(\hat{\mathbf{k}}, t) \simeq \lim_{k \to 0} \frac{1}{2V} \sum_{ij} \sum_{\mathbf{q}} A_{ij}^{\alpha}(\mathbf{k}, \mathbf{q}) \left[\overline{A}_{ij}^{\beta}(\mathbf{k}, \mathbf{q})\right]^* \exp\left\{-i\left[z_i(\mathbf{q}) + z_j(\mathbf{k} - \mathbf{q})\right]t\right\} , \qquad (11)$$

where the summations are extended over all collective modes and over all values of the wave vector $\mathbf{q}$ of these modes. The mode-coupling amplitudes $A_{ij}^{\alpha}(\mathbf{k}, \mathbf{q})$ are given by

$$A_{ij}^{\alpha}(\mathbf{k}, \mathbf{q}) = \frac{1}{V} < \left[Q\hat{\mathbf{k}} \cdot \mathbf{j}_\alpha(\mathbf{k})\right]^* a_i(\mathbf{q}) a_j(\mathbf{k} - \mathbf{q}) > . \qquad (12)$$

The mode-coupling amplitudes $\overline{A}_{ij}^{\beta}(\mathbf{k}, \mathbf{q})$ are defined analogously, with adjoint modes $\overline{a}_i$ and $\overline{a}_j$. Since the plasmon modes and the gyro-plasmon modes contain a term $q_v(\mathbf{k})/k$ the mode-coupling amplitudes for these modes can be divergent in the long-wavelength limit. Indeed, one finds:

$$\frac{1}{V} < \left[\hat{\mathbf{k}} \cdot \mathbf{j}_\epsilon(\mathbf{k})\right]^* \frac{q_v(\mathbf{q})}{q} g(\mathbf{k} - \mathbf{q}) > = \frac{q_v}{q\beta^2}(\hat{\mathbf{k}} - \hat{\mathbf{k}} \cdot \hat{\mathbf{q}}\hat{\mathbf{q}}) , \qquad (13)$$

for small values of the wavenumbers.

Slowly decaying contributions to the mode-coupling expression for the time correlation function $F_{\alpha\beta}(\hat{\mathbf{k}},t)$ arise if both modes i and j are undamped for small wavenumber. For an unmagnetized ionic mixture the generalized plasmon modes can therefore be excluded from the sum over the modes in (11). The contribution with the slowest decay arises from the coupling of the currents to a viscous mode and to a mixed heat-diffusion mode. Since the mode-coupling amplitudes for this coupling are of zeroth order in the wavenumber the resulting long-time behaviour of $F_{\alpha\beta}(\hat{\mathbf{k}},t)$ is proportional to $t^{-3/2}$, so that the Fourier transform $F_{\alpha\beta}(\hat{\mathbf{k}},z)$ is finite for $z \to i0$. Hence, the transport coefficients occurring in the frequencies of the heat-diffusion modes are all finite for a general unmagnetized ionic mixture. If the mixture is well-poised, however, the above reasoning is not conclusive, since in that case the plasmon modes are no longer damped for small wavenumbers. The coupling of the energy current to a viscous mode and a plasmon mode is characterized by a mode-coupling amplitude that diverges for small wavenumber, as follows from (13). As a consequence the time correlation function $F_{11}(\hat{\mathbf{k}},t)$ has a tail proportional to $t^{-1/2}\cos(\omega_\mathrm{p} t + \theta)$, with θ a phase factor. Owing to the oscillating factor its Fourier transform is still finite as z goes to $i0$.

If a magnetic field is present the picture changes. For a general mixture the dominant contributions to the tails of the time correlation functions stem from the coupling of the currents to the heat-diffusion modes. The resulting tails are found to be proportional to $t^{-5/2}$. However, for the well-poised mixture in a magnetic field the situation is completely different. The gyro-plasmon modes are no longer damped. The coupling of the energy current to two gyro-plasmon modes in such a way that the zeroth-order mode frequencies in the exponent of (11) compensate each other leads to a slowly decaying tail in $F_{11}(\hat{\mathbf{k}},t)$ proportional to $t^{-1/2}$, without an accompanying oscillating factor. Hence, the transport coefficients appearing in the frequencies of the heat-diffusion modes for a well-poised magnetized ionic mixture are divergent. The conclusion is that the presence of several species of particles with different charge-mass ratios is necessary for the validity of magnetohydrodynamics, at least if dissipative effects are to be included.

REFERENCES

1) L. G. Suttorp and A. J. Schoolderman, Physica 141A (1987) 1.

2) M. C. Marchetti, T. R. Kirkpatrick and J. R. Dorfman,
J. Stat. Phys. 46 (1987) 679, 49 (1987) 871.

3) L. G. Suttorp and A. J. Schoolderman, Physica 143A (1987) 494.

4) A. J. Schoolderman and L. G. Suttorp, Physica 144A (1987) 513.

5) A. J. Schoolderman and L. G. Suttorp, J. Stat. Phys. 53 (1988) 1237.

6) A. J. Schoolderman and L. G. Suttorp, Physica A 156(1989)795.

7) A. J. Schoolderman and L. G. Suttorp, in print.

Chapter VI:
Quantum Electron Liquids in Strong Coupling

DENSITY FUNCTIONAL THEORY OF SUPERCONDUCTORS REGARDED AS TWO-COMPONENT PLASMAS

Walter KOHN

Department of Physics, University of California Santa Barbara
Santa Barbara, California 93106, U.S.A.

The recent density functional formulation of superconductivity is discussed as a theory of a 2-component plasma. The two "fluids" are described by the total electron density, $n(r)$, and the pair-function, $\Delta(r)$. Connections with the following theories are discussed: Density functional theory (DFT) of a one-component plasma without and with an external magnetic field, at T=0 and T>0; simple two-component plasmas; the Bogoliubov Gennes equations; Ginsburg-Landau theory; and the phenomenological 2-fluid model of superconductivity. The DFT formulation may be especially appropriate for high T_c materials, because of their strong normal many body effects and short coherence lengths.

1. BASIC DENSITY FUNCTIONAL THEORY.

Density functional theory (DFT) was originally developed a system of N interacting, non-magnetic electrons —a one-component plasma— under the influence of a static external potential, $v(r)$, in its quantum mechanical ground state.[1] It was shown that a ground state density distribution, $n(r)$, was consistent with only <u>one</u> external potential, $v(r)$. Thus, in principle, $n(r)$ determined the Hamiltonian of the system and hence all the properties derivable from this Hamiltonian, such as the pair correlation function, $g(r_1,r_2)$, excited state energies, E_n, etc. The theory was formulated in terms of a functional for the ground state energy, in which the density was the independent variable,

$$E_{v(r)}[n(r)] \equiv \int v(r)n(r)dr + F[n(r)], \tag{1}$$

where $F[n(r)]$ is a universal functional of the density $n(r)$ (not explicitly dependent on $v(r)$). This functional attains its minimum value —which is the ground state energy— when $n(r)$ is the correct ground state density corresponding to the potential $v(r)$. The functional $F[n(r)]$ is, of course, not exactly known, but useful approximations exist.

Soon afterwards it was shown[2] that the variational principle, $\delta E_v[n(r)]/\delta n(r)=0$, could be cast into the form of (in principle) exact, self-consistent single particle equations, structually analogous to the approximate Hartree equations,

$$\left(-\frac{1}{2m}\nabla^2 + v_{eff}(r) - \varepsilon_j\right)\varphi_{j,\sigma}(r) = 0,$$

$$n(r) = \sum_{\sigma=+,-}\sum_{j=1}^{N} |\varphi_{j,\sigma}(r)|^2,$$

$$v_{eff}(r) = (r) + \int \frac{n(r')}{|r-r'|}dr' + \frac{\delta E_{xc}[n(r)]}{\delta n(r)}. \quad (2)$$

Here the exchange-correlation functional, $E_{xc}[n(r)]$, like $F[n(r)]$, is a universal functional of the density, $n(r)$. These so-called Kohn-Sham (KS) equations, with suitable approximations for the functional $E_{xc}[n(r)]$, have been the basis of the great majority of applications of DFT to ground states of atoms, molecules and especially perfect and imperfect solids.

2. GENERALIZATIONS

Subsequently the theory was generalized in many directions, including the following[3]: multicomponent plasmas, like the electron-hole droplets in Si and Ge; finite temperature ensembles; electrons with spin magnetism and —rather recently— orbital magnetism.[4] The generalization for superconductors,[5,6] reported here, builds on this earlier work.

In discussing the spin-magnetism of electrons in a magnetic field B(r), taken in the z-direction, we have only one kind of particle —electrons— but it is appropriate to regard the system as a <u>two</u>-component plasma, the two fluids being spin up and spin down electrons with densities $n_+(r)$ and $n_-(r)$, as equivalently the total density, $n(r) = n_+(r) + n_-(r)$, and the magnetization density, $m(r) = n_+(r) - n(r)$, The energy functional now takes the form

$$E_{v(r),B(r)}[n(r),m(r)] = \int v(r)n(r)dr - \int B(r)m(r)dr + F_{xc}[n(r),m(r)], \quad (3)$$

where F_{xc} is a universal functional of its arguments and the correct densities, $n(r)$ and $m(r)$ minimize $E_{v,B}[n,m]$ and yield the ground state energy. There is here a one-to-one correspondence between the two external "potentials" and the two densities,

$$\{v(r), B(r)\} \leftrightarrow \{n(r), m(r)\}. \quad (4)$$

In analogy with the KS-equations (3) for a one-component plasma one had obtained a set of coupled one–particle equations for spin up and spin-down orbitals, $\varphi_{j+}(r)$ and $\varphi_{j-}(r)$, involving two effective potentials, $v_{eff}(r)$ and $B_{eff}(r)$.

For finite temperatures, T, there exists a functional for the grand potential,

$$\Omega^\beta_{v(r)-\mu,B(r)}[n(r),m(r)] \equiv \int v(r)n(r)\, dr - \int B(r)m(r)\, dr$$
$$+ F_{xc}^\beta[n(r),m(r)], \tag{5}$$

where $\beta \equiv (kT)^{-1}$, μ is the chemical potential, and $F_{xc}^\beta[n(r),m(r)]$ is a universal functional of its arguments, depending parametrically on β. The functional (5) is minimized by the correct $n(r)$, $m(r)$, yielding the correct value of Ω. This variational principle can also be cast into a set of temperature-dependent KS equations, involving temperature-dependent effective potentials, $v_{eff}^\beta(r)$ and $B_{eff}^\beta(r)$.

We add one pertinent remark. A system, like solid Fe below its Curie temperature, with $B(r) \equiv 0$, which is spontaneously magnetized, can in principle, be exactly treated in two different ways: (a) As a one component plasma with the correspondence $v(r) \leftrightarrow n(r)$; or (b) as a two-component plasma in a magnetic field $B(r)$, with the correspondence of Eq.(4), in the limit $B(r) \to 0$. However, whereas, in approach (b) the basic physics of spontaneous magnetization can be represented by rather simple forms of $F_{xc}^\beta[n,m]$, much more sophisticated forms of $F_{xc}^\beta[n]$ are needed to obtain the same accuracy using approach (a). Thus (b) is greatly preferable.

3. SUPERCONDUCTORS

We turn now to superconductors. The technical details are described in two recent papers.[5,6] Here we shall emphasize the ideas.

First of all, we limit ourselves to purely electronic pairing mechanisms,[7] which are believed to be principally responsible for the superconductivity in the so-called high-T_c superconductors. (The inclusion of electron-phonon pairing mechanisms is certainly possible in principle, but I have at this time no reason to believe that DFT is a useful formalism for their discussion.) The formal structure of the DFT for superconductors is independent of the details of the pairing mechanism and depends only on such general characteristics as to whether the pairing is s-like singlet or p-like triplet etc. For definiteness we shall assume the former.

As is well known, for singlet superconductors, (in the absence of a magnetic field), not only does the density operator have a finite expectation value,

$$\left\langle \sum_{\sigma=+,-} \varphi_\sigma^+(r)\varphi_\sigma(r) \right\rangle \equiv n(r), \tag{6}$$

but also the pair destruction operator,

$$\langle \varphi_+(r)\, \varphi_-(r) \rangle \equiv \Delta(r); \tag{7}$$

$\Delta(r)$ is called the pair function. Its non-local generalization is

$$<\varphi_+(r)\varphi_-(r')> \equiv \Delta(r,r').[8] \tag{8}$$

The pair function, $\Delta(r)$ (or its non-local generalization $\Delta(r,r')$) is the hallmark of superconductivity, just as $m(r)$ is the hallmark of spontaneous magnetization. In analogy with the magnetic field, $B(r)$, we therefore introduce a singlet pairing field, $D(r)$, (or its non-local version, $D(r,r')$). Such a field does have physical reality in the proximity effect (a superconductor generates a physical pairing field in an adjacent normal metal). But here we introduce it as a formal device, like the infinitesimal $B(r)$ for spin magnets, and eventually let it approach zero.

$$H_{v,D} \equiv \int \varphi^+(r) [-\frac{\nabla^2}{2} + v(r) - \mu]\varphi(r) \, dr +$$
$$+ \frac{1}{2} \int \varphi^+(r) \varphi^+(r') \frac{1}{|r-r'|} \varphi(r')\varphi(r) \, drdr'$$
$$- \int [D^*(r) \varphi_+(r)\varphi(r) + h.c.] \, dr. \tag{9}$$

The two "fluids" in superconductors are thus described by $n(r)$ and $\Delta(r)$. The density functional formulation, for finite temperatures, leads to the grand potential functional

$$\Omega^\beta_{v,D}[n(r),\Delta(r)] \equiv \int n(r)v(r)dr - \int [D^*(r) \Delta(r) + h.c.] \, dr$$
$$+ F^\beta[n(r),\Delta(r)] , \tag{10}$$

where $F^\beta[n(r), \Delta(r)]$ is as universal functional of its arguments, <u>which incorporates all normal and superconducting many-body effects</u>. Minimization of Eq.(10) with respect to $n(r)$ and $\Delta(r)$ leads to their correct functional forms and the correct ground potential Ω^β. In the physical limit $D(r) \to 0$, but, below the critical temperature, $\Delta(r)$ remains finite.

Extensions to include spin magnetic and diamagnetic effects (such as the Meissner effect, current vortices etc) have been carried out. The superconducting extension of the KS equations (2) have the form of the Bogoliubov Gennes equations but include all normal and superconducting many body effects, in principle exactly. We have also obtained the connections with the Ginzburg-Landau equations.[9]

It remains to learn as much as possible about the universal functional $F^\beta[n(r), \Delta(r)]$, or about $F_{xc}^\beta[n(r),\Delta(r)]$, the analog of $E_{xc}[n(r)]$ in Eq.(2). Some progress can be made purely formally. Thus the smallness of $\Delta(r)$ compared to $n(r)$ leads to a truncated expansion up to $O(\Delta^2)$. The possible symmetry characteristics of $E_{xc}^\beta[n(r),\Delta(r)]$ can also be obtained from formal considerations. However, the specific functional form of E_{xc}^β will require use of detailed experiments, on an atomic scale, and/or a microscopic theory.

We believe that the above DFT may be especially pertinent to high-T_c superconductor. Their superconducting coherence lengths are only a few times a typical atomic spacing. Thus spatial inhomogeneities, on an atomic scale, are significant and DFT is intrinsically a theory of spatially inhomogeneous systems. Secondly, the high-T_c materials, in their normal state, are known to have strong many-body effects which obviously are also present in the superconducting state. DFT can, in principle, incorporate these.

Finally we remark that this theory may be regarded as the microscopically exact version of the phenomenological 2-fluid model of superconductivity.

Acknowledgements

This research was supported in part by the National Science Foundation through Grant No. DMR 97-03434 and the U.S. Office of Normal Research under contract No. N00014-84-K-0548.

REFERENCES
1) P. Hohenberg and W. Kohn, Phys. Rev. **136**, B864 (1964).
2) W. Kohn and L. J. Sham, Phys. Rev. **140**, A 1133 (1965).
3) For a review see W. Kohn and P. Vashishta, in *Theory of the Inhomogeneous Electron Ones,* edited by S. Lundquist and N. March (Plenum, New York, 1983), p.79
4) G. Vignale and M. Rasolt, Phys. Rev. Lett. **59,** 2360 (1987).
5) L. N. Oliveira et al., Phys. Rev. Lett., **60**, 2430 (1988).
6) W. Kohn et al., Journ de Physique, to appear Oct. 1989.
7) W. Kohn and J. M. Luttinger, Phys. Rev. Lett. **15**, 524 (1965), and much recent work on resonating valence bonds, spin-bags, etc.
8) The dependence on (r'-r), when Fourier transformed, describes the momentum dependence of Δ.
9) See, for example, P.G. de Gennes, Superconducivity of Metals and Alloys (Benjamin, New York, 1966)

GREEN'S FUNCTION AND DYNAMIC CORRELATIONS OF ELECTRONS IN METALS

Aiichiro NAKANO and Setsuo ICHIMARU

Department of Physics, University of Tokyo, Bunkyo-ku, Tokyo 113, Japan

A dynamic hypernetted-chain theory of quantum electron liquids is developed, in which a closed set of equations for Green's functions, self-energies, density- (and spin-) fluctuation excitations and dynamic local-field corrections are obtained through functional derivatives of time-dependent multiparticle correlation functions and response functions. Dynamic structure factor is evaluated in accord with the sum rules and asymptotic behavior for the dielectric function as well as for a vertex correction describing short range effects. Nearly exact Green's functions for valence electrons, obtained from a self-consistent solution to the resulting Dyson equations, predict the quasiparticle properties accurately. We show how such a Green's function formalism can account for the experimental photoemission bandwidths and electron mean free paths.

1. FUNCTIONAL DERIVATIVES

For the valence electrons in simple metals such as Al and Na, effects of crystal potentials are weak so that an electron liquid model is applicable for the description of quasiparticle properties.[1] It is a model in which the electrons are immersed in a uniform neutralizing backgrouned of positive charges. The electron liquids in metals are strongly coupled[2] in that the Coulomb coupling parameter, $r_s = (3/4\pi n)^{1/3} me^2/\hbar^2$, takes on a value greater than unity (typically, $2 < r_s < 6$). Here m and e denote the mass and electric charge of an electron, and n refers to the number density of the electrons.

We begin with defining the single particle Green's function (GF) as[3]

$$\mathcal{G}_\sigma(1,1') = -i\langle 0|T[\psi_{H\sigma}(1)\psi^\dagger_{H\sigma}(1')\mathcal{S}]|0\rangle / \langle 0|\mathcal{S}|0\rangle . \tag{1}$$

Here, $1 = (\vec{r}_1, t_1)$ denotes a four-dimensional coordinate, T is the time-ordering operator, $\psi^\dagger_{H\sigma}(1)$ and $\psi_{H\sigma}(1)$ are the creation and annihilation Heisenberg operators for an electron with spin σ, and $|0\rangle$ refers to the ground state. The scattering matrix, $\mathcal{S}$ is given by

$$\mathcal{S} = T\exp[-\frac{i}{\hbar}\sum_\sigma \int d1 \int d1' \psi^\dagger_{H\sigma}(1)\phi_\sigma(1,1')\psi_{H\sigma}(1')] , \tag{2}$$

where $\phi_\sigma(1,1')$ is an external field nonlocal in space and time.

The ν-body response functions $\chi^{(\nu)}_{\sigma\tau...\upsilon}(1,1';...;\nu,\nu')$ are then defined in terms of the functional derivatives of the GF with respect to $\phi_\mu(j,j')$:

$$\chi^{(\nu)}_{\sigma\tau...\upsilon}(1,1';...;\nu,\nu') = -i\frac{\delta^{\nu-1}}{\delta\phi_\upsilon(\nu,\nu')...\delta\phi_\tau(2,2')}\mathcal{G}_\sigma(1,1'). \tag{3}$$

In particular $\chi_{\sigma\tau}(1,2) = \chi_{\sigma\tau}^{(2)}(1^+,1;2^+,2)$ is the density response function between spin components σ and τ, where $1^+ = (\vec{r}_1, t_1+0)$ and 0 means a positive infinitesimal, $\chi(1,2) = \sum_{\sigma\tau}\chi_{\sigma\tau}(1,2)$ is the density response function, and $\zeta(1,2) = \sum_{\sigma\tau}(2\delta_{\sigma\tau}-1)\chi_{\sigma\tau}(1,2)$ is the spin response function.

Using equations of motion, we derive the following hierarchal equations[3] in paramagnetic cases:

$$\mathcal{G}(k) = [k_4 + \mu/\hbar - \hbar k^2/2m - \Sigma_\sigma(k)]^{-1}, \tag{4}$$

$$\Sigma_\sigma(k) = -\sum_\tau \int \frac{d^4q}{(2\pi)^4} v(q) \chi_{\sigma\tau}^{(2)}(k+q/2, q) \mathcal{G}_\sigma^{-1}(k), \tag{5}$$

$$\chi(k) = \chi_L(k)/\{1 - v(k)[1 - G(k)]\chi_L\}, \tag{6}$$

$$G(k) = -\frac{i\hbar}{2n}\sum_{\sigma\tau\upsilon} \int \frac{d^4q}{(2\pi)^4} \frac{\vec{k}\cdot\vec{q}}{|\vec{q}|^2} \chi_{\sigma\tau\upsilon}^{(3)}(k-q,q)\chi_{\upsilon\tau}^{-1}(k), \tag{7}$$

$$\zeta(k) = \chi_L(k)/[1 + v(k)J(k)\chi_L(k)], \tag{8}$$

$$J(k) = -\frac{i\hbar}{2n}\sum_{\sigma\tau\upsilon} \int \frac{d^4q}{(2\pi)^4} \frac{\vec{k}\cdot\vec{q}}{|\vec{q}|^2}(2\delta_{\sigma\tau}-1)\chi_{\sigma\tau\upsilon}^{(3)}(k-q,q)\chi_{\upsilon\tau}^{-1}(k), \tag{9}$$

where $k = (\vec{k}, k_4)$ is a four dimensional wavevector, μ is the chemical potential, $v(k) = 4\pi e^2/|\vec{k}|^2$ is the Coulomb potential, and $\chi_L(k)$ is the Lindhard polarizability.[1,4]

The functions $\mathcal{G}(k), \chi(k)$, and $\zeta(k)$ retain many-body effects in the self-energy $\Sigma(k)$, and the local-field corrections (LFC's) $G(k)$ and $J(k)$; the latter functions in turn depend on the two- and three-body response functions. If these response functions are expressed in terms of $\mathcal{G}, \chi$, and ζ, truncation is completed; we thus obtain a closed set of equations.

2. CONVOLUTION APPROXIMATION AND DYNAMIC HYPERNETTED-CHAIN SCHEME

For analyses of the multiparticle response functions, we find it instructive to consider ν-body correlation potentials[3] defined as the functional derivatives of Σ with respect to $\mathcal{G}$:

$$\Xi_{\sigma\tau...\upsilon}^{(\nu)}(1,1';...;\nu,\nu') = \frac{\delta^{\nu-1}}{\delta\mathcal{G}_\upsilon(\nu,\nu')\cdots\delta\mathcal{G}_\tau(2,2')}\Sigma_\sigma(1,1'). \tag{10}$$

We then introduce a local approximation,[3] such that

$$\Xi_{\sigma\tau...\upsilon}^{(\nu)}(1,1';...;\nu,\nu') = \tilde{\Xi}_{\sigma\tau...\upsilon}^{(\nu)}(1,...,\nu)\delta(1,1')\cdots\delta(\nu,\nu'). \tag{11}$$

The short-rangedness of the correlation potentials as compared with the long-rangedness of the bare Coulomb interaction provides a justification for the local approximation (11). In Eq. (11), $\delta(1,1')$ refers to a four-dimentional δ function. In the local approximation, two- and three-body response functions can be expressed as[3]

$$\chi^{(2)}(p,k) = -i\hbar^{-1}\sum_\sigma \mathcal{G}_\sigma(p-k/2)\mathcal{G}_\sigma(p+k/2)\Gamma(k)/\epsilon(k), \tag{12}$$

$$\chi_{\sigma\tau\upsilon}^{(3)}(k,q) = \sum_\eta \chi_{0\eta}^{(3)}(k,q)\epsilon_{\sigma\eta}^{-1}(k)\epsilon_{\tau\eta}^{-1}(q)\epsilon_{\eta\upsilon}^{-1}(k+q)$$

$$- \hbar\sum_{\eta\theta\iota}\tilde{\Xi}_{\eta\theta\iota}^{(3)}(k,q)\chi_{\sigma\eta}(k)\chi_{\tau\theta}(q)\chi_{\iota\upsilon}(k+q), \tag{13}$$

where $\epsilon(k)$ is the dielectric function, and $\Gamma(k)$ is the vertex function defined as

$$\Gamma(k) = \chi_0^{-1}(k)[v(k) + \chi^{-1}(k)]^{-1}. \qquad (14)$$

Here we define

$$\chi_0(k) = \sum_\sigma \chi_{0\sigma}(k),$$

$$\chi_{0\sigma}(k) = -i\hbar^{-1} \int \frac{d^4p}{(2\pi)^4} \mathcal{G}_\sigma(p - k/2)\mathcal{G}_\sigma(p + k/2),$$

$$\tilde{\epsilon}_{\sigma\tau}^{-1}(k) = \chi_{\sigma\tau}(k)/\chi_{0\sigma}(k),$$

$$\chi_{0\sigma}^{(3)}(k) = -i\hbar^{-2} \int \frac{d^4p}{(2\pi)^4} [\mathcal{G}_\sigma(p)\mathcal{G}_\sigma(p + k)\mathcal{G}_\sigma(p - k) + (k \leftrightarrow q)].$$

Substitution of Eq. (12) in Eq. (5) yields

$$\Sigma_\sigma(k) = \frac{i}{\hbar} \int \frac{d^4q}{(2\pi)^4} \frac{v(q)}{\epsilon(q)} \Gamma(q)\mathcal{G}_\sigma(k + q). \qquad (15)$$

If we ignore the last term of Eq. (13) and substitute the remainder in Eqs. (7) and (9), we obtain[3]

$$G(k) = -\frac{i\hbar}{n\chi_0(k)} \int \frac{d^4q}{(2\pi)^4} \frac{\vec{k} \cdot \vec{q}}{|\vec{q}|^2} \chi_0^{(3)}(k - q, q)\tilde{\epsilon}^{-1}(k - q)\tilde{\epsilon}^{-1}(q), \qquad (16)$$

$$J(k) = -\frac{i\hbar}{n\chi_0(k)} \int \frac{d^4q}{(2\pi)^4} \frac{\vec{k} \cdot \vec{q}}{|\vec{q}|^2} \chi_0^{(3)}(k - q, q)\xi^{-1}(k - q)\tilde{\epsilon}^{-1}(q), \qquad (17)$$

where $\chi_0^{(3)}(k, q) = \sum_\sigma \chi_{0\sigma}^{(3)}(k, q)$, $\tilde{\epsilon}(k)^{-1} = \Gamma(k)/\epsilon(k) = \chi(k)/\chi_0(k)$, and $\xi^{-1}(k) = \zeta(k)/\chi_0(k)$. Hierarchies in the many-body correlations and response functions are thus truncated at the second stage involving density- and spin-fluctuations. Consequently, Eqs. (4), (6), (8), and (15)–(17) constitute a closed set of equations for $\mathcal{G}$, χ, and ζ.

The physical contents in the resulting set of equations have been elucidated.[3] First the neglect of the last term in Eq. (13) reduces in the classical, static limit to the *convolution approximation* to the three-body correlation function. Significance of the convolution approximation for the long-ranged Coulombic systems and its relation to the hypernetted chain approximation have been well investigated;[5,6] it ensures perfect screening and thereby maintains the short-rangedness of the correlation potentials. We may thus regard Eqs. (16) and (17) as the LFC in the dynamic hypernetted-chain approximation. Second, equations for $\mathcal{G}$ reduce in the classical limit ($\hbar \to 0$) to the Fokker-Plank equations; they are consistent with the notion of single particles scattered by field fluctuations. Finally, Eq. (16) reduces in the weak-coupling limit ($r_s \to 0$) to the LFC derived by Toigo and Woodruff[7] by the analysis of lowest-order proper-polarization diagrams; hence, it correctly describes the exchange effect in the weak-coupling limit.

Formalisms analogous to those described above may be applicable to superconductors as well: Starting with the Luttinger-Ward variational principle[8] extended to superconducting states, we have derived[9] a thermodynamic-potential functional $\Omega[\Delta]$ of the gap function $\Delta(\vec{r}_1, \vec{r}_1') = \langle \psi_\uparrow(\vec{r}_1)\psi_\downarrow(\vec{r}_1')\rangle$, the expansion coefficients of which consist solely of the normal-state GF's and the direct functions

$$c^{(2\nu)}(1,1';2,2';...;2\nu-1,2\nu'-1;2\nu,2\nu')$$
$$= \frac{\delta^{2\nu}\Phi}{\delta\mathcal{F}^\dagger(1,1')\delta\mathcal{F}(2,2')\cdots\delta\mathcal{F}^\dagger(2\nu-1,2\nu'-1)\delta\mathcal{F}(2\nu,2\nu')}, \quad (18)$$

where Φ is the interaction part of the thermodynamic potential[8] and $\mathcal{F}(1,1')$ and $\mathcal{F}^\dagger(1,1')$ are the usual anormalous GF's.[10] In the local approximation, $c^{(2)}$ may be expressed in terms of the normal-state GF and the density- and spin-response functions alone, so that the knowledge of the latter functions is suffiecient for the expansion of Ω up to quadratic terms in Δ.

3. SUM RULES AND ASYMPTOTIC BEHAVIOR FOR THE DIELECTRIC FUNCTION AND VERTEX CORRECTION

Dyson equation (4) contains the self-energy as the interaction part of GF. For its solution we hereafter normalize the wavenumber and energy in units of $k_F = (3\pi^2 n)^{1/3}$ and $E_F = \hbar^2 k_F^2/2m$.

For an accuracy of formulation, we regard it essential that the following exact sum rules and boundary conditions on the dielectric functions[1,2] are properly taken into consideration:

(i) Compressibility sum rule in the static long-wavelength limit.
(ii) Frequency-moment sum rules in the high-frequency limit.
(iii) Kimball relation in the short-wavelength limit.[11]
(iv) Niklasson condition[12]

$$\lim_{k\to\infty} \frac{1}{\epsilon(\vec{k},\omega)} = 1 - v(k)\chi_L(\vec{k},\omega) + \frac{2}{3}[1-g(0)][v(k)\chi_L(\vec{k},\omega)]^2, \quad (19)$$

where $g(0)$ is the value of the radial distribution function at the origin. Special attention should be paid to the classical limit of Eq. (19): it has been shown[13] that the large k limit and the small $\hbar$ limit are not interchangeable. If we take the classical limit first, the coefficient 2/3 on the right-hand side of Eq. (19) must be replaced by unity.

(v) Integral of Im $\epsilon^{-1}(\vec{k},\omega)$ over wavenumber and frequency is proportional to the interaction enrgy.
(vi) The plasmon damping in the long wavelength limit is related to the strength of multipair exitation in this domain.

We present here an analytic expression for $\epsilon(\vec{k},\omega)$ which satisfies all of the conditions cited above, on the basis of a microscopic theory by Utsumi and Ichimaru.[14] The explicit formula in ref. 14 contains long- and short-time relaxation functions, $\tau(k)$ and $\Omega(k)$, the static LFC

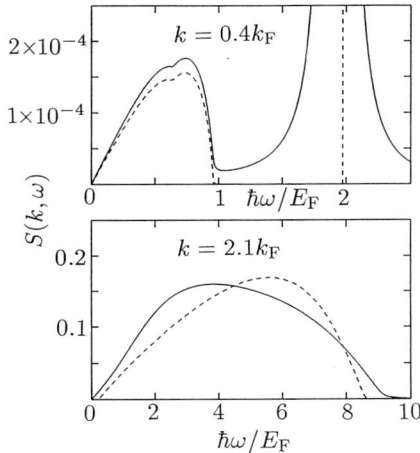
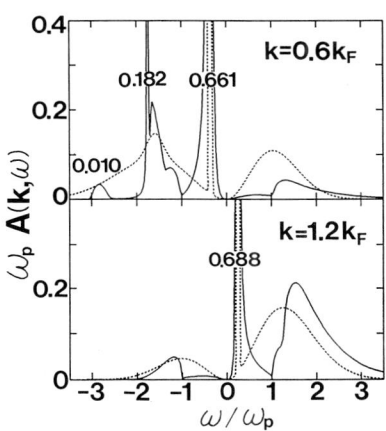

FIGURE 1
Dynamic structure factor for valence electrons at $r_s = 3.93$ appropriate for Na. The solid curves are the present parametrization; the dashed curves, in the random-phase approximation.

FIGURE 2
Parametrized values (dashed curves) of $A(k,\omega)$ and the corresponding imaginary parts (solid curves) of Eq. (4) for valence electrons in Na. The numbers on the figure denote the relative strengths under the solid curves.

$G(k)$, and the correlation function $I(k)$. Here, $\Omega(k)$ is written as $\Omega_0/[1+(k/2)^2]$, $G(k)$ is given by the Ichimaru-Utsumi formula[15] with $1 - g(0)$ multiplied by the factor $2/3 + (\pi/2)^{1/2}/6$, and $I(k)$ is evaluated by the use of the analytic formula[15] for the static LFC. Then, $\tau(k)$ is calculated from these functions through Eq. (3.19) of ref. 14. In the calculation, we use the Monte Carlo (MC) values[16] for the correlation energy, Yasuhara's formula[17] for $g(0)$, and the plasmon damping calculated by Utsumi and Ichimaru.[18] Figure 1 depicts the dynamic structure factor $S(k,\omega)$ so calculated for $r_s = 3.93$ appropriate to Na.

There is another important sum rule concerning the short-range scattering. The two-particle response function must satisfy[19,20] the following relation:

$$\lim_{k\to\infty} \frac{3\pi k^4}{8\lambda r_s}[1 + 3\pi \int_0^\infty d\omega \mathrm{Im} \int\int \frac{d^3 p\, dx}{(2\pi)^4} \chi^{(2)}(\vec{p}, x; \vec{k}, \omega)] = g(0), \qquad (20)$$

where $\lambda = (4/9\pi)^{1/3}$. This sum rule imposes a constraint on the vertex function through Eq. (12).

In the well known GW approximation,[21] Γ is set equal to unity. It then follows that right-hand side of Eq. (20) becomes identically unity. Since $g(0) \leq 1/2$, the strong repulsive correlations in short-range scattering events are not properly taken into account in such a GW approximation. Equation (12) with Γ given by Eq. (14), on the other hand, satisfies[19] Eq. (20). Recently, Mahan and Sernerius[22] adopted, in their calculation of the self-energy, another vertex correction $\Gamma(\vec{k},\omega) = [1 + v(k)G(k)\chi_L(\vec{k},\omega)]^{-1}$, where $G(k)$ is the static approximation for the LFC. This expression likewise satisfies the sum rule (20).

4. NEARLY EXACT GREEN'S FUNCTIONS FOR VALENCE ELECTRONS

Many of the existing theories on the single-particle Green's function are classified by the ways that the Dyson equations (4) and (5) are solved. Hedin[21] in his pioneering work calculated the self-energy in a GW approximation (i.e., $\Gamma = 1$) with ϵ in the random-phase approximation[1] (RPA) and with the free-fermion Green's function: it may thus be called a lowest-order perturbative (LP) solution. In recent angle-resolved photoemission experiments,[23,24] however, bandwidth of valence electrons much narrower than those predicted in these solutions were observed. Static local-field effects beyond the RPA were considered by Northrup et al.[25,26] and by Lyo and Plummer[24] in GW calculations of the self-energy; these authors were thereby able to account for the observed large reduction in the bandwidths. Mahan and Sernelius[22] included both the static local-field effect and the vertex correction, within confines of LP solutions. The resulting bandwidths were nearly identical to the LP solutions[21] in GW approximation with the RPA ϵ. An inclusion of Γ in the calculation of Σ in Eq. 15 shall be called a GWΓ scheme.

Rietschel and Sham[27], on the other hand, obtained a self-consistent (SC) iterative solution to Eqs. (4) and (15) for $\mathcal{G}$ in a GW approximation where the RPA was retained for ϵ. Their results implied a widening of the bandwidth at metallic densities, contrary to experimental indications. The present authors[19,20] took account of the static and dynamic local-field effects in ϵ and a vertex correction in the calculation of Σ. The resulting GWΓ equations were then solved self-consistently. The dielectric function is expressed in terms of the dynamic structure factor $S(\vec{k}, \omega)$ as[1,2]

$$\frac{1}{\epsilon(\vec{k},\omega)} = 1 + \frac{v(k)}{3\pi^2}\int_0^\infty dx \frac{S(\vec{k},x) - S(\vec{k},-x)}{\omega - x + i0\mathrm{sgn}(\omega)}.$$

The vertex correction is accounted for in a simplified form[19,20]

$$\Gamma(\vec{k},\omega) = \epsilon(\vec{k},\omega) + \frac{k^2 g(0) + k_s^2}{k^2 + k_s^2}[1 - \epsilon(\vec{k},\omega)], \quad (22)$$

which satisfies both the long- and short-wavelength boundary conditions including Eq. (20).

We solved Eqs. (4) and (15) for $\mathcal{G}(\vec{k},\omega)$ in terms of its spectral function $A(\vec{k},\omega)$, which was expressed as a summation of three distinct contributions:[19] the quasiparticle peak $A_{\mathrm{qp}}(\vec{k},\omega)$; resonant states between plasmons and holes, $A_{\mathrm{ph}}(\vec{k},\omega)$; and the background $A_{\mathrm{bg}}(\vec{k},\omega)$. The special shapes and strength of those excitations were characterized by ten parameters, which were determined by an iteration method. The iteration continued until the relative change in each of the ten parameters between two successive steps fell below 10^{-3}. Figure 2 exhibits the parametrized solutions (the dashed curves) thus obtained and $A(\vec{k},\omega)$ (the solid curves) calculated from the imaginary parts of Eq. (4) for $r_s = 3.93$, where ω_p is the plasma frequency.

In Fig. 3, we compare the values of the exchange correlation part of the chemical potential $\mu_{\mathrm{xc}} = \mu - 1$, in the present theory with those μ_{MC} in the MC calculations.[16] In LP solutions to GW equations, as we observed in refs. 24–28, the local-field effect acts to enhance the scattering rates of quasiparticles, resulting in an increase of $|\mu_{xc}|$ over those predicted with the RPA.[2] This overestimate is now corrected by the inclusion of the vertex correction in the SC solutions to the GWΓ equations; as a consequence we observe a good agreement with MC values ($\leq 2\%$ for $r_s \leq 10$). The LP solutions in the GWΓ scheme by Mahan and Sernerius[22] (solid squares) also achieved a good agreement with the MC values. However, their results become worse for the larger values of r_s, because of the lowest-order perturbation approximation. Both the vertex correction and the self-consistency are indispensable in a correct description of quasiparticle properties.

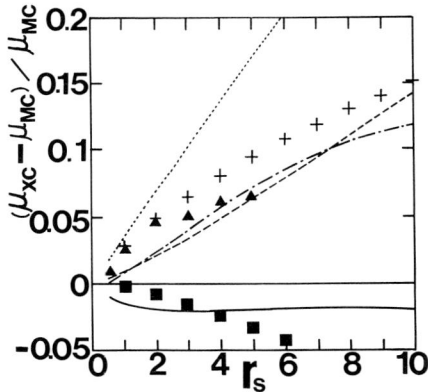

FIGURE 3
The exchange-correlation chemical potential μ_{xc} as a function of r_s; μ_{MC} denotes the MC values of μ_{xc} in ref. 16. The solid curve is the SC calculation in the GWΓ scheme; the dot-dashed curve, SC in the GW scheme; the dashed curve, LP calculation in the GWΓ scheme; the dotted curve, LP in the GW scheme. The crosses denote the values in ref. 21; the filled triangles, ref. 27; the filled squares, ref. 22.

Figure 4 shows the values of the renormalization constant, $Z(1) = [1 - \partial\Sigma(1,0)/\partial\omega]^{-1}$, at the Fermi surface. The open circles represent the values of $Z(1)$ calculated by the variational method[28] based on the Fermi hypernetted-chain (FHNC) approximation. The present solution (solid curve) agrees well with the FHNC results, especially at large r_s values. The calculations by Northrup et al.[26] (open triangles) are almost identical to the LP solution to GW equations, and overestimate the correlation effects.

In Fig. 5, we plot the values of the effective-mass ratio, $m_*/m = [1 - \partial\Sigma(1,0)/\partial\omega]/[1 + (1/2)\partial\Sigma(1,0)/\partial k]$, calculated in various theoretical schemes. The present theory (the solid curve) predicts a monotonic decrease in m_* as a function of r_s, in a way similar to recent calculations.[27,29]

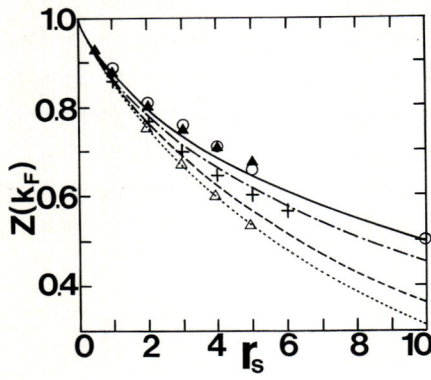 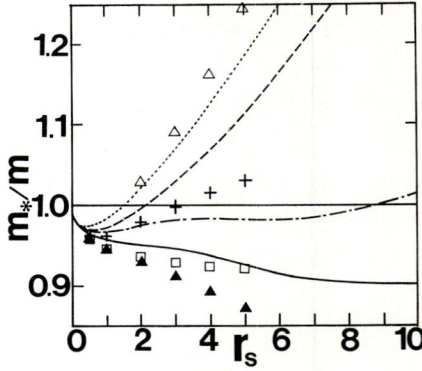

FIGURE 4
The normalization constant at the Fermi surface as a function of r_s. The open circles denote the values in ref. 28; the open squares, ref. 26; other symbols correspond to the cases in ref. 3.

FIGURE 5
The effective-mass ratio as a function of r_s. The open squares denote the values in ref. 29; other symbols correspond to the cases in Figs. 3 and 4.

5. PHOTOEMISSION BANDWIDTHS AND ELECTRON MEAN FREE PATHS

The many-body effects on the valence-band structures are represented in the self-energy correction, $\Delta\mathrm{Re}\Sigma(k) = E_{\mathrm{qp}}(k) - k^2 + 1$ where $E_{\mathrm{qp}}(k)$ denotes the peak position of $A_{\mathrm{qp}}(\vec{k},\omega)$. In Figs. 6(a) and 6(b), we depict the values (the dashed curves) of $\Delta\mathrm{Re}\Sigma(k)$ and $\Delta\mathrm{Re}\Sigma(0)$ in the present theory together with the corresponding photoemission data (the open circles[24] and the open square[30]). The two sets of values differ diametrically, in that the experimental data indicate a substantial narrowing of the bandwidths while the computed results imply a widening. The bandwidth narrowing has been observed universally in simple metals [cf. Fig. 6(b)], while its theoretical widening is related closely to the decrease in m_*/m (cf. Fig. 5).

This discord may be resolved with the aid of a relaxation-time argument in the following way:[19,20] A typical photoelectron involved in the study of the valence band has an energy nearly 18eV above the Fermi surface with a lifetime of $\tau \sim 2.5 \times 10^{-16}$s. The characteristic time for plasmons to respond to and equilibrate with a valence electron is estimated approximately as the plasma time, $2\pi/\omega_p \sim 6.9 \times 10^{-16}$s, and that for electron-hole pair is the Fermi time $2\pi/E_F \sim 1.3\times 10^{-15}$s, either of which is greater significantly than τ; hence, little time is available for establishment of $A_{\mathrm{ph}}(\vec{k},\omega)$ and $A_{\mathrm{bg}}(\vec{k},\omega)$ in photoemission experiments. The relaxation-time argument mentioned above may lead to consideraion of a nonequilibrium GF in which $A_{\mathrm{ph}}(\vec{k},\omega)$ and $A_{\mathrm{bg}}(\vec{k},\omega)$ are partially suppressed. Here, we study a limiting case where the nonequilibrium GF consists solely of $A_{\mathrm{qp}}(\vec{k},\omega)$. The SC solution to GWΓ equations now predicts a narrowing of the bandwidths to an extent observed experimentally, as the solid curves in Figs. 6(a) and 6(b) demonstrate.

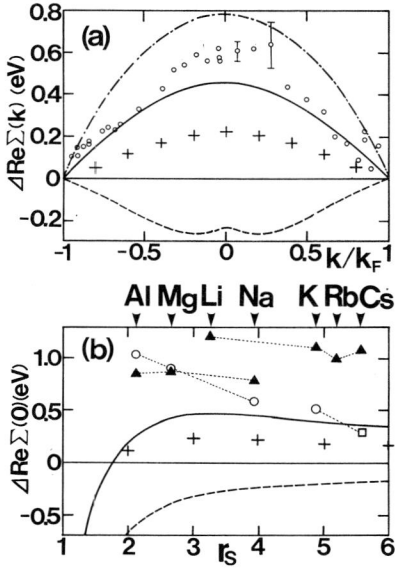

FIGURE 6
Comparison of the experimetal and theoretical self-energy correction for Na (a) and bandwidth correction for simple metals (b). The dashed curves refer to the SC equilibrium calculations in the GWΓ scheme; the solid curves, SC nonequilibrium in the GWΓ scheme; the dot-dashed curve (a) and the filled triangles (b), SC nonequilibrium with the core-electron effects. The crosses are from ref. 21. The open circles (ref. 24) and the open squares (ref. 30) represent the experimental data.

Northtrup et al.[25,26] and Lyo and Plummer[24] accounted for the observed reduction in the bandwidths, by using a GF consisting solely of $A_{\rm qp}(\vec{k},\omega)$ without a justification. As we remarked earlier, however, such an ansatz for the GF, when used in equilibrium situations, leads to predictions contradictory to the known quasi-particle properties.

Finally we take into account the core-electron effects[19,20] by replacing Γ/ϵ by

$$\tilde{\epsilon}_c^{-1}(\vec{k},\omega) = \Gamma(\vec{k},\omega)/\epsilon_c\epsilon(\vec{k},\omega) + [\omega_{\rm pc}^2/2\omega_c(k)]\{[\omega - \omega_c(k) + i0]^{-1} - [\omega + \omega_c(k) - i0]^{-1}\}, \quad (23)$$

where ϵ_c, $\omega_{\rm pc}$, and $\omega_c(k) = \omega_c + k^2$ are determined by the use of the observed plasmon energy[31] and the f-sum rule including core electrons. The results of the nonequilibrium GF calculations are described by the dot-dashed curve in Fig. 6(a) and by the solid triangles in Fig. 6(b).

The mean free paths of an electron with energy $E_{\rm qp}(k)$ and momentum k are calculated as $k/|{\rm Im}\Sigma(k, E_{\rm qp}(k))|$. In Fig. 7, we plot the results of equilibrium GF calculation with (solid curves) and without (dashed curves) core electron effects for Na and Al and compare them with the experimental data. Inclusion of the core-electron effects clearly brings about a significantly improved account of experiments,[32-34] to the extent of predicting and explaining kinks at $E \sim 50$eV (130eV) for Na (Al) associated with an onset of core-electron excitations. Predicted also are separate kinks at $E \sim 6$eV (20eV) arising from the plasmon excitations.

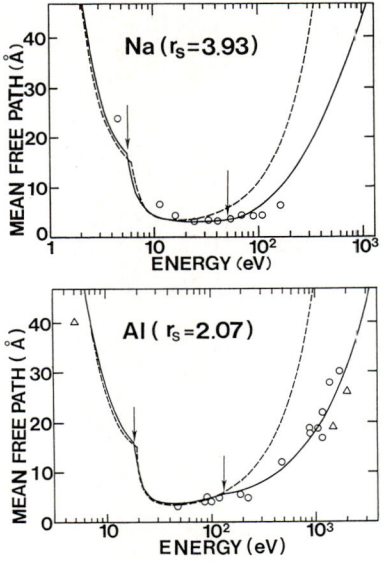

FIGURE 7
Comparison of the experimental and theoretical mean-free paths for Na and Al. SC calculations in the GWΓ scheme with (solid curves) and without (dashed curves) core-electron effects are plotted. The open circles (ref. 33) and open triangles (ref. 32) for Al and the open circles (ref. 34) for Na are the experimental values. The arrows indicate the kink positions mentioned in the text.

REFERENCES

1) D. Pines and P. Nozières, *The Theory of Quantum Liquids* (Benjamin, N.Y., 1964).

2) S. Ichimaru, Rev. Mod. Phys. **54** (1982) 1017.

3) A. Nakano and S. Ichimaru, Phys. Rev. B **39** (1989) 4930; further references may also be found here.

4) J. Lindhard, K. Dan. Vidensk. Selsk. Mat.-Fys. Medd. **28**(8) (1954) 1.

5) H. Iyetomi, Prog. Theor. Phys. **71** (1984) 427.

6) S. Ichimaru, H. Iyetomi, and S. Tanaka, Phys. Rep. **149** (1987) 91.

7) F. Toigo and T. O. Woodruff, Phys. Rev. B **2** (1970) 3958.

8) J. M. Luttinger and J. C. Ward, Phys. Rev. **118** (1960) 1417; see also G. Baym, Phys. Rev. **127** (1962) 1319.

9) A. Nakano, Doctoral thesis, University of Tokyo, 1989 (unpublished).

10) A. L. Fetter and J. D. Walecka, *Quantum Theory of Many-Particle Systems* (McGraw-Hill, N.Y., 1971).

11) J. C. Kimball, Phys. Rev. A **7** (1973) 1648.

12) G. Niklasson, Phys. Rev. B **10** (1974) 3052.

13) A. Nakano and S. Ichimaru, Phys. Lett. **136** (1989) 227.

14) K. Utsumi and S. Ichimaru, Phys. Rev. B **22** (1980) 1522.

15) S. Ichimaru and K. Utsumi, Phys. Rev. B **24** (1981) 7385.

16) D. M. Ceperley and B. J. Alder, Phys. Rev. Lett. **45** (1980) 566; for parametrization, see S. H. Vosko, L. Wilk, and M. Nusair, Can. J. Phys. **58** (1980) 1200.

17) H. Yasuhara, Solid State Commun. **11** (1972) 1481.

18) K. Utsumi and S. Ichimaru, Phys. Rev. B **23** (1981) 3291.

19) A. Nakano and S. Ichimaru, Phys. Rev. B **39** (1989) 4938.

20) A. Nakano and S. Ichimaru, Solid State Commun. **70** (1989) 789.

21) L. Hedin, Phys. Rev. **139** (1965) A796.

22) G. D. Mahan and B. E. Sernerius, Phys. Rev. Lett. **62** (1989) 2718.

23) E. Jensen and E. W. Plummer, Phys. Rev. Lett. **55** (1985) 1912.

24) I.-W. Lyo and E. W. Plummer, Phys. Rev. Lett. **60** (1988) 1558.

25) J. E. Northrup, M. S. Hybertsen, and S. G. Louie, Phys. Rev. Lett. **59** (1987) 819.

26) J. E. Northrup, M. S. Hybertsen, and S. G. Louie, Phys. Rev. B **39** (1989) 8198.

27) H. Rietschel and L. J. Sham, Phys. Rev. B **28** (1983) 5100.

28) L. J. Lantto, Phys. Rev. B **22** (1980) 1380.

29) H. Yasuhara and Y. Ousaka, Solid State Commun. **64** (1987) 673.

30) S. A. Lindgren and L. Wallden, Phys. Rev. Lett. **61** (1988) 2894.

31) H. Raether, *Excitation of Plasmon and Interband Transitions by Electrons* (Springer, Berlin, 1980).

32) H. Kanter, Phys. Rev. B **1** (1970) 522; **1** (1970) 2357.

33) J. C. Tracy, J. Vac. Sci. Tech. **11** (1974) 280; see also C. J. Powell, Surf. Sci. **44** (1974) 29.

34) R. Kammerer, J. Barth, F. Gerken, C. Kunz, S. A. Flodstrom, and L. I. Johansson, Phys. Rev. B **26** (1982) 3491.

FREQUENCY-DEPENDENT LOCAL-FIELD FACTOR $G(k,\omega)$ FOR A TWO-DIMENSIONAL ELECTRON GAS

K.S. SINGWI
Department of Physics and Astronomy, Northwestern University
Evanston, Illinois 60208

1. INTRODUCTION

The problem of electron correlations in the intermediate and strong coupling regimes is a long standing one. During the last few years with the discovery of heavy-fermion compounds and of high-T_c copper oxides, there has been a renewed interest in this problem. I shall here confine myself to the specific problem of frequency dependence of the exchange-correlation effects in a 2D homogeneous electron gas. In the mean-field approach exchange and correlation effects are generally lumped into the "so-called" local-field factor G defined through the following equation:[1]

$$\chi(k,\omega) = \frac{\chi^0(k,\omega)}{1-v(k)[1-G(k,\omega)]\chi^0(k,\omega)}, \qquad (1)$$

where $\chi(k,\omega)$ in the density-density response function of an interacting electron gas and $\chi^0(k,\omega)$ is the corresponding function in the noninteracting case. $v(k)$ is the Fourier transform of the bare coulomb potential. Equation (1) is a formal definition of G and does not explicitly contain the effects of self-energy and quasi-particle renormalization. In situations where the latter effects are small, Eq. (1) may serve as a good approximation. For liquid ^{3}He it is definitely a poor approximation.

In the Random-Phase Approximation (RPA), $G = 0$. The RPA is a remarkably simple and successful approximation. In the interesting region of metallic densities (i.e., $r_s \sim 2\text{-}6$) it runs into certain difficulties. For example, the pair-correlation function becomes negative at small inter-particle separation and the compressibility sum rule is violated. Physically, the RPA does not account for the fact that the true electron distribution seen by one particular electron is different from the average distribution. In fact, each electron, because of the Pauli principle and the coulomb repulsion, tends to exclude other electrons from its neighborhood, thus surrounding itself by an "exchange correlation" hole. This effect modifies at short range the average RPA field, and the modification is usually written as an extra term of the form - $v(q)G(q,\omega)\delta n(q,\omega)$. With this effect included the response function takes the form of Eq. (1).

Techniques of the many-body theory are used to calculate the function G. The Feynman graphical method although very systematic has not proved very useful because of the mathematical complexities involved in evaluating the higher-order graphs. The kinetic equation method offers a natural way to determine the local-field correction. One arrives at a very simple expression[2] for the static part of G.

$$G(k) = -\frac{1}{n}\int \frac{d\vec{q}}{(2\pi)^3} \frac{\vec{k}\cdot\vec{q}}{q^2}\left[S(|\vec{k}-\vec{q}|)-1\right]. \quad (2)$$

(STLS). The price that one pays for this simplicity is in the uncontrolled nature of the approximation. Its merit is to be judged by the results it yields. I believe that the full potential of the kinetic equation method has not yet been fully exploited.

If one uses the Hartree-Fock approximation for the structure factor $S(q)$ in Eq. (2), one finds the lowest-order approximation to $G(k)$ very close to the Hubbard form[3]

$$G_H(k) = \frac{1}{2}\frac{k^2}{k^2+k_F^2}, \quad (3)$$

which was one of the first attempts to improve upon the RPA. Fully self-consistent solutions for $G(k)$ have been obtained numerically, and various useful parametrizations exist in literature. G vanishes as k^2 for $k \to 0$, and tends to a constant for $k \to \infty$.

The correlation energy as a function of electron density (i.e. r_s) calculated with the fully self-consistent local-field factor of Eq. (2) both in the 3D and 2D electron gas agrees remarkably well with the numerical simulation results of Ceperley and co-workers.[4] This gives us some confidence in the use of Eq. (2).

Very recently, Senatore and Pastore[5] using the density-functional theory of freezing have studied Wigner crystallization of the quantum electron liquid. The energy difference between the solid and the liquid phase can be written as

$$\Delta E = T_0[n(r)] - \frac{3}{5}NE_F + \frac{V}{2}\sum_{\vec{K}\neq 0} v(\vec{K})[1-G(\vec{K})]|n_{\vec{K}}|^2, \quad (4)$$

where $n(\vec{r})$ is the density of the solid phase and is given by

$$n(\vec{r}) = n_0 + \sum_{\vec{K}\neq 0} n_{\vec{K}} e^{i\vec{K}\cdot\vec{r}} \quad (5)$$

and where the $\vec{K}$'s are the reciprocal lattice vectors of the chosen structure (BCC). $T_0[n(r)]$ is the ideal kinetic energy at density $n(r)$. Since the ideal kinetic energy difference is always positive and, therefore, in the second-order theory the crystalline phase may become stable as seen from Eq. (4) if and only if $G(\vec{K}) > 1$ for some $\vec{K}$ and the contribution from such vectors is dominant in the potential energy sum on the right hand side of Eq. (4). Thus the key ingredient in the theory is the value of the local-field factor $G(\vec{K})$. Using the numerical values of $S(K)$ of Ceperley and Eq. (2), Senatore and Pastore find that ΔE of Eq. (4) vanishes at the r_s value of 128. This theoretical value of the critical r_s is in remarkably good agreement with the value $r_s = 100 \pm 29$ obtained by Ceperley from quantum Monte-Carlo calculations. In Fig. 1 are shown the plots of $G(q)$ for various values of r_s as calculated by Senatore and Pastore using Eq. (2) and $S(q)$ of Ref. 4. It is evident from Fig. 1 that the fluid is progressively softening with increasing r_s against modulation by density waves with wave vectors in the range q_f to $2q_F$. There are other interesting implications of this observation into which I cannot go here.

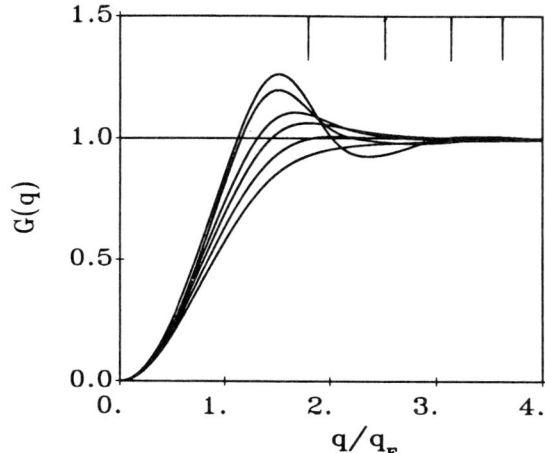

FIGURE 1
Local-field factor $G(q)$ for the spin-polarized electron fluid at increasing values of the coupling strength. The various curves refer to $r_s = 2, 5, 10, 10, 20, 50, 100$ in order of increasing peak height. The bars at the top of the figure show the positions of the RLV of the BCC lattice.

Exhaustive discussion of the static local-field factor and its properties can be found in the review articles of Ref 1.

2. FREQUENCY DEPENDENCE OF G

So far local-field correction has been treated in the static approximation. In general it is a complex function and depends on frequency, thereby implying that the exchange-correlation hole is dynamic in nature. No satisfactory expression for the frequency dependent G exists so far, although several attempts have been made in this direction.[6-8] A detailed analysis[6] of the first-order corrections to the screened response function leads to a complex $G(k,\omega)$ which has unphysical singularities. Besides, it does not lead to plasmon damping since the $ImG(k,\omega) = 0$ in the ω-region beyond the particle-hole continuum. At least second-order corrections are needed to get the damping of the collective mode. These leading corrections were studied in a systematic manner by Glick and Wong.[9] These authors have derived a closed form expression for the imaginary part of the dielectric function $\varepsilon(k,\omega)$ in the limit of high frequency and small wave number within the two-particle-hole-pair excitation approximation. In a recent paper[10] Holas and Singwi have extended the validity of the result of Glick and Wong and also generalized it to an arbitrary potential $v(k)$ in three-, two-, and one- dimensional space. In the following we shall use these results to derive an interpolation formula for the complex local-field factor $G(k,\omega)$ in a 2D electron gas which then immediately leads to an expression for the ω- dependent exchange-correlation potential - a result analogous to that derived by Gross and Kohn[11] in the 3D case. At the present state of our knowledge the interpolation procedure is the most satisfactory one since it is based on our exact knowledge of the behavior of $ImG(k,\omega)$ for both small ω and large ω; and the parameters of the interpolation formula are determined through the use of sum rules.

(a) ω-dependence of ImG in the 2D case. From Ref. 10 (Eq. 2.15) we have

$$\lim_{\omega \to \infty} \varepsilon_2(k,\omega) = \frac{11\pi}{64} \frac{1}{(k_F a_B)^3} \frac{(k/k_F)^3}{(\hbar\omega/2E_F)^5}, \qquad (6)$$

where $\hbar k_F$ is the Fermi momentum. In the 3D case the wave number and frequency dependence of ε_2 is, respectively, k^2 and $\omega^{-11/2}$ (See. Eq. (2.9) of Ref. 10). From the definition of G (Eq. 1), it follows that:

$$\lim_{\omega \to \infty} ImG(k,\omega) = Im\,\varepsilon(k,\omega) \left[\frac{\omega}{\omega_p(k)}\right]^4 \left[1+0\left(\frac{1}{\omega^2}\right)\right], \qquad (7)$$

where $\omega_p(k)$ is the plasmon frequency and is given by

$$\omega_p^2(k) = \frac{2\pi n e^2 k}{m} - \frac{r_s}{\sqrt{2}} \left(\frac{k}{k_F}\right) \left(\frac{2E_F}{\hbar}\right)^2 \tag{8}$$

Using (6) and (8) in (7), we have

$$\mathrm{Im}G(k,\omega) = \frac{11\pi}{64} \left(\frac{r_s}{\sqrt{2}}\right) \left(\frac{k}{k_F}\right) \left(\frac{\hbar\omega}{2E_F}\right)^{-1} \tag{9}$$

The corresponding result in the D=3 case is

$$\mathrm{Im}G(k,\omega) = \frac{23}{60} \left(\frac{4}{9\pi}\right)^{1/3} r_s \left(\frac{k}{k_F}\right)^2 \left(\frac{\hbar\omega}{2E_F}\right)^{-\frac{3}{2}} \tag{10}$$

(b) ω-dependent exchange-correlation potential:

The function $f_{xc}^h(k,\omega)$, introduced by Gross and Kohn and which is related to the ω-dependent exchange-correlation potential, is given by

$$f_{xc}^h(k,\omega) = -v(k)G(k,\omega) \tag{11}$$

From (10) and (11) we have

$$\mathrm{Im}f_{xc}^h(k,\omega) = -\frac{11\pi^2}{32} \frac{\hbar^3}{m^2 a_B^2} \omega^{-1} \tag{12}$$

for large ω. In the linear response regime, the ω-dependent exchange-correlation potential is given by[11]

$$V_{xc}(\vec{r},\omega) = f_{xc}^h(k=0, \omega; n_o) n_1(r,\omega), \tag{13}$$

$n_1(r,\omega)$ is the induced density in the presence of the external potential $v_1(\vec{r},\omega)$.

(c) $G(k,\omega)$ in D = 2 case.

Since we know the behavior of ImG for both small and large ω, we write

$$\mathrm{Im}G(k,\omega) = \frac{\omega_1(k)\omega}{\omega_2^2(k)+\omega^2}, \tag{14}$$

where $\omega_1(k)$ and $\omega_2(k)$ are two parameters to be determined. The full $G(k,\omega)$ is analytic in the upper half of the complex ω plane and can be written as

$$G(k,\omega) = G(k,\infty) + i \frac{\omega_1(k)}{\omega+i\omega_2(k)} \tag{15a}$$

$$= G(k,\infty) + \frac{\omega_1(k)\omega_2(k)}{\omega^2+\omega^2(k)} + i \frac{\omega\,\omega_1(k)}{\omega^2+\omega^2(k)}, \tag{15b}$$

$\omega_2(k) > 0$.

From (15b) we have for the static G

$$G(k,0) = G(k,\infty) + \frac{\omega_1(k)}{\omega_2(k)}, \qquad (16)$$

and

$$\lim_{\omega\to\infty}\left[\omega\mathrm{Im}G(k,\omega)\right] = \omega_1(k) \qquad (17)$$

Comparing (17) with (9) we have

$$\omega_1(k) = \frac{11\pi}{64}\left(\frac{r_s}{\sqrt{2}}\right)\left(\frac{k}{k_F}\right)\left(\frac{2E_F}{\hbar}\right) \qquad (18)$$

From (16) we have

$$\omega_2(k) = \frac{\omega_1(k)}{G(k,0)-G(k,\infty)} \qquad (19)$$

$G(k,\infty)$ can be determined from the 3rd moment sum rule. It is given by[8]

$$G(k,\infty) = I_d(k) - \frac{3}{4}\frac{\sqrt{2}}{r_s}\frac{k}{k_F}\delta_{kin}, \qquad (20)$$

where

$$\delta_{kin} = \left[<E_{kin}> - <E_{kin}{}_o>\right]\Big/<E_{kin}{}_o> \qquad (21)$$

and

$$I_d(k) = \frac{1}{2\pi k^3}\int d^2q \frac{[\vec{k}\cdot\vec{q}]^2}{q}\left[S(q) - S(|\vec{k}+\vec{q}|)\right] \qquad (22)$$

$<E_{kin}>$ is the exact kinetic energy of the interacting electron system and $<E_{kin}{}_o>$ that of the noninteracting system. $<E_{kin}>$ can be expressed in terms of ε_c - the correlation energy per particle, and is given by[12]

$$<E_{kin}> = <E_{kin}{}_o> - \frac{d}{dr_s}\left[r_s\varepsilon_c(r_s)\right] \qquad (23)$$

Thus $G(k,\infty)$ is known if we know $S(q)$ and $\varepsilon_c(r_s)$. The latter two quantities are known from Ceperley's calculations. Thus the parameters $\omega_1(k)$ and $\omega_2(k)$ in (15) are determined.

In the limit of small k, $G(k,\infty)$ has a simple form. It can be shown that[10]

$$\lim_{k\to 0}\left[G(k,\infty)\frac{k_F}{k}\right] = \frac{5}{6\pi} + \frac{7}{8\sqrt{2}}r_s\varepsilon_c + \frac{19}{16\sqrt{2}}r_s^2\frac{d\varepsilon_c}{dr_s} \qquad (24)$$

where ε_c is in rydbergs.

Also from Ref. 12, Eq. (3.6c), we have

$$\lim_{k \to 0} \left[G(k,0)\frac{k_F}{k} \right] = \frac{1}{\pi} + \frac{1}{8\sqrt{2}} r_s^2 \frac{d\varepsilon_c}{dr_s} - \frac{1}{8\sqrt{2}} r_s^3 \frac{d^3\varepsilon_c}{dr_s^2} \qquad (25)$$

Thus a knowledge of ε_c from the recent work of Tanatar and Ceperley[13] as a function of r_s enables one to calculate $\lim k \to 0 G(k,\omega)$ and hence $\lim k \to 0 f_{xc}^h(k,\omega)$ fairly precisely. A knowledge of the latter may prove useful in time-dependent-density-functional calculations in the 2D case. In future it would be interesting to subject the parametrized form (15) of G to experimental test.

3. ACKNOWLEDGEMENTS

My thanks are due to Drs. Holas and Senatore for many interesting discussions.

REFERENCES

1. See for example: K.S. Singwi and M.P. Tosi in Solid State Physics 36, 177 (1981), edited by M. Ehrenreich, F. Seitz and D. Turnbull (Academic, New York); Setsuo Ichimaru, Rev. Mod. Physics 54, No. 4 1017 (1982).

2. K.S. Singwi, M.P. Tosi, R.H. Land and A. Sjölander, Phys. Rev. 176, 589 (1968).

3. J. Hubbard, Proc. Roy. Soc. A240, 539 (1957); A243, 336 (1958).

4. D.M. Ceperley, Phys. Rev. B18, 3126 (1978), D.M. Ceperley and B.J. Alder, Phys. Rev. Lett. 45, 567 (1980).

5. G. Senatore and G. Pastore, Phys. Rev. Lett. (to be published).

6. A. Holas, P.K. Aravind and K.S. Singwi, Phys. Rev. B20, 4912 (1979).

7. F. Brosens, J.T. Devreese and L.F. Lemmens, Phys. Stat. Sol. (b) 80, 99 (1977); 81, 551 (1977); 82, 117 (1977); J.T. Devreese, F. Brosens and L.F. Lemmens, Phys. Rev. B21, 1349 (1980); B21, 1363 (1980).

8. A. Czachor, A. Holas, S.R. Sharma and K.S. Singwi, Phys. Rev. B25, 2144 (1982)

9. A.J. Glick and W.F. Long, Phys. Rev. B4, 3455 (1971).

10. A. Holas and K.S. Singwi, Phys. Rev. B40, 158 (1989).

11. E.K.U. Gross and W. Kohn, Phys. Rev. Lett. 55, 2850 (1985).

12. N. Iwamoto, Phys. Rev. A30, 3289 (1984).

13. B. Tanatar and D.M. Ceperley, Phys. Rev. B39, 5005 (1989).

VARIATIONAL THEORY OF ELECTRON LIQUID

Yasutami TAKADA

Institute for Solid State Physics, University of Tokyo
7-22-1 Roppongi, Minato-ku, Tokyo 106, Japan

We review the effective-potential expansion (EPX) method which was invented with the purpose of combining the advantages of perturbation-theoretic and variational approaches, while avoiding their disadvantages. We present the motivation to introduce the EPX method to study the electron liquid, describe the formulation in its most advanced form, and show the calculated results of the physical quantities such as the correlation energy, the one-body momentum distribution function, and the pair correlation function. Future problems about the EPX method are also discussed.

1. INTRODUCTION

The electron liquid is a system consisting of a large number of electrons embedded in a uniform positive back ground. The electrons interact with one another through the Coulomb interaction. Thus the Hamiltonian is written in second quantization as

$$H = H_0 + V, \qquad (1)$$

where

$$H_0 = \sum_{k\sigma} \varepsilon_k C_{k\sigma}^+ C_{k\sigma}, \qquad (2)$$

and

$$V = \frac{1}{2} \sum_{q \neq 0} \sum_{k\sigma} \sum_{k'\sigma'} V(q)\, C_{k+q\sigma}^+ C_{k'-q\sigma'}^+ C_{k'\sigma'} C_{k\sigma}, \qquad (3)$$

with $\varepsilon_k = \hbar^2 k^2/2m$ and $V(q) = 4\pi e^2/q^2$. As usual, $C_{k\sigma}$ is the annihilation operator of an electron specified by wave vector $\mathbf{k}$ and spin σ. In the following, we measure momenta and energies in units of the Fermi momentum $\hbar k_F$ and Rydberg $me^4/2\hbar^2$, respectively. Then the system can be described by only one parameter r_s, defined by $r_s = me^2/\alpha\hbar^2 k_F$ with $\alpha = (4/9\pi)^{1/3} = 0.521$.

Until now, many discussions[1] have been done on the correlation effects in the electron liquid and they elucidate the following three aspects as the magnitude of the transferred wave vector **q** changes.

(1) For small-q processes, there is a strong correlation among very many electrons due to the long-range nature of the Coulomb interaction. This correlation effect can be included completely and conveniently by the infinite sum of the ring diagrams [the Random-Phase Approximation (RPA)] in the usual perturbation-theoretic method.[2]

(2) When q is of the order of k_F, the Pauli principle working between the electrons of parallel spins reduces the effect of the ring diagrams considerably.[3] In the perturbation-theoretic and related methods, this exclusion effect, or the exchange contribution is included by the introduction of the so-called local-field correction $G(q)$. Several sum rules are employed to provide a useful $G(q)$,[1] but the Pauli principle cannot be satisfied exactly by such an approximate method. It is satisfied only when direct and exchange diagrams are included in pairs and treated faithfully. Besides, $G(q)$ obtained for the discussion of the normal-state properties cannot be used directly for the pairing potential of the Cooper pair, because $G(q)$ appears in the electron-hole scattering channel while the pairing potential is defined in the electron-electron scattering channel. Thus we need a more serious consideration as to what types of the vertex corrections are actually included in $G(q)$, once we go beyond the study of the normal-state properties of the electron liquid and want to investigate a possibility of superconductivity in it.

(3) For $q > 2k_F$, the multiple scattering between two electrons of anti-parallel spins become important, in particular, for large r_s. These processes can be treated by the sum of the ladder terms.[4] But they can also be described well by the variational wave function of the Bijl-Dingle-Jastrow form[5-7]

$$|\Phi_0\rangle = \prod_{i<j} f(r_{ij}) |0\rangle , \qquad (4)$$

where $|0\rangle$ is the state defined by a plane-wave Slater determinant and $f(r)$ is a two-particle correlation function to be determined variationally.

It is, of course, a difficult problem to construct a theory satisfying these requirements at different q. In the perturbation-theoretic approach, much effort has been paid to combine the ring and the ladder diagrams together with the exchange diagrams in the form of the dielectric function. An apparent problem is that the theories in this approach are not self-contained. They need to draw information from external sources to obtain an appropriate dielectric function in order to have quantitatively good results. This situation indicates that those theories cannot give an immediate answer for the system with the bare interaction $V(\mathbf{q})$ in Eq.(2) which is different from the purely Coulombic interaction as is often the case in real solids having a **q**-dependent inter-band dielectric constant.

The variational approach is, on the other hand, self-contained. So many attempts[8-17] have been done with the trial function of the type (4) until the Fermi hypernetted-chain (FHNC) theory is established in which the variational parameter f(r) is determined by the solution of the Euler-Lagrange-type equation[18,19]. As reviewed by Krotscheck[20], this FHNC theory is good enough to provide the correlation energy ε_c in excellent agreement with the results obtained by the Green's function Monte Carlo (GFMC) method[21] for any r_s. However, if we look at the FHNC theory more closely, we come to know that the Pauli principle does not hold order by order in the FHNC infinite sum. Besides, although the term "RPA" is also used in the FHNC theory, the meaning of the ring diagrams is different. The energy denominators which appear in the perturbation expansion are totally ignored in the theory. This makes ε_c deviate more from the exact value as r_s decreases. (The error becomes more than 8% for $r_s \ll 1$.[19]) The absence of the energy denorminators may be rather insensitive to the quantities like ε_c which are determined by the contribution of all the electrons, but it will cause a serious problem in such physical quantities as determined mostly by the electron near the Fermi surface at which the energy denominators vanish. In fact, it is known well that the one-body momentum distribution function n(k), defined by

$$n(k) \equiv <\Phi_0|C_{k\sigma}^+ C_{k\sigma}|\Phi_0> / <\Phi_0|\Phi_0>, \qquad (5)$$

shows a totally unphysical behavior near the Fermi surface in the one-dimensional Hubbard model[22] if the Gutzwiller-type trial function[23], a special form of Eq.(4), is used.[24] Namely, n(k) does not decrease but increases monotonically with a big jump at $k = k_F$ as k increases. This leads to an unphysically large renormalization factor z_F at the Fermi surface. (It is known that the theories including the energy denominators give a physically reasonable result for n(k).) Therefore, even in the three-dimensional electron liquid with a Jastrow form, a careful examination is necessary for the physical quantities determined at the Fermi surface like z_F.

In 1983, the present author proposed a new variational approach[25] which can potentially satisfy all the three requirements at the same time. In the approach, an effective potential $\tilde{V}$ is introduced to define a variational many-body wave function and all physical quantities are expanded in terms of $\tilde{V}$ rather than the bare one V. The expansion series is very similar to that in the perturbation-theoretic approach. Thus the summation of the ring and ladder terms can be done exactly with the energy denominators included. The Pauli principle is satisfied completely order by order by taking direct and exchange terms in pairs. The theory is self-contained, because $\tilde{V}$ is determined variationally. Further, the theory can be easily extended to the study of superconductivity. Since $\tilde{V}$ plays an essential role, this new approach is named the method of effective-potential expansion (EPX).

In this review paper, we give a brief account of the EPX method in its most advanced form [26] in Sec.2. The calculated results of the physical quantities like ε_c, $n(k)$, z_F, the spin-dependent static structure factor $S_{\sigma\sigma'}(q)$, the pair correlation function $g_{\sigma\sigma'}(r)$, and the spin-antiparallel zeroth component of the Landau's Fermi-liquid parameter are given in Sec.3. Some of these results were published in the past.[27] In Sec.4, we discuss the future problems relating on the EPX method as well as the electron liquid itself.

2. EPX METHOD

In the EPX method, we do not start with the trial function (4) but with a more complicated form, given by

$$|\Phi_0\rangle = \sum_{n=0}^{\infty} \frac{1}{n!} \left[\sum_{m=1}^{\infty} U_m\right]^n |0\rangle . \tag{6}$$

The correlation operator U_m is defined with the use of the long- and short-range parts of the effective potential, $\tilde{V}_l$ and $\tilde{V}_s$, as

$$U_m = \frac{1}{(i\hbar)^m m!} \int_{-\infty}^{0} \exp(0^+ t_1) dt_1 \ldots \int_{-\infty}^{0} \exp(0^+ t_m) dt_m \, T[\tilde{V}_l(t_1)\ldots\tilde{V}_l(t_m)]_L$$

$$+ \frac{1}{(i\hbar)^m (m-1)!} \int_{-\infty}^{0} \exp(0^+ t_1) dt_1 \ldots \int_{-\infty}^{0} \exp(0^+ t_m) dt_m \, T[\tilde{V}_s(t_1)\tilde{V}_l(t_2)\ldots\tilde{V}_l(t_m)]_L . \tag{7}$$

Here, $\tilde{V}(t)$ is defined as

$$\tilde{V}(t) = \exp[i(H_0 - \mu N)t/\hbar] \, \tilde{V} \, \exp[-i(H_0-\mu N)t/\hbar], \tag{8}$$

with H_0 in Eq.(2), μ the chemical potential, and N the total number operator. The symbol T and the subscript L represent, respectively, the T product and the instruction to collect only terms described by linked graphs. Note also that each cluster enclosed by the T operator is defined not to link directly to other clusters in Eq.(6). For $\tilde{V}_l$ and $\tilde{V}_s$, we choose the same form as Eq.(3) for V with the replacement of $V(q)$ by $\tilde{V}_l(q)$ or $\tilde{V}_s(q)$.

The energy expectation value E_0 with respect to $|\Phi_0\rangle$ can be expanded in power series of the effective potential with the use of the linked-cluster theorem. If we write the series in powers of $\tilde{V}_s$ as

$$E_0 \equiv \langle \Phi_0 | H | \Phi_0 \rangle / \langle \Phi_0 | \Phi_0 \rangle$$

$$= \sum_{n=0}^{\infty} E^{(n)} , \tag{9}$$

$E^{(n)}$ is the nth-order term in $\tilde{V}_s$ but it can still be expanded in $\tilde{V}_I$ up to infinite order. In order to treat the infinite sum of $\tilde{V}_I$, we assume the form for $\tilde{V}_I(q)$ as

$$\tilde{V}_I(q) = V(q) / [\frac{1}{2} + \frac{1}{2} \exp(q^2/q_c^2)], \qquad (10)$$

where q_c is a variational parameter to be explored in the region of $0 < q_c < 0.3 k_F$. In this choice, $\tilde{V}_I(q)$ is not zero only for small-q processes. Thus the ring diagrams with respect to $\tilde{V}_I(q)$ are predominantly important. All of them are included in each $E^{(n)}$. Then the actual expansion parameter in Eq.(9) is not $\tilde{V}_s(q)$ but $\overline{V}_s(q)$, defined by

$$\overline{V}_s(q) = \tilde{V}_s(q) / [1 + \tilde{V}_I(q) \Pi_0(q,0)]^2 \qquad (11)$$

in the static approximation, where $\Pi_0(q,i\omega)$ is the polarization function for the imaginary frequency $i\omega$ in the RPA, given by

$$\Pi_0(q,i\omega) = 2\sum_{k\sigma} \frac{n_{k\sigma}(1-n_{k+q\sigma})(\varepsilon_{k+q} - \varepsilon_k)}{\omega^2 + (\varepsilon_{k+q} - \varepsilon_k)^2}, \qquad (12)$$

with the distribution function of a free-electron system $n_{k\sigma} = \theta(k_F - |k|)$. In addition to the ring diagrams, we have also taken the ring-exchange diagrams, self-energy diagrams, and its exchange partners with respect to $\tilde{V}_I$ into account in $E^{(0)}$ and $E^{(1)}$. It is necessary to include them in order to satisfy the Pauli principle order by order.

For large q, $\overline{V}_s(q)$ is at most $V(q)$ and is small. For small q, $\overline{V}_s(q)$ is also small because of the screening factor in Eq.(11). Thus $\overline{V}_s(q)$ is a good expansion parameter for the whole value of q. Therefore we first neglect the terms higher than second order in Eq.(9) but include all the lower-order terms to determine $\overline{V}_s(q)$ by the numerical solution of the Euler-Lagrange-type equation

$$\delta E_0 / \delta \overline{V}_s(q) = 0. \qquad (13)$$

We then estimate the effects of the higher-order terms by evaluating them with the use of $\overline{V}_s(q)$ thus obtained. Note that Eq.(13) reduces to the Bethe-Salpeter equation in the ladder approximation for large q. The obtained equation is nothing but the one discussed by Yasuhara.[4]

Once we obtain $\overline{V}_s(q)$ and q_c variationally, we can calculated many physical quantities. For example, the one-body momentum distribution function $n(k)$ can be calculated from Eq.(5) with an expansion similar to that for E_0 in Eq.(9). The spin-dependent static structure factor $S_{\sigma\sigma'}(q)$ is also given by

$$S_{\sigma\sigma'}(q) = \frac{2}{N} \sum_{kk'} <\Phi_0|C_{k\sigma}^+ C_{k+q\sigma} C_{k'\sigma'}^+ C_{k'-q\sigma'}|\Phi_0> / <\Phi_0|\Phi_0>. \quad (14)$$

The pair correlation function $g_{\sigma\sigma'}(q)$ is given by the Fourier transform of $S_{\sigma\sigma'}(q)$. The Landau's Fermi-liquid function can be calculated by the second functional derivative of E_0 in Eq.(9) with respect to $n_{k\sigma}$;

$$f_{k\sigma,k'\sigma'} = \frac{\delta^2 E_0}{\delta n_{k\sigma} \delta n_{k'\sigma'}}. \quad (15)$$

3. NUMERICAL RESULTS

The correlation energy is defined as the difference between the total energy E_0 and that in the Hartree-Fock approximation. Calculated results for the correlation energy per electron ε_c in Rydbergs are given in Table 1 in which $\varepsilon_c^{(2)}$ represents the result with all the terms in Eq.(9) up to second order, while $\varepsilon_c^{(4)}$ and $\varepsilon_c^{(6)}$ are, respectively, those with the ring diagrams, their exchange partners, and the self-energy direct and exchange diagrams in fourth and sixth orders with respect to $\bar{V}_s(q)$ determined from Eq. (13) in addition to the terms in $\varepsilon_c^{(2)}$. For comparison's sake, we have also shown the data from other methods such as the variational Monte Carlo (VMC) of Ceperley,[16] the GFMC of Ceperley and Alder,[21] the FHNC of Lantto,[18] Zabolitzky,[19] and Krotscheck,[20] the perturbation-theoretic approach of Suehiro, Ousaka, and Yasuhara,[28] and the coupled-cluster formalism of Emrich and Zabolitzky.[29]

r_s	q_c/k_F	$\varepsilon_c^{(2)}$	$\varepsilon_c^{(4)}$	$\varepsilon_c^{(6)}$	VMC	GFMC
1	0.156	-0.117	-0.121	-0.121	-0.120	-0.120
2	0.148	-0.085	-0.089	-0.090	-0.087	-0.090
3	0.144	-0.069	-0.073	-0.073	-0.072	—
4	0.140	-0.059	-0.062	-0.063	-0.062	—
5	0.138	-0.052	-0.055	-0.056	-0.055	-0.056

r_s	L	Z	K	SOY	EZ
1	-0.143	-0.114	-0.134	-0.121	-0.122
2	-0.099	-0.086	-0.094	-0.091	-0.090
3	-0.079	-0.071	-0.075	-0.076	-0.074
4	-0.067	-0.061	-0.064	-0.066	-0.063
5	-0.058	-0.054	-0.056	-0.059	-0.056

TABLE 1
Correlation energy per electron in Ry units. Note that $\varepsilon_c^{(2)}$, $\varepsilon_c^{(4)}$, and $\varepsilon_c^{(6)}$ are the present results with the terms up to second, fourth, and sixth orders in $\bar{V}_s(q)$ in Eq.(9). Other data indicated by VMC, GFMC, L, Z, K, SOY, and EZ are, respectively, taken from Ref.16, 21, 18, 19, 20, 28, and 29. The optimum value for q_c in Eq. (10) to define $\bar{V}_l(q)$ is also shown here.

An example of the calculated results for the one-body momentum distribution function n(k) is given in Fig.1 for $r_s=5$. Except for k near k_F, our result is very close to that in the FHNC method of Lantto.[18] Near the Fermi surface, however, our result deviates significantly from it and goes closer to that in the GFMC.[21] Due to the absence of the energy denominators in the trial function (4), n(k) becomes inaccurate in the FHNC method for k near k_F. Thus the result for the renormalization factor at the Fermi surface z_F becomes unreasonably large in the FHNC as shown in Table 2 in which the result in the perturbation-theoretic approach of Rice[30] is also shown for comparison's sake.

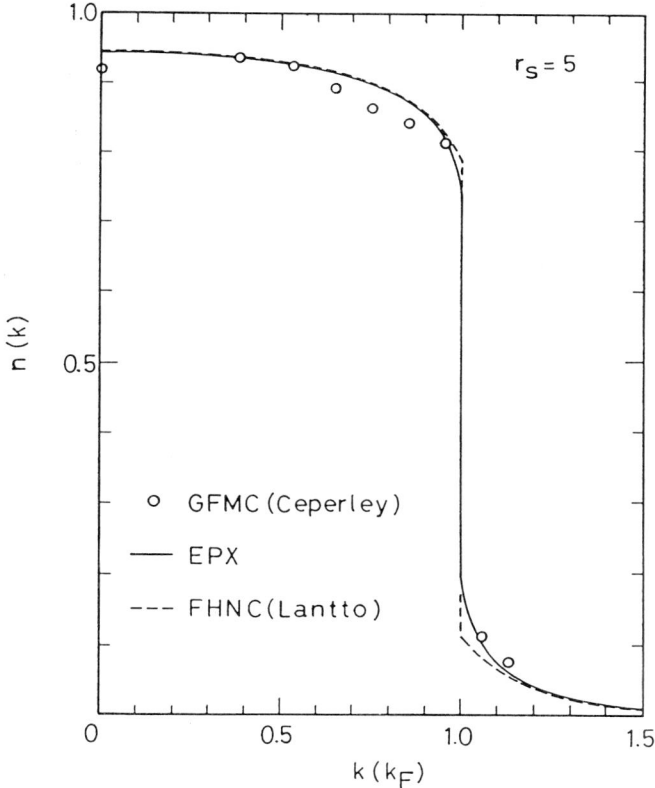

FIGURE 1
One-body momentum distribution function n(k) for $r_s=5$. The result of Lantto[18] in the FHNC and that in the GFMC[21] are also shown by the dashed curve and the open circles, respectively.

r_s	$z_F^{(2)}$	$z_F^{(4)}$	$z_F^{(6)}$	R	L	K
1	0.88	0.87	0.87	0.86	0.89	0.89
2	0.78	0.77	0.77	0.77	0.81	0.82
3	0.69	0.68	0.68	0.70	0.76	0.76
4	0.61	0.60	0.61	0.65	0.71	0.72
5	0.55	0.53	0.54	0.60	0.66	0.69

TABLE 2
Renormalization factor at the Fermi surface. As in Table 1, $z_F^{(2)}$, $z_F^{(4)}$, and $z_F^{(6)}$ are the results with the terms up to second, fourth, and sixth orders in $\bar{V}_s(q)$, respectively. Other data indicated by R, L, and K are, respectively, obtained by the perturbation-theoretic approach of Rice[30] and by the FHNC of Lantto[18] and Krotscheck.[20]

The spin-dependent static structure factor calculated up to sixth order is shown in Fig.2 together with the result of Suehiro, Ousaka, and Yasuhara[28] in the perturbation-theoretic approach. The overall feature is the same between these two results. As for the spin-parallel pair correlation function, $g_{\uparrow\uparrow}(0)$ vanishes exactly because the Pauli

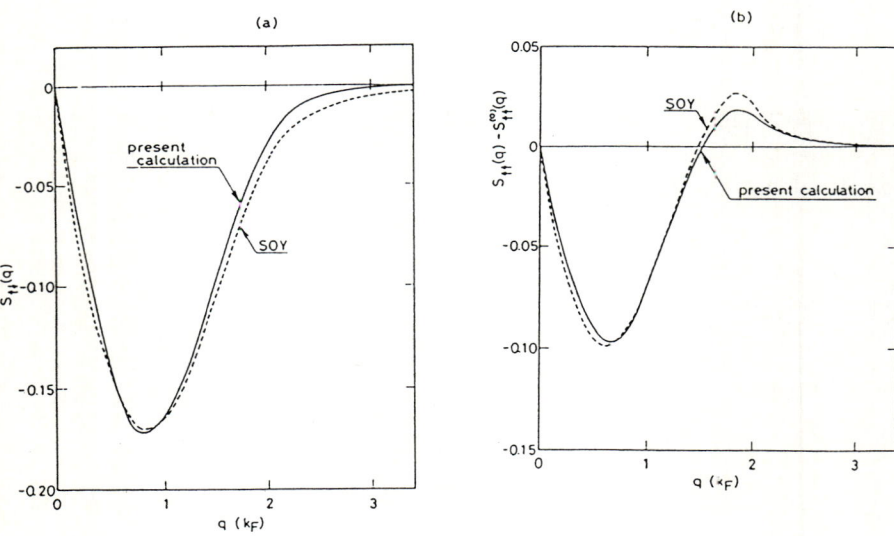

FIGURE 2
Spin-dependent static structure factor $S_{\sigma\sigma'}(q)$ for $r_s=4$. The results of Ref.28 in the perturbation-theoretic approach are given by the dashed curves. In (a) and (b), $S_{\uparrow\downarrow}$ and $S_{\uparrow\uparrow}-S_{\uparrow\uparrow}(0)$ are given, respectively, where $S_{\uparrow\uparrow}(0)$ is the static structure factor in the Hartree-Fock approximation.

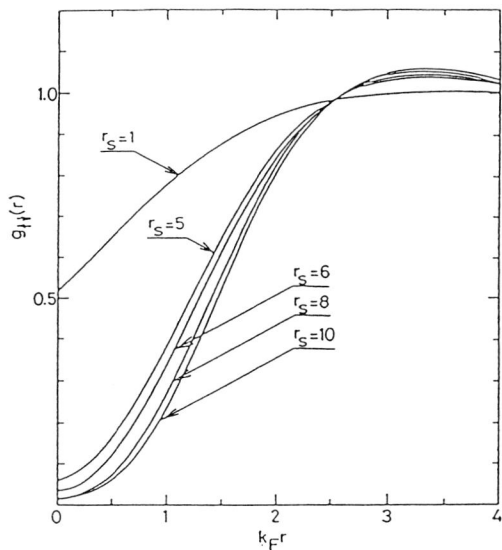

FIGURE 3
Spin-antiparallel pair correlation function $g_{\uparrow\downarrow}(r)$ for $r_s=1,5,6,8$ and 10.

principle is strictly imposed order by order in our approach. The spin-antiparallel part $g_{\uparrow\downarrow}(r)$ is always positive as shown in Fig.3. At $r = 0$, it is very close to that of the approximate formula of Yasuhara[4]

$$g_{\uparrow\downarrow}(0) = \{ 2\,(\alpha r_s/\pi)^{1/2}\, I_1[4\,(\alpha r_s/\pi)^{1/2}] \}^2 \,, \qquad (16)$$

for any r_s, where I_1 is the first-order modified Bessel function of the first kind.

Finally, as an example of the calculated results for the Fermi-liquid parameters, the difference between the compressibility ratio κ_0/κ and the spin susceptibility ratio χ_0/χ obtained by other methods[31,32] is compared in Table 3 with $F_{\uparrow\downarrow}$, defined by

$$F_{\uparrow\downarrow} = \sum_{k'} \delta(\varepsilon_{k'}-\mu)\, f_{k\sigma,k'-\sigma}\,. \qquad (17)$$

According to the Landau's Fermi-liquid theory, we obtain

$$F_{\uparrow\downarrow} = \frac{1}{2}\left(\frac{\kappa_0}{\kappa} - \frac{\chi_0}{\chi}\right), \qquad (18)$$

r_s	$F_{\uparrow\downarrow}$	$[\kappa_0/\kappa - \chi_0/\chi]/2$
1	-0.022	-0.02
2	-0.070	-0.08
3	-0.136	-0.14
4	-0.214	-0.20
5	-0.301	-0.27

TABLE 3
$F_{\uparrow\downarrow}$ defined in Eq. (17) obtained by the calculation of the terms up to second order in $\bar{V}_s(q)$. It is compared with the result of $[\kappa_0/\kappa - \chi_0/\chi] / 2$ obtained by the calculation of the ground-state energy (Refs. 31 and 32).

where κ_0 and χ_0 are, respectively, the compressibility and the spin susceptibility of a free-electron system. Although our calculation for $F_{\uparrow\downarrow}$ is done only up to second order in $\bar{V}_s(q)$ by now, the agreement shown in Table 3 is quite remarkable.

4. FUTURE PROSPECTS

The EPX method has been proven to provide independent and reliable data on the properties of the electron liquid in the normal state. It is very advantageous that the EPX method can give progressively and systematically better results with the progress of computer technology by the systematic inclusion of higher-order terms in Eq.(9).

Since we have obtained a self-contained and reliable method for the many-body problem, we should calculate the physical quantities for a system with an arbitrary $V(q)$ in order to explore the conditions for $V(q)$ to provide interesting properties such as superconductivity with a high transition temperature T_c. A study along this line has already been initiated by the calculation of T_c in the system with $V(q)$ slightly different from the purely Coulombic potential. A new insight into a pairing potential for high-T_c superconductivity has been obtained from such a study.[33] We have pointed out a conceptual difference between superconductivity originating from a repulsive potential and superconductivity from an attractive potential like a BCS superconductor with the aid of phonons. This type of study should be done more extensively not only for the ordered states like superconductivity but also for the normal state. From such a study, we may obtain a new aspect as to the heavy-electron system.

Extensions of the EPX method will also be interesting and important. A finite-temperature version of the EPX method has already been formulated.[34] Extensions to treat dynamical properties and inhomogeneous electron liquids are other two important subjects in the future. From the viewpoint of formulation, the latter problem is much easier, but a considerably big progress of computer technology is needed for the actual implementation of the "band-structure"-like calculation in the EPX method. Progress along this line will contribute to the improvement on the variational theory based on the trial function (4) which has already treated the inhomogeneous systems such as metal surfaces[35,36] and bulk materials like diamond.[37]

ACKNOWLEDGEMENTS

The author wishes to thank A.W. Overhauser for the suggestion of a formalism leading to the EPX method and for the encouragement at the early stage of this study. He also acknowledges many useful discussions with T. Ando, H. Fukuyama, M. Kohmoto, W. Kohn, H. Shiba, and H. Yasuhara on the many-body problem.

REFERENCES
1) For recent reviews, S. Ichimaru, Rev. Mod. Phys. **54** (1982) 1017. Since we are concerned with the variational approach to the electron liquid here, most of the works in the perturbation-theoretic approach, namely, the Green's function method and its variations are not referred in this paper, but those works can be found in this review article.
2) M. Gell-Mann and K. Brueckner, Phys. Rev. **106** (1957) 364.
3) J. Hubbard, Proc. Roy. Soc. London **A243** (1958) 336.
4) H. Yasuhara, Solid State Commun. **11** (1972) 1481.
5) A. Bijl, Physica (Utrecht) **7** (1940) 869.
6) R.B. Dingle, Philos. Mag. **40** (1949) 573.
7) R. Jastrow, Phys. Rev. **98** (1955) 1479.
8) T. Gaskell, Proc. Phys. Soc. London **77** (1961) 1182; **80** (1962) 1091.
9) M.S. Becker, A.A. Broyles, and T. Dunn, Phys. Rev. **175** (1968) 224.
10) R. Monnier, Phys. Rev. **A6** (1972) 393.
11) D.K. Lee and F.H. Ree, Phys. Rev. **A6** (1972) 1218.
12) G. Keiser and F.Y. Wu, Phys. Rev. **A6** (1972) 2369.
13) F.A. Stevens and M.A. Pokrant, Phys. Rev. **A8** (1973) 990.
14) J.D. Talman, Phys. Rev. **A10** (1974) 1333; **A13** (1976) 1200.
15) S. Chakravarty and C.W. Woo, Phys. Rev. **B13** (1976) 4815.
16) D. Ceperley, Phys. Rev. **B18** (1978) 3216.
17) P. Horsch and P. Fulde, Z. Phys. **B36** (1979) 23.
18) L. J. Lantto, Phys. Rev. **B22** (1980) 1380.
19) J. G. Zabolitzky, Phys. Rev. **B22** (1980) 2353.

20) E. Krotscheck, Ann. Phys. (N.Y.) **155** (1984) 1.
21) D.M. Ceperley and B.J. Alder, Phys. Rev. Lett. **45** (1980) 566.
22) J. Hubbard, Proc. Roy. Soc. London **A276** (1963) 238.
23) M. C. Gutzwiller, Phys. Rev. **134** (1964) A923.
24) W. Metzner and D. Vollhardt, Phys. Rev. Lett. **59** (1987) 121.
25) Y. Takada, Phys. Rev. **A28** (1983) 2417.
26) Y. Takada, Phys. Rev. **B37** (1988) 155.
27) Y. Takada, Phys. Rev. **B30** (1984) 3882; **B35** (1987) 6923.
28) H. Suehiro, Y .Ousaka, and H. Yasuhara, J. Phys. **C19** (1986) 4263.
29) K. Emrich and J. G. Zabolitzky, Phys. Rev. **B30** (1984) 2049.
30) T.M. Rice, Ann. Phys. (N.Y.) **31** (1965) 100.
31) S.H. Vosko, L. Wilk, and M. Nusair, Can. J. Phys. **58** (1980) 1200.
32) Y. Kawazoe, H. Yasuhara, and M. Watabe, J. Phys. **C10** (1977) 3293.
33) Y. Takada, Phys. Rev. **B39** (1989) 11575.
34) T. Kita and Y. Takada, unpublished.
35) X. Sun, M. Farjam, and C.W. Woo, Phys. Rev. **B28** (1983) 5599.
36) E. Krotscheck, W. Kohn, and G.X. Qian, Phys. Rev. **B32** (1985) 5693.
37) S. Fahy, X. W. Wang, and S. G. Louie, Phys. Rev. Lett. **61** (1988) 1631.

LANDAU INTERACTION FUNCTION AND EFFECTIVE MASS OF AN ELECTRON LIQUID

Hiroshi YASUHARA and Yumi OUSAKA†

College of General Education, Tohoku University, Sendai 980, Japan
†Sendai National College of Technology, Sendai 989-31, Japan

We have evaluated the effective mass m^* of an electron liquid from a systematic consideration of diagrams for the Landau interaction function $f^{\sigma\sigma'}(\mathbf{p},\mathbf{p}')$. It is found that the mass ratio m^*/m is a slowly decreasing function of r_s and remains smaller than unity throughout the whole region of metallic densities. A deviation of m^*/m from unity is as small as 8% even at the lowest metallic densities.

The long-range correlations characteristic of an electron liquid can adequately be described by means of the RPA[1]. The application of the RPA is valid for high densities where long-range parts of electron correlations are dominant. Strongly correlated features of an electron liquid at metallic densities are very difficult to clarify. In 1968 Singwi and coworkers[2] managed to construct a self-consistent theory(STLS theory) which treated successfully with short-range correlations at metallic densities. The study of an electron liquid at metallic densities has since been advanced by a large number of authors using a wide variety of methods[3]. At present, the values of the correlation energy calculated from different methods agree within 0.5mRy per electron throughout the whole region of metallic densities. A clear understanding of short-range correlations in diagrammatic terms of standard many-body perturbation theory was obtained by one of the authors (H.Y)[4] in 1972. A series of particle-particle ladder diagrams are indispensable for the adequate description of the spin-antiparallel pair distribution function $g^{\uparrow\downarrow}(r)$ at short distances. It has been proved that only the contributions of the static structure factor $S(q)$ from the particle-particle ladder parts of energy diagrams give its asymptotic form of order q^{-4} for large q and as a consequence of the cancellation between direct and exchange contributions $S^{\uparrow\downarrow}(q)$ (the spin-antiparallel part of $S(q)$) has the exact asymptotic form as follows: $S^{\uparrow\downarrow}(q) = -(4/3)(\alpha r_s/\pi)(p_f/q)^4 g^{\uparrow\downarrow}(0) + ...$, $\alpha = (4/9\pi)^{1/3}$. The success of the STLS theory probably can be ascribed to the fact that the self-consistent determination of the local field correction in the formulation amounts approximately to solving the integral equation for the particle-particle ladder interactions. The idea of interpolating between small and large momentum transfer interactions that was initially proposed by Hubbard and Nozieres-Pines in the late fifties can be developed into a theory[5] valid for the region of metallic densities if one takes account of not only the second order exchange diagram but also higher order particle-particle ladder diagrams and their exchange counterparts as the local field correction to the RPA.

In spite of all these developments over the past two decades the study of the effective mass m^* at metallic densities still remains unsettled; the high density expansion of the effective mass ratio m^*/m $(= [1 - (\alpha r_s/2\pi)(2 + ln(\alpha r_s/\pi)) + ...]^{-1})$ is a decreasing function of r_s for $r_s \lesssim 1.0$ but in almost all the previous calculations[7] m^*/m increases with r_s throughout the region of metallic densities and even becomes larger than unity. The evaluation of m^*/m requires a more detailed knowledge of electron correlations, compared with thermodynamical quantities such as the ground state energy, the compressibility and the spin-susceptibility. According to Landau's formulas, the mass ratio m^*/m, the compressiblity ratio κ_f/κ and the spin-susceptibility ratio χ_f/χ are related to the Landau interaction function $f^{\sigma\sigma'}(\mathbf{p},\mathbf{p}')$ $(=f_0(\mathbf{p},\mathbf{p}')+\delta_{\sigma\sigma'}f_e(\mathbf{p},\mathbf{p}'))$ as follows: $m/m^* = 1 - \int_{-1}^{1}dx\,x f(x)$, $\kappa_f/\kappa = m/m^* + \int_{-1}^{1}dx\,f(x)$, and $\chi_f/\chi = m/m^* + \int_{-1}^{1}dx\,f_e(x)$ where x is the direction cosine of the angle between $\mathbf{p}$ and $\mathbf{p}'$ and $f_0(x), f_e(x)$ and $f(x)(=2f_0(x)+f_e(x))$ denote the corresponding reduced functions like $f(\mathbf{p},\mathbf{p}') = (8\varepsilon_f/3N)f(x)$. As for κ_f/κ and χ_f/χ, the precise value of these two quantities is available without a knowledge of $f^{\sigma\sigma'}(\mathbf{p},\mathbf{p}')$. The following fitting formulas obtained from thermodynamical calculations are useful; $\kappa_f/\kappa = 1-(\alpha r_s/\pi)-(1-ln2)(\alpha r_s/\pi)^2+0.171(\alpha r_s/\pi)^3$, $\chi_f/\chi = 1-(\alpha r_s/\pi)-(\alpha r_s/\pi)^2 ln(\alpha r_s/\pi)/1.481$. From a systematic consideration of diagrams for $f^{\sigma\sigma'}(\mathbf{p},\mathbf{p}')$ we have determined[8] an approximate expression for $f^{\sigma\sigma'}(\mathbf{p},\mathbf{p}')$ with the help of the information about κ_f/κ and χ_f/χ. $f^{\sigma\sigma'}(\mathbf{p},\mathbf{p}')$ is given by the functional derivative of the self-energy $\Sigma(p,\varepsilon[G])$ for $\varepsilon = E_p$(quasiparticle's pole) with respect to quasiparticle's distribution function n_p. $\Sigma(p,\varepsilon[G])$ depends on n_p only through its constituent G (the Green function). G depends on n_p in two different ways; directly and indirectly through the self-energy implicitly included in G. We have considered the diagrams for $f^{\sigma\sigma'}(\mathbf{p},\mathbf{p}')$ that come from the direct dependence(see Fig.1). The omission of the indirect contributions may, for the most part, be compensated by taking the 1st iterative solution $\Sigma(p,\varepsilon_p[G])$ ($\varepsilon_p = \hbar^2 p^2/2m$) instead of $\Sigma(p, E_p[G])$. From a summation of the diagrams shown in Fig.1(a) and (b) we have obtained the following approximate expression: $f^{\uparrow\uparrow}(\mathbf{p},\mathbf{p}') = -\Omega^{-1}4\pi e^2/\{(\mathbf{p}-\mathbf{p}')^2\tilde{\varepsilon}(\mathbf{p}-\mathbf{p}',0)\}$, $\tilde{\varepsilon}(\mathbf{p}-\mathbf{p}',0) = 1 + v(\mathbf{p}-\mathbf{p}')\pi_0(\mathbf{p}-\mathbf{p}',0)\Gamma(\mathbf{p}-\mathbf{p}')$ and $\Gamma(\mathbf{p}-\mathbf{p}') = (1-G(\mathbf{p}-\mathbf{p}'))^2/\{1+v(\mathbf{p}-\mathbf{p}')G(\mathbf{p}-\mathbf{p}')(1-G(\mathbf{p}-\mathbf{p}'))\pi_0(\mathbf{p}-\mathbf{p}',0)\}$, where Ω is the volume of the system, $v(q) = 4\pi e^2/q^2$, $\pi_0(q,0)$ is the static Lindhard function and $G(q)$ is the local field factor defined by the ratio of the irreducible particle-hole interaction to $v(q)$(see Fig.2(a)). If one takes the limiting value of $\pi_0(\mathbf{p}-\mathbf{p}',0)$, $G(\mathbf{p}-\mathbf{p}')$ and $\Gamma(\mathbf{p}-\mathbf{p}')$ as $\mathbf{p}' \to \mathbf{p}$ in the above expression, $f^{\uparrow\uparrow}(\mathbf{p},\mathbf{p}')$ can be reduced into a physically appealing form; i.e., $f^{\uparrow\uparrow}(\mathbf{p},\mathbf{p}') = -(1/\Omega)4\pi e^2/\{(\mathbf{p}-\mathbf{p}')^2+k_s^2\}$, $k_s^2 = k_{TF}^2/\{1+(1-\kappa_f/\kappa)\}$ where k_{TF} is the Thomas-Fermi screening wavenumber;in the process of the calculations we have used Ward's identities relating the vertex part ($G(q)$ in our approximation) to the compressiblity κ. Note that such a simple modification of the Thomas-Fermi screened interaction with a scaling of the screening factor is valid for the region of

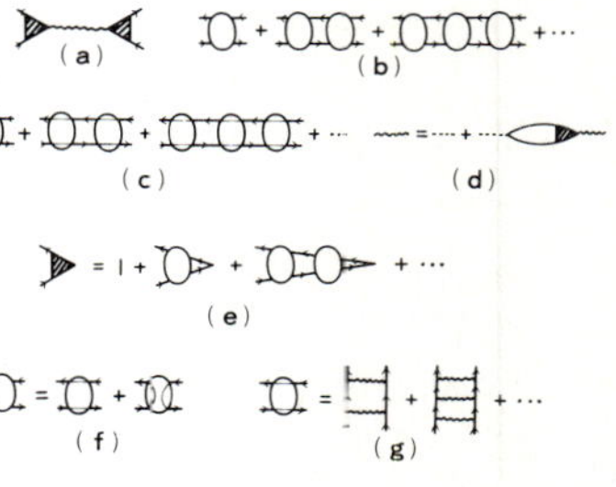

Fig.1

(a),(b):diagrams for $f^{\uparrow\uparrow}$; (c):diagrams for $f^{\uparrow\downarrow}$; (d): the screened Coulomb interaction; (e): the vertex part; (f): the irreducible particle-hole interaction; (g): the particle-particle ladder interactions

metallic densities; the scaling factor $1 + (1 - \kappa_f/\kappa)$ is reduced to unity in the limit $r_s \to 0$ and increases with r_s up to two at $r_s \simeq 5.3$ where κ_f/κ vanishes. A comparison between three screening factors is made in Fig.3; the Wigner-Seitz model gives the strongest screening; we have defined the Wigner-Seitz screening wavenumber, k_{WS} by putting the $\mathbf{q} = 0$ Fourier component of the Wigner-Seitz potential equal to $4\pi e^2/k_{\mathrm{WS}}^2$. The previous calculations of $f^{\uparrow\downarrow}(\mathbf{p}, \mathbf{p}')$ are not only overestimated in magnitude but also show the logarithmic divergence in the neighborhood of $\mathbf{p}' = -\mathbf{p}(x = -1)$ for lack of higher order particle-particle ladder interactions shown in Fig.1(g). With two parameters and the value of χ_f/χ we have devised an approximate expression for $f^{\uparrow\downarrow}(\mathbf{p}, \mathbf{p}')$ corresponding to the diagrams shown in Fig.1(c), in which the logarithmic divergence is suppressed. The one parameter has been determined so that the spin-antiparallel RPA contribution to m^*/m should be reproduced from $f^{\uparrow\downarrow}(\mathbf{p}, \mathbf{p}')$ in its RPA limit and the other so that the value of κ_f/κ should be reproduced from $f^{\uparrow\uparrow}(\mathbf{p}, \mathbf{p}')$ and $f^{\uparrow\downarrow}(\mathbf{p}, \mathbf{p}')$ through Landau's formula. The resulting $f^{\uparrow\downarrow}(\mathbf{p}, \mathbf{p}')$ is shown in Fig.2(b). The mass ratio m^*/m calculated from $f^{\uparrow\uparrow}(\mathbf{p}, \mathbf{p}')$ and $f^{\uparrow\downarrow}(\mathbf{p}, \mathbf{p}')$ is shown in Fig.4. Since the effective mass enhancement caused by $f^{\uparrow\downarrow}(\mathbf{p}, \mathbf{p}')$ is rather small, the overall features of m^*/m shown in Fig.4 can well be represented by the following analytic formula that can easily be calculated

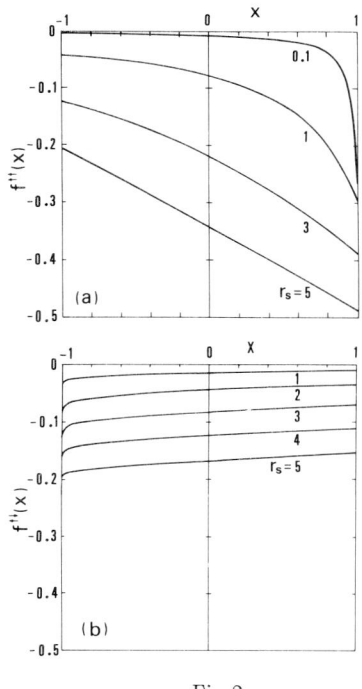

Fig.2
(a): $f^{\uparrow\uparrow}(x)$; (b): $f^{\uparrow\downarrow}(x)$

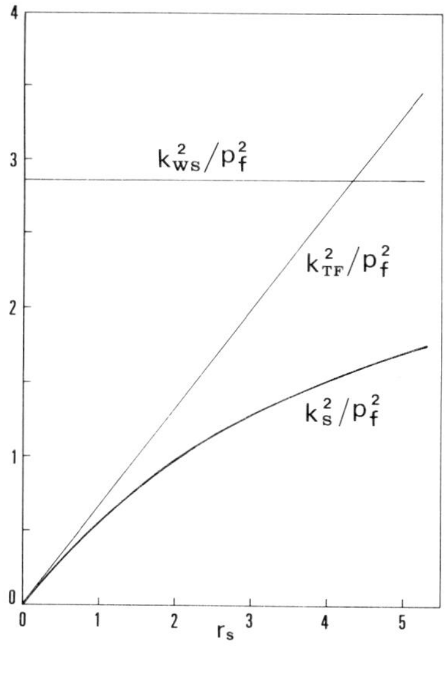

Fig.3
A comparison between k_{TF}^2/p_f^2, k_s^2/p_f^2 and k_{WS}^2/p_f^2

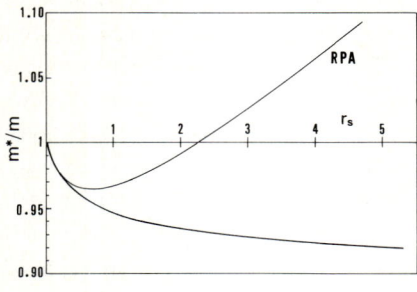

Fig.4
r_s-dependent feature of m^*/m

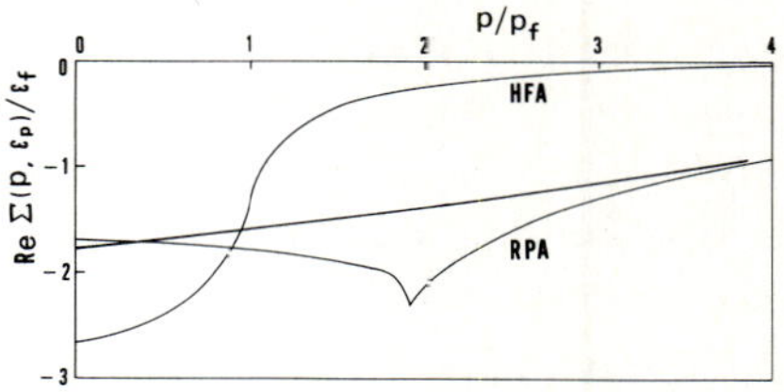

Fig.5
$Re\Sigma(p,\varepsilon_p)$ as a function of p/p_f

only from the simplified version of $f^{\uparrow\uparrow}(\mathbf{p},\mathbf{p}')$; $(m^*/m)^{-1} = 1 - (\alpha r_s/\pi)[1 + (1/2)\{1 + 2(\alpha r_s/\pi)\Gamma(0)\}ln\{(\alpha r_s/\pi)\Gamma(0)/(1+(\alpha r_s/\pi)\Gamma(0))\}]$, $\Gamma(0) = \{1+(1-\kappa_f/\kappa)\}^{-1}$. It is reduced to the high density expansion if one puts $\Gamma(0) = 1$. Finally, in Fig.5 we have shown the probable value of $Re\Sigma(p,\varepsilon_p)$ predicted from m^*/m and μ_{xc} (the exchange and correlation contribution to the chemical potential).

REFERENCES

1) D. Pines, The Many-Body Problem (BENJAMIN, London, 1962).

2) K. S. Singwi, M. P. Tosi, R. H. Land and A. Schölander, Phys. Rev. 176 (1968) 589.

3) K. S. Singwi and M. P. Tosi, Solid State Physics, Vol.36, eds. H. Ehrenreich, F. Seitz and D. Turnbull(ACADEMIC PRESS, New York, 1981) pp. 177-266.
S. Ichimaru, Rev. Mod. Phys. 54 (1982) 1017.

4) H. Yasuhara, Solid State Commun. 11 (1972) 1481; J. Phys. Soc. Japan 36 (1974) 361; Physica 78 (1974) 420. See also B. B. Hede and J. P. Carbotte, Can. J. Phys. 50(1972)4512; D. N. Lowy and G. E. Brown, Phys. Rev. B12 (1975) 2138; J. G. Kimball, Phys. Rev. A7 (1973) 1648; G: Niklasson, Phys. Rev. B10 (1974)3052.

5) Y. Ousaka, H. Suehiro and H. Yasuhara,J. phys. C17 (1985) 4951; ibid. 18 (1985) 4471; ibid. 19 (1986) 4247; ibid. 19 (1986) 4263; ibid. 21 (1988) 4045.
H. Suehiro, Doctor Thesis of Hokkaido University (1987).

6) M. Gell-Man, Phys. Rev. 106 (1957) 369.

7) T. M. Rice, Ann. Phys. NY 31 (1965) 100; L. Hedin, Phys. Rev. 139A (1965) 796; A. W. Overhauser, Phys. Rev. B3 (1971) 1388.
L. Hedin and S. Lundqvist, Solid State Physics, Vol. 23, eds. F. Seitz, D. Turnbull and H. Ehrenreich(ACADEMIC PRESS, New York, 1969) pp. 1-181.

8) H. Yasuhara and Y. Ousaka, Solid State Commun. 64 (1987) 673.

RPA, VERTEX CORRECTION AND SUPERCONDUCTIVITY IN TWO-DIMENSIONAL MODELS

Kenji YONEMITSU

Department of Physics, University of Tokyo, Bunkyo-ku, Tokyo 113, Japan

Superconductivity is investigated, in a CuO_2 plane which includes the hopping integral t between copper d and oxygen p orbitals, the energy difference Δ of the two orbitals, and the on-site Coulomb repulsion U on the d orbital. It is assumed that spin fluctuations, which are determined self-consistently in RPA, mediate pairing near the antiferromagnetic transition. When Δ/t is large, the model is effectively reduced to the Hubbard model. In this model, the effect of vertex correction is studied and found to increase the transition temperature by several times. When Δ/t is small, the lower band which involves the Fermi surface becomes p-like and d holes become heavy with increasing U. However, superconducting states are found neighboring antiferromagnetic states when U is not too large.

1. INTRODUCTION

Pairing mechanism of high-T_c superconductivity in oxides[1] has been studied with a lot of efforts. Some kind of spin fluctuations has been considered to give rise to a pairing interaction because the superconductivity appears near an antiferromagnetic state.[2]

The oxides have CuO_2 planes in common, which contain copper $3d_{x^2-y^2}$ orbitals and oxygen $2p_x$, $2p_y$ orbitals. The two-dimensional Hubbard model has been frequently used for simplicity. According to the idea that spin fluctuation estimated in RPA mediates the pairing, perturbation calculations have been performed to show that d-wave superconductivity is realized in this model.[3]

In this paper, superconductivity is investigated in a CuO_2 plane, where only the lowest band is partly occupied by holes. When the energy difference Δ between d and p orbitals is large enough, compared with the hopping integral t between them and the on-site Coulomb repulsion U on the d orbital, the lowest band is d-like and this system is reduced to the Hubbard model. The effect of vertex correction is studied in the Hubbard model. The correction enhances scatterings near momentum transfers $(\pm\pi, \pm\pi)$, which cause the gap function to have a d-wave symmetry. Though it increases the effective mass at the same time, it leads to an increase of the transition temperature by several times.

When Δ is comparable to or smaller than t, the lowest band has a mixed character of d-like and p-like. As U increases, the character becomes p-like and the effective mass of d holes becomes heavy. Consequently, spin fluctuations are suppressed when U is very large. However, d-wave superconductivity appears neighboring to an antiferromagnetic state when U is not too large.

2. SELF-CONSISTENCY EQUATIONS

The model of a CuO_2 plane, which is called dp model hereafter, is given by the Hamiltonian

$$\mathcal{H}_{dp} = -\Delta \sum_{i,\sigma} d_{i\sigma}^\dagger d_{i\sigma} - t \sum_{<i,j>,\sigma} [(d_{i\sigma}^\dagger a_{j\sigma} + d_{i\sigma}^\dagger b_{j\sigma}) + \text{h.c.}] + U \sum_i n_{di\uparrow} n_{di\downarrow} \quad , \tag{1}$$

where $n_{di\sigma} = d_{i\sigma}^\dagger d_{i\sigma}$, $d_{i\sigma}^\dagger$, $a_{i\sigma}^\dagger$, and $b_{i\sigma}^\dagger$, creates holes with spin σ at site i in the copper $d_{x^2-y^2}$, oxygen p_x, and p_y orbitals, respectively. An oxygen site is located on each bond between neighboring copper sites, while the copper sites are on a square lattice. The nearest-neighbor sites are denoted by $<i,j>$. With $\Delta = t_H/\eta^2$, $t = t_H/\eta$, $U = U_H$, and the chemical potential $\mu = \mu_H - 4t_H - t_H/\eta^2$ in the limit $\eta \to 0$, the dp model is reduced to the Hubbard model on a square lattice with the hopping integral t_H, the on-site Coulomb repulsion U_H, and the chemical potential μ_H. In the self-energy part $\Sigma(p)$, only the dd-component $\Sigma_{dd}(p)$ is non-vanishing because the interaction exists only between d holes.

The normal state self-energy part, which is represented diagrammatically in Fig. 1, is determined self-consistently up to the linear term in $i\omega_n$

$$\Sigma_{dd}(p) = \chi(\vec{p}) + i\omega_n(1 - z(\vec{p})) \ , \ \text{with} \ \chi(\vec{p}) = \Sigma_{dd}(\vec{p}, i\omega_n = 0) \ , \ z(\vec{p}) = 1 - \frac{\partial \Sigma_{dd}(\vec{p}, i\omega_n)}{\partial i\omega_n}\Big|_{i\omega_n=0} \ . \quad (2)$$

It is evaluated at zero temperature to be decoupled from the anomalous self-energy part. The dd- and pp-components of the normal state Green function are given by

$$G_{dd}(p) = \frac{1}{z(\vec{p})} \sum_{\gamma=\pm} w_{+\gamma}(\vec{p}) \frac{1}{i\omega_n - \xi_\gamma(\vec{p})} \ , \quad G_{pp}(p) = \sum_{\gamma=\pm} w_{-\gamma}(\vec{p}) \frac{1}{i\omega_n - \xi_\gamma(\vec{p})} \ , \quad (3)$$

where p is a linear combination of a and b holes and orthogonal to the nonbonding combination. The weight of the d orbital in the γ band is denoted by

$$w_\gamma(\vec{p}) = \frac{\xi_\gamma(\vec{p}) - (-\mu)}{\xi_\gamma(\vec{p}) - \xi_{-\gamma}(\vec{p})} \ . \quad (4)$$

The one-particle energy $\xi_\pm(\vec{p})$ is a function of $\chi(\vec{p})$, $z(\vec{p})$, μ and the bare bonding and antibonding energies $\epsilon_\pm(\vec{p})$. The irreducible bubble diagram is determined by

$$\chi_d(q) \equiv -\sum_k G_{dd}(k)G_{dd}(k+q) = \sum_{\vec{k},\nu\mu} \frac{w_\nu(\vec{k})w_\mu(\vec{k}+\vec{q})}{z(\vec{k})z(\vec{k}+\vec{q})} \frac{f_\mu(\vec{k}+\vec{q}) - f_\nu(\vec{k})}{\xi_\nu(\vec{k}) - \xi_\mu(\vec{k}+\vec{q}) + i\omega_q} \ , \quad (5)$$

$f_\nu(\vec{k})$ being the Fermi distribution function at zero temperature of the quasihole with energy $\xi_\nu(\vec{k})$.

The transition temperature T_c and the gap function $\Delta(\vec{p})$ at $T = T_c$ are obtained through transformation of the gap equation into

$$\Delta(\vec{p}) = -\sum_{\vec{k}} \frac{R(\vec{p},\vec{k})}{z(\vec{p})z(\vec{k})} [\sum_\gamma u_\gamma(\vec{k}) \frac{\tanh(\xi_\gamma(\vec{k})/2T_c)}{2\xi_\gamma(\vec{k})}] \Delta(\vec{k}) \ , \quad (6)$$

$$u_\gamma(\vec{p}) \equiv \frac{\xi_\gamma(\vec{p})^2 - (-\mu)^2}{\xi_\gamma(\vec{p})^2 - \xi_{-\gamma}(\vec{p})^2} \ , \quad (7)$$

whose kernel depends on momentum variables only. The procedure has been frequently used in the electron gas.[4] Here the kernel is determined by

$$R(\vec{p},\vec{k}) \simeq R^{(0)}(\vec{p},\vec{k}) \equiv \mathcal{S}_{n(\vec{p})} \mathcal{S}_{m(\vec{k})} \Gamma_d(p,k) \ , \quad (8)$$

where the vertex part $\Gamma_d(p,k)$ for the particle-particle channel is represented diagrammatically in Fig. 2 and averaging over a frequency variable is performed by

$$\mathcal{S}_{n(\vec{p})} \cdots = \int_{-\infty}^\infty \frac{d\omega}{2\pi} \sum_\gamma \frac{u_\gamma(\vec{p})}{2|\xi_\gamma(\vec{p})|} \frac{2|\xi_\gamma(\vec{p})|}{\omega^2 + \xi_\gamma(\vec{p})^2} \cdots / \sum_\nu \frac{u_\nu(\vec{p})}{2|\xi_\nu(\vec{p})|} \ . \quad (9)$$

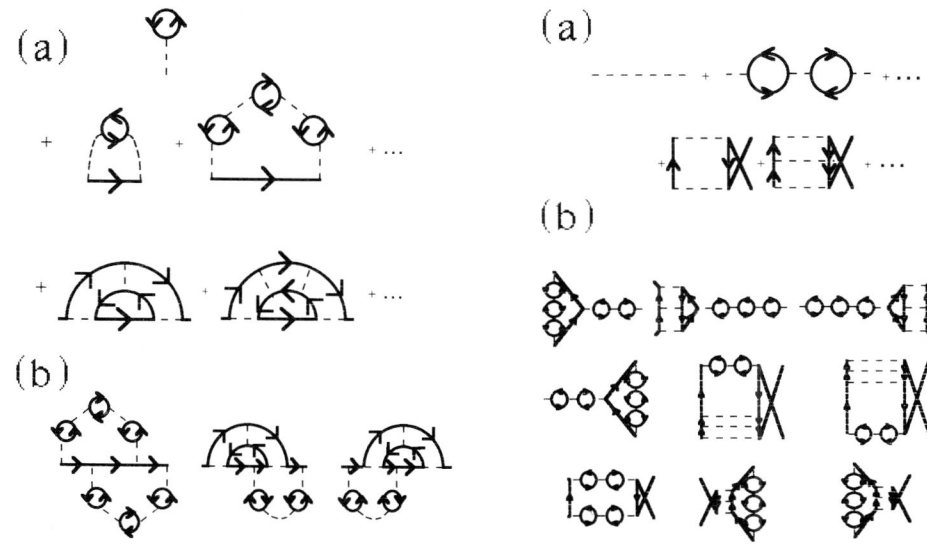

FIGURE 1
(a) The RPA normal state self-energy part with the Hartree term. (b) Its lowest order corrections.

FIGURE 2
(a) The RPA vertex part for the particle-particle channel. (b) Its lowest order corrections.

3. VERTEX CORRECTION IN THE HUBBARD MODEL

In RPA, the kernel $R(\vec{p}, \vec{k})$ has a large amplitude when both $\vec{p}$ and $\vec{k}$ are close to the Fermi line. Where momentum transfers $\vec{q} = \vec{p} \pm \vec{k}$ are close to (π, π), it has large narrow peaks, leading to a d-wave symmetry of the gap function. The ω independent term $\chi(\vec{p})$ of the normal state self-energy part behaves in such a way that the electronic density of states decreases near the Fermi line. The effective mass $z(\vec{p})$ has a large value near the Fermi line. The gap function has a large amplitude near the Fermi line.

The correction to RPA enhances these properties in RPA. It enhances strong scatterings with momentum transfers $\vec{q} \simeq (\pm\pi, \pm\pi)$ by the same order of magnitude as the RPA kernel $R(\vec{p}, \vec{k})$. The ω independent term $\chi(\vec{p})$ decreases the electronic density of states near the Fermi line further. The effective mass $z(\vec{p})$ near the Fermi line increases by a few percent. The gap function $\Delta(\vec{p})$ has a larger amplitude near the Fermi line than that in RPA.

The corrections are regarded as to renormalize the normal and anomalous self-energy parts by multiplying factors which are larger than unity and smooth functions of momenta. Consequently, the transition temperature T_c is raised. It is also found that the range of the hole concentration becomes wide with which the superconductivity takes place.

4. SUPERCONDUCTIVITY IN THE CuO_2 PLANE

In this section, the dp model is studied in RPA. When Δ/t is large and U/t is small, the lowest band is d-like ($w_-(\vec{p}) \simeq 1.0$) (Fig. 3), and superconductivity has similar properties to that in the Hubbard model.

When Δ is comparable to or smaller than t, the lowest band has a mixed character of d-like and p-like. Its weight $w_-(\vec{p})$ of the d orbital becomes small as U increases (from $w_-(\vec{p}) \geq 0.5$ ($U = 0$)

to $w_-(\vec{p}) \simeq 0.0$ ($U \gg t$)), mainly due to the Hartree term. The situation differs from that in the Hubbard model, where the contribution from the Hartree term can be abscrbed into the chemical potential. Then the increase of the d hole density is suppressed upon doping (Fig. 3). Because the effective mass is raised only for d holes by the interaction, the d hole density has a much smaller jump at the Fermi line than the p hole density. Therefore, the low-lying excitations are p-like quasiholes.

The irreducible bubble $\chi_d(q)$ has large contribution from intraband excitations in the lower band, so that it is small when the weight $w_-(\vec{p})$ is small and the effective mass $z(\vec{p})$ of d holes is large, as shown in (5). Spin fluctuations are determined by the irreducible bubble $\chi_c(q)$ multiplied by U in RPA. When U is large but $\chi_d(q)$ is small, RPA does not have its validity. Tentative results with $t = 1.0$, $\Delta = 1.0$, and $U = 100$ show that spin fluctuations do not become so large as to give rise to antiferromagnetic or superconducting states.

On the other hand, a superconducting state and an antiferromagnetic state appear as shown in Fig. 4, at least when $t = 1.0$, $\Delta = 1.0$, and $1.0 \leq U \leq 10.0$. The antiferromagnetic state is denoted by hatching in the figure. All the superconducting states which are found have a d-wave symmetry and are neighboring antiferromagnetic states.

When U is much larger than t and Δ, holes are expected to occupy mainly d orbitals until n becomes unity, and mainly p orbitals afterwards. Meanwhile, in the present Fermi-liquid approach, the curves of the d and p hole densities do not have singularity, but they are smooth near $n \simeq 1.0$. However, tendency up to intermediate U would be reflected in the present results, where d-wave superconductivity takes place near the antiferromagnetic transition, as in the Hubbard model.

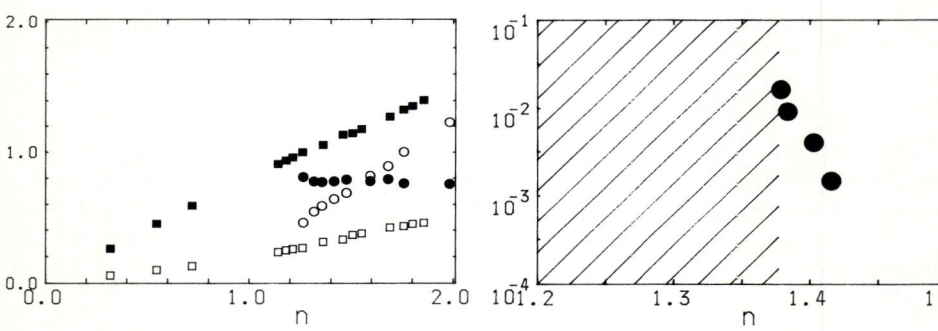

FIGURE 3
The d hole density (closed symbols) and the p hole density (open symbols), as functions of the total hole density n, when $t = 1.0$, $\Delta = 3.0$, $U = 1.0$ (squares) and when $t = 1.0$, $\Delta = 1.0$, $U = 8.0$ (circles).

FIGURE 4
The transition temperature T_c, as a function of the total hole density n. $t = 1.0$, $\Delta = 1.0$, $U = 8.0$.

REFERENCES

1] G. Bednorz and K. A. Müller: Z. Phys. B **64** (1986) 189.

2] G. Shirane, Y. Endoh, R. J. Birgeneau, M. A. Kastner, Y. Hidaka, M. Oda, M. Suzuki, and T. Murakami: Phys. Rev. Lett. **59** (1987) 1613.
Y. Tokura, J. B. Torrance, T. C. Huang, and A. I. Nazzal: Phys. Rev. B **38** (1988) 7156.

3] D. J. Scalapino, E. Loh, Jr., and J. E. Hirsch: Phys. Rev. B **34** (1986) 8190.
N. E. Bickers, D. J. Scalapino, and R. T. Scalettar: Int. J. Mod. Phys. B **1** (1987) 687.
H. Shimahara and S. Takada: J. Phys. Soc. Jpn. **57** (1988) 1044.
N. E. Bickers, D. J. Scalapino, and S. R. White: Phys. Rev. Lett. **62** (1989) 961.

4] D. A. Kirzhnits, E. G. Maksimov, and D. I. Khomskii: J. Low Temp. Phys. **10** (1973) 79.

ABSENCE OF EXPONENTIAL SCREENING IN QUANTUM MECHANICAL PLASMAS

A. ALASTUEY * and Ph.A. MARTIN **

(*) Laboratoire de Physique Théorique et Hautes Energies, Université de Paris-Sud, F-91405 Orsay Cedex, France
(**) Institut de Physique Théorique, Ecole Polytechnique Fédérale de Lausanne, CH-1015 Lausanne, Switzerland

1. INTRODUCTION

One of the most fundamental property of a system of charged particles in thermal equilibrium is the screening of the Coulomb potential $\phi(r) = 1/r$. The following familiar and widely accepted picture is part of the standard background knowledge. A particle is surrounded by a screening cloud of opposite charge having an extension λ, the screening length. The charge distribution of the particle together with its cloud produces, in the medium, an effective potential $\phi_{\text{eff}}(r) \sim \exp(-r/\lambda)/r$, which is exponentially screened. For classical systems, this picture has been first introduced by Debye and Hückel on the basis of a mean-field theory, and has been firmly established in recent years through rigorous proofs valid in the weak-coupling regime[1], and in two-dimensional solvable models[2]. For quantum systems, the mean-field treatments, like the Thomas-Fermi theory or the random phase approximation, also predict an exponential screening. However, the rigorous results are still scarce, and even doubts have been raised on the possible exponential fall-off of the quantum charge-charge correlations[3]. The point of this paper is to present strong evidences that the equilibrium correlations of quantum mechanical charges do not cluster exponentially fast, irrespective of the value of the density and of the temperature. Thus the Debye screening does not exist, strictly speaking, in real matter.

Let us give a qualitative understanding for this lack of exponential clustering. In the classical case, the fall-off of the correlations faster than any inverse power implies that all the multipoles of any given particle configuration are shielded by the corresponding screening clouds[4]. This perfect organization is destroyed by the quantum fluctuations. In the quantum system, there is a minimal screening which ensures that the monopoles, i.e. the total charges, vanish (there is no bare Coulomb potential seen in equilibrium matter), but the higher order multipoles do not vanish in general. The latter generate then multipole forces which in turn induce algebraic tails in the correlations. These multipole forces are different from the Van der Waals forces due to complex polarizable entities, atoms or molecules, as it is examplified in the case of the one-component plasma (OCP) where no binding occurs. In fact, our study shows that these Van der Waals forces cannot be exponentially screened by the free quantum charges and thus also contribute to the algebraic tails.

Our strong evidences for the absence of exponential screening rely on the following two approaches. First, we investigate the large-distance behaviour of the quantum corrections to the classical correlations for both an OCP and a multicomponent system described by Maxwell-Boltzmann statistics (section 2). This investigation gives algebraic lower bounds for the decay of the quantum correlations (for instance, the particle-particle correlations should not decay faster than $1/r^6$ in a multicomponent system). Second, we study a simplified model where only two quantum charges are immersed in a classical plasma (section 3). We then determine the exact asymptotic behavior of the correlations of these two charges, which is found as $1/r^6$ when $r \to \infty$. A preliminary[5] and a detailed[6] account of this work are published elsewhere, while part of it is reviewed[4].

2. SEMI-CLASSICAL EXPANSIONS

First, we consider the case of the OCP. The OCP is an ensemble of identical point particles with charge e and mass m, embedded in a uniform neutralizing background with charge density $-e\rho$ if ρ is the mean-particle density. We have calculated the $\hbar^4$-term in the Wigner-Kirkwood (WK) expansion of the equilibrium quantum charge-charge correlation $S^{qm}(r)$, with Maxwell-Boltzmann statistics ($\hbar$ is the Planck constant). We find

$$S^{qm}(r) = S(r) + \hbar^2 S^{(2)}(r) + \hbar^4 S^{(4)}(r) + \ldots \tag{1}$$

where $S(r)$ is the corresponding classical function, and $S^{(2)}(r)$, $S^{(4)}(r)$, ..., are also expressible in terms of the classical correlation functions of the OCP. The latter are known to cluster exponentially fast in the Debye screening regime[1]. So does the second-order term. The fourth-order term can be written as

$$S^{(4)}(r) = \frac{2}{5}\left(\frac{\beta^2 e}{24m}\right)^2 \int d\underline{r}_1\, d\underline{r}_2\, S(|\underline{r}-\underline{r}_1|)\, f(|\underline{r}_1-\underline{r}_2|)\, S(r_2) + \text{exponential terms} \tag{2}$$

where $f(r)$ behaves as the square of the dipole potential for large r,

$$f(r) \sim \sum_{ij}^{3} \left(\partial_i \partial_j \frac{1}{r}\right)^2 \tag{3}$$

One finds that the asymptotic behavior of (2) is algebraic, i.e.

$$S^{(4)}(r) \sim \frac{7}{16\pi^2}\left(\frac{\beta e}{m}\right)^2 \frac{1}{r^{10}}, \quad \text{when } r \to \infty. \tag{4}$$

We have extended the above analysis to all the terms of the WK expansion of the equilibrium correlations of a multicomponent quantum system, with a regularized Coulomb potential and Maxwell-Boltzmann statistics. Starting from the functional integration formalism, we introduce a diagrammatic-expansion scheme which generates the WK expansion in a systematic way. This scheme allows a qualitative analysis of the large-distance behaviors of the quantum correlations, up to any order in $\hbar$. All the $\hbar^{2n}$-terms ($n \geq 2$) decay algebraically with powers depending on the nature of the considered correlations. For instance, one finds $1/r^6$, $1/r^8$ and $1/r^{10}$ for respectively the particle-particle, particle-charge and charge-charge correlations; slower decays like $1/r^3$ are found for higher-order correlations when two groups of two or more particles are separated by a large distance r.

The functional integration formalism enlights the essential role played by the intrinsic quantum fluctuations in the appearance of the algebraic tails. As illustrated in section 3, a quantum particle can be considered as a classical random charged filament carrying multipoles. Once the averages over quantum fluctuations have been performed, these "ghost" multipoles give raise to effective algebraic potentials of the type (3) which are ultimately responsible for the algebraic tails in the correlations. Since this mechanism is intrinsic to quantum mechanics, algebraic decay should occur at any value of the thermodynamic parameters. In fact, the power laws computed in the semi-classical expansion can reasonably be viewed as lower bounds on the decay of the quantum correlations. We have checked that these bounds are compatible with the upper bounds found through an asymptotic analysis of the evolution equations for the imaginary time Green functions. This analysis, which is non-perturbative and valid for Fermi statistics, makes extensive use of the Kubo-Martin-Schwinger equilibrium condition and relies on reasonable a priori assumptions (existence of the thermodynamic limit, monotonous decay).

As far as the statistics is concerned, the exchange terms in the OCP are expected to be exponentially small with $\hbar$, while for multicomponent systems, the short-range regularization of the Coulomb potential should not drastically alter the large-distance behaviors in the semi-classical regime. The quantum algebraic tails are expected to have a little practical influence for systems where the quantum effects are small. For instance, in the cases of sodium chloride electrolyte at room temperature, and of the interior of a white dwarf made of C^{12} nuclei embedded in a non-polarizable degenerate electron gas, we have checked that the cross-over distance above which the algebraic tails dominate the classical exponential behavior, is forty to sixty times the classical screening length. Of course, in systems where the quantum effects are essential (like the electrons in a metal), the algebraic tails should play a more important role.

3. THE HYDROGEN ATOM IN A CLASSICAL PLASMA

The model consists of two quantum mechanical particles of charges e_1, e_2 and masses m_1, m_2 embedded in a classical plasma and in thermal equilibrium with it. The hamiltonian of these two quantum charges is

$$H(C) = \frac{|p_1|^2}{2m_1} + \frac{|p_2|^2}{2m_2} + e_1 e_2 \phi(|r_1-r_2|) + \sum_{i=1}^{2} e_i \int dr \, \phi(|r_i-r|) Q(r, C) \tag{5}$$

with $Q(r, C)$ the charge density of the classical particles in the configuration C. The corresponding equilibrium distribution is defined by

$$\rho(r_1, r_2) = Z_0^{-1} \int dC \, (r_1, r_2 | \exp[-\beta H(C)] | r_1, r_2) \exp[-\beta V_0(C)] \tag{6}$$

where $V_0(C)$ is the Coulomb energy of the classical gas and Z_0 is the corresponding partition function. With the help of the Feynman-Kac formula, one can represent the kernel in (6) as a functional integral on closed Brownian paths $\xi_i(s)$, $0 \leq s \leq 1$, with gaussian measures $D\xi_i$

$$(r_1, r_2 | \exp[-\beta H(C)] | r_1, r_2) = (2\pi \lambda_1 \lambda_2)^{-3} \int D\xi_1 \int D\xi_2 \exp[-\beta E(r_1 \xi_1, r_2 \xi_2; C)] \tag{7}$$

$$E(r_1 \xi_1, r_2 \xi_2; C) = e_1 e_2 \int_0^1 ds \, \phi(|r_1 - r_2 + \lambda_1 \xi_1(s) - \lambda_2 \xi_2(s)|)$$

$$+ \sum_{i=1}^{2} \int dr \int_0^1 ds \, \phi(|r_i + \lambda_i \xi_i(s) - r|) Q(r, C) \tag{8}$$

and $\lambda_i = (\beta \hbar^2/m_i)^{1/2}$ are the thermal de Broglie lengths.

It is useful to consider the quantum particles as random charged filaments at r_1 and r_2 with charge densities

$$e_i n_i(r) = e_i \int_0^1 ds \, \delta(r_i + \lambda_i \xi_i(s) - r), \qquad i = 1, 2 \tag{9}$$

and to compare the quantum energy (8) with the corresponding classical interaction energy $U_{cl}(r_1 \xi_1, r_2 \xi_2; C)$. These two energies differs obviously by the term (with $r = r_1 - r_2$)

$$W(r, \xi_1, \xi_2) = e_1 e_2 \int_0^1 ds_1 \int_0^1 ds_2 \left(\delta(s_1-s_2) - 1\right) \phi(|r + \lambda_1 \xi_1(s) - \lambda_2 \xi_2(s)|) \tag{10}$$

which behaves as a dipole-dipole potential as $r \to \infty$

$$W(r, \xi_1, \xi_2) \sim \frac{e_1 e_2}{2} \int_0^1 ds_1 \int_0^1 ds_2 \left(\delta(s_1-s_2) - 1\right) \left[(\lambda_1 \xi_1(s_1) - \lambda_2 \xi_2(s_2)) \cdot \nabla\right]^2 \frac{1}{r} \tag{11}$$

Introducing the decomposition $E = U_{cl} + W$ in (7) and noting that

$$\int dC \exp\left[-\beta(V_0(C) + U_{cl}(r_1 \xi_1, r_2 \xi_2; C))\right] = Z(r_1 \xi_1, r_2 \xi_2)$$

is the partition function of the classical gas in presence of the two additional external filamentous charges (9), the distribution (6) takes the form

$$\rho(\underline{r}_1, \underline{r}_2) = (2\pi\,\lambda_1\lambda_2)^{-3} \int D\xi_1\, D\xi_2\, \exp\left[-\beta\bigl(W(\underline{r}_1 - \underline{r}_2, \xi_1, \xi_2) + F(\underline{r}_1\,\xi_1, \underline{r}_2\,\xi_2)\bigr)\right] \quad (12)$$

where $F(\underline{r}_1\,\xi_1, \underline{r}_2\,\xi_2)$ is the excess free energy when the two charges (9) are immersed in the classical gas. Finally, the dimensionless truncated correlation of the quantum charges $h(|\underline{r}_1-\underline{r}_2|) = \rho(\underline{r}_1)^{-1}\,\rho(\underline{r}_2)^{-1}\,\rho(\underline{r}_1, \underline{r}_2) - 1$ ($\rho(\underline{r}_i)$ are the one point distributions) takes the simple form in the thermodynamic limit

$$h(r) = \int \tilde{D}\xi_1 \int \tilde{D}\xi_2 \left\{\exp\left[-\beta(\phi_{eff}(\underline{r}, \xi_1, \xi_2)) + W(\underline{r}, \xi_1, \xi_2)\right] - 1\right\} \quad (13)$$

where $\tilde{D}\xi_1$ are renormalized Brownian paths measures. In (13) $\phi_{eff}(\underline{r}, \xi_1, \xi_2)$ is the classical effective potential between the two charged filaments.

Assuming that the classical plasma is in the Debye-Hückel regime, external charges are perfectly screened, implying that $\phi_{eff}(\underline{r}, \xi_1, \xi_2)$ decays exponentially fast as $r \to \infty$. It is therefore the quantum-mechanical potential W that governs the asymptotic behavior of $h(\underline{r})$. Neglecting ϕ_{eff}, expanding $\exp(-\beta W)$ in (13) and using (11), we can generate an asymptotic inverse-power expansion for $h(r)$

$$h(r) = -\beta \int \tilde{D}\xi_1 \int \tilde{D}\xi_2\, W(\underline{r}, \xi_1, \xi_2) + \frac{\beta^2}{2}\int \tilde{D}\xi_1 \int \tilde{D}\xi_2\, (W(\underline{r}, \xi_1, \xi_2))^2 + \ldots \quad (14)$$

As a consequence of rotation invariance and of the harmonicity of the Coulomb potential, it is easily seen that the first order term does not contribute, while one finds that the second order term in (14) behaves as B/r^6, B a strictly positive constant. The higher order terms have a faster decay, thus $h(r) \sim B/r^6$, $r \to \infty$. This $1/r^6$ behavior originates from the averaged square of the random dipole-dipole like interaction W. It is in agreement with the findings of the perturbative analysis of section 2.

REFERENCES

1) D. Brydges and P. Federbush, Comm. Math. Phys. 73 (1980) 197; J. Imbrie, Comm. Math. Phys. 87 (1983) 515.

2) B. Jancovici, Solvable Models, this volume

3) D. Brydges and E. Seiler, J. Stat. Phys. 42 (1986) 405

4) Ph. A. Martin, Rev. Mod. Phys. 60 (1988) 1075

5) A. Alastuey and Ph. A. Martin, Europhys. Lett. 6 (1988) 385

6) A. Alastuey and Ph. A. Martin, Phys. Rev. A, in print.

Chapter VII:
Metallic Systems

NATURE OF PHONONS, ISOTOPE EFFECT, AND SUPERCONDUCTIVITY IN $Ba_{1-x}K_xBiO_3$

Macros H. DEGANI,[1] Rajiv K. KALIA and P. VASHISHTA

Materials Science Division, Argonne National Laboratory, Argonne, IL 60439, USA

Phonon density-of-states of insulating $BaBiO_3$ in orthorhombic phase and superconducting $Ba_{0.6}K_{0.4}BiO_3$ in cubic phase are studied using the molecular dynamics (MD) method. The MD results are compared with the recent inelastic neutron scattering experiments and electron tunneling experiments. The exponent of the oxygen isotope effect is calculated from the first moment of the phonon density of states (the weak coupling limit) and from the solution of Eliashberg gap equations. Results are compared with isotope effect experiments. Evidence based on inelastic neutron scattering, tunneling, and isotope effect experiments when combined with the MD calculations suggest that this material is a normal weak coupling BCS superconductor with strong coupling of the carriers to the high energy oxygen phonons.

1. INTRODUCTION

1.1 General Remarks

Recent discovery of superconductivity in $Ba_{1-x}K_xBiO_3$ at 30K is of great interest because of the difference in the properties of this material and other high-T_c oxide superconductors. $Ba_{1-x}K_xBiO_3$ is cubic in the superconducting phase, $0.35<x<0.5$, whereas the other high-T_c materials have a distinctly planar (Cu-O) structure. This material has no copper and displays none of the antiferromagnetism common to other high-T_c materials. Measurements by Hinks et al.[1] on $Ba_{0.6}K_{0.4}BiO_3$ indicate a substantial oxygen isotope effect, $\alpha=0.4$, unlike the $YBa_2Cu_3O_{7-\delta}$ which shows a negligible isotope effect. Batlogg et al.[2] find a smaller value $\alpha=0.22$, whereas in a more recent experiment Kondoh et al.[3] have determined $\alpha=0.35$ in $Ba_{0.6}K_{0.4}BiO_3$. Infrared measurements[4] reveal a superconducting gap ($2\Delta=8.7$ meV) with $2\Delta/k_BT_c=3.5$. Since there is no evidence for

---

[1]Permanent Address: Instituto de Física e Química de São Carlos, USP 13560 - São Carlos - S.P. - Brasil

magnetic fluctuations and the transition temperature is not so high that one has to invoke an exotic mechanism, it is quite possible that phonon coupling between carriers is responsible for the superconductivity in $Ba_{1-x}K_xBiO_3$.[5] Attempts to carry out electron tunneling experiments in superconductor-insulator-superconductor (S-I-S) junctions in $YBa_2Cu_3O_{7-\delta}$ have not been very successful due to the very small coherence length (~ 10Å). In $Ba_{1-x}K_xBiO_3$, however, electron tunneling experiments on S-I-S junctions have recently been carried out by Zasadzinski et al..[6] This is due in part to the fact that the coherence length in this material is of the order of ~50Å. Even though it has not been possible to invert the Eliashberg gap equations using the tunneling data, it is quite clear that there are images of excitations in the tunneling data at high energies in the range of 60 meV. This observation of high energy structure in the tunneling data has important consequences as we shall discuss in the following sections.

The pairing theory[7] (BCS) has given an accurate account of the properties of conventional superconductors as well as superfluid ^{3}He. With the discovery of high-T_c oxide superconductors, the question has been raised whether the pairing theory continues to be applicable but with an exceptionally strong pairing interactions or whether a totally different framework must be developed. It is easier to address some of these questions on cubic, nonmagnetic 30K oxide superconductors $Ba_{1-x}K_xBiO_3$ than on highly complex materials like $YBa_2Cu_3O_{7-\delta}$ with antiferromagnetic phases lingering in the vicinity of superconducting phase.

1.2 Structural Information and T_c versus Composition

Using neutron diffraction technique Shiyou Pei et al.[8] have determined the structure of five phases in the $Ba_{1-x}K_xBiO_3$ system for potassium concentrations in the range 0.0<x<0.5 below 473K. Only the perovskite phase which exists for x>0.37 exhibits bulk superconductivity. Fig. 1 shows the crystal structures in $Ba_{1-x}K_xBiO_3$, as a function of temperature and potassium concentration, as well as the behavior of the superconducting transition temperature.

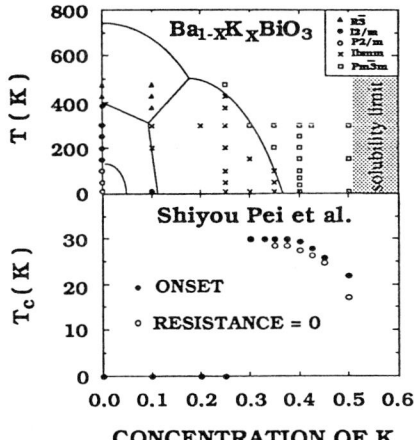

FIGURE 1
Phase diagram for $Ba_{1-x}K_xBiO_3$ and T_c versus concentration.

2. MOLECULAR DYNAMICS SIMULATION

2.1 Interaction Potentials

Effective interparticle interactions were used in the molecular dynamics simulations. The potentials[9] include steric repulsions between ions, Coulomb interactions due to charge transfer effects, and charge-dipole interactions due to large electronic polarizability of O^{--} ions. For $BaBiO_3$ there are six interaction potentials, whereas for $Ba_{0.6}K_{0.4}BiO_3$ there are ten.[10]

2.2 Molecular Dynamics Method

The molecular dynamics[11] calculations were done with 540 particle orthorhombic $BaBiO_3$ and 625 particle cubic $Ba_{0.6}K_{0.4}BiO_3$ at the experimental number densities. The unit cells of the two systems are shown in Fig. 2. The structural parameters used in the MD simulations are given in Table I.

Table I   Experimental data used in the MD simulations.

	$BaBiO_3$	$Ba_{0.6}K_{0.4}BiO_3$
Mass density	7.88 g/cm^3	7.33 g/cm^3
Lattice Structure	orthorhombic a=6.2000Å b=6.1561Å c=4.3474Å	cubic a=4.3160Å

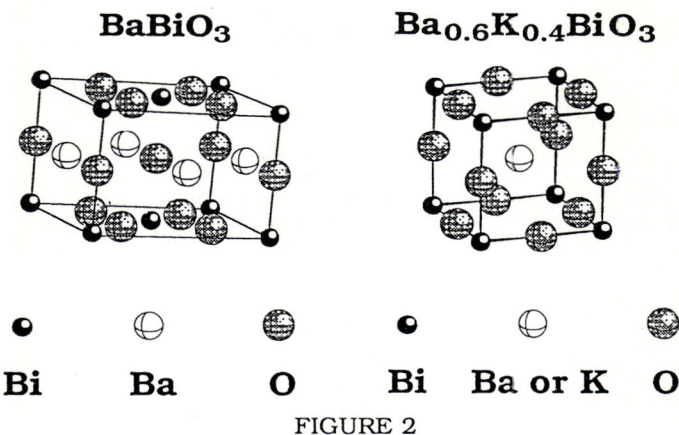

FIGURE 2
Unit cell for $BaBiO_3$ and $Ba_{1-x}K_xBiO_3$.

The $Ba_{0.6}K_{0.4}BiO_3$ system was obtained from $BaBiO_3$ by randomly replacing 40% of the Ba atoms with K atoms. Before calculating the phonon DOS, it was ensured that the systems were dynamically stable in the appropriate symmetries. The phonon DOS was calculated using three methods: (1) velocity auto-correlation functions, (2) the equation of motion method,[12] and (3) direct diagonalization of the dynamical matrix. The results of all these three calculations are in good agreement with one another.

2.3 Dynamical Stability of the Systems

To establish the dynamical stability of $BaBiO_3$, the system was put in the orthorhombic structure in a fixed volume MD cell. The partial pair distribution function and bond angle distributions were calculated to verify the bond lengths and coordination numbers etc. The system was slowly heated to 600K and thermalized for several thousand time steps. The system was run uninterruptedly for more than 30,000 time steps and various structural correlations were calculated to examine the symmetry. The system at 600K was slowly cooled and then subjected to a steepest descent quench[13](SDQ) which is a mathematically well defined method of examining the underlying mechanically stable structures. The pair correlation functions and bond angle distributions were calculated again to ascertain the symmetry of the MD system. After performing the above mentioned procedure on the 540 particle $BaBiO_3$ system it was determined that the resulting final symmetry was indeed the same as that of the starting orthorhombic structure. The 625 particle cubic $Ba_{0.6}K_{0.4}BiO_3$ system was obtained by randomly replacing 40% of the Ba atoms with K atoms. This system was subjected to the same procedure to ensure the dynamic stability of $Ba_{0.6}K_{0.4}BiO_3$ system.

3. PHONON DENSITY OF STATES

3.1 Methods of Calculation

Before calculating the phonon density of states it was ensured that the systems were dynamically stable. The phonon density of states was calculated using three different methods.

The first method involves calculating the velocity autocorrelation functions for each species of atoms from the molecular dynamics trajectories. The partial densities of states are obtained by fourier transforming the time correlations.

$$F_N(\omega) = \sum_\beta \left(4\pi b_\beta^2 \right) C_\beta F_\beta(\omega) / \sum_\beta 4\pi b_\beta^2 C_\beta \qquad (1)$$

$$F(\omega) = \sum_\beta C_\beta F_\beta(\omega) \qquad (2)$$

$$F_\beta(\omega) = \int_0^\tau Z_\beta(t) \cos(\omega t) \, e^{-\lambda(t/\tau)^2} \, dt \qquad (3)$$

$$Z_\beta(t) = \frac{\left\langle \sum_{i(\beta)} \mathbf{v}_i(t) \cdot \mathbf{v}_i(0) \right\rangle}{\left\langle \sum_{i(\beta)} \mathbf{v}_i^2(0) \right\rangle} \qquad (4)$$

where $Z_\beta(t)$ is the normalized velocity autocorrelation function for β^{th} species and its fourier transform $F_\beta(\omega)$ is the partial density of states normalized to the total number of particles, N, of the system. The total density of states is obtained by summing the concentration weighted partial densities of states. Additional weighting with the coherent neutron cross-sections is required to obtain the neutron density of states, $F_N(\omega)$, for comparison with the generalized density of states, $G(\omega)$, obtained in the neutron scattering experiment.[5]

The second method involves calculating the displacement autocorrelation function using the equation of motion method.[12] To implement this method it is essential to bring the system to the local minimum by carrying out the steepest descent quench which guarantees that the force and the velocity for each particle is zero. Each particle is then given a random displacement and the system is allowed to evolve according to the classical equations of motion.

$$F(\omega) = \frac{4}{\pi \delta_m^2} \int_0^\tau f(t) \cos(\omega t) \, e^{-\lambda(t/\tau)^2} \, dt \qquad (5)$$

$$f(t) = \sum_{i\mu} \delta r_{i\mu}(t) \, \delta r_{i\mu}(0) \qquad (6)$$

$$\delta r_{i\mu}(t) = r_{i\mu}(t) - r_{i\mu 0} \qquad (7)$$

$$\delta r_{i\mu}(0) = \delta_m \cos\theta_{i\mu} \qquad (8)$$

where $r_{i\mu}(0)$ and $r_{i\mu}(t)$ are the μth (x, y, or z) component of the displacement of the ith particle, $r_{i\mu 0}$ are the equilibrium positions of the atoms, δ_m is the amplitude of the initial displacement, and $\theta_{i\mu}$ are random angles distributed uniformly between 0 and 2π.

Finally, the phonon density of state can be calculated by direct diagonalization of the dynamical matrix. As with the equation of motion method it is essential to bring the system in the correct symmetry with zero forces on each particle. This is accomplished by applying the SDQ to the MD configuration. Since we know the interaction potential functions under the influence of which the system is dynamically stable, we can calculate the elements of the dynamical matrix by numerically calculating the x, y, and z derivatives of the force on each particle.

$$F_{j\nu} = -\frac{\partial \phi}{\partial r_{j\nu}} \qquad (9)$$

$$D_{ij}^{\mu\nu} = -\frac{1}{\sqrt{M_i M_j}} \frac{\partial F_{j\nu}}{\partial r_{i\mu}} \qquad (10)$$

$$\omega^2 u_{i\mu}^{(0)} = \sum_{j=1}^{N} \sum_{\nu=1}^{3} D_{ij}^{\mu\nu} u_{j\nu}^{(0)}. \qquad (11)$$

Using standard numerical methods the 1620x1620 dynamical matrix for $BaBiO_3$ and the 1875x1875 matrix for $Ba_{0.6}K_{0.4}BiO_3$ was diagonalized to obtain the eigenvectors and eigenvalues from which the partial and total density of states can be obtained.

3.2 Comparison with Neutron Scattering Experiments

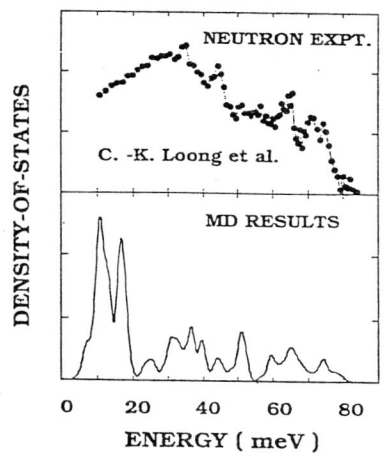
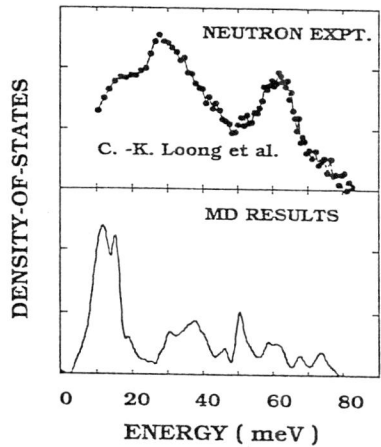

FIGURE 3
Phonon DOS for BaBiO$_3$.
Experiment and MD simulation.

FIGURE 4
Phonon DOS for Ba$_{0.6}$K$_{0.4}$BiO$_3$.
Experiment and MD simulation.

For BaBiO$_3$, Fig. 3 display the phonon density of states from neutron scattering and MD simulation, respectively. The experiment shows prominent peaks at 35, 43, 63, and 71 meV. In addition, there is an indication of a shoulder at 24 meV and weak features in the region of 50-58 meV. In the MD results, we find peaks at 25, 32, 37, 45, 51, 60, 66, and 74 meV. Because of limited resolution of the neutron measurements, the MD peaks at 32 and 37 meV are observed experimentally as a single peak at 35 meV. Similarly, the MD peaks at 60 and 66 meV appear as a single peak at 63 meV in the neutron experiment. Note that there are two additional peaks at 11 and 16 meV in the MD results. These peaks are not observed in the neutron measurements because of the low-energy cutoffs.

For Ba$_{0.6}$K$_{0.4}$BiO$_3$ the phonon density of states are shown in Fig. 4. Neutron measurements reveal that K doping broadens the peaks due to disorder caused by random substitution of K on Ba sites and also shifts the density of states toward lower energies. Because of broadening, only two bands at 30 and 60 meV remain in $Ba_{0.6}K_{0.4}BiO_3$. An overall broadening of the peaks and a shift toward lower energies are also evident from the MD results shown in Fig. 3 and Fig. 4: (1) a band extending from 25 to 37 meV. (2) a peak around 51 meV, (3) a band between 54 and 65 meV, and (4) small peaks at 67 and 73 meV. In addition, there are small features at 25 and 46 meV. The simulation also shows peaks at 11 and 15 meV, whereas only a shoulder at 16 meV is visible experimentally. Higher-resolution neutron measurements may reveal the additional features observed in the simulation.

3.3 Partial Density of States

To understand the origin of the peaks in the density of states, we examine the MD results for the partial phonon density of states for $BaBiO_3$ shown in Fig. 5. There is a clear delineation in the peaks associated with Ba and Bi on one hand and O on the other. For Ba there is only one main peak at 11 meV, whereas for Bi there are two peaks at 12 and 17 meV. Clearly, in the total density of states the peak at 11 meV is due to both Ba and Bi and the peak at 15 meV is due to Bi alone. Above 20 meV the entire spectrum arises from oxygen vibrations.

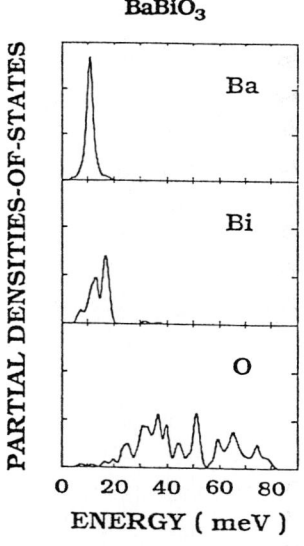

FIGURE 5

Molecular dynamics results for partial DOS for $BaBiO_3$.

For $Ba_{0.6}K_{0.4}BiO_3$, the partial phonon density of states from simulation are shown in Fig. 6. As in the case of $BaBiO_3$, the peaks above 20 meV are due to oxygen vibrations. It is clear from the partial density of states for oxygen in $BaBiO_3$ and $Ba_{0.6}K_{0.4}BiO_3$ that there is an overall broadening and softening of the peaks in $Ba_{0.6}K_{0.4}BiO_3$. The main peak in the K partial density of states is at 20 meV and it strongly overlaps with the second peak at 16 meV in the Bi partial density of states. In the total density of states, the contributions from Ba, K, and Bi give rise to peaks around 11 and 15 meV.

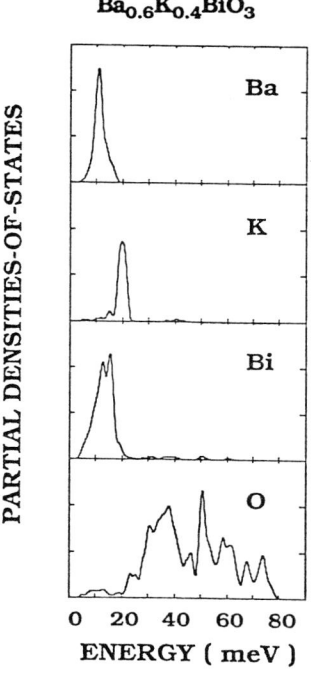

FIGURE 6

Molecular dynamics results for partial DOS for $Ba_{0.6}K_{0.4}BiO_3$.

4. ELECTRON TUNNELING

Electron tunneling experiments on S-I-S and S-I-N junctions were carried out by Zasadzinski et al..[6] All significant features in the MD density of states and neutron generalized density of states are seen in the tunneling measurements. MD calculations were also carried out to determine the energies of symmetric breathing modes for oxygen vibrations around Bi (~

35 meV), and Ba and K (~ 60 meV). Neutron and tunneling experiments together with MD simulations suggest coupling of carriers to oxygen breathing modes around 35 and 60 meV.

5. ISOTOPE EFFECT DUE TO ^{16}O TO ^{18}O SUBSTITUTION

5.1 Reference value of the isotope effect exponent, α_r

The isotope effect is manifested in the vibrations of a lattice when the mass of one or more of the species is changed by isotopic substitution without changing interatomic forces. Let us define the first frequency moment of the phonon density of states as,

$$<\omega> = \int \omega F(\omega)d\omega \; / \int F(\omega)d\omega \; . \tag{12}$$

For a monoatomic system in which M is the mass of each atom, $<\omega>$, behaves as, $<\omega> \approx M^{-1/2}$. For a multicomponent systems such as oxide superconductors, we define the oxygen isotope effect exponent as,

$$<\omega> \approx M_O^{-\alpha_r} \; , \tag{13}$$

where M_O is the oxygen mass and α_r is the reference isotope effect exponent which is generally smaller than 1/2 depending on the masses of other atoms in the unit cell.

When superconductivity is due to electron-phonon coupling and the strong coupling effects are included, the isotope effect of the lattice is reflected through the superconducting transition temperature,

$$T_c \equiv <\omega> e^{-1/N(0)V} \equiv <\omega> e^{-1/(\lambda - \mu^*)} \equiv <\omega> e^{-f(\lambda, \dots, \mu^*)} \; . \tag{14}$$

In an oxide superconductor the oxygen isotope effect in T_c can be expressed as,

$$T_c \approx M_O^{-\alpha}, \quad \alpha = \alpha_r - \delta\alpha, \tag{15}$$

where α_r arises from $<\omega>$ and $\delta\alpha$ from f in the exponent. Clearly, for a monoatomic system α_r is 1/2 and $\delta\alpha$ is a measure of strong coupling corrections. For most oxide superconductor α_r is around 0.4.[10] This is due to heavy masses of other atomic species when compared to the oxygen mass. For $Ba_{0.6}K_{0.4}BiO_3$ the reference isotope effect exponent was calculated from $^{16}<\omega>$ and $^{18}<\omega>$. Results of these calculations are summarized in Table II.

Table II. MD results for the first moment, in meV, and α_r, for $Ba_{0.6}K_{0.4}BiO_3$.

$^{16}<\omega>$	$^{18}<\omega>$	α_r
33.50	31.91	0.41

5.2 Isotope effect exponent, α, from Eliashberg gap equation.

Implications of the strong coupling effects are studied within the Eliashberg theory[14] with a model of $\alpha^2(\omega)F(\omega)$ for ^{16}O, using $\lambda=1$ and $^{16}T_c=29.5K$, which is consistent with the electron tunneling experiments.[6] Using the $^{18}F(\omega)$ and the same model for $\alpha(\omega)$ and the same value of $\mu^*(=0.12)$ as for ^{16}O, the gap equations were solved to obtain $^{18}T_c$. Results of these calculations are summarized in Table III.

Table III. α from T_c for $Ba_{0.6}K_{0.4}BiO_3$ using Eliashberg equations.

$^{16}T_c$ (K)	$^{18}T_c$ (K)	α
29.5	28.2	0.38

6. CONCLUSIONS

MD simulations in conjunction with the neutron, tunneling, infrared, and isotope effect measurements reveal that $Ba_{0.6}K_{0.4}BiO_3$ is a weak coupling superconductor in which the carriers have a strong matrix element to high frequency oxygen vibrations.

Acknowledgements

This work was supported by the U.S. DOE Basic Energy Sciences-Materials Science, under contract No. W-31-109-ENG-38. MD simulations were done on the Energy Research Cray Supercomputer at the National MFE Computing Center(Livermore). M.H.D would like to thank Fundação de

Amparo à Pesquisa do Estado de São Paulo (FAPESP), Brazil for a research fellowship.

REFERENCES

1) D. G. Hinks, D. R. Richards, B. Dabrowski, D. T. Marx, and A. W. Mitchell, Nature **335** (1988) 419.

2) B. Batlogg, R. J. Cava, L. W. Rupp, Jr., A. M. Mujsce, J. P. Remeika, W. F. Peck, Jr., A. S. Cooper, and G. P. Espinosa, Phys. Rev. Lett. **61** (1987) 1670.

3) S. Kondoh, M. Sera, Y. Ando and M. Sato, Physica C **157** (1989) 469.

4) Z. Schlesinger, R. T. Collins, J. A. Calise, D. G. Hinks, A. W. Mitchell, Y. Zheng, and B. Dabrowski, N. E. Bickers, and D. J. Scalapino, submitted to Phys. Rev. B.

5) C.-K. Loong, P. Vashishta, R. K. Kalia, M. H. Degani, D. L. Price, J. D. Jorgensen, D. G. Hinks, B. Dabrowski, A. W. Mitchell, D. R. Richards, and Y. Zheng, Phys. Rev. Lett. **62** (1989) 2628.

6) J. F. Zasadzinski, N. Tralshawala, D. G. Hinks, B. Dabrowski, A. W. Mitchell, D. R. Richards, Physica C **158** (1989) 519.

7) J. Bardeen, L. N. Cooper, and J. R. Schrieffer, Phys. Rev. **108** (1957) 1175.

8) S. Pei, J. D. Jorgensen, B. Dabrowski, D. G. Hinks, D. R. Richards, A. W. Mitchell, J. M. Newsam, S. K. Sinha, D. Vaknin, and A. J. Jacobson, submitted to Phys. Rev. B.

9) P. Vashishta, and A. Rahman, Phys. Rev. Lett. **40** (1978) 1337, and P. Vashishta, R. K. Kalia, and I. Ebbsjö, Phys. Rev. B **39** (1988) 6034.

10) M. H. Degani, R. K. Kalia, and P. Vashishta, to be published.

11) A. Rahman, and P. Vashishta, in The Physics of Superionic Conductors, edited by J. W. Perram (Plenum, New York, 1983), and J. P. Hansen, and I. R. McDonald, Theory of simple liquids (Academic Press, 1976).

12) D. Beeman and R. Alben, Adv. in Phys. **26** (1977) 339.

13) R. Fletcher, Practical Methods of Optimization (Wiley, New York, 1980), and F. H. Stillinger, and T. A. Weber, Science **225** (1984) 983.

14) G. M. Eliashberg, Zh. Eksp. Teo. Fiz. **38** (1960) 966.

MICROSCOPIC DERIVATION OF LANDAU-GINZBURG FREE ENERGY FOR AN ION-ELECTRON TWO-COMPONENT PLASMA

Kuniyoshi EBINA and Makoto KABURAGI

College of Liberal Arts, Kobe University, Nada, Kobe 657, Japan

We present a general formulation to derive the Landau-Ginzburg free energy microscopically for an ion-electron two-component plasma with the ion density as an order parameter. We utilize the outcome of the structural expansion theory. We discuss the importance of the third-order invariant in the Landau-Ginzburg free energy due to the three-body interaction through electron medium.

1. INTRODUCTION

The Landau theory provides a powerful framework for studying continuous as well as first-order phase transitions.[1] It has been generalized to the case where the order parameter is a spatially varying function. The free energy as a functional of the order-parameter field is called the Landau-Ginzburg Hamiltonian and provides a unifying starting point for the modern theory of critical phenomena.[2] Originally, the "Hamiltonian" is introduced phenomenologically from the symmetry considerations mainly interested in the criticality of the transition. If the free-energy functional is properly given microscopically, it may as well be used to determine quantitatively the phase diagrams. Most generally, the free energy functional can be derived[2] by introducing, in the microscopic Hamiltonian, a spatially varying external field which linearly couples to the dynamical variable whose expectation value constitutes the order parameter, then calculating the relevant thermodynamic potential as a functional of the external field, and finally performing a functional Legendre transformation of the thermodynamic potential in order to change the independent variables from the external field to the order parameter field. This procedure provides not only a conceptual basis to the theory but probably also a method to actually construct an approximate free-energy functional.

The density functional theory, which deals with the quantum electron systems as well as classical systems (see a review[3] for strongly coupled plasma), can provide a microscopic basis to the Landau theory for problems where the particle density is a relevant quantity. Recently it has extensively applied to freezing of classical liquids.[4] For the freezing problem the third-order invariant in the Landau expansion may be very important[5] in explaining the special stability of bcc phase just under melting line, but its kernel has almost never been calculated microscopically. On the other hand the structural expansion theory[6-9], which has been used to investigate the structural stability of simple metals such as metallic hydrogen, can also provide a method for constructing the free energy through the procedure mentioned above. However, as far as the present authors know, this possibility has not yet been worked out.

In this paper we present a general formulation to construct the Landau free energy for the two-component plasma (TCP) of ions of charge Ze and electrons $-e$ with the ion density as an order parameter by eliminating the electron degrees of freedom. We want to utilize fully the outcome of the structural expansion theory. We base our discussion on the thermodynamic formulation of quantum many-particle systems by Luttinger and Ward.[10] We discuss the consequences from the resulting free-energy functional in the adiabatic approximation by utilizing the knowledge of the structural expansion. In particular we emphasize the importance of the third-order invariant coming from the three-body interaction between ions through electron medium.

2. GENERAL FORMULATION

Following Luttinger and Ward[10-12], we write the thermodynamic potential Ω of the TCP with the ion and the electron chemical potentials μ_i and μ_e as a functional of the ion and the electron thermal Green's functions $G_i(x,x';\zeta_l)$ and $G_e(x,x';\zeta_l)$, where ζ_l is the Matsubara frequency. Because of the global charge neutrality, μ_i and μ_e must have an implicit relation. We assume for definiteness that the ions are Fermions though the statistics are irrelevant in the Boltzmann regime. Introducing the external potential $w(x)$ which couples linearly to the ion density $\rho_i(x) = \beta^{-1} \sum G_i(x,x; \zeta_l)$, we have

$$\Omega[G_i, G_e] = \beta^{-1} \sum_l \mathrm{tr}\{\ln G_i(\zeta_l) + [G_i^{-1}(\zeta_l) - G_i^{(0)-1}(\zeta_l)]G_i(\zeta_l)\}e^{x_l 0+}$$
$$+ \beta^{-1} \sum_l \mathrm{tr}\{\ln G_e(\zeta_l) + [G_e^{-1}(\zeta_l) - G_e^{(0)-1}(\zeta_l)]G_e(\zeta_l)\}e^{x_l 0+}$$
$$+ \Phi[G_i, G_e] + \beta^{-1} \sum_l \mathrm{tr}[G_i(\zeta_l) w], \qquad (1)$$

where G_i, G_e and w are matrices whose x, x' components are $G_i(x,x';\zeta_l)$, $G_e(x,x';\zeta_l)$ and $w(x)\delta(x-x')$ respectively, and $\Phi[G_i, G_e]$ is the contribution of all closed linked skeleton diagrams with the thin lines replaced by the thick lines such that its functional derivatives with respect to G_i and G_e are respectively the ion and electron self-energies $\Sigma_i(x',x;\zeta_l)$ and $\Sigma_e(x',x;\zeta_l)$. The Dyson equations derived from the variational principle, $\delta\Omega/\delta G_i = 0$ and $\delta\Omega/\delta G_e = 0$, relates the Green's functions and the self-energies. Expanding the self-energies and the Green's functions in terms of $w(x)$[9,12], we can solve the set of Dyson equations order by order with a suitable approximation for $\Phi[G_i, G_e]$. Inserting the obtained expressions for G_i and G_e into eq.(1), we shall get the thermodynamic potential $\Omega[w(x)]$ as a functional Taylor series expansion in $w(x)$:

$$\Omega(\mu_i, [w(x)]) = \Omega^{(0)}(\mu_i) + \int \Omega^{(1)}(x) w(x) + \frac{1}{2}\int \Omega^{(2)}(x,x') w(x) w(x')$$
$$+ \frac{1}{6}\int \Omega^{(3)}(x_1,x_2,x_3) w(x_1) w(x_2) w(x_3) + \cdots, \qquad (2)$$

where the kernels

$$\Omega^{(n)}(x_1,\cdots,x_n) = \left.\frac{\delta^{(n)}\Omega}{\delta w(x_1)\cdots\delta w(x_n)}\right|_{w(x)=0} \qquad (3)$$

can be calculated from the unperturbed Green's functions $G_i^{(0)}$ and $G_e^{(0)}$. We note here that $\Omega^{(1)} = \delta\Omega/\delta w(x) = \rho_i(x)$. Now we restrict ourselves to globally uniform systems. Then we can assume that the spatial average of $w(x)$ is zero; the uniform component of $w(x)$ can be absorbed into $-\mu_i$. Thus introducing $\Delta\rho_i(x) = \rho_i(x) - \bar{\rho}_i$ with $\bar{\rho}_i$ being an average ion density, and performing a Legendre transformation

$$F(\bar{\rho}_i, [\Delta\rho_i(x)]) = \Omega(\mu_i, [w(x)]) + \int \bar{\rho}_i \mu_i - \int \Delta\rho_i(x) w(x), \qquad (4)$$

we get

$$F(\bar{\rho}_i, [\Delta\rho_i(x)]) = F^{(0)}(\bar{\rho}_i) + \frac{1}{2}\int F^{(2)}(x, x')\Delta\rho_i(x)\Delta\rho_i(x')$$

$$+ \frac{1}{6}\int F^{(3)}(x_1, x_2, x_3)\Delta\rho_i(x_1)\Delta\rho_i(x_2)\Delta\rho_i(x_3) + \cdots, \qquad (5)$$

where

$$\int F^{(2)}(x, x'')\Omega^{(2)}(x''; x') = -\delta(x - x'), \qquad (6)$$

$$F^{(3)}(x_1, x_2, x_3) = \int \Omega^{(3)}(x_1', x_2', x_3') F^{(2)}(x_1, x_1') F^{(2)}(x_2, x_2') F^{(2)}(x_2, x_2'), \qquad (7)$$

Here we note that in eq.(2) the average density $\bar{\rho}_i$ generally changes when we change $w(x)$ even if the uniform component $-\mu_i$ is fixed. Therefore we must fine-tune μ_i to keep $\bar{\rho}_i$ fixed. In the freezing problem, where the chemical potential for solid is the same as that of liquid, eq.(5) includes terms due to the shift of the average density.[13,14] If we fix the particle density and expand eq.(5) further in terms of the shift of chemical potential, there appear correction terms.[15] For a globally uniform system we expand $\Delta\rho_i(x)$ in a Fourier series $\Delta\rho_i(x) = \sum \rho_q e^{iq \cdot x}$ and then we get the Fourier transformed version of eq. (5)-(7).

3. ADIABATIC APPROXIMATION

Except probably for the situations where the electron localization couples with the ion motion, we can take the trace of electron degrees of freedom first with a fixed ion configuration giving the effective pair, three-body, etc. potentials among ions. Then the free-energy functional may be decomposed as

$$F[\rho_q] = F_{id}[\rho_q] + F_{pot}[\rho_q] + F_c[\rho_q]. \qquad (8)$$

Here $F_{id}[\rho_q]$ is the free-energy functional for the ideal ion system, which can be calculated by retaining only the first line and the last term in eq.(1). The kernels can be obtained by the Legendre transformation from the many-point ring diagrams $\Pi_{in}^{(0)}(q_1, q_2, \cdots, q_n)$. The properties of these diagrams for the Fermion case in the quantum limit has been worked out in the structural expansion theory[6-9]. In the Boltzmann limit it reduces to the usual ideal gas term. The mean-field potential energy $F_{pot}[\rho_q]$ for ions is an analog of the Hartree term in the electron system, which includes here the effective ion-ion interaction screened by the electron medium

$$F_{pot}^{(2)}[\rho_q] = \sum_q Z^2 \tilde{v}(q) \rho_q \rho_{-q}, \qquad (9)$$

where $\tilde{v}(q) = v(q)/\varepsilon_e(q)$ with $v(q) = 4\pi e^2/q^2$ and $\varepsilon_e(q)$ the static dielectric function of the uniform electron gas, and the effective three-body interaction

$$F_{pot}^{(3)}[\rho_q] = \sum_q Z^3 u_3(\boldsymbol{q}_1, \boldsymbol{q}_2, \boldsymbol{q}_3) \rho_{q_1} \rho_{q_2} \rho_{q_3}, \tag{10}$$

where

$$u_3(\boldsymbol{q}_1, \boldsymbol{q}_2, \boldsymbol{q}_3) = \Pi_{e_3}(\boldsymbol{q}_1, \boldsymbol{q}_2, \cdots, \boldsymbol{q}_n) \tilde{v}(\boldsymbol{q}_1) \tilde{v}(\boldsymbol{q}_2) \tilde{v}(\boldsymbol{q}_3) \tag{11}$$

with $\Pi_{e_3}(\boldsymbol{q}_1, \boldsymbol{q}_2, \cdots, \boldsymbol{q}_n)$ standing for the sum of all three-point irreducible diagrams for the uniform electron gas, and so on. Finally the correlation term $F_c[\rho_q]$, which takes care of everything else, is obtained by considering the ion correlation with many-body interactions.

4. DISCUSSIONS

The third-order term $F_{pot}^{(3)}[\rho_q]$ above corresponds to the third-order energy in the structural expansion for electron systems. This energy plays an important role in the structural stability of metallic hydrogen[16] ($Z = 1$). This is because the kernel (10) has a strong wave-vector dependence, which cannot properly be treated in the square-gradient type theory. This brings a special stability of the structure in which as many equilateral triangles with $q=\sqrt{3}k_F$ appear in the reciprocal lattice. It requires an anisotropic structure or a periodic modulation of the lattice for an atomic phase of hydrogen. Another possibility is to form a crystal with two ions in a unit cell (molecular phase). In any case, the atomic phase of bcc structure would not be realized just under the melting line as log as the third order term in $F_c[\rho_q]$ does not suppress the effect. We shall investigate this problem numerically in the near future.

REFERENCES

1) L.D. Landau, Phys. Z. Sowjetunion 11 (1937) 26.
2) e.g., D.J. Amit, Field theory, the Renormalization Group and Critical Phenomena (World Scientific, Singapore, 1984).
3) S. Ichimaru, H. Iyetomi and S. Tanaka, Phys. Rep. 149 (1987) 91.
4) A.D.J. Haymet and D.W. Oxtoby, J. Chem. Phys. 84 (1986) 1769.
5) S. Alexander and J.P. McTague, Phys. Rev. Lett. 41 (1978) 702.
6) P. Lloyd and C.A. Sholl, J. Phys. C 1 (1968) 1620.
7) E.G. Brovman and Y.M. Kagan, Usp. Fiz. Nauk 112 (1974) 369.
8) J. Hammerberg and N.W. Ashcroft, Phys. Rev. B 9 (1974) 409.
9) T. Nakamura, H. Nagara and H. Miyagi, Prog. Theor. Phys. 63 (1980) 368.
10) J.M. Luttinger and J.C. Ward, Phys. Rev. 118 (1960) 1417.
11) J.M. Luttinger, Phys. Rev. 119 (1960) 1153.
12) K. Ebina, Prog. Theor. Phys. 69 (1983) 1686.
13) T.V. Ramakrishnan and M. Yussouf, Phys. Rev. B 19 (1979) 2775.
14) A.D.J. Haymet and D.W. Oxtoby, J. Chem. Phys. 74 (1981) 2559.
15) W. Kohn and J.M. Luttinger, Phys. Rev. 118 (1960) 41.
16) K. Ebina and H. Miyagi, Anisotropic structures of metallic hydrogen, to be published (1989).

THERMODYNAMIC PROPERTIES OF A LIQUID METAL USING A SOFT-SPHERE REFERENCE SYSTEM

M. HASEGAWA[*], I. KONDO[†], M. WATABE[†] and W. H. YOUNG[§]

[*]Faculty of Engineering, Iwate University, Morioka 020, Japan
[†]Faculty of Integrated Arts and Sciences, Hiroshima University, Hiroshima 730, Japan
[§]Physics Department, College of Science, Sultan Qaboos University, P.O.Box 32486, AL-Kohd, Oman

Some structural and thermodynamic properties of liquid Na and Al are calculated by applying the Gibbs-Bogoliubov method to interatomic potentials obtained using pseudopotential theory, the reference systems being soft-sphere fluids interacting with inverse-power potentials. It is pointed out that the reference fluids with intermediate softness enjoy certain advantages over the extreme hard-sphere and one-component plasma cases.

1. INTRODUCTION

The Gibbs-Bogoliubov (GB) variational method has served well for calculating the structural and thermodynamic properties of liquid metals. A hard-sphere (HS) reference system has often been invoked (see, e.g. ref.1), using the hard-sphere diameter, or equivalently the packing fraction η, variationally. More recently, the merits of a coulombic reference system (the one-component plasma or OCP) have been investigated.[2,3] Here, the variational parameter is the effective charge.

Now, hard-sphere and coulombic interactions are merely the extreme members of a large class of inverse-power potentials, and the question arises of the efficacy of the intermediate ones (to be called soft-sphere (SS) potentials hereafter) in the GB calculation of the properties of liquid metals. This problem has been investigated previously,[2,4] using the model due to Ross[5] which is a modification of the HS system to simulate the properties of the inverse 12-th power potential system. Some general discussions on this problem have also been given by Gray and Young.[6] We have investigated the problem further; by resolving first the problems of technicality for preparing precise enough information about the properties of the SS systems to be used in a convenient way as reference systems, we have implemented successfully the GB procedure with the SS reference systems. Some of the results for liquid Na and Al will be presented below.

2. THE GB METHOD AND SOFT-SPHERE REFERENCE FLUIDS

We wish to calculate the thermodynamic and structural properties of a single component fluid at temperature T and mean density $\rho=N/\Omega$, here N being the total number of atoms present and Ω being the total volume. The interatomic potentials are supposed to be pairwise, known and given by v(r). In relation to this system, we specify a reference fluid, identical in all respects except that the pair potential is $v_n(r) = \varepsilon(\sigma/r)^n$. It is essential for practical convenience that this fluid is characterised by the single coupling parameter $\Gamma_n = (\varepsilon/k_BT)(\sigma/a)^n$, where $a = (3/4\pi\rho)^{1/3}$.

Let us denote the excess Helmholtz free energy per atom of the reference system by $F_n^{ex}(\Gamma_n)$ and the radial distribution function by g_n. Then, by the GB principle, the upper-bound of the structure-dependent part of the true excess free energy (per atom) of the actual fluid is given by

$$F^{ex} = F_n^{ex} + \frac{1}{2}\rho\int d\mathbf{r}[v(r) - v_n(r)]g_n(r)$$

If we minimise this expression with respect to Γ_n, then the minimum value becomes our free energy estimate and, furthermore, with the optimising Γ_n inserted, $g_n(r)$ becomes our approximation to the radial distribution function. With this same value of Γ_n used, it may be shown[6] that the appropriate estimate of the excess entropy is given by $S_n^{ex}(\Gamma_n)$.

For implementing the above procedure, we need F_n^{ex} and g_n. For F_n^{ex}, we may use accurate interpolation formulae provided by DeWitt[7] from Monte Carlo (MC) calculations.[8] The interpolation of whole radial distribution functions $g_n(r)$, each of which must then be used repeatedly in numerical calculation, is more difficult to achieve. In the present work this problem is solved successfully by using the structural functions obtained by the modified HNC (MHNC) method due to Rosenfeld and Ashcroft[9] to fit and interpolate the MC generated data. The MHNC results we obtained are in excellent agreement with MC calculations.[10]

3. RESULTS FOR Na AND Al

We chose Na and Al for detailed study, these being known to have relatively soft and hard core ions respectively. For computing the effective interionic potentials we invoke pseudopotentials obtained by the method of Hasegawa and Watabe[11] and screening functions in the approximation due to Ichimaru and Utsumi.[12] Using these potentials for v(r), the GB method as described in section 2 can be used to obtain structural and thermodynamic results on the basis of the SS reference systems.

The variation of F^{ex} calculated at the respective melting point of Na and Al, as the coupling constant Γ_n is valid, is shown in Fig. 1, together with the results for the HS and OCP cases. Fig. 1(a) demonstrates the known fact for Na that the OCP is more favourable than the HS reference system. But, in addition, it shows that the SS cases provide significantly lower energy bounds than either. Furthermore, once this point is noted, there is little further to choose in energy among this subgroup. Actually, we have found that, among the optimising SS structures, the one for the n=6 case provides the most favourable comparison with experiment. Our structure calculations have also confirmed the superiority of the OCP over the HS reference system, already known for this metal, while the OCP and the SS reference systems are quite competing with each other.

For Al, we obtain the known results, in Fig.1(b), that hard spheres provide a better reference system than the OCP. But, as for Na, the intermediate cases are better (in the GB sense) than either. This time, however, of the cases calculated, the n=12 result is significantly the lowest. The results of the optimising structural functions have shown that, around the principal peak, the HS structure factor compares better with experiment than the n=12 one but, at higher arguments, the reverse is true, while, in real space, the n=12 result, i.e. $g_{12}(r)$, is very superior, giving excellent agreement with experiment.

We have also calculated the entropies. For the excess entropies at melting, the

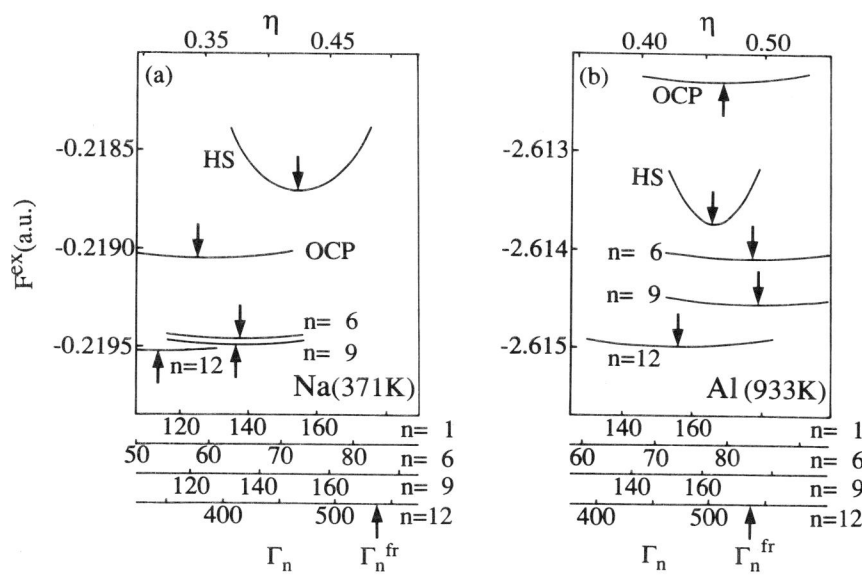

FIGURE 1

Variation of F^{ex} with coupling parameter Γ_n for n=1(OCP), 6, 9, 12 and ∞ (hard-spheres). The packing fraction η is used for Γ_∞. The arrows indicate the positions of the minimising Γ_n and η. Γ_n^{fr} is the coupling constant at which the inverse n-th power liquid freezes : (a) Liquid Na at 371K; (b) Liquid Al at 933K.

extreme (OCP and HS) reference systems provide better results than the intermediate ones for Na, but the situation is reversed for Al. The results for the variation of the entropy with temperature, obtained by repeating the GB procedure with the effective ion- ion potential recalculated at each temperature, imply that the n=6 and n=12 reference systems respectively for Na and Al give the specific heats in excellent agreement with experiment and beyond what a HS reference system can achieve.

4. CONCLUSIONS

By resolving the problems of the technicalities, especially the interpolation procedure for the structural functions, it has now become possible to utilise the SS fluid as reference systems in the GB method as conveniently as the HS system and the OCP. As shown in the present work, the SS fluids enjoy certain advantages over the two extreme (HS and OCP) cases as reference systems for liquid metals; they yield better total free energies (in the GB sense) and, on the whole, although there is a degree of ambiguity, the structure factors and entropies are improved also.

REFERENCES

1) N.W. Ashcroft and D. Stroud, Theory of the thermodynamics of simple liquid metals, in; Solid State Physics, Vol.33, eds. H. Ehrenreich, F. Seitz and D. Turnbull (Academic , New York, 1978) pp. 1-81.

2) M. Ross, H.E. DeWitt and W.B. Hubbard, Phys. Rev. **A24** (1981) 1016.

3) K.K. Mon, R. Gan and D. Stroud, Phys. Rev. **A24** (1981) 2145.

4) J.A. Moriarty, D.A. Young and M. Ross, Phys. Rev. **B30** (1984) 578.

5) M. Ross, J. Chem. Phys. **71** (1979) 1567.

6) P. Gray and W.H. Young, Phys. Chem. Liq. **13** (1983) 159.

7) H.E. DeWitt, Equilibrium statistical mechanics of strongly couled plasmas by numerical simulation, in; Strongly Coupled Plasms, eds. G. Kalman and P. Carini (Plenum, New York, 1978) pp. 82-115.

8) W.G. Hoover, S.G. Gray and K.W. Johnson, J. Chem. Phys. **55** (1971) 1128.

9) Y. Rosenfeld and N.W. Ashcroft, Phys. Rev. **A20** (1979) 1208.

10) J.P. Hansen and J.J. Weis, Mol. Phys. **23** (1972) 853.
    J.P. Hansen and D. Schiff, Mol. Phys. **25** (1973) 1281.

11) M. Hasegawa and M. Watabe, J. Phys. **C16** (1983) L29.

12) S. Ichimaru and K. Utsumi, Phys. Rev. **B24** (1981) 7385.

ELECTRON-ION STRONG COUPLING EFFECTS IN DENSE HYDROGEN PLASMAS I. EQUATION OF STATE AND ELECTRIC CONDUCTIVITY

Shigenori TANAKA,[†] Xin-Zhong YAN,[‡] and Setsuo ICHIMARU[†]

[†]Department of Physics, University of Tokyo, Bunkyo-ku, Tokyo 113, Japan

[‡]Institute of Physics, Chinese Academy of Sciences, Beijing 100080, P. R. China

Theory of strong correlations between electrons and ions in hydrogen plasmas near metal-insulator boundaries is developed. Novel features include appearance of "Rydberg states" in the electron-ion short-range correlation, the equation of state and the electric conductivity.

In an earlier series of investigations,[1,2] which will be referred to as IMTY, the present authors and Mitake developed a strong coupling theory of dense hydrogen plasmas appropriate to interior of the main-sequence stars and to final stages of inertial-confinement fusion plasmas. In a dense plasma, the strong Coulomb/exchange coupling between the charged particles beyond the random-phase approximation[1] becomes essential. In the density-response formalism, on which the IMTY theory is based, such a strong-coupling effect can be treated rigorously in terms of the local-field corrections[1] (LFC's), $G_{\mu\nu}(k)$, where $\mu, \nu = 1, 2$ with $1 =$ electrons and $2 =$ ions.

When the density and/or temperature of the plasma are lowered toward the conditions for the onset of pressure and/or thermal ionization, as expected in the interior of Jovian planets and possibly in the brown dwarfs,[3] the Coulomb coupling between the electrons and the ions become particularly pronounced; a trend toward an incipient formation of bound pairs (i.e., neutral atoms) should be revealed in the features of the electron-ion correlations. A major shortcoming of the IMTY theory lies in its inaccuracy in treating such an effect of strong electron-ion coupling as the plasma approaches the metal-insulator boundaries. They approximately expressed the radial distribution function (RDF) between electrons and ions, $g_{12}(r)$, in terms of linear response of the electrons against the ions; as a consequence, $G_{12}(k)$ vanished identically in their calculation. The predicted equation of state[2] (EOS) did not show a tendency toward incipient bound pairs; the calculated values[2] of the conductivities remained relatively high near the metal-insulator boundaries.

We here present[4] a strong coupling theory of dense hydrogen plasmas applicable in the vicinity of the metal-insulator boundaries. Strong Coulomb/exchange correlations are analyzed through an integral equation approach, which adopts the hypernetted-chain (HNC) approximation[1] for the classical ion-ion correlations and the modified convolution approximation[5] (MCA), justified both in the classical plasmas[6] and in the quantum electron liquids,[7] for the quantum-mechanical electron-electron and electron-ion correlations. The resulting HNC/MCA equations are solved self-consistently for $g_{\mu\nu}(r)$ and $G_{\mu\nu}(k)$. The EOS thus calculated reveals emergence of "Rydberg states" in the metallic phase, implying physically an approach toward an insulator phase. The Rydberg states are found to modify the electron-ion correlations remarkably and thereby act to reduce the electric and thermal conductivities.

The strong coupling effects in dense plasmas have been studied theoretically by a number of investigators.[1] Earlier in 1982, Dharma-wardana and Perrot[8] (referred to as DP) presented a density-functional theory of hydrogen plasmas: In this theory, $g_{22}(r)$ was analyzed in the HNC approximation; $g_{11}(r)$, however, was treated in the density-functional formalism with the local-density approximation (LDA); essential k-dependent effects in $G_{11}(k)$ were thus ignored. Furthermore, they neglected the contributions of exchange-correlation potential altogether in their LDA treatment of $g_{12}(r)$; $G_{12}(k) = 0$ was assumed at the onset. Whenever feasible (Figs. 2 and 3 below), the results of the present HNC/MCA study are compared with those of DP, although degrees of approximations involved are significantly different.

The Coulomb-coupling parameter[1] for a hydrogen plasma at temperature T with n electrons (and ions) in a unit volume is $\Gamma = e^2/ak_BT$, where $a = (3/4\pi n)^{1/3}$ is the ion-sphere radius. The density and degeneracy parameters for the electrons are $r_s = a/a_B$ and $\theta = k_BT/E_F$, where a_B is the Bohr radius and E_F is the Fermi level of the electrons in the ground state. We take the condition for pressure ionization in hydrogen approximately at $E_F = $ Ry, where Ry $= me^4/2\hbar^2 = 13.6$eV, the ionization energy of a hydrogen atom in the ground state. Analogously the condition for thermal ionization is taken approximately at $k_BT = $ Ry. Hydrogen is in an ionized, metallic state, when $E_F > $ Ry or $k_BT > $ Ry.

For an isolated hydrogen atom in the ground state, the value of the joint probability function between an electron and an ion at $r = 0$ is $g_{12}^0(0) = 4r_s^3$; the electron-ion interaction energy in such an atom is -2Ry. As hydrogen plasmas approach the metal-insulator boundaries, coupling between electrons and ions becomes so pronounced that these features of bound states emerge in their correlation characteristics. Such will be called "Rydberg states" in hydrogen plasmas, which imply incipient bound states.

In Fig. 1, we show the results for $G_{\mu\nu}(k)$ at $\theta = 1$ and $\Gamma = 1$ calculated in the HNC/MCA scheme. Note that $G_{11}(k) \geq 0$, $G_{22}(k) \geq 0$, and $G_{12}(k) \leq 0$. This fact implies that the differences between the effective potentials, $v_{\mu\nu}(k)[1-G_{\mu\nu}(k)]$, and the bare potentials, $v_{\mu\nu}(k)$, are *always* negative (i.e., attractive). In the strong-coupling regime, the magnitude of $G_{12}(k)$ remarkably increases in the large-k domain, leading to enhancement of the effective electron-ion attraction at short distances and thereby to incipient bound states.

Figures 2 and 3 show the RDF's at $r_s = 1$ and $\Gamma = 2$. The results with $G_{12}(k) = 0$ and those of DP are also shown there for comparison. We observe that the HNC/MCA values of $g_{\mu\nu}(r)$ exceed those with $G_{12}(k) = 0$, resulting from the enhanced electron-ion attraction. Apparent difference in $g_{22}(r)$ between HNC/MCA and DP may be attributed to the neglect

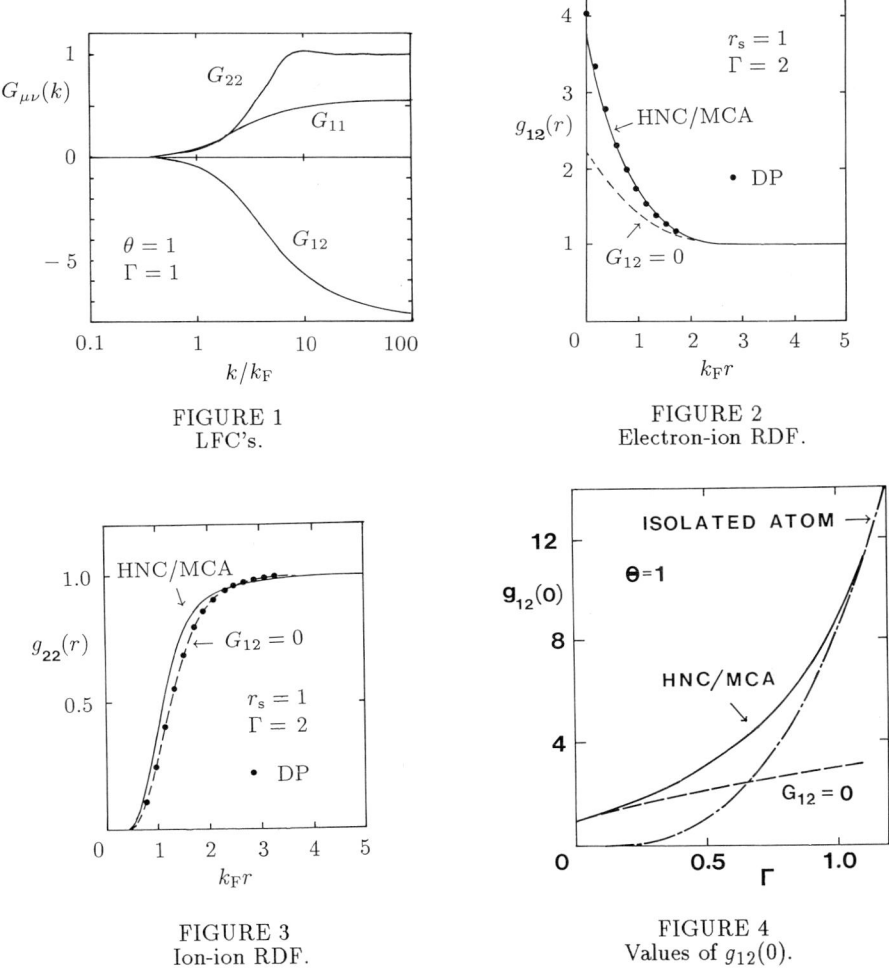

FIGURE 1
LFC's.

FIGURE 2
Electron-ion RDF.

FIGURE 3
Ion-ion RDF.

FIGURE 4
Values of $g_{12}(0)$.

of the attractive, electrton-ion correlation potential in the latter.

Figure 4 shows the values of $g_{12}(0)$ predicted in various theoretical schemes. The HNC/MCA value asymptotically approaches the isolate-atom value of $g_{12}^0(0)$ as Γ increases, pointing to an emergence of incipient bound states or Rydberg states.

Another evidence for the Rydberg states are observed in the calculated values of the interaction energy in Fig. 5. The value of E_{int} in the HNC/MCA scheme approaches the isolated-atom value of $-2\mathrm{Ry}$ as Γ increases.

The generalized Coulomb logarithm[1,2] for the electric conductivity σ is calculated as

$$L = \frac{3\sqrt{\pi}\theta^{3/2}}{4} \int_0^\infty \frac{dk}{k} f(k/2) \frac{S_{22}(k)}{|\tilde{\varepsilon}(k)|^2}[1 - G_{12}(k)],$$

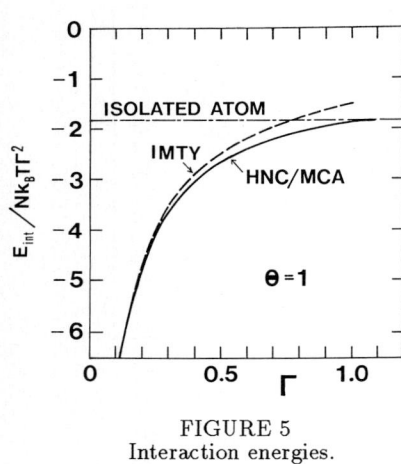
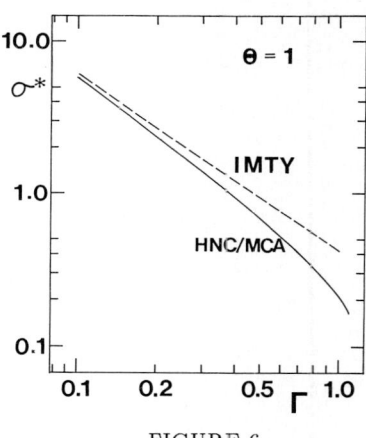

FIGURE 5
Interaction energies.

FIGURE 6
Normalized electric conductivity.

where $f(k)$ is the Fermi distribution function, $S_{22}(k)$ is the structure factor of the ions, $\tilde{\epsilon}(k) = 1 - v(k)[1 - G_{11}(k)]\chi_1^{(0)}(k)$, and $\chi_1^{(0)}(k)$ is the free-electron polarizability. In Fig. 6, we plot the computed values of the conductivity, normalized by the plasma frequency of the electrons. The steep decrease of the HNC/MCA conductivity in the strong-coupling regime implies an approach to an insulator phase, pointing once again to incipient bound states.

Details of the theory will be published elsewhere.[4]

REFERENCES

1) For a review, S. Ichimaru, H. Iyetomi, and S. Tanaka, Phys. Rep. **149** (1987) 91.

2) S. Ichimaru, S. Mitake, S. Tanaka, and X.-Z. Yan, Phys. Rev. A **32** (1985) 1768; S. Ichimaru and S. Tanaka, Phys. Rev. A **32** (1985) 1790; S. Tanaka and S. Ichimaru, Phys. Rev. A **32** (1985) 3756.

3) W. B. Hubbard, Plasma thermodynamics and the evolution of brown dwarfs and planets, this volume.

4) S. Tanaka, X.-Z. Yan, and S. Ichimaru, to be published.

5) K. Tago, K. Utsumi, and S. Ichimaru, Prog. Theor. Phys. **65** (1981) 54.

6) X.-Z. Yan and S. Ichimaru, J. Phys. Soc. Jpn. **56** (1987) 3853.

7) S. Tanaka and S. Ichimaru, Phys. Rev. B **36** (1989) 1036.

8) M. W. C. Dharma-wardana and F. Perrot, Phys. Rev. A **26** (1982) 2096.

DENSITY FUNCTIONAL APPROACH TO PARTICLE CORRELATIONS AND ELECTRONIC STRUCTURE IN DENSE PLASMAS

Chandre DHARMA-WARDANA

Division of Physics, National Research Council of Canada, Ottawa, Canada K1A 0R6

The self-consistent calculation of the internal electronic structure of ions in hot plasmas, and the evaluation of the mean electron- and ion-distributions around a given ion in the plasma, are most effectively carried out using density functional theory. The density functional model in the context of recent developments is discussed. We introduce electron-ion and ion-ion "pseudopotentials" derived from the density functional calculation itself and consider how the internal structure of the ions as well as ion-configuration fluctuations in the plasma could be treated to give a more realistic account of plasma properties. Impurity-ion-ion correlations near an impurity relevant to, eg., ion-microfield calculations and to the plasma mapping of the excitations in the fractional quantum Hall effect, are also discussed.

1. INTRODUCTION

At the most basic level any plasma or liquid metal can be considered as a system of electrons and nuclei interacting via the Coulomb potential, the ions forming a disordered but correlated system. The electrons may occupy continuum states or bound states associated with individual ions, and possibly clusters of ions. This brings us directly against the major difficulty involved in any basic theory of plasmas - namely, the self-consistent determination of the electronic and ionic configurations of the system, at a given temperature and density starting from, say, the total hamiltonian of the system. A very fruitful approach to this type of problems is provided by density functional theory (DFT) which uses the one-particle densities of the system as the variational functionals of an exact many-body theory.[1]

However, at the very outset we remark that DFT in its present form provides no rigorous answer to the calculation of excited states and related properties (eg.. optical properties). Unfortunately, other rigorous methods (eg.. Greens function methods[2]) also remain on the whole computationally intractable and hence we shall focus on the application of DFT in its present form to plasma problems since here we have at least a possibility of a rigorous treatment of static properties of the system. A dynamical theory can then be built up with the static information as essential inputs defining appropriate basis sets for the electrons, equilibrium density distributions for the ions and so forth.[3,4]

2. THE DENSITY FUNCTIONAL MODEL

Several approaches to the formulation of a density functional model of plasmas are possible. A computationally oriented new approach due to Car and Parrinello[5] has not been applied to plasmas so far. Chihara[6] starts from the coupled hypemeted chain (HNC) equations for the ion-ion, ion-electron and electron-electron pair correlations and invokes DFT to construct the direct correlation functions appearing in the HNC-equations, or in the relevant response functions for his quantal-

HNC formulation. In our applications of DFT to plasmas[7] we proceed from the Hohenberg-Kohn Mermin result that the thermodynamic potential Ω is a unique functional of the electron and ion densities n(r), ρ(r), and that $\Omega[n,\rho]$ is a minimum for the true density distributions. This leads to the coupled functional variational equations

$$\delta\Omega[n,\rho]/\delta n = 0 \qquad (2.1)$$

$$\delta\Omega[n,\rho]/\delta\rho = 0 \qquad (2.2)$$

In order to obtain a tractable form for (1) and (2) we consider a fixed nucleus of charge Z_0 at the origin and study the plasma in the external potential of this nucleus. Particle correlations die off after a certain characteristic distace R such that all the pair-distribution functions (PDF), i.e, g(r) become essentially unity. Hence we consider a "correlation sphere "of radius R. Then the system under study consists of the central nucleus interacting with the plasma made up of a distribution of electrons n(r) and ions ρ(r) distributed around it. For r > R, n(r) and ρ(r) will tend to the mean electron and ion densities $\bar{n}$ and $\bar{\rho}$ respectively. The ions in the distribution will be called *field ions* to distinguish them from the central ion. In effect, the central ion can be one of the field ions which has been singled out for detailed study. The field ions can be treated in a simplified manner to calculate a detailed picture of the electronic bound states, and continuum states, associated with the central ion. The detailed information about the central ion obtained from such a calculation can be used to construct a better model of the the field ions to be used in a new calculation, etc. Instead of using a single ion as the generator of the external potential we may use a cluster of ions but the problem becomes too difficult unless a round about approach is used (Sec. 6). Here we restrict ourselves to the single ion model and use atomic units (e = m = $\hbar$ = 1).

The one-body ion distribution about the central ion is given by

$$\rho(r) = \Sigma' \; \delta(\vec{r} - \vec{r}_i) \qquad (2.3)$$

$$\text{---> } \bar{\rho} \; g_{0i}(r) \qquad (2.4)$$

In (4) we have introduced the PDF $g_{0i}(r)$ for the probability of finding a field ion at a distance r from the central nucleus which is excluded from the primed sum. If we are considering a homogeneous plasma the central nucleus is equivalent to the field nuclei and $g_{0i}(r)$ becomes identical with $g_{ii}(r)$. Going from (3) to (4) paves the way for replacing a discontinuous distribution by a smoothed-out average distribution g(r). Such a smoothed-out distribution is the average over all the ionic configurations and is valid only for processes involving time scales greater than ion-relaxation times. Schemes for transcending this shortcoming will be discussed in Section 6.

The use of a smoothed-out distribution removes the deep Coulomb potentials associated with the nuclear positions of the field ions However, we only need a simplified picture of the field ions and hence we could attempt to allow for the deep Coulomb potential by the use of pseudopotentials (Sec..3). Thus we introduce $V_{ei}(r)$ as the pseudopotential defining the interaction of an electron with a field ion, and similarly $V_{ii}(r)$ as the interaction between two field ions. The interaction between an electron and the central nucleus is simply the Coulomb potential $-Z_0/r$. Similarly, the interaction between a field ion and the central nucleus is $-Z_0 V_{ei}(r)$. We write the pseudopotentials in a form which retains the analogy with the Coulomb potentials.

$$V_{ei}(r) = -\zeta_1(r)/r \quad \text{and} \quad V_{ii}(r) = \zeta_2(r)/r \qquad (2.5)$$

The electron distribution around the central nucleus, denoted by $n(r)$, consists of a bound electron distribution $n_b(r)$ and a free electron distribution $n_f(r)$. Thus we write

$$n(r) = n_b(r) + n_f(r), \quad n_f(r) = \bar{n} + \Delta n_f(r)$$

$$N_b = \int n_b(r)\, dr^3 \qquad (2.6)$$

If there are bound states associated with each field ion, then this enters into the formulation of the pseudopotentials. Thus the point-ion model would be to take $\zeta_1 = \bar{Z} = Z - N_c$ where N_c is the number of bound electrons associated with each field ion. N_c is close to the number N_b contained in the distribution $n_b(r)$ but not *equal* to the latter since $n_b(r)$ contains the bound electrons of the central nucleus *and* the distribution $\rho(r)$.

The electron density $n(r)$ may be thought of as arising from a superposition of individual spherical charge distributions of each ion. Thus the free electron density displacement $\Delta n_f(r)$ and the bound density $n_b(r)$ are re-expressed as

$$\Delta n_f(r) = \Delta n_f^{(0)}(r) + \sum_i{}' \Delta n_f^{(1)}(|r - r_i|)$$

$$(2.7)$$

$$n_b(r) = n_b^{(0)}(r) + \sum_i{}' n_b^{(1)}(|r - r_i|)$$

The primes on the summations exclude the origin. Here $\Delta n^{(1)}_f(r)$ is the displaced free electron distribution associated with a single *field ion*, while $\Delta n^{(0)}_f(r)$ is that associated with the central nucleus. Similarly $n^{(0)}_b(r)$, $n^{(1)}_b(r)$ refer to bound electron distributions. The electron distribution associated with a single field ion is thus $n^{(1)}(r) = n^{(1)}_b(r) + \Delta n^{(1)}_f(r)$. If the nuclear charges Z_0 and Z are identical then $n^{(0)}(r)$ is equal to $n^{(1)}(r)$. Then $n^{(1)}(r)$, viz, the electron density associated with an ion can be determined for a given ion distribution $g_{ii}(r)$ using a Fourier transformation.

We should note that DFT does not provide a variational principle for the free-electron and bound-electron distributions separately, but deals with the total distribution. Which electrons are free or bound is merely a matter of energy scales and hence should be irrelevant for the complete problem. But since the latter is beyond our reach model construction becomes necessary. Keeping these remarks in mind we write the thermodynamic potential of the plasma as

$$\Omega = T[n,\rho] + F_e + F_{ei} + F_i - \mu_i \int \rho(r) d\vec{r} - \mu_e \int n(r) d\vec{r} \qquad (2.8)$$

where

$$\Omega = T[n,\rho] + F_e + F_{ei} + F_i - \mu_i \int \rho(r) d\vec{r} - \mu_e \int n(r) d\vec{r} \qquad (2.9)$$

$$F_e = -\int \frac{Z_0}{r} n(r) d\vec{r} + \frac{1}{2} \int \frac{n(r) n(r')}{|\vec{r}-\vec{r}'|} d\vec{r} d\vec{r}' + \int F^e_{xc}[n,\rho] d\vec{r} \qquad (2.10)$$

$$F_{ei} = -\int \frac{\zeta_1(\vec{r}-\vec{r}') \rho(r) n(r')}{|\vec{r}-\vec{r}'|} d\vec{r} d\vec{r}' + \int F^{ei}_c[n,\rho] d\vec{r} \qquad (2.11)$$

$$F_i = \int Z_0 \frac{\zeta_1(r) \rho(r)}{r} + \int \frac{\zeta_2(\vec{r}-\vec{r}') \rho(r) \rho(r')}{|\vec{r}-\vec{r}'|} d\vec{r} d\vec{r}' + \int F^i_c[n,\rho] d\vec{r} \qquad (2.12)$$

Here $T[n,\rho]$ is the kinetic energy functional for a non-interacting system, having the true interacting density distributions $n(r)$, $\rho(r)$. The free energy contributions and chemical potentials are denote by F and μ. The first term in (10) gives the Coulomb interaction between the nucleus at the origin ("external potential") and the electron distribution, while the second term is the electron-electron interaction. The exchange-correlation functional $F^e_{xc}[n,\rho]$ and similarly the correlation functionals F^{ei}_c, F^i_c appearing in Eqs.. (11) and (12) bring in the many-body effects. When Eq..(8) is used in (1) and (2) the variational equations take the form

$$\left[\delta F^0/\delta n + V^e(r) - \bar{\mu}_e\right] \delta n = 0 \qquad (2.13)$$

$$\left[\delta F^0/\delta \rho + V^i(r) - \bar{\mu}_i\right] \delta \rho = 0 \qquad (2.14)$$

These are equations obeyed by non-interacting particles moving in the effective potentials $V^e(r)$ and $V^i(r)$ such that

$$V^e(r) = -[Z_0/r + V^e_P(r)] + V^e_{xc}(r) - V^e_{xc}(R) \qquad (2.15)$$

$$V^i(r) = [Z_0 \zeta_1(r)/r + V^i_p(r)] + V^i_c(r) - V^i_c(R) \tag{2.16}$$

where

$$V^e_p(r) = \int d\vec{r}\,' [\zeta_1(\vec{r}-\vec{r}\,')\rho(r')-n(r')]/|\vec{r}-\vec{r}\,'| \tag{2.17}$$

$$V^i_p(r) = \int d\vec{r}\,' [\zeta_2(\vec{r}-\vec{r}\,')\rho(r')-\zeta_1(\vec{r}-\vec{r}\,')n(r')]/|\vec{r}-\vec{r}\,'| \tag{2.18}$$

When the field-ions are replaced by point ions of effective charge $\bar{Z}$ then $\zeta_1 \longrightarrow \bar{Z}$ and $\zeta_2 \longrightarrow \bar{Z}^2$. Then $V^e p(r)$ and $V^i p(r)$ reduce to the usual Poisson potentials used in the earlier version of the theory.[7] The new Eqs..(17)-(18) reflect the complications due to the internal structure in the field-ions. However, a proper treatment of internal structure of the field ions is necessary if we are to treat electron-ion correlations in a systematic way.

Eqs..(15) and (16) contain the exchange and correlation potentials arising from the free energy contributions F^e_{xc}, F^{ei}_c, F^i_c.

$$V^e_{xc}(r) = \delta F^e_{xc}[n,\rho]/\delta n + \delta F^{ei}_c[n,\rho]/\delta n \tag{2.19}$$

$$V^i_c(r) = \delta F^i_c[n,\rho]/\delta\rho + \delta F^{ei}_c[n,\rho]/\delta\rho \tag{2.20}$$

These arise from correction terms of the type $\langle n(r)n(r')\rangle - \langle n(r)\rangle\langle n(r')\rangle$, $\langle n(r)\rho(r')\rangle - \langle n(r)\rangle\langle\rho(r')\rangle$, and $\langle\rho(r)\rho(r')\rangle - \langle\rho(r)\rangle\langle\rho(r')\rangle$ and bring in the many-body effects. In our previous studies we neglected $F^{ie}_c[n,\rho]$ along with the internal structure of the field ions which were modeled as point particles of effective charge $\bar{Z}$. The correlation potentials for the electron subsystem were taken in the local density approximation from the known exchange correlation potential of a uniform electron gas at finite temperatures[8]. The correlation potential for the ions was modeled as a sum of HNC graphs[9]. As for the F^{ei}_c contributions, Chihara has shown how they may be related to the direct correlation functions and response functions of the system[10]. In ref. 11 it is argued that Chihara's approach is best applied to the free-electron part of the electron distribution and not to the bound electrons.

Eq..(13) applies to electrons and it is just the Kohn-Sham Schrodinger equation. Thus

$$[-\nabla^2/2 + V^e(r)]\phi_v(r) = \varepsilon_v \phi_v(r) \tag{2.21}$$

The Kohn-Sham orbitals and eigenvalues cannot be rigorously identified with the single particle spectrum of the central ion *plus* the surrounding particle distributions. However recent studies[12] of the Dyson equation in the "GW"-approximation for solid semi-conductors seem to suggest that although the spectrum is poorly rendered by the eigen*values* (cf. the band gap problem), the eigen*states* are reasonably well represented by the Kohn-Sham orbitals. What ever the short-comings of the Kohn-Sham spectrum, they have the property that the one-particle density

$$n(r) = \sum_v |\phi_v(r)|^2 f(\varepsilon_v) \tag{2.22}$$

is correctly given. In (22) $f(\varepsilon_v)$ is the Fermi factor for the occupational probability of the Kohn-Sham state v, which stands for (n,l,m) for bound states, and (k,l,m) for continuum states with $\varepsilon_v = k^2/2$. In actual calculations n(r) is not estimated as in (22) but written as $n(r) = n_b(r) + \Delta n_f(r) + \bar{n}$ where $n_b(r)$ is obtained from the discrete finite sum over bound states. The bound states have exponentially decaying boundary conditions and negative energies. The displaced density $\Delta n_f(r)$ is obtained from an integral over the continuum states which are characterized by phase shifts. These phase shifts should satisfy the finite temperature version of the Friedel sum rule[9] at self-consistency.

Eq.. (14) applies to the ion-subsystem and gives the ion-density distribution:

$$\rho(r) = \bar{\rho}\, g_{ii}(r) = \bar{\rho}\, \exp[-\beta V^i(r)] \tag{2.23}$$

Given a plasma at a temperature T, free electron density $\bar{n}$ and nuclear charge Z, a density functional calculation would proceed by starting off with trial solutions n(r) and ρ(r) for solving the coupled equations (22) and (23). In the first stage of the calculations one would use a point-ion model with $\zeta_1(r)$ and $\zeta_2(r)$ replaced by $\bar{Z}$ and $\bar{Z}^2$ respectively. The first stage of the calculation would yield $g_{ii}(r)$, $n_b^{(1)}(r)$, $\Delta n_f^{(1)}(r)$ etc for constructing the pseudopotentials needed in an improved set of calculations.

3. THE PSEUDOPOTENTIALS

In metals physics the electrons are degenerate and only a narrow band of energies at the Fermi energy plays a significant role in the interactions. Also, the thermal energy is much smaller than the typical electronic bound state energies and hence the ions with their inner cores can be specified in a relatively simple manner, at least for non-transitional metals. Hence the construction of pseudopotentials for electron-ion interactions in terms of states orthogonalized to the core states is a viable procedure. In a plasma such an approach is not useful. Instead we use the picture of a "mean-ion" or "average atom" that comes out of the DFT calculation as the building block for the pseudopotentials. The equations (2.6) and (2.7) enable us to picture each ion in the plasma as carrying a bound electron distribution $n^{(1)}{}_b(r)$ and a free-electron displacement $\Delta n^{(1)}{}_f(r)$ centered on its nucleus of charge Z. Within this picture we can write, for an electron at r,

$$V_{ei}(r) = -\zeta_1(r)/r = -Z/r + \int d\vec{r}_1 \, n_b^{(1)}(r_1)/|\vec{r}-\vec{r}_1| \qquad (3.1)$$

For two field ions of nuclear charges Z_1 and Z_2, separated by r and carrying bound-electron distributions $n^{(1)}{}_b(r_1)$ and $n^{(1)}{}_b(r_2)$ centered on each nucleus, we have

$$\begin{aligned}V_{ii}(r) = \zeta_2(r)/r = Z_1 Z_2/r &- Z_1 \int d\vec{r}_2 \, n_b^{(1)}(r_2)/|\vec{r}+\vec{r}_2| \\ &- Z_2 \int d\vec{r}_1 \, n_b^{(1)}(r_1)/|\vec{r}-\vec{r}_1| \\ &+ \int d\vec{r}_1 d\vec{r}_2 \, n_b^{(1)}(r_1) n_b^{(2)}(r_2)/|\vec{r}+\vec{r}_2-\vec{r}_1| \end{aligned} \qquad (3.2)$$

The bound charge distributions remain frozen during each stage of the DFT calculation. Initially we do a "zeroth order" DFT-calculation using the point-ion model for the field ions. This gives us the first estimate of $n_b{}^{(1)}(r)$ to formulate the pseudopotentials for the next round of DFT-calculations, etc. In effect, these "pseudopotentials" have to be calculated for each plasma and are not "transferable" quantities. However, they can be used for other calculations, eg.. the electrical conductivity from V_{ei}. Note that in this model the effective ionic charge $\bar{Z}$ emerges as a meaningful quantity only as a limit of $\zeta_1(r)$ for large r, ie., when the ions are probed from far away.

4. SIMPLIFIED DENSITY FUNCTIONAL MODELS

Most "average atom" models are restricted to a calculation of the electronic structure of a single ion confined in a cell [14], in jellium, in a step like potential -well[15] or in a Debye profile[16]. These are DFT models where simplified field-ion configurations are assumed. An interesting variant has been recently studied by Ofer, Nardin and Rosenfeld[17] where they replace the electronic problem by the Thomas-Fermi approximation but solve for the average ionic configuration using the HNC equations. These authors show that they obtain $g_{ii}(r)$ which are in good agreement with the more sophisticated calculations of Dharma-wardana and Perrot, and of Chihara. Their work suggests that the ion -correlations are not sensitive to the details of the potential imposed by the electronic structure. This is consistent with the experience from liquid metals where even hard sphere models are used to obtain the $g_{ii}(r)$. However, the details are needed for an accurate description of S(k) especially for small-k.

5. CORRELATIONS IN IMPURITY PLASMAS

An important class of plasma problems arises where the properties of an impurity ion placed in the plasma become relevant. This is typical of many plasma spectroscopy problems.

Another celebrated application arises in the fractional quantum Hall effect [18] (FQHE) since Laughlin's model can be mapped into that of a classical plasma. In an impurity plasma we need to consider (a) $g^0_{ii}(r)$ which defines the ion-ion correlations in the uniform plasma without the impurity at the origin, (b) $g_{0i}(r)$ where subscript 0 indicates the impurity (c) $g_{ii}(r)$ which defines the field ions in the inhomogeneous plasma. Both (a) and (b) can be calculated from the DFT procedure outlined above. Thus (a) is obtained from a calculation where the central ion is identical to the field ions, while (b) is obtained from a calculation where the central ion of charge Z_0 is the impurity. However, $g_{ii}(r)$ of the inhomogeneous plasma is really a three-particle problem, viz, $g(\vec{r}_1, \vec{r}_2 | 0)$ since the ion-ion correlations are needed in the presence of the impurity (usually the "radiator" in plasma spectroscopy) held at the origin. A standard approach is to use the Kirkwood decomposition

$$g(1,2 | 0) = g(0,1)g(0,2)g^0(1,2) \qquad (5.1)$$

where $g(0,1)$ and $g(0,2)$ are simply $g_{0i}(r)$ while $g^0(1,2)$ is $g_{ii}^0(r)$. The use of the homogeneous $g^0(r)$ in (5.1) is an approximation which needs to be improved, as seen from our calculations[19] of microfields and from FQHE studies. The added correlations embodied in $\Delta h(1,2 | 0) = g(1,2) - g^0(1,2)$ have been named impurity-plasma-plasma corrections (ipp-corrections[19]) and are essentially those referred to as "non-central" correlations by Iglesias et al[20]. The idea of retaining the product form with a modified $g(1,2)$ has also been examined[21] in the context of triplet correlations in homogeneous plasmas but the present problem is in a sense simpler. The simplest approach[22] to the present problem is to consider a two-component plasma (TCP) where one of the components (impurity) has a vanishingly small concentration. Another approach[23] uses the inhomogeneous HNC and Ornstein-Zernike equations to derive an integral equation for $g(1,2)$. If there are N particles in the correlation sphere of volume Ω_c then quantities of the order of $1/N$ have to be retained since the impurity density is also of the order of $1/N$. Some of the essential differences in the calculated excitation energies in the FQHE are probably related to such inconsistencies.

In the TCP model the plasma is made up of plasma ions of density ρ_p and impurity ions of density ρ_i (note change of notation, ie., now the object of the calculation is $g_{pp}(r) = 1 + h_{pp}(r)$, and the ipp-correction is $\Delta h_{pp}(1,2 | 0)$ etc.). Let the homogeneous plasma density $\bar{\rho}$ be explicitly denoted by ρ^0, with N particles in Ω_c. The Ornstein-Zernike (O-Z) relation is

$$h^0_{pp}(\vec{r}_1, \vec{r}_2) = C^0_{pp}(\vec{r}_1, \vec{r}_2) + \rho^0_p \int C^0_{pp}(\vec{r}_1, \vec{r}_3) h^0_{pp}(\vec{r}_3, \vec{r}_2) d\vec{r}_3 \qquad (5.2)$$

We remove one of the plasma particles and introduce the impurity. The new densities are $\rho_p = (N-1)/\Omega_c$, $\rho_i = 1/\Omega_c$. The new O-Z relations are for a TCP but without terms involving C_{ii} since there is only a single impurity.

$$h_{pp}(\vec{r}_1,\vec{r}_2) = C_{pp}(\vec{r}_1,\vec{r}_2) + \rho_p \int C_{pp}(\vec{r}_1,\vec{r}_3) h_{pp}(\vec{r}_3,\vec{r}_2) d\vec{r}_3$$
$$+ \rho_i \int C_{pi}(\vec{r}_1,\vec{r}_3) h_{ip}(\vec{r}_3,\vec{r}_2) d\vec{r}_3 \qquad (5.3)$$

$$h_{ip}(\vec{r}_1,\vec{r}_2) = C_{ip}(\vec{r}_1,\vec{r}_2) + \rho_p \int C_{ip}(\vec{r}_1,\vec{r}_3) h_{pp}(\vec{r}_3,\vec{r}_2) d\vec{r}_3$$

The TCP is translationally invariant and hence we have $h_{pp}(\vec{r}_1,\vec{r}_2) = h_{pp}(|\vec{r}_1 - \vec{r}_2|)$. Since $\rho_p = \rho^0_p - \rho_i$ we have, from Eq. (5.3),

$$h_{pp}(\vec{r}_1,\vec{r}_2) = C_{pp}(\vec{r}_1,\vec{r}_2) + \rho^0_p \int C_{pp}(\vec{r}_1,\vec{r}_3) h_{pp}(\vec{r}_3,\vec{r}_2) d\vec{r}_3$$
$$+ \rho_i \int [C_{pi}(\vec{r}_1,\vec{r}_0) h_{ip}(\vec{r}_0,\vec{r}_2)$$
$$- C_{pp}(\vec{r}_1,\vec{r}_0) h_{pp}(\vec{r}_0,\vec{r}_2)] d\vec{r}_0 \qquad (5.4)$$

We have used r_0 instead of r_3 in the last term in square brackets. If we write the above as

$$h_{pp}(\vec{r}_1,\vec{r}_2) = h^0_{pp}(\vec{r}_1,\vec{r}_2) + \rho^i_p \Delta h_{pp}(\vec{r}_1,\vec{r}_2) \qquad (5.5)$$

we see that $h_{pp}(\vec{r}_1,\vec{r}_2) \longrightarrow h^0_{pp}(\vec{r}_1,\vec{r}_2)$ as $\rho_i \longrightarrow 0$. The corrections to leading order in ρ_i to h^0_{pp} are hence contained in Δh_{pp} evaluated using zeroth order quantities. Note that $\Delta h_{pp}(\vec{r}_1,\vec{r}_2)$, an integral over the impurity position $\vec{r}_0$ appears in the FQHE. But microfield calculations[19] require $\Delta h_{pp}(\vec{r}_1,\vec{r}_2|\vec{r}_0)$ prior to the $\vec{r}_0$ integration. This is given by

$$\Delta h(\vec{r}_1,\vec{r}_2|\vec{r}_0) = \rho_i [C_{pi}(\vec{r}_1,\vec{r}_0) h_{ip}(\vec{r}_2 - \vec{r}_0) - C_{pp}(\vec{r}_1,\vec{r}_0) h_{pp}(\vec{r}_0 - \vec{r}_2)] \qquad (5.6)$$

Owing to the convolution structure of the O-Z equations Eq. (5.6) has to be symmetrized in r_1 and r_2, although this is not necessary if r_0 is to be integrated over. The various published calculations for the FQHE do not seem to have included all the terms presented in Eq. (5.6). Corrections which are second order in Δh are generated on iterating the O-Z equations. They are also conveniently calculable from the O-Z equations of an inhomogeneous system. We shall not discuss them here due to limitations of space.

6. TREATMENT OF ION-CONFIGURATION EFFECTS

The DFT-model discussed so far solves the problem of determining the ion-electron and ion-ion PDFs to good accuracy. It also provides an "average-atom" or "mean ion" type description of the electronic states of the ions in the plasma. It also provides phase shifts, pseudopotentials, $\bar{Z}$ etc which are relevant to a complete description of "mean-ions".

In this "mean-ion" picture two types of fluctuations are ignored: (a) fluctuations in the electronic configuration, (b) fluctuations in the ionic environment. The first of these, viz, (a) arises from the use of mean occupation factors $f(\varepsilon_v)$ for the energy levels v even though at any instant they should be just 0 or 1. One approach to the correction of this involves detailed calculations for many specific electronic configurations.[24] Approaches based on statistical fluctuation-models, and using many-body methods[4] have been examined. In this paper we will not consider this aspect of the problem in detail.

The fluctuations in the ionic environment involve time scales τ_i which are much longer than electronic time scales τ_e. Hence the electrons see the "instantaneous" ionic environment and *not* the ion distribution represented by $\bar{\rho}\, g_{ii}(r)$. These fluctuations will have little or no effect on the deep electronic states but will affect the higher-lying bound states of the ions. One way of dealing with ion-configuration fluctuations is to use ion-microfield distributions. Instead of using ion-microfields, a better way would be to use the fluctuating ion-potentials themselves. The advantage here is that the potential acting on the radiating electron at some location $\vec{r}$ can be used rather than the (approximate) use of the field at the nucleus (r = 0) for all values of $\vec{r}$. Also, microfields only take account of *non-spherical* ion density fluctuations, completely ignoring spherical density fluctuations.

6.1 Probability distribution for fluctuating ion-potentials

Let us assume that the instantaneous configuration of the field-ions is $\{\vec{r}_i\} = \vec{r}_1, \vec{r}_2, \ldots \vec{r}_N$. Then the potential felt by a test charge at some arbitrary location $\vec{r}$ is given by

$$v_t(\vec{r}, \{\vec{r}_i\}) = \sum_i{}' v(\vec{r} - \vec{r}_i) \tag{6.1}$$

$$v(\vec{r} - \vec{r}_i) = \frac{\zeta_1(\vec{r} - \vec{r}_i)}{|\vec{r} - \vec{r}_i|} - \int \frac{\Delta n_f^{(1)}(|\vec{r}_2 - \vec{r}_i|)}{|\vec{r} - \vec{r}_2|} d\vec{r}_2 \tag{6.2}$$

This is the potential from the ion at $\vec{r}_i$ having a bound charge cloud (hence the ζ_1) as well as the free electron-displacement $\Delta n^{(1)}{}_f(r)$ associated with it. Let the value of this potential at the origin, ie.., r = 0 be denoted by $v(r_i)$. The microfields are evaluated with $V_0 = \Sigma' v(r_i)$. If $v(r_i)$ is known its value at any $\vec{r}$ is known assuming we know the functional form (6.2). The probability of occurrence of the potential (6.1) is simply the probability of occurrence of the configuration $\{\vec{r}_i\}$. Just as in the theory of ion-microfields, or in the virial cluster expansion we look for the probability $W(V_0)$ of finding the potential V_0 at $r = 0$. We define a Fourier transform such that

$$W(V_0) = \int e^{-ik_0 V_0} W(k_0) \, dk_0/2\pi$$

Then it is easy to show that

$$W(k_0) = \exp[\bar{\rho} w_1 + \frac{\bar{\rho}^2}{2} w_2 + ..]$$

where

$$w_1 = \int \phi_1(\vec{x}_1) g_{ii}(\vec{x}_1) d\vec{x}_1$$

and $\phi_1(x_1) = [e^{-ik_0 v(x_1)} - 1]$.

Hence, for a given set of values $\{V^s{}_0\}$, ie. values of the potential at the origin, a set of probabilities for the potentials, viz, $\{W(V^s{}_0)\}$ can be calculated. This set of probabilities is applicable not just to $r = 0$ but to all $\vec{r}$ given by the functional form used to calculate V_0. Hence, given the set of potentials $\{V^s(r)\}$ and their probabilities we can use the Kohn-Sham equation and obtain a corresponding set of eigenvalues $\varepsilon^s{}_v$. We solve Eq.. (2.21) where $V^e(r)$ is given by Eq..(2.15) but with the $V^e{}_P(r)$ of Eq..(2.17) replaced by $V^s(\vec{r})$. Such a calculation will provide a probability distribution for each eigenvalue of energy ε_v. This method is in principle superior to the microfield approach, directly applicable to strongly coupled situations, and uses the $g_{ii}(r)$, $\zeta_1(r)$ etc obtained from the mean-ion model to build in the effect of ion-configuration fluctuations on the atomic physics. A similar approach can be used for dealing with electron fluctuation effects as well. Also, a many-body approach involving the calculation of self-energies etc can be implemented for each energy $\varepsilon^s{}_v$. However, since these ideas have not yet been tested at the computational level we will not consider them in any greater detail.

7. ACKNOWLEDGEMENTS

Many valuable discussions with François Perrot (CEA - France) are gratefully acknowledged. The author wishes to thank Carl Moser (CECAM) and François Grimaldi (CEA) for fostering the collaborations leading to much of the work reviewed here.

REFERENCES

1. P. Hohenberg and W. Kohn, Phys. Rev. B 136, (1984) 864; N.D. Mermin, Phys. Rev. A 177 (1965) 1441; W. Kohn and L.J. Sham, Phys. Rev. A140 (1965) 1133

2. For reviews see W.D. Kraeft, D. Kremp, W. Ebeling, and G. Röpke, Quantum Statistics of Charged Particles (Plenum 1986), also M.W.C. Dharma-wardana, JQSRT 27 (1982) 315, M.W.C. Dharma-wardana, F. Grimaldi, A. Lecourt, J.-L. Pelissier, Phys. Rev. A 21, (1980) 379 and references there-in.
3. For example, F. Nardin, G. Jacucci, and M.W.C. Dharma-wardana, Phys. Rev. A 37 (1988) 1028, F. Grimaldi, A. Grimaldi-LeCourt and M.W.C. Dharma-wardana, Phys. Rev. A 32 (1985) 1063.
4. F. Perrot and M.W.C. Dharma-wardana, Phys. Rev. A 29 (1986) 1378.
5. R. Car and M. Parinello, Phys. Rev. Let. 55 (1985) 2471.
6. J. Chihara, J. Phys. C 17 (1986) 1633, also see Strongly Coupled Plasma Physics, eds. F.J. Rogers and H.E. DeWitt (Plenum, New York (1986) p. 315.
7. M.W.C. Dharma-wardana in Strongly Coupled Plasmas, eds. F.J. Rogers and H.E. DeWitt (Plenum, New York 1986) p. 275.
8. F. Perrot and M.W.C. Dharma-wardana, Physical Rev. A 30 (1984) 2619, and S. Ichimaru, H. Iyetomi, and S. Tanaka, Phys. Reports 149 (1987) 91.
9. M.W.C. Dharma-wardana and F. Perrot, Physical Rev. A 26 (1982) 2096.
10. J. Chihara, Phys. Rev. A 33 (1986) 2575.
11. F. Perrot, Y. Furutani, and M.W.C. Dharma-wardana, Phys. Rev. A (submitted).
12. M.S. Hybertsen and S.G. Louis, Phys. Rev. B 34 (1986) 5390, see also Ref. 4 for computations using Dyson's equation for plasmas.
13. W.A. Harrison, Pseudopotentials in the Theory of Metals (Benjamin, New York 1966).
14. B.F. Rozsnyai, Phys. Rev. 145 (1972) 1137.
15. D.A. Liberman, JQSRT 27 (1982) 335.
16. J. Davis and M. Blaha, JQSRT 27 (1982) 307.
17. D. Ofer, E. Nardi, Y. Rosenfeld, Phys. Rev. A 38 (1988) 5801.
18. The Quantum Hall Effect, eds R.E. Prange and S.M. Girvin (Springer-Verlag, New York 1987).
19. F. Perrot and M.W.C. Dharma-wardana, Phys. Rev. A (submitted).
20. C.A. Iglesias and C.F. Hooper, Phys. Rev. A 25 (1982) 1049.
21. J.-L. Barrat, J.-P. Hansen and G. Pastore, Phys. Rev. Lett. 58 (1987) 2075 and references there-in.
22. R.B. Laughlin, Surf. Sci. 142 (1984) 163.
23. H.A. Fertig and B.I. Halperin, Phys. Rev. B 36 (1987) 6302.
24. e.g. see F. Perrot in Strongly Coupled Plasma Physics, eds F.J. Rogers and H.E. DeWitt (Plenum, New York 1986) p. 293 and Physica A 150 (1988) 357.
25. F. Grimaldi and A. Grimaldi-LeCourt, JQSRT 27 (1982) 373.

EFFECT OF THE ELECTRON-ION CORRELATION POTENTIALS ON THERMODYNAMIC FUNCTIONS IN DENSE H AND He PLASMAS

François PERROT

Centre d'Etudes de Limeil-Valenton, BP 27, 94195 Villeneuve St Georges, France

Yoichiro FURUTANI

Department of Electrical and Electronic Engineering, Okayama University, Okayama 700, Japan

Chandre DHARMA-WARDANA

National Research Council, Ottawa K1A OR6, Canada

The electron-ion correlation effect is examined, using electron-ion correlation potentials, expressed in terms of a short-range part of the electron-ion correlation function. The influence of electron-ion correlations on relevant thermodynamic quantities is discussed.

1. INTRODUCTION

We use Density Functional Theory (DFT), to study the effect of the electron-ion (e-i) correlation potentials (CP) W^{ei} expressed in terms of the short-range part of the e-i direct correlation function (DCF). We introduce in Section 2 the two CP's W_e^{ei} and W_i^{ei}, acting separately on electrons and ions, which improves on earlier work[1,2]. Section 3 is devoted to numerical results for H and He plasmas. Two sets of calculations were performed, viz., *with* and *without* W^{ei}, to investigate the influence of the e-i DCF on the static structure factor, the degree of ionization and the ion-ion (i-i) interaction energies.

2. THE e-i CORRELATION IN DFT

The DFT states that the exact electron density $n(r)$ obeys a Kohn-Sham equation and the ion density $\rho(r)$ a HNC-type equation. These equations contain the potentials

$$U_e(\vec{r}) = -U_H(\vec{r}) + W_e^{ee}(\vec{r}) + W_e^{ei}(\vec{r}) \tag{1}$$

$$U_i(\vec{r}) = \bar{Z}U_H(\vec{r}) + W_i^{ii}(\vec{r}) + W_i^{ei}(\vec{r}) \tag{2}$$

where $U_H(\vec{r}) = \dfrac{Z}{r} - \dfrac{1}{r} * (\Delta n - \bar{Z}\Delta\rho)$. \hfill (3)

$U_H(\vec{r})$ is the Hartree potential. $\Delta n(\vec{r})[=n(\vec{r})-\bar{n}]$ and $\Delta\rho(\vec{r})[=\rho(\vec{r})-\bar{\rho}]$ are the displaced densities with, e.g., $\bar{n}=n(r\to\infty)$ and the asterisk denotes the convolution. $\bar{Z}$ is the effective charge of non-central ions. When it differs from Z, they carry bound charges. W_e^{ee}, W^{ei} and W_i^{ii} are the electron-electron (e-e) exchange correlation potential, the e-i and i-i CP's, respectively. We also use the local density approximation (LDA) to construct W_e^{ee}, using the free energy of a homogeneous electron gas at temperature T. We thus write

$$W_e^{ee}(r) = \frac{\delta F_{xc}^{ee}[n(\vec{r}),T]}{\delta n(\vec{r})} \qquad (4)$$

Next, W_i^{ii} can be given, within the HNC approximation, by

$$W_i^{ii}(\vec{r}) = -\frac{1}{\beta}(h_{ii}+\beta U_i)*\bar{\rho}h_{ii}, \qquad h_{ii} = \Delta\rho(\vec{r})/\bar{\rho} \qquad (5)$$

where h_{ii} is the i-i radial correlation function. Eq.(5) can be rewritten in terms of the short-range part of the i-i DCF, i.e., $\tilde{c}_{ii}(\vec{r}) = c_{ii}(\vec{r}) + Z\bar{Z}/r$. Thus

$$W_i^{ii}(\vec{r}) = -\frac{1}{\beta}\tilde{c}_{ii}*\bar{\rho}h_{ii} \qquad (6)$$

In our earlier work[1,2], W^{ei} was neglected. This RPA-type approximation requires improvement for light ions involving core states. To construct W^{ei}, we first consider the classical TCP, using the HNC approximation and obtain

$$W_e^{ei}(\vec{r}) = -\frac{1}{\beta}\tilde{c}_{ei}*\bar{\rho}h_{ii} \quad \text{and} \quad W_i^{ei}(\vec{r}) = -\frac{1}{\beta}\tilde{c}_{ei}*\bar{n}h_{ei} \qquad (7,8)$$

with $h_{ei} = \Delta n/\bar{n}$. These expressions are correct in the quantum regime, provided $\tilde{c}_{ei}$ is defined, following Chihara[3], by

$$\tilde{c}_{ei}(q) = -\beta\bar{n}\frac{h_{ei}(q)}{\chi_e^0(q)} \qquad (9)$$

with $\chi_e^0(q)$ the density response function of a non-interacting electron gas.

3. NUMERICAL RESULTS AND DISCUSSION

Self-consistent calculations were carried out for H and He plasmas in a range of electron densities and temperatures going from weak to strong coupling regimes with $\Gamma_e (=\beta/r_s$, where r_s is the electron sphere radius) which varies between 0.07 and 4.35. In Table 1, we show the effective charge $\bar{Z}$ of non-central ions, the ionic structure factor $S(q=0)$ at the long wavelength limit and the 1s bound level. When $\bar{Z}$ differs from $Z(=1$ for H), $\bar{\rho} \neq \bar{n}$ and $S(0) - \Delta n(0) = (1-\bar{Z})\Delta\rho(0)$, which is not zero. The 1s level exists at small densities (e.g., $r_s=4$ in Table 1) and disappears at higher densities ($r_s=0.5$). A detailed discussion of the temperature dependence of relevant thermodynamic functions will be given elsewhere[5]. In Figs.1 and 2, the radial profile of W_e^{ei} and W_i^{ei} are illustrated for $r_s=1$ and for different temperatures. Another important effect of e-i correlation appears

Table 1. Results for H plasmas. ε_{1s} is the 1s bound level in hartrees, $\bar{Z}$ the effective charge and S(0) the i-i structure factor at q=0.

				with no e-i CP			with e-i CP		
r_s	T(eV)	T/T_F	Γ_e	ε_{1s}	$\bar{Z}$	S(0)	ε_{1s}	$\bar{Z}$	S(0)
4	12.5	3.99	0.54	-0.222	0.740	0.571	-0.208	0.749	0.619
4	25	7.98	0.27	-0.283	0.915	0.532	-0.272	0.916	0.536
4	50	16.0	0.14	-0.325	0.971	0.519	-0.320	0.971	0.520
4	100	31.9	0.07	-0.364	0.984	0.519	-0.362	0.984	0.523
0.5	12.5	0.06	4.35	-	1.0	0.125	-	1.0	0.109
0.5	25	0.12	2.18	-	1.0	0.205	-	1.0	0.191
0.5	50	0.25	1.09	-	1.0	0.314	-	1.0	0.305
0.5	100	0.50	0.54	-	1.0	0.417	-	1.0	0.413

in the i-i interaction energies, viz., E_{ii} for a pair of bare ions and E_{ii}^* for a pair of dressed ions defined, respectively, by

$$E_{ii} = \frac{1}{2} \beta Z^2 \bar{\rho} \int d\vec{r} \, \frac{1}{r} h_{ii}(r) \qquad (10)$$

and $E_{ii}^* = \frac{1}{2} \beta \bar{\rho} \int d\vec{r} \, \phi_{ii}^*(r) \, h_{ii}(r) \qquad (11)$

where $\phi_{ii}^*(r)$ is the pair correlation potential which produces the same h_{ii} for a TCP. Results are shown in Table 2. E_{ii} and E_{ii}^* hardly vary at high densities, where the strong i-i correlation makes the e-i one less effective. Finally, results for He plasmas are tabulated for a few sets of parameters in Table 3. General features are similar to those for H plasmas.

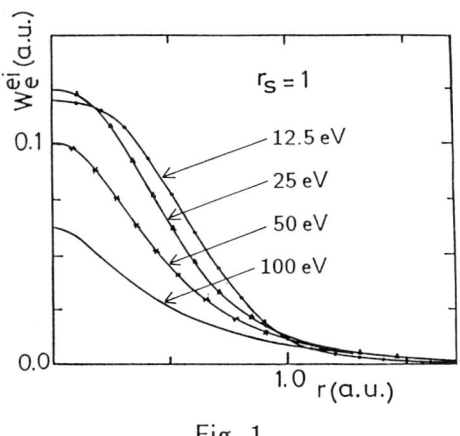

Fig. 1

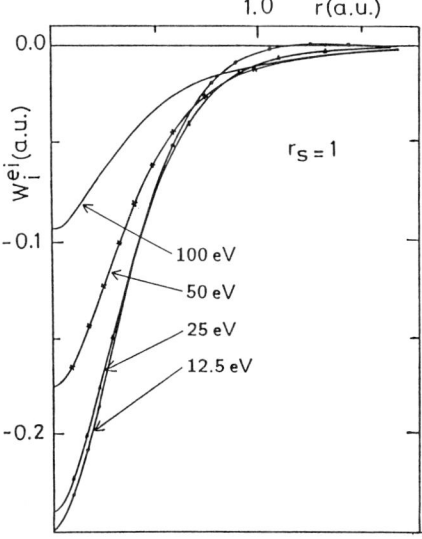

Fig. 2

Radial profile of the two CP's, W_e^{ei} and W_i^{ei}, versus r in Figs. 1 and 2, respectively.

Table 2. Energy integrals for H plasmas in units of $k_B T$.

		with no e-i CP		with e-i CP	
r_s	T(eV)	E_{ii}	E_{ii}^*	E_{ii}	E_{ii}^*
4	12.5	0.513	0.0485	0.129	0.0257
4	25	0.0677	0.0303	0.0658	0.0277
4	50	0.0269	0.0140	0.0266	0.0136
4	100	0.0101	0.0056	0.0101	0.0056
0.5	12.5	2.860	1.346	2.890	1.382
0.5	25	1.224	0.580	1.232	0.583
0.5	50	0.493	0.237	0.494	0.234
0.5	100	0.190	0.094	0.190	0.093

Table 3. DFT calculations for He plasmas, including the e-e xc CP in the LDA. The bound levels are in hartrees, while the interaction energies are in units of $k_B T$.

		with no e-i CP				with e-i CP			
r_s	T(eV)	ε_{1s}	ε_{2s}	$\bar{Z}$	S(0)	ε_{1s}	ε_{2s}	$\bar{Z}$	S(0)
2	50	-1.003	-	1.716	0.363	-0.9204	-	1.727	0.345
2	100	-1.210	-0.0079	1.906	0.349	-1.166	-0.0058	1.908	0.341
1	50	-0.1554	-	1.439	0.275	-0.0979	-	1.520	0.267
1	100	-0.3650	-	1.662	0.306	-0.3061	-	1.680	0.295

		with no e-i CP		with e-i CP	
		E_{ii}	E_{ii}^*	E_{ii}	E_{ii}^*
2	50	0.3567	0.1376	0.3493	0.1199
2	100	0.1504	0.0753	0.1497	0.0732
1	50	0.9226	0.2874	0.8789	0.2393
1	100	0.3820	0.1550	0.3763	0.1449

4. CONCLUDING REMARK

Within the framework of DFT, the effect of the e-i CP on the static correlational properties of H and He plasmas was examined in terms of the short-range part of the e-i DCF. Though our approach parallels the "QHNC" formalism due to Chihara, we emphasize that the present approximation is, in principle, limited to the e-i correlation between *free* electrons and ions, and is likely unable to describe the effect of deeply *bound* electrons. More complete theory of this effect deserves a future study.

REFERENCES

1) M.W.C. Dharma-wardana and F. Perrot, Phys. Rev. A26 (1982) 2096
2) F. Perrot and M.W.C. Dharma-wardana, Phys. Rev. A29 (1984) 1378
3) J. Chihara, Phys. Rev. A33 (1986) 2575
4) F. Perrot and M.W.C. Dharma-wardana, Phys. Rev. A30 (1984) 2619
5) F. Perrot, Y. Furutani and M.W.C. Dharma-wardana, submitted to Phys. Rev. A.

ENERGY LOSS OF CHARGED PARTICLES IN LIQUID AND AMORPHOUS METALS

Fukuo YOSHIDA

Research Reactor Institute, Kyoto University, Osaka 590-04, Japan

In the energy loss of heavy low-energy particles by conduction electrons in metals we show the existence of a dynamical part representing ionic effects. Contribution of this dynamical part is discussed and compared with conventional results.

1. INTRODUCTION

The energy loss of charged particles has been a subject of basic importance in atomic and condensed matter physics[1,2]. There are, as is well known, two approaches to study the energy loss problem, the binary collision theory and the response or dielectric function approach. By the latter we can naturally incorporate collective effects, and dispense with cut-off parameters. Most of previous works dealing with the energy loss problems in plasmas rely on the random-phase approximation(RPA) or its generalized versions, and to our knowledge there are few works paying attention to the dynamical contribution[3-7].

In this paper we derive a general formula containing a dynamical contribution for the energy loss of low-energy heavy particles based on the response function method. It is pointed out that in liquid and amorphous metals this contribution is analogous to the Ziman formula[8] for the electrical conductivity, and becomes relatively non-negligible when the electron mean free path is of the order of the interatomic distance.

2. FORMULATION

We consider a system composed of charged particles with mean number density n and particle's mass m at absolute temperature T, under the presence of opposite charges in the background. When a particle of mass M with energy E is impinging upon the system, the energy loss rate dE/dx of this projectile is defined by

$$dE/dx = -(nm/4\pi hE) \int dq d\omega \ |V(q)|^2 (\omega q) S(q,\omega) \qquad (1)$$

where $q=|k-k_1|$ and $\omega=(E-E_1)/\hbar$ with $k(k_1)$ being the wavevector of the projectile before(after) scattering. In the above equations $V(q)$ represents the interaction potential of the projectile and a particle in the system, and $S(q,\omega)$ is the dynamical structure factor. In eq.(1) the integration range for q is from 0 to ∞ and that for ω is given by the kinematical condition as

$$\omega_+(q) = vq - \omega_r(q) \geq \omega \geq \omega_-(q) = -vq - \omega_r(q), \quad \text{with} \quad \omega_r(q) = \hbar q^2/2M \quad (2)$$

where $v=(2E/M)^{1/2}$ is the velocity of the projectile. The quantity dE/dx is also related to the imaginary part $\chi''(q,\omega)$ of the response function $\chi(q,z)$ via the relation

$$S(q,\omega) = (\hbar/\pi)\{1-\exp(-\beta\hbar\omega)\}^{-1} \chi''(q,\omega), \quad \text{with} \quad \beta=(k_BT)^{-1}. \quad (3)$$

For heavy projectiles ($M \gg m$) the recoil term $\omega_r(q)$ in eq.(2) is negligibly small. When the velocity of the projectile is sufficiently small under this condition, we can expand $S(q,\omega)$ around $\omega=0$ in eq.(1), to obtain

$$dE/dx = -v(1/6\pi\hbar)(m/M)n \int dq \, |V(q)|^2 q^4 [\partial S(q,\omega)/\partial \omega]_{\omega=0} \quad (4)$$

3. DYNAMICAL EFFECTS IN ENERGY LOSS FOR LOW ENERGY PROJECTILES

The response function $\chi(q,z)$ can be generally written in the form[9]

$$\chi(q,z) = \chi_0(q,z)/[1-v(q,z)\chi_0(q,z)] \quad (5)$$

where $v(q,z)$ is a frequency-dependent effective potential. The quantity $\chi_0(q,z)$ is the response function for a reference system, which is usually taken as a non-interacting system, i.e., $\chi_0 = \chi_f$. It is noted that the local field correction factor $G(k,z)$ has the one-to-one correspondence to $v(q,z)$ as $v(q,z) = nv(q)[1-G(q,z)]$. Equation (5) enables us to write the imaginary part $\chi''(q,\omega)$ in terms of that of $\chi_0(q,z)$ and $v(q,z)$, and from this we obtain the derivative of $S(q,\omega)$ at $\omega=0$ in terms of $\xi(q)$;

$$\xi(q) = \xi_0(q) + \xi_d(q) \quad (6)$$

$$\xi_0(q) = \lim_{\omega \to 0}[\chi_0''(q,\omega)/\omega], \quad \xi_d(q) = -\chi_0(q)^2 \lim_{\omega \to 0}[v''(q,\omega)/\omega]. \quad (7)$$

Substituting it into eq.(4), we arrive at the formula

$$dE/dx = -v(m/M)(1/6\pi^2)n \int dq \, |V^*(q)|^2 q^4 \xi(q) = [dE/dx]_s + [dE/dx]_d \quad (8)$$

where $V^*(q)$ is the screened potential given by

$$|V^*(q)|^2 = |V(q)|^2/D(q,0) = |V(q)|^2/[1+v'(q,0)\chi_0(q)]^2. \quad (9)$$

In the above equation the value of the real part $v'(q,\omega)$ of $v(q,z)$ at $\omega=0$ is denoted as $v'(q,0)$. The static part $[dE/dx]_s$ and the dynamical part $[dE/dx]_d$ in eq.(8) correspond to $\xi_0(q)$ and $\xi_d(q)$, respectively. The former represents the kinetic contribution from $\chi_0(q,z)$, and the latter a dynamical contribution from the ω-dependence of $v(q,z)$. In the RPA and similar static approximations $\xi_d(q)=0$ or $\xi(q)=\xi_0(q)$.

The static part corresponds to treating opposite charges as uniform, as in the OCP. Using the relation $\xi_\theta(q)=\pi\beta S_\theta(q,\omega=0)$ and introducing the mean free path ℓ_s and the relaxation time τ_s, we may rewrite it as

$$[dE/dx]_s = -E/\ell_s \quad , \quad \ell_s = \ell_s(v) = \tau_s v \tag{10}$$

$$1/\tau_s = (1/3\pi)(m/M)^2 \beta (n/m) \int dq \; q^4 |V^*(q)|^2 S_\theta(q,\omega=0). \tag{11}$$

For the dynamical part we make use of the expression

$$v(q,z) = [1/\chi(q) - 1/\chi_\theta(q)] + (m/q^2)zR_2(q,z) \tag{12}$$

where $\chi(q)$ is the static susceptibility and $R_2(q,z)$ is the interaction part of a second-order memory function $M_2(q,z)$ for the density fluctuation[9]. Substituting eq.(12) into the second equation of eq.(7) we obtain the similar result for $[dE/dx]_d$ as eq.(10) in terms of $\ell_d = \ell_d(v) = \tau_d v$ with τ_d given by

$$1/\tau_d = (1/3\pi^2)(m/M)^2 n \int dq \; q^2 |V^*(q)|^2 \chi_\theta(q)^2 R_2'(q,\omega=0). \tag{13}$$

4. ENERGY LOSS IN LIQUID AND AMORPHOUS METALS

We apply the results obtained above to metals. The static susceptibility $\chi_\theta(q)$ for free electrons decreases rapidly like q^{-2} for large q, and thus the contribution of $M_2'(q,\omega=0)$ from the large q-region is less important. It is convenient in the present case to adopt $R_2'(q,\omega=0) \simeq M_2'(q=0,\omega=0)$. If the q-dependence of $R_2'(q,\omega=0)$ was not significant compared with other factors in eq.(13) it would be a quite good approximation. We then obtain

$$1/\tau_d = (1/\tau')(1/3\pi^2)(m/M)^2 n \int dq \; q^2 |V^*(q)|^2 \chi_\theta(q)^2 . \tag{14}$$

The quantity $1/\tau' = M_2'(q=0,\omega=0)$ is the relaxation time for the electrical conductivity, $\sigma = (ne^2/m)\tau'$, as in our previous work[9]. The relative importance of the dynamical part is measured by the ratio $\tau_s/\tau_d = \ell_s/\ell_d = \Lambda^*/\Lambda$, where $\Lambda = v_F \tau'$ is the electrical mean free path and Λ^* is introduced by

$$\Lambda^* = (mv_F/\beta\pi)\int dq \; q^2 |V^*(q)|^2 \chi_\theta(q)^2 \; / \int dq \; q^4 |V^*(q)|^2 S_\theta(q,\omega=0). \tag{15}$$

The quantity Λ^* is determined solely by the properties of conduction electrons.

The quantity $\Lambda(A)$ is related to the electrical resistivity $\rho(\mu\Omega.cm)$ by $\Lambda \simeq 10^2 r_s^2/\rho$, where r_s is the dimensionless characteristic radius. The values of ρ are in fact roughly of the order of 10 in liquid metals, but these tend to be as large as 10^2 for amorphous metals and Λ is comparable to the interatomic distance[10]. We have actually evaluated eq.(15) by using reasonable local field correction factors. It is found that the values of Λ^* are about 1-2A for r_s = 2-5 and not so sensitive to the local field correction compared with τ'. This result implies that the dynamical contribution due to ions tends to become a large correction when the electron mean free path is as small as the interatomic distance.

5. CONCLUDING REMARKS

We have derived the general formula, eq.(8), for the energy loss rate of low-energy heavy particles. In addition to the conventional static term it contains the dynamical one. We have found that in liquid and amorphous metals it represents ionic correlations, analogous to the Ziman formula, and qualitatively discussed the possibility about its non-negligible contribution.

REFERENCES

1) J.D. Jackson, Classical Electrodynamics, 2nd edition(Wiley, New York, 1976) chap.13.

2) R.M. More, Atomic physics in inertial confinement fusion, in: Applied Atomic Collision Physics, vol.2(Academic Press, New York,1984).

3) E. Nardi, E. Peleg and Z. Zinamon, Phys. Fluids 21(1978) 574.

4) N.R. Arista and W. Brandt, Phys. Rev. A23(1981) 1898.

5) G. Maynard and C. Deutsch, Phys. Rev. A26(1982) 665.

6) Yu.S. Sayarov, Z. Phys. A313(1983) 9.

7) S. Tanaka and S. Ichimaru, J. Phys. Soc. Jpn. 54(1985) 2537.
   X.Z. Yan, S. Tanaka, S. Mitake and S. Ichimaru, Phys. Rev. A32(1985) 1785.

8) J.M. Ziman, Phil. Mag. 6(1961) 1013.
   T.E. Faber and J.M. Ziman, Phil. Mag. 11(1965) 153.

9) S. Takeno and F. Yoshida, Prog. Theor. Phys. 60(1978) 1304.
   F. Yoshida and S. Takeno, Phys. Rep. 173(1989) 301.

10) P.J. Cote and L.V. Meisel, Electrical transport in glassy metals, in: Glassy Metals 1, eds. H.J. Guntherodt and H. Beck(Springer, Berlin, 1981) pp. 141-166.

STUDIES OF A STRONGLY COUPLED PLASMA PRODUCED IN A CAPILLARY DISCHARGE

J. F. BENAGE, Jr., L. A. JONES, R. J. TRAINOR, Jr., W. R. SHANAHAN, R. L. SHEPHERD*, and D. P. NOTHWANG

Los Alamos National Laboratory, Physics Division, P.O. Box 1663, MS E526
Los Alamos, New Mexico 87545 USA

We have carried out an experiment measuring the properties of a strongly coupled ($\Gamma \sim 0.8$) plasma produced in a capillary discharge. Using an analytical model and some 1-D MHD calculations along with the measurements, we were able to determine the density, temperature, strong coupling parameter Γ, and the resistivity of this plasma. We then compared our resistivity measurements to various theories.

For the past years we have been developing an experimental program to measure the electrical resistivity of a strongly coupled plasma which is produced in a capillary discharge. This work was begun when it was realized, utilizing the ideas of McCorkle[1], that one could produce a high density, low temperature plasma using a small diameter capillary. This led to an initial set of experiments done by Shepherd, et al.[2] We report here the continuation of these experiments and a slightly different analysis of the results based on an analytical model we have developed and also on some MHD calculations which predict the dynamics of the capillary discharge.

In the experiment, the plasma is created by passing a large current, 550 kA, through a 20 μm hole in a block of polyurethane. The current breaks down the inside wall of the capillary and the plasma formed fills the center rapidly. The high pressure of this plasma produces a shock which compresses and moves through the plastic. Behind the shock, the hot plasma ablates material from this shock compressed region and thus widens the plasma channel.

To determine the resistivity of this plasma, several quantities must be measured. We must measure the voltage across the plasma, the current through the plasma, and the size of the conducting channel. We must also measure the temperature and density to compare our results to theory. At present, we use a capacitive voltage probe and a Rogowski coil to measure the voltage and current, and we use a filtered x-ray diode array to determine the temperature. We cannot yet measure the size of the conducting channel or the density. We therefore must rely on our model and on measurements of the expanding shock, which we measure using a visible framing camera. Together these measurements

---

* Permanent address: Lawrence Livermore National Laboratory

and the model allow us to compute the resistivity of the plasma.

We can briefly describe the model in the following way. We assume the discharge can be divided into three uniform regions as shown in Figure 1. These regions are separated by a shock front and an ablation front. The properties of these regions can be determined by solving a system of ten equations. These equations are the conservation of mass, energy, and momentum across the fronts, the equation of state for the plasma and shock compressed region, a relation between the velocity of the ablation front and the temperature of the plasma, and an equation for determining the fluid velocity of the plasma region.

These equations can be written as

$$\left. \begin{array}{l} \rho_0 v_s^2 = \rho_1(v_s^2 - v_p^2) \\ P_1 = \rho_0 v_s v_p \\ P_1 = \tfrac{1}{2}\rho_0 v_p^2 + \rho_0 C_1 T_1 \end{array} \right\} \text{Conservation equations at the shock front}$$

$$\left. \begin{array}{l} \rho_1(v_A^2 - v_p^2) = \rho_2 v_A^2 \\ P_1 - P_2 = \rho_2 v_A(v_p - v_2) \\ P_1 - P_2 + \dfrac{Q}{\pi r_A^2 l} = \rho_2(C_2 T_2 - C_1 T_1) + \tfrac{1}{2}\rho_2(v_2^2 - v_p^2) \end{array} \right\} \text{Conservation equations at the ablation front}$$

$$P_2 = \dfrac{\rho_2 N_A k T_2}{\overline{A}}[1 + \overline{Z}]$$

$$v_A = v_p + v_R \quad \text{where} \quad v_R = \dfrac{\sigma T^3}{\rho_1 C_2}$$

and

$$v_2 = 0$$

where v_s, v_p, v_A, v_2 are the shock, particle, ablation, and plasma velocity, P, C, and T are the pressure, specific heat, and temperature for region 1 or 2, Q is the energy supplied to the plasma, r_A is the radius of the ablation front, l is the length of the plasma, N_A is Avagadro's number, $\overline{A}$ is the average atomic mass, and σ is Stefan-Boltzmann's constant. We have left out the equation of state for the shock compressed region, but this equation along with the first three will determine the Hugoniot curves. Thus we can substitute these curves for the equations. To solve this system, we plug in the energy supplied to the system and numerically iterate over time.

The results of this model will be compared to the experiment and to the MHD calculations. For now we just consider the graph in Figure 2, which shows the density profile as calculated by the code. We note the densities are fairly uniform and divided into three distinct regions, exactly as we assumed.

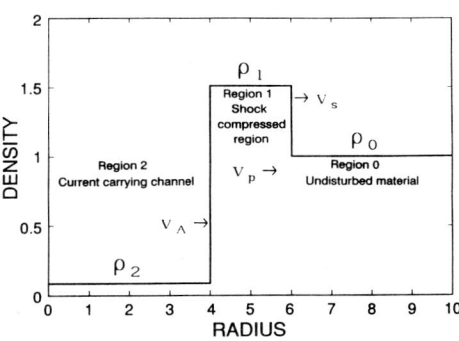

FIGURE 1
Assumed radial density profile for the capillary discharge.

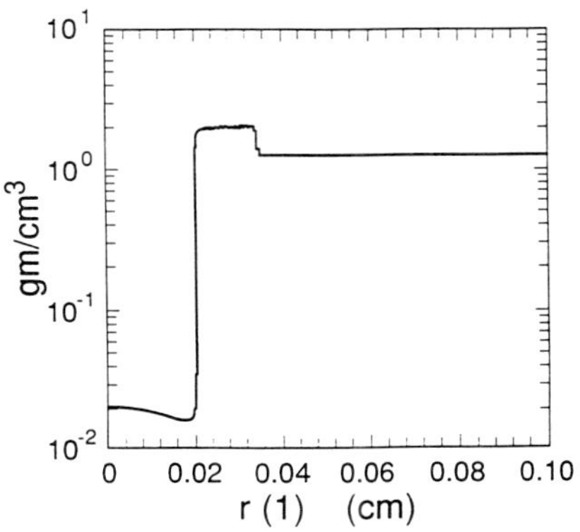

FIGURE 2
Radial density profile calculated by 1-D MHD code.

The experimental results are shown in Figures 3-5. In Figure 3, we compare the shock velocity as measured with the velocities predicted by the model and the 1-D code. We can see that both match the experiment fairly well, but the code does slightly better. Figure 4 contains a graph showing the temperature profile as measured compared to the calculations. We can see that the model and 1-D code don't do as well here. However, their values are almost within the error bars and the differences could be accounted for by changing the specific heat and by taking into account any non-uniformities in the temperature in the plasma.

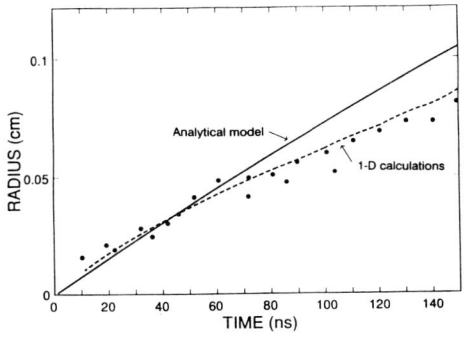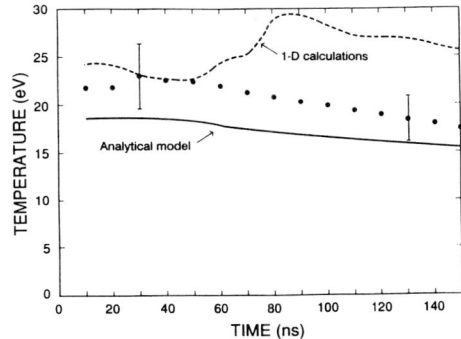

FIGURE 3
Comparison of the calculated and measured shock front.

FIGURE 4
Comparison of the calculated and measured temperature profile.

From these figures it is clear the model and 1-D code do an adequate job of modeling the discharge. We therefore are justified in utilizing these calculations to determine the density, strong coupling Γ, and the radius of the ablation front which is used in calculating resistivity. From the code we estimate $n_e = 4 \times 10^{21}$ cm^{-3} and $\Gamma \sim 0.8$ for our plasma. The velocity of the ablation front averages 3.8×10^5 cm/s and we use this value to determine the size of the conducting plasma. We can now use these values to determine the resistivity and compare to several theories.

Figure 5 shows the resistivity as a function of temperature for the experiment and several theories. We note here that Spitzer[3] does nearly as well as the Rinker[4] values even though the coulomb logarithm is < 2. The Rinker values do the best, fitting almost within the error bars. The surprising result is that Ichimaru's theory doesn't fit the data well. We have no explanation for this.

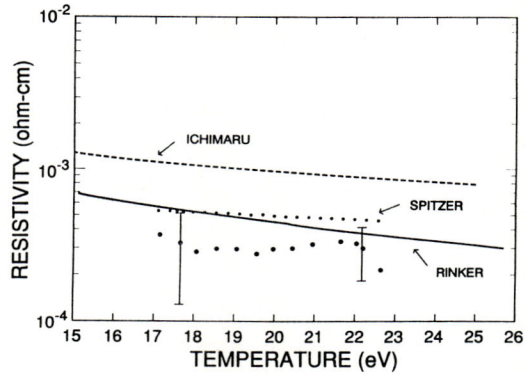

FIGURE 5
Comparison of theoretical and measured resistivities as a function of temperature.

In conclusion, we have measured the resistivity of a strongly coupled plasma and the results fit the theory by Rinker best. In the future we plan more experiments and hope to cover a wider range of densities and temperatures.

REFERENCES

1) R. A. McCorkle, Appl. Phys. A 26 (1981) 261-270.
2) R. L. Shepherd, D. R. Kania, and L. A. Jones, Phys. Rev. Lett. 61, (Sept. 12, 1988) 1278.
3) L. Spitzer, Jr. and R. Harm, Phys. Rev. 89 (1953) 977.
4) G. Rinker, Los Alamos National Laboratory Report, LA-10608-MS, February 1986.
5) S. Ichimaru and S. Tanaka, Phys. Rev. A 32, (Sept. 1985) 1790.

THE MEASUREMENT OF TRANSPORT PROPERTIES IN STRONGLY COUPLED PLASMAS

R.L. SHEPHERD, D.R. KANIA, L.A. JONES*, D.H. SCHNEIDER and R.E. STEWART

Physics Department, Lawrence Livermore National Laboratory
Livermore, CA 94550

*Physics Division, Los Alamos National Laboratory,
Los Alamos, NM 87545

Transport properties of strongly coupled plasmas are of interest in solid state physics, astrophysics, and high energy-density plasma physics. Many theoretical models have been presented to describe transport phenomena in strongly coupled plasmas, however few experiments have been performed to compare with calculations. We present the results of an experiment which measure the low-frequency electron resistivity in a partially degenerate ($1.2 < \theta < 1.9$), strongly coupled ($0.6 < \Gamma < 1.8$) plasma. In addition, we suggest experiments to measure the stopping power and the energy transport in strongly coupled plasmas.

1. INTRODUCTION

Transport phenomena can play a significant role in determining the temporal and spatial evolution of plasmas. The transport parameters are expressed as analytic expression when the collisions in the plasma are binary (no particle correlations). A useful parameter in determining the importance of correlations is the comparison of the potential energy of the particles to the kinetic energy of the particles, or (for electron-electron correlations),

$$\Gamma = \frac{e^2}{k_B T}\left(\frac{4\pi n_i}{3}\right)^{1/3} \qquad (1)$$

e is the elementary charge, k_B is the Boltzmann constant, T is the ion temperature, and n_i is the ion density. When Γ (the strong coupling parameter) is of order one, the plasma is said to be strongly coupled. Also effecting the transport is the particle degeneracy. The Fermi degeneracy parameter is used to measure the level of degeneracy and is expressed as,

$$\theta = \frac{2 m_e k_B T}{h^2}(3\pi^2 n_e)^{-2/3} \qquad (2)$$

where m_e is the electron mass, and n_e is the electron density. When θ is much less than one the plasma is degenerate. Although transport properties have been calculated for a wide range of densities and temperatures, until recently, few measurements have been made in the strong coupling regime. In this paper we present the results from an experiment which measured the low frequency resistivity in a partially degenerate, strongly coupled plasma. In addition, we propose measuring the stopping power in plasmas of similar parameters and suggest a possible method of measuring radiative thermal conductivity.

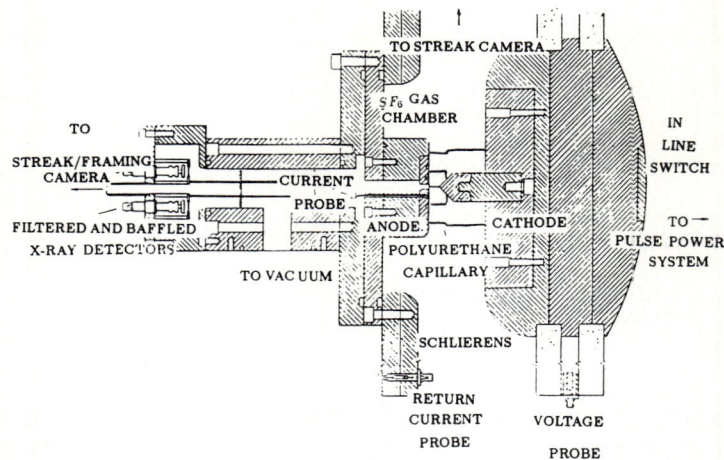

Figure 1: Experimental Apparatus

2. RESISTIVITY EXPERIMENT

Numerous models have been presented to describe the electrical conduction in cold, dense matter. However, few experiments have been conducted to verify existing theories. The experiments described here were used to measure the resistivity in a partially degenerate, strongly coupled plasma. The plasmas were created by pulsing 600 kV across a 1.7 cm long, 20 μm diameter hole (capillary) in polyurethane[1] (see Figure 1). When the pulsed power system is discharged, the inner walls of the capillary are heated to conduction temperatures, accreting material from the walls and filling the initial void with plasma. As the current rises, the plasma expands at a velocity of 6×10^5 cm/s by ionizing the surrounding cold material, producing a cold ($\approx$ 10 eV), dense ($\approx 6 \times 10^{22}$ e-/cm^3) plasma. During the first 100 ns of the discharge, the plasma characteristics are determined and the resistivity is inferred.

The diagnostics were designed to measure four macroscopic parameters of the plasma column; temperature, radial column size, current, and voltage. The temperature was measured by assuming a blackbody spectral distribution and applying the two foil method[2] to data collected using an X-ray diode (XRD) array.

Radial column size measurements were made axially and radially using framing and streak photography, respectively. Schlieren photographs were used to show the interface of the critical density of the probe laser with the cold polyurethane. Based on the size and temperature measurements, the free stream velocity of particles out the anode of the capillary was calculated and found to be small. Thus, once the plasma volume is large compared to the initial capillary volume, the plasma is essentially at solid density.

The voltage drop across the capillary was measured by capacitively coupling a voltage probe to the capillary cathode, while the current was measured at the face of the capillary by using a localized Rogowski coil.

2.1 Inferred Resistivity

The resistivity is inferred by combining the above measurements. The voltage supplied by the external circuit (V_p) is

$$V_p = I_p J_0 \eta_0 + \frac{4\pi l_p}{c^2} \frac{d}{dt}\left[\frac{r_1^2 J_0}{4} + \frac{r_1^2 J_0}{2} \ln\left(\frac{r_0}{r_1}\right)\right] \quad (3)$$

where J_0 is the current density, η_0 is the resistivity, r_0 is the chamber radius, r_1 is the plasma radius, and l_p is the plasma length. Assuming an isotopic, homogeneous current density, multiplying by current and solving for the resistivity we find

$$\eta_0 = \frac{\pi r_1^2}{I^2 l_p}\left[IV_p - \frac{l_p}{c^2}\left[\frac{I\,dI}{dt}\left(1 + 2\ln\left(\frac{r_0}{r_1}\right)\right) - 2I^2 \frac{dr_1}{dt}\right]\right] \quad (4)$$

A plot of nine resistivities as a function of time are displayed in Figure 2. The error in the resistivity measurement is characterized by the spread in the data displayed in Figure 2 and the error propagated in the calculation by other measured quantities. Combining the systematic and random errors, the total error was estimated[3] and the *true* resistivity was found to be $\approx$ 2.5 x 10^{-3} Ohm-cm with values as larges as 6 x 10^{-3} Ohm-cm and as small as 7.5 x 10^{-4} Ohm-cm being possible. The assumption of uniform current distribution was shown to be self-consistently valid by estimating the skin depth and comparing it to the size of the plasma.

The experimental measurement has been compared to four theories; 1) Spitzer, 2) a modified Spitzer, 3) a semi-analytic expression by Ichimaru[4], and a numerical calculation by Rinker[5]. The Spitzer and modified Spitzer differ by the method of calculating the Coulomb logarithm. The Spitzer calculation uses a classical Coulomb logarithm while the modified Spitzer utilizes a molecular dynamics calculation of the Coulomb logarithm. The Ichimaru calculation is a quantum-statistical approach based on fully ionized hydrogen. The Rinker calculation is based on the Ziman formula. The resistivity of the individual species is calculated and combined using a technique successfully employed in the study of liquid metals. The computed resistivities are compared to the measurement in Table 1.

In summary, the plasmas created in the capillary experiment have a strong coupling parameter (Γ) between approximately 0.6 and 1.8 and a Fermi degeneracy parameter (Θ) between 1.9 and 1.2. This makes the plasmas partially degenerate and in the intermediate to strongly coupled regime. Three theories

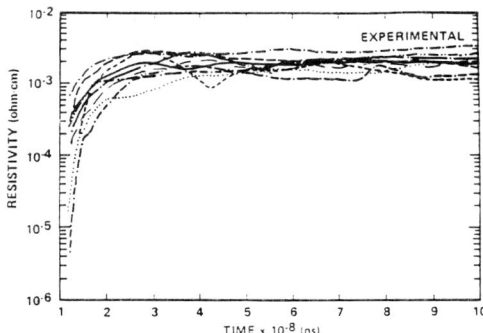

Theory	Resistivity (Ohm-cm)	Measured/Theory
Spitzer	9.8 x 10^{-6}	256
Modified Spitzer	5.9 x 10^{-5}	42
Ichimaru	2.5 x 10^{-4}	10
Rinker	1.8 x 10^{-3}	1.4
Measured value	2.5 x 10^{-3}	—

Table 1: Resistivity Comparison

Figure 2: Resistivity versus time for nine shots

are used for comparison; Spitzer, Ichimaru, and Rinker. The Spitzer and modified Spitzer calculations are approximately a factor of 250 and 42 less than the measured resistivity, respectively. This, however, is to be expected because the classical assumptions break down in plasmas in this density and temperature regime. Calculations based on the Ichimaru theory is approximately a factor of 10 less than the measured value. The Ichimaru calculation corrects the resistivity for the level of degeneracy and strong coupling in the plasma. However, the assumption of a fully ionized plasma is inapplicable to the capillary plasmas which have bound states. This fact is suspected to be the cause of the poor agreement between the Ichimaru calculation and the measured value. The calculation by Rinker, (which includes ionic structure) agrees *best* with the data.

3. OTHER TRANSPORT PROPERTY MEASUREMENT

The need of further experimental research into the transport properties of strongly coupled plasmas is great. However, the difficulties involved in doing transport experiments are considerable. Because strongly coupled plasmas are usually created in a transient state, the plasma parameters have to be determined in a time dependent fashion. Furthermore, the transport property must be measured without perturbing the plasma significantly. The following sections are intended as suggestion for measuring two transport properties; the stopping power and the radiative thermal conductivity.

3.1 Stopping Power

The stopping power of a plasma can be broken into three main contributions: a) stopping due to collision with the nucleus, b) stopping due to collision with the bound electrons and c) stopping due to collision with the free electrons. For low Z projectiles (say Z=6), the nuclear contribution to the stopping power is small except at extremely low projectile energies ($E_{proj.} \leq 20$ keV) The stopping power due to the electrons can be expressed by the Bethe formula

$$\left(\frac{dE}{dx}\right)_{electronic} \approx -\left(\frac{Z^* e}{\upsilon_{ion}}\right)^2 \left[\frac{4\pi(Z^* - \overline{Z})e^2}{m_e} \ln \Lambda_b + \omega_p^2 \ln \Lambda_f\right] \quad (5)$$

where ω_p is the plasma frequency, (Z^*) is the effective charge of the ion beam, $\overline{Z}$ is the average charge of the plasma ions, υ_{ion} is the ion velocity, and $\ln\Lambda_b$ and $\ln\Lambda_f$ is the Coulomb logarithm for the bound and free electrons, respectively. The value of the parameters in front of the bound and free electron Coulomb logarithms are approximately the same. Thus the relative magnitudes of the free and bound stopping powers are determined by the respective Coulomb logarithms. For a 100 keV ion beam impinging on a 2 eV carbon target, one can show that when $n_e \ll 3.7 \times 10^{22}$ e-/cm^3, the stopping power is dominated by collisions with free electrons.

Stopping power measurements could be made by utilizing the energy of the Auger electrons emitted from an ion beam after passing through a plasma. Auger electrons will be produced by the relaxation of metastable states created by inner-shell excitation collisions. The energy of an Auger electron (in the laboratory frame at zero degrees) emerging from the ion beam after the decay of the metastable state is

$$E_{A_0} = [E_0^{1/2} \pm E_{A_0}^{'1/2}]^2 \quad (6)$$

where E_0 is the electron-ion mass ratio times the ion beam energy ($E_0 = E_{Ion} \times m_e/M_{Ion}$), and E'_{A_0} is the energy of the Auger electron in the center of mass frame. Thus, the Auger electron energy is dependent on the ion beam energy and when the ion beam has passed through a target, the Auger electron will be emitted at a lower energy, corresponding to the energy loss by the ion beam. An electron spectrometer could be positioned to collect and measure the energy spectrum of the Auger electrons after the ion beam has passed through the target.

3.2 Radiative Energy Transport (Opacity)

Another transport parameter which would be of interest if measured is the radiative conductivity. In general, the conductivity is,

$$\frac{1}{K} = \frac{1}{K_r} + \frac{1}{K_c} \qquad (7)$$

where K_r is the radiative conductivity and K_c is the electron thermal conductivity. When $K_c \gg K_r$, the conductivity is approximately K_r and the energy transport is

$$\underline{Q} = \underline{S} = \frac{16\sigma l T^3}{3} \underline{\nabla} T \qquad (8)$$

where $\underline{Q}$ is the heat flux, $\underline{S}$ is radiant energy flux, σ is the Stefan-Boltzmann constant, $\underline{\nabla}T$ is the temperature gradient, and l is the Rosseland mean free path. Thus, by measuring the radiant energy flux and the temperature gradient, the radiative conductivity, $16\sigma l T^3/3$, can be inferred. A scenario for such an experiment would be a shock heated laser plasma. The radiant intensity emitted from planar surfaces could be measured, while the temperature gradient was measured by spatially resolving the emission perpendicular to the shock. This measurement would be difficult, but is feasible.

4. CONCLUSIONS

In conclusion, the resistivity has been measured in a partially degenerate strongly coupled plasma. Future experiments could include stopping power measurements and energy transport measurements. These experiments are currently being considered but are recognized as conceptually difficult.

This work was performed under the auspices of the U.S. Dept. of Energy by Lawrence Livermore National Laboratory under contract No. W-7405-ENG-48.

REFERENCES

1) Polyurethane is 62.1% carbon, 24.2% oxygen, 9.0% hydrogen, and 4.8% nitrogen.

2) R. H. Huddlestone and S. L. Leonard, editors, Plasma Diagnostic Techniques. (Academic Press, New York, 1965).

3) R. L. Shepherd, "The Measure of the Resistivity and Study of the Dynamics of a Cold, Dense Plasma Created in a Capillary Discharge, PhD. thesis, University of Michigan, 1987.

4) S. Ichimaru and S. Tanaka, Phys. Rev. 32A (1985) 1790.

5) G. Rinker, "Transport Coefficients for Carbon, Hydrogen, and the Organic Mixture C_2H_3, technical report LA-10608-MS, Los Alamos National Laboratory, 1986.

ELECTRICAL RESISTIVITY OF STRONGLY COUPLED PLASMAS IN INTENSE FIELDS*

Robert CAUBLE and Forrest J. ROGERS

Lawrence Livermore National Laboratory, Livermore California 94550 USA

Wojtek ROZMUS

Theoretical Physics Institute, Department of Physics, University of Alberta, Edmonton, Alberta T6G 2J1 Canada

It is now possible to measure the electrical resistivity, or alternatively the electron collision frequency, in strongly coupled plasmas through the use of very fast laser pulses. In such experiments, the effect of the intense laser field needs to be included in the calculation of material transport coefficients. We derive a form for the electrical resistivity which includes the effects of strong correlations, as well as an external electric field. Our results are compared with other theories and a recent set of experiments.

1. INTRODUCTION

The advent of intense, subpicosecond laser pulses has opened a new window in the study of very high density plasmas[1]. When such a pulse is focused on a solid, much of the laser energy absorbed by the ionizing electrons is rapidly conducted away by cold matter below an absorption skin depth. The result is a cold very dense plasma. The theoretical interpretation of these experiments relies mainly on the electron-ion collision frequency, which governs laser light absorption through inverse bremsstrahlung, and on the electron thermal conductivity, which controls plasma cooling. Given these conditions and the plasma environment, the form of the collision frequency must be appropriate to strongly coupled plasmas in the presence of a strong electromagnetic pump source. Below we take into account both major effects by calculating the electrical resistivity or conductivity, in strongly coupled plasma including the presence of a strong electric field.

2. GENERAL THEORY

The electron collision frequency, ν, electrical conductivity, σ, and electrical resistivity, ρ, share the following relations:

---

* Work performed under the auspices of the US Department of Energy at Lawrence Livermore National Laboratory under Contract No. W-7405-ENG-48 and with the support of the Natural Sciences and Engineering Research Council of Canada.

$$\sigma = \omega^2_{pe}/(4\pi\nu) \text{ and } \rho = 4\pi\nu/\omega^2_{pe} \text{ or } \rho = 1/\sigma,$$

where ω_{pe} is the electron plasma frequency. Forms for the plasma dc[2] and ac[3] resistivity in the case of weak coupling are well known. The bases for these formulations are inappropriate in the presence of a strongly coupled plasma[4-7]. Attempts to find an appropriate form for these transport coefficients in strong coupling must consider strong electron-ion and ion-ion correlations, electron degeneracy, quantum effects, and completeness of the solution (since most of the final forms are Chapman-Enskog polynomial expansion types). For this calculation, we concentrate on classical statistics in order to elucidate a more complete solution (and include an external electric field). Quantum effects are contained entirely in an effective electron-ion potential and degeneracy is ignored altogether.

We start by writing the kinetic equation for the phase space particle-particle distribution functions, $C_{ab}(r_a, r_b, p_a, p_b, t)$, in the disconnected approximation[8,9], which contains a memory function form of the collision operator; $a,b = i,e$ for point ions and electrons. "Lower order" (analytic) solutions of C_{ab} must be used in the collision operator to obtain a solution[10].

The next step is to "solve" the kinetic equation by expanding in terms of an infinite set of orthogonal momentum polynomials. We assume separability of momentum and configuration spaces, i.e., classical statistics. The transport coefficients are easily recognized by separating the hydrodynamic momentum powers from the remainder of the set. The number of terms retained beyond the hydrodynamical subspace dictates the completeness of the solution. Most of the correction is included by using two additional polynomials[11] corresponding to the so-called two Sonine polynomial approximation. For dc resistivity, the details of this process can be seen in Ref 12.

For ac resistivity, the time dependence of the analytic C_{ab}'s must be retained. The simplest correlation propagator is the free streaming form[10]; we use this for ion motion. But for electron transport, we take the free streaming correlation modified by the presence of an external electric field. For the two Sonine case where the electric field, E, is strong the result is

$$\rho_2(\omega) = \rho_1(\omega) - \Theta(\omega),$$

where ρ_1 is the one Sonine result, given by

$$\rho_1(\omega) = \frac{n}{\omega^2_{pe}\sqrt{2\pi kT}} \int_{-1}^{1} \mu^2 d\mu \int_0^{\infty} dl\, l^3\, c_0^{ei}(l)\, V^{ei}(l)\, \Delta(l)\, \Omega(l,\omega)$$

where

$$\Delta(l) = S^{ee}(l) S^{ii}(l) - S^{ei}(l) S^{ie}(l)$$

and

$$\Omega(l,\omega) = \exp\left\{\frac{-m}{2kTl^2}\left(\omega + \frac{eEl}{m\omega}\mu\right)^2\right\}.$$

The term, $\Theta(\omega)$, represents corrections to the one Sonine case and contains formulae similar to these. Above, S^{ab} and c_D^{ei} are static correlation functions and V^{ei} is the bare electron-ion potential. In weak coupling, ρ_2/ρ_1 is a known ratio.

RESULTS

We gear our results to recently published values of the electrical resistivity of solid density aluminum[13]. The plasmas were created by 400 fsec laser pulses. Values of the temperature were derived and presented as a function of the laser intensity. A reproduction of the data appears in Fig. 1.

Our static correlation functions were found in the hypernetted chain (HNC) approximation using a pseudopotential developed by an activity expansion of the partition function[14]. Results for the aluminum resistivity are presented in Fig. 1, along with results from another, completely different method[15]. Lee and More[15] (L&M) evaluated ρ by obtaining parameters from the scattering cross section found by a partial wave solution to the Schrödinger equation with the Thomas-Fermi potential. Connections were made to low temperatures. Other calculations, Ichimaru and Tanaka[16] and Boercker, Rogers, and DeWitt[17], both using Coulomb potentials (not pseudopotentials) and a high temperature limit of the Fermi-Dirac distribution, produce much smaller values of ρ. [The latter results are equivalent to the one Sonine approximation and those values in the figure have been reduced by the appropriate ratio.[16,17]]

We note that there is little distinction between our results and L&M over the range we have calculated, although the curves do not fit the data very well. The closeness is somewhat remarkable since our approach is very different to that of L&M. For the cases here, the degeneracy parameter is greater than one for all but the lowest temperatures, so our calculation should not suffer greatly from nonclassical effects. We have not calculated a pseudopotential for temperatures less than 10 eV. For comparison, we have approximated the correlations by their Debye forms and the potential by a screened Thomas-Fermi-like form, which allows an analytic solution.

For most of the data here, there is approximately a 10-20% increase in the value of ρ due to assumed intensities of electric fields. We predict that higher laser field intensities can increase the resistivity by as much as 50%. Here the presence of the non-zero laser frequency is more important. We note that, when the aluminum is maintained at solid density, none of the calculations predict a ρ as large as the data around 40 eV ($\approx$ 200 $\mu\Omega$-cm).

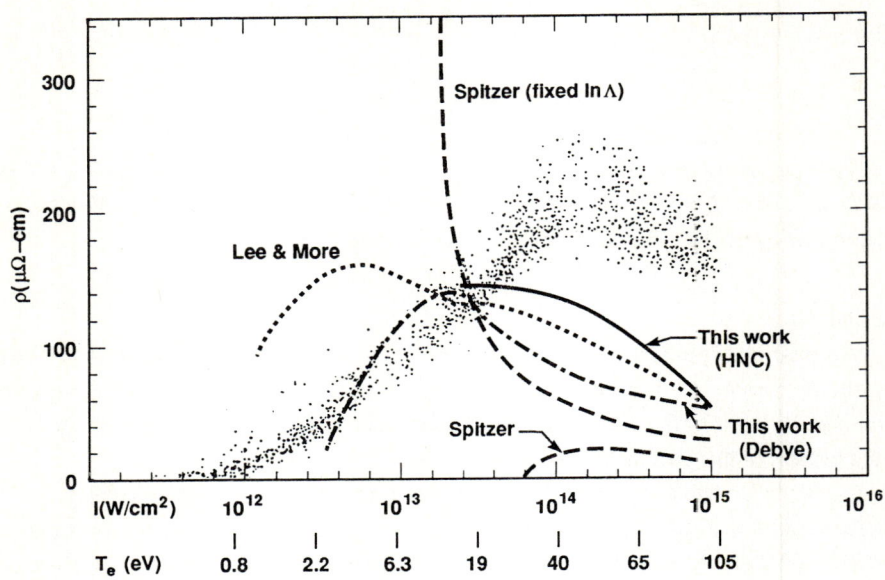

FIGURE 1
Electrical resistivity data from Milchberg[13] compared with theories of Lee and More[15] and this work with HNC and Debye correlations, as well as Spitzer[2].

REFERENCES

1) M. M. Murnane, H. C. Kapteyn, and R. W. Falcone, Phys. Rev. Lett. 62 (1989) 155.

2) L. Spitzer, Jr., Physics of Fully Ionized Gases (Interscience, New York, 1962).

3) J. M. Dawson and C. Oberman, Phys. Fluids 5 (1962) 517.

4) J. P. Hansen and I. R. McDonald, Phys. Rev. A23 (1981) 2041.

5) M. Baus, J. P. Hansen, and L. Sjögren, Phys. Lett. 82A (1981) 180.

6) F. J. Rogers, H. E. DeWitt and D. B. Boercker, Phys. Lett. 82A (1981) 331.

7) R. Cauble and D. B. Boercker, Phys. Rev. A28 (1983) 944.

8) G. F. Mazenko, Phys. Rev. A9 (1974) 360; G. F. Mazenko and S. Yip, in Statistical Mechanics, Part B, ed. B. J. Berne (Plenum, New York, 1974).

9) J. Wallenborn and M. Baus, Phys. Rev. A18 (1978) 1737.

10) R. Cauble and W. Rozmus, Phys. Fluids 28 (1985) 3387.

11) R. H. Williams and H. E. DeWitt, Phys. Fluids 12 (1969) 2326.

12) R. Cauble and W. Rozmus, J. Plas. Phys. 37 (1987) 405.

13) H. M. Milchberg, R. R. Freeman, S. C. Davey, and R. M. More, Phys. Rev. Lett. 61 (1988) 2364.

14) F. J. Rogers, Phys. Rev. A29 (1984) 868.

15) Y. T. Lee and R. M. More, Phys. Fluids 27 (1984) 1273.

16) S. Ichimaru and S. Tanaka, Phys. Rev. A32 (1985) 1790.

17) D. B. Boercker, F. J. Rogers, and H. E. DeWitt, Phys. Rev. A25 (1982) 1623.

GENERATION OF A STRONGLY COUPLED PLASMA WITH ELECTRON TEMPERATURE AROUND 4.2 K IN CRYOGENIC HELIUM GASES

Kazuo MINAMI, Keizo KATO, Akira SUGAWARA and Takahiro NOMURA

Department of Electrical Engineering, Niigata University, Niigata 950-21 Japan

A novel experimental attempt for generating a strongly coupled plasma is described; we observe a decaying plasma in its late stage created by pulsed discharges in helium gases at a temperature near 4.2 K. Coulomb coupling constant with a value ranging from 0.0056 to 0.12 is obtained. Prospect of research on strongly coupled plasmas in this way is examined.

1. INTRODUCTION

The Coulomb coupling constant is given by $\Gamma = 2.69 \times 10^{-3} n^{1/3}$ $(cm^{-3}) Te^{-1}(K)$, where the notations are standard. Plasmas with Γ not much less than unity are called strongly coupled or non-ideal plasmas. Such plasmas, high densities or low temperatures, can be realized in laboratory experiments by laser fusion, shock waves caused by explosives or laser cooling of ions confined by electrostatic and magnetic fields. In these experiments, Γ of the order of unity have been attained. In this paper, an novel attempt is made to generate a strongly coupled plasma in a different way from previous works; we observe a very late stage of a decaying plasma initiated by energetic pulsed discharges with high density and temperature in helium gases at a temperatures near 4.2 K. Goldan and Goldstein measured collision cross section of the electrons in a decaying plasma in cryogenic helium gases in a glass tube immersed in a liquid helium at 4.2 K. Although their work disclosed various interesting results, they were not interested in the state of strongly coupled plasma and did not give the practical values of Γ. The new point in our experiment is that an attempt is made to observe a very late afterglow plasma in which Γ may become a value not much less than unity.

In 2. Experimental setup and procedure are described. Experimental results and discussion are given in 3.

2. EXPERIMENTAL SETUP AND PROCEDURE

The stainless steel cylindrical plasma container shown in FIG. 1(a) is divided into two parts by a plane of thin wire grid. The decaying plasma is initiated at a gap between a pair of electrodes in the lower part of the cylinder by a pulsed discharge of voltage 7 kV, current 750 A and time duration 1 μsec. The measurements are made in the upper part of the cylinder which is TE_{011} mode cavity at the resonant frequency of 2.83 GHz. The loaded Q value without plasmas is 5300. Plasma parameters, density and temperature, are measured in a single shot by detecting the phase change in addition to the absolute value of transmission coefficient T of small signal microwaves. The block diagram of the microwave circuit is shown in FIG. 1(b). It is a microwave interferometer including the plasma in the cavity. Plasma parameters $(\omega_{pe}/\omega)^2$ and ν_e/ω are given by the relations derived by one of the present authors[7], where ω_{pe} is the average plasma frequency related to plasma density, ν_e is the electron collision frequency with helium atoms.

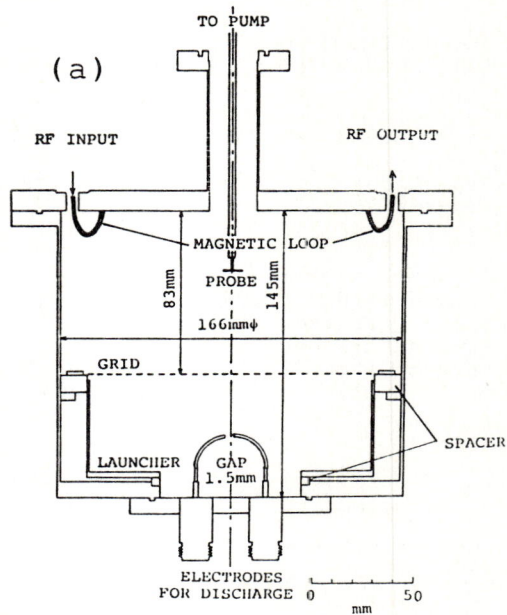
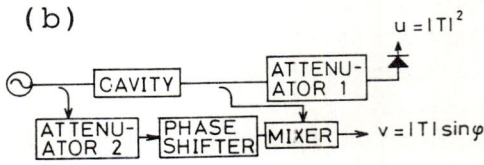

FIGURE 1
(a) Cylindrical plasma container with TE_{011} cavity.
(b) Block diagram of microwave interferometer including plasma cavity.

3. EXPERIMENTAL RESULTS AND DISCUSSION

Examples of changes in time of the density and ν_e/ω are shown, respectively, in FIG. 2(a) and (b). Note that the decay times of the density in cryogenic helium gas are hundred times longer than those in gases at room temperature for a given pressure. The observed ν_e is shown to be roughly constant in time and to be proportional to the gas pressure. These fact suggest that the observed ν_e is the electron neutral collision frequency which is given by $\nu_e = n_g \sigma v_{th}$. Here, n_g, σ and v_{th} are, respectively, the gas number density, the momentum transfer cross section and the electron thermal velocity. Since the magnitude of σ in cryogenic helium gases is known[6] to be $1.0-1.9 \times 10^{-15}$ cm^2, the electron temperature T_e can be estimated by the measured collision frequency. The results are shown in FIG. 3. For relatively high gas pressures of the order of 100 Pa, T_e is roughly at 4.2 K, whereas for low gas pressures, T_e rises to the order of 50 K. This is because the cooling by the helium gases through collisions is less effective for low pressures.

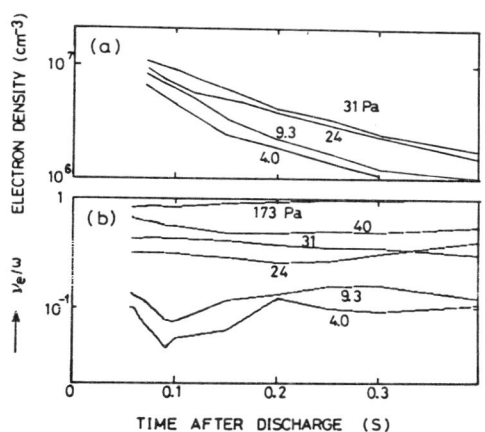

FIGURE 2
Change in time of (a) plasma density and (b) electron collision frequency.

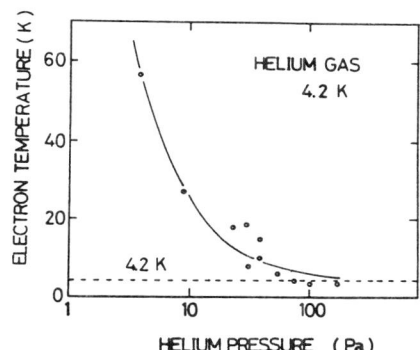

FIGURE 3
Observed electron temperatures as a function of helium gas pressure at 4.2 K.

Next, we carry out measurement of propagation of diffusion waves[8] in our decaying plasma. In fact, the plasma container shown in FIG. 1(a) is similar to the double plasma device.[9] The inner vessel in the lower part of the cylinder is insulated and a wave packet of AC voltage up to 25 V peak to peak with frequencies 0.3 - 2.0 kHz is applied at a late period of the decaying plasma. The vessel is used as a wave launcher and the plane of wire grid is kept at the ground potential. The density perturbation propagates upward in the plasma cavity and is detected by a receiving probe which can be moved on the axis. The time delay and attenuation of the density diffusion wave are measured and are compared with a theoretical analysis assuming cryogenic electron temperatures. The dispersion relation for angular frequency ω and wave number k of one dimensional diffusion waves in a uniform quasi-neutral plasma with $\Gamma \ll 1$ is derived. The electron temperatures estimated from the dispersion relation are found to be of the order of 4.2 K.

In conclusion, we have attempted to generate a strongly coupled plasma by observing a late afterglow plasma in cryogenic helium gases. The density from 10^7 to 10^6 cm^{-3} and the electron temperature from 55 to 4.2 K were observed. These plasma parameters correspond to the Coulomb coupling constants from 5.6×10^{-3} to 0.12. The values of Γ of the order of unity may be attainable for gas temperatures less than 1 K. In such plasmas, electron cooling due to ambipolar diffusion is considered to prevail the heating due to three body recombination.

ACKNOWLEDGEMENTS

The authors are indebted to Prof. Setsuo Ichimaru, University of Tokyo, for his invaluable discussion and suggestions. Thanks are also due to Prof. Mitsuru Awano, Tokyo Engineering University for his supporting and interest in the present work.

REFERENCES
1) S. Ichimaru, Rev. Mod. Phys. 54 (1982) 1017.
2) H. Motz, The Physics of Laser Fusion (Academic Press, London, 1979) pp. 224.
3) Yu.V. Ivanov, V.B. Mintsev, V.E. Fortov and A.N. Dremin, Zh. Eskp. Teor. Fiz (1976) 216 [Sov. Phys. JETP 44 (1976) 112].
4) J.M. Malmberg, T.M. O'Neil, A.W. Hyatt and C.F. Driscoll, Bull. Am. Phys. Soc. 28 (1983) 1155.
5) J.J. Bollinger and D.J. Wineland: Phys. Rev. Lett. 53 (1984) 348.
6) P.D. Goldan and L. Goldstein, Phys. Rev. 138-A (1965) 39.
7) K. Minami, H. Mitera and S. Takeda, Jpn. J. Appl. Phys. 26 (1987) 1153
8) M. Kando and S.Takeda: J. Phys. Soc. Jpn. 36 (1974) 579.
9) R.J. Taylor, D.R. Baker and H. Ikezi: Phys. Rev. Lett 24 (1970) 206.

MEASUREMENT OF THE DYNAMIC FORM FACTOR AT LOW FREQUENCIES FOR A PLASMA WITH $\Gamma = .06$

A. W. DeSILVA and Y. Q. ZHANG

Laboratory for Plasma Research, University of Maryland, College Park, MD 20742-3511[1]

We have measured, at low frequencies (ion acoustic and below), the dynamic form factor $S(k,w)$ for argon plasmas having electron density of $1 \times 10^{17} \text{cm}^{-3}$, and a temperature of about 2 eV, giving a coupling parameter Γ of about 0.06. We present the results of these measurements and discuss the possibilities for extending such measurements to plasmas having coupling parameters closer to unity.

The frequency spectrum of light scattered by a plasma reflects the frequency spectrum $S(k,w)$ of fluctuations of electron density at a wavenumber $\mathbf{k} = \mathbf{k}_s - \mathbf{k}_i$, where $\mathbf{k}_s$ and $\mathbf{k}_i$ refer, respectively, to the scattered and incident light wavenumbers. Thus, both the $\mathbf{k}$ and w dependence are accessible to study by light scattering.

Experiments were performed in plasmas created in a simple pulsed arc of 100 μsec duration in argon or helium gas. The helium discharges were not reproducible enough to yield useful data. Plasma conditions were measured by standard spectroscopic means, and the argon plasma composition was deduced from the Saha equation to be $ne : n0 : n1 : n2 = 1 : .01 : .75 : .12$, ($n0$ is the neutral atom density, $n1$ the first ionized state, etc.).

The light source for scattering was a CO_2 laser having a pulsed output of 100 watts at 10.6 μ, with 100 μsec duration. Light from the laser was focussed into the plasma and scattered light was observed at angles of from 3 to 10 degrees to the forward direction, giving a factor of over three in variation of k.

The light detector used, a Ge:Cu detector cooled to 4K, had a bandwidth of over 400 MHz. The scattered light was mixed with an optical local oscillator, and the heterodyne output from the detector was recorded by a transient digitizer at 1.4 Gsample/sec., for an interval of 7 μsec on each firing of the arc. As the signal-to-noise ratio of the system was less than unity, about 1500 shots were recorded, and the results Fourier analyzed and averaged. The resulting spectra are displayed in Figs. 1–3. A local oscillator frequency different from that of the probe was used to shift the signal out of the detector's 1/f noise region, so the zero frequency fluctuations appear at 53 MHz in the figures.

---

[1] Work supported by the U.S. National Science Foundation

The data have been compared with several theories: the kinetic theory of Linnebur and Duderstadt,[1,2] which uses modeled relaxation coefficients, a kinetic theory using a BGK collision term, and a two-fluid theory. The latter uses Braginskii's theory[3] to compute the conductivity, and the fluctuation-dissipation theorem is applied to obtain the spectrum. This result is superposed on the data of Figs. 2 and 3. The results of the BGK theory are superposed on Fig. 1, and Fig. 2 shows also the LD theory. In the case of the LD theory, the peaked structure seen in the experiment is absent. Boercker has shown that this structure is recovered in a one-component version of the LD theory and suggests that the two-component theory would also show the correct structure with suitable modification of the damping coefficients.[4]

In the case of the BGK theory, the relative size of the peaks is not well modeled, theory predicting the central peak to be too large. A modification to the theory in which the ion-electron collisional interaction is explicitly introduced produced a better fit, shown in Fig. 1.

The applicability of this technique to measurement of the form factor of plasmas with coupling parameters approaching unity is limited by absorption of the probe light beam by the plasma. Absorption mechanisms include both free-free and free-bound interactions. The free-free absorption coefficient α_{FF} may be easily calculated,[5] and the results are shown in Fig. 4. $\alpha_{FF}(w)$ scales as the square of the electron density and falls approximately as the square of the frequency. Plasmas having density high enough to approach the strongly coupled regime thus suffer large absorption of the probing radiation, forcing one to ever smaller plasma dimensions to maintain optical transparency. Use of shorter wavelength probing radiation can alleviate the problem, but here one may encounter problems with free-bound absorption.

Free-bound absorption is dependent upon the density of atoms or ions in excited states high enough to be within reach of the continuum limit for the photons of the illuminating laser beam. This is quite sensitive to the particular conditions of the discharge, influenced not only by ions within the scattering volume, but also by the boundary layer through which both incident and scattered light must pass. The free-bound absorption coefficient α_{FB} for a hydrogen plasma with 1.5 eV temperature is also shown in Fig. 4. α_{FB} scales linearly as the ground state ion density. As the radiation frequency rises, successively lower energy levels become accessible for ionization, and the absorption coefficient jumps at these points. In the case of plasmas of other species, the structure becomes more complex, but the general features are the same, namely, that as the photon energy becomes large enough to excite the lowest energy levels of the ion, large steps appear in $\alpha_{FB}(w)$, and it becomes critical to pick a radiation frequency carefully, in order to avoid absorption peaks.

Absorption, besides restricting the size of the plasma, also modifies the plasma conditions through its heating effect. Heating may be minimized by use of suitably small probe radiation power, at the expense of signal-to-noise ratio. This may, however, be recovered

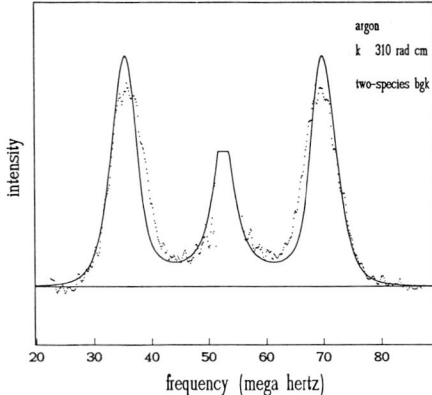

Fig. 1. Fluctuation spectrum at $k = 310$ cm^{-1}. Solid line is BGK theory including collisional ion-electron interaction.

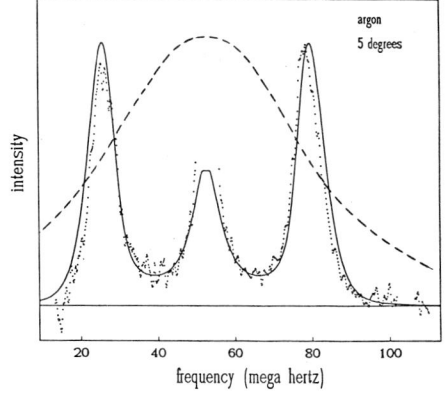

Fig. 2. Fluctuation spectrum at $k = 517$ cm^{-1}. Solid line is 2-fluid theory, dashed line LD theory (see text).

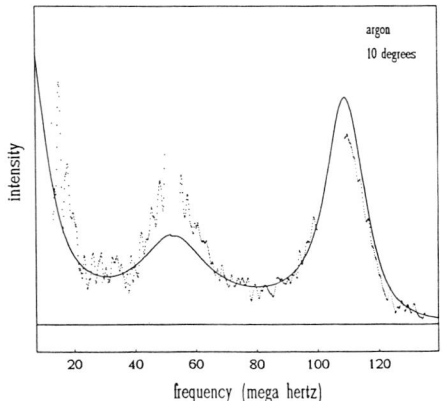

Fig. 3. Fluctuation spectrum at $k = 1033$ cm^{-1}. Solid line is 2-fluid theory.

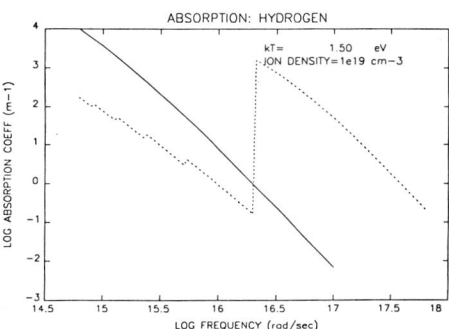

Fig. 4. Absorption coefficient for 1.5 eV, n = 10^{19} cm^{-3} plasma. Solid line: free-free; dashed line: free-bound (hydrogen).

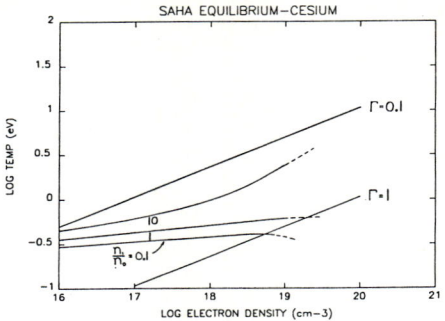

Fig. 5. Saha equilibrium of cesium plasma. Parameter is ratio of ion to neutral density.

through longer observation time. The dense plasmas of interest are most likely to be short-lived, so one is forced into requiring large numbers of shots and excellent reproducibility.

In the foregoing, the question of production of a suitable plasma for study has been ignored. In fact, for all ordinary gases, at the temperatures and densities that would yield $\Gamma = 1$, the equilibrium degree of ionization would be $< 1\%$. In order to raise the degree of ionization it would be necessary to use a species having low ionization potential, such as cesium. Figure 5 shows the Saha equilibrium conditions for cesium, including lines of constant Γ. It appears that a cesium plasma with $n_e = 10^{19} \text{cm}^{-3}$ and $T_e = 1eV$ would be about 70% ionized and have $\Gamma = 0.5$. The FF absorption depth for this plasma would be about 1 cm, and FB absorption would be negligible if the boundary layers were kept thin.

REFERENCES

1) A. N. Mostovych and A. W. DeSilva, Phys. Rev. A 34, 3238 (1986).

2) E. J. Linnebur and J. J. Duderstadt, Phys. Fluids 16, 665 (1973).

3) S. I. Braginskii, Reviews of Plasma Physics, Vol. 1, ed. M. A. Leontovich, (Consultants Bureau, NY, 1964).

4) D. B. Boercker, private communication.

5) G. Bekefi, Radiation Processes in Plasmas, (Wiley, NY, 1966).

Chapter VIII:
Metal-Insulator Transition

THERMODYNAMIC AND STRUCTURAL PROPERTIES OF FLUID METALS IN THE
METAL-INSULATOR TRANSITION RANGE

Friedrich HENSEL

Institute of Physical Chemistry and Materials Science Center,
Philipps-University, Hans-Meerwein-Straße, D-3550 Marburg, FRG

Experimental results for fluid metals near the liquid-vapor critical point show that strong variation in the electronic structure due to the metal-nonmetal transition manifests itself in a correspondingly strong thermodynamic-state-dependence of the effective interparticle interaction. The existence of this state-dependence noticeably influences the thermodynamic and structural features of expanded fluid metals.

1. INTRODUCTION

A great deal of effort has been devoted to the study of the thermophysical properties of liquid metals expanded by heating toward the liquid-vapor critical point. Much of the activity is motivated by the large number of current and potential applications of fluid metals as high temperature working fluids for advanced energy technologies. The selection of a particular metal and the design and operation of engineering processes requires the knowledge of reliable thermodynamic data of metals at the proposed operating conditions. Moreover, the essential need for safety analysis and risk assessment requires us to extend our knowledge of those properties up to conditions far above the proposed temperatures and pressures of normal operation, preferably up to and beyond the liquid-vapor critical point.

From another point of view, that of fundamental science, fluid metals are also of special interest because they are fundamentally distinct from normal insulating fluids in that the interparticle potentials are not quantities related to intrinsically atomic properties, but rather depend strongly on the nature of the electron gas which screens the Coulombic interaction. Moreover near the metal-nonmetal transition, which occurs in the liquid when it is made to expand by heating to critical conditions, one might expect that the nature of the interparticle interaction changes dramatically, from metallic to a van der Waals type interaction. Such a strong thermodynamic-state-dependence of the "effective" interparticle interactions would be expected to influence considerably any property of fluid metals, in particular the thermodynamic and kinetic features of the liquid-vapor phase transition.

Several attempts have been made during the past two decades to explore the problem experimentally, but the subject has remained elusive until quite recently.

Part of the difficulty arises from the fact that the high cohesive energies of metals place the critical region at temperatures and pressures too high for easy experimental investigation. It is therefore not surprising that relatively accurate experimental results are available for only the three metals with the lowest critical temperatures: Hg (T_c = 1478° C, p_c = 1673 bar, ρ_c = 5.8 g/cm^3) [1], Cs (T_c = 1651° C, p_c = 92.5 bar, ρ_c = 0.38 g/cm^3) [2], and Rb (T_c = 1744° C, p_c = 124.5 bar, ρ_c = 0.29 g/cm^3) [2]. The experimental information for these metals is accurate enough to permit determination of the asymptotic behavior of the properties near the critical point.

2. EXPANDED ALKALI METALS

A great deal of effort has been devoted to the experimental[3-6] and theoretical[7-12] investigation of expanded fluid alkali metals because as elemental, monovalent metals they closely resemble the expanded crystals of hydrogen or alkali atoms considered by Mott in his original discussion of the metal-nonmetal transition. Much of this effort has focussed on cesium because of its relatively low critical temperature and pressure. The experimental evidence shows that the electrical conductivity σ [3] of expanded liquid cesium decreases greatly before a metal-nonmetal transition occurs close to the critical point. The combined analysis[7] of magnetic susceptibility[4] and Knight shift[5] measurements reveals that precursors of the metal-nonmetal transition, i.e. many - electron correlations of the type invoked in the theory of the Mott - Hubbard transition [11,13] play an essential role in determining the electronic properties of expanded metallic liquid cesium for densities $\rho \leq 1.4$ g/cm^3 (the density at the normal melting point is 1.84 g/cm^3). As the density of the liquid metal decreases, two effects are expected to occur: the average coordination number decreases and the average near-neighbor distance increases. In general, to describe the properties of a liquid metal at different densities, these two factors have to be taken into consideration. The central importance of these effects has stimulated neutron diffraction measurements of the structure factors S(Q) of fluid Cs and Rb up to the critical points[6,14]. Figure 1 displays as a typical example the Fourier transform of S(Q), i.e. the pair correlation function g(R), of Cs at various temperatures and densities ranging from the melting point up to the critical region. The following noteworthy changes in g(R) occur with decreasing density or increasing temperature. The intensity of the main peak of g(R) is strongly reduced and broadened, whereas the peak position R_1 shifts only slightly toward higher R. Now the pair correlation function g(R) is related to the radial distribution function n(R) = $4\pi R^2 n g(R)$, which determines the number of neighboring atoms n(R)dR in a spherical shell of radius R and thickness dR centered on a

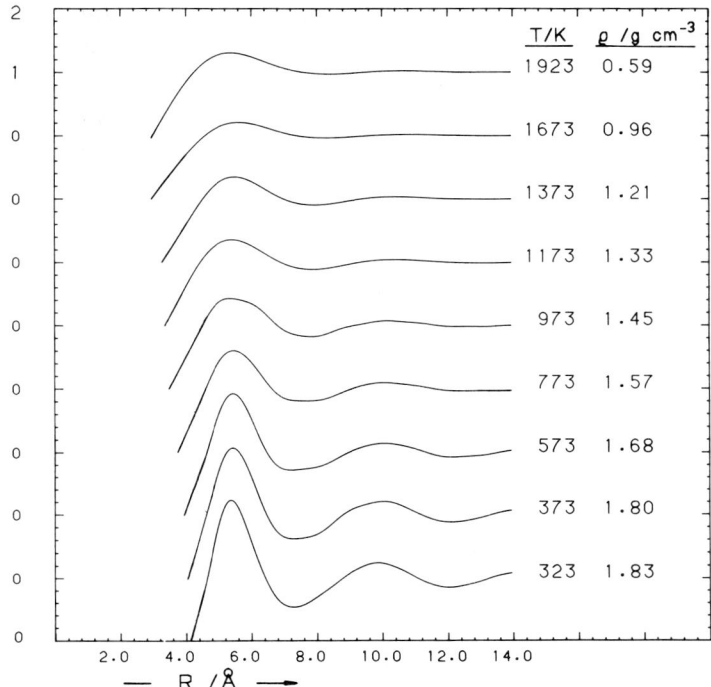

FIGURE 1
Pair correlation function g(R) of expanded liquid cesium at conditions near the liquid-vapor coexistence

particular atom of interest. Detailed analysis[6] of data such as those displayed in Figure 1 shows that for Cs and Rb the average coordination number N_1, defined by the area under the first peak in n(R), tends to decrease approximately linearly as the density is decreased by thermal expansion. Because there is usually considerable overlap between the first- and second-neighbor peaks, experimental values of N_1 depend sensitively on the method employed to define and integrate the first neighbor peak. For reasons of reliability and consistency we used the method of "symmetrical main maximum" (see inset in Figure 2) for all data shown in Figure 2. It is evident that the dominant effect of thermal expansion on the structure of Cs is a reduction of the average coordination number N_1 rather than an increased neighbor distance. Data for the nonmetal argon[15] exhibit the same trend. This simply reveals the important point that termal expansion of a liquid reorganizes the short-range order; it does not only increase the interatomic distances as it would in crystals. The knowledge of this feature has been particularly important for theoretical attempts to model the electronic

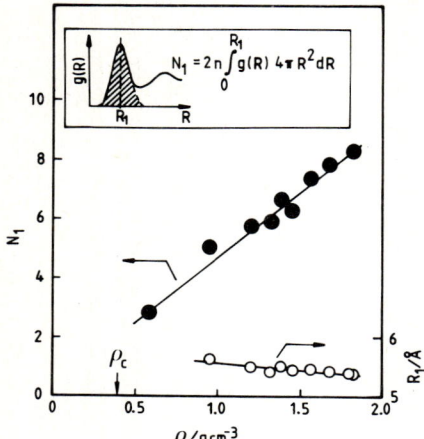 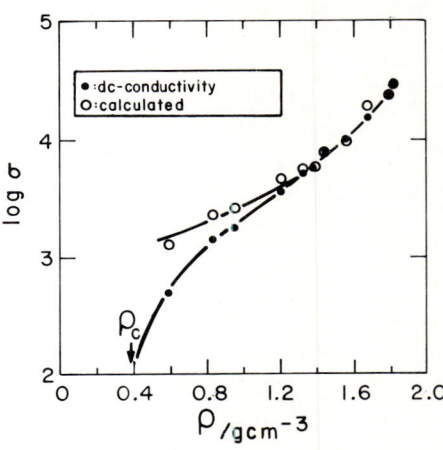

FIGURE 2
Average number of nearest neighbors N_1 for expanded liquid cesium together with the average next neighbor distance R_1 of cesium as a function of density ρ

FIGURE 3
Measured and calculated (NFE-model) electrical conductivity σ of expanded liquid cesium as a function of density at conditions near the liquid-vapor coexistence

structure of expanded metals by the use of band structure considerations[7].

The structural data are especially suited to explore the limitations of the nearly free electron (NFE) approach for the expanded alkali metals because the electrical transport is correlated with the variation of $g(R)$. In the NFE-model the electrical conductivity is described by the Ziman formula[16]:

$$\sigma^{-1} = \frac{3\pi m^2 \Omega}{4\hbar^3 e^2 k_F^6} \int_0^{2k_F} S(Q) V(Q)^2 Q^3 \, \Omega \tag{2}$$

where $V(Q)$ denotes the screened ion preudopotential, k_F the wavenumber of the electrons at the Fermi surface and Ω is the atomic volume. For $V(Q)$ the Ashcroft empty core potential[17] was used combined with a density dependent dielectric function $\varepsilon(Q)$ which takes into account exchange and correlation effects[18]. The only parameter in this model potential is the radius r_c which was chosen to match the measured value of the electrical conductivity near the melting point. In Figure 3 the results of these calculations are compared with the experimental values of the conductivity[3]. The agreement is satisfactory for high densities. However, it is obvious from Figure 3 that the applied formalism breaks down at the relatively high density of about 1.4 g cm^{-3}. This failure of the NFE-model for densities lower than 1.4 g/cm^3 is very close to the range where the onset of the magnetic susceptibility enhancement shows that electron-electron correlation

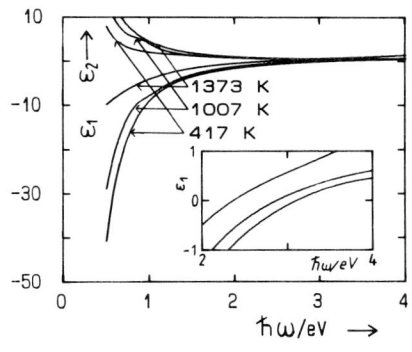
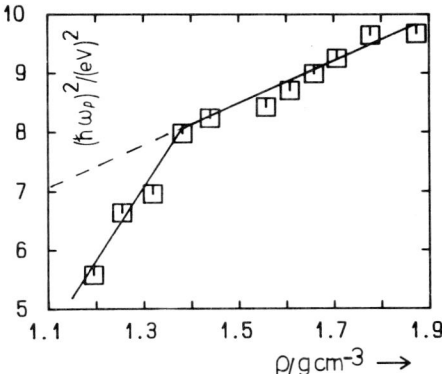

FIGURE 4
Examples for $\varepsilon_1(\hbar\omega)$ and $\varepsilon_2(\hbar\omega)$ of liquid cesium at different constant temperatures or densities. The inset displays the energy range around $\varepsilon_1(\hbar\omega) = 0$.

FIGURE 5
Density dependence of the plasma frequency $\hbar\omega_p$ of liquid cesium.

become important. The theory of the susceptibility enhancement by correlation [13] assumes that the majority of sites in the correlated metal is instantaneously singly occupied leaving only a small fraction f doubly occupied. The effective mass in the highly correlated metal is enhanced according to $m^*/m_e = 1/2\,f$. This must also be important in the electrical conductivity, but in fact a much more direct measurement of the effective mass may be made from the plasma frequency

$$\omega_p^2 = \frac{4\pi Ne^2}{m^*} \qquad (2).$$

In Figure 4 we show selected examples of the real and imaginary parts of the complex dielectric function $\varepsilon = \varepsilon_1 + i\,\varepsilon_2$, derived from reflectivity measurements[19], at different temperatures or densities. The ε_1-curves are of special interest because near the plasma resonance, $\hbar\omega_p$, ε_2 is very small so that ω_p is determined quite accurately by the condition that $\varepsilon_1(\omega_p) = 0$ (see the inset in Figure 4). These ω_p-values are plotted in Figure 5 in the form ω_p^2 versus density. It can be seen at a glance from Figure 5 that the density dependence of ω_p can be closely approximated by equation (2) with a constant effective optical mass m^* for densities larger than 1.4 g/cm^3, but for densities smaller than 1.4 g/cm^3, i.e. in the range where the magnetic data indicate the presence of large correlation effects, an increasing strong enhancement of the effective optical electron mass m^* is observed.

The presence of large correlation effects implies that the effective interparticle interaction must suffer marked changes as the density of the fluid is decreased, and therefore changes in structural and thermodynamic properties may well appear. Useful information about this change stems from recent neutron diffraction investigations[20] which were undertaken with the intention of determining the liquid structure factor S(Q) and its isothermal pressure derivative $(\partial S(Q)/\partial p)_T$ of liquid cesium over ranges of temperature and pressure sufficiently wide to include conditions close to the critical point where the metallic properties disappear. Selected experimental results for the pressure dependence of S(Q) of liquid cesium are presented in Figure 6 for five different subcritical temperatures. In the high-density metallic region of cesium at 343 K the dominant pressure effect on S(Q) (or equivalently on the pair correlation function g(R)) is a slight shift of the first peak positions reflecting a corresponding slight change in the mean interatomic distance while otherwise little change is seen. This behavior is in contrast to that of normal nonconducting liquids for which pressure change mainly leads to a change in the number of nearest neighbors. But it is in agreement with the uniform fluid model (UFM)[21] which can be characterized in the following way: it is generally agreed that at high densities the alkali-metal effective interionic pair potential V(R) has a soft repulsion for small distances R and a weak attractive minimum at intermediate R. At larger distances the potential oscillates, taking on the asymptotic form

$$V(R) = \cos(2k_f R)/R^3 \qquad R \to \infty \qquad (3)$$

where $k_f = (3\pi^2 n)^{1/3}$ represents the Fermi wave vector and n is the number density of the conduction electrons. The UFM assumes that the structure of the fluid remains relatively fixed. The effect of density changes is simply to scale all distances with the density dependence of k_f ($k_f \propto \rho^{1/3}$). Hence g(R), the radial distribution function is a universal function g(X), where $X = R\rho^{1/3}$, and consequently S(Q) is a universal function S(Y), where $Y = Q\rho^{-1/3}$. This leads to a prediction for the density derivative as follows:

$$(\partial S(Q)/\partial \rho)_T = (-Q/3\rho)(\partial S(Q)/\partial Q)_{\rho,T} \qquad (4)$$

Egelstaff et al.[22] verified experimentally that the liquid metal rubidium satisfies eq. (4) in the region of the first maximum of S(Q), the expected region of validity of the UFM. It has also been shown that the UFM does not agree with data for the density dependence of S(Q) for liquid neon[33] or the Lennard-Jones model fluid, because the number of nearest neighbors changes with density in this case.

The contrast was critically analyzed[24] in a number of papers which use a selection of model fluids (the hard-sphere fluid, the Yukawa fluid, the Lennard-Jones fluid, and the Price-model potential of liquid metals) each of which is solvable analytically within the mean spherical aooproximation or within the Percus-Yevick approximation. The conclusion which can be drawn from the results

Thermodynamic and structural properties of fluid metals

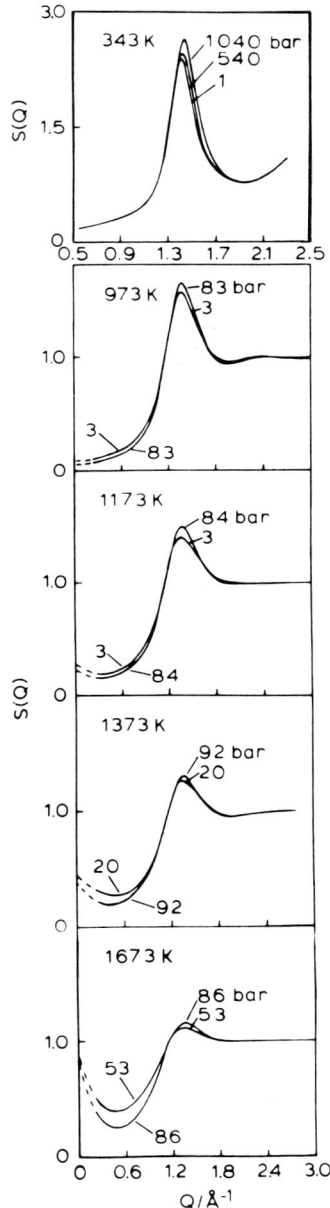

FIGURE 6
Structure factors S(Q) of expanded liquid cesium as a function of temperature and pressure

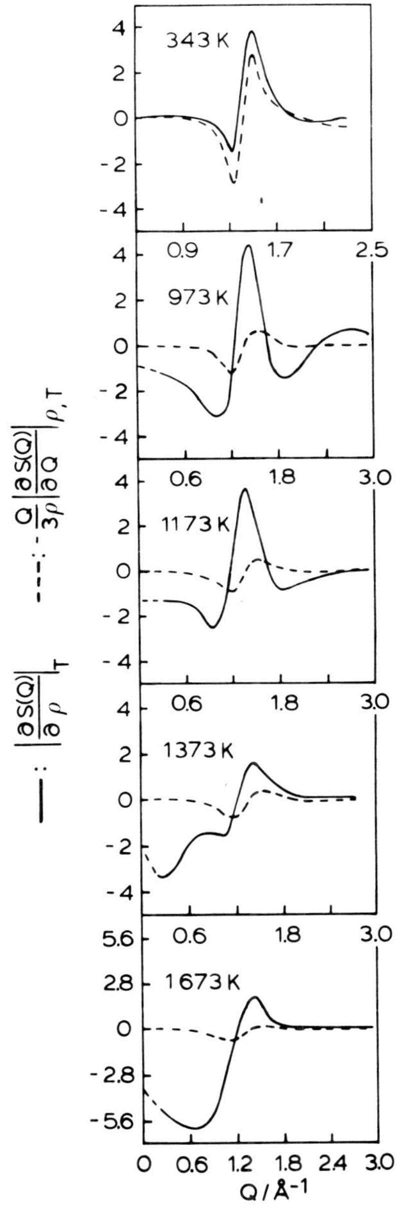

FIGURE 7
The density (thick curves) and wavenumber (thin curves) derivatives of the structure factor

is that two characteristic features of the effective interionic potential V(R) of metals are mainly responsible for the experimentally observed UFM behavior: the softness of the repulsion and the presence of the Friedel oscillations in the tail of the potential.

The prediction of the UFM (eq. (4)) is compared to the experimental data on cesium in Figure 7. The UFM fits reasonalby well the experimental data for the high-density metallic region of cesium (the average density ρ is about 1.87 g/cm^3 at T = 343 K). However, it is obvious from Fig.7 that its validity breaks down with increasing temperature corresponding to a decrease in ρ (at T = 973 K ρ = 1.46 g/cm^3, at T = 1173 K ρ = 1.34 g/cm^3, at T = 1373 K ρ = 1.22 g/cm^3, and at T = 1673 K ρ = 0.97 g/cm^3). The data for T = 973 and 1173 indeed do not agree with eq. (4) but show characteristics which are qualitatively very similar to those expected for neon and Lennard-Jones fluids. This observation seems to be consistent with the behavior of the conductivity σ. As already noted σ has suffered marked changes at these temperatures and densities, and the electron mean free path calculated from σ (assuming nearly free electron behavior) has approached a value not very different from the mean interatomic distance. The finite electronic mean free path, however, corresponds to a blurring of the Fermi surface, and this is obviously not included in calculations of the typical ion-ion potential. A crude estimate for a point ion model[25] shows that such a blurring damps out the oscillatory bahavior in the ion-ion interaction and makes at the same time its repulsive part harsher. This result is consistent with the observed correlation between the mean free path of electrons in expanded liquid cesium and the behavior of the structure factor data. The most surprising feature of the data in Figure 6 and 7 are the shapes of S(Q) and its density derivative $(\partial S(Q)/\partial \rho)_T$ in the low-Q region for T = 1373 K and T = 1673 K (which are at reduced temperatures $\Delta T/T_c$ of 0.29 and 0.13, respectively). At such large distances from the critical point insulating fluids (e.g., Ar) do not show such a marked enhancement in S(Q) for small Q. One explanation of such an effect, which has been discussed by Kahl and Hafner[26], is that it reflects the strong density dependence of the attractive part of the effective interionic interaction if screening is reduced as the metal-nonmetal transition is approached. Interestingly, the region of density where these enhancements occur is the same as that in which magnetic, optical and conductivity data indicate the presence of many-electron correlation effects.

3. EXPANDED DIVALENT MERCURY

The large amount of work devoted over the last two decades to the experimental and theoretical investigation of the metal-nonmetal transition in fluid mercury has been fully reviewed [27,28] and need not detain us here. It is

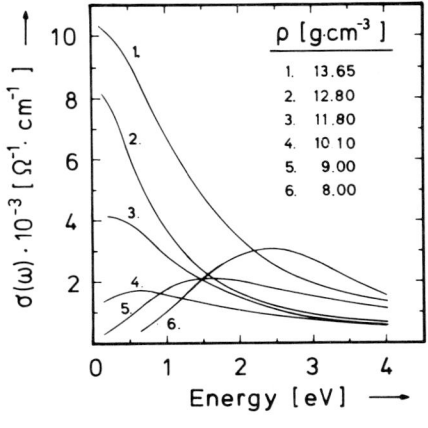 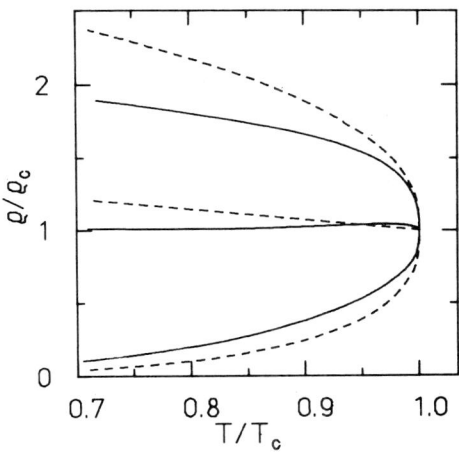

FIGURE 8
Optical conductivity $\sigma(\omega)$ of expanded mercury at conditions near the liquid-vapor coexistence; the corresponding densities ρ are given in the figure

FIGURE 9
The reduced densities of the coexisting vapor and liquid of Xenon (outer dotted curves) and mercury (inner solid curves)

sufficient to summarize a few noteworthy observations. Measurements such as those of the conductivity, the termoelectric power, the Knight shift, the optical reflectivity, the Hall effect and the equation of state show that for densities down to about 11 g/cm^3 the properties of mercury can be described by the NFE-theory[16] of metals but, with further decreasing density, a rather gradual metal-semiconductor transition occurs in the range ρ = 9 to 8 g cm^{-3}. Figure 8 shows the optical conductivity $\sigma(\omega)$ measured by reflectivity along the coexistence line of liquid mercury at various densities and temperatures. The change in the shape of the $\sigma(\omega)$-curves nicely illustrates the gradual diminution of metallic properties with decreasing density. In the nearly-free-electron range, ρ is larger than 11 g cm^{-3} and the low frequency optical conductivity is Drude-like. A gradual change from metallic to nonmetallic behavior occurs between 11 and 9 g cm^{-3}. For still smaller densities, the shape of the $\sigma(\omega)$-curves is consistent with the view that for densities lower than 9 g cm^{-3}, either a true energy gap opens or a range of energy exists that is so thinly populated with states that their contribution to the optical properties is negligibly small. The opening of this gap or "pseudogap", means that mercury changes macroscopically to a nonmetal at the same density. Hence the high electrical conductivity nearer to the liquid-vapor critical point density ρ_c = 5.8 g/cm^3 is simply a consequence of the presence of an ionization equilibrium. A detailed analysis of the experimentally determined equation of state[29] in that region demonstrates that this volume-dependent thermal electronic excitation leads to a nonnegligible constribution to the total pressure of the fluid. It is not suprising, therefore, that the liquid-vapor phase behavior of mercury differs strongly from that of normal molecular fluids as demonstrated by the comparison of the coexisting liquid (ρ_L) and vapor (ρ_V) densities of Hg, and the inert element Xe, in Figure 9. The data to be compared are represented in dimensionless form by the critical constants. Poor reduced correlation between Hg and Xe is observed. The reduced diagram in Figure 9 demonstrates a second interesting consequence of the presence of the ionization equilibrium in mercury as the critical region is traversed. Fluid mercury voilates the hundred year old empirical law of rectilinear diameter over surprisingly large temperature range. By cortrast, the deviations from this law are extremely (often immeasurably) small for the coexistence curves of essentially all normal, insulating one-component fluids[30]. The law states that the locus of the tie-line midpoints $\rho_d = 1/2(\rho_L + \rho_V)$ is a linear function of T. Since both ρ_L and ρ_V approach the limiting density, ρ_c, at the liquid-vapor critical point, the law can be written

$$\rho_d - \rho_c = D_1 \tau \tag{5}$$

where $\tau = T_c - T/T_c$ and D_1 is a constant. Theoretical arguments based on the penetrable-sphere model and on the renormalization-group analysis have raised doubts about the validity of the law. They predicted the existence of a singular diameter which has instead the expansion

$$\rho_d - \rho_c = D_0 \tau^{(1-\alpha)} + D_1 \tau + \ldots \tag{6}$$

where $\alpha = 0.11$ is the same exponent that describes the behavior of the constant volume specific heat c_V. Since $(1-\alpha) = 0.89$ is not very different from unity, the true singularity is difficult to separate out from the analytic temperature term. The coefficient D_1 does not even have to be much larger than D_0 for the analytic term to dominate over the entire range accessible to experimentation. The latter causes the invisibility of the $(1-\alpha)$-singularity for most nonmetallic fluids (see e.g. Xe in Figure 9).

It is seem at a glance from Figure 9 that the shape of the diameter of mercury is completely different. Indeed, outside the asymptotic critical region the diameter actually slopes toward higher densities, opposite to the behavior seen for Xenon. Inside the critical region, there does indeed appear to be a $(1-\alpha)$-singularity as predicted by eq. (6). This finding strongly supports the hypothesis, that the strong thermodynamic-state-dependence of the effective interparticle interactions, caused by the presence of the ionization equilibirum, is responsible for the strength of the singular term in the diameter of the liquid-vapor coexistence curve of mercury.

4. PHASE EQUILIBRIA IN HG-HE MIXTURES

Whilst considerable progress has been made in understanding the phase equilibria in binary fluid mixtures of similar components such as two molecular fluids, the characteristics of phase diagrams for binary mixtures of components of different types such as metallic and molecular fluids are less well understood. In the following we give a brief description of the first results of phase equilibria experiments on the binary mixture of a metallic (Hg) and a molecular (He) component[31]. This system was chosen because its study can contribute to an understanding of the interaction of neutral He with the itinerant states in dense degenerate plasma. The solubility of He in metallic Hg at normal density is extremely small. In view of the close interrelation between the density induced ionization transition and the vapor-liquid transition, it can be expected that this behavior extends up to the supercritical region. This is confirmed by the measurements which were performed by the so-called synthetic method: Known amounts of Hg and He were filled into a high-temperature (T)-high-pressure (p) cell. Subsequently, at constant volume V, T is increased and recorded simultaneously with p. The p-T-curves at constant V have disontinuities in the slope at the transition from a two-phase to a homogeneous one-phase system. Combinations of such points give "isopleths", phase equilibrium curves for constant composition on the three-dimensional phase boundary surface. Figure 10 shows some of these results. To the left of the isopleths is the two phase region. The critical curve is an envelope of the isopleths. The corresponding critical curve ascends by about 10 K with a pressure of 640 bar. Two distinct fluid phases are shown to exist in Hg-He-mixtures at temperatures and pressures higher than the critical temperature and pressure of pure mercury.

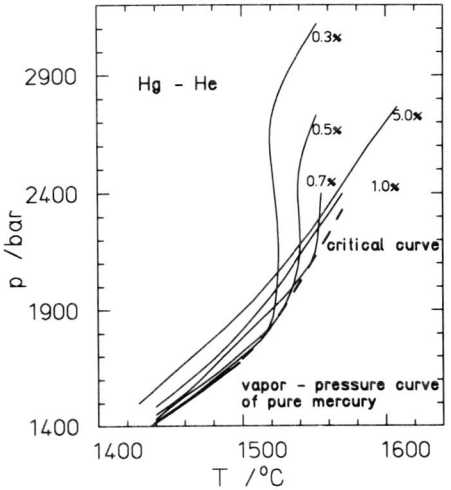

FIGURE 10
Curves of constant He-content "isopleths" and the critical curve as the high temperature limit of the heterogenous two phase region[31].

REFERENCES

1) W.Götzlaff, G.Schönherr and F.Hensel, Z.Phys.Chem. N.F. 156 (1988) 219
2) S.Jüngst, B.Knuth and F.Hensel, Phys.Rev.Lett. 55 (1985) 2160.
3) F.Hensel, S.Jüngst, F.Noll and R.Winter, in: Localisation and Metal Insulator Transitions, eds. D.Adler, and H.Fritsche (Plenum Press, London, 1985)
4) W.Freyland, Phys.Rev.B 20 (1979) 5140.
5) W.El-Hanany, G.F.Brennert and W.W.Warren, Phys.Rev.Lett. 50 (1983) 540.
6) R.Winter, T.Bondensteiner, W.Gläser and F.Hensel, Ber.Bunsenges. Phys.Chem. 91 (1987) 1327.
7) W.W.Warren, Jr. and LF.Mattheis, Phys.Rev.B 30 (1984) 3103.
8) J.R.Franz, Phys.Rev.B 29 (1984) 1565.
9) R.E.Goldstein and N.W.Ashcroft, Phys.Rev.Lett. 55 (1985) 2164.
10) J.H.Rose, Phys.Rev.B 23 (1981) 552.
11) N.F.Mott, Metal-Insulator Transitions (Taylor and Francis, London, 1974).
12) J.P.Hernandez, Phys.Rev.A 34 (1986) 1316.
13) W.F.Brinkmann and T.M.Rice, Phys.Rev.B 2 (1970) 4302.
14) G.Franz, W.Freyland, W.Gläser, F.Hensel and E.Schneider, J. De Phys. 41 (1980) C8-192.
15) C.J.Pings, in: Physics of Simple Liquids, eds. H.N.V.Temperley, J.S. Rowlinson, G.S.Rushbrooke (North-Holland, Amsterdam, 1968) p. 387.
16) T.E.Faber, Introduction to the Theory of Liquid Metals, (Cambridge University Press, Cambridge, 1972).
17) N.W.Ashcroft, Phys.Lett. 23 (1966) 48.
18) V.Heine and I.Abarenkov, Phil.Mag. 9 (1964) 451.
19) B.Knuth, Doctoral Thesis, University of Marburg, 1989.
20) R.Winter, F.Hensel, F.Bodensteiner and W.Gläser, J.Phys.Chem. 92 (1988) 7171.
21) P.A.Egelstaff, I.D.Page and C.R.T.Heard, J.Phys.C: Solid State Phys. 4 (1971) 1453.
22) P.A.Egelstaff, J.B.Suck, W.Gläser, R.McPherson and A.Teistma, J. De Phys. 41 (1980) C8-222.
23) P.A.Egelstaff and S.S.Wang, Can.J.Phys. 50 (1972) 2461.
24) P.T.Cummings and P.A.Egelstaff, J.Phys.F: Met.Phys. 12 (1982) 233.
25) T.Gaskell and N.H.March, Phys.Lett. 7 (1963) 169.
26) G.Kahl and J.Hafner, Phys.Rev.A 29 (1984) 3310.
27) F.Yonezawa and T.Ogawa, Suppl.Prog.Theor.Phys. 72 (1982) 1.
28) F.Hensel and H.Uchtmann, Annu.Rev.Phys.Chem. 40 (1989) 61.
29) W.Götzlaff, Doctoral Thesis, University of Marburg, 1988.
30) R.E.Goldstein, A.Parola, N.W.Ashcroft, M.W.Pestak, M.H.W.Chen, J.R.de Bruyn and D.A.Balzarin, Phys.Rev.Lett. 58 (1987) 41.
31) G.Schäfer, Docotoral Thesis, University of Marburg, 1988.

THEORETICAL STUDY OF ATOMIC AND ELECTRONIC STRUCTURES IN MICROCLUSTERS OF POTASSIUM AND MERCURY

Fumiko YONEZAWA and Shoichi SAKAMOTO

Department of Physics, Faculty of Science and Technology, Keio University, Hiyoshi 3-14-1, Kohoku-ku, Yokohama 223, Japan

By making use of the molecular-orbital method, we evaluate the total energy, the ionization energy IP and the level spacing $\Delta\varepsilon$ near the highest-occupied molecular orbital of potassium(K_N) and mercury(Hg_N) microclusters up to size $N=48$. For K_N, the atomic configurations composed of pairs of K atoms (i.e., K_2) tend to be stable, and the size dependences of IP and $\Delta\varepsilon$ indicate that K_N is metallic even when N is as small as 3. For Hg_N, on the other hand, the atomic structures with high coordination numbers are energetically favourable, and the behaviours of IP and $\Delta\varepsilon$ imply that the transition from non-metallic to metallic nature takes place around $20 \leq N \leq 33$.

§1. INTRODUCTION

The physics of metallic microclusters has recently attracted much attention from various points of view.[1)-3)] In particular, we could pick up, among other things, the following two aspects as the most interesting problems. That is:

(1) What kind of the atomic configuration yields the most stable structure for a microcluster with a given number N of atoms?

(2) Approximately at which size does a microcluster undergo a transition from an insulating to metallic behaviour?

In connection with the former problem, the condensed-matter physics has not been successful so far to tell, from the first principle, what kind of a bulk structure is energetically the most favourable for a given set of atoms under a given set of environmental conditions such as temperature and pressure. The difficulty originates in the fact that the essentially many-body interactions must be treated. Our wishful thinking is that the subtlety might be relaxed when we deal with microclusters in which the numbers of constituent atoms are not enormously large.

In order to cope with the latter problem, we first have to specify the definition of a metal. This is rather straightforward in the framework of the band theory. To microclusters of a finite size, however, the arguments based upon the band theory no longer apply, and a question arises; 'what is a metal anyway?'

Another intriguing issue is whether or not the solutions to the above problems vary from material to material. Do microclusters of simple metals such as alkali metals behave in a manner similar to that of, say, mercury in which p electrons are expected to play an important role in the binding of atoms?

With these points in mind, it is the purpose of the present article to study the electronic and atomic structures of potassium K (as an example of alkali metals) and of mercury Hg. In §2, we mention what is known from experimental data, and in §3 we explain our method for theoretical calculations. Some atomic configurations of microclusters which we investigate are illustrated. Our results for K and Hg are presented in §4, and discussion is given in §5.

§2. INFORMATION FROM EXPERIMENTS

Now, let us face the question of 'what is a metal?' A material is regarded as a metal if it activates a metal detector located at an airport which one has to go through before getting on board. The detector is designed such that it responds when the material passing there contains movable electrons. Actually, the existence of movable or non-tightly-bounded electrons is the underlying assumption for the jellium model employed to the analyses of the electronic structures in microclusters. Judging from the fact that the jellium model has succeeded in accounting for the magic numbers and so forth[1], the concept of non-tightly-bounded electrons seems to be plausible for microclusters of reasonably large sizes. But, it is not clear at all that the validity of this assumption can be automatically extended to microclusters of very small sizes.

It has been put forward that the size dependence of the ionization potential IP of microclusters serves as a criterion for testing whether a system is metallic or not. The experimental data due to Saunders[4] are shown in Fig.1(a) where the IP's of K_N are depicted as a function of $N^{-1/3}$. Among out of all IP values, the results from microclusters with closed shells are denoted by open circles, which we can observe to change linearly, satisfying the relation

$$IP = WF + Ae^2/R \qquad (2.1)$$

where WF is the work function for a bulk material, A a constant whose magnitude is about 0.5, e the charge of an electron, and R the radius of a microcluster under consideration. The relation $R \propto N^{-1/3}$ is assumed to hold well for a microcluster with a closed shell. Since the second term of Eq.(2.1) describes the energy required to charge a metallic sphere, it has been asserted that the system is metallic when the IP of closed-shell microclusters is linear with $1/R$. Figure 1(a) suggests that K_N, when classified by this criterion, behaves like a metal even when N is as small as 8.

As shown in Fig.1(b), the experimental IP of Hg microclusters[5] exhibits the property qualitatively different from that of K microclusters. The IP data for $N \leq 17$ or so stay on a branch which Hensel et al[5)-7)] have asserted as being the non-metallic (NM) branch of van der Waals' type. They also argue that the transition occurs in the region of $17 \leq N \leq 70$ from this NM branch to a metallic (M) branch which is linear in $1/R$, here again following Eq.(2.1), and which extrapolates to the *WF* of a bulk Hg in the limit of *N* being infinite.

One of our aims in this article is to clarify, from theoretical considerations, the physical meaning of the criterion using the *IP* to examine whether a system is metallic or not.

§3. METHOD

We evaluate the eigen values and functions of electrons in a microcluster by making use of the Hartree-Fock-Roothaan (HFR) method[8] for the molecular-orbital (MO) calculations. The wave function $\psi_i(r)$ of the *i*-th MO is described in terms of the atomic orbital $\phi_\mu(r) = \phi(r - R_\mu)$ associated with an atom at R_μ as

$$\psi_i(r) \equiv \sum_\mu C_{\mu i} \phi_\mu(r) \tag{3.1}$$

where $C_{\mu i}$ is the coefficient to be determined self-consistently by solving the HFR equation.

For K, we approximate $\phi_\mu(r)$ by a modified Gauss function of the form $\chi_\mu(r) \propto r^2 \exp(-\alpha_s r^2)$, in which the parameter $\alpha_s^{-1/2}$ gives the maximum position of $\chi_\mu(r)$. In our self-consistent calculations of the energy levels of electrons, we choose α_s such that the *IP* of an isolated K is equal to the value obtained experimentally. The atomic wave function $\chi_\mu(r)$ with α_s thus defined shows good agreement, in some essential aspect, with that due to Herman and Skillman.[9]

For Hg, we express $\phi_\mu(r)$ in terms of four Gauss functions corresponding to s, p_x, p_y and p_z atomic orbitals, in which appear two parameters α_s and α_p, the former being determined so that the *IP* of an isolated Hg is reproduced and the latter being determined so that the modification of the MO due to the electron correlations is suitably taken into account.

It may sound very crude to say that we approximate the atomic orbitals by using only one or a few bases. The important point to make here is that we are not after the extraordinarily accurate energy levels, but we rather want to grasp the rough idea about what is going on when the number *N* of atoms in a microcluster is increased from 1 to up about 50.

Armed with the method as described here, we are now ready to calculate the energy levels ε_i of electrons in a microcluster and the total energy E_{tot} of the whole cluster for any atomic configuration.

Our task at this stage is to find out which atomic configuration gives the minimum of E_{tot} for a given number *N* of atoms in a cluster. On the basis of the conjecture that the E_{tot} would become the smallest for a configuration with high symmetry, we pick up several configurations of relatively high symmetry for each *N*. Some examples of atomic configurations we study are illustrated in Fig.2.

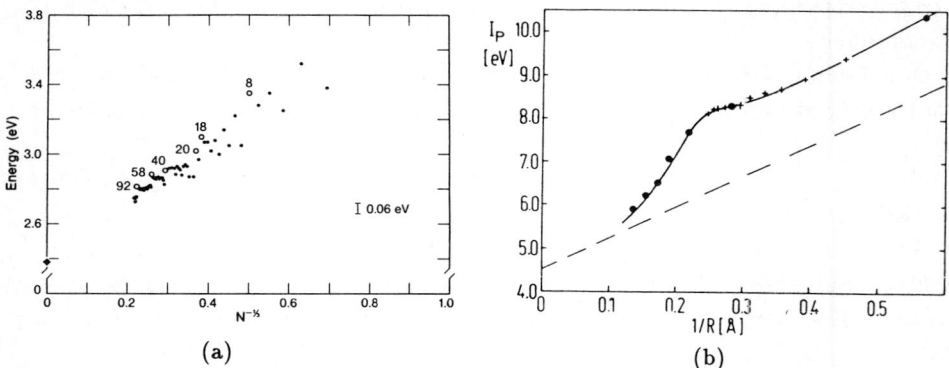

Fig.1 Experimental results of IP vs. $N^{-1/3}$. (a) K_N due to Walt et al.[4] (b) Hg_N due to Hensel et al.[5]

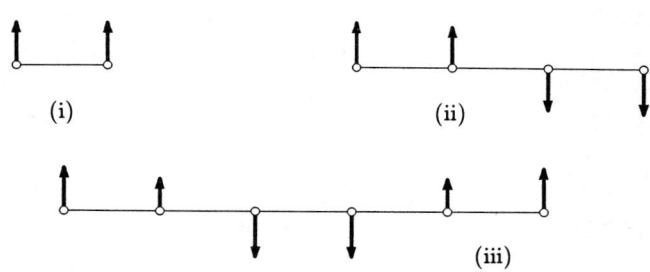

Fig.2 Some atomic structures of microclusters studied in this paper.
(a) *Linear chains with equal atomic distances.*
 Arrows indicate the coefficients $C_{\mu i}$ of the respective basis μ which form the wave function of the HOMO. Details are given in §4. (i) $N=2$; (ii) $N=4$; and (iii) $N=6$.

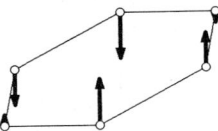

Fig.2 (continued)

(b) *Polygons — triangle, square, pentagon, hexagon, etc.*
 A regular hexagon is given as an example where $C_{\mu i}$'s are also illustrated.

When a specific configuration is chosen, we still are left with one more freedom, that is, we can always enlarge or reduce the configuration without changing the geometrical structure. Therefore, we denote a configuration by a characteristic length such as the interatomic distance R, carry out the calculations of ε_i and E_{tot} for various R's of a certain geometrical structure and search for R at which E_{tot} becomes minimum.

§4. RESULTS FOR POTASSIUM AND MERCURY

4.1. Stable Structure

According to the prescriptions as explained in the preceding section, E_{tot}'s per atom are derived for several structures with $1 \leq N \leq 48$, and a few examples for K_N are presented in Fig.3 as functions of R, the structures employed being a pair $(N=2)$, a hexagon $(N=6)$ and a cube $(N=8)$. Here, we recognize two remarkable features; (1) $E_{tot}(N)$ measured from $E_{tot}(N=1)$ is negative except for a region of very small values of R and has a minimum; and (2) E_{tot} at the minimum decreases as N increases. Similar results are obtained for Hg as shown in Fig.4 for a pair $(N=2)$, a tetrahedron $(N=4)$, an fcc $(N=8)$ and a polyhedron with icosahedral symmetry $(N=33)$.

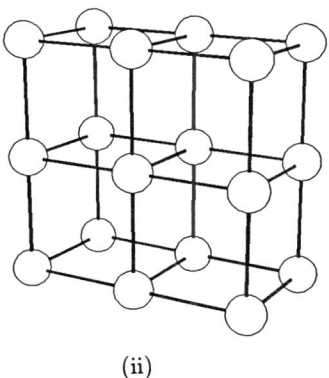

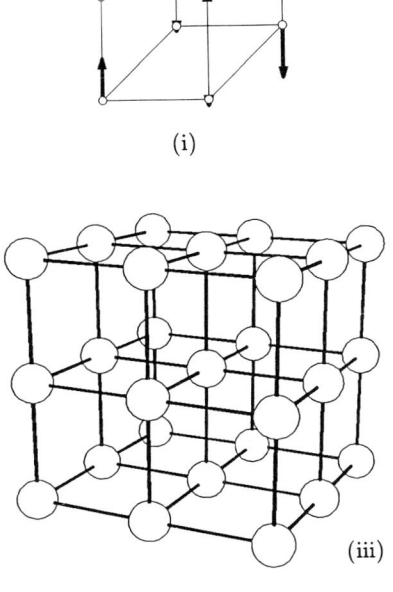

Fig.2 (continued)

(c) *Polyhedra with cubic symmetry.*
  (i) A regular cube with $C_{\mu i}$'s $(N=8)$.
  (ii) A collection of 4 cubes $(N=18)$.
  (iii) A collection of 8 cubes $(N=27)$.

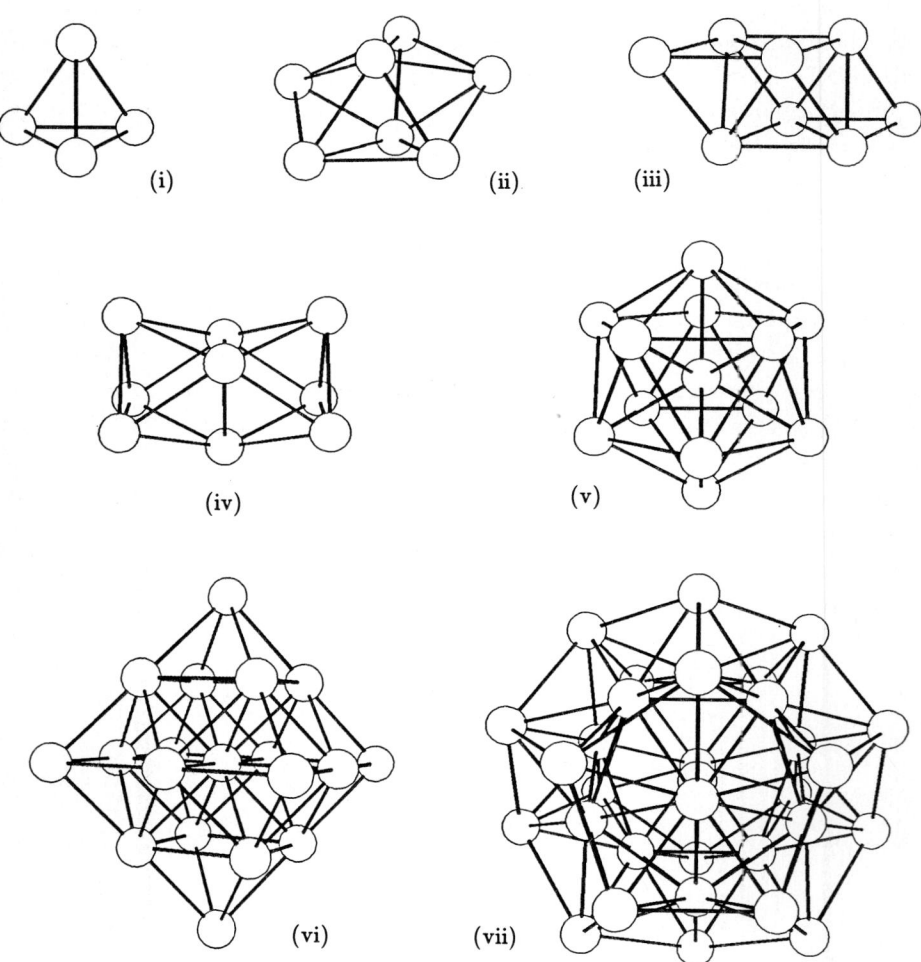

Fig.2 (continued)

(d) *Polyhedra with high coordination numbers.*
   Each of the following figures contains a lot of triangles formed by three nearest-neighbour atoms.
   (i) A regular tetrahedron ($N=4$).
   (ii) A decahedron ($N=7$).
   (iii) An fcc stacking of A(1)B(3)C(3)A(1) ($N=8$).
   (iv) An hcp stacking of A(3)B(3)A(3) ($N=9$).
   (v) An icosahedron ($N=13$).
   (vi) An fcc cluster with one central atom, 12 nearest-neighbour atoms and 6 second-neighbour atoms ($N=19$).
   (vii) A polyhedron with icosahedral symmetry where a tetrahedron is attached onto each face of an icosahedron ($N=33$).

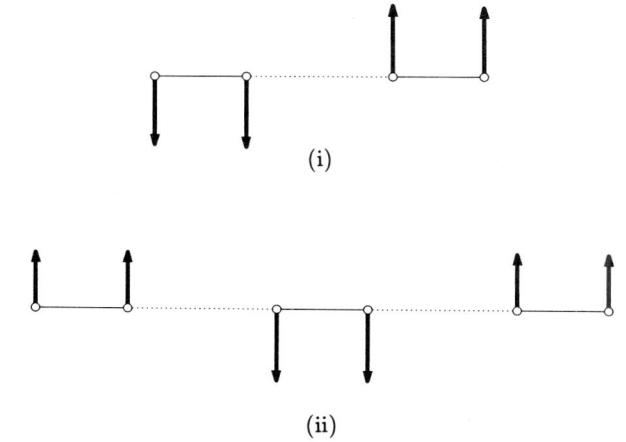
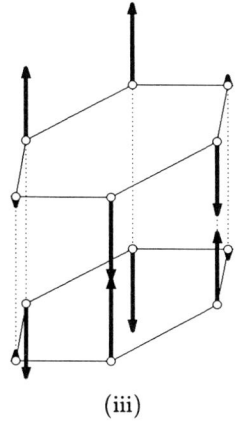
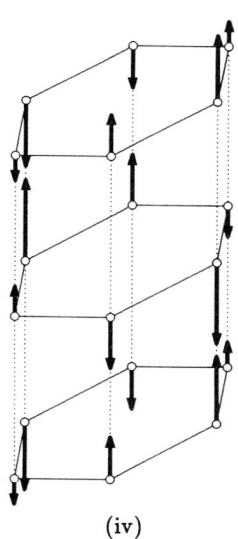
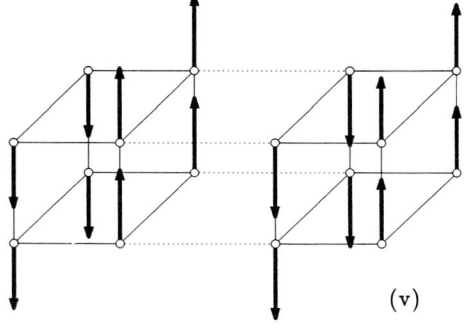

Fig.2 (continued)

(e) *Collections of smaller units*

Chains composed of 2 pairs (i) and 3 pairs (ii) of atoms. Stackings of 2 hexagons (iii) and 3 hexagons (iv). A collection of 2 cubes (v). In all of these figures, the coefficients $C_{\mu i}$'s are described by arrows.

It must be noted however that $E_{\text{tot}}(N)$ does not always decrease when N is increased because the atomic configuration used for the calculation of $E_{\text{tot}}(N)$ may not be energetically the most favourable structure for that value of N. The situation is clearly observed in Fig.5(a) and 5(b) in which E_{tot} for various structures are shown as functions of N for K_N and Hg_N respectively. A necessary condition for a structure to be stable is for $E_{\text{tot}}(N)$ to be smaller than $E_{\text{tot}}(N')$ with $N'<N$. Unless this condition is fulfilled, that particular structure of size N would fall apart.

The structures satisfying this necessary condition are denoted by fulfilled circles in Fig.5(a) and 5(b). For K_N, they are a pair ($N=2$), two pairs ($N=4$) as depicted in Fig.2(e-i), a hexagon ($N=6$), a cube ($N=8$), a stacking of two hexagons ($N=12$) as shown in Fig.2(e-iii), a collection of two cubes ($N=16$) as given in Fig.2(e-v) and a stacking of three hexagons ($N=18$) as illustrated in Fig.2(e-iv).

The aspects characteristic of these structures are that the atoms in each structure tend to form a pair with a length nearly equal to the distance between two atoms in an isolated K_2. In the figures mentioned above, the arrows indicate the values of the coefficients $C_{\mu i}$ for $i=$HOMO (the highest-occupied molecular orbital). From a careful analysis of the magnitudes and directions of these arrows in a structure, we can study the properties of the HOMO and get a clue to the elucidation of the reasons why their structures have relatively low energies. For instance, by comparing Fig.2(e-i) with Fig.2(a-ii), we learn the difference between the behaviours of these two wave functions.

One interesting conclusion reached immediately therefrom is that the N-dependence of E_{tot} forms distinct series for even- and odd-number clusters. This is because, in an odd-number cluster, there exists an extra atom which cannot find a partner to form a pair, and this extra atom, failing to recognize binding energy, adds an extra amount of energy to E_{tot}. In fact, in Fig.5(a), the filled circles for the even-number clusters as described above form one series while the open circles corresponding to a chain of size 3, 5, 7 and 9 form another.

For Hg_N, the structures denoted by filled circles in Fig.5(b) are a pair ($N=2$), a triangle ($N=3$), a tetrahedron ($N=4$), a decahedron ($N=7$) as illustrated in Fig.2(d.ii), an fcc cluster ($N=8$) as denoted in Fig.2(d.iii), an extended fcc cluster ($N=19$) as shown in Fig.2(d.vi) and a polyhedron with icosahedral symmetry ($N=33$) as depicted in Fig.2(d.vii). In contrast with the case of K, there is no such thing as an even- or odd-number series.

The feature common to all these structures is that, when each pair of nearest-neighbour atoms are connected by a bond in a structure, there exist many regular or almost regular triangles. In a structure with many triangles of bonds, the coordination number (i.e., the number of the nearest-neighbour atoms) is high. Since the p components of the bases $\chi_\mu(\mathbf{r})$ works to favour the bonding orbitals energetically, the higher the coordination number, the lower the total energy E_{tot}.

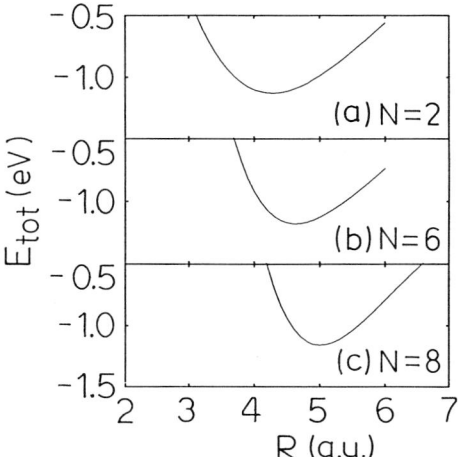

Fig.3 $E_{tot}(N)$ — the total energy per atom — of K_N. The origin of the energy is taken to be $E_{tot}(N=1)=0$. (a) $N=2$; (b) $N=6$ (hexagon); and (c) $N=8$ (cube).

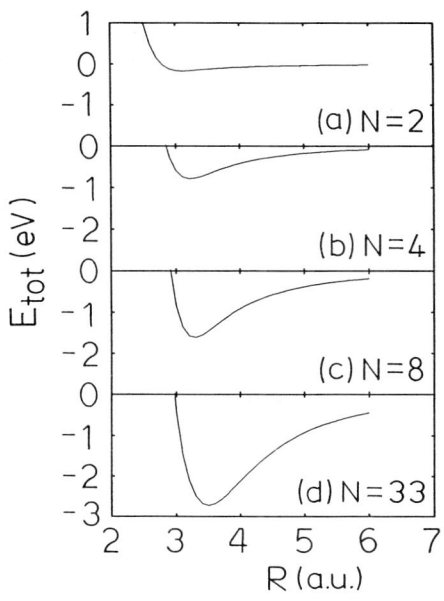

Fig.4 $E_{tot}(N)$ — the total energy per atom — of Hg_N. (a) $N=2$; (b) $N=4$ (tetrahedron); (c) $N=8$ (fcc); (d) $N=33$ (poly-icosahedron).

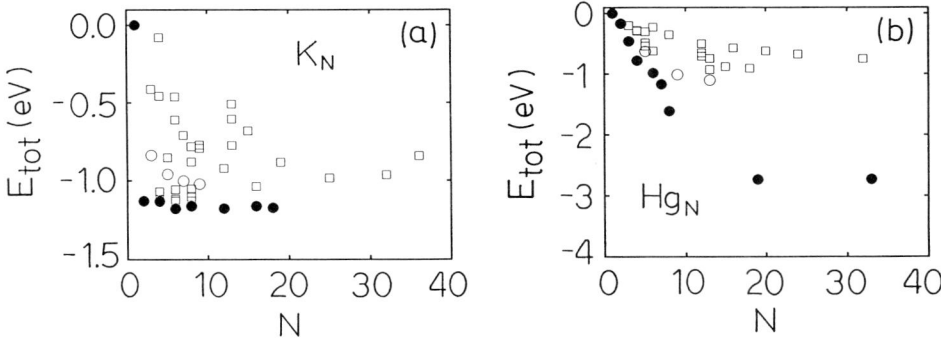

Fig.5 $E_{tot}(N)$ vs N. (a) K_N; (b) Hg_N.

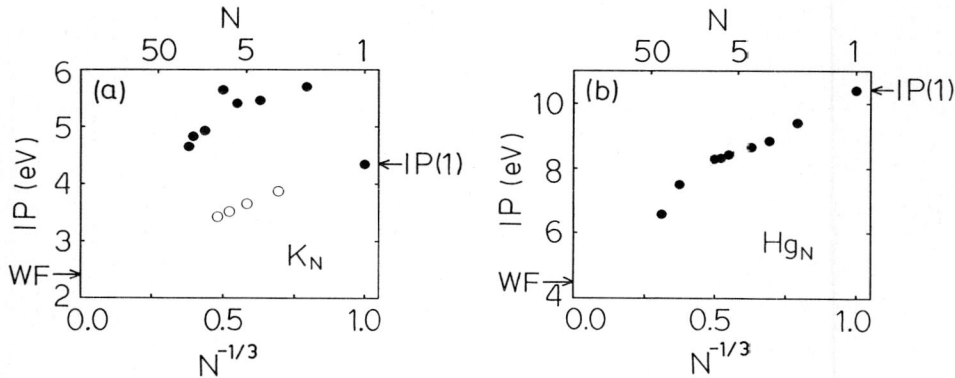

Fig.6 $IP(N)$ vs $N^{-1/3}$. (a) K_N ; (b) Hg_N.

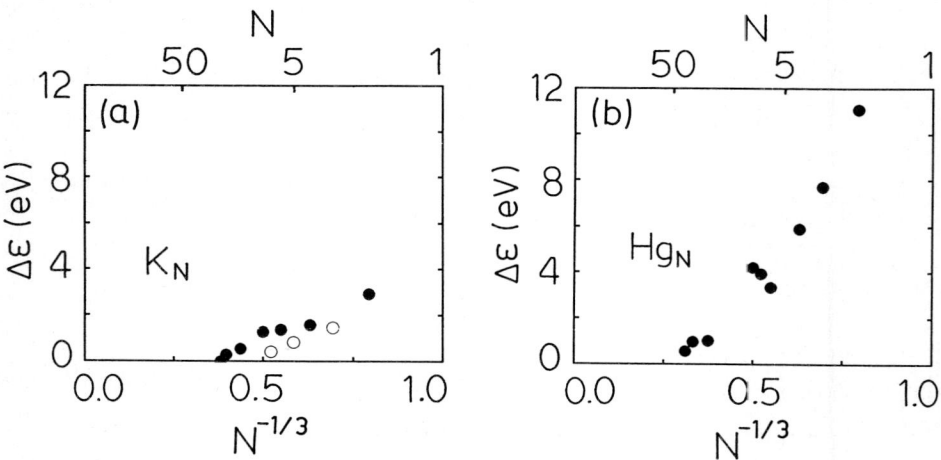

Fig.7 $\Delta\varepsilon(N)$ vs $N^{-1/3}$. (a) K_N ; (b) Hg_N.

4.2. Ionization Potential and Level spacing

The ionization potential IP of a microcluster is defined by

$$IP(N) = E_{\text{tot}}(N;R_0;\text{ion}) - E_{\text{tot}}(N;R_1;\text{neutral}) \qquad (4.1)$$

in which $E_{\text{tot}}(N;R_0;\text{neutral})$ indicates the total energy per atom for a structure of size N for a neutral microcluster at the length R_0 where E_{tot} becomes minimum. On the other hand, $E_{\text{tot}}(N;R_1;\text{ion})$ corresponds to an ionized microcluster. Based upon the assumption that the atomic configuration does not drastically change during the ionization process, we take the same structure for both terms on the rhs of Eq.(4.1).

The IP thus enumerated are presented in Fig.6(a) for K_N. Here again, we can observe two distinct series for even- and odd-number clusters denoted respectively by filled and open circles. When extrapolated to the limit of infinite N, both series point the WF of the bulk K. In either case, the IP is linear and obeys Eq.(2.1), thus implying that the system is metallic for $N \geq 3$.

For Hg_N, the IP are demonstrated in Fig.6(b). It is worth mentioning that the IP's of both even- and odd-number clusters (shown by filled and open circles respectively) stay on the same curve. The IP's start departing from the upper branch when N exceeds 20 or so. This tendency is in good agreement with the experimental results, and therefore the transition from the NM to M branch seems to be ascertained theoretically.

It is interesting to examine the energy-level spacing $\Delta \varepsilon$ near the HOMO as a function of $N^{-1/3}$, which is given in Fig.7(a) for K_N and in Fig.7(b) for Hg_N. It is clearly seen from these figures that $\Delta \varepsilon$ for K_N is small enough for $N \geq 3$, indicating the metallic nature of a cluster. On the other hand, $\Delta \varepsilon$ for Hg_N becomes reasonably small only when N is larger than 20 or so, thus suggesting that clusters of sizes less than 20 or so are not metallic.

This behaviour of $\Delta \varepsilon$ provides a physical interpretation of the criterion by means of the IP for judging whether a microcluster is metallic or not.

§5. DISCUSSION

By making use of the molecular-orbital calculations, we have derived $E_{\text{tot}}(N)$ and $IP(N)$ of K_N and Hg_N microclusters, and we also have made analyses of the atomic structures of clusters. The conclusions we have obtained are summarized as follows.

[1] Between the physical properties of K_N and Hg_N microclusters, there exist differences not only in the electronic properties but also in the atomic configurations, the latter originating in the former.

[2] *Results for K_N microclusters:*

 [A] Atoms prefer to combine in pairs, and as a consequence, microclusters consisting of even atoms have total energies per atom, E_{tot}, lower than those of microclusters composed of odd atoms. This is related to the fact that each atom supply only one valence electron of s-nature just like other alkali metals. Therefore, the two electrons in K_2 occupy the relatively low HOMO, thus making K_2 stable.

[B] The $E_{tot}(N)$ and $IP(N)$ of K_N as functions of N form two distinct series each corresponding to even- and odd-number clusters. This feature again comes from the above-mentioned characteristics of electrons.

[C] The behaviours of the IP and $\Delta \varepsilon$ of K_N as functions of N indicate that K_N clusters are metallic when $N \geq 3$.

[3] *Results for* Hg_N *microclusters*

[A] For Hg_N clusters, the atomic configurations with high coordination numbers tend to be stable. This is due to the fact that the electrons with p-nature work to construct bonds between neighbouring atoms, thus saving binding energies.

[B] The $E_{tot}(N)$ and $IP(N)$ as functions of N show no distinct differences for even- and odd-number clusters. This is because the electronic structure does not favour the formation of atomic pairs in Hg.

[C] The behaviours of the IP and $\Delta \varepsilon$ of Hg_N as functions of N imply that the transition from non-metal to metal takes place in the region of $20 \leq N \leq 33$.

[4] The level spacing $\Delta \varepsilon$ provides a physical interpretation for metallic and non-metallic nature in a system as small as microclusters.

References

1) W. A. de Heer, W. D. Knight, M. Y. Chou and M. L. Cohen, *Solid States Physics* (ACADEMIC PRESS, INC. 1987) 94.
2) S. Sugano, Y. Nishina and S. Ohnishi eds. *Microclusters* (Springer Series in Material Science 4, 1987).
3) G. Benedek, T. P. Martin and G. Pacchioni eds. *Elemental and Moleculer Clusters* (Springer Series in Material Science 6, 1988).
4) W. A. Saunders, Ph.D. Thesis, University of California, Berkeley (1986).
5) K. Rademann, B. Kaiser, U. Even and F. Hensel, Phys. Rev. Lett. **59** ,2319 (1987).
6) F. Hensel and H. Uchtmann, Annu. Rev. Phys. Chem. **40** ,61 (1989).
7) K. Rademann, Ber. Bunsenges. Phys. Chem. **93** ,653 (1989).
8) B. K. Rao and P. Jena, Phys. Rev. B **32** ,2058 (1985).
9) F. Herman and S. Skillman, *Atomic Structure Calculations* (Prentice Hall, Englewood Cliff, N.J. 1963).
10) S. Sakamoto and F. Yonezawa, to be submitted to Phys. Rev. B.

IONIZATION EFFECTS IN A MODEL FLUID*

John P. HERNANDEZ

Department of Physics and Astronomy, University of North Carolina, Chapel Hill, NC 27599-3255 USA

A free energy model is considered for a system consisting of neutral atoms, ions, molecules, and molecular ions, neutralized by a Fermi gas of electrons. The ionization balance is determined by extremizing the system free energy with respect to the density of either sign charges. The system exhibits a vapor-liquid phase transition and an insulator-metal transition. The dependence of the results on material parameters is discussed.

1. INTRODUCTION

Materials which have a liquid metal regime require theoretical treatments in which their electronic and thermodynamic properties are treated in a unified manner. At low enough densities such materials cannot be metallic and the structure of a fluid is intimately affected by its electronic properties. Thus, the insulator-metal transition is interlinked to the vapor-liquid transition.

Model treatments for this problem were undertaken by the author. The coulombic Hartree interactions and screening were first explored[1]. Charge-neutral interactions have also been considered from the point of view that the ionization energy for atoms is changed in a medium and that the binding of molecules and molecular ions also affects the energetics of the system[2-4]. It is the purpose of this note to unify previous work by discussing the parameter dependences. The work is still based on a simple model in order to attempt to extract the basic physics rather than delving into the complexities required for a rigorously complete treatment of the problem.

2. MODEL

The model is based on an approximation to the free energy of the system. It consists, at first, of a mixture of ideal gases of the species which may exist in the material, then it is supplemented by interactions. Thus, a chemical approach is taken. The thermodynamic properties are, of course, derivable from the system free energy.

---

*Partially supported by the NATO Scientific Affairs Division and the UNC Research Council

Neutral atoms and molecules, ions and molecular ions contribute to the free energy with ideal gas terms

$$kTN_i \ln(N_i/ez_i).$$

In which k is Boltzmann's constant, T is temperature, N_i the number corresponding to the ith species, e is the base of natural logarithms, and z_i the partition function. This last term, for an atom, would be given by

$$s_o (MkT/2\pi\hbar^2)^{3/2} [\Omega(1 - \rho\nu)],$$

where s_o is a spin degeneracy factor, M the mass, $\hbar$ Plank's constant divided by 2π and the square bracket gives the system free volume. For an ion, this type of term needs to be modified by $\exp(- I/kT)$ in which I is the ionization potential. For more complicated species, the generalizations are straightforward.

The unbound electrons can be treated as a Fermi gas with a free energy contribution

$$kT \int_0^\infty dE\ g(E) \left[fE/kT + f \ln f + (1 - f) \ln(1 - f) \right]$$

in which a density of states g(E) is incorporated along with the Fermi function

$$f = \left[\exp\left(\frac{E - \mu}{kT}\right) + 1 \right]^{-1},$$

with the chemical potential (μ) being related to the electron density

$$n_e(\mu) = \int_0^\infty dE\ g(E)\ f/\Omega.$$

The simplest model would use a free particle density of states.

Interactions are then added to the free energy. Screened Hartree terms can be incorporated by considering the introduction of an external charged particle at the origin and the linear resonse of the charges in the system, to set up an electrostatic potential $\phi(r)$. Then, for example, the system becomes inhormogeneous with z for the ions modified via

$$z_+(r) = z_+ \exp(- e\ \phi(r)/kT)$$

and the electron density varying with a functional relation

$$n(r) = n_e(\mu + e\ \phi(r)).$$

Equilibrium would require equality of the electrochemical potentials for the charged species at some r to the values far away from the introduced external charge, for example,

$$\frac{\partial F(r)}{\partial n(r)} = \frac{\partial F(\infty)}{\partial n}$$

This requirement yields the position dependent densities as a function of $\phi(r)$.

On the other hand, Poisson's equation yields $\phi(r)$ as the screened coulomb potential.

For the system with atoms, mono-positive ions and electrons, the square of the inverse screening length, in linear response, is

$$k_o^2 = \frac{4\pi n e^2}{kT} \left[\frac{kT}{n} \frac{\partial n_e}{\partial \mu} + \frac{1}{n} \left[\frac{\Omega}{kT} \frac{\partial^2 F}{\partial N_+^2} \right]^{-1} \right],$$

in which a fraction $(kT/n)(\partial n_e/\partial \mu)$ of the unbound electrons screen and the last term gives the fraction of the positive ions, ruled by excluded volume, which participate in the screening. The Debye-Huckel and Fermi-Thomas limits are obtained as appropriate. The fiction of the external charge is now replaced with the actual charges in the system to obtain the interaction free energies due to the screened Hartree terms.

Charge-neutral interactions can also be included. For example, in a homogenous system in which the neutral atom density is ρ, s-wave scattering of electron by the neutrals shifts the ionization potential from the free atom value (I) to $I + \hbar^2 4\pi\rho a/2m$, for small electron-atom s-wave scattering length (a). For a net attraction between an electron and a neutral (a < 0), the effective ionization potential in the medium is lowered. This could be incorporated, in the spirit of the previous linear response treatment, so that the neutral atom density also becomes position dependent. Such a treatment has not been carried out, up to now, except in an average sense.

3. RESULTS

With the system free energy now augmented by interactions, the electronic behavior can be probed by examining the electron density and chemical potential (μ) as a function of the thermodynamic variables T and ρ (the latter being the atomic density, neutrals plus ions and also including molecules and molecular ions). If the unbound electron density of states is taken to be zero below some E = 0, the system would be metallic if $\mu > 0$ (μ in the band) or insulating for $\mu < 0$ (μ in the gap). There do exist conditions in which there is substantial ionization but $\mu < 0$, so that the system is a non-metallic plasma. Also, the vapor-liquid behavior in mean field can be examined, for example, by examining the isobars as a function of ρ and T with

$$P = - \frac{\partial F}{\partial \Omega}\bigg|_{T,N}.$$

In general, the vapor-liquid instability ($\partial P/\partial \Omega|_{T,N} > 0$) as density is increased, below the critical temperature, arises from the attractive Hartree terms which favor density increases. The restabilization is mainly due to screening. The critical temperature for the model increases with atomic ionization potential, as would be expected. Hence, electron-neutral interactions, if attractive, reduce the critical temperature as they decrease the effective ionization potential in the medium. The formation of molecules generally favors ionization because of a decreased ionization potential. The metallic regime is at high densities and low temperatures.

It is worthwhile emphasizing that although the insulator-metal and vapor-liquid transitions are strongly linked, there is no strong reason to have a coincidence between metalization and the vapor-liquid critical point, ie the insulator-metal transition need not coincide with that of vapor to liquid.

4. SUMMARY

A simple model for a fluid can be constructed based on free energy considerations. The model treats electronic and thermodynamic effects on the same basis and thus can lead to linked insulator-metal and vapor-liquid transitions. Although the model yields reasonable phenomena dependence on material parameters, it is not entirely satisfactory. The chemical nature of the model is strongly biased to proceeding from low density properties. The model fails to account for any liquid structure (aside from Hartree effects) and indeed would be expected to fail at high densities given its nature, a mixture of ideal gases augmented by interactions. Perhaps a more satisfactory model, within the spirit of this work, would be a lattice model in which atoms and vacancies may occupy the lattice positions. Such a model would remove the bias we have introduced and yield a more satisfactory treatment of geometric structure. Alloying with vacancies ameliorates the rigidity of introducing the lattice and includes knowledge, from neutron scattering, that the nearest neighbor distance has only a weak density dependence[5].

REFERENCES

1) John P. Hernandez, Phys. Rev. A $\underline{34}$, 1316 (1986)

2) J. P. Hernandez, G. Schönherr, W. Götzlaff and F. Hensel, J. Phys. C $\underline{17}$, 4421 (1984)

3) John P. Hernandez, Phys. Rev. A $\underline{31}$, 932 (1985)

4) John P. Hernandez, Phys. Rev. Lett. $\underline{57}$, 3183 (1986)

5) G. Franz, W. Freyland, W. Glaser, F. Hensel and E. Schneider, J. de Phys. $\underline{41}$, C8-194 (1980)

THE INSULATOR-METAL TRANSITION IN DENSE PLASMAS

Helmut HESS

GDR Academy of Sciences, Central Institute of Electron Physics, Hausvogteiplatz 5 - 7, Berlin 1086, GDR

Insulator-metal transitions will occur in a variety of materials if the mean interatomic distance is decreased due to changes in pressure, temperature, or, in the case of mixtures, in concentration[1]. In atomic or molecular gases which are dielectric at standard conditions, and where an insulator-metal transition has been found at low temperatures and high pressures, such a transition is predicted also at higher temperatures leading from a more or less ionised plasma to a metallic fluid and ending in a second critical point. An experimental verification of this so-called plasma phase transition (PPT) is still pending.

1. INTRODUCTION

One dozen years ago, we were concerned with preparations for experiments in a mighty device - a so-called ballistic compressor. As we expected, a strongly coupled plasma should be produced by this machine due to a nearly isentropic compression. To limit surprises to a friendly extent, we calculated plasma compositions in a wide range of parameters, especially for xenon[2].

As we soon learned, the use of a density-dependent lowering of ionisation energy alone is sufficient for producing solutions, e.g. in a p-v plane (p - pressure, v - specific volume) which looks like Van der Waals loops well separated from the usual gas-liquid transition. We realised[3] the connection between this phenomenon and the dielectric-metal transition discussed as early as in 1943 by Landau and Zeldovich[4,5] also for higher temperatures. Further we understood that there exist many different groups of systems showing an insulator-metal transition:
- metal vapours
- metal ammonia solutions
- metal methylamine solutions
- films of alkali and rare-gas atoms
- doped semiconductors
- electron-hole systems
- molecular and atomic gases

Only for the last group the plasma phase transition could up to now not yet be verified by experiments. Just this group will be our main subject in the following.

2. THE INSULATOR-METAL TRANSITION AT LOW TEMPERATURE
2.1. Herzfeld's Criterion

In a medium which shows dielectric behaviour at normal conditions, a transition to a metallic state will be expected if it is sufficiently compressed. As early as in 1927 Bridgman[6] wrote: "... that at ordinary temperatures sufficiently high pressures are capable of breaking down the quantum structure of atoms, reducing matter to an electrical gas of electrons and protons." So he introduced what was later called pressure ionisation.

In the same volume of Physical Review Herzfeld[7] brought out a criterion for determining when an element will show metallic conductivity. This will occur when the molar volume V_M becomes smaller than the molar refractivity R:

$$V_M < R \qquad (1)$$

where R is connected with the polarisability α by $R = 4\pi\alpha L/3$. L is the Avogadro number. The fulfilment of eq. (1) is often called "polarisation catastrophe".

Ross[8] interpreted the Herzfeld criterion from a geometrical point of view. Taking into account that R is equal to the actual volume occupied by the atoms themselves, the available volume V_M becomes too small and a change to another state with a smaller (atomic) volume is necessary. As Herzfeld was already able to

Species	$\alpha/\text{Å}^3$	R/cm^3	V_M^{tr}/cm^3	V_M/R
He	0.204	0.514	17.25*	33.6
Ne	0.392	0.988	13.98	14.1
Ar	1.63	4.11	24.62	5.99
Kr	2.465	6.22	29.66	4.77
Xe	4.01	10.11	37.09	3.67

Table 1: Molar refractivity and molar volume of the rare gases. The molar volume is taken for the solid at the triple point; the molar refractivity is calculated from the static polarisability α.   *) at 4 K.

show, a large number of metals and nonmetals obeys this simple criterion.

In Table 1 corresponding data are given for the rare gases[9]. As can be seen, V_M is always greater than R what corresponds to the dielectric character of the rare gas solids at low pressure. The last column further gives the factor by which the solid has to be compressed to fulfil the Herzfeld criterion. On the other hand, the metallic alkalis show the expected opposite behaviour (Table 2). They already fulfil the Herzfeld criterion at low pressure: V_M taken at 20 °C and atmospheric pressure is smaller than the molar refractivity R.

Species	$\alpha/\text{Å}^3$	R/cm^3	V_M/cm^3	V_M/R
Li	24.3	61.3	13.1	0.214
Na	23.6	59.5	23.7	0.398
K	43.4	109	45.4	0.417
Rb	47.2	119	56.2	0.472
Cs	59.5	150	71.1	0.474

Table 2: Molar refractivity[10] and molar volume of the alkali metals. The molar volume is taken for the solid at 20 °C and atmospheric pressure.

2.2. Mott's Criterion

Often the insulator-metal transition is connected with the name of Mott. In extending a model of Wilson[11] he pointed out that there must be a discontinuous transition from a dielectric to a metal at T = 0 K if the lattice parameter becomes smaller than a critical value[12,13]. An estimate was given for the corresponding transition density

$$n_{cr}^{M\ 1/3} a_H \approx 0.2...0.3 \qquad (2)$$

with a_H the generalised "Bohr radius" $a_H = \varepsilon \hbar^2/me^2$. Here ε is the dielectric constant.

Taking into account the already mentioned connection between the polarisability and the atomic volume[8], the Herzfeld criterion (1) and the Mott criterion (2) become identic within a factor of the order of unity.

The Mott criterion was initially derived for one-electron atoms, and therefore hydrogen and the alkalis should be used as examples (Table 3). Here the critical Mott densities are compared with the densities at the critical point for the liquid-gas transition - the lowest density a liquid can reach.

Species	$a_H/\text{Å}$	n_{cr}^{M}/cm^{-3}	n_{cr}^{L-G}/cm^{-3}
H	0.67	5.84 + 22	1.89 + 22
Li	1.58	4.46 + 21	9.05 + 21
Na	1.82	2.92 + 21	7.81 + 21
K	2.22	1.61 + 21	2.91 + 21
Rb	2.43	1.22 + 21	2.03 + 21
Cs	2.66	1.00 + 21	1.71 + 21

Table 3: Critical Mott densities for the transition insulator-metal compared with the critical densities for the liquid-gas transition. The atomic radii are from[14]. The constant in eq. (2) is 0.26[15].

Despite the relatively high critical temperatures of the alkalis the criterion works well. The densities in the critical point are always higher than the Mott densities reflecting the fact that the liquid state of the alkalis is always metallic. In hydrogen, on the contrary, the Mott density is higher what means that compression is necessary to reach a metallic state.

In the last years, efforts were made to calculate the conditions under which hydrogen and the rare gases as well as their isoelectronic neighbours become metallic at low temperature. All these transitions are located at pressures of about 100 GPa and higher. Meanwhile some of them are confirmed by experiments. The insulator-metal transition in xenon, for instance, was found by Nelson and Ruoff in 1979[16]. It is very impressive to follow up the interaction between experimentalists and theorists in determining finally the pressure reached at the transition[17,18]. This procedure gives convincing evidence for the confidence the theorists have in their methods.

It should be mentioned that the calculations as well as the experimental verifications achieved up to now are in remarkably good agreement with the predictions of the simple criteria due

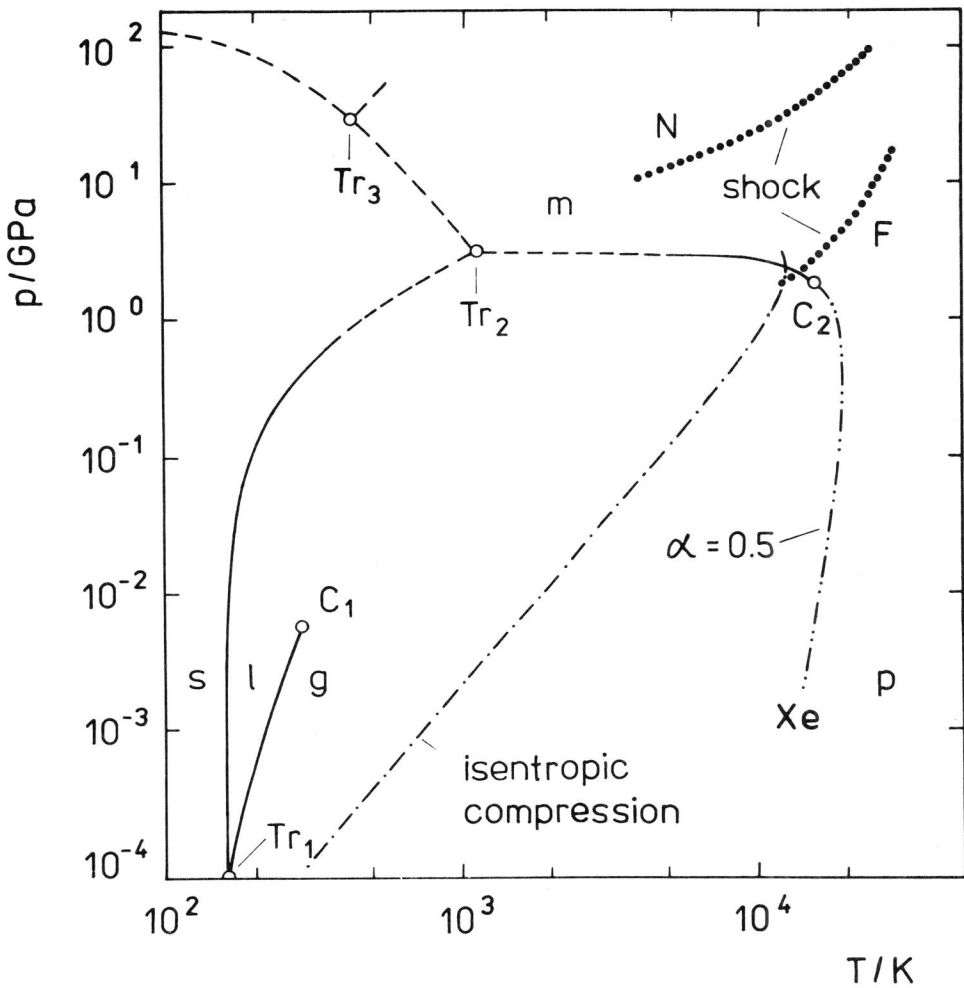

FIGURE 1

Pressure-temperature diagram of xenon. The full lines in the lower left are transition lines for the different phases: s - solid, l - liquid, g - gas. Tr_1 is the corresponding triple point, C_1 the critical point of the liquid-gas transition. The full line in the upper right is the (calculated) transition line of the plasma phase transition ending in a critical point C_2[40,41]. The dashed lines as well as the triple points Tr_2 and Tr_3 are speculative only (m - metallic). The dotted curves are results from shock-wave experiments: N - Nellis[22,23]; F - Fortov[24,25]. The dashed-dotted curve is an isentropic compression which will be possible in the future by our ballistic compressor. The dashed-double-dotted curve gives a degree of ionisation of 50 % (p - plasma).

to Herzfeld and Mott.

3. THE INSULATOR-METAL TRANSITION AT HIGHER TEMPERATURES - THE PLASMA PHASE TRANSITION (PPT)

A pressure-temperature diagram for xenon shall serve for introducing the dense-plasma problem (Fig. 1). Beside the well-known transitions solid-liquid (melting) and liquid-gaseous (vaporisation) with the triple point Tr_1 and the critical point C_1, the zero-temperature insulator-metal transition can be seen in the upper left.

Introduced as a vague idea by Landau and Zeldovich in 1943[19] and worked out in more detail by Norman and Starostin in 1968[20,21], a transition from a more or less ionised gas to a metallic fluid was proposed which should be of first order below a certain temperature corresponding to another critical point C_2. It was called plasma phase transition (PPT). It seems reasonable that the PPT transition line ends in a second triple point Tr_2 on the melting line and further that there exists a third triple point Tr_3 where the transition line between the solid and the fluid metallic state is starting from. Up to now, however, the dashed lines in this region are speculative only.

The slope of the transition line is controlled by the Clausius-Clapeyron equation: changes of entropy and volume in the same direction give a positive slope, changes in different directions a negative slope. Further, the environment of a triple point is controlled by the "180° rule"[26] which states that in a phase diagram with thermodynamically proper independent variables, no phase occupies more than 180° of angle around a triple point.

3.1. The PPT in Hydrogen

Till now, the most work has been done for hydrogen mainly due to the relevance of the corresponding parameters for astrophysical objects as, for instance, the heavier planets. The results of more recent theoretical work has been summarised in Figure 2. The conventional transitions are shown in the lower left, the zero--temperature insulator-metal transition in the upper left. A continuous transition from a plasma to a highly degenerated quantum fluid occurs from the bottom to the top at the right hand side (at high temperatures), and in the intermediate range transition lines for the PPT are shown due to different authors[27-32].

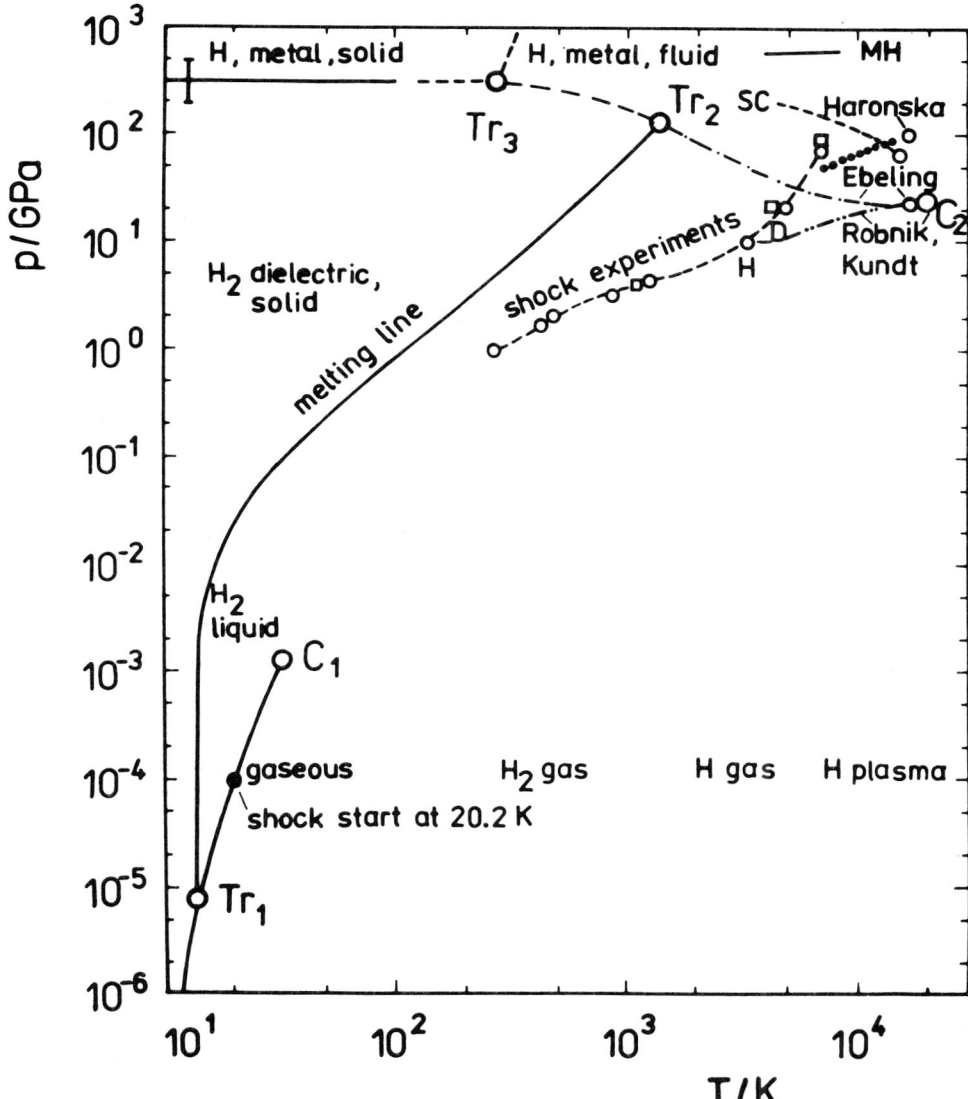

FIGURE 2
p-T diagram of hydrogen. In the upper right different calculated lines of coexistence and critical points C_2 of the PPT are given the authors of which are indicated: Robnik and Kundt[27], Ebeling[28], Haronska[29], SC[30], MH[32]. Further shock-wave experiments[34,35] in hydrogen(H) and deuterium (D) are shown.

The slopes of the transition lines are positive[27,29,32] as well as negative[28,30,31]. From a phenomenological point of view (simultaneous increase of density and entropy) the negative slope seems to be more probable. The critical temperature calculated in the more recent work does not vary so much, whereas the critical pressures differ by more than one order of magnitude. The reason for this behaviour can be understood using scaling relations proposed by Hess[33]. According to this proposal the critical temperature is connected with the well-known ionisation energy for the undisturbed ion, whereas the critical pressure depends upon the repulsive parts of the potentials of the ions and atoms (the "diameters" of these particles) where very different approximations are used for.

In the most recent work[30,31], the low degree of ionisation beyond the transition line (in the "metallic" phase) is in contrast to the earlier work and to the idea of an ionisation catastrophe. The reason for this difference is not yet clear.

The shock-wave experiments shown in Figure 3 were done in Livermore[34,35]. The values beyond the transition line[28] were obtained in double-shocked deuterium. The PPT line of coexistence of deuterium should not be very different from that of hydrogen but the maximum possible shock pressure increases proportional to the atomic mass (that is one reason for using deuterium).

Generally, an Hugoniot undergoes an unsteadiness in passing a phase transition line examples of which can be found in [36]. For recognising this effect, many points around and on the transition are necessary. In the case of the deuterium experiments related to the transition calculated in [28], practically only two points at both sides of the transition line exist (initial and final state of the double-shocked deuterium). This is not sufficient for deciding whether there is a phase transition or not.

Two earlier papers on hydrogen should be mentioned, both from 1972. Kerley[37] did not find any phase transition in the fluid region. The transition from the neutral gas to the plasma is continuous in the whole range. It is not clear due to which assumptions he could avoid the ionisation catastrophe in his calculations. Nevertheless, his triple point Tr_2 is one basis for our estimation of this point.

The other work is from Grigoryev and co-workers[38,39]. It is of high interest because all those features which come out from recent calculations can already be found there. That is true for the second triple point and for the negative slope of the coexistence line in a p-T diagram. Further there can be seen for the first time a crossing of isotherms and a cooling during a compression process.

Grigoryev's group was the first who claimed having made metallic hydrogen. They used an isentropic compression by explosives and measured the density with the help of γ rays.

3.2. The PPT in Xenon

In comparison with hydrogen, the PPT in xenon is distinguished by a virtually lower critical pressure ($p_c \approx$ 2 GPa compared with 20...500 GPa; cf. Figure 1[40,41]) what makes the probability higher for an early experimental verification. As in the case of hydrogen, the shock-wave experiments until now give no evidence for the existence of a phase transition. On the other hand the ballistic compressor experiments done in our institute reached only about 1 GPa. We are now on the way to enlarge the length of our small-cross-section compressor with the goal to reach pressures above 2 GPa.

For representing some details of the thermodynamic behaviour, a p-n diagram of xenon around the PPT is shown in Figure 3[40,42]. The region of coexistence is encircled by a dotted line (schematically). The critical point C_2 is at the bottom of this region which do not show the usual sinusoidal form. Due to a density increase on both sides away from the critical point, the region of coexistence has a scythe-like shape. As can be seen near the critical point C_2, there are crossing isotherms what means that each point in the p-n plane can be described by two different temperatures.

Recently, shock-wave experiments in Livermore on nitrogen[43] revealed features very similar to our PPT calculations. May be that there a PPT has already been found although the authors speak of a smooth transition.

Up to now, only onefold ionisation has been taken into account. At least in the rare gases however, the subsequent ionisation stages overlap the earlier ones. The problem of multiple ionisation has only recently been tackled and the impression has been risen that the ionisation catastrophe is much more radical than earlier thought.

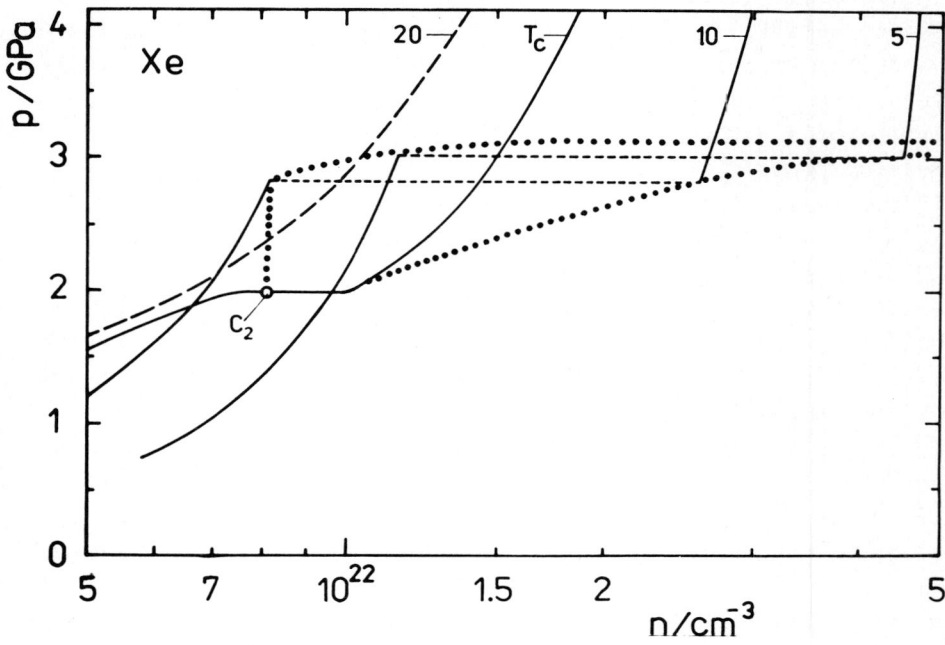

FIGURE 3
p-n diagram of xenon around the PPT. C_2 - critical point of the PPT. The temperatures along the isothermes are given in thousands of K. The region of coexistence is encircled by a dotted line (schematically).

4. CONCLUSIONS

During the preceding Workshop on Strongly Coupled Plasma Physics three years ago there was only one talk[44] on plasma phase transitions. The growing interest in this subject is documented by at least four corresponding talks during this conference. In the future, the theory has to be improved concerned the potentials at small distances, and beyond the hydrogen issue, multiple ionisation has to be considered. The experimental verification has to be tackled, and there is a good chance by shock waves as well as by ballistic compressors to decide whether a PPT exists or not.

REFERENCES

1) F. Yonezawa and T. Ogawa, Suppl. Progr. Theor. Phys. 72 (1982) 1

2) H. Dienemann and H. Hess, Proc. 5th Int. Conf. Gas Discharges Liverpool 1978, IEE Conf. Publ. No 165, p. 135

3) H. Dienemann, G. Clemens and W. D. Kraeft, Ann. Phys. (Leipzig) 37 (1980) 544

4) L. Landau and J. Zeldovich, Acta Physicochimica U.R.S.S. 18 (1943) 194

5) J. Zeldovich and L. Landau, JETF 14 (1944) 32

6) P.W. Bridgman, Phys. Rev. 29 (1927) 188

7) K.F. Herzfeld, Phys. Rev. 29 (1927) 701

8) M. Ross, J. Chem. Phys. 56 (1972) 4651

9) G. A. Cook (ed.), Argon, Helium and the Rare Gases (Interscience Publishers, New York 1961), Vol. I, pp. 151 and 351-355

10) B. Fricke, J. Chem. Phys. 84 (1986) 862

11) A.H. Wilson, Proc. Roy. Soc. A 133 (1931) 458

12) N. F. Mott, Proc. Phys. Soc. A 62 (1949) 416

13) N.F. Mott, Rev. Mod. Phys. 40 (1968) 677

14) W. Finkelnburg, Quantum Theory of Matter (Springer, New York 1968)

15) P. P. Edwards and M. J. Sienko, Phys. Rev. B 17 (1978) 2575

16) D.A. Nelson and A.L. Ruoff, Phys. Rev. Lett. 42 (1979) 383

17) M. Ross and A.K. McMahan, Phys. Rev. B 21 (1980) 1658

18) K.S. Chan, T.L. Huang, T.A. Grzybowski, T.J. Whetten, and A.L. Ruoff, Phys. Rev. B 26 (1982) 7116

19) L. Landau and J. Zeldovich, Acta Physicochimica U.R.S.S. 18 (1943) 194

20) G. E. Norman and A.N. Starostin, Teplofiz. vys. temp. 6 (1968) 410

21) G.E. Norman and A.N. Starostin, Teplofiz. vys. temp. 8 (1970) 413

22) R.N. Keeler, M. van Thiel and B.J. Alder, Physica 31 (1965) 1437

23) W.J. Nellis, M. van Thiel, and A.C. Mitchell, Phys. Rev. Lett. 48 (1982) 816

24) V.B. Mintsev and V.E. Fortov, Pisma v ZhETF 30 (1979) 401

25) Yu.B. Zaporozhets, V.B. Mintsev, V.E. Fortov, and O.M. Batovskij, Pisma v ZhTF 10 (1984) 1339

26) J.C. Wheeler, J. Chem. Phys. 61 (1974) 4474

27) M. Robnik and W. Kundt, Astron. Astrophys. 120 (1983) 227

28) W. Ebeling and W. Richert, Phys. Lett. 108 A (1985) 80

29) P. Haronska, Thesis, Univ. Rostock 1986

30) D. Saumon and G. Chabrier, Phys. Rev. Lett. 62 (1989) 2397

31) G. Chabrier and D. Saumon, Pressure Ionization in Fluid Hydrogen, this volume

32) M.S. Marley and W.B. Hubbard, Icarus 73 (1988) 536

33) H. Hess, High Pressure Research 1 (1989) 203

34) M. van Thiel, L.B. Hord, W.H. Gust, A.C. Mitchell, M. D'Addario, K. Boutwell, E. Wilbarger, and B. Barret, Phys. Earth Planet. Inter. 9 (1974) 57

35) W.J. Nellis, A.C. Mitchell, M. van Thiel, G.J. Devine, and R.J. Trainor, J. Chem. Phys. 79 (1983) 1480

36) S.B. Kormer, Uspekhi fiz. nauk 94 (1968) 641

37) G.I. Kerley, Phys. Earth Planet. Inter. 6 (1972) 78

38) F.V. Grigoryev, S.B. Kormer, O.L. Mikhailova, A.P. Tolochko, and V.D. Urlin, Pisma v ZhETF 16 (1972) 286

39) F.V. Grigoryev, S.B. Kormer, O.L. Mikhailova, A.P. Tolochko, and V.D. Urlin, ZhETF 69 (1975) 743

40) H. Hess and T. Kahlbaum, Proc. XIth AIRAPT Conf. on High Pressure Science and Technology, Kiev 1987, Kiev Naukova Dumka 1989, p. 238

41) T. Kahlbaum and H. Hess, Proc. XXVth EHPRG Meeting on High Pressure Geosciences and Material Synthesis, Potsdam 1987, Akademie-Verlag Berlin 1988, p. 90

42) A. Förster, W. Ebeling, and W. Richert, Proc. XXVth EHPRG Meeting on High Pressure Geosciences and Material Synthesis, Potsdam 1987, Akademie-Verlag Berlin 1988, p. 90

43) H.B. Radousky, W.J. Nellis, M. Ross, D.C. Hamilton, and A.C. Mitchell, Phys. Rev. Lett. 57 (1986) 2419

44) H. Hess and W. Ebeling, Proc. Workshop on Strongly Coupled Plasma Physics, Santa Cruz 1986, eds. F.J. Roger and H.E. DeWitt (Plenum Press, New York, 1987), p. 185

PRESSURE IONIZATION IN FLUID HYDROGEN

[1]G. CHABRIER, [2]D. SAUMON

[1]ENS Lyon, 46 Allée d' Italie, 69364 Lyon Cedex 07, France
[2]Department of Physics & Astronomy, University of Rochester, Rochester, NY 14627, USA

I. INTRODUCTION

During the last decade, the substantial developments in high-pressure technology and the wealth of information gathered on giant planets have generated a lot of interest for the physics of dense hydrogen. The recent discovery of very promising brown dwarf candidates[1] will allow comparison of observations with theoretical models. Models of these cool, compact objects require a detailed knowledge of the effects of the strongly non-ideal behavior o f dense hydrogen. Not mentioning the fact that the problem of partial ionization, in particular, is of special importance to stellar astrophysics. On the other hand since Wigner and Huntington[2] suggested the possibility of metallizing hydrogen at high pressure, there has been considerable interest in the insulator metal phase transition of hydrogen. Most of the extensive work has focussed on the zero-temperature isoth erm where it is now widely recognized that this transition may be accomplished by dissociation of the molecular phase into a monoatomic metal or by closure of the conduction gap which leads to a conducting molecular state[3]. Estimated transition pressures are predicted in the 1–5 Mbar range, approached by recent compression experiments[4]. In the fluid phase, theoretical calculations have suggested that the transition occurs via a first order phase transition, the so-called Plasma Phase Transition (PPT), which ends on a critical point around 0.23 Mbar and between 16000K and 19000K[5,6]. At $T \simeq 6000K$, shock-wave experiments have confirmed the stability of the molecular phase up to 0.8 Mbar[7].

In view of the substantial improvements in the statistical physics of dense fluids and plasmas over the past few years, we computed a new, detailed equation of state (EOS) for fluid hydrogen at high density, which covers the domain of partial ionization and gives new, extensive results for the PPT[8]. The model free energy is discussed in section II and the results for pressure dissociation, pressure ionization and the related PPT, are presented in section III. In section IV we consider briefly the application of this new EOS to static models of low-mass brown dwarfs. This new EOS can be applied formally to pure helium and hydrogen-helium mixtures.

II. DESCRIPTION OF THE MODEL FREE ENERGY

In our free energy model we adopt a "chemical picture", in the sense that we assume the existence of independent, bound configurations such as H atoms and H_2 molecules, interacting with pair potentials. At densities corresponding to pressure ionization, such a scheme is erroneous as the concept of individual pair potential fails[9], requiring the use of quantum-statistical many-body theory, i.e. a "physical picture" where only fundamental particles (electrons and nuclei) exist. Although formally exact, the physical picture involves approximations in solving the quantum mechanical problem of three or more particles[10] and is based, in practice, on expansions which converge only at low density and high temperature[11]. The calculation of an EOS for practical applications is rendered nearly impossible by the formidable complexity of these theories. In view of these difficulties, the chemical picture remains a very powerful method.

Our EOS consists of a neutral and a fully ionized model, which represent, respectively, the low-density, low-temperature and the high-density and/or high temperature limits of a general model applied in the partial ionization zone. While the EOS of the fully ionized plasma is now well understood, considerable uncertainty remains in the dense atomic-molecular phase. This uncertainty is reflected in estimates of the PPT transition pressure. We therefore developed a very detailed treatment of the microphysics in this regime.

II. 1) Model for Neutral Hydrogen

At low density ($\rho < 1 \text{gcm}^{-3}$), low temperature ($T < 10^4 \text{K}$), hydrogen is adequately described as a neutral mixture of H atoms and H_2 molecules. The concentrations of ions H^- and H_2^+ are found to be negligible[12] ($< 10^{-3}$) and they are ignored in our EOS. We assume factorizability of the partition function in translational (F_{trans}), configurational (F_{conf}) and internal (F_{int}) contributions. The free energy reads:

$$F(N_H, N_{H_2}, V, T) = F_{\text{trans}} + F_{\text{conf}} + F_{\text{int}} + F_{\text{qm}} \qquad (1)$$

II. 1a) *The Configuration Term* F_{conf}

The configuration term represents the interactions between the different particles in their ground states. Computation of these interactions in the binary mixture requires the knowledge of three interaction potentials $\phi_{H_2 H_2}$, ϕ_{HH_2} and ϕ_{HH}. For $\phi_{H_2 H_2}(r)$ we use an effective potential derived from shock tube experiments[7] which implicitly includes many-body effects. Since no similar experimental data exist for $\phi_{HH}(r)$ and $\phi_{HH_2}(r)$, we have used ab initio potentials[13]. We treat the spin dependence of the H-H interaction by averaging the interaction potentials of the singlet and triplet states; the resulting $\phi_{HH}(r)$ has no bound states. The three potentials have been fitted by generalized Morse potentials. The long-range Van der Waals attractive part ($\sim 1/r^6$) is poorly represented by exponential functions but this has no consequence on the thermodynamics of the system at the temperatures of interest.

The configuration free energy F_{conf} is derived from these interaction potentials in the framework of the WCA fluid perturbation expansion[14]. In this theory, the interaction potential $\phi(r)$ is split into a repulsive reference potential $\phi^{\text{ref}}(r)$, and a weak, attractive, perturbation potential $\phi^{\text{pert}}(r)$. We approximate the free energy of the reference system by that of a hard sphere fluid, which is known analytically[15], whereas the contribution of the perturbation potential is given by the first term of the free energy expansion (High Temperature Approximation):

$$F_{\text{conf}}(N,V,T) = F_{\text{HS}}(N_1,N_2,\sigma_1,\sigma_2,V,T) + \frac{1}{2V}\sum_{\alpha,\beta=1}^{2} N_\alpha N_\beta \int \phi_{\alpha\beta}^{\text{pert}}(r) g_{\alpha\beta}^{\text{HS}}(r) d\vec{r} \quad (2)$$

Here the $g_{\alpha\beta}(r)$ are the hard sphere pair correlation functions[16] and σ_1 and $\sigma_2 (1 \equiv H; 2 \equiv H_2)$ are the *density- and temperature- dependent hard sphere diameters* determined thermodynamically by the WCA criterion[14]. The failure of the WCA expansion scheme at high density has been corrected by using a novel potential separation which preserves the additivity of the reference system, i.e. $\sigma_{12} = 1/2(\sigma_{11} + \sigma_{22})$, as required by the hard sphere EOS, through fulfilling the WCA criterion [17]. Comparison of the excess internal energy and pressure derived from this expansion scheme with Monte Carlo (MC) simulations for the density and temperature range of interest agree within 3%. This assesses the validity of the configuration energy for the H-H_2 mixture for a given set of potentials.

II. 1b) *The Internal Free Energy* F_{int}

The effect of near-neighbor interactions on the internal structure of bound species is essential to a correct description of pressure dissociation and ionization. We have used a new approach based on an occupation probability formalism[18]. In this formalism, the internal free energy reads:

$$F_{\text{int}} = -kT \ln \sum_{\alpha=1}^{2} N_\alpha \sum_i w_{\alpha i} g_{\alpha i} e^{-\epsilon_{\alpha i}/kT} \quad (3)$$

where i runs over all internal states of species α and $w_{\alpha i}, g_{\alpha i}$ and $\epsilon_{\alpha i}$ are respectively the occupation probability, the multiplicity and the unperturbed energy of state i. The probability that a given state is destroyed by interaction with neighboring particles is $1 - w_{\alpha i}$ ($0 \le w_{\alpha i} \le 1$). The *density dependent* $w_{\alpha i}$ are computed from the configuration term F_{conf} in the free energy. This ensures consistency of both the interactions and their effects on the internal partition function. It also provides a smooth cutoff of the internal partition function, and therefore a plausible pressure dissociation/ionization effect in the H/H_2 mixture. Moreover the present method uses unperturbed energy eigenvalues and does not invoke hypothetical energy level shifts of doubtful validity. As a matter of fact, such shifts have been shown to be too small to be significant[19]. Calculations based on this approach are in excellent agreement with emissivity measurements of hydrogen plasmas[20].

In practice, however, one must resort to a linearization of F_{conf} to compute the occupation probability (see Ref. 18 for details). In addition our occupation probabilities include neutral particle interactions only (excluded volume). The effect of the charged species, mainly Stark ionization, requires a knowledge of the plasma microfield distribution which is too complex to be tractable in the context of our calculation.

In our treatment of the H_2 partition function, we have included all rotational and vibrational levels of each bound state of the molecule[21].

The term F_{qm} in equation (1) is the quantum correction to the free energy at high density, calculated for the ground states to the first non-vanishing order in the Wigner-Kirkwood $\hbar^2$ expansion.

II. 2. Model for Fully Ionized Hydrogen

For $kT \gtrsim 1Ry$ or $\rho \gtrsim 2gcm^{-3}$ (corresponding to $r_s \sim 1$ where r_s is the mean interionic spacing in units of the Bohr radius), the fluid is a fully ionized electron-proton plasma.

II. 2a) *The high-density regime* ($\rho > 2gcm^{-3}$)

At these densities the mean electron-ion potential energy $E_{ie} = e^2/r_s a_0$ ($a_0 \equiv$ Bohr radius) is smaller than the electron Fermi energy E_F so that the ion-electron interaction is sufficiently weak for the plasma to be described as the superposition of a screened ionic fluid and a rigid electron background[22]. Under these conditions the free energy of the plasma can be written in the form:

$$F = F_{id} - NkT \ln \int e^{-\beta U_{eff}} d\vec{r} + F_{xc} + F_{qm} \qquad (4)$$

where F_{id} denotes the ionic and electronic perfect gas contributions. The term F_{xc} is the exchange and correlation free energy of the electron gas at finite temperature, evaluated with an accurate fit[23]. The second term on the r.h.s. of equation (4) is the free energy of the screened ionic fluid, calculated in the framework of the hypernetted-chain (HNC) theory for the *temperature- and density- dependent screened Coulomb potential* $U_{eff}(k, V, T) = 4\pi e^2/k^2 \epsilon(k, V, T)$[24]. The dielectric function ϵ is the finite temperature Lindhard function corrected with a local field correction for the short range interactions between electrons[25]. This model free energy shows excellent agreement with existing Monte Carlo calculations at finite and zero temperature[26] and with more detailed models using non-adiabatic theories[27] in the domain of low degeneracy ($\theta > 1$) where the finite-temperature effects of the electrons are expected to be important[28].

II. 2b) *The Low-Density Regime* ($\rho < 2gcm^{-3}$)

As the density of the electron gas decreases, the ion-electron interaction can no longer be treated within the linear response approximation. At very low densities ($\log \rho \lesssim -3$), the electrons behave almost classically and the plasma may be treated as a pseudo-classical two-component plasma (TCP) in which the protons and electrons interact with pseudo- potentials[29]. We calculated the thermodynamics of this TCP in the framework of the HNC theory.

At intermediate densities the free energy is interpolated smoothly between the high density- and low-density models described above. The quantum correction F_{qm} for the ions is again calculated to leading order in $\hbar^2$, using a Wigner-Kirkwood expansion for the screened potential U^{eff}.

II. 3) Model for the Partial Ionization Zone

The two models described above can be combined to provide a description of partial ionization. In this general model, we treat the interaction between charge and neutral particles through a *polarization potential*[11]. This potential is approximated by a hard core interaction inside the atomic or molecular radius and by a screened potential outside the core. The hard core radii are chosen to be the Bohr radius for H and an equivalent radius for H_2. This hard core contribution effectively reduces the volume available for the ionic and electronic ideal terms by a factor $(1 - \eta)$ where η is the hard core packing fraction of H and H_2. The second contribution introduces an additional polarization term F_{pol} to the free energy, given by:

$$F_{pol} = 4kT \frac{N_{e^-}}{V} \sum_{\alpha=1}^{2} N_\alpha B_{\alpha i} \tag{5}$$

The $B_{\alpha i}$ denote the virial coefficients of particles of species α calculated from the screened potential U^{eff} (section II. 2a).

The general model free energy for the hydrogen fluid finally reads:

$$\begin{aligned} F(V, T, N_{H_2}, N_H, N_{H^+}, N_{e^-}) = & F_{id}(V, T, N_{H_2}, N_H) + F_{id}((1-\eta)V, T, N_{H^+}, N_{e^-}) \\ & + F_N^{ex}(V, T, N_{H_2}, N_H) + F_I^{ex}(V, T, N_{H^+}, N_{e^-}) \\ & + F_{pol}(V, T, N_{H_2}, N_H, N_{H^+}, N_{e^-}) \end{aligned} \tag{6}$$

where the subscript "*id*" denotes the ideal contribution whereas F_N^{ex} and F_I^{ex} stand for the non-ideal contributions of the neutral model and the fully ionized model developed in section II1 and II2 respectively. Having imposed the electroneutrality condition, we minimize the free energy (6) at fixed total density and temperature to obtain the chemical equilibrium of the four component mixture (H_2, H, H^+, e^-).

III. RESULTS AND DISCUSSION

The only high-temperature, high density data available for hydrogen comes from shock tube experiments[30]. Single- and double-shock points for deuterium are shown in Fig. 1 along with theoretical Hugoniots derived from our model after including modifications appropriate for deuterium. In view of the experimental uncertainty, the agreement is excellent. Points along the Hugoniot curves are given in Table 1 where we note that molecular dissociation is substantial at high density. At low density, relative abundances and pressure are in good agreement with activity expansion calculations carried up to 0.03gcm^{-3} [31].

We calculated the limit of stability of the model free energy (6) as a function of the density along several isotherms and found a first order phase transition in the regime of pressure ionization. Characteristics of the transition are given in Table 2*. Since temperature ionization does *not* occur via a phase transition, the phase transition line must end in a critical point. We find the parameters of this point to be $P_c = 0.61 \text{Mbar}, T_c = 15300 \text{K}, \rho_c = 0.35 \text{gcm}^{-3}$. This transition line is shown in Fig. 2 where it is compared with existing experimental results and

---

* The small differences with Table 1 of Ref. 8 originate in the correction of an error in the former calculation.

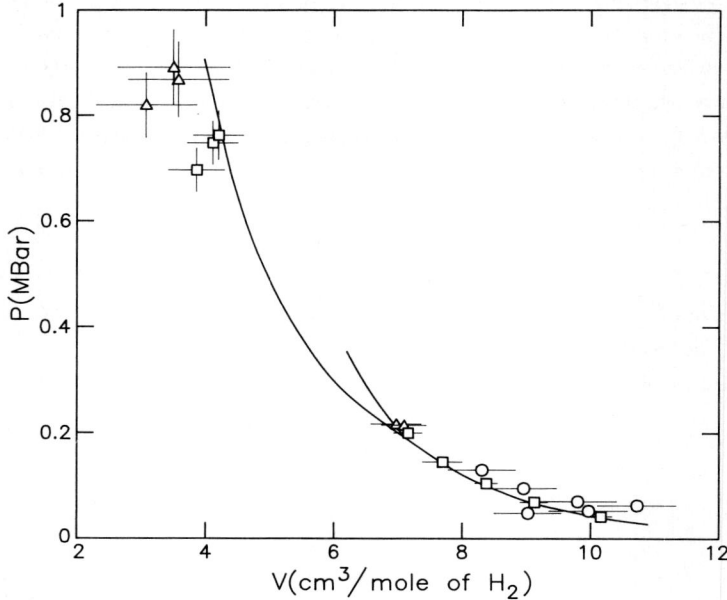

Figure 1: Principal and reflected Hugoniot of fluid deuterium. Experimental data points include the work of Van Thiel *et. al.* ($\triangle$; Ref. 30c.), Dick and Kerley (o; Ref. 30b.), and Nellis *et. al* ($\square$; Ref. 30a).

T (K)	V (cm^3/mole)	P (MBar)	χ_H (%)
20.23	23.600	1.57×10^{-3}	0
978	9.870	4.75×10^{-2}	0
1674	8.856	7.83×10^{-2}	1.2×10^{-4}
2881	7.836	0.132	0.08
4280	7.100	0.198	1.41
4781	6.895	0.222	2.50
6500	6.393	0.307	8.14
7500	6.218	0.354	12.0
Double shock, Reflected from 7.10 cm^3/mole			
5000	5.386	0.394	3.06
6600	4.222	0.773	8.38

Table 1: Points along the Hugoniots of Fig. 1, computed from our model. The last column gives the concentration of H atoms, $\lambda_H = N_H/(N_H + N_{H_2})$.

other theoretical estimates. We note that the slope $\partial P_c/\partial T$ is negative, consequence of the cumulative effect of pressure ionization *and* temperature ionization as temperature increases. The range of predicted transition pressures is of about one order of magnitude, as can be seen in Fig. 2. This can be understood in the light of the rather crude approximations used in previous efforts. The transition line of Marley and Hubbard[32] (MH) is obtained by imposing the phase transition conditions ($T_1 = T_2$, $P_1 = P_2, \mu_1 = \mu_2$) between a fully ionized plasma and a molecular fluid. Hydrogen atoms are excluded from their treatment, and plasma and molecules are not allowed to coexist (except along the coexistence curve). By performing a similar calculation with the two models described in II. 1 and II. 2, we obtain a somewhat lower transition pressure, but still in good agreement with MH considering the remaining differences in our respective models.

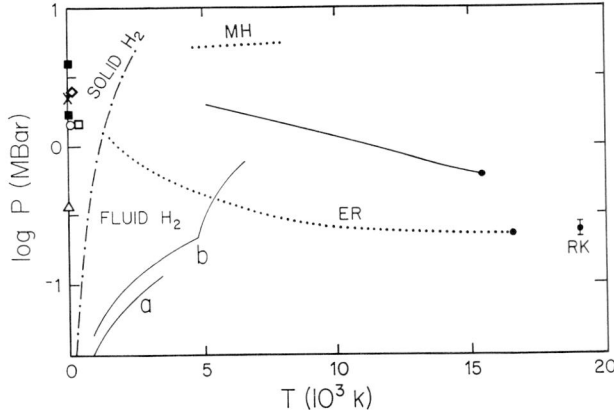

Figure 2: P-T diagram for hydrogen in the pressure-ionization regime. Heavy solid line: the PPT and the critical point. Other theoretical estimates for the PPT and the critical point are labeled MH (Ref. 32), RK (Ref. 5) and ER (Ref. 6). The zero temperature calculations of Ref 3a (X) and 3b (□) for the phase transition are indicated. Dash-dotted curve: theoretical melting curve of H_2 from Ref 7. The curves labeled a and b are respectively experimental single- and double-shock Hugoniots of H_2 and D_2 (Refs 30a and 7). Open symbols indicate the highest pressures reached in static compression experiments; □ (Ref. 4a), ○ (Ref. 4b), △ (Ref. 4c), and ◊ (Ref. 4d).

On the other hand, Robnik and Kundt[5] (RK) neglect molecules altogether but allow for a mixture of atoms, protons and electrons. Both Ebeling and Richert[6] (ER) and RK model the neutral interactions with unrealistically harsh hard sphere potentials (with *fixed* hard sphere diameters). Consequently, the atomic phase becomes unstable at lower pressures, probably underestimating the transition pressure, which depends sensitively on the choice of hard sphere diameters. Hard sphere potentials also induce very sudden ionization, an important qualitative

T (K)	P (MBar)	ρ^I gcm^{-3}	ρ^{II}	χ^I	χ^{II}	ΔS 10^7erg K^{-1}g^{-1}
5010	2.14	0.75	0.92	1.4×10^{-3}	0.48	5.07
6020	1.95	0.70	0.88	2.1×10^{-3}	0.50	4.87
7240	1.62	0.64	0.80	3.0×10^{-3}	0.50	4.49
8710	1.39	0.58	0.74	5.1×10^{-3}	0.51	4.19
10470	1.13	0.51	0.65	8.8×10^{-3}	0.52	3.83
12590	0.895	0.43	0.55	0.020	0.50	3.47
15140	0.631	0.35	0.38	0.17	0.33	1.17
15310	0.614	0.35	0.35	0.18	0.18	0

Table 2: The characteristics of the plasma phase transition. For each temperature T, we give the pressure P, the density and the ionization fraction of each phase, (ρ^I, ρ^{II}) and (χ^I, χ^{II}), repectively, and the change in entropy, $\Delta S = S^{II} - S^I$.

difference with our results (see below). Both works ignore the internal structure of the bound species as well as the charged-neutral coupling. We note that the ER coexistence curve crosses the experimental double-shock Hugoniot of deuterium (Fig. 2). It is generally agreed that there is no evidence for the PPT in the available data.

Figure 3 shows the concentration of molecules and charged particles as a function of density for a few isotherms. We draw the following conclusions:

(i) The system undergoes a first-order phase transition from a neutral phase ($\chi_{e^-} \leq 10^{-2}$ for $T < T_c$) to a partially ionized phase ($\chi_{e^-} \leq 0.5$) as ρ increases. (ii) At the transition pressure, the degree of ionization increases discontinuously whereas the concentration of molecules drops drastically, indicating that *molecular dissociation and pressure ionization occur at almost the same density*. (iii) Above the critical density, the system reaches complete ionization *very gradually*. This points out the qualitative difference found when treating pressure ionization with realistic, albeit flawed, potentials and with pure hard sphere interactions. (iv) Molecules are the dominant neutral species at high density. Even though our model for the neutral species is highly questionable above the transition density, we believe these qualitative features to be physically realistic.

A critical aspect of our model, when applied to pressure ionization, is that it is based on pair interaction potentials which are virtually unknown and even lose their physical meaning at these high densities. In addition the linearization inherent to the application of the occupation probability formalism becomes doubtful at these densities. The introduction of Stark ionization by the microfield distribution of the plasma should favor lower transition pressures. The apparition of band structure in the neutral fluid was also ignored. We have verified by explicit *ex post facto* calculations that the band structure in the H_2 molecule has little effect on the PPT:

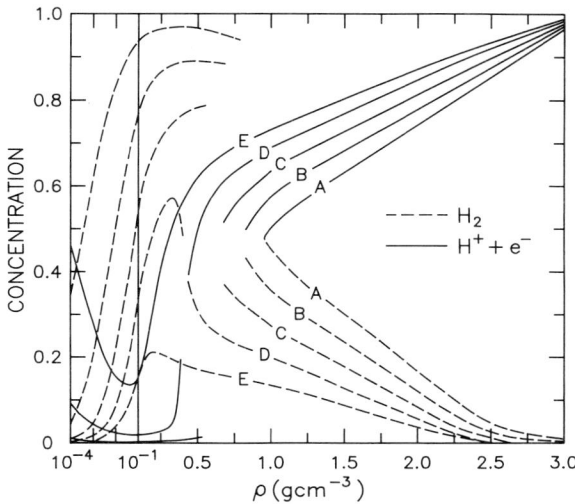

Figure 3: Concentration of H_2 and of charged particles ($H^+ + e^-$). Temperatures are: A, 5010K; B, 7240K; C, 10470K; D, 15140K; and E, 21880K. The left panel shows the low-density behavior on a log-density scale.

using the density dependent gap calculated at zero temperature[3a], we estimated the fraction of thermally excited electrons to vary from less than 10% at 8000K to 2% at the critical point. However, we cannot rule out the possibility that pressure ionization occurs smoothly by closure of the conduction gap.

IV. EFFECT OF THE PPT ON THE THERMAL STRUCTURE OF LOW-MASS BROWN DWARFS

The PPT predicted by our model occurs at densities and temperatures characteristic of giant planets and low-mass brown dwarfs. It is of great interest to estimate its effect on the structure of these objects. Real stars contain other elements besides hydrogen but we expect a pure hydrogen model to illustrate the effects to be expected from more realistic compositions. To keep this calculation as simple as possible, we simply computed adiabats in our EOS table. The approximation of adiabatic structure is justified for low mass objects, as they are expected to be fully convective. Since they are also degenerate, their mechanical structure is not affected by the PPT. It is a simple matter to estimate the effect of the PPT on the thermal structure of low mass objects by looking at the surface to core temperature relation. This is done by comparing the core temperatures of the true adiabat (constant entropy throughout the object, i.e. no PPT) and a discontinuous adiabat with the same surface entropy but with the central entropy raised by the corresponding ΔS of the PPT.

We present the result for two adiabats, one with the same specific surface entropy as Jupiter (S/k=7.56 per proton) and one characteristic of a slightly subcritical brown dwarf (S/k=10.32 per proton). We find that the PPT *raises* the central temperature by 33% and 16% respectively, for a given surface boundary condition. Using a very elementary cooling theory, it can be shown that the age-luminosity relation will be affected by the same amounts. That is, for a given surface boundary condition, models with a PPT would be 33% and 16% older, respectively.

V. CONCLUSION

We have developed an improved EOS, based on realistic interparticle potentials and a self-consistent treatment of the internal level, for fluid hydrogen at high density. In the partial ionization zone we generated a complete model of four interacting neutral and charged species. This model *predicts* a first order Plasma Phase Transition for which we give the characteristics and the critical point. This work does not solve the problem of pressure ionization but we believe the very detailed treatment of the microphysics included in this EOS to represent a significant improvement over existing models. By applying this model to adiabats characteristic of low-mass brown dwarfs and giant planets, we have shown that the PPT can have important consequences on the thermal structure, and thus the evolution, of these objects.

ACKNOWLEDGEMENTS

We wish to thank H. M. Van Horn for his continuous interest in this work. This work was supported in part by NSF Grants No. AST-87-06711 and No. PHY-88-08146 and by NASA Grant No. NAGW-1476 through the University of Rochester. D. S. acknowledges financial support from the Link Foundation.

REFERENCES

1. W. J. Forrest, these proceedings.

2. E. Wigner and H. B. Huntington, J. Chem. Phys. **3**, 764 (1935).

3a.) C. Friedli and N. W. Ashcroft, Phys. Rev. **B16**, 662 (1977); b.) B. I. Min, J. F. Jansen and A. J. Freeman, Phys. Rev. **B33**, 6383 (1988).

4a.) H. K. Mao, P. M. Bell and R. J. Hemley, Phys. Rev. Lett. **55**, 99 (1985); b.) R. J. Hemley and H. K. Mao, ibid **61**, 857 (1988); c.) J. Van Straaten and J. F. Silvera, Phys. Rev. **B37**, 1989 (1988); d.) H. K. Mao and R. J. Hemley, Science **244**, 1462 (1989).

5. M. Robnik and W. Kundt, Astron. Astrophys. **120**, 227 (1983).

6. W. Ebeling and W. Richert, Phys. Lett. **108A**, 80 (1985).

7. M. Ross, F. H. Ree, and D. A. Young, J. Chem. Phys. **79**, 1487 (1983).

8. D. Saumon and G. Chabrier, Phys. Rev. Lett. **62**, 2397 (1989).

9. S. Chakravarty, J. H. Rose, D. Wook, and N. W. Ashcroft, Phys. Rev. **B24**, 1624 (1981).

10. F. Rogers, Phys. Rev. **A29**, 868 (1984) and references therein.

11. W. D. Kraeft, D. Kremp, W. Ebeling and G. Röpke, "Quantum Statistics of Charged Particle Systems", Plenum, New York (1986).

12. D. Mihalas, private communication.

13a.) W. Kolos and L. Wolniewicz, J. Chem. Phys. **43**, 2429 (1965); b.) R. N. Porter and M. Karplus, J. Chem. Phys. **40**, 1105 (1964).

14. J. D. Weeks, D. Chandler, and H. C. Andersen, J. Chem. Phys. **54**, 5237 (1971).

15. G. A. Mansoori, N. F. Carnahan, K. E. Starling, and T. W. Leland, Jr., J. Chem. Phys. **54**, 1523 (1971).

16. E. W. Grűndke and D. Henderson, Mol. Phys. **24**, 269 (1972).

17. D. Saumon, G. Chabrier and J. J. Weis, J. Chem. Phys. **90**, 7395 (1989).

18. D. G. Hummer and D. Mihalas, Astrophys. J. **331**, 794 (1988).

19. W. L. Wiese, D. E. Kelleher, D. R. Paquette, Phys. Rev. **A6**, 1132 (1972).

20. W. Dăppen, L. Anderson, D. Mihalas, Ap.J **319**, 195 (1987).

21. K. P. Huber and G. Herzberg, "Molecular Spectra and Molecular Structure", Van Nostrand Reinhold Company, New York, (1979).

22. N. W. Ashcroft and D. Stroud in "Solid State Physics", H. Ehrenreich, F. Seitz, D. Turnbull eds, Academic, New York (1978).

23. S. Ichimaru, H. Iyetomi and S. Tanaka, Phys. Rep. **149**, 93 (1987).

24. H. Iyetomi and S. Ichimaru, Phys. Rev. **A34**, 433 (1986).

25. K. Utsumi and S. Ichimaru, Phys. Rev. **A26**, 603 (1982).

26a.) W. B. Hubbard and W. Slattery, Astrophys. J. **168**, 131 (1971); b.) H. Totsuji and K. Tokami, Phys. Rev. **A30**, 3175 (1984).

27. S. Tanaka, S. Mitake, X. Z. Yan, S. Ichimaru, Phys. Rev. **A32**, 1779 (1985).

28. G. Chabrier, Phys. Lett. A134, 275 (1989).

29a.) H. Minoo, M. H. Gombert, and C. Deutsch, Phys. Rev. **A23**, 924 (1981); b.) J. P. Hansen and I. R. McDonald, Phys. Rev. **A23**, 2041 (1981).

30a.) W. J. Nellis, A. C. Mitchell, M. Van Thiel, G. J. Devine and R. J. Trainor, J. Chem. Phys. **79**, 1480 (1983); b.) R. D. Dick a nd G. I. Kerley, J. Chem. Phys. **73**, 5264 (1970); c.) M. Van Thiel, L. B. Hord, W. H. Gust, A. C. Mitchell, M. D'addario, K. Bautwell, E. Wilbarger, and B. Barret, Phys. Earth Planet. Inter. **9**, 57 (1974).

31. F. J. Rogers, private communication.

32. M. S. Marley and W. B. Hubbard, Icarus **73**, 536 (1988).

THERMODYNAMICS AND TRANSPORT IN DENSE PARTIALLY IONIZED PLASMAS

Wolf-Dietrich KRAEFT[a], Manfred Klaus KILIMANN[b], and Dietrich KREMP[b]

Ernst-Moritz-Arndt-Universität, Greifswald, 2200 DDR (a),

Wilhelm-Pieck-Universität, Rostock, 2500 DDR (b)

It is shown on the basis of Green functions technique that macroscopic properties of partially ionized plasmas are essentially determined by the Coulomb interaction and by the symmetry postulate. Especially the many particle effects such as screening, self energy and modifications of the two particle spectrum lead to drastic changes of the plasma features as compared with those in dilute systems.

1. INTRODUCTION

Dense plasmas are of interest both from the technical and the scientific point of view. The Coulomb interaction and the symmetry postulate lead to a number of interesting features especially in the case of partially ionized plasmas.

Effects which should be mentioned are the dynamical screening which is of interest, e.g., for the study of dispersion relations and the propagation of waves[1]. Screening is a typical many particle effect; as a result of many particle effects, the single particle energy of charge carriers shows an essential deviation from the ideal kinetic energy. The resulting self energy represents a shift and a broadening of the single particle spectrum; especially the latter is connected with the finite life time of single particle states.

The density dependence of the energy and the finite life time of the two particle states has interesting consequences. Here we have a strong interplay between screening, self energy and the Pauli principle; see chapter 2. The consequences of the deviation of the two particle bound state behaviour from that in dilute plasmas are nonideality effects which lead to modifications of the chemical equilibrium, i.e., of the mass action law , and of the recombination and ionization processes under nonequlibrium conditions.

Especially the resulting nonlinear behaviour of physical quantities leads to a variety of physical phenomena, such as discussed in chapters 4-6. Generally one can mention that the simple laws of thermodynamics and transport have to be modified. In particular we will see that the change of the single particle behaviour and that of two particle states coincide with certain drastic changes of macroscopic properties which are mainly connected with drastic changes of the fraction of bound states in connection with changes of the number density.

We mention still that the existence of bound states is, for a given system of charge carriers, restricted to a " corner of correlation " in the density-temperature plane[2], see figure 1. Here we remark only that we have, outside of this region, practically a nondegenerate highly ionized ideal system (high temperatures), or an ideal degenerate completely ionized system (high densities), respectively. A detailed discussion of the different regions is given, e.g., in [2] .

2. MANY PARTICLE EFFECTS. GREEN FUNCTION TECHNIQUE

2.1 General remarks. Basic equations

The Green functions technique is known to be a very powerful tool in order to describe the macroscopic properties of strongly coupled plasmas. Here we outline only main ideas and refer the reader to the literature; see , e.g., [2-5].

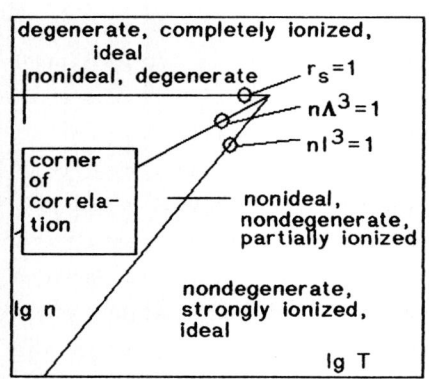

Fig.2.1 Density (n)- temperature (T)-plane. Λ- De Broglie-, I- Landau-length

A key quantity is the single particle Green function, which is defined by equation (2.1) as certain mean value of a time ordered product of creation and annihilation operators. From the quantity defined thus we get the polarization function Π or the two particle Green function G_2 with the help of functional derivatives with respect to the effective external potential or simply the external potential according to eq.(2.2) or (2.3), respectively. The screened potential V_s is connected with Π and the bare potential via equation (2.4). The single particle Green function itself must be determined from the Dyson equation (2.5), in which the many particle effects are condensed in the self energy Σ. We write

$$G_1(11') = (1/i\hbar)\langle T\{\psi(1)\psi^+(1')\}\rangle . \qquad (2.1)$$

Here T is the time ordering symbol, and $1 \equiv \{r_1,t_1\}$ and

$$\Pi = - \delta G_1 / \delta U^{eff} , \qquad (2.2)$$

where U^{eff} is the effective external potential, which includes an external one and the Hartree potential (Hartree self energy) as well. Further we have

$$[G_2(121'2') - G_1(11')G_1(22')] = \delta G_1/\delta U \qquad (2.3)$$

with U being the external potential. Dropping all arguments, we may write

$$V^s = V/(1 - \Pi V^s) . \qquad (2.4)$$

The Dyson equation reads
$$G_1 = G_1^o + G_1^o \Sigma G_1, \qquad (2.5)$$
where G_1^o is the single particle Green function of free particles.

Of course, one has to do certain approximations in order to get explicit results. In the following sections we will give expressions for the self energy in the random phase approximation and discuss the behaviour in the nondegenerate limit. Furthermore one uses Padè-approximations, which are valid in a wide range. In practical applications it is often useful to use, instead of momentum dependent quantities, rigid shifts of the single particle energies, which correspond to mean values, which are equal to the corresponding chemical potentials [2,6].

2.2 Self energy of charged particles

In the so-called V^s-approximation[2] the self energy reads
$$\Sigma_a(12) = iV_{aa}^s(12)G_a(12), \qquad (2.6)$$
which is usually split up into a Hartree-Fock contribution
$$\Sigma_a^{HF}(12) = iV_{aa}(12)G_a(12) \qquad (2.7)$$
and a correlation part (Montroll-Ward), which reads
$$\Sigma_a^{corr}(12) = i(V_{aa}^s(12) - V_{aa}(12))G_a(12). \qquad (2.8)$$
Equation (2.6) reads in the momentum Matsubara frequency representation
$$\Sigma_a^{HF}(\mathbf{p}) = - \int d\mathbf{k} \, e_a^2 \, f_a(\mathbf{p}-\mathbf{k})/[(2\pi)^3 \, \varepsilon_0 k^2] \qquad (2.9)$$
and is a real quantity. In the nondegenerate limit we get
$$\Sigma_a^{HF}(\mathbf{p}) = -\{2e_a^2/[\varepsilon_0 (2s_a+1)]\} n_a \Lambda_a^2 \, {}_1F_1(1,3/2; -\hbar^2 p^2/(2m_a kT)), \qquad (2.10)$$
where ${}_1F_1$ is the confluent hypergeometric function. The evaluation of (2.8) is much more complicated and will be given below by numerical integration.

The mean value of Σ is sometimes a sufficient approximation (rigid shift approximation [6]) and is related to the interaction part of the chemical potential by
$$\langle \Sigma_a \rangle = \Delta_a = \mu_a^{int}.$$
Using (2.8) in the nondegenerate limit we obtain the simple expression (Debye approximation)
$$\Delta_a = -e_a^2 \varkappa/(8\pi\varepsilon_0). \qquad (2.11)$$
The inverse Debye radius $\varkappa = r_D^{-1}$ is given in the nondegenerate limit by
$$\varkappa^2 = 2ne^2/(\varepsilon_0 kT) \qquad (2.12)$$
and is in the general case connected with the fugacities via (electrons only)
$$\varkappa^2 = 2e^2 \exp(\beta\mu_e)/(\varepsilon_0 kT \Lambda^3). \qquad (2.13)$$
The fugacities or chemical potentials, respectively, are connected with the densities by the relation

$$\beta\mu_a = I_{1/2}^{-1}(\, n_a \Lambda_a^3 \,/(2s_a+1)), \tag{2.14}$$

where $\Lambda_a^2 = h^2/(2\pi m_a kT)$ and I^{-1} is the inverse Fermi integral.

For the numerical evaluation of (2.8) it is useful to take the following expression for the real part

$$\text{Re}\,\Sigma_a(k,\omega) = -\hbar P \int dp'd\omega'\, (2\pi)^{-4}\, 2V_{aa}(p)*\{1-f_a(p'+k)+n_B(\omega')\}$$

$$* \text{Im}\,\varepsilon^{-1}(p',\omega'+io)/[\hbar(\omega-\omega')-\hbar^2(p'+k)^2/2m_a]. \tag{2.15}$$

Equation (2.15) was evaluated for the dispersion $\hbar\omega = \hbar^2 k^2/2m_a$ in the nondegenerate limit using the RPA dielectric function [2]. The result is shown in fig. 2.2, see also [7]. The single particle energy has then the following shape

$$E = p^2/2m + \text{Re}\,\Sigma + \text{Im}\,\Sigma\,; \tag{2.16}$$

the damping may be omitted in simple cases, and the real part of Σ is included usually in the shape of a rigid shift (chemical potential) or by introducing an effective mass [2].

2.3 Two particle states

The investigation of the behaviour of two particle states is one of the most interesting tasks in dense plasma physics. Especially the shift of energy levels and the formation and destruction of bound states leads to a number of rather interesting features of the optical, thermodynamic and transport properties. The starting point is here the equation for the two particle Green function [2,6,8-11]. Equations of such type are usually referred to as Bethe Salpeter equations. The corresponding homogeneous equation is a generalized Schrödinger equation and plays the role of an effective wave equation. As compared with the usual Schrödinger equation of an isolated pair of charge carriers, essentially four additional effects are taken into account, namely (i) the Pauli blocking, (ii) the Hartree-Fock self energy, (iii) the effectice dynamical screening and (iv) the effective dynamical self energy. As discussed in [2,9], there is an extensive compensation between the different effects, e.g., between the Pauli blocking, which is due to symmetry of the elementary particles and leads to a restriction in

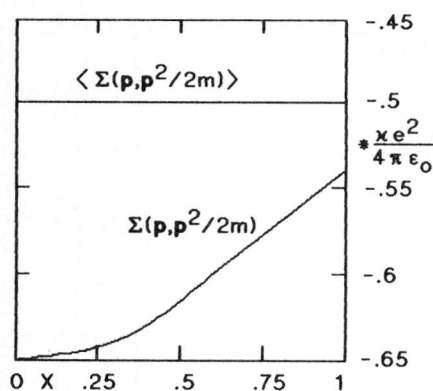

Fig. 2.2 Self energy after eq. (2.15). Mean value eq. (2.11). $\hbar\omega_p/kT=.64$
$X^2 = p^2/(2m\hbar\omega_p)$; ω_p - plasma frequency

the occupation of the phase space, and the Hartree Fock self energy. An equivalent compensation takes place at the level of screening.

In this paper, we give the numerical evaluation for two model situations. (i) We consider electrons and ions (protons) having Hartree Fock self energy only which are suject to Pauli blocking, and (ii) we take additionally a rigid Debye shift and the statically screened Debye potential. For the connection between the screening parameter and the chemical potential occuring in the Fermi function we use eq.(2.14).

Our starting point is an effective wave equation of the following shape

$$\{E(\mathbf{p}_1) + E(\mathbf{p}_2) - \hbar z\} \psi(\mathbf{p}_1\mathbf{p}_2) + [1 - f(E(\mathbf{p}_1)) - f(E(\mathbf{p}_2))] * (2\pi)^{-3}$$
$$* \int V^s(\mathbf{p}_2' - \mathbf{p}_2) \delta(\mathbf{p}_1' + \mathbf{p}_2' - \mathbf{p}_1 - \mathbf{p}_2) \psi(\mathbf{p}_1'\mathbf{p}_2') \, d\mathbf{p}_1' d\mathbf{p}_2' = 0. \qquad (2.17)$$

In the above mentioned case (i) we take instead of V^s simply the Coulomb potential, while the single particle energy is taken to be

$$E(\mathbf{p}) = \hbar^2 p^2/2m + \Sigma^{HF}(\mathbf{p}) \,, \qquad (2.18)$$

where the self energy Σ is taken according to (2.9) or (2.10). As a simplification we consider only the relative motion with respect to the center of mass. The result is shown in fig. 2.3 [12]. It can be seen, that there is a strong compensation between self energy and Pauli blocking acting for the bound state energy up to the "border" between degeneracy and nondegeneracy, i.e., near $\beta\mu = 0$. This compensation does not act for the two particle states of the continuum, so that there is a larger density dependence. At higher densities, the energies of bound states and continuum states come together asymptotically, however there is apparently no crossing.

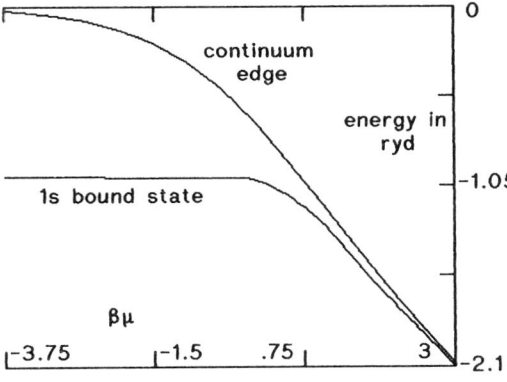

Fig. 2.3 Bound state energy and continuum state energy in Hartree-Fock approximation. 1s-state, hydrogen, kT=1 ryd

In the second example we take a rigid self energy shift according to (2.11), and the screened potential V^s is taken to be the statically screened Coulomb potential with $\varkappa$ to be connected with $\beta\mu$ according to (2.13). The Fermi functions were dropped. The result is given in fig. 2.4[12-14]. The behaviour is similar to that of fig. 2.3, however now the bound state energy and the energy of continuum states come close together already at lower densities. Again there is no crossover of the curves. In fig 2.5 both self energy contributions are taken into account, and as screened potential we have now only the difference between the Coulomb and the Debye potentials. Quantitatively, there is a difference between figs. 2.5 and 2.4, however the Debye quantities give the main contribution at low densities [12,13].

It is interesting to discuss the behaviour of the wave functions. An isolated two particle bound state is localized in the position space and thus relatively extended in momentum space. With increasing density, the influence of the surrounding medium becomes more and more essential, and so the states become less localized, and the wave function becomes sharply peaked in momentum space. This behaviour is in correspondence with that of continuum states, so that, above certain density, there is no essential difference between a bound state and the continuum state. The behaviour of a bound state wave function is shown in fig. 2.6 for very different densities [12].

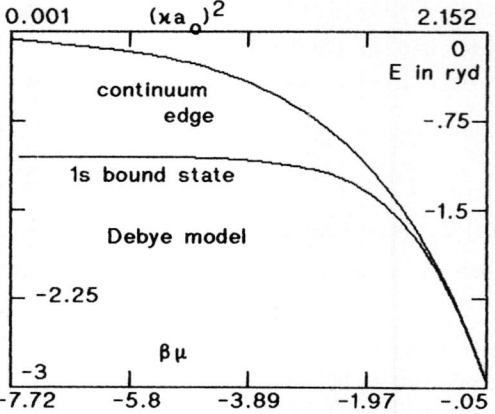

Fig. 2.4 Bound state energy and continuum for hydrogen. kT =1 ryd. a_0-Bohr radius

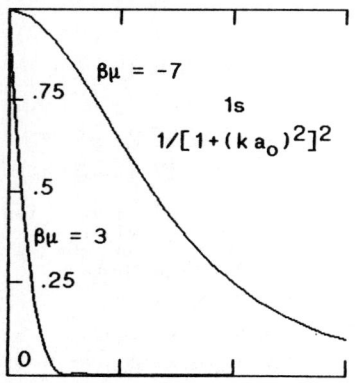

Fig. 2.6 Wave functions $\psi(k)$ corresponding to fig. 2.5 for different densities, maximum is taken to be unity. For $\beta\mu=-7$-unperturbed

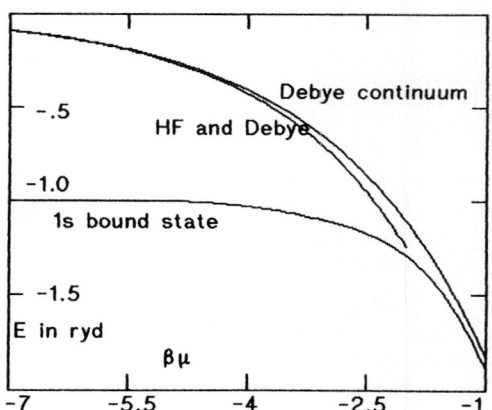

Fig.2.5 Bound state energy and continuum edge for hydrogen. kT= 1 ryd. Hartree-Fock and Debye self energy, Pauli blocking, Debye potential

In the following chapters we will briefly discuss the question how the self energies and the two particle properties influence the thermodynamic and the transport properties. Especially we will see that the self energy shifts and the effective ionization energies, i.e., the difference between the bound state energy and the continuum, give rise to essential deviations from the ideal behaviour.

3. THERMODYNAMIC EQUILIBRIUM

In connection with nonideality effects, the processes of ionization and recombination rather interesting ones. These questions are dealt with in Chapters 4-6. However, already under equilibrium conditions we have interesting questions. So we may discuss the mass action law and the question of stability in connection with phase transitions and critical points [15]. For a detailed discussion see[16].

An appropriate starting point for the discussion of thermodynamic problems is the formula for the density as a function of the chemical potential

$$n_a(\mu T) = \int f_a(\omega) A_a(p\omega) d\omega dp/(2\pi)^4 , \qquad (3.1)$$

where the spectral function A of the single particle Green function contains the interaction with the surrounding medium via the self energy Σ to any order[2]. Starting from (3.1) it is possible to derive a mass action law of the shape given in eq. (4.9)[17].

4. KINETIC EQUATIONS FOR NONIDEAL REACTING PLASMAS

Usually the nonequlibrium properties of many particle systems may be described by the Boltzmann kinetic equation. However, in the case of nonideal reacting plasmas, this equation must be generalized in many directions[18,19]. Using the powerful method of nonequilibrium Green functions[3], it is possible to take into account many particle effects (nonideality effects) such as self energy, screening, Pauli blocking and two particle energy shifts[20,21]. The result of such considerations is the following kinetic equation for the distribution function of the electrons

$$\{ \partial/\partial T + \partial E_a/\partial \mathbf{p}_a \, \partial/\partial \mathbf{R}_a - \partial E_a/\partial \mathbf{R}_a \partial/\partial \mathbf{p}_a \} f_a(\mathbf{p} \, R \, T) = \sum_b I_{ab} + \sum_{bc} I_{abc} . \qquad (4.1)$$

Here E_a is the quasiparticle energy

$$E_a = \mathbf{p}^2/2 + \Delta(n \, \mathbf{p}) \qquad (4.2)$$

in the "rigid shift" approximation. I_{ab} is the usual Boltzmann collision term given by

$$I_{ab} = (2\pi \hbar)^3/(V\hbar) \int d\mathbf{p}_b d\bar{\mathbf{p}}_a d\bar{\mathbf{p}}_b \, |\langle \mathbf{p}_a \mathbf{p}_b | T_{ab} | \bar{\mathbf{p}}_b \bar{\mathbf{p}}_a \rangle|^2 \, 2\pi \, \delta(E_{ab} - \bar{E}_{ab})$$

$$*\{f_a f_b (1 \pm \bar{f}_a)(1 \pm \bar{f}_b) - (1 \pm f_a)(1 \pm f_b) \bar{f}_a \bar{f}_b\} \qquad (4.3)$$

with $E_{ab} = E_a + E_b$.

I_{abc} is the three particle collision integral containing the reaction terms. With the atomic distribution function $F_n(P)$, n- internal quantum number, P- center of mass momentum, this contribution has the form

$$I_{abc} = (2\pi\hbar)^6/(V\hbar) \sum_c \sum_x \{\int dp_b dp_c dp \; |\langle P_a P_b P_c | T^{0x} | xp\rangle|^2 \; 2\pi\delta(E_o - E_x)$$

$$*(f_x \bar{N}_o - N_x \bar{f}_a \bar{f}_b \bar{f}_c) + (2\pi\hbar)^{-3} \sum_{n_{bc}} \int dp \; dP_{bc} |\langle p \; n_{bc} P_{bc} | T^{1x} | xp\rangle|^2$$

$$*2\pi\delta(E_1 - \bar{E}_x)(f_x \bar{N}_1 - N_x \bar{F}_{n_{bc}} \bar{f}_a)\}. \tag{4.4}$$

Here $|px\rangle$ are the asymptotic three particle states. The quantum number x gives the classification of the states with respect to the asymptotic initial states (channels) of three particle scattering. p denotes all the other observables. Further explanations are given in the following table.

x (channel)	$\|xp\rangle$	E_x	f_x	N_x
a + b + c	$\|p_a\rangle\|p_b\rangle\|p_c\rangle$	$E_a + E_b + E_c$	$f_a f_b f_c$	$(1 \pm f_a)(1 \pm f_b)(1 \pm f_c)$
a + (b+c)	$\|p_a\rangle\|n_{bc}P_{bc}\rangle$	$E_a + E_{n_{bc}}$	$f_a F_{n_{bc}}$	$(1 \pm f_a)(1 + F_{n_{bc}})$
b + (a+c)	$\|p_b\rangle\|n_{ac}P_{ac}\rangle$	$E_b + E_{n_{ac}}$	$f_b F_{n_{ac}}$	$(1 \pm f_b)(1 + F_{n_{ac}})$
c + (a+b)	$\|p_c\rangle\|n_{ab}P_{ab}\rangle$	$E_c + E_{n_{ab}}$	$f_c F_{n_{ab}}$	$(1 \pm f_c)(1 + F_{n_{ab}})$

The possibility of chemical reaction leads to the condition

$$\sum_{bc} \int dp_a [I_{abc}] = W_a \neq 0, \tag{4.5}$$

where W_a is a source function.

Let us now consider macrophysical consequences of the kinetic equation. We introduce the densities and the mean velocities by

$$n_a(R\,T) = \int dp_a/(2\pi\hbar)^3 f_a(p_a RT);$$

$$u(RT) = 1/n_a \int p_a/m_a \; f_a(p_a RT) * dp_a/(2\pi\hbar)^3.$$

Then we obtain from the kinetic equation the following reaction diffusion equation

$$\partial n_a/\partial T + \nabla_a n_a u_a = W_a(n_1 \ldots n_f) = \partial U/\partial n_a. \tag{4.6}$$

In the case of a three component system a, b, (a b) and neglecting exchange reactions, the source function has the simple shape

$$W_a = \sum_{c=a,b} (\alpha_c n_{ab} n_c - \beta_c n_a n_b n_c), \tag{4.7}$$

where the rate coefficients α and β must be determined from (4.5). Using the energy conservation it is easy to show that one gets from (4.5) the following relation between α and β [22]:

$$\beta = \alpha \Lambda_a^3 \Lambda_b^3 / \Lambda_{ab}^3 \exp[-\beta(E_n - \Delta_a - \Delta_b + \Delta_o)],$$

where E_n is the binding energy of an isolated two particle complex, and the Δ's are the relevant self energy shifts of single particles and pairs. Further we obtain for α for a system of electrons (e), ions (i) and atoms (A)

$$\alpha = \alpha_{id} \Lambda_e^3 \exp[-\beta(\Delta_e + \Delta_i - \Delta_A)]. \tag{4.8}$$

In the case of thermodynamic equilibrium, we have $W_a = 0$, and therefore

$$n_{ab}/n_a n_b = \Lambda_a^3 \Lambda_b^3 / \Lambda_{ab}^3 \exp[-\beta(E_n + \Delta_{ab} - \Delta_a - \Delta_b)]. \tag{4.9}$$

This is just the Saha equation for nonideal plasmas.

5. NONIDEALITY AND NONLINEAR IONIZATION KINETICS IN PLASMAS

We consider now a partially ionized plasma consisting of electrons, ions and atoms. Further we use the simple model of ambipolar diffusion. Then we obtain the following reaction diffusion equation

$$\partial n_e / \partial T + D \nabla_e^2 n_e = W_e. \tag{5.1}$$

Due to the nonideality effects, W_e is depedent on the density in a nonlinear way. Therefore we may expect nonlinear phenomena in dense plasmas such as [23] ionization fronts, dissipative structures and phase separation.

Obviously the behaviour and the properties of the solutions of (5.1) are essentially determined by the zeroes $c_m(nT)$ of $W_e(nT)$. Especially the c_m determine the behaviour of the solutions of the stationary equation (5.1) and thus of the stationary states of the homogeneous problem. One can see easily that $c_m = 0$ is one of the zeroes, and that further zeroes are given by the solutions of the Saha equation (4.9) or from Fig. 5.1, in which we have presented the solution of the equation

$$\mu_e + \mu_i = \mu_A .$$

A special property of the function $c_m(nT)$ is the following one. For $T < T_k$ and $n_1(T) < n_e < n_2(T)$ we obtain 3 zeroes $c_1 < c_2 < c_3$, where c_1 and c_3 represent stable stationary states, while c_2 corresponds to an unstable one.

Therefore we have a bistable behaviour; Fig. 5.2 represents the bifurcation diagram of Fig. 5.1, and n_1 and n_2 are the bifurcation points. At these points we observe the transition from the bistable to the monostable behaviour [24,25].

This behaviour has many interesting consequences. For example, in an infinite system withe the boundary conditions $c(-\infty) = c_1$, $\nabla c(-\infty) = 0$, $c(\infty) = c_3$, $\nabla c(\infty) = 0$, we obtain under the condition

$$\int_{c_3}^{c_2} W_a \, dc = \int_{c_2}^{c_1} W_a \, dc$$

the so-called kink-solutions which describe a phase separation between a phase with the degree of ionization $n_e/n = c_1$ and a phase with $n_e/n = c_3$ [24,25].

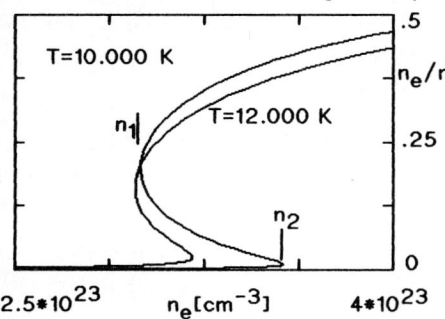

In finite systems, we obtain in the bistable region the evolution of an initial profile to a stationary front solution, which describes an ionization front[25]. Such an evolution is shown in Fig. 5.3.

Fig. 5.1 Degree of ionization in the bistability region. n_1, n_2 — bifurcation points

6. ELECTRICAL CONDUCTIVITY OF A NONIDEAL PLASMA

A very interesting property of partially ionized plasmas is the electrical conductivity[26]. This quantity is determined by the scattering processes included in equation (4.1). Here especially the many particle effects should be mentioned, which lead, e.g., to substantial changes in the ionization equilibrium, see eq. (4.9).

For simplification we neglect in eq. (4.1) the three body scattering of free particles and inelastic and reaction processes as well. Then we have in the homogeneous stationary case a kinetic equation, which accounts for charge-charge and charge-neutral scattering

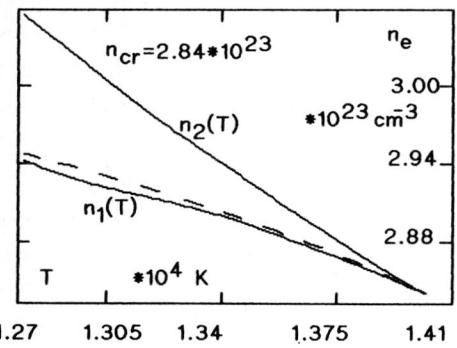

$$\partial E_a/\partial R_a \, \partial/\partial p_a \, f_a(pT) = I_{ee} + I_{ep} + I_{eH},$$

where I_{eH} is given by the elastic free-bound scattering contribution of I_{abc}. Together with eq. (4.9) the distribution function may be determined in the following manner:

(i) electron - electron, electron - ion and electron - atom scattering is taken into account;

Fig. 5.2 Region of bistability for H-plasma Padè-approximation for self energy shift, e-A interaction included. Bifurcation at n_1 and n_2

(ii) the phase shifts η_l for electron-electron and electron-ion scattering are determined from solutions of the radial Schrödinger equation with static Debye potential; (iii) for the determination of the phase shifts of the electron-atom scattering we use the close coupling equation in first order perturbation theory including exchange effects; (iv) we solve the kinetic equation for the stationary case using the well known expansion with respect to Sonine polynomials. Results for the electrical conductivity are given in Figs. 6.1 and 6.2 [27]. A characteristic feature of the figures is the minimum and the subsequent strong increase of the conductivity. This effect is produced by the increasing importance of the e-A scattering which is due to the formation of bound states with increasing density. The increase is due to the pressure ionization near the Mott density. As one may expect the influence of the formation of atoms on the conductivity is larger at lower temperatures. The individual properties are accounted for by the potentials.

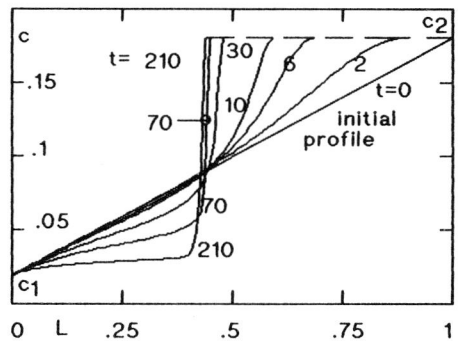

Fig 5.3 Evolution of an ionization front profile: degree of ionization versus length of the system. c_1, c_2 - stable zeroes of the source function W. T=13.000 K, $n=2.95*10^{23} m^{-3}$, timestep 10^{-6}.

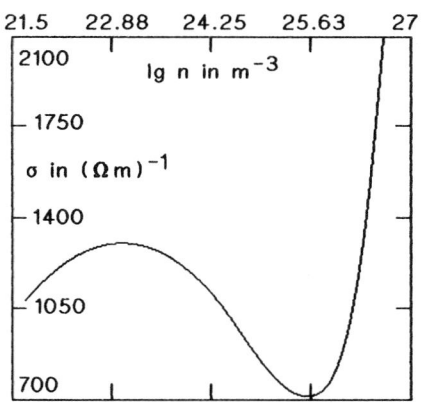

Fig. 6.1 Electrical concuctivity of a partially ionized Cs plasma at 4000 K as a function of the total electron density

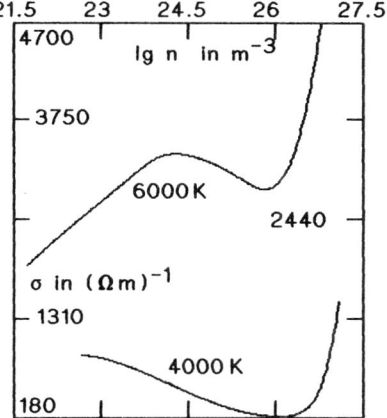

Fig. 6.2 Electrical conductivity of a partially ionized Na plasma as a function of the total electron density

The problem of the electrical conductivity of strongly coupled plasmas was also dealt with in [28].

REFERENCES

1) G. Kalman and K.I. Golden, Phys. Rev. A29 (1984) 844; G. Kalman and X.-Y. Gu, Phys. Rev. A36 (1987) 3999, and references quoted therein
2) W.D. Kraeft, D. Kremp, W. Ebeling and G. Röpke, Quantum Statistics of Char-

ged Particle Systems (Plenum, London and New York, 1986; Akademie-Verlag, Berlin,1986; Mir, Moscow, 1988)
3) L.P. Kadanoff and G. Baym, Quantum Statistical Mechanics (Benjamin, Memlo Park, 1962)
4) F.J. Rogers and H.E. DeWitt, Strongly Coupled Plasma Physics (Plenum, New York, 1987)
5) W. Ebeling, W.D. Kraeft and D. Kremp, Theory of Bound States and Ionization Equlibrium in Plasmas and Solids (Akademie-Verlag, Berlin, 1976)
6) R. Zimmermann, Many Particle Theory of Highly Excited Semiconductors (Teubner, Leipzig, 1988)
7) W.D. Kraeft et al. (submitted)
8) K. Kilimann, W.D. Kraeft and d. Kremp, Phys. Letters, 61A (1977) 393
9) R. Zimmermann et al., phys. stat. sol. (b)90 (1978) 175
10) K. Kilimann, Doctor thesis II (Rostock University, 1978)
11) G. Röpke, K. Kilimann, D. Kremp and W.D. Kraeft, Phys. Letters 68A (1978) 329. For level shifts of multiply charged ions see e.g., X.-Z. Yan and S. Ichimaru, Phys. Rev. A34 (1986) 2173; W. Ebeling and K. Kilimann, Z. Naturforsch. 44a (1989) 519
12) W.D. Kraeft, K. Kilimann and D. Kremp (submitted)
13) W.D. Kraeft, W. Ebeling and K. Kilimann (submitted)
14) F.J. Rogers, H.C. Graboske and D.J. Harwood, Phys. Rev. A1 (1970) 1577; F.J. Rogers and H.E. DeWitt, Phys. Rev. A8 (1973) 1061
15) W. Ebeling, W.D. Kraeft, D. Kremp and G. Röpke, Physica 140A (1986) 160
16) W.D. Kraeft, W. Ebeling, D. Kremp and G. Röpke, Ann. Physik 45 (1988) 429
17) D. Kremp, W.D. Kraeft and A.J.D. Lambert, Physica 127A (1984) 72
18) Yu.L. Klimontovich and D. Kremp, Physica 109A (1981) 517 ; Yu.L. Klimontovich, D. Kremp and W.D. Kraeft, Theory of Chemically Reacting Gases and Plasmas, Adv. Chem. Phys. 58 (1987) 175
19) J.A. McLennan, J. Stat. Phys. 28 (1982) 521
20) D. Kremp, M. Schlanges and T. Bornath, J. Stat. Phys. 41 (1985) 661
21) M. Schlanges, Doctor thesis II (Rostock University, 1985), T. Bornath, Doctor thesis I (Rostock University, 1986)
22) M. Schlanges, T. Bornath and D. Kremp, Phys. Rev. A38 (1988) 2174
23) W. Ebeling, Strukturbildung bei irreversiblen Prozessen (Teubner, Leipzig,1970; Mir, Moscow, 1976)
24) W. Ebeling, A. Förster, D. Kremp and M. Schlanges, Physica (in press)
25) M. Bonitz, C. Kremp and D. Kremp, Wiss. Zeitschr. Univ. Rostock (in press)
26) D. Kremp, G. Röpke and M. Schlanges, in (W. Ebeling et al., Eds.) Transport Properties of Dense Plasmas (Akademie-Verlag, Berlin, 1984; Birkhäuser, Boston, Basel, 1984)
27) F. Bialas, Doctor thesis I (Rostock University, 1989)
28) S. Ichimaru and S. Tanaka, Phys. Rev. A32 (1985) 1790; F. Sigeneger et al., Physica A152 (1988) 365; S. Arndt et al., Contr. Plasmaphys., to appear 1989

Chapter IX:
Atomic and Molecular States and Radiation

GENERALIZED SCHRODINGER EQUATIONS FOR SHIFTS, WIDTHS, AND WAVE FUNCTIONS OF ATOMIC AND MOLECULAR STATES IN DENSE MATTER

M.D. GIRARDEAU

Department of Physics and Institute of Chemical Physics,
University of Oregon, Eugene, OR 97403, U.S.A.

There are a variety of approaches to the definition and calculation of wave functions of atoms and molecules immersed in a perturbing medium. These include (1) the self-consistent field method, (2) the microfield method, (3) the Bethe-Salpeter equation, (4) determination of the discrete eigenfunctions of an appropriate reduced density matrix, and (5) variational methods based on the existence of bound-state poles of appropriate Green's functions and the dependence of the positions of these poles on the bound state wave functions. After brief review of approaches (1)-(4) with emphasis on their limitations, the approach (5), which leads to generalized Schrodinger equations for simultaneous determination of shifts, widths, and wave functions of medium-perturbed bound states, will be described in more detail. This approach makes use of Liouville-space methods and is sufficiently general to encompass both equilibrium and nonequilibrium states of the medium. Some explicit results are obtained for a partially ionized hydrogen plasma.

1. SCF METHOD

A system consisting of an atom or molecule immersed in a medium such as a dense plasma is described by a Hamiltonian

$$H(X,Y) = H_a(X) + H_m(Y) + V_{am}(X,Y) \qquad (1.1)$$

where X denotes the set of all configurational (position and spin) variables of the constituents (electrons and nuclei) of the atom or molecule, $H_a(X)$ is the corresponding free-atom or free-molecule Hamiltonian, Y and $H_m(Y)$ are the corresponding objects for the medium in the absence of the atom or molecule, and $V(X,Y)$ is the atom-medium interaction. (From now on the word "atom" will be used for definiteness, but the discussion will be general and includes molecules as well.) Here the constituents of the atom are assumed to be distinguishable from those of the medium. This is actually not the case since both contain electrons and usually also the same species of nuclei. This restriction will be lifted in Sec. 5 by use of second quantization, but is convenient here to avoid extraneous complications. The pure-state SCF method is a quantum-mechanical variational method based on trial states

$$\psi_{trial}(X,Y) = \psi_a(X)\psi_m(Y) \qquad (1.2)$$

Necessary conditions for a minimum of the energy expectation value with respect to functional variation of ψ_a and ψ_m subject to the constraints

$$\langle\psi_a|\psi_a\rangle = \langle\psi_m|\psi_m\rangle = 1 \qquad (1.3)$$

are the generalized Hartree-Fock SCF equations

$$[H_a(X) + V_a(X) - \lambda_a]\psi_a(X) = 0$$
$$[H_m(Y) + V_m(Y) - \lambda_m]\psi_m(Y) = 0 \qquad (1.4)$$

where λ_a and λ_m are the Lagrange multipliers for constraints (1.3) and where the SCF potentials are

$$V_a(X) = \int V_{am}(X,Y) |\psi_m(Y)|^2 dY$$
$$V_m(Y) = \int V_{am}(X,Y) |\psi_a(X)|^2 dX \qquad (1.5)$$

Although it is natural to interpret λ_a as the perturbed atomic energy, there are some problems with this interpretation. In fact, any atomic state has only a finite lifetime due to bound-bound and bound-continuum transitions induced by the perturbing medium, and these finite-lifetime effects are omitted from the SCF approach.

The SCF method can be generalized to nonzero temperature by use of the Gibbs-Bogoliubov variational principle for the free energy[1], which states that

$$F \leq F_{trial} = F_o + \langle H-H_o\rangle_o \qquad (1.6)$$

Here F is the Helmholtz free energy and F_o is that of some "trial Hamiltonian" H_o. Variational parameters or functions are introduced into a judicious joice of H_o and the right side of (1.6) is minimized with respect to variation of these parameters or trial functions. For the sake of simplicity let us again restrict ourselves to the case of a single atom. Then the following choice of H_o is natural:

$$H_o = \sum_\alpha \epsilon_\alpha P_\alpha + \sum_\mu \epsilon_\mu P_\mu \qquad (1.7)$$

Here $\psi_\alpha(X)$ are orthonormal bound (discrete) atomic trial functions, ϵ_α are their energy expectation values $\langle\psi_\alpha|H_a|\psi_\alpha\rangle$, and P_α and P_μ are the projection operators onto ψ_α and ψ_μ. This leads[1] to generalized Schrodinger equations for ψ_α and ψ_μ containing nonzero-temperature generalizations of the SCF potentials. These equations have the same difficulty of interpretation as in the pure-state case since the atomic energy levels are real and therefore omit finite-lifetime effects.

2. MICROFIELD METHODS

Microfield methods are one of many approaches to the theory of spectral line shapes of atoms in gases and plasmas[2-6]. One calculates (in some approximation) Stark shifts of atomic levels due to the instantaneous electric fields of neighboring ions, then averages over some distribution of ion positions to obtain level profiles. Such an approach could in principle be applied to calculation of plasma-perturbed atomic wave functions, but in practice this would be very difficult due to the unsymmetrical instantaneous atomic environment. Furthermore, this method shares with the SCF method the omission of true level width (finite lifetime) effects, which can become large for the highest bound states.

3. BETHE-SALPETER EQUATION

This equation, originally developed to treat the relativistic bound state problem[7,8] and based on the spectral representation of the two-particle Green's function, can be generalized to systems of nonzero density and temperature by use of appropriate thermal Green's functions. This approach has been used to calculate energies and wave functions of atoms and plasmas. The development is complex and the original work[9,10] should be consulted for details. The method does take into account finite-lifetime effects but the physical significance of the atomic wave functions is not transparent. Indeed, this is already the case for the simpler zero-temperature, zero-density Bethe-Salpeter wave functions of relativistic quantum field theory[7,8].

4. DISCRETE ATOMIC EIGENFUNCTIONS OF THE REDUCED DENSITY MATRIX

The thermal equilibrium one-atom reduced density matrix $\rho(X,X')$ is defined by

$$\rho(X,X') = \frac{\sum_i e^{-\beta E_i} \int \psi_i(X,Y)\psi_i^*(X',Y)\, dY}{\sum_i e^{-\beta_i}} \tag{4.1}$$

where the notation is the same as in Sec. 1 (X for atomic coordinates, Y for those of the medium) and ψ_i are the orthonormal states of the full Hamiltonian (1.1) with eigenvalues E_i. $\rho(X,X')$ is a hermitian kernel and hence possesses orthonormal eigenfunctions ψ_α and eigenvalues λ_α satisfying

$$\int \rho(X,X')\, \psi_\alpha(X')\, dX' = \lambda_\alpha\, \psi_\alpha(X) \qquad (4.2)$$

In cases where there exist bound states of the atom immersed in the medium, the eigenvalue spectrum $\{\lambda_\alpha\}$ will contain both a discrete and continuum part, and the medium-perturbed atomic wave functions may be defined as the discrete eigenfunctions ψ_α.

This definition of perturbed bound-state wave functions has been used, for example, in the theory of superconductivity, where the Cooper pair wave function is an eigenfunction of the two-electron reduced density matrix[11]. It has the advantage of being a fundamental approach but also several disadvantages:(a) the reduced density matrix of a system as complicated as a partially-ionized plasma can only be evaluated in crude approximation;(b) the λ_α, being real, do not incorporate information regarding finite-lifetime effects.

5. VARIATIONAL METHOD FOR SHIFTS, WIDTHS, AND WAVE FUNCTIONS

This method is an outgrowth of the standard treatment of quasiparticles in condensed matter via the one-particle Green's function. An atom immersed in a plasma is regarded as a "quasiatom" of finite lifetime whose energy shift and width (inverse lifetime) and perturbed wave function can be determined in a unified way from a variational principle which leads to a generalized Schrodinger equation[12].

Like standard Green's function methods, the development is most conveniently carried out using second quantization. Denote the second-quantized states of the medium by $|\phi_\mu\rangle$. These are essentially the second-quantized versions of the medium wave functions $\psi_\mu(Y)$ of Sec. 1, Eq. (1.10) ff. Now, however, I will use the symbol ϕ for states, reserving ψ for the quantized field operators. The use of second quantization allows the $|\phi_\mu\rangle$ to contain many bound atoms and ions as well as free electrons, as is the case for a real, partially-ionized plasma. They play the role of a medium if one more atom is added to the system. Denote the discrete wave functions of the added atom (or molecule) by $\phi_\alpha(X)$, the same as $\psi_\alpha(X)$ of Sec. 1. Let n be the number of constituents and let $x_1 \cdots x_n$ be their individual coordinates. Then the operator $\hat{A}_\alpha^\dagger$ which creates one atom with wave function ϕ_α is

$$\hat{A}_\alpha^\dagger = N \int dx_1 \cdots dx_n \, \phi_\alpha(x_1 \cdots x_n) \, \hat{\psi}_1^\dagger(x_1) \cdots \hat{\psi}_n^\dagger(x_n) \tag{5.1}$$

Here $\hat{\psi}_j^\dagger(x_j)$ is the quantized field creation operator (Fermi operator for an electron, Bose or Fermi operator for a nucleus depending on nuclear spin) for the jth constituent. $\phi_\alpha(x_1 \cdots x_n)$ means the same thing as $\phi_\alpha(X)$ but with coordinates written more explicitly. Assuming the ϕ_α orthonormal, the state vectors $\hat{A}_\alpha^\dagger |0\rangle$ will be also provided that the normalization constant N is chosen properly. A state with one more atom with wave function ϕ_α added to the system is represented in second quantization by $\hat{A}_\alpha^\dagger |\phi_\mu\rangle$. For application to a plasma one should use a statistical ensemble of such states with weights w_μ which are normalized Boltzmann factors in thermal equilibrium, but may be more general in order to allow treatment of nonequilibrium systems.

The one-atom retarded Green's function is

$$G_\alpha(t) = -i \langle \hat{A}_\alpha(t) \hat{A}_\alpha^\dagger(0) \rangle, \quad t \geq 0 \tag{5.2}$$

where $\langle \cdots \rangle$ denotes the ensemble average

$$\langle \cdots \rangle = \sum_\mu w_\mu \langle \phi_\mu | \hat{A}_\alpha(t) \hat{A}_\alpha^\dagger(0) | \phi_\mu \rangle \tag{5.3}$$

$\hat{A}_\alpha(t)$ is the Heisenberg operator

$$\hat{A}_\alpha(t) = e^{it\hat{H}} \hat{A}_\alpha e^{-it\hat{H}} \tag{5.4}$$

$\hat{H}$ is the second-quantized Hamiltonian for the whole system including all medium-medium and medium-atom interactions, and units with $\hbar=1$ are assumed.

Introducing the Laplace transform

$$\tilde{G}_\alpha(z) = \int_0^\infty G_\alpha(t) e^{izt} \, dt \tag{5.5}$$

defined for z in the upper half complex plane, as well as the Liouville superoperator $\hat{L}$ defined, for any operator $\hat{A}$, by

$$\hat{L}\hat{A} \equiv [\hat{A}, \hat{H}] = \hat{A}\hat{H} - \hat{H}\hat{A} \tag{5.6}$$

one finds

$$\hat{A}(t) = e^{-it\hat{L}} \hat{A} \tag{5.7}$$

and

$$\tilde{G}_\alpha(z) = \langle [(z - \hat{L})^{-1} \hat{A}_\alpha] \hat{A}_\alpha^\dagger \rangle \tag{5.8}$$

The eigenvalue spectrum of $\hat{L}$ consists of all differences of eigenvalues of $\hat{H}$ and (for a macroscopic system) is dense on the real axis, constituting a cut of the function $\tilde{G}_\alpha(z)$ which is analytic in the upper half plane. When analytically continued into the lower half plane it will develop a pole z_α whose physical significance is the complex perturbed atomic energy $z_\alpha = \epsilon_\alpha + \xi_\alpha - i\gamma_\alpha$ where ϵ_α is the energy of the isolated atom, ξ_α is the energy shift, and γ_α the width. Clearly the z_α depend on the wave functions $\phi_\alpha(X)$. This dependence can be used to formulate a variational principle for simultaneous determination of the z_α and ϕ_α. One might, for example, minimize $\epsilon_\alpha + \xi_\alpha$ with respect to variation of ϕ_α, leading to a generalization of the SCF approach of Sec.1 to include correlation effects. Another natural criterion is maximization of the lifetime (minimization of γ_α), an atom in a medium being "most likely" to be found in a perturbed state of maximal lifetime. These two criteria are combined naturally by requiring that z_α be stationary under variation of ϕ_α subject to orthonormality:

$$[\delta/\delta\phi_\alpha^*(X)] \, [z_\alpha - \sum_\beta \lambda_{\alpha\beta}(\phi_\alpha|\phi_\beta)] = 0 \qquad (5.9)$$

$\lambda_{\alpha\beta}$ being the Lagrange multipiers for the orthonormality constraint.

In order to obtain more explicit results one must introduce an appropriate decomposition of the Liouvillian into a diagonal part $\hat{L}_0$ and off-diagonal part $\hat{L}'$ and define a corresponding "Liouvillian self energy" which will give the level shifts ξ_α and widths γ_α. If one retains the description in terms of the operators $\hat{A}$ and $\hat{A}^\dagger$ then one encounters difficulties in effecting such a separation because (a) these operators satisfy complicated nonelementary commutation relations[13,14], both with themselves and with the annihilation and creation operators for the atomic constituents (electrons and nuclei); (b) this set of operators is not linearly independent, the $\hat{A}_\alpha^\dagger$ being expandable in terms of products of constituent creation operators via (5.1), equivalent to the expandability of bound state wave functions in terms of plane-wave products. These difficulties can be circumvented by effecting a change of representation via a unitary transformation[13,14] which transforms the single-atom composite-particle states $\hat{A}_\alpha^\dagger |0\rangle$ into elementary-particle states $\hat{a}_\alpha^\dagger |0\rangle$ where the $\hat{a}_\alpha$ and $\hat{a}_\alpha^\dagger$ satisfy elementary Bose or Fermi commutation or anticommutation relations and commute or anticommute with the constituent field operators $\hat{\psi}_i$ and $\hat{\psi}_i^\dagger$ of Eq. (5.1). The importance of treating bound composites "as if" elementary (in some respects) in formulating the statistical mechanics of partially ionized plasmas has been stressed before[15,16].

One replaces the composite-particle one-atom Green's function (5.2) by an elementary-particle one-atom Green's function[12,17,18]

$$g_\alpha(t) = -i \langle \hat{a}_\alpha(t) \hat{a}_\alpha^\dagger(0) \rangle \quad , \quad t \geq 0 \qquad (5.10)$$

and correspondingly

$$\tilde{g}_\alpha(z) = \langle [(z - \hat{L})^{-1} \hat{a}_\alpha ; \hat{a}_\alpha^\dagger] \rangle \qquad (5.11)$$

$\hat{L}_o$ and $\hat{L}'$ are defined relative to a suitable linearly independent operator basis $\{\hat{B}_i\}$ such that the $\hat{B}_i$ are "eigenoperators" of $\hat{L}_o$ in the sense $\hat{L}_o \hat{B}_i = \lambda_i \hat{B}_i$ with λ_i the corresponding eigenvalue of $\hat{L}_o$. One then expands the Liouvillian Green's operator $(z-\hat{L})^{-1}$ as a geometric series in $\hat{L}'$, the Liouvillian Born series:

$$(z - \hat{L})^{-1} = \hat{G}_o(z) + \hat{G}_o(z) \hat{L}' \hat{G}_o(z) + \cdots \qquad (5.12)$$

where $\hat{G}_o(z) = (z-\hat{L}_o)^{-1}$, the unperturbed Liouvillian Green's operator.

In order to present the general ideas without extraneous complications I will consider here only the case of a partially-ionized atomic hydrogen plasma. Then the ϕ_α are of the form $\phi_\alpha(Xx)$. It is convenient to work instead with the momentum wave function $\tilde{\phi}_\alpha(Kk)$, the Fourier transform of ϕ_α. In Fock-Tani representation the Hamiltonian decomposes into a diagonal part $\hat{H}_o$ and nondiagonal part $\hat{H}'$ where[13,14]

$$\hat{H}_o = \sum_\alpha \epsilon_\alpha \hat{a}_\alpha^\dagger \hat{a}_\alpha + \sum_K \epsilon_K \hat{p}_K^\dagger \hat{p}_K + \sum_k \epsilon_k \hat{e}_k^\dagger \hat{e}_k \qquad (5.13)$$

Here $\hat{p}_K$, $\hat{p}_K^\dagger$, $\hat{e}_k$, $\hat{e}_k^\dagger$ are Fermi annihilation and creation operators for free protons and electrons in plane-wave states with energies $\epsilon_K = K^2/2M_p$, $\epsilon_k = \frac{1}{2}k^2$ (atomic units). The bound atomic annihilation and creation operators $\hat{a}_\alpha$, $\hat{a}_\alpha^\dagger$ are Bose operators commuting with the $\hat{p}_K$, $\hat{p}_K^\dagger$, $\hat{e}_k$, $\hat{e}_k^\dagger$, and the corresponding unperturbed atomic energies $\epsilon_\alpha = (\alpha|H_{pe}|\alpha)$ are the diagonal elements of the matrix

$$(\alpha|H_{pe}|\alpha') = \sum_{Kk} (\epsilon_K + \epsilon_k) \tilde{\phi}_\alpha^*(Kk) \tilde{\phi}_{\alpha'}(Kk)$$
$$+ \Omega^{-1} \sum_{Kkq} \tilde{v}_{pe}(q) \tilde{\phi}_\alpha^*(K+q, k-q) \tilde{\phi}_{\alpha'}(Kk) \qquad (5.14)$$

The index α stands for the quantum number set $\{\vec{p}, n\ell m\}$, $\vec{p}$ being the translational wave vector and $\{n\ell m\}$ the usual quantum numbers of the relative wave function. The Coulomb interactions in momentum space are denoted by $\tilde{v}_{pp}(q)=\tilde{v}_{ee}(q)=-\tilde{v}_{pe}(q)$. One may include static screening by taking these to be, e.g., $4\pi e^2/(q^2+q_s^2)$. Long-range collective effects appear in collective plasmon variables which contribute to some atomic processes[19] but are omitted here for simplicity. The interaction Hamiltonian is[13,14]

$$\hat{H}' = \sum_{\alpha\alpha'}{}' \hat{a}_\alpha^\dagger (\alpha|H_{pe}|\alpha') \hat{a}_{\alpha'}$$
$$+ \sum_{\alpha Kk} [\hat{p}_K^\dagger \hat{e}_k^\dagger (Kk|H_{pe}|\alpha)\hat{a}_\alpha + \hat{a}_\alpha^\dagger (\alpha|H_{pe}|Kk)\hat{e}_k\hat{p}_K] + \cdots \quad (5.15)$$

with bound-continuum transition matrix element

$$(Kk|H_{pe}|\alpha) = (\epsilon_K+\epsilon_k)\phi_\alpha(Kk) + \Omega^{-1}\sum_q \tilde{v}_{pe}(q)\phi_\alpha(K+q,k-q)$$
$$- \sum_{K'k'} \Delta(Kk,K'k')[(\epsilon_{K'}+\epsilon_{k'})\phi_\alpha(K'k') + \Omega^{-1}\sum_q \tilde{v}_{pe}(q)\phi_\alpha(K'+q,k'-q)] \quad (5.16)$$

Here Δ is the bound state kernel

$$\Delta(Kk,K'k') = \sum_\alpha \phi_\alpha(Kk)\phi_\alpha^*(K'k') \quad (5.17)$$

In the absence of the medium the ϕ_α reduce to unperturbed eigenfunctions $\phi_\alpha^{(0)}$ satisfying

$$(\epsilon_K+\epsilon_k)\phi_\alpha^{(0)}(Kk) + \Omega^{-1}\sum_q \tilde{v}_{pe}(q)\phi_\alpha^{(0)}(K+q,k-q) = \epsilon_\alpha^{(0)}\phi_\alpha^{(0)}(Kk) \quad (5.18)$$

Then by orthonormality $(\alpha|H_{pe}|\alpha')$ of Eq. (5.14) becomes diagonal and the bound-continuum matrix element (5.17) vanishes identically by cancellation. This is the mathematical representation of the stability of unperturbed bound states in the absence of perturbation by the medium. However, the optimal variational solutions in the presence of the medium have finite lifetimes, undergoing both bound-bound transitions via $(\alpha|H_{pe}|\alpha')$ and bound-continuum transitions via $(Kk|H_{pe}|\alpha)$ and $(\alpha|H_{pe}|Kk)$.

The natural choice of operator basis $\{\hat{B}_i\}$ is such that each $\hat{B}_i$ is a normally ordered product of creation and annihilation operators. Defining $\hat{L}_0$ and $\hat{L}'$ by

$$\hat{L}_0\hat{A} \equiv [\hat{A},\hat{H}_0] \quad , \quad \hat{L}'\hat{A} \equiv [\hat{A},\hat{H}'] \quad (5.19)$$

one finds that each such $\hat{B}_i$ is an eigenoperator of $\hat{L}_0$ with eigenvalue equal to the sum of energies of annihilation operators minus the sum of energies of creation operators[12]. A basis element $\hat{\Pi}^\dagger\hat{\Pi}\hat{a}_\alpha$ will have exactly the same eigenvalue ϵ_α as does $\hat{a}_\alpha$ provided that $\hat{\Pi}$ is any product of annihilation operators. The repeated occurrence of such terms in $\hat{L}'$ causes a "pileup of divergences" from repeated factors $(z-\epsilon_\alpha)^{-1}$ in (5.11), (5.12) which can be summed to give a self-energy denominator $[z-\epsilon_\alpha-\Sigma_\alpha(z)]^{-1}$. This is more general than the usual definition[20] of self energy since "quasidiagonal" diagrams (involving energy denominators from basis elements $\hat{\Pi}^\dagger\hat{\Pi}a_\alpha$) contribute as well as those involving $\hat{a}_\alpha$ itself. Σ_α is the sum of contributions of all "irreducible α-self-energy" diagrams whose definition[12] will be illustrated by evaluating some leading contributions. The energy Green's function is representable in the form

$$\mathcal{G}_\alpha(z) = [z - \epsilon_\alpha - \Sigma_\alpha(z)]^{-1}[1 + f_\alpha + \sigma_\alpha(z)] \qquad (5.20)$$

where f_α is the atomic distribution function $\langle \hat{a}_\alpha^\dagger \hat{a}_\alpha \rangle$ ($\ll 1$ for a nondegenerate plasma) and σ_α is a sum of diagrams which are irreducible but not quasidiagonal. The pole z_α entering into the variational equation (5.10) depends only on Σ_α, being the solution of

$$z_\alpha - \epsilon_\alpha - \Sigma_\alpha(z_\alpha) = 0 \qquad (5.21)$$

The variational equation for $\mathcal{P}_\alpha$ can be written in the form

$$(\epsilon_K+\epsilon_k)\mathcal{P}_\alpha(Kk) + \Omega^{-1}\sum_q \tilde{v}_{pe}(q)\mathcal{P}_\alpha(K-q,k+q) + \delta\Sigma_\alpha(z_\alpha)/\delta\mathcal{P}_\alpha^*(Kk)$$
$$= [1 - \Sigma'_\alpha(z_\alpha)]\sum_\beta \lambda_{\alpha\beta}\mathcal{P}_\beta(Kk) \qquad (5.22)$$

where $\Sigma_\alpha'(z)$ is the derivative of $\Sigma_\alpha(z)$ with respect to z. Taking the inner product of (5.22) with $\mathcal{P}_\alpha$ and using (5.21) together with orthonormality yields the useful identity

$$[1 - \Sigma_\alpha'(z_\alpha)]\lambda_{\alpha\alpha} = z_\alpha \qquad (5.23)$$

which can be used to rewrite (5.22) in the form

$$(\epsilon_K+\epsilon_k)\mathcal{P}_\alpha(Kk) + \Omega^{-1}\sum_q \tilde{v}_{pe}(q)\mathcal{P}_\alpha(K-q,k+q)$$
$$+ \sum_{\beta K'k'} V_{\alpha\beta}(Kk,K'k')\mathcal{P}_\beta(K'k') = z_\alpha\mathcal{P}_\alpha(Kk) \qquad (5.24)$$

where $V_{\alpha\beta}$ is a nonlocal, complex, nonlinear optical potential matrix defined implicitly by

$$\delta\Sigma_\alpha(z_\alpha)/\delta\mathcal{P}_\alpha^*(Kk) - [1-\Sigma'_\alpha(z_\alpha)]\sum_{\beta(\neq\alpha)}\lambda_{\alpha\beta}\mathcal{P}_\beta(Kk)$$
$$= \sum_{\beta K'k'} V_{\alpha\beta}(Kk,K'k')\mathcal{P}_\beta(K'k') \qquad (5.25)$$

The first-order contributions to Σ_α are depicted in Fig. 1.

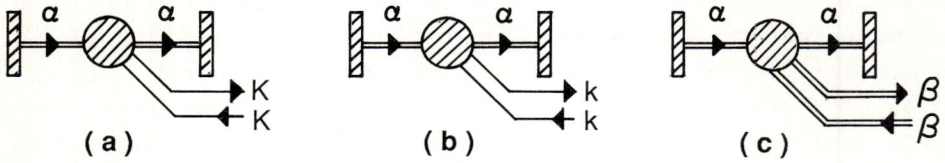

Fig. 1. First-order contributions to atomic self energy.

Diagram (a) represents the contribution of elastic scattering. Reading from left to right, the right-oriented line with label α stands for the operator $\hat{a}_\alpha$ on which $\hat{G}_0 \hat{L}' \hat{G}_0$ operates in (5.11), (5.12), the line-termination "anchor" on the left indicating that this is not a free line (α not summed). The vertex (cross-hatched circle) represents action of $\hat{L}'$, in particular that term yielding terms $\hat{p}_K^\dagger \hat{p}_K \hat{a}_\alpha$ when applied to $\hat{a}_\alpha$, the three lines to the right of the vertex standing for the three factors in this product (right-oriented lines for annihilation operators, left-oriented line for the creation operator. The proton lines (label K) are free (K to be summed) and hence unterminated on the right, while the atom line (label α) is not free and hence has a line-termination "anchor" on the right. Diagrams (b) and (c) are similar but for the atom-electron and atom-atom interactions. Each diagram can be interpreted as the result of a scattering process in which the lines directed toward the vertex represent initial-state particles and those directed away from the vertex represent final-state particles. The sum of these three processes gives, by the diagram rules[12], the first-order self energy

$$\Sigma_\alpha^{(1)}(z) = \sum_K (\alpha K|H|\alpha K) f_K + \sum_k (\alpha k|H|\alpha k) f_k + \sum_\beta [(\alpha\beta|H|\alpha\beta) + (\alpha\beta|H|\beta\alpha)] f_\beta \quad (5.26)$$

where f_K, f_k, and f_β are the proton, electron, and atom distribution functions:

$$f_K = \langle \hat{p}_K^\dagger \hat{p}_K \rangle, \quad f_k = \langle \hat{e}_k^\dagger \hat{e}_k \rangle, \quad f_\beta = \langle \hat{a}_\beta^\dagger \hat{a}_\beta \rangle \quad (5.27)$$

The necessary atom-proton, atom-electron, and atom-atom matrix elements are given in the literature[13,14].

The expression for the first-order optical potential matrix $V_{\alpha\beta}^{(1)}$ follows by substitution of (5.27), (5.29) into (5.26). $\Sigma_{\alpha'}^{(1)}$ vanishes since $\Sigma_\alpha^{(1)}$ is independent of z, and there are no off-diagonal $\lambda_{\alpha\beta}$ since $V_{\alpha\beta}^{(1)}$ is hermitian, ensuring that the z_α are real (no level widths) in first order and that $\tilde{\phi}_\alpha$ belonging to different eigenvalues are automatically orthogonal without the necessity of off-diagonal $\lambda_{\alpha\beta}$. One finds

$$V^{(1)}_{\alpha\beta}(Kk,K'k') = -\delta_{\alpha\beta}\delta_{KK'}\delta_{kk'}\Omega^{-1}\left[\sum_{K''}f_{K''}\tilde{v}_{pp}(K''-K)\right.$$
$$\left.+\sum_{k''}f_{k''}\tilde{v}_{ee}(k''-k)\right]$$
$$-\tfrac{1}{2}\delta_{\alpha\beta}\Omega^{-1}\sum_{q}\tilde{v}_{pe}(q)(f_{K}+f_{K+q}+f_{k}+f_{k-q})\delta_{K',K+q}\delta_{k',k-q}+\cdots \quad (5.28)$$

where "···" stands for more complicated terms quadratic in wave functions arising from atom-atom interaction [third term in (5.26)]. $V^{(1)}_{\alpha\beta}$ is a generalized SCF potential.

Lack of space forbids exhibiting details of the second-order contributions here, but I will show one of the self-energy diagrams contributing to atomic level width and nonhermitian terms in $V_{\alpha\beta}$:

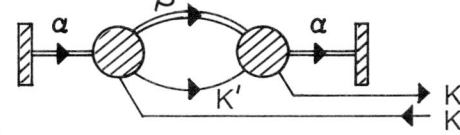

Fig. 2. Example of a second-order contribution.

The corresponding second-order contribution to $\Sigma_\alpha(z_\alpha)$ is

$$\sum_{\beta KK'}\frac{|(\alpha K|H|\beta K')|^2 f_{K'}}{\epsilon_\alpha-\epsilon_\beta-\epsilon_{K'}+\epsilon_K+i\eta} \quad (5.29)$$

Since we are calculating here only to second order, z_α in (5.29) has been approximated by its value to zero order, which is $\epsilon_\alpha+i\eta$, the positive imaginary infinitesimal $i\eta$ occurring because the value just above the real-axis cut of $\mathcal{\tilde{g}}_\alpha(z)$ is required. Application of the Dirac identity $1/(x+i\eta) = P/x - i\pi\delta(x)$ leads to a contribution to the level shift from the principal value term and to the level width from the Dirac delta function term. There are a number of other second-order terms which behave similarly.

REFERENCES

1) M.D. Girardeau and R.M. Mazo, Variational methods in statistical mechanics, in: Advances in Chemical Physics, Vol. 24, eds. I. Prigogine and S.A. Rice (Academic Press, New York, 1974) pp. 187-255.

2) H. Griem, Spectral Line Broadening by Plasmas (Academic Press, New York, 1974).

3) C.F. Hooper, Phys. Rev. 149 (1966) 77, 165 (1968) 215.

4) H. Pfennig and E. Treffitz, Z. Naturforsch. A21 (1966) 697.

5) R. Stamm, Simulation studies of the dynamics of plasma microfields, in: Spectral Line Shapes, Vol. 2, ed. K. Burnett (Walter de Gruyter, Berlin, 1983) pp. 3-29.

6) C. Iglesias and J. Dufty, Ion-electron correlations in spectral line shape theory, in: Spectral Line Shapes (ibid.) pp. 55-69.

7) M. Gell-Mann and F. Low, Phys. Rev. 84 (1951) 350.

8) E.E. Salpeter and H.A. Bethe, Phys. Rev. 84 (1951) 1232.

9) K. Kilimann, D. Kremp, and G. Ropke, Teor. Mat. Fiz. 55 (1983) 448.

10) W.-D. Kraeft, D. Kremp, W. Ebeling, and G. Ropke, Quantum Statistics of Charged Particle Systems (Plenum Press, New York, 1986), pp. 118 ff.

11) C.N. Yang, Revs. Mod. Phys. 34 (1962) 694.

12) M.D. Girardeau, Liouvillian propagator technique for perturbed wave functions, level shifts and broadenings of composite particles in a many-body medium, in: Lecture Notes in Physics, Vol. 142, Recent Progress in Many-Body Theories, eds. J.G. Zabolitzky, M. de Llano, M. Fortes, and J.W. Clark (Springer, Berlin, 1981) pp. 355-363.

13) M.D. Girardeau, Int. J. Quantum Chem. 17 (1980) 25 and references therein.

14) J.C. Straton and M.D. Girardeau, Phys. Rev. A (in press).

15) W. Ebeling, Physica 73 (1974) 573.

16) W.-D. Kraeft, D. Kremp, W. Ebeling, and G. Ropke, ibid.[10], pp. 85 ff., 118 ff., 184 ff.

17) Y. Soulet and A. Gomes, J. Statis. Phys. 25 (1981) 695.

18) R. Fleckinger and Y. Soulet, Physica 119A (1983) 243.

19) M.D. Girardeau and F.A. Gutierrez, Phys. Rev. A 38 (1988) 1624.

20) R.D. Mattuck, A Guide to Feynman Diagrams in the Many-Body Problem (McGraw-Hill, London, 1967) Chaps. 10, 14.

DYNAMICS OF ELECTRIC FIELDS IN STRONGLY COUPLED PLASMAS

James W. DUFTY and Lorena ZOGAIB

Department of Physics, University of Florida, Gainesville, FL 32611, USA

The equilibrium joint probability density for electric fields at two different times is considered for both neutral and charged points. The behavior of this distribution function is discussed in the Gaussian, short time, and high field limits. An approximate global description is proposed using an independent particle model as an extension of corresponding approximations for the single time field distribution.

1. INTRODUCTION

The force on a charged particle or charge distribution (e.g., dipole) in a plasma is determined by the electric field due to all other particles. The equilibrium distribution of field values, $Q(\varepsilon)$, is well known from Monte Carlo, molecular dynamics, and theoretical studies.[1] More generally, it is of interest to know how the electric field distribution changes in time for specified initial states. For example, if the electric field is measured to have some value at t = 0 in an equilibrium plasma, what is the probability distribution for its values at some later time? A wide class of dynamical properties can be calculated from knowledge of this equilibrium joint distribution function for two field values at two times, $f(\vec{\varepsilon}, t; \vec{\varepsilon}', t')$.

In spite of its importance for describing both radiative and transport properties of atoms and ions in plasmas,[2,3] there has been only limited study of this distribution, either by simulation[4] or by theory.[5] In this brief review, emphasis is placed on limiting behavior that can be extracted (formally) without restriction on the plasma conditions. In general the discussion is applicable to classical or quantum, single or multi-component, weakly or strongly coupled plasmas. The limits considered are the Gaussian limit (small field values at highly charged points), the short time limit, and the high field limit. In addition, the weak coupling independent particle (mean field) model is renormalized to yield an approximate joint distribution for all fields and times that is consistent with the known limits described here. Some difficulties and alternative approaches to this problem are described in the last section.

2. DEFINITIONS AND GENERAL PROPERTIES

Consider a system of particles with overall charge neutrality and identify a specific particle with charge, position, and momentum $(Q_o, \vec{R}_o, \vec{P}_o)$. In the following we will be interested in the ion electric field, $\vec{E}$, at the specified neutral or positive particle, due to all other ions of the system. Each configuration leads to a field value, $\vec{\varepsilon}$, and the distribution of such values in equilibrium is defined by,

$$Q(\varepsilon) = \langle \delta(\vec{\varepsilon}-\vec{E}) \rangle = (2\pi)^{-3} \int d\vec{\lambda}\, e^{-i\vec{\lambda}\cdot\vec{\varepsilon}}\, e^{I(\lambda)} \qquad (2.1)$$

The brackets, $\langle \cdots \rangle$, denote an equilibrium ensemble average and the second equality defines the associated generating function, $I(\lambda)$. The latter is introduced as a more suitable object for theoretical analysis. The equilibrium distribution is time independent due to stationarity of the equilibrium state. Dynamical properties of the fields can be determined from the joint distribution function,

$$f(\vec{\varepsilon},t;\vec{\varepsilon}',t') = \langle \tfrac{1}{2}[\delta(\vec{\varepsilon}-\vec{E}(t)),\delta(\vec{\varepsilon}'-\vec{E}(t'))]_+ \rangle$$

$$= (2\pi)^{-6} \int d\vec{\lambda}d\vec{\lambda}'\, e^{-i\vec{\lambda}\cdot\vec{\varepsilon}-i\vec{\lambda}'\cdot\vec{\varepsilon}'}\, e^{G(\vec{\lambda},\vec{\lambda}';t-t')} \qquad (2.2)$$

where $G(\vec{\lambda},\vec{\lambda}';t-t')$ is the corresponding generating function. The brackets $[,]_+$ denote the anti-commutator so it is a real, positive distribution. The notation for G indicates that f depends on t and t' only through their difference, as follows from stationarity. In all of the following we set $t' = 0$, without loss of generality. Several important properties of f and G follow from their definitions and the symmetries of the equilibrium state (time reversal invariance, stationarity, and mixing):

$$f(\vec{\varepsilon},0;\vec{\varepsilon}',0) = \delta(\vec{\varepsilon}-\vec{\varepsilon}')Q(\varepsilon) \quad ; \quad G(\vec{\lambda},\vec{\lambda}';0) = I(|\vec{\lambda}+\vec{\lambda}'|) \qquad (2.3)$$

$$f(\vec{\varepsilon},t;\vec{\varepsilon}',\infty) = Q(\varepsilon)Q(\varepsilon') \quad ; \quad G(\vec{\lambda},\vec{\lambda}';\infty) = I(\lambda)+I(\lambda') \qquad (2.4)$$

$$f(\vec{\varepsilon},t;\vec{\varepsilon}',0) = f(\vec{\varepsilon},-t;\vec{\varepsilon}',0) = f(\vec{\varepsilon}',t;\vec{\varepsilon},0) \quad ;$$

$$G(\vec{\lambda},\vec{\lambda}';t) = G(\vec{\lambda},\vec{\lambda}';-t) = G(\vec{\lambda}',\vec{\lambda};t) \qquad (2.5)$$

In addition, rotational invariance of the equilibrium ensemble implies that $f(\vec{\varepsilon},t;\vec{\varepsilon}',0)$ depends on the field variables only through their magnitudes and the angle between them; a similar result applies for $G(\vec{\lambda},\vec{\lambda}';t)$.

Most of the analysis below refers to the generating function, G. A useful formal representation for it is given by,

$$G(\vec{\lambda},\vec{\lambda}';t) = \ln\{<\frac{1}{2}[e^{i\vec{\lambda}\cdot\vec{E}(t)},e^{i\vec{\lambda}'\cdot\vec{E}}]_+>\} \quad (2.6)$$

Equations (2.3)-(2.4) show that for long and short times $G(\vec{\lambda},\vec{\lambda}';t)$ is simply related to the corresponding (known) generating function for the static distribution, I. A convenient alternative representation is therefore,

$$G(\vec{\lambda},\vec{\lambda}';t) = [1-\alpha(\vec{\lambda},\vec{\lambda}';t)][I(\lambda)+I(\lambda')] + \alpha(\vec{\lambda},\vec{\lambda}';t)I(|\vec{\lambda}+\vec{\lambda}'|) \quad (2.7)$$

where $\alpha(\vec{\lambda},\vec{\lambda}';t)$ is a function that varies from 1 to 0 as t goes from 0 to ∞,

$$\alpha(\vec{\lambda},\vec{\lambda}';t) = [G(\vec{\lambda},\vec{\lambda}';t)-G(\vec{\lambda},\vec{\lambda}';\infty)]/[G(\vec{\lambda},\vec{\lambda}';0)-G(\vec{\lambda},\vec{\lambda}';\infty)] \quad (2.8)$$

3. GAUSSIAN LIMIT

If the generating function, $G(\vec{\lambda},\vec{\lambda}';t)$, is expanded in powers of $\vec{\lambda}$ and $\vec{\lambda}'$ the leading contribution is quadratic in these variables, yielding a Gaussian distribution for $f(\vec{\varepsilon},t;\vec{\varepsilon}',0)$,

$$f(\vec{\varepsilon},t;\vec{\varepsilon}',0) = [1-\alpha^2(t)]^{-3/2} Q_g(\frac{\vec{\varepsilon}-\alpha(t)\vec{\varepsilon}'}{[1-\alpha^2(t)]^{1/2}}) Q_g(\varepsilon') \quad (3.1)$$

where $\alpha(t) \equiv \alpha(\vec{0},\vec{0};t) = <\frac{1}{2}[E_i(t),E_i]_+>/<E^2>$ is the equilibrium electric field autocorrelation function, and $Q_g(\varepsilon)$ is the Gaussian limit of $Q(\varepsilon)$,

$$Q_g(\varepsilon) = [3/2\pi<E^2>]^{3/2} \exp[-3\varepsilon^2/2<E^2>] \quad (3.2)$$

The electric field autocorrelation function is simply related to the Fourier transformed charge density autocorrelation function, $S(\vec{k},\vec{k}',t)$, by,

$$<\frac{1}{2}[E_i(t),E_i]_+> = (2\pi)^{-6}\int d\vec{k}d\vec{k}' \; \vec{f}^*(\vec{k})\cdot\vec{f}(\vec{k}')S(\vec{k},\vec{k}',t) \quad (3.3)$$

Here $\vec{f}(\vec{k})$ is the form factor representing the Fourier transform of the single particle field amplitude ($\sim \vec{k}/k^2$ for Coulomb fields). Since good approximations to $S(\vec{k},\vec{k}',t)$ are available for strongly coupled plasmas,[6] the function $\alpha(t)$ can be considered known and (3.1) is completely determined. It may be shown that this result satisfies the conditions (2.3) - (2.5). This limit is only applicable for electric field distributions at charged points for which the second moment, $\langle E^2 \rangle$, exists. Furthermore, it requires that configurations corresponding to weak fields dominate to justify small $\vec{\lambda} \cdot \vec{E}$. This restricts the result to small $\vec{\varepsilon}$, $\vec{\varepsilon}'$ and large charge on the particle at which the field is calculated.

Equivalently, Eq. (3.1) can be expressed as the solution to a non-Markovian Fokker-Planck equation,

$$\frac{\partial}{\partial t} f(\vec{\varepsilon}, t; \vec{\varepsilon}', 0) = \frac{\partial}{\partial \varepsilon_i} \gamma(t) [\frac{\partial}{\partial \varepsilon_i} + 3\varepsilon_i / \langle E^2 \rangle] f(\vec{\varepsilon}, t; \vec{\varepsilon}', 0) \qquad (3.4)$$

where $\gamma(t) = -\langle E^2 \rangle \dot{\alpha}(t) / 3\alpha(t)$.

The Gaussian limit can be improved by the ad hoc replacement of Q_g by Q in Eq. (3.1). In this way the dependence on the exact static distribution is preserved for long and short times. A related approximation has been considered by Smith, et al.[4] as a "weak diffusion" model. A more systematic extension in the same spirit is obtained by expanding only $\alpha(\vec{\lambda}, \vec{\lambda}'; t)$ in the representation (2.8). Then the joint distribution function is found to be,

$$f(\vec{\varepsilon}, t; \vec{\varepsilon}', 0) = \int d\vec{\varepsilon}_1 Q(\varepsilon_1; \alpha(t)) Q(|\vec{\varepsilon} - \vec{\varepsilon}_1|; 1 - \alpha(t)) Q(|\vec{\varepsilon}' - \vec{\varepsilon}_1|; 1 - \alpha(t)) \qquad (3.5)$$

where $Q(\varepsilon; x)$ is simply related to the generating function for $Q(\varepsilon)$ in Eq. (2.1) by,

$$Q(\varepsilon; x) = (2\pi)^{-3} \int d\vec{\lambda} \, e^{-i\vec{\lambda} \cdot \vec{\varepsilon}} \, e^{x I(\lambda)} \qquad (3.6)$$

The result (3.5) reproduces the exact static distributions in the limits $t \to 0$ and $t \to \infty$, and reduces to (3.1) if Q is replaced by Q_g. In all these cases the joint distribution is completely determined by the static electric field distribution and the time dependent electric field autocorrelation function.

4. SHORT TIME LIMIT

The exact short time limit of $G(\vec{\lambda}, \vec{\lambda}'; t)$ is easily obtained from the representation (2.6), with the leading term given by,

$$G(\vec{\lambda}, \vec{\lambda}'; t) = I(|\vec{\lambda} + \vec{\lambda}'|) + \frac{1}{2} \lambda_i \lambda_j' F_{ij}(|\vec{\lambda} + \vec{\lambda}'|) \, t^2 + (\text{order } t^4) \qquad (4.1)$$

where $F_{ij}(\lambda)$ is related to the conditional average for the initial time derivative of the electric field, $\dot{E}_i$,

$$F_{ij}(\lambda) \, e^{I(\lambda)} = \int d\vec{\varepsilon} \, Q(\varepsilon) \, \langle \dot{E}_i \dot{E}_j \rangle_\varepsilon \, e^{i\vec{\lambda} \cdot \vec{\varepsilon}} \qquad (4.2)$$

$$\langle \dot{E}_i \dot{E}_j \rangle_\varepsilon = \langle \dot{E}_i \dot{E}_j \delta(\vec{\varepsilon} - \vec{E}) \rangle / Q(\varepsilon) \qquad (4.3)$$

To obtain the expression (4.2) and (4.3) explicit use has been made of the assumption that the ion center of mass degrees of freedom can be treated classically.

To calculate $f(\vec{\varepsilon}, t; \vec{\varepsilon}', 0)$ from (4.1) a change of variables is made to $\vec{\lambda} + \vec{\lambda}'$ and $(\vec{\lambda} - \vec{\lambda}')/2$. The integration over $\vec{\lambda} + \vec{\lambda}'$ is then performed to leading order in t with the result,

$$f(\vec{\varepsilon}, t; \vec{\varepsilon}', 0) = Q(a)(2\pi)^{-3} \int d\vec{\ell}\, e^{-i\vec{\ell} \cdot \Delta\vec{\varepsilon}} \exp\{t^2 [A_3(a) - \ell_i \ell_j \langle \dot{E}_i \dot{E}_j \rangle_a]/2\}$$

$$= Q(a)\, e^{A_3(a)t^2/2} (2\pi t^2)^{-3/2} A_1^{-1} A_2^{-1/2} \exp\{-(\Delta\varepsilon)_\perp^2 / 2A_1 t^2$$

$$- (\Delta\varepsilon)_\parallel^2 / 2A_2 t^2\} \qquad (4.4)$$

Here $\Delta\vec{\varepsilon} = \vec{\varepsilon} - \vec{\varepsilon}'$ and $\vec{a} = (\vec{\varepsilon} + \vec{\varepsilon}')/2$; also, $\Delta\vec{\varepsilon}_\perp$ and $\Delta\vec{\varepsilon}_\parallel$ are the components of $\Delta\vec{\varepsilon}$ perpendicular and parallel to $\vec{a}$, respectively. The coefficients $A_m(a)$ are related to $\langle \dot{E}_i \dot{E}_j \rangle_a$ by,

$$A_1(a) = \frac{1}{2} \langle [\dot{E}^2 - (\hat{a} \cdot \dot{\vec{E}})^2] \rangle_a, \quad A_2(a) = \langle (\hat{a} \cdot \dot{\vec{E}})^2 \rangle_a$$

$$A_3(a) = [4Q(a)]^{-1} \frac{\partial^2}{\partial a_i \partial a_j} Q(a) \langle \dot{E}_i \dot{E}_j \rangle_a = [4Q(a)]^{-1} \frac{\partial}{\partial a_i} Q(a) \langle \ddot{E}_i \rangle_a \qquad (4.5)$$

Equation (4.4) is asymptotically exact as $t \to 0$. It satisfies the conditions (2.3) and (2.5), but does not yield the long time limit (2.4). It is Gaussian in the field difference, $\Delta\vec{\varepsilon}$, but has a more complex dependence on the field $\vec{a}$ through the conditional correlation function $\langle \dot{E}_i \dot{E}_j \rangle_a$. This may be calculated in an approximation suitable for strongly coupled plasmas as follows. First, it is expressed in terms of the conditional pair correlation function as,

$$\langle \dot{E}_i \dot{E}_j \rangle_a = k_B T \sum_\alpha m_\alpha n_\alpha \int d\vec{r}\, \left(\frac{\partial E_{\alpha i}(\vec{r})}{\partial r_\ell}\right)\left(\frac{\partial E_{\alpha j}(\vec{r})}{\partial r_\ell}\right) g_\alpha(\vec{r}; \vec{a}) \qquad (4.6)$$

Here, $n_\alpha g_\alpha(\vec{r}; \vec{a})$ is the pair correlation function for a particle of species α at a position $\vec{r}$ relative to the field point, given that the field there has the value $\vec{a}$. Also, $\vec{E}_\alpha(\vec{r})$ is the field due to that particle. This quantity can be calculated from a suitable functional derivative of the generating function, $I(\lambda)$, for the static distribution. Since accurate approximations for $I(\lambda)$ are known,[1] suitable for strongly coupled plasmas, they may be used to determine $g_\alpha(\vec{r};\vec{a})$ as well. The details of such a calculation for various approximations to $I(\lambda)$ are discussed in reference 7. In particular, for the APEX approximation to $I(\lambda)$,[8] Eq. (4.6) reduces to,

$$\langle \dot{E}_i \dot{E}_j \rangle_a = k_B T \sum_\alpha m_\alpha n_\alpha \int d\vec{r}\, \left(\frac{\partial E_{\alpha i}(\vec{r})}{\partial r_\ell}\right)\left(\frac{\partial E_{\alpha j}(\vec{r})}{\partial r_\ell}\right) g_\alpha(\vec{r}) Q(\vec{a}-\vec{E}^*_\alpha(\vec{r}))/Q(\vec{a}) \qquad (4.7)$$

where $g_\alpha(\vec{r})$ is the usual equilibrium radial distribution function for a particle of species α at a distance r from the field point, and $\vec{E}^*_\alpha$ is the APEX effective field for that particle. The short time distribution (4.4) can therefore be calculated to a good approximation for strongly coupled plasmas from knowledge of $Q(\varepsilon)$ and the radial distribution function. This short time result has been evaluated in detail for the special case of a neutral point and independent particles.[9]

An alternative short time description is given by the Fokker-Planck equation,

$$\frac{\partial}{\partial t^2} f(\vec{\varepsilon},t;\vec{\varepsilon}',0) = \frac{\partial}{\partial \varepsilon_i} D_{ij}(\varepsilon)[\frac{\partial}{\partial \varepsilon_j} - V_j(\varepsilon)] f(\vec{\varepsilon},t;\vec{\varepsilon}',0) \qquad (4.8)$$

$$D_{ij}(\varepsilon) = \tfrac{1}{2}\langle \dot{E}_i \dot{E}_j \rangle_\varepsilon, \qquad V_i(\varepsilon) = \frac{\partial}{\partial \varepsilon_i} \ln[Q(\varepsilon)] \qquad (4.9)$$

It can be shown that the solution to this equation, with initial condition (2.3) is asymptotically exact as $t \to 0$ and has the properties (2.4) and (2.5) as well. For finite times the short time forms (4.4) and (4.8) are not equivalent. Equation (4.8) has the advantage of approaching the correct long time limit and therefore may extrapolate approximately to longer times than (4.4). On the other hand, we have an explicit form for (4.4) whereas we have not been able to construct the solution to (4.8).

5. HIGH FIELD LIMIT

The behavior of the distribution function for very large values of the fields is expected to be governed by the fields of one particle close to the field point at $t = 0$ and one particle nearby at time t. To extract this contribution it is convenient to introduce a cluster expansion for $G(\vec{\lambda},\vec{\lambda}';t)$ which is a generalization the Baranger-Mozer expansion for the static distribution.[1] Define the functions, $\phi_\alpha(\vec{r},\lambda)$, by,

$$e^{i\vec{\lambda}\cdot\vec{E}_\alpha(\vec{r})} = 1 + \phi_\alpha(\vec{r},\lambda) \tag{5.1}$$

Substitution of this expression in Eq. (2.6) gives G as a functional of ϕ, i.e., $G = G[\phi]$. A functional Taylor series expansion in ϕ then has the leading terms,

$$G[\phi] = I[\phi(\lambda)] + I[\phi(\lambda')]$$

$$+ \sum_\alpha \sum_{\alpha'} \int d\vec{r}d\vec{r}' \phi_\alpha(\vec{r},\lambda)\phi_{\alpha'}(\vec{r}',\lambda') C_{\alpha\alpha'}(\vec{r},\vec{r}';t) + \ldots \tag{5.2}$$

The first two terms on the right side denote the usual Baranger-Mozer expansions for the static generating functions $I(\lambda)$ and $I(\lambda')$. The last term is the leading dynamical contribution, bilinear in the function ϕ, and the dots indicate other dynamical contributions higher order in ϕ. The density time correlation function for species α, α' is defined by,

$$C_{\alpha\alpha'}(\vec{r},\vec{r}';t) = \langle n_\alpha(\vec{r},t)[n_{\alpha'}(\vec{r}')-\langle n_{\alpha'}(\vec{r}')\rangle]\rangle \tag{5.3}$$

where $n_\alpha(\vec{r})$ is the number density for species α.

For particles near the field point the field is approximately Coulomb. This suggests a scaling in (2.2) according to $\lambda\varepsilon \to \lambda$, $r^2\varepsilon \to r^2$, which shows that each factor of the function ϕ introduces a factor of $\varepsilon^{-3/2}$ or $\varepsilon'^{-3/2}$. Formally, therefore, the higher order terms not shown in (5.2) become negligible for sufficiently large ε,ε'. Furthermore the exponential of G in (2.2) also can be expanded to quadratic order in ϕ. The terms contributing for large ε, ε' are then found give,

$$f(\vec{\varepsilon},t;\vec{\varepsilon}',0) \to Q(\varepsilon)Q(\varepsilon')$$

$$+ \sum_\alpha \sum_{\alpha'} \int d\vec{r}d\vec{r}' \delta(\vec{\varepsilon}-\vec{E}_\alpha(\vec{r}))\delta(\vec{\varepsilon}'-\vec{E}_{\alpha'}(\vec{r}'))C_{\alpha\alpha'}(\vec{r},\vec{r}';t) \tag{5.4}$$

where $Q(\varepsilon)$ and $Q(\varepsilon')$ are given by their high field limits,

$$Q(\varepsilon) \to \sum_\alpha n_\alpha \int d\vec{r}\, \delta(\vec{\varepsilon}-\vec{E}_\alpha(\vec{r}))g_\alpha(r) \tag{5.5}$$

For the large fields considered we may replace $\vec{E}_\alpha(\vec{r})$ in (5.4) by Coulomb fields and perform the integration to give,

$$f(\vec{\varepsilon},t;\vec{\varepsilon}',0) \to Q(\varepsilon)Q(\varepsilon')$$

$$+ \frac{1}{4}(\varepsilon\varepsilon')^{-9/2} \sum_\alpha \sum_{\alpha'} (Z_\alpha Z_{\alpha'} e^2)^{3/2} C_{\alpha\alpha'}(\hat{\varepsilon}\sqrt{Z_\alpha e/\varepsilon},\ \hat{\varepsilon}'\sqrt{Z_{\alpha'} e/\varepsilon'};t) \quad (5.6)$$

where $C_{\alpha\alpha'}(\vec{r},\vec{r}';t)$ is the species density correlation function defined in (5.3). The high field limit is seen to be determined from radial distribution functions, $g_\alpha(r)$, and $C_{\alpha\alpha'}(\vec{r},\vec{r}';t)$. The explicit form of the second term on the right side of (5.6) has been evaluated and discussed for the case of a neutral field point and independent particles.[2,4]

The interpretation of (5.6) as the single field contribution can be made more apparent by rewriting it in the equivalent form,

$$f(\vec{\varepsilon},t;\vec{\varepsilon}',0) \to \sum_\alpha \sum_{\alpha'} \sum_i \sum_j \langle \delta(\vec{\varepsilon}-\vec{E}_\alpha(\vec{r}_i(t)))\delta(\vec{\varepsilon}'-\vec{E}_{\alpha'}(\vec{r}_j)) \rangle \quad (5.7)$$

where the summations range over all particles. Equation (5.7) represents all possible ways to produce the fields, $\vec{\varepsilon}'$ at $t = 0$ and $\vec{\varepsilon}$ at t, by a single particle (not necessarily the same). This limit also preserves the properties (2.3)-(2.5). We note that for short times and large fields the second term on the right side of (5.6) dominates the first term, i.e., both fields are due to the same single particle.

6. INDEPENDENT PARTICLE MODEL

The above discussion has focused on limiting behavior of the joint distribution function for electric fields. To obtain a more global description for all times and fields, suitable approximations are required. In this section we propose an independent particle model that generalizes those used for calculating the static distribution, $Q(\varepsilon)$. An appropriate starting point is the cluster expansion (5.2). In the extreme weak coupling limit it is possible to show that only the leading static and dynamic terms in the series survive,

$$G[\phi] \to \sum_\alpha n_\alpha \int d\vec{r}\ [\phi_\alpha(\vec{r},\lambda) + \phi_\alpha(\vec{r},\lambda')]g_\alpha(\vec{r})$$

$$+ \sum_\alpha \sum_{\alpha'} \int d\vec{r} d\vec{r}'\ \phi_\alpha(\vec{r},\lambda)\phi_{\alpha'}(\vec{r}',\lambda')C_{\alpha\alpha'}(\vec{r},\vec{r}';t) \quad (6.1)$$

where $g_\alpha(\vec{r})$ and $C_{\alpha\alpha'}(\vec{r},\vec{r}';t)$ are calculated in the random phase approximation. Models of the form (6.1), i.e. retaining only the structure of the first two terms in the functional expansion, will be referred to as independent particle models. For noninteracting ions and neutral field points the independent particle model is exact. However, even for noninteracting ions there are higher order terms in the functional expansion if the field point is charged. The result (6.1) should properly be understood as resulting instead from the weak coupling limit (mean field, quantum Vlasov dynamics).

The independent particle model assumes the form (6.1), but replaces $g(r)$ and $C_{\alpha\alpha'}(\vec{r},\vec{r}';t)$ by their exact values. Furthermore, an effective field, $\vec{E}_\alpha^*(\vec{r})$, and an associated function, $\phi_\alpha^*(\vec{r},\lambda)$ (analogous to (5.1)) are introduced. It is straightforward then to express ϕ, and consequently also G, as a functional of functional of ϕ^*. The details are described in reference 10 for the static case and extention to the dynamic case is straightforward. This renormalized independent particle model then has a form similar to (6.1),

$$G[\phi] \equiv G[\phi^*] \to \sum_\alpha n_\alpha \int d\vec{r} [\phi_\alpha^*(\vec{r},\lambda) + \phi_\alpha^*(\vec{r},\lambda')] R_\alpha(r) g_\alpha(r)$$

$$+ \sum_\alpha \sum_{\alpha'} \int d\vec{r} d\vec{r}' \phi_\alpha^*(\vec{r},\lambda) \phi_{\alpha'}^*(\vec{r}',\lambda') R_\alpha(r) R_{\alpha'}(r') C_{\alpha\alpha'}(\vec{r},\vec{r}';t) \quad (6.2)$$

Here, R_α is the ratio of the field magnitudes E_α/E_α^*. The effective field $\vec{E}_\alpha^*$ is now chosen such that the independent particle model approximation for $Q(\varepsilon)$ yields the exact second moment, $<E^2>$ (or related condition for neutral points). With such a choice it is possible to show that the joint distribution function calculated from (6.2) yields the exact time dependent second moments, $<E_i(t)E_j>$. The fundamental symmetries (2.3)-(2.5) are also preserved.

7. DISCUSSION

The limits described in sections 3 - 5 provide in most cases explicit and practical expressions for calculating the joint distribution function from knowledge of the static microfield distribution, $Q(\varepsilon)$, the pair distribution functions, $g_\alpha(r)$, and the electric field autocorrelation function. It appears that good approximations for these quantities already exist for a wide range of conditions, including strong coupling. Problems remain even at this level, however, for more complex plasma conditions (e.g., strong coupling with degenerate electrons and internal ionic states). The limits explore the parameter space of time and field values for small fields at all times, short time for all fields, and large fields at all times.

The simplest expression for $f(\vec{\varepsilon},t;\vec{\varepsilon}',0)$ is (3.1), the Gaussian limit. This describes a diffusive behavior in each of field variable, from an initial delta function to a final Gaussian distribution. This result applies only for the charged field point case (and small field values); the small field behavior for neutral field points is not known. The approach to the Gaussian limit is not simple, however, since the field autocorrelation function $\alpha(t)$ changes sign as a function of t.[3] It is not clear from the formal analysis here what is the range of validity of this approximation. Based on analysis of the corresponding limit for $Q(\varepsilon)$ it is expected to improve with increased charge on the field point, smaller field values, and strong coupling.

The short time limit (4.4) is also Gaussian, but only in one of the two independent field values. This result is applicable to both neutral and charged points at arbitrary plasma coupling. It describes the broadening, shift, and attenuation of the initial delta function on a time scale less than an appropriate inverse plasma frequency. The dependence of the coefficients $A_m(a)$ on $\vec{a} \equiv (\vec{\varepsilon}+\vec{\varepsilon}')/2$ has been investigated thus far only for the independent particle model. This should be accurate more generally in the limit of large a. For the OCP, neutral field point, and Coulomb fields, it is found that $A_3 = 0$ and both A_1 and A_2 have an algebraic growth for large a.[9] This represents enhanced diffusion for large initial field values.

The high field limit (5.6) or (5.8) associates the total field at the field point with a single particle at each time. The corresponding limit for $Q(\varepsilon)$, Eq. (5.5), agrees with the high field Holtzmark and nearest neighbor form. Also, the short time limit of (5.6) or (5.8) agrees with the high field limit of (4.4). This latter comparison indicates that the high field limit for $f(\varepsilon,t;\varepsilon',0)$ applies for field values comparable to those for which $Q(\varepsilon)$ approaches its large field limit. The relationship in (5.6) to the charge density autocorrelation function suggests the possibility of studying some of the long time properties outside the Gaussian limit. Of course, the independent particle model for the OCP, neutral point, and no interactions can be evaluated immediately to give,

$$f(\vec{\varepsilon},t;\vec{\varepsilon}',0) \to Q(\varepsilon)Q(\varepsilon')$$

$$+ \frac{n}{4}(Ze)^3(\pi u^2 t^2)^{-3/2}(\varepsilon\varepsilon')^{-9/2}\exp\left\{-(\frac{\hat{\varepsilon}}{\sqrt{\varepsilon}} - \frac{\hat{\varepsilon}'}{\sqrt{\varepsilon'}})^2 Ze/u^2 t^2\right\} \qquad (7.1)$$

where $u^2 = 2k_BT/m$.

The independent particle model defined by (6.2) is expected to be a good approximation uniformly across field values and time. The basis for this expectation is the validity of the corresponding model for $Q(\varepsilon)$. The renormalization procedure leading to the latter is understood as a partial resummation of the Baranger-Mozer cluster series such that the leading term (independent particle term) is exact through terms quadratic in $\vec{\lambda}$, as well as yielding the correct large $\vec{\lambda}$ limit. The dynamical model described here has the similar property of interpolating between the large and small $\vec{\lambda}$ limits (it is again exact through quadratic terms in $\vec{\lambda}$, $\vec{\lambda}'$). Although the statistical mechanics in this description has been reduced to known quantities, it suffers the drawback of considerable numerical difficulty to perform the remaining integrations. While Monte Carlo methods can be applied, the calculation is inefficient since the function being calculated depends on four variables (two field values, their relative direction, and time) plus state conditions.

In light of the last comment above, we consider it prudent to pursue in parallel the possibility of more phenomenological, but possibly more practical descriptions. The stochastic model of Brissaud and Frisch[5] is an example of such an approach. They assume that the electric field is a stochastic variable governed by a Kangaroo process (a generalization of the discrete Poisson and continuous Kubo-Anderson processes). It has the advantage of admitting exact solutions and the flexibility to contain the exact static distribution $Q(\varepsilon)$ and electric field autocorrelation function, $\langle\vec{E}(t)\cdot\vec{E}\rangle$. However, it has unphysical features such as a remnant of the initial delta function singularity for all times. An opposite extreme is the Fokker-Planck process given by (3.1) generalized as described in section 3 to include the exact static distribution and electric field autocorrelation function. Again, this has the advantage of allowing practical calculations, but it seems to be in disagreement for high field values with the limited computer simulation results.[4] We have studied a stochastic equation for $f(\vec{\varepsilon},t;\vec{\varepsilon}',0)$ that combines both Fokker-Planck diffusive behavior for small fields and Kangaroo behavior for large fields.[11] We hope to have a preliminary evaluation of this approach shortly. In the mean time, it would be useful to have one or more "benchmark" simulations of $f(\vec{\varepsilon},t;\vec{\varepsilon}',0)$ to test such theoretical models.

ACKNOWLEDGEMENTS

This research was supported by National Science Foundation grant PHY 8822581. The research of L. Z. was also supported by a fellowship from the DGAPA of the Universidad National Autónoma de México.

REFERENCES

1) For a recent review with references see J. W. Dufty, Electric microfield distributions, in Strongly Coupled Plasma Physics, eds. F. J. Rogers and

H. E. DeWitt (Plenum, NY, 1987).

2) J. W. Dufty, The microfield formulation of spectral line broadening, in Spectral Line Shapes, eds. B. Wende (W. de Gruyter, NY, 1981).

3) D. B. Boercker, C. A. Iglesias, and J. W. Dufty, Phys. Rev. A$\underline{36}$ (1987) 2254.

4) E. Smith, R. Stamm, and J. Cooper, Phys. Rev. A$\underline{30}$, (1984) 454.

5) A. Brissaud and U. Frisch, JQSRT $\underline{11}$, (1971) 1767 ; J. Math. Phys. $\underline{15}$, (1974) 524.

6) For a recent review with references see S. Ichimaru, Rev. Mod. Phys. $\underline{54}$ (1982) 1017.

7) F. Lado and J. W. Dufty, Phys. Rev. A$\underline{36}$, (1987) 2333.

8) C. Iglesias, J. Lebowitz, and D. Mac Gowan, Phys. Rev. A$\underline{28}$, (1983) 1667.

9) L. Zogaib and J. W. Dufty, Electric field dynamics, in Proceedings of the 9th International Conference on Spectral Line Shapes, (Nicholas Copernicus University Press, Torun, 1988).

10) J. W. Dufty, D. B. Boercker, and C. A. Iglesias, Phys. Rev. A$\underline{31}$, (1985) 1681.

11) L. Zogaib, PhD thesis, UNAM (in preparation).

ELECTRON-ION STRONG COUPLING EFFECTS IN DENSE HYDROGEN
PLASMAS II. ELECTRIC LEVELS OF IMPURITY IONS

Shigenori TANAKA and Setsuo ICHIMARU

Department of Physics, University of Tokyo, Bunkyo-ku, Tokyo 113, Japan

Energy levels of impurity ions Ne^{9+} and C^{5+} embedded in dense hydrogen plasmas are analyzed by adopting specific model descriptions for those radiator atoms. The average polarizations of surrounding plasmas is described through a solution to the HNC/MCA integral equations which take account of correlations between the radiators and the plasma particles. We pay special attention to the electron-ion and the electron-radiator strong coupling effects beyond the RPA, leading to strongly attractive forces between them. We further take account of the effects of plasma particles penetrating inside the bound-electron orbitals. Through comparison between model calculations, we find that the atomic potentials felt by the bound electrons are elevated when the strong coupling effects and the penetration effects are explicitly taken into account; the resulting energy levels become shallower and higher levels disappear.

1. INTRODUCTION

The study on electronic energy levels of impurity ions embedded in dense hydrogen plasmas is important in application to diagnose those high-density plasmas found in the inertial-confinement fusion experiments or in the interior of main-sequence stars. Such a method of analysis is essential in the theoretical predictions on opacities and ionization transitions in a dense material.

In this paper we work with the static screening model of the surrounding plasma particles, as has been adopted by a number of investigators.[1-3] We investigate the extent to which the electronic levels are modified by the electron-ion and electron-radiator strong coupling effects beyond the RPA, and by the effects of plasma particles penetrating inside the bound-electron orbitals.

2. MODELS

We consider a radiator atom with nuclear charge $Z_N e$ immersed in a hydrogen plasma with number density n and temperature T. We assume that the radiator retains a single bound electron so that its electric charge is given by $Z_R e$ with $Z_R = Z_N - 1$. We are thus concerned with the hydrogenic energy levels of the bound electron. The effective potential felt by the bound electron may then be expressed as

$$V_e(r) = -\frac{Z_N e^2}{r} + V_p(r), \quad (1)$$

where $V_p(r)$ is the polarization potential produced by the surrounding plasmas.

In model I, we neglect the polarization potential altogether; the bound electron thus feels the potential produced by the nucleus alone.

We proceed to take into account the screening effects of the plasmas through the radiator-electron and radiator-ion correlation functions.[3] The polarization potential is then calculated in accord with the following three schemes.

In model II, we regard the radiator as a point particle with electric charge $Z_R e$ in the calculation of the radiator-plasma correlation functions, as ref. 3 did. The correlation functions are then obtained through solution to integral equations for the radiator-electron-ion three-component system. Only the radiator-ion, ion-ion and electron-electron strong-coupling effects are taken into consideration in this model with the classical HNC approximations.

In model III, we still assume a point-particle approximation to the radiator. In this model, however, we take additional account of the electron-ion and electron-radiator strong-coupling effects beyond the RPA, as we did in ref. 4. These effects lead to strongly attractive forces between those opposit charges. All the correlation functions are thus calculated through a solution to the HNC/MCA integral equations.[4]

We remark here that the point-particle approximation is valid only when magnitude of the nuclear charge is so large and the principal quantum number under consideration is so small that the radius of the bound-electron orbital is significantly smaller than the mean spacing between plasma particles. An improvement of such an approximation is therefore called for when a high-density material is treated.

The model IV, the most accurate scheme in this work, considers a possibility that the plasma particles may penetrate inside the bound-electron orbitals. We thus construct the effective field felt by the bound electron imposing the boundary conditions near and far from the nucleus as

$$V_e(r) \xrightarrow[r \to 0]{} -\frac{Z_N e^2}{r} + V_p(r; Z_N), \tag{2}$$

$$V_e(r) \xrightarrow[r \to \infty]{} -\frac{Z_R e^2}{r} + V_p(r; Z_R). \tag{3}$$

Here $V_p(r; Z)$ is obtained by regarding a radiator as a point particle with an electric charge Ze. We then interpolate between the near and far fields by introducing a characteristic length distinguishing between the two limiting behaviors; this parameter in turn is determined self-consistently by requiring that the characteristic length be equal to the radius of the bound-electron orbital obtained through a solution to the Schrödinger equation.

3. RESULTS

We take up the cases of impurity ions Ne^{9+} and C^{5+} embedded in dense hydrogen plasmas. We calculated the energy levels for 5 combinations of plasma parameters. In Figure 1 we show a result for the effective potential felt by the 1s bound electron in the case of Ne impurity at $T = 1.71 \times 10^7 K$ and $n = 2.56 \times 10^{26} cm^{-3}$; the resulting energy levels (expressed in units of eV) are listed in Table 1 and are compared with the results in the Debye-Hückel (DH) and ion-sphere (IS) models.

We find in the figure that the atomic potentials felt by the bound electron are elevated as the strong coupling efects and the penetration effects are explicitly taken into account. The resulting energy levels accordingly become shallower and the higher levels vanish in model IV, as seen in Table 1. We remark in this connection that the radius r_{nl} of 1s orbital calculated in the model IV is comparable with the interparticle spacing of the surrounding plasmas, which implies the significance of the penetration effects.

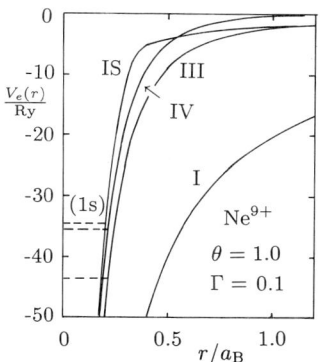

FIGURE 1
Effective potentials.

TABLE 1
Energy levels.

n	l	I	II	III	IV (r_{nl}/a_B)	DH	IS
1	0	-1360.58	-678	-592.90	-484.14 (0.175)	-593.42	-474.50
2	0	-340.15	-12.9	-11.30	——	-11.15	-8.42
2	1	-340.15	-3.65	-3.37	——	-3.56	——
3	0	-151.18	-3.24	-2.91	——	-3.04	-2.63
3	1	-151.18	-1.59	-1.44	——	-1.57	-1.51
3	2	-151.18	-1.51	-1.37	——	——	——

In light of the results with other parametric combinations as well, we have found that an inclusion of strong coupling effects elevates each energy level and that those effects become more pronounced as the plasma coupling gets stronger. The penetration effects manifest themselves when the interparticle spacing of the plasma particles becomes on the same order of magnitude as the radius of the bound-electron orbital.

REFERENCES

1) S. Skupsky, Phys. Rev. A **21** (1980) 1316.

2) U. Gupta and A. K. Rajagopal, Phys. Rep. **87** (1982) 259.

3) X.-Z. Yan and S. Ichimaru, Phys. Rev. A **34** (1986) 2173.

4) S. Tanaka, X.-Z. Yan, and S. Ichimaru, Electron-ion strong coupling effects in dense hydrogen plasmas I. Equation of state and electric conductivity, this volume.

EQUATION OF STATE AND OPACITY OF DENSE PLASMAS

F.J. ROGERS, and C.A. IGLESIAS

Physics Department, Lawrence Livermore National Laboratory
Livermore, CA 94550

Recent research on the equation of state and opacity of dense plasmas has been strongly driven by discrepancies between calculation and observation of a wide variety of astrophysical quantities such as, the solar neutrino flux rate, the period ratio of Cepheid variable stars, and lithium depletion in the Hyades stellar cluster. A summary of some recent work based on a many-body activity expansion of the grand-partition function is given. An integral part of this work involves methods for calculating atomic data and occupation numbers for excited states in dense plasmas.

1. INTRODUCTION

The properties of partially-ionized plasmas have been of interest for a long time, in large part due to their occurrence in stars. Until recently it was thought to be sufficient to model these properties with simple approximations, e.g., the Saha ionization equilibrium equation, hydrogenic scaling of photoabsorption processes, etc. However, measurements of the solar neutrino flux rate from the Davis Cl^{37} experiment[1], the development of Helioseismology[2], observation of period ratios in Cepheid variable stars[3] and the observed low abundance of lithium in the Hyades cluster[4], for example, have led to discrepancies between theory and observation that may be in part due to the simplistic modeling of partially-ionized plasmas, particularly the opacity. In addition to stars, partially ionized plasmas are now routinely produced and studied in the laboratory (see papers by Fortov[5], Popovich[6], Mostovych[7], Benage et. al[8] this volume) and in practical applications.

In the present paper we give a review of our work on the equation of state and opacity of partially-ionized multi-component plasmas. Due to the fundamental importance of these plasma properties, as mentioned above, several other groups are independently pursuing similar programs.[9,10,11] Recent work that applies Green's function methods to obtain the equation of state and transport properties of partially-ionized plasmas is given in the review by Kraeft.[12]

Construction of an opacity code is a three part process: 1) Equation of state; 2) Atomic physics; and 3) Photoabsorption processes. Our approach for each of these parts is briefly summarized in the following section. The emphasis is placed on the equation of state since it is most closely connected to this subject of the volume and it also forms the physical basis of our opacity calculations.

2. EQUATION OF STATE

A common method to obtain the equation of state is through average atom density function methods, particularly in the Thomas-Fermi approximation.[13,14] This method works best at high density while detailed statistical mechanical methods are more suitable for low to moderate density. This is especially true when opacity is also to be calculated since lowly occupied states, far removed from the dominate species, may strongly influence the photoabsorption cross-section.[15] These detailed statistical mechanical approaches are commonly characterized as being in the "chemical picture" when a free energy expression is explicitly minimized. Recent work in the chemical picture has been carried out by Mihalas and his collaborators[16-19], Sevastyanenko[20], Ebeling and his collaborators[21-23] and by Saumon and Chabier.[24]

The starting point in the chemical approach is a free energy expression allowing for all nuclear types and bound ionic species that can be formed from them. This requires that one assert a knowledge of the effect of the full system on the internal states of atoms and ion. This is true of even the recent work of Hummer and Mihalas, i.e. they appeal to experiment to conclude that there is no significant plasma effect on the bound energy levels.

On the other hand, we use the so-called "physical picture" approach starting from the grand canonical ensemble of a classical system of electrons and nuclei interacting through the Coulomb interaction. Quantum mechanics is inserted only in the final expression after a completely global classical result, free of classical Coulomb divergencies, has been obtained. The effect of the plasma environment on the internal states is obtained directly from the statistical mechanical analysis. Our approach is based on the fact that many particle correlations are highly classical and, consequently, it is restricted to regions where the de Broglie wavelength is less than the plasma screening length. We have, however, developed methods for systematically introducing quantum mechanics into the many-particle correlations.

Initially, we treat the plasma as though it is an ordinary gas with short range interaction. The first step is, thus, to develop the logarithm of the grand partition function into an activity expansion, i.e.

$$\frac{1}{v} \ln \Xi = \beta P (\{z_i\},\{b_j\}) \qquad (1)$$

where

$$z_i = \lambda_i^{-3} e^{-\beta \mu_i / kT} \quad (2)$$

is the activity for species i, i ranges over electrons and ions, and the b_i are species dependent cluster coefficients.

The next step is to eliminate the activity in favor of the density to obtain a multi-component virial expansion, i.e. carryout the transformation

$$\beta P(\{z_i\},\{b_i\}) = \beta P(\{\rho_i\},\{S_i\}) \quad (3)$$

where ρ_i is the number density of species i and the B_i are species dependent virial coefficients. Eq (3) works very well for the fluid phase of neutral gases when the B_i are accurately known. Frequently, due to the difficulty of measuring or calculating the B_i, they are calculated approximately from a pair potential. In this approximation it is much easier, instead, to use modified integral equation methods.[25,26] A similar observation can be made for plasmas, except, it is necessary to introduce a pseudopotential to treat the electron-ion interaction.[27]

The virial expansion given by Eq (3) is insufficient at low temperature for short-ranged attractive interactions, due to the formation of dimers, trimers, etc. This is because the B_i become exponentially large due to the presence of bound states and it is necessary to associate these terms with new variables for composite particles.[28] Due to the physical connection of the cluster coefficients with bound states it is natural to carryout this renormalization in the activity expansion. In the case of plasmas, the virial expansion given by Eq (3) is insufficient even for full ionization, since the B_i are all divergent, and it must be diagrammatically redeveloped to remove these divergencies[29], i.e.

$$\beta P(\{\rho_i\},\{B_i\}) = \beta P(\{\rho_i\},\{S_i\}) \quad (4)$$

where the S_n are the Abe functions. The function S_1 occurring in Eq. (4) is defined to be the summation over ring diagrams and the higher S_i are closely related to the virial coefficients of the Debye potential. For example the S_2 function is given by

$$S_2 \equiv \sum_{ij} \rho_i \rho_j s_{ij} \quad (5)$$

where

$$s_{ij} = -B_{ij}(T,\lambda_D) + \Phi_{ij} \qquad (6)$$

$$\Phi_{ij} = 2\pi \int_0^\infty r^2 dr \left(\beta v_{ij} - \frac{\beta(v_{ij})^2}{2} \right), \qquad (7)$$

λ_D is the Debye length and V_{ij} is the Debye-Hückel potential between particles of type i and j.

The Abe series is a divergence free expansion of the pressure for fully ionized plasmas but it also suffers from the exponential increase at low temperature that the virial series encounters for associating neutral gases; thus requiring a renormalization of the equivalent activity series. To obtain the plasma activity expansion we could go all the way back to Equation (1) but, in fact, an easier route is available. That is, since the Abe procedure has already solved the classical divergence problem for the canonical ensemble, go from it to the equivalent grand canonical result. This gives[30]

$$\beta P(\{\rho_e\},\{S_i\}) = \beta P(\{z_i\},\{\hat{o}_i S_j(\{z_i\})\}) \qquad (8)$$

where $\hat{o}_i$ is a differential operator acting on the S_i. For a one component plasma the result is explicitly

$$\beta P = z + S + \sum_{m=2}^{\infty} \frac{z}{m!} \left(\frac{\partial}{\partial z} z\right)^{m-2} \left(\frac{\partial S}{\partial z}\right)^m \qquad (9)$$

where

$$S = -\sum_{j=2}^{\infty} \frac{z^j \beta_j}{j-1} \qquad (10)$$

is the basic definition for a short-ranged potential and

$$S = \sum_{j=1}^{\infty} S_j \qquad (11)$$

for the redeveloped form appropriate for the Coulomb interaction. In Eq (11) S_1 is the ring sum, S_2 is given by Eq (5) with ρ replaced by z. The sum equivalent to Eq (1) occurring in the canonical ensemble is just the negative excess free energy, whereas, Eq (1) does not appear to have a simple physical meaning.

The pressure expression given by Eq (8) has an additional complication that must be removed before it can be renormalized to introduce additional activity variables for composite particles. That is, it involves the function S which is expressed in terms of virial coefficients, whereas, to renormalize it to account for composite particles we need to work with cluster coefficients having direct physical meaning. The next step is thus to carry out the transformation

$$\beta P(\{z_i\},\{\delta_{\underline{n}} S_{\underline{n}}\}) = \beta P(\{z_i\},\{C_{\underline{i}}\}) ,\qquad(12)$$

where $C_{\underline{1}}$ is like $S_{\underline{1}}$ in that it is a sum over ring diagrams but in addition it involves sums over various type of linked ring diagrams. The $C_{\underline{i}}$ for $\underline{i} \leq 2$ are built up from the $S_{\underline{i}}$ similar to expressing the $b_{\underline{i}}$ in terms of the $B_{\underline{i}}$. The first few $C_{\underline{i}}$ are given by

$$C_{\underline{2}} \equiv S_{\underline{2}} \qquad(13)$$

$$C_{\underline{3}} \equiv S_{\underline{3}} + \frac{z}{2}\left(\frac{\partial S_{\underline{3}}}{\partial z}\right)^2 .\qquad(14)$$

We now replace the classical $b_{\underline{i}}$ for the screened Coulomb potential appearing in the $C_{\underline{i}}$ with their quantum mechanical counterparts, i.e. replace integrals with traces.

The quantum mechanical version of Eq (1) is next renormalized to account for the formation of composite particles, i.e., we carryout the transformation[31]

$$\beta P(\{z_i\},\{C_{\underline{i}}\}) = \beta P(\{z_i\},\{C'_{\underline{i}}\}) ,\qquad(15)$$

where C'_j ranges over electrons, nuclei, and composite particles. The C'_j, occurring is Eq (15), do not include strong bound state contributions since they have been used to create new activity variables.

A detailed description of the equation and methods used to calculate the equation of state of partial ionized plasmas, allowing for strong coupling of the heavy ions, is given in reference (31). A major outcome of the analysis is that strong bound states are unshifted by the plasma while the continuum moves down with increasing Coulomb coupling. The Debye-like screening that was originally in the bound state parts of the C_j, associated with the nuclear index i, has been used to create new activity variables for composite particles such that i → i'. When the continuum is lowered to the vicinity of a bound state, that state is no longer used to define a new composite particle variable. Instead it remains explicitly in the C_j along with the scattering contributions. The contributions to C'_j are screened by a Debye-like potential. In other words, the number of states is determined by screened potential, but their energies are not.

3. ATOMIC PHYSICS

The result that bound states, except possibly those near the plasma continuum edge, are unshifted allows great simplification in atomic data calculations. It means that we can use isolated atom bound state energy levels and standard line broadening theory to calculate bound-bound and bound free absorption cross-sections. This is fortuitous, since, in order to calculate the equation of state and opacity of an astrophysical mixture of twelve or more materials, requires a knowledge of a very large number of energy levels and oscillator strengths. One possibility for doing this is to create a large data base and develop sophisticated algorithms for data retrieval. We have instead opted to develop a procedure for calculating the data online from prefitted effective potentials.[32]

For conceptual purposes, when discussing the effective potentials, it is convenient to define the electron configurations as having two components. The first component is a "parent" configuration consisting of all the electrons in a given configuration except one. The excluded electron defines the second component or "running" electron. The parent configuration defines the effective potential for all the subshells available to the running electron. Our analytic effective potentials are a sum of Yukawa terms, one for each occupied shell in the parent configuration.

$$V_{ei}(r) = -\frac{2}{r}\left[(Z-\nu) + \sum_{n=1}^{n^*} N_n e^{-\alpha_n r}\right] \quad (16)$$

(in Rydgergs), where

$$\nu = \sum_{n=1}^{n^*} N_n \quad (17)$$

is the number of electrons for the parent ion, N_n the number of electrons in the shell with principal quantum number n, n* the maximum value of n for the parent configuration, and α_n the screening parameter for electrons in shell n. For continuum states the long-range tail is expontentially screened by the plasma.

The screening parameters are obtained from an iterative solution of a spin averaged Dirac equation by matching the experimental one-electron ionization energies. The screening parameters, thus obtained for each isoelectronic sequence, are fitted to simple polynominals. For each parent configuration the potential $V_{ei}(r)$ is also used to calculate the configuration-averaged energies for all relevant excited states available to the running electron.

Opacity calculations require atomic data corresponding to numerous excited parent configurations, including core excitations. Screening parameters for these excited parent configurations are obtained from scaling laws that adjust the ground-state parent screening parameters. In this procedure the Dirac equation is first solved for states of the running electron in the ground state parent potential. These results then are used to rescale the screening parameters to account for promotions of electrons within the parent configuration. Term splitting of the levels is introduced through perturbation theory.[33]

4. OPACITY

In this section we discribe the OPAL opacity code being developed at LLNL. At present the code is restricted to temperatures sufficiently high (above 1 eV) that molecular absorption is negligible and photon energies low enough (less than about 10 kev) that relativistic effects are small. The code is thus well suited to computing opacities for laser produced plasmas and stellar interiors. In the parameter range currently being treated the dominant absorption processes are electronic transitions in the field of ions; i.e. line transitions (bound-bound); photoionization (bound-free), and inverse bremsstrahlung (free-free). At high temperatures, when the ions are highly striped, photon scattering from free electrons is also important. A brief description of how these cross-sections are calculated follows.

As described in the equation of state section, many-body statistical mechanics indicates that bound state energies are unscreened. Bound-bound transitions thus depend on the unscreened line locations and associated oscillator strengths. Using the parametric potential we can obtain 1% or better accuracy for the configuration averaged energies when compared to experiment for ions relevant to the solar interior. In order to match real experiments, it is necessary to include the configuration term splitting. The angular momentum coupling is done using standard perturbation theory methods and includes either Russell-Saunders or intermediate coupling depending on the levels and nuclear charge. The term splitting results are not as accurate as the configuration averages but are better than 10% when compared to experiment. The strong oscillator strengths are also in the order of 10% accuracy, but very little experimental data are available. Of course, the comparisons are done for isolated ions and it must be emphasized that little is known of how ions behave under the extreme conditions of stellar interiors. Not all transitions are computed with term splitting. Transitions from lower bound levels with quantum number greater than 5 are considered in the configuration average only. If in the future this "switch" is insufficient (perhaps in the much colder regions in the photosphere) it is easily changed in the code to some higher value.

In nature, spectral lines experience broadening, for example, Doppler effects, collisions, and natural lifetimes. In OPAL lines from single electron ions are treated with standard linear Stark theory[34]. These line shapes are in good agreement with experiment. In the near future we will include similar calculations for Helium-like and Lithium-like ions which have also compared well with experiments. For all other transitions we use Voigt profiles where the Gaussian width is due to doppler broadening and the lorentz width is due to estimated natural plus electron impact collisions.[34] The latter is computed using the second order dipole approximation. These are not scaled hydrogenic results but use the same wavefunctions as in the atomic data calculations.

Bound-free cross sections are computed explicitly for all levels with angular momentum quantum number less than 5. For levels with principal quantum number greater than 5, the configuration term structure is neglected. The resulting bound-free cross sections have compared well with experiment even for neutral atoms[35] except when configuration interaction effects are important. Such effects, however, are expected to be very small for solar interior calculations. For levels with $\ell \geq 4$ we use scaled hydrogenic cross sections.

Since the parametric potential model provides good results for bound-free cross sections where both bound and scattering states are required, we assume that the parametric potential method will also be valid for free-free calculations where only scattering states are necessary. We compute explicitly the dipole matrix elements except in some limiting regions where simpler approximations are valid. For example, for small photon energies elastic scattering cross sections are useful and easy to compute. Plasma screening effects are introduced into the electron-ion interaction. These effects are described by Rogers[35] and reduced to the Debye-Hückel result for weakly coupled plasmas. This approach provides two improvements over calculations using Coulomb Gaunt factors. The first is corrections due to the plasma screening at small photon energies. The second is corrections at large photon energies where the scattering electron can penetrate bound electron orbits and see a higher effective charge.

The treatment of photon scattering from free electrons follows Boercker.[36] There, the transport cross section is computed including the many electron effects; that is,

$$\sigma_{sc}(k) = \sigma_t \int_{-1}^{1} d(\cos\theta)[1+\cos^2\theta][1-\cos\theta] \, S(k) \tag{18}$$

where

$$k = \frac{4\pi}{\lambda} \sin\left(\frac{\theta}{2}\right) \tag{19}$$

$$S(k) = 1 + h_{RPA}(k) + h_\chi(k) \tag{20}$$

Presently we consider conditions such that, collisions between the plasmas consitutents are sufficiently frequent that the plasma can be assumed to be in local thermodynamic equilibrium (LTE). Since the mean free path of an average photon is small compared to the scale of the matter temperature gradients inside a star, the photons are in equilibrium with the matter and have a black body spectrum at the material temperature. The transport of photons will then be well described by the diffusion approximation with the diffusion constant given by the Rosseland mean opacity, K_R,

$$\frac{1}{K_R} = \int_0^\infty du \, \frac{1}{\tilde{K}(u)} \, \frac{\partial B(u,T)}{\partial T} \tag{21}$$

where the weighting function

$$\frac{\partial B(u,T)}{\partial T} = \frac{15}{4\pi^4} \, \frac{u^4 e^u}{(1-e^u)^2} \tag{22}$$

peaks at u (=photon energy/T)=4 with T the matter temperature in units of energy. The extinction coefficient is defined by

$$\tilde{K}(u) = K_{abs}(u)(1-e^{-u}) + K_{sc}(u) \tag{23}$$

where K_{abs} is the absorption coefficient and K_{sc} is the scattering cross section,

$$K_{abs}(u) = \sum_{ijk} N_{ijk} \, \sigma_{ijk}(u)$$

$$K_{sc}(u) = N_e \, \sigma_{sc}(u) \, . \tag{24}$$

Here, N_{ijk} is the number of ions of charge j in electronic level k of element species i, $\sigma_{ijk}(u)$ is the absorption cross section for photons with energy u by those levels, and N_e the free electron number density with $\sigma_{sc}(u)$ the photon scattering cross section. The Rosseland mean opacity, given by Eq (21), is the usual input for astrophysical calculations.

5. OPACITY CALCULATIONS

The classical Population I Cepheids have been important in astrophysics as distance indications and as tests of stellar evolution theories. However, present stellar models give period ratios larger than those observed for the double-mode Cepheids.[37] Following a suggestion by Simon[38] we examined the Rosseland mean opacities for astrophysical mixtures at $\rho=10^{-5}$ gm/cm^3 and T=20 and 60 eV. We compared our results with the Los Alamos Astrophysical Opacity Library[39] used in many stellar model calculations. No significant differences were seen at the higher temperature, but a factor of 2.2 increase in the opacity was computed at the lower temperature. It is interesting that Andreasen[3] found that such an increase in the opacity goes a long way towards removing the discrepancies between theoretical and experimental period ratios.

Several studies[40,41] suggest that a 10-20% increase in the opacity of solar matter just below the convection zone can yield very good agreement between computed and observed low degree P-mode oscillations. Our recent calculations[7] along a ρ-T tract from the solar center to conditions above the convection zone boundary gives such an increase in opacity.

This work was performed under the auspices of the U.S. Department of Energy by Lawrence Livermore National Laboratory under contract No. W-7405-ENG-48.

REFERENCES

1) J.N. Bahcall, R.K. Ulrich, Rev. Mod. Phys. 60 (1988) 297.
2) J. Christensen-Dalsgard, D. Gough, and J. Toomre, Science 229 (1985) 923.
3) G.K. Andreasen, Astron. Astrophys. 201 (1988) 72.
4) P.H. Bodenheimer Ap. J 142 (1965) 451.
5) V.E. Fortov, Physics of strongly coupled plasma in dynamic experiments, this volume.
6) M.M. Popovic', Interprethation of experimental values of DC conductivity and spectral lineshape, this volume.
7) A.N. Mostovych, Laser produced optically thin strongly coupled plasmas, this volume.
8) J.F. Benage, L.A. Jones, R.J. Trainor, W.R. Shanahan, R.L. Shepherd, and D.P. Nothwong, Study of strongly coupled plasma produced in capillary discharge, this volume.
9) M. Seaton, J.Phys. B: Atom. Molec. Phys. 20 (1987) 6363.
10) R.E.H. Clark and A.L. Merts, J. Quant. Spectrosc. Radiat. Trans. 38, (1987) 287.
11) B.F. Rosznyai, Ap. J. 341, (1989) 414.
12) W.D. Kraeft, Thermodynamics and transport in partially ionized plasmas, this volume.

13) C. Dharma-Wardona, Density functional approach to particle correlations and electronic structure in dense plasmas, this volume.

14) S. Eliezer, A. Ghatak and H. Hora, An introduction to equations of state: Theory and applications (Cambridge Univ. Press, London, 1986).

15) V.V. Dragalov and V.G. Novikov, High temperature 26 (1989) 660.

16) W. Däppen, L.S. Anderson, and D. Mihalas, Ap. J. 319 (1987) 195.

17) D.G. Hunner and D. Mihalas, Ap. J. 331 (1988) 794.

18) D. Mihalas, W. Däppen, and D.G. Hummer, Ap. J. 331 (1988) 815.

19) W. Däppen, D. Mihalas, H.G. Hummer, and B. Mihalas, Ap. J. 332 (1988) 261.

20) V. Sevastyanenko, Beitr Plasmphys. 25 (1985) 151.

21) W. Ebeling, Equation of state and ionization of dense plasmas, in: Inside the Sun, eds. G. Berthomieu and M. Cribier (Klawer Academic Publ. Dordrecht, 1989).

22) W. Ebeling and Kilmann, Z. Naturforsch 449 (1989) 519.

23) W. Ebeling, Contrib. Plasmas Phys. 29 (1989) 165.

24) D. Saumon and G. Chabrier, Phys. Lett. 62 (1989) 2397.

25) G. Zerah and J.P. Hansen, J. Chem Phys. 84 (1986) 2336.

26) F.J. Rogers and D.A. Young, Phys. Rev. A30 (1984) 999.

27) F.J. Rogers, Phys. Rev. A29 (1984) 868.

28) W. Ebeling, W.D. Kraeft, and D. Kremp, Theory of Bound States and Ionization Equilibrium in Plasmas and Solids (Akademie - Verlag, Berlin 1977).

29) R. Abe, Prog. Theor. Phys. 22 (1959) 213.

30) F.J. Rogers and H.E. DeWitt, Phys. Rev. 8 (1973) 1061.

31) F.J. Rogers, Phys. Rev. A24 (1981) 1531.

32) F.J. Rogers, B.G. Wilson, and C.A. Iglesias, Phys. Rev. A38 (1988) 5007.

33) R.D. Cowan, The theory of atomic structure (University of California, Berkeley, 1981).

34) R.W. Lee, J. Quant. Spectras. Radiat. Transfer (1988) 561; H.R. Griem, Spectral line Broadening by plasmas, (Academic Press, New York, 1974).

35) F.J. Rogers, Ap. J. 310 (1986) 723.

36) D.B. Boercher, Ap. J. Lett. 316 (1987) 316.

37) A.N. Cox, Ann. Rev. Astr. Ap. 18 (1980) 15.

38) N.R. Simon, Ap. J. 260 (1982) L87.

39) W.F. Huebner, A.L. Merts, N.H. Magee, M.F. Argo, Los Alamos Scientific Report LA-6760-M.

40) S.G. Korzennik and A.K. Ulrich, Ap. J. 339 (1989) 1144.

41) A.N. Cox, J. Gusih, and R. Kidman, Ap. J. 342 (1989) 342.

42) C.A. Iglesias and F.J. Rogers, Astrophysical opacities at LLNL, in: Inside the Sun, eds. G. Berthomieu and M. Cribier (Klawser Academic Publ., Dordrecht, 1989).

SOME INTERPRETATION OF EXPERIMENTAL VALUES OF DC ELECTRICAL CONDUCTIVITY AND SPECTRAL LINE SHAPE

Marko M. POPOVIC, Yves VITEL[†] and Anatolij A. MIHAJLOV

Institute of Physics—Belgrade, Maksima Gorkog 118, 11080 Zemun, Yugoslavia
[†]Univeresite "P. et M. Curie", Paris, France

Pulsed arc in which relevant measurements of electroconductivity and the shapes of spectral lines could be conducted has been chosen. The results of the measurements of electroconductivity and the spectral lines width are given. The disagreements with the usual calculations are explained by an appearance of some other particles interactions in dense nonideal plasmas.

1. INTRODUCTION

Electroconductivity of plasma can be determined properly by the precise measurements of the basic measurable parameters such as: volt-ampere characteristics, electrical field strength and electrical current, axis temperature and radial temperature profiles.[1] In order to get a more complete knowledge of plasma, it is helpful to know the types and quantities of materials which are converted to plasma (initial gas pressure in the case of gaseous plasma), total pressure in the measuring time and quantity of particles on which the pressure depends, especially the concentration of electrons. These parameters are measurable in some experiments and, with the knowledge of plasma geometry, they enable the proper and precise determination of plasma electroconductivity mentioned above.[2]

Theoretical calculation of the electroconductivity of multi component plasmas of higher but not extreme densities—those which deviate from an ideal plasma but still are not a strongly coupled plasma, is rather complicated and has never been conducted starting from the fundamental principals but always some form of modeling has been used.[3] Also, there is a discrepancy between the calculated and measured values of electroconductivity. In numerous cases the measured value is lower than the calculated one, although in some cases it exceeds the so called "Spitzer" value.[8,9] Calculation using modeling gives somewhat lower values than the experimental ones.[10] One of the essential problems that appears in the calculations is the choice of an effective potential of the electron-ion interaction.[3,11] This paper will give a short review of the experimental possibilities for obtaining a plasma well defined in space and time. We will analyse parameters measurable directly and their influence in determining plasma parameters and its electroconductivity. The second part of the paper we will dedicate to a more detailed analysis of the cases in which an electroconductivity value experimentally determined is reliable so that it could be used to prove theoretical calculations. We will also discuss a possible contribution of elelctron scattering on atoms both in the ground and excited states. We will especially point out the possibility of electrons scattering on oscillating microfields in plasma causing the lower values determined experimentally.[12]

The radiation of plasmas of higher densities—which includes weakly nonideal plasmas, is dominantly a continuum with the lines from neutral and ionized atoms superimposed on it. The determination of the real profiles and shapes of spectral lines is more difficult due to selfabsorption in the plasma. It is even more difficult in the cases of high density plasmas which are optically thick and the line spectrum originates from the boundary layers of the plasma. In these cases neither the geometry nor the parameters of the plasma from which the radiation appears are not determined precisely, so the measured values can be indicative only. Only measurements in which above difficulties can be eliminated could be submitted to a serious interpretation.[13-17]

In the third part of our paper, besides the various important experimental details which ensure correct measurement of the spectral line shapes and intensities in weakly nonideal plasmas, we will give the available experiental results and compare it with existing calculations. We will try to explain the disagreement of the experimental and calculated values of the spectral line widths and shifts.

In the last part of the paper we will suggest further work on the experimental determination of DC electrical conductivity and spectral line shape.

2. PRODUCTION AND DIAGNOSTICS OF HIGH DENSITY PLASMAS

The detailed description of different ways for the production of high density plasmas, in the range from weakly nonideal plasma to the plasma of extreme parameters, have been given in review articles and monographies published before.[18-21] We do not have enough space to analyse of all of them here. We will give only some criteria which we consider to be fulfilled in experiments from which one expects to get proper conclusions about electroconductivity and spectral characteristics of plasmas. We take into account that the experiments in which nonideal plasma is produced are mostly conducted with some other, at first practical, technological requirements and that plasma usually is just either an instrument for conducting some technological process or a difficulty encountered in such a process. Thus, a better knowledge about plasma characteristics—primarily thermodynamical, sometimes kinematical or optical, is desirable in order to better conduct a process which can be empirically settled even without that knowledge.

In an experiment in which we want to determine DC electroconductivity of plasma, and that is usually a stationary discharge or stationary discharge with additional pulses or a pulsed electrical discharge in gas (shock tube, furnace with high pressure or some other ways for plasma production are less convenient for this purpose), we can have a known initial pressure—and thus the number of particles in the process also. The duration of plasma is many orders of magnitude longer than any given process in the plasma. Plasma has a geometry given in advance, which can be controlled during the duration of the plasma. Pulsed arcs in a quartz tube have an advantage due to axial homogeneity of plasma and the same value of the electrical field by the whole positive column, while in stationary arcs it is rather difficult to differentiate between plasma in the "metallic ring" of the cascade arc and free positive column between cascades. That is the reason why in our work we are mostly concentrated to the study of plasma produced in a pulsed arc in flash-lamps charged with inert gases and/or hydrogen.[14,21,23,24] Regarding the diagnostics and possibilities of precise measurements of all the parameters necessary for the determination of electroconductivity see Ref. 2. Here we will just shortly give which control and diagnostic components such a typical experiment contains.[12] The setup of the experiment is given at the Fig. 1.

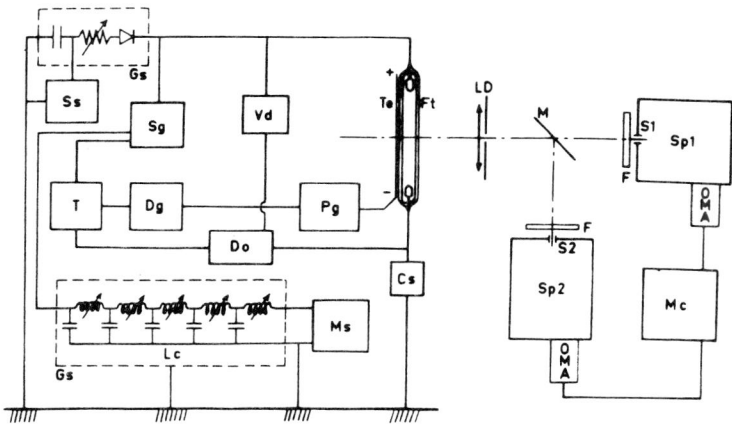

FIGURE 1
The experimental set-up. Cs: low resistance coaxial shunt, D: diaphragm, Dg: delay generator, Do: digital oscilloscope, F: high order rejecting filter, Ft: flashtube, Gs: grounded shield, L: lens, Lc: LC cells with tunable L, M: movable mirror, Mc: microcomputer, Ms: main supply, OMA: optical multichannel analyzer, Pg: pulse generator, Sg: spark-gap, Sp1, Sp2: spectrographs, Ss: simmer supply, S1, S2: entrance slits, T: trigger, Te: trigger electrode, Vd: voltage divider.

Not included in the figure above is a laser interferometer with three mirrors or a modified Mach-Zehnder interferometer.[25]

In this way, it is possible to determine independently the temperature of plasma from the continuum or from the optically thick infrared atomic line. The temperature profile can be determined by either side-on or the end-on measurements. The difference between the values measured (Fig. 2) is within the experimental error limits.[26] In the end-on measurements a special L-shape flashtube is used. Then the measurements are conducted under the high optical thickness and the plasma emits like the blackbody (Fig. 3).

The electron density profiles $n_e(r)$ are calculated using the radial distribution of the continuous emission coefficient, obtained through an Abel inversion in the ultraviolet part of

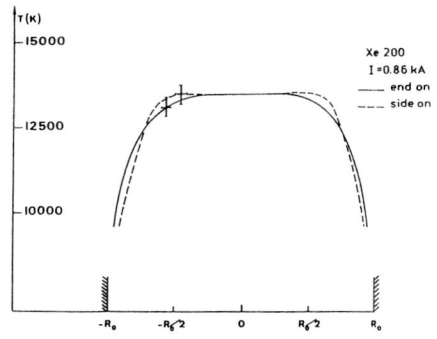

FIGURE 2
Temperature profile of the plasma column of pulse dischage in Xenon. R_0–tube wall, 0–axis, $\frac{\Delta T}{T} \leq 3\%$.

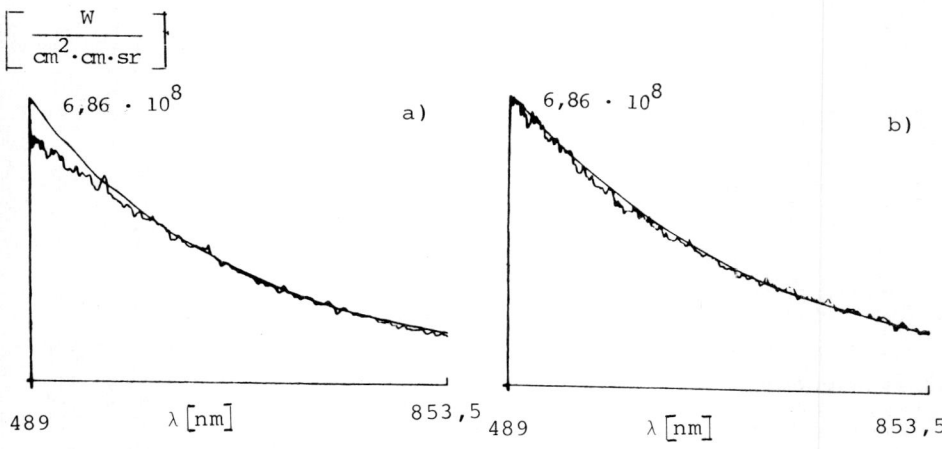

FIGURE 3
End-on emitted radiation from L-Shaped pulse-arc compared with blackbody radiation at axis temperature; a) Kr, $T = 14.900$K, b) Xe, $T = 23.300$K. It is espied that in the case of Kr plasma only at higher wave lengths are emitted as a blackbody while in the case of Xe plasma they are emitted as a blackbody in the whole interval of the wave lengths measured.

the spectrum, where plasma is optically thin. This can be checked in two ways. The first one is by comparison with the mean value obtained by laser interferometry, and the other one deducted from the width or the well known hydrogen lines adding 1–2% of hydrogen to the basic gas. The profile $n_e(r)$ is given in the Fig. 4.

It is certain that when different and independent methods are used the electron density has a 5% precision and the temperature is known with a 3% precision.[12]

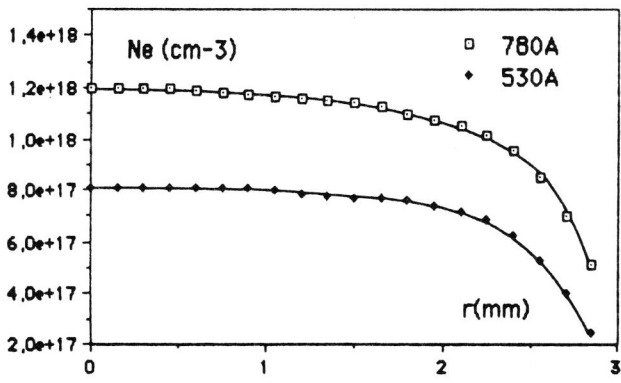

FIGURE 4
Electron density profiles in Xe plasmas for various currents.

The current intensity is directly measured by a low resistance coaxial shunt and electrical field strength by means of a precise voltage divider. The precision in both the cases is better than 1%. The electrode voltage drop can be eliminated introducing auxiliary electrodes.[27]

3. DC ELECTRICAL CONDUCTIVITY

The local values of electrical conductivity in a cyllindrically symmetric plasma are derived from Ohm's law procedure described in Ref. 1. This is a consequent procedure with well defined and controlled numerical accuracy but with high demand in accuracy of basic experimental information. For the pulse arc described in section 2, we consider these requirements fulfilled and the electroconductivity values determined experimentally are of 10% precision.

In Table 1, the measured and Spitzer's values of electroconductivity for Argon and Xenon are given. Differently from representing the results up to now, where different values of both the temperature and the concentration corresponded to the same measured value of electroconductivity, we have succeeded to select 3–4 points with the same temperature and different concentration from numerous measurements. As it has been espied with that, σ_{sp}/σ_{exp} relation increased with the increase of conceantration. The difference can not be explained by the electron-atom collisions. We estimate the neutral contribution to the conductivity always less than 20%. So, we conclude that electrons should be disturbed in their movement. Using the idea of Kurilenkov and Valuev[28] we can see that this disturbance can originate from the scattering of electrons by the oscillating microfields.

TABLE 1
Measured and calculated values of electroconductivity for Argon and Xenon.

	n_a 10^{25} m^{-3}	n_e 10^{23} m^{-3}	σ_{sp} Ω^{-1} m^{-1}	σ_{exp} Ω^{-1} m^{-1}	σ_{sp}/σ_{exp}
	0.35	5.9	9.100	8.300	1.09
Argon	0.71	10.1	9.900	7.900	1.25
$T = 16.400$K	1.42	13.1	10.300	7.600	1.35
	2.13	15.4	10.600	6.400	1.65
Xenon	0.71	6.2	6.900	4.640	1.48
$T = 12.400$K	1.42	11.2	7.570	4.380	1.73
	2.13	12.5	7.720	4.110	1.88
Xenon	0.71	7.2	7.120	4.800	1.48
$T = 12.600$K	1.42	12.6	7.860	4.630	1.70
	2.13	14.0	8.010	4.350	1.84

4. PROFILE OF THE SPECTRAL LINES

In the experiments with a pulsed discharge in inert gases, in spite of strong continuum and very large lines, it is possible to select a few lines convenient for study.[14] The selfabsorption must be controlled very strictly. It is controlled by the twofold optical path, with a mirror of known reflection installed behind the arc. The line studied has to be set apart and the detection set adjusted to correct automatically removing the continuum contribution to the radiation intensity. The experimental line profile of ArII 480,6nm is given as an example in Fig. 5. The profile seems to be ideal until the concentration reaches a value of 5×10^{23} m^{-3}. With higher concentrations it seems that on the ideal profile is superposed some oscillations. This experimental fact needs a more detailed systematical analysis.

Experimental values of half-widths and shifts of lines show the dependence on charged particle concentrations in plasma (Fig. 6).

In Figure 6, the measured values of the spectral line half-widths and shifts for ArII 480,6nm and ArII 484,7nm are compared with the calculated values using Grim's semiempirical formula.[29] It is obvious in the Figure that all the values calculated and measured are in agreement till the value of 5×10^{23} m^{-3}. With the increase of the concentration the disagreement is higher and higher. We suppose it is caused by the interacting electron perturbing the nearby ions and atoms. However, when Debye radius becomes of the same order of magnitude as the interparticle distance, the interacting electron can not influence other atomic particles, except that one with whom it interacts. If only this is taken into account, in the case of ionic line, the calculated values should be halved.

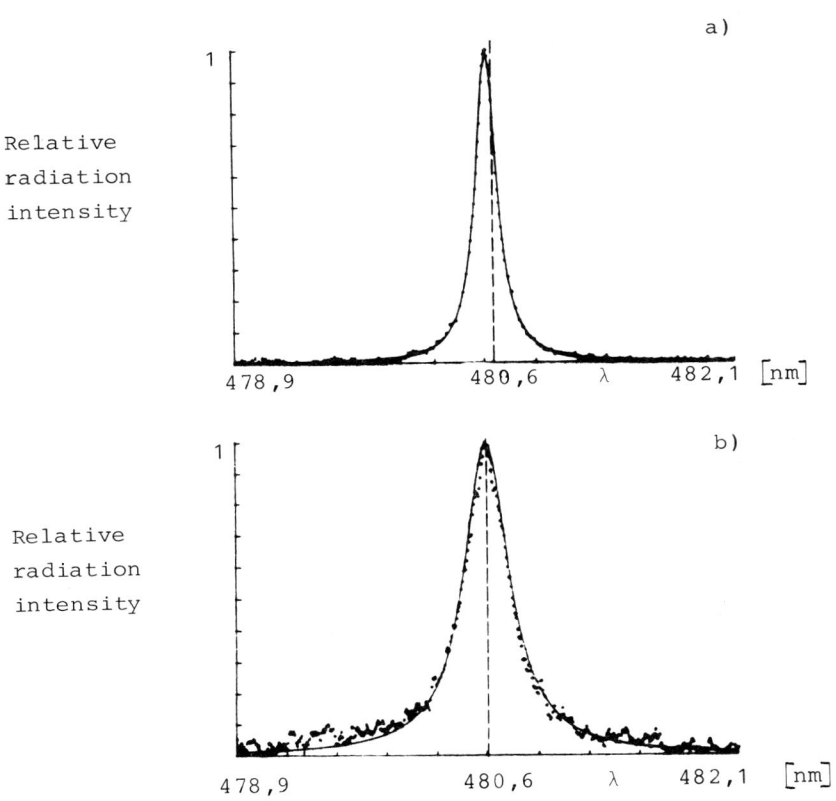

FIGURE 5
Line profile of ArII 406,8nm: a) $N_e = 6 \times 10^{17}$ cm^{-3}, $T = 16.500$K, $\Delta\lambda = 0.16$nm, b) $n_e = 14.9 \times 10^{17}$ cm^{-3}, $T = 18.000$K, $\Delta\lambda = 0.34$nm.

5. CONCLUDING REMARKS

We have analysed in detail the conditions which have to be fulfilled for the experimental determination of DC electroconductivity. A pulsed discharge in inert gases is most convenient for fulfilling these conditions. The values of electroconductivity obtained experimentaly with increasing concentrations of electrons differ more and more from the values obtained in classical calculations. This difference can not be explained by introducing a correction for electron collisions with atoms. The decreased electroconductivity could be caused by electrons scattering in fluctuations of a plasma microfield.

Nonideal plasma radiation is characterised by a continuum with a line spectrum superimposed. With careful treatment it is possible to measure isolated and non-absorbed line widths. The halfwidths measured are not increased with the increasing concentration as it would be

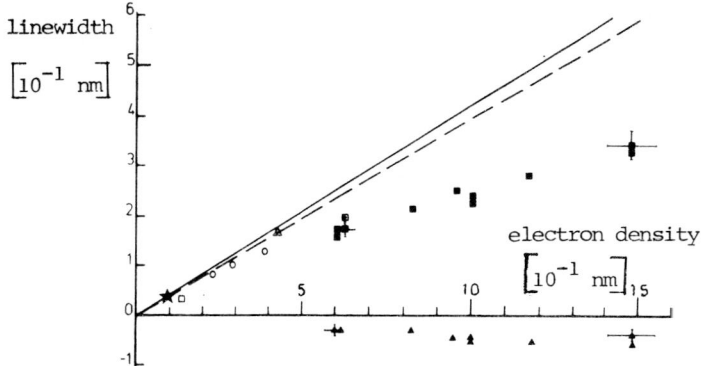

FIGURE 6
Linewidth and shift variation against the electron density for ArII 480,6nm and ArII 484,7nm lines: solid square, linewidth; solid triangle, shift; astarisk, Chapell et al;[30] open circle, Konjevic et al;[31] open square with dot, Labat et al;[32] triangle, Roberts;[33] open square, Nick and Helbig.[34]

obtained in the usual calculations. We consider a detailed analysis of the calculation reliability necessary.

ACKNOWLEDGEMENT

This paper has been supported in part by the Yugoslav Scientific fund, Contract P-169 and by CRNS, URA 1096, France. It is a result of Yugoslav–French scientific collaboration.

REFERENCES

1) M. M. Popovic and S. S. Popovic, Strongly-coupled plasma diagnostics of DC electrical conductivity, in *Strongly Coupled Plasma Phycics*, eds. F. J. Rogers and H. E. DeWitt (Plenum, New York, 1986) pp. 99-108.

2) M. M. Popovic, Radiative and Transport properties of weakly nonideal plasmas, in XIX ICPIG papers, eds. V. Zigman (Belgrade, 1989) in press.

3) A. A. Mihajlov, D. Djordjevic, M. M. popovic, T. Meyer, Z. Luft. and W. D. Kraeft, Contrib. plasma physics (1989) in print.

4) C. Goldbach, G. Nollez, S. S. Popovic, and M. M. Popovic, Z. Naturforch. **39**a (1978) 11.

5) K. Günther, M. M. Popovic, S. S. Popovic, and R. Radtke, J. Phys. D **9** (1976) 1139.

6) R. L. Shepherd, D. R. Kania, and L. A. Jones, Phys. Rev. Lett. **61** (1988) 1278.

7) Yu. V. Ivanov, V. B. Mintsev, V. E. Fortov, and A. N. Dremin, Zh. Eksp. Teor. Fiz. **71** (1976) 216.

8) C. J. Timmermans, G. M. W. Kroesen, P. M. Vallinga, and D. C. Schram, Z. Naturforch, **43**a (1988) 806.

9) H. H. Brouwer and P. P. J. M. Schram, Physica **114**A (1987) 589.

10) I. I. Tovstronajt-Polin and C. A. Triger, Teplofizika, Visokih Temperatur **26** (1988) 417.

11) S. Ichimaru, S. Mitake, S. Tanaka, and X.-Z. Yan, Phys. Rev. A **32** (1985) 1768, and S. Ichimaru and S. Tanaka, Phys. Rev. A **32** (1985) 1790.

11) Y. Vitel, A. Mokhtari, and M. Skowronek, J. Phys. B (1989) in print.

13) Y. Vitel, J. Phys. B **20** (1987) 2327.

14) Y. Vitel and M. Skowronek, J. Phys. B **20** (1987) 6477 and 6493.

15) Y. Vitel, M. Skowronek, M. S. Dimitrijevic, and M. M. Popovic, Astron. Astrophys. **200** (1988) 285.

16) N. I. Uzelac and N. Konjevic, Phys. Rev. A **33** (1986) 1349.

17) M. S. Dimitrijevic and M. M. Popovic, Astron. Astrophys. **217** (1989) 201.

18) M. M. Popovic, Measured Porperties of Nonideal Plasmas, in IX SPIG, eds. R. K. Janev (Dubrovnik, 1978) pp. 481–500.

19) P. P. Kulik, V. A. Rjablji, and N. V. Ermohin, Neidealnaja Plasma (Moskva, 1984).

20) K. Günther and R. Radtke, *Electric Properties of Weakly Nonideal Plasmas*, (Berlin, 1984).

21) V. E. Fortov and I. T. Iakubov, Fizika Neidealnoi Plazmi (Chernogolovka, 1984).

22) M. M. Popovic, S. S. Popovic, and S. M. Vukovic, Fizika **6** (1974) 29.

23) M. M. Popovic, Electric Properties of Dense Plasmas in High Current Pulsed Discharge, in VI SPIG, eds. M. V. Kurepa (Split, 1972) pp. 651–666.

24) K. Günther and R. Radtke, J. Phys. E **8** (1975) 371.

25) D. Djordjevic, K. Günther, R. Radtke, and R. Ulbricht, Exp. Techn. Phys. **25** (1977) 433.

26) Y. Vitel, M. Skowronek, and M. M. Popovic, Flashtubes as secondary radiation standard, in *Gas Dischage Application* (Venezia, 1988) p. 701.

27) R. Radtke and K. Günther, J. Phys. D **9** (1976) 1131.

28) Yu. K. Kurilenkov and A. A. Valuev, Beitr. Plasmaphys. **24** (1984) 161.

29) H. R. Griem, Phys. Rev. **165** (1968) 258.

30) J. Chapelle, F. Cabannes, and J. Blandin, J. Q. S. R. T. **8** (1968) 1201.

31) N. Konjevic, J. Labat, Li. Cirkovic, and J. Puric, Z. Physik **235** (1970) 35.

32) J. Labat, S. Djenize, Li. Cirkovic, and J. Puric, J. Phys. B **7** (1974) 1174.

33) D. E. Roberts, J. Phys. B **1** (1968) 53.

34) K. P. Nick and V. Helbig, Physica Scripta **33** (1986) 55.

EXPERIMENTAL STUDY OF OPTICAL PROPERTIES OF STRONGLY COUPLED PLASMAS

V.E. FORTOV, V.E. BESPALOV, M.I. KULISH, S.I. KUZ

Institute of High Temperatures USSR Academy of Sciences, Moscow
Institute of Chemical Physics USSR Academy of Sciences, Chernogolovka, Moscow Region

In an explosion-type generator, nonideal argon, argon-hydrogen and xenon plasmas have been generated. The electron densities have been varied in a wide range. Emission profiles of spectral lines of neutral argon and Balmer alpha line in dense argon plasmas have been measured. The radiation of strongly nonideal xenon and argon plasmas in the range $\Gamma \leq 4.6$ has been investigated.

1. INTRODUCTION

Optical properties of strongly coupled plasmas are of great physical interest, enabling us to observe the effect of interparticle interaction on the dynamics and the energy spectrum of electrons in a dense disordered medium. The plasma emission provides information on temperature and concentration of the particles, as well as on photoionization and photorecombination processes[1]. The theoretical concepts that have been developed pertain to rarefied plasmas, where the elementary processes are easily separated and the influence of the plasma environment reduces to the broadening of the spectral lines and the shift of the photorecombination thresholds, which have been registered in numerous experiments on rarefied plasmas. With the density increasing, the optical consequences of nonideality become evident prior to corresponding modifications of the thermodynamic and transport properties. Further growth of the nonideality leads to a considerable reconstruction of the energy structure for electrons. This results in diminishing the population of the upper part of energy levels[2], disappearance ("nonrealization") of some of the highly excited states[3] and deformation of the electron spectrum in compressed plasmas[4]. Experimental studies of nonideal plasmas of argon, xenon and argon-hydrogen mixture obtained by the hydrodynamic heating resulting from viscous energy dissipation in shock-wave fronts[5] are presented.

2. PLASMA GENERATION

The generation of dense plasmas was carried out by the explosion-type generators[1] that produce square shock waves (fig 1). The plasma parameters depend on the initial state of a gas and on the velocity of the target expansion into a gas. When the mass of an explosive is constant, the mass velocity U of the target is mainly determined by the thickness Δ of aluminium strikers. This dependence is presented in fig.2, for the expansion of the aluminium target into the argon (60%)-hydrogen (40%) mixture at 1 bar. The mass velocity was registered by the pins.

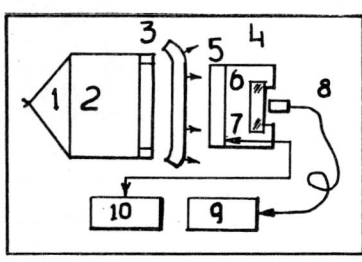
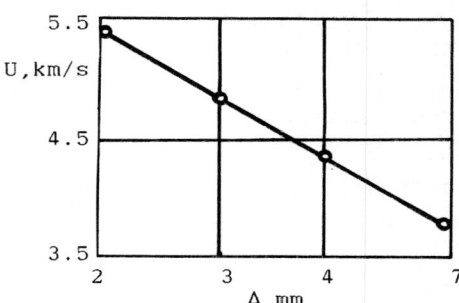

FIGURE 1
Set-up of experiments with the explosive compression. 1)Detonation lens; 2)explosive charge;3)metal striker; 4)test assembly;5)target;6)test gas; 7)protected pins;8)quartz fibre with a diaphragm;9)spectrograph;10)oscillograph.

FIGURE 2
Aluminium target mass velocity U versus striker thickness Δ.

The front velocity in the gas was measured by the optical baseline method. However, for the argon-hydrogen mixture, unlike rare gases, this method gives wrong results because of the time delay (≅0.8μs) between the shock wave creation in the gas and the plasma emission growth.

The calculation of the plasma parameters was carried out within the Debye theory in the grand canonical ensemble of statistical mechanics[1].

The observed form of optical signals in our experiments is described by the formula[4]:

$$I(t) = I_0 (1 - e^{-k \times t}), \qquad (1)$$

with k independent of t. This kind of dependence corresponds to the linear growth of the plasma layer thickness.

3. TEMPERATURE MEASUREMENT TECHNIQUES

The temperature of the plasma is determined from the plasma radiation, the thickness of the plasma layer being estimated from the velocity measurments. The radiation is calibrated in absolute value against a standart tungsten ribbon lamp[5]. To avoid an error connected with relatively small temperature (2800 K) of reference source, another temperature standart was aslo used - the radiation of a shock wave of velocity 6.25 km/s at initial pressure 1 bar. The temperature of the argon plasma (18600 K) is in the range of measured values and makes it possible to produce temperature measurement with the accuracy of ±300K.

4. STARK SHIFT OF ArI LINES

An investigation of ArI radiation in dense plasma appear to be of interest due to the data[4] on free-free and bound-free transitions in the range $n_e \leq 10^{20}$ cm^{-3}, and those on the Stark broadening and shift[8,9] for $n_e \leq 4 \times 10^{17}$ cm^{-3}.

In our work the set of lines of neutral argon was studied for $n_e \leq 10^{19}$ cm^{-3}. When the electron concentration is $> 2 \times 10^{19}$ cm^{-3} the group of lines under investigation displayes the effect of "dissolution".

The experimental techniques which included a spectrograph with gating electron-optical converter and made it possible to measure the plasma radiation with the resolution $\tau = 50$ ns, was described in[10] earlier. The thermodynamic parameters of shock-compressed plasmas, calculated for the velocity of shock front 6.25 km/s are displayed in table 1.

Table 1. The thermodynamic parameters of argon plasmas.

	P_0, bar	$n_e, 10^{18}$ cm^{-3}	$n_a, 10^{19}$ cm^{-3}	$T, 10^4$ K	P, bar	Γ
1	.08	3.7	1.3	1.65	46	.32
2	.15	6.6	2.4	1.70	85	.4
3	.2	8.3	3.2	1.73	113	.43
4	.25	10.0	3.8	1.75	140	.47
5	.3	12.0	4.7	1.77	170	.5

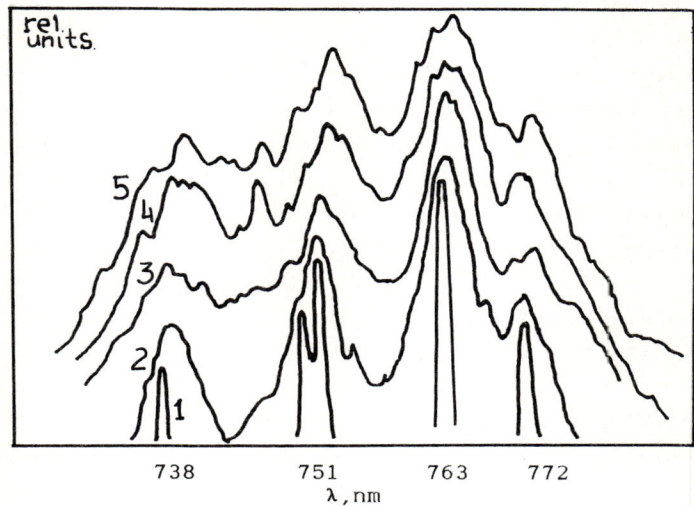

FIGURE 3
ArI lines. 1) Reference sourse; 2), 3), 4) and 5) experiments for plasma parameters displayed in lines 2, 3, 4 and 5, table 1, correspondingly.

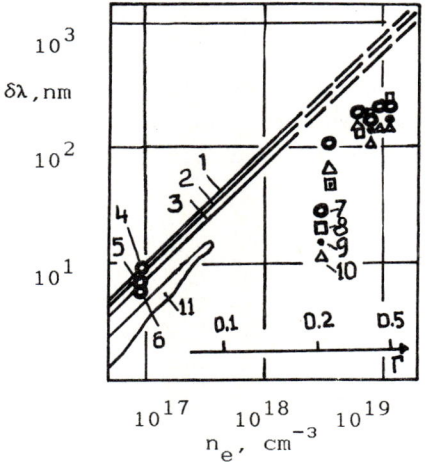

FIGURE 4
ArI 4s(4s')-4p(4p') lines shifts versus concentration of electrons. 1)-3) calculation by Griem; 4)-6) measurements[9] for 750.4, 751.5, 772.4 nm; 7)-10) our points for 750.4, 738.4, 763.5, 772.4 nm; 11) region of experimental points[8] for 763.5 nm.

The microphotograms presented in fig.3 were developed and

spectral lines shifts were obtained. For all spectral lines a red shift was observed as well as in the experiments at $n_e \leq 4 \times 10^{17} cm^{-3}$. However, under the discussed condition this shift stands out in stronger relief and reaches a few nanometers. The calculation of the Stark broadening and shift[11,12] in the region $\Gamma \geq 0.2$ (for $T=1.6 \times 10^4$ K, $n_e = 2 \times 10^{18} cm^{-3}$) under the condition of strong interparticle interaction is extrapolative.

For $n_e > 5 \times 10^{18} cm^{-3}$ fourfold or fivefold smaller shifts (fig.4) were observed. The analogous tendency was revealed in the experiments[13] with adiabatically compressed xenon plasma ($n_e = 3 \times 10^{18} cm^{-3}$).

5. INVESTIGATION OF THE BALMER SERIES LINES

The spectral region near the photoionization threshold was investigated in a set of experimental studies (see ref. 1) to confirm the disappearence ("nonrealization") of some of the highly excited states as a result of the microfild effect in the plasma. In the articles[14,15,16] the study of Balmer lines in plasmas generated in a flashtube is presented. The detailed account[15,16] of the H_α shift and the broadening is presented for the range of electron densities from 6×10^{17} to $10^{18} cm^{-3}$ in an argon-hydrogen mixture. These densities are in a region where the ion-radiator dynamic effects[16] are relatively important.

In our work the test assembly is filled with the argon-hydrogen mixture in the proportion 60% - 40% at the initial pressures of 1 or 2 bar. The composition and impurities were controlled by the mass-spectrograph. An argon-hydrogen mixture was chosen since a shock wave of velocity ≈ 6 km/s in pure hydrogen produces low temperatures ($T \approx 4000$ K) and the electron cocentrations of $n_e \approx 10^{13} cm^{-3}$. In a pure argon with the addition of hydrogen ≤ 5% at this conditions the electrons concentration[4] is $\approx 10^{20} cm^{-3}$ at $T \approx 20000$ K. However, the absorption coefficient connected with free-free transitions is so high, that it hinders the hydrogen lines to be measured. Therefore in the present experiments the initial mixture composition was chosen to cause the electron concentrations of $5 \times 10^{17} - 8 \times 10^{18}$ cm^{-3}.

The measurement of the H_α line profile in this range was performed on the 400 mm diffraction spectrograph with the electron-optical converter and 600 or 300 lines/mm graiting. The radiation from the experimental set-up was directed to entrance slit of the spectrograph through a quartz fibre ≈ 30 m long and 200 μm in diameter. In front of the photocathode of the electron-optical converter three step nonselective filters for the densitometric development were installed. In the rest the scheme of the experiment did not differ from that described in part 4. This technique of intensity measurments by the film density gives the accuracy of less than 20 %.

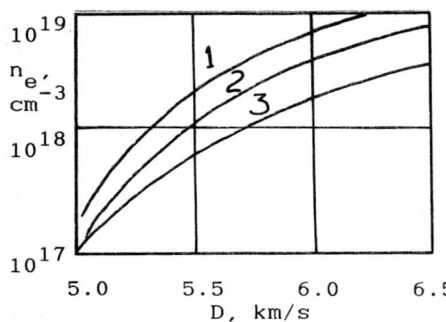
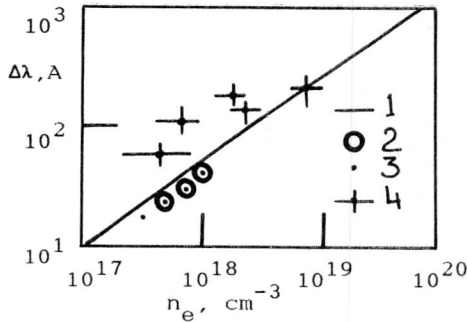

FIGURE 5
Composition of argon-hydrogen plasma versus front velocity D. 1) electron concentration; 2) Ar^+ concentration; 3) H^+ concentration.

FIGURE 6
Full width at half maximum of Balmer alpha line. 1) calculation by Griem, 2) experiment[15]; 3) calculation[16]; 4) our experiments.

Fig 5. presents calculated curves of the plasma composition versus the shock velocity, D. For the velocities D≤5km/s the variation of n_e on D is significant and this determines the main experimental error.

The full widths H_α line measurement are displayed in fig.6. The calculation[12] is presented for the mean value of the experimental temperatures range, since the temperature dependence is weak. The experimental results of paper[15] and calculated points[9,16] are described by the density dependence $\simeq n_e^{2/3}$. The experimental points of this work approach to this dependence in the region of $n_e > 5 \times 10^{18} cm^{-3}$, and for smaller densities layers above. This kind of dependence is likely to be due to the influence of neutral particles, concentration of which (see table 2) in these experiments is practically constant and varies only with the variation of the initial pressure of the mixture.

Table 2. The composition of the argon-hydrogen mixture plasmas.

	P_0, bar	T, K	$n_e, 10^{18} cm^{-3}$	Ar+H, $10^{20} cm^{-3}$	D, km/s
1	1.0	15600	8.0	1.8	6.43
2	1.0	11500	0.5	1.8	5.12
3	2.0	11700	0.8	3.6	5.12
4	1.0	13200	2.6	1.8	5.67
5	1.0	13000	2.3	1.8	5.62
6	2.0	13200	3.6	3.6	5.63

To measure the plasma radiation in a wide spectral range of 400-1100 nm another technique was used which differes by the registration of radiation in spectral channels. In the focal plane of the diffraction spectrograph with dispersions 2,4, or 8 nm/mm (three replaced graitings) five-channel photosensitive receiver was set. The spectral width of each channel is 2,4, or 8 nm. The photoreceivers were the fast p-i-n photodiodes with spectral sensitivity within the range of 400-1100 nm. In the case of the 600 lines/mm graiting, the distance between the channels was 30 nm. The photoelectrical signal through the amplifier with the dynamic range $\simeq 10^4$ entered the analog-to-digital converter with the sampling rate up to the 20 MHz. The polychromator calibration was produced against tungsten lamp and the shock-compressed argon as described above. The ribbon lamp signals in the "green" region are weak and to increase the accuracy the manifold measuring was carried out.

The absorption coefficients x' were determined from the radiation growth with the increase in the plasma layer thickness according to formula (1), the difference between the shock-wave and the mass velocities being known. The experimental data (fig. 7) were compared with those calculated according to[2], where the contribution of the levels above the boundary one (4p, 4p' for argon) was taken into account as integral. The H_α line was calculated without the account of the plasma microfields influence on the levels population. The calculated and measured by two methods half widths coincided within the experimental error. The experimental full intensity of the line appeared to be an order less than the calculated one, because the decrease in the upper levels population was not taken into account.

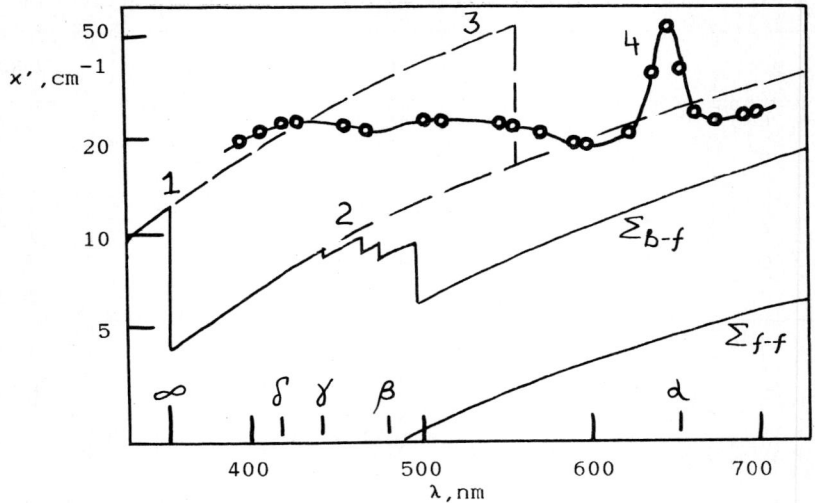

FIGURE 7
Absorption coeffisient x' of argon-hydrogen plasma. 1) photorecombination threshold of hydrogen; 2) photorecombination thresholds of argon 4p,4p'; 3) Inglis-Teller photorecombination threshold shifts; 4) our experiment.

6. Ar AND Xe DENSE PLASMAS IN THE REGION OF LARGE Γ.

The method described above was also applied to study the nonideal Ar and Xe plasmas at $n_e \leq 6.1 \times 10^{21} cm^{-3}$ and $\Gamma \leq 4.6$. As it was shown in[1], the plasmas generation conditions in this range can be realized by the use of the simple linear scheme of the shock wave exitation at the increase in the initial pressure up to 10's bar. The measurements were performed in the IR region (900-1100nm), where the metallization effects[1,17] influence on the plasma emission is quite probable at $n_e \geq 10^{21}$.

The increase in the xenon initial pressure caused the decrease in the xenon radiation down to one tenth as compared with that of the blackbody (fig.9). The stationary value of the radiation intensity was attained for ≅100 ns. The polichromator resolution time determined by the sampling time (50 ns) did not allow us to follow the initial site of the curve in detail. However, the qualitative description of its behavior is based on the absorption layer formation in xenon before the shock-wave front for the period which is longer than that of the optical thick layer of the shock-compressed xenon. The "gap" position is n_e-independent. Similar bands of the inert gas absorption in the IR region induced by electron beam pulse, which caused the formation of excited diatomic molecules were found in the paper[17]. The formation of Xe_2^* molecules in our experiments was facilitated by the action of the strong UV radiation from the shock-wave front.

In the experiments with argon a two-layer (steel-aluminium) target was used, whose expansion produced a shock wave of a higher initial velocity and determined the shock-wave profile decreasing in time. In the argon plasma radiation the absorption, though in the 997±2 nm channel which is much closer to the absorbing band center, increased gradually with the narrowing of the unperturbed argon layer. There was no absorption observed at <1 mm thickness (fig.8).

The radiation absorption by the Xe_2^* molecules hindered the measurements of the plasma optical properties in the resonant $\omega \sim \omega_p \sim (4 \cdot \pi \cdot n_e^2 / m)^{1/2}$ frequency range. It should be noted that the measurements of the reflection coefficient of xenon dense plasma were carried out at the plasma layer growth for <50 ns.

The argon plasma radiation does not deviate from that produced by the blackbody at the frequency lower than the plasma one (except for the Ar_2^* absorption described above).

Apparently, a more fast measurement techniques and a different approach are needed for a more thorough study of the plasma optical properties in the resonant region.

Table 3. The composition of xenon and argon plasmas.

	P_0, bar	n_e, 10^{21} cm^{-3}	n_a, 10^{21} cm^{-3}	P, 10^4 bar	T, 10^4 k	Γ	D, km/s
Xenon							
1	30	6.1	2.2	5.6	3.2	4.6	5.5
2	16	3.0	0.9	2.5	3.1	3.4	5.6
3	6	1.2	0.3	1.0	2.8	2.3	5.6
4	3	0.56	0.2	0.45	2.7	1.8	5.6
5	1	0.2	0.07	0.14	2.5	1.2	5.6
Argon							
6	50	2.2	7.0	4	2.75	3.5	7.4 (max)

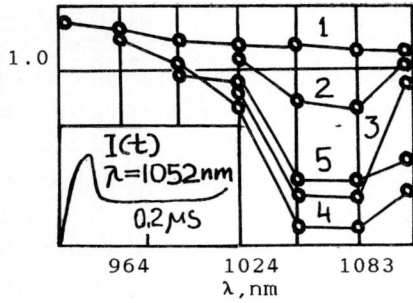
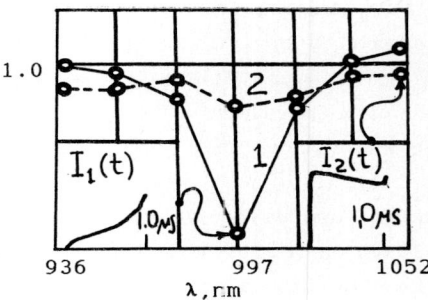

FIGURE 8
Emission of xenon plasma normalized to the black-body emission. 1), 2), 3), 4) and 5) refered to 5, 4, 3, 2 and 1 lines of table 3.

FIGURE 9
Emission of argon plasma normalized to the black-body emission. 1) initial moments after creation shock wave in a gas; 2) thickness of unperturbed gas before the front <1 mm. The plasma parameters are in line 6 of table 3.

ACNOWLEGEMENT

The autors are indebted to V.K. Gryaznov for the calculation of plasma parameters presented here.

REFERENCES

1) V.E. Fortov and I.T. Iakubov, Fizika Neidealnoj Plazmy (USSR Academy of Sciences, Chernogolovka, 1984).

2) V.G. Sevast'yanenko, Preprints 30, 32, (ITPM, USSR Academy of Sciences,Novosibirsk, 1980).

3) G.A. Kobzev, Yu.K. Kurilenkov and G.E. Norman, Teplofiz. Vys. Temp. 15, (1977) 647.

4) V.E. Bespalov, V.K. Gryaznov, and V.E. Fortov, Zh. Eksp. Teor. Fiz. 76, (1979) 140.

5) V.E. Fortov, Usp. Fiz. Nauk 138, (1982) 361.

6) Ya.B. Zel'dovich and Yu.P. Raizer, Fizika Udarnykh Voln i Vysokotemperaturnykh Gidrodinamicheskikh Yavlenii (Nauka, Moscow, 1966).

7) V.I. Malyshev, Vvedenie v Eksperimentalnuju Spektroskopiju (Nauka, Moscow, 1979).

8) L. Bober and R.S. Tankin, J. Quant. Spectrosc. Rad. Transfer, 10, (1970) 991.

9) Dowglas W. Jonnes, K. Musiol and W.L.Wiese, in: Spectral Line Shapes, Vol. 2, eds. Walter de Gruyter and Co. (N. Y.,1983) p 125.

10) V.E. Bespalov, M.I. Kulish and V.E. Fortov, Teplofiz. Vys. Temp. 24, (1986) 995.

11) H.R. Griem, Plasma Spectroscopy (Mc Graw-Hill, New York, 1964).

12) H.R. Griem, Spectral line broadening by plasmas (Acad.Press, New York and London, 1974).

13) H. Hess, L. Hitzchke , Metzke et al., Pros. XII SPIG (Sibenik,1984) p.453.

14) V.E. Gavrilov, T.V. Gavrilova and T.N. Fedorova, Opt. i Spectr., 58, (1985) 1228, 59, (1985) 518.

15) Y. Vitel, J. Phys. B, At. Mol. Phys., 20, (1987) 2327.

16) Dipak H. Oza and Ronald L. Green, J. Phys. B; At. Mol. Opt. Phys.,21, L5-L8 (1988).

17) Yu.B. Zaporozhets, V.B. Mintsev and V.E. Fortov, VII Conf. Fizika Nizkotemperaturnoj Plasmy (Tashkent 1989) pp. 45-46.

18) Shigeyoshi Arai, Takefumi Oka, Masuhiro Kogoma and Masashi Imamura, J. Chem. Phys. 68, (1978) 4595.

MANY-ELECTRON EFFECTS ON DYNAMIC PROCESSES IN DENSE MATTER

Stephen M. YOUNGER

X-Division, Los Alamos National Laboratory, Los Alamos, New Mexico 87545, USA

Large scale calculations quantum mechanical calculations of dynamic processes in dense helium suggest that quasi-molecular effects may play an important role in determining the transport properties of dense plasma.

1. INTRODUCTION

Recent experiments suggest that many-electron processes may be important in determining both the transport and radiative properties of strongly interacting plasma. In x-ray measurements of the high density implosion of an argon filled laser fusion capsule Hooper et al.[1] observed a prominent high energy feature associated with the heliumlike 1s-2p transition. This feature occurred only near peak compression when the plasma was *both* hot and dense. Although it is typical to observe a rich satellite structure at lower photon energies, structures at high energy are unusual, and are difficult to interpret in terms of a single-atom model of the emitter. We have suggested that such features may be associated with the formation of transient *molecular* complexes resulting from stochastic thermal motion within the plasma, i.e., encounters close enough to allow significant hybridization of the atomic orbitals. If an electron collision occurs while the ions are in such a configuration, the resulting photon emission will occur at the instantaneous *molecular* transition energy, which can be shifted to higher energy as the ion pair approaches the united atom limit.

At lower temperatures, two stage gas gun measurements of the shock Hugoniot of helium at densities up to 0.6 g/cc and inferred temperatures of 2 eV have been performed by Nellis et al.[2] Modeling of these experiments suggests that the effective interatomic interaction at high density is substantially softer than one would expect from accurate theoretical calculations or measurements of the binary He-He interaction.

These experiments suggest that the dynamics of matter at high density may be more complex than previously anticipated. Indeed, both of the above results indicate that many-center interactions are important in determining the properties of dense matter. Theories which include only a single atom in an "effective" medium may omit important dynamic correlations wherein the forces within and between atoms are determined by the instantaneous configuration of its neighbors. We have recently performed a series of calculations aimed at addressing this point.

2. SELF-CONSISTENT FIELD MOLECULAR DYNAMICS

In order to explicitly study the effects of many-atom interactions at high densities we have developed a new theory of dynamic matter based on a synthesis of techniques from quantum chemistry and molecular dynamics. Our method consists of the repetition of three steps. First, we use a molecular orbital Hartree-Fock approximation to compute the electronic structure of a large cluster of atoms. Using this wavefunction to derive a charge density, we calculate the net electrostatic forces acting on each of the nuclei. Finally, we allow the nuclei to move slightly in response to the forces. By repeating these three steps we are able to generate time histories of the atomic motion, allowing later postprocessing of quantities such as correlation functions and transport coefficients. This is an adiabatic approximation propagating a ground state configuration in the Born-Oppenheimer approximation.[3]

3. DYNAMIC SCREENING

3.1. Static Calculations.

Before describing the results of the full SCFMD calculations, it is worthwhile to examine the effect of many-atom screening on a small configuration of atoms. Figure 1 shows the effect of neighbor atoms on the binary interaction force.. If the total force on atom B were simply a superposition of pair potentials the ratio in Figure 1 would always be one, since the neighbor atoms are symmetric about the test atom. The reduction in the force when neighbors are present is the result of the self-consistent redistribution of electronic charge density throughout the cluster. The addition of the nuclear forces of atoms B,C, and D results in a net potential which is deepest at the location of atom B. This deep potential draws in electronic charge density, which serves to screen atom B from the effect of atom A.

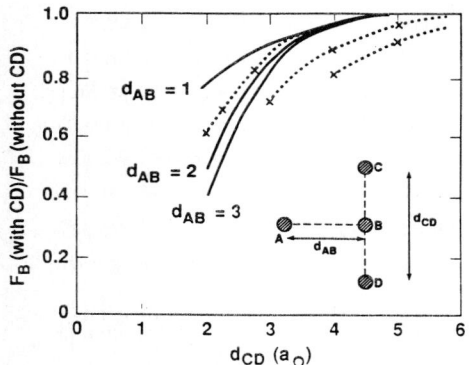

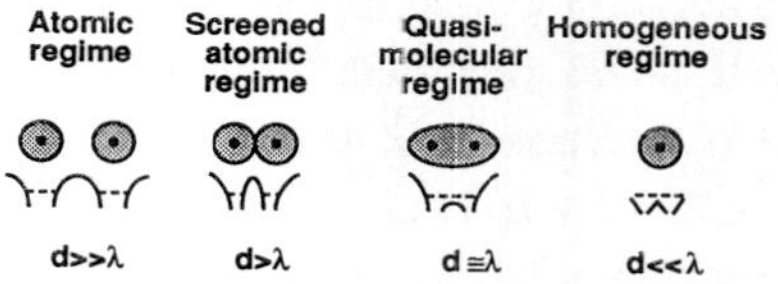

FIGURE 2
Four regimes in dense plasma. d is the internuclear distance, λ is a characteristic electron wavelength.

FIGURE 1
Effect of near neighbors on interatomic force.
Solid: electrostatic forces. Dashed: energy gradient forces.

In a related study[4], calculations were performed on a body-center-cubic cluster of nine atoms using a finite temperature Hartree-Fock-Slater approximation. As the lattice spacing was decreased, i.e. the density increased, the effective number of bound electrons at first decreased reflecting the onset of pressure ionization. At higher densities, however, the bound population *increased* as the electrons were attracted to the deep and extended potential well created by the superposition of the nine nuclear potentials.

These calculations led us to propose that dense plasma can be divided into four regimes, classified by the relationship of the interionic spacing to the characteristic wavelength of the electrons. As shown in Figure 2, at low densities the electrons occupy separate potential wells and the kinetics can be modeled as sums of binary encounters. As the density increases the tails of the atomic wavefunctions begin to overlap, resulting in screening of the nuclear potentials and a rise in atomic eigenvalues. This reduction in binding is associated with pressure ionization. As the density increases still further, however, it is possible for clusters of nuclei to *collectively* bind electrons. As in a covalent bond, the electrons are shared among several centers. Since the effective potential in the center of the cluster is deeper than that associated with a single atom, the electrons are more tightly bound and the bound atomic population increases. At very high densities the superposition of atomic potentials results in a dense packing of narrow wells which are singly and collectively unable to bind electrons. This is the limit of a homogeneous electron gas.

Regular crystalline lattices do not exist in a dense plasma. It is possible, however, to form transient clusters of ions due to the stochastic motion of atoms at high temperature. Such clusters are extremely important for the dynamics of the ensemble, since it is in close collisions that the largest momentum transfer occurs. Hence for energy transport in dense matter it is important to self-consistently treat the electronic and ionic motion. This was the purpose of our self-consistent field molecular dynamics calculations.

3.2. Dynamic Calculations

Self-consistent field molecular dynamics calculations on helium at densities between 0.1 and 3.0 g/cc and temperatures of 1 and 5 eV were performed on samples of 23 and 30 atoms. Periodic boundary conditions were imposed to mimic the effects of extended media and to minimize surface effects. The wavefunction was expanded in a Gaussian basis set of two or three functions attached to each of the nuclei.

Figure 3 illustrates the normalized velocity autocorrelation function $Z(t)=<\mathbf{v}(t)\cdot\mathbf{v}(0)>/<\mathbf{v}(0)\cdot\mathbf{v}(0)>$ where $\mathbf{v}$ is the velocity and the brackets indicate averages over the atoms as well as the initial time. In addition to the SCFMD results, shown by solid lines, we also give the results of classical molecular dynamics calculations performed with several different pair interactions potentials. At low density all of the calculations agree save the one which used the high density potential of Young et al. This is not surprising since this potential *assumes* the presence of nearest neighbors which produce screening. At low density and high temperature it is possible for atoms to approach one another without any neighbors nearby. As the density increases the SCFMD velocity autocorrelation function assumes a progressively longer decay time than those computed from binary pair potentials. This is due to the formation of transient quasi-molecular clusters which dynamically screen the interatomic interactions.

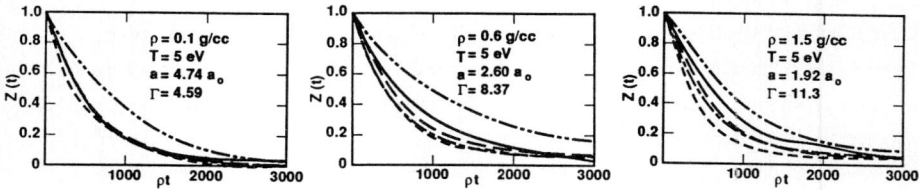

FIGURE 3
Velocity autocorrelation functions for dense helium.——SCFMD; ······Aziz et al., Hartree-Fock short range potential[5];— — Aziz et al. quantum Monte Carlo short range potential[6];—··—Young et al.[7] derived from high density linear muffin tin calculations which assume the constant presence of several near neighbors;—·— pair potential derived from the same atomic basis set used in our SCFMD calculations.

4. SUMMARY

Self-consistent field molecular dynamics calculations suggest that transient quasi-molecular configurations may have an important effect on dynamic processes in dense matter. Although not explicitly discussed in this paper, we expect that such effects may be especially pronounced for electron transport.

ACKNOWLEDGEMENT

It is a pleasure to acknowledge helpful interactions with Gayle Sugiyama and Alan Harrison of LLNL. This work performed under the auspices of the U.S. DOE by the Lawrence Livermore National Laboratory under Contract No. W-7405-ENG-48.

REFERENCES

1) C.F. Hooper, D.P. Kilcrease, R.C. Mancini, L.A. Woltz, D.K. Bradley, P.A. Jaanimagi, and M.C. Richardson, Phys. Rev. Lett. 63 (1989) 267.
2) W.J. Nellis, N.C. Holmes, A.C. Mitchell, R.J. Trainor, G.K. Governo, M. Ross, and D.A. Young, Phys. Rev. Lett. 53 (1984) 1248.
3) S.M. Younger, A.K. Harrison, and G. Sugiyama, Phys. Rev. A, in print.
4) S.M. Younger, A.K. Harrison, K. Fujima, and D. Griswold, Phys. Rev. Lett 61 (1988) 962.
5) R.A. Aziz, V.P.S. Nain, J.S. Carley, W.L. Taylor, and G.T. McConville, J. Chem. Phys. 70 (1979) 4330.
6) R.A. Aziz, F.R.W. Mc Court, and C.C.K. Wong, Mol. Phys. 61 (1987) 1487.
7) D.A. Young, A.K. McMahon, and M. Ross, Phys. Rev. B 24 (1981) 5119.

Chapter X:
Shock-Compressed Plasmas and Inertial-Confinement-Fusion Plasmas

LASER PRODUCED OPTICALLY-THIN STRONGLY COUPLED PLASMAS

A. N. MOSTOVYCH, K. J. KEARNEY, and J. A. STAMPER

Laser Plasma Branch, Plasma Physics Division
U.S. Naval Research Laboratory, Washington, D.C. 20375*

Introduction

Although strongly coupled plasmas are common in astrophysical environments and have been the subject of many theoretical and numerical investigations,[1] very little is known about them experimentally. Astrophysical observations help to test theoretical models, such as the ones for stellar evolution or the luminosity of brown dwarfs, but they offer little hope for a detailed comparison to strongly coupled plasma theory. These observations are generally convoluted with many physical effects and the state of the plasmas from which they originate are not well known. On the other hand, well characterized laboratory experiments are generally limited to the weakly coupled regime. The few experiments with strong coupling, i.e. $\Gamma \geq 1$, are limited to one component plasmas,[2,3] to capillary discharges,[4] or to strongly shocked solids and gases.[5,6,7] The experiments with one component plasmas cannot address the issues relevant to two-component plasmas, while the capillary discharge and shock wave experiments are often difficult to diagnose and are generally optically thick such that studies of their optical properties are not possible.

We have developed a technique by which optically-thin strongly coupled plasmas with $n_e \approx 10^{20}$ cm^{-3}, $T_e \approx$ 2-6 eV, and Γ = .3-1.5 (where $\Gamma = (\Gamma = e^2 Z^{5/3} (3N_e/4\pi)^{1/3}/T$) is the ion coupling parameter) can be produced under controlled conditions. These plasmas are produced by laser ionization and heating of a dense Al vapor and the plasma temperature and density are independently measurable. As a result, experiments with these plasmas are suitable for straightforward tests of theoretical predictions. In this review, we will introduce this technique, discuss the kind of plasmas that can be produced, compare our results to the Saha equation of state, and present measurements of plasma opacities from which we determine the Coulomb logarithm Ln(Λ). Surprisingly, our measured value for Ln(Λ) agrees almost exactly with that of the classical result based on Debye shielding.

---

*This work is supported by the U.S. Office of Naval Research.

Plasma Production

A laser produced plasma is used as a source for these experiments. The plasma is produced by a sequence of steps which are very different from the usual laser-plasma experiments, where a blow-off plasma is produced by the direct interaction of an intense laser beam with a solid target. Instead, a thin film of metal is first explosively vaporized with a moderate intensity (10^{11} watts/cm^2) laser beam, the metallic vapor is shaped by a set of slits, and the vapor is then heated and ionized by a second laser. By this process the plasma density and temperature become independent variables and may be controlled at will.

The details of this process are shown in Figure 1 and 2. A glass substrate is coated with a metallic film (Al in this case) and a laser beam is used to vaporize this metal film by irradiation through the glass. The intensity of the laser is below the threshold for glass breakdown but is sufficiently high to permit almost instantaneous vaporization of the metal. The exact details of the laser-metal-glass interaction are not completely understood at this time, however, the end result is that the metal film vaporizes (most likely by shock heating) in a sufficiently short time (2-5ns; short compared to the transverse acoustic time) such that the expansion of the metal vapor from the glass substrate is predominantly 1-dimensional and very uniform. By this method, first described by Mayer and Bush,[8] it is possible to generate metallic vapors with densities up to solid density and temperatures of less than 1 eV. A sample interferogram of a dense Al vapor is shown in Figure 2. For the most part the Al vapor is optically thick, expands at a constant velocity, and is highly supersonic as is indicated by the very small transverse expansion following shearing with a sharp knife edge. From the amount of transverse flow immediately following the knife edge it is possible to estimate[9] that the temperature of the Al vapor is less than 0.3 eV. Measurements of black-body emission intensities indicate similar

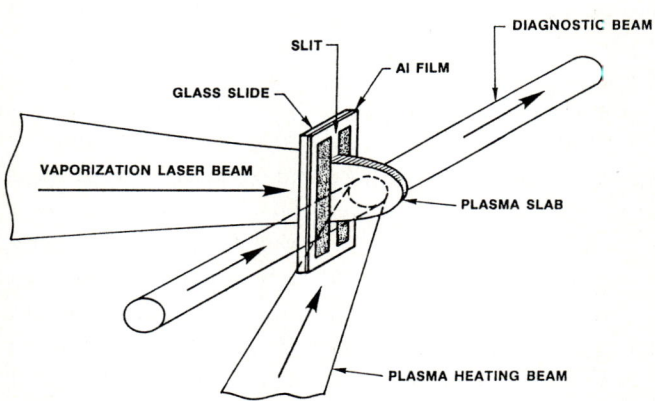

Fig. 1   Schematic of technique used to produce optically-thin strongly coupled plasmas.

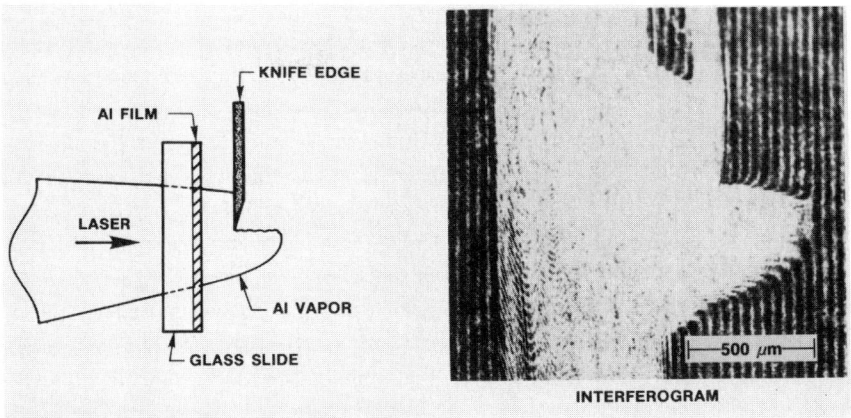

Fig. 2   Interferogram of a dense Al vapor expansion. Transverse flow past knife edge is used to estimate temperature.

temperatures. At these low temperatures, substantial ionization of the Al vapor is not expected. Interferometry data from the Al vapor expansions show no evidence of plasma formation. In a plasma the index of refraction decreases with increasing density. However, in Figure 2 the curvature of the the interference fringes is consistent with an increasing index of refraction - indicating a negligible ionization fraction.[10]

For the self-similar 1-dimensional model of Mayer and Bush[8] the vapor density is linearly related to the initial solid density by the ratio of the vapor expansion length to the initial thickness of the metal film. For the example in Figure 2 the vapor density is on the order of $n_v = 10^{20}$ cm^{-3}. Since the vapor density falls monotonically with the continuing expansion, a choice of densities up to solid density is available in each expansion.

For interesting SCP experiments it is necessary to ionize the Aluminum vapor and make it optically thin. Given the supersonic nature of the vapor flow it is easy to control the transverse dimensions of the vapor cloud by a set of slits, Figure 3a. The Al vapor flows through the slits to form a thin slab ($\Delta x \approx 100 \mu$m) of vapor which is optically thin. After the vapor slab has reached the appropriate expansion length for the desired initial density a second laser of moderate intensity ($10^{10} - 10^{11}$ W/cm^{-3}) is used to heat and ionize the Al vapor. Typically, for initial densities of 10^{20}cm^{-3}, the heating laser is fired at about 50-80ns after the vaporization of the Al film. The heating occurs over a time of about 8-10ns and during the heating pulse fringe reversal is seen in the interferograms, Figure 3b. This is direct evidence of plasma formation while the visibility of the fringes throughout the plasma volume proves that the plasma is optically thin.

Plasma Characteristics

The production of an optically thin plasma has the unique advantage of permitting the use of optical diagnostic techniques to measure the plasma properties. This is very important because, as a result, measurements can be made independently and accurately. A sample of some of the diagnostics deployed on this experiment is shown in Figure 4. A mode-locked glass laser (1.054μm) serves as a primary optical probe. This laser has a sufficiently short pulse-length (300-500ps) to freeze the plasma motion and it can be doubled (0.527μm), tripled (0.3513μm), or Raman shifted (0.6258μm) to permit multiple wavelength interferometry or wavelength dependent absorption measurements. An intensity calibrated 3/4m monochromenter is used for black-body and absolute intensity measurements while a 1/3m spectrometer with an Optical-Multichannel-Analyzer (OMA) is used to measure atomic spectral emission from the plasmas.

The plasma electron density is determined from interferograms, produced by a polarization wavefront interferometer[11] and the mode-locked laser probe. The probe laser timing is adjustable such that it can be fired at any time during the heating laser pulse and the interferograms are recorded on film with a magnification of about 5x. An example of such an interferogram is shown in Figure 3b. Interferometry measures the cumulative phase change, which is proportional to $\int n_e(x)dx$ but not to $n_e(x)$ directly. As a result, the plasma thickness must be known accurately to infer the average transverse density from the interferograms. The slab geometry of this experiment has the advantage that the plasma thickness is primarily determined by the slit size even though some correction for the thermal expansion of the heated plasma is necessary. Corrections to the plasma thickness are made on the basis of an ion-acoustic expansion velocity during the course of the heating pulse.

Typical transverse and axial density profiles are shown in Figure 5. Peak densities of 10^{20}cm^{-3} or larger are easily accessible. For the experiments where simultaneous two-wavelength interferometry was implemented, the interferograms from the the two different wavelengths gave identical density profiles as long as only the plasma dispersion relation $\omega^2 = k^2c^2 + \omega_p^2$ is used to invert the interferograms. As a result, the the heated plasmas are well ionized and whatever neutral atoms that may be in the plasma do not affect the electron density measurement.

The plasma temperature is measured by two separate methods. One is based on the ratio of atomic line intensities while the other is due to the application of Kirchoff's law to simultaneous plasma opacity and absolute intensity measurements. At temperatures of about 2eV or less the ratio of the Al III 3605.4 and 3709.2 Ang. lines is very sensitive to the plasma temperature. A calculation of this ratio, including all multiplets, is displayed in Figure 6a. These lines are from the same ionization stage, thus the ratio does not depend on the knowledge of the equation of state but only on the assumption that the atomic levels are distributed according to the Boltzman

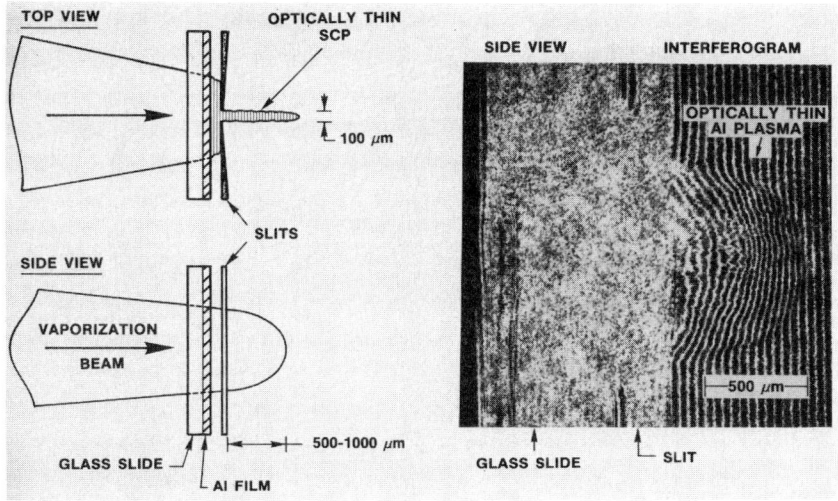

Fig. 3 a) Schematic of Al-film target with slits. b) Interferogram of typical plasma after laser heating.

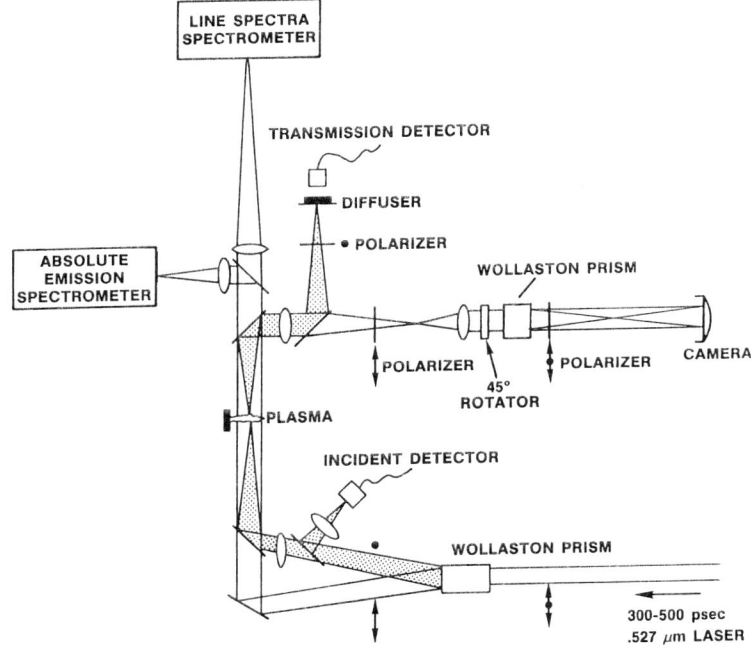

Fig. 4 Diagram of diagnostics set-up.

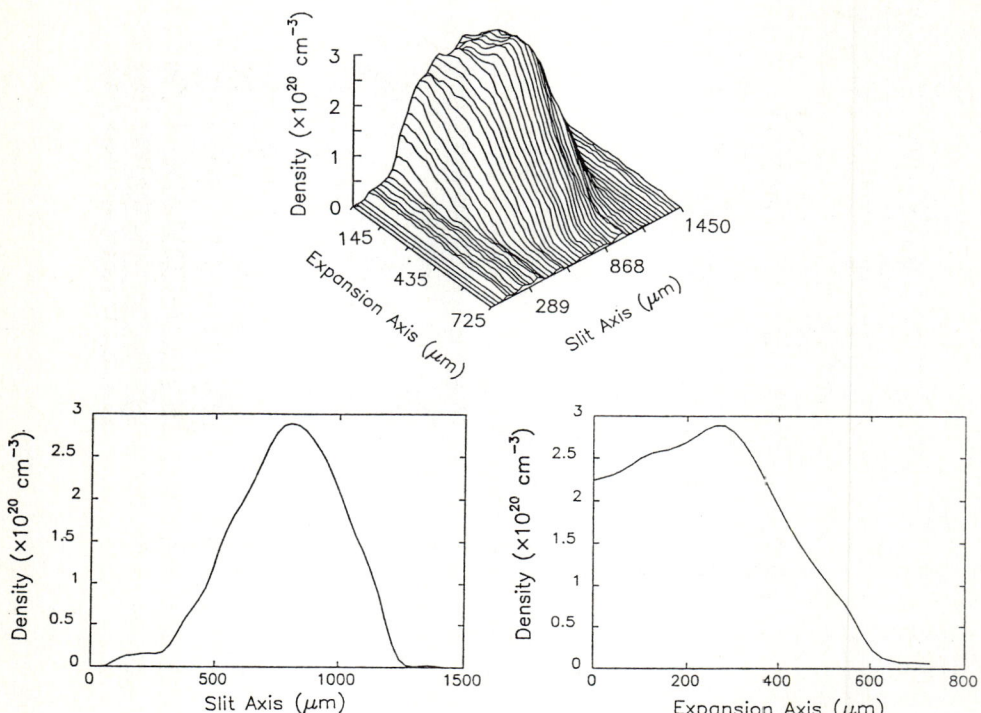

Fig. 5 a) 2-dimensional density profile of a typical SCP plasma. Inverted from interferogram in Fig. 3b. b) Transverse density profile. c) Axial density profile.

distribution. For $T_e \approx 2eV$ and $N_e \approx 10^{20} cm^{-3}$ the conditions[12] for using the Boltzman assumption, i.e. LTE, are well satisfied. Similarly, the atomic levels of these lines are sufficiently lower than the effective ionization level of the Al III ion ($E_{ionization} - E_{level} \leq 2\Delta E$; where ΔE is reduction of the ionization energy due to plasma screening effects) such that the bound-state energy should be an accurate representation of the bound electron energy in the Boltzman distribution. Sample OMA time-integrated spectra of these lines for a heated plasma and an unheated vapor are shown in Figure 6b. The ratio of these lines is plotted along with the calculations in Figure 6a. The plasma temperature for this data is about 2eV. In spite of the time-integrated nature of the spectra this temperature measurement corresponds to the peak temperature of the plasma. This is shown by the dashed curve in Figure 6a which corresponds to the calculated integrated-line-intensity ratios (gaussian time dependence, similar to heating laser) as a function of peak temperature. The amplitude of both of the Al III lines falls so rapidly with temperature that the low temperature components of the emission contribute negligibly to the line intensities.

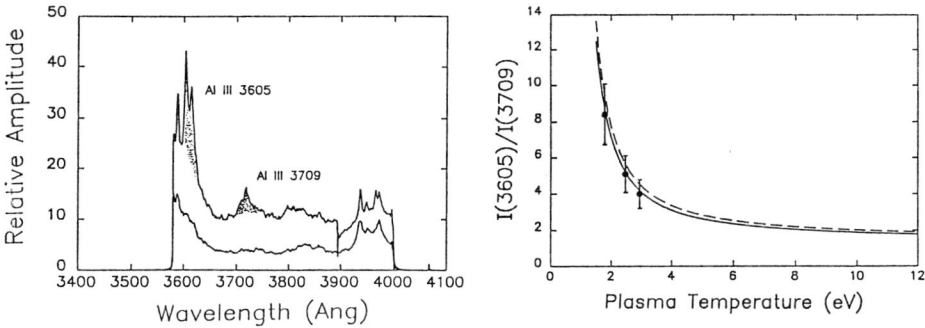

Fig 6. a) Typical Al spectra with and without laser heating. Lower spectrum is the no heating case. b)Electron temperature as a function of the 3605 Ang. and 3709 Ang. line ratios. Data points due to measured line ratios.

The plasma temperature is also determined from the measured plasma opacity and absolute emission intensity. These quantities are related to the temperature through Kirchoff's Law. The temperature determination from this method is time resolving and it is not restricted to low temperatures, i.e. $T \leq 2eV$. In practice, a portion of the probing laser is focused onto the plasma slab, as in Figure 4, and the degree to which this beam is absorbed is monitored by a set of fast ($\tau \approx 350ps$) photodiodes. The absorption beam and the interferometer beam have orthogonal polarizations such that polarizing optics can be used to make the two measurements independent but simultaneous. Similarly, the absolute emission of the plasma is measured at the same time with a 3/4 meter monochrometer. The monochrometer is tuned near the wavelength of the probe and it images the same area of the plasma as is sampled by the absorption beam. The spatial resolution of this measurement is on the order of 50µm while the time resolution of the monochrometer PM-tube is about 2ns. For a slab emitter with uniform emission and absorption properties the emission intensity is related to the temperature through $I(\nu,T) = I_p(\nu,T)(1-I/I_o)$. were $I_p(\nu,T)$ is the Planck distribution, I/I_o is the transmission fraction of the probing beam, and ν is the frequency of the absorbing radiation. This is equivalent to black-body emission with an emissivity given by $(1 - I/Io)$, i.e., a "gray body". Thermal smoothing of the density gradient L_n at the edges of the slab insures that $L_n >> \lambda_1$; thus the effect of reflection from the plasma slab boundary is negligible.[13]

The absolute emission technique for measuring temperature is more difficult experimentally; however, by not being restricted to low temperatures it permits a wider choice of plasma conditions for SCP experiments. In Figure 7a, the easily accessible plasma conditions for 1.06µm laser heating of Al vapors (0.5µm Al film targets, 100µm slits, 60 and 80 ns expansion times) are displayed. The temperature and density were measured at the same time and at the same position in the plasma slab. The curves shown with the data are from a Saha calculation for a range of initial Al vapor densities. Within experimental error the Saha equation fits the

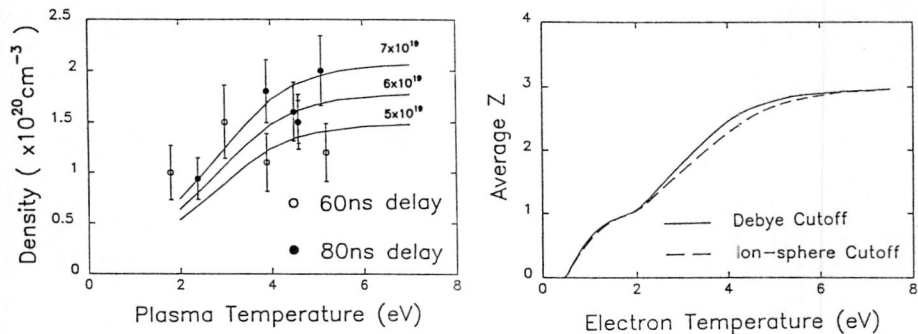

Fig. 7   a) Electron density in plasma center compared to a Saha equation calculation. b) Saha equation calculation of $<Z>$. At $T_e \approx$ 5-8 eV all of the ions are triply ionized.

data. The initial vapor densities which are needed for a good fit are the same as expected from a simple volumetric expansion of the metal film. In this calculation of the Saha equation the reduction of the ionization energy is calculated on the basis of Debye screening. The partition functions are calculated from tabulated atomic tables for all principle quantum numbers n for which the tables are complete, otherwise the hydrogenic approximation is used.[12] This calculation is not very sensitive to the model used for the reduction of the ionization energy as is illustrated by the plot of the average ionization Z (Figure 7b). This is especially true around the singly- and triply-ionized stages of Al where distinctive ionization shelves are observed and the two calculations are identical. It appears that the gap in the ionization energy between the Al II and Al III stages and the Al IV and Al V stages is sufficiently large that it dominates over the details of the calculation. As a result, for the data around 4 to 6 eV, where the density is no longer increasing with temperature, it is possible to conclude that $Z \approx 3$ even though the equation of state may not be known exactly.

The accessible ion coupling parameters ($\Gamma = e^2 Z^{5/3} (3N_e/4\pi)^{1/3}/T$) for the 1.06μm heating experiments are plotted in Figure 8. The average ionization Z is taken from the Saha calculation as in Figure 7. Ion coupling increases with T_e to a maximum of about $\Gamma \approx 1.4$ at $T_e \approx$ 4-5eV. The increase is due to an increase in N_e and Z with temperature; however, when the ionization saturates at Z = 3 continued increases in T_e produce a decreases in Γ. To our knowledge these plasmas have the highest Γ achieved in a two-component optically thin plasma. In addition, the plasmas are in full local equilibrium considering that the temperature equilibration times ($\tau_{eq} \sim$ .2ns) and the interparticle collision times ($\tau_i \sim 10^{-3}$ns, $\tau_e \sim 10^{-6}$ns) are much shorter than the 8-10ns plasma heating time. Similarly the the mean-free path between collisions (l $\sim 10^{-3}$μm) is much shorter than the smallest plasma dimension (l $\sim$ 100μm). These properties make this plasma source ideal for studying the effects of strong coupling in experimental plasmas.

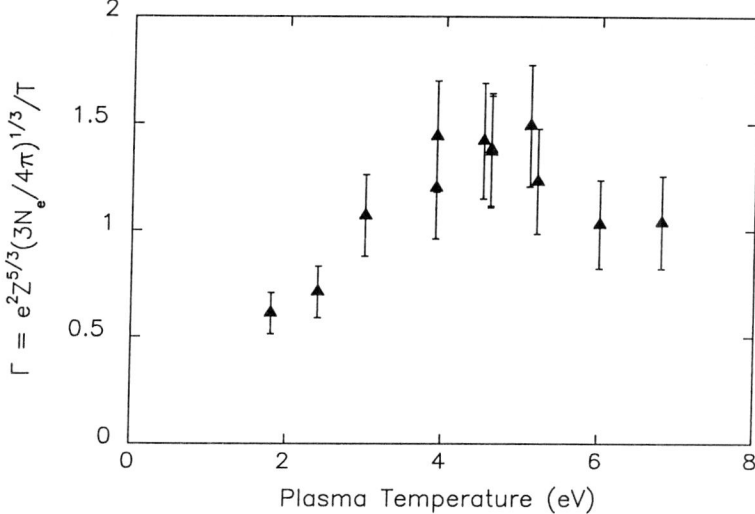

Fig. 8   Accessible ion coupling parameters in the experiment.

Plasma Opacity

In general the plasma opacity is due to absorption of electromagnetic radiation by bound-bound, bound-free and free-free electronic transitions. Bound-bound transitions have the largest absorption cross sections due to the resonant nature of that interaction, however for wavelengths far from resonance they are not important and bound-free absorption may become dominant. For the conditions of the experiments described in this work the situation is even simpler: Bound-free absorption is negligible because visible photons are not energetic enough to excite electrons into the continuum from the well populated low lying states and because the plasma is sufficiently coupled such that the closely spaced high statistical weight levels near the ionization limit no longer form bound states. For example, for $N_e \approx 10^{20} cm^{-3}$ and $T_e \approx 2eV$ Al II is the dominant stage of ionization and the calculated[14] optical depth for $0.527 \mu m$ light due to bound-free absorption is about $2000 \mu m$ - in comparison to about $100 \mu m$ for inverse bremsstrahlung. As a result, measurements of plasma opacity at the second harmonic ($0.527 \mu m$) of the probe laser directly give the absorption coefficient of the plasma for inverse bremsstrahlung absorption without the complication of atomic effects. This is an interesting regime for opacity studies in strongly coupled plasmas because by isolating the absorption to only inverse bremsstrahlung it is possible to measure the Coulomb logarithm $Ln(\Lambda)$ directly. Since the coulomb logarithm contains the details of the coulomb interaction between particles it is the main source of uncertainty in calculations of inverse bremsstrahlung in the strongly coupled regime.

The coefficient for inverse bremsstrahlung absorption is given by[15]

$$\kappa = 1.89 \times 10^6 T_e^{1/2} Z^2 N_e N_i \nu^{-3} (1 - e^{-h\nu/KT_e})(1 - \omega_p^2/\omega^2)^{-1/2} \text{Ln}(\Lambda) \text{ cm}^{-1} \quad (1)$$

where Z is the charge state of the ion, ν is the frequency of the incident photons, and Ln(Λ) is the coulomb logarithm. The $1 - \omega_p^2/\omega^2$ term represents a correction for field swelling at high densities and $1 - e^{h\nu/KT_e}$ accounts for stimulated emission by the plasma. For the conditions of these experiments $(\omega_p/\omega)^2$ is small and $h\nu/KT_e$ is of order unity. Consequently, the density correction may by ignored but the complete form of the stimulated emission correction must be retained.

Inverse bremsstrahlung measurements are made using the same optical set-up, Figure 4, as was used for the opacity component of the "gray body" temperature measurements. The temperature, density, and opacity are measured at the same time and at the same position in space. Conveniently, a small amount of the absorption beam can be allowed to image onto the interferogram film, thus, the position at which the opacity is measured is recorded along with the interferogram on every shot. The measured absorption coefficient for the plasma conditions in Figure 8 ranges between 50 and 150 cm^{-1}. The average optical depth $l \approx 1/\kappa$ is about 100μm.

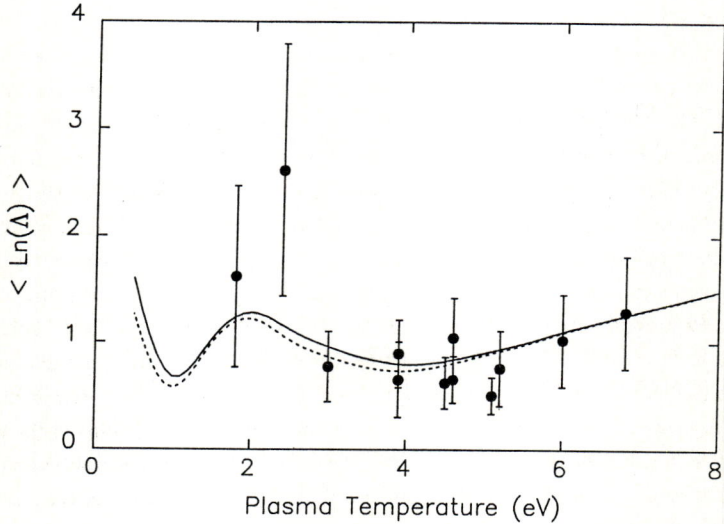

Fig. 9 Experimental values for the average Coulomb logarithm along with the predictions of classical theory.

For a plasma composed of several stages of ionization the $Z^2N_iLn(\Lambda)$ product in the absorption coefficient is replaced with $\Sigma\ Z_i^2N_iLn(\Lambda)$. The sum is over all of the ionization stages and it is assumed that the inverse bremsstrahlung absorption process is independent for each stage of ionization. It is convenient to define the average of the Coulomb logarithm as

$$<Ln(\Lambda)> = \frac{\Sigma\ Z_i^2\ N_i\ Ln(\Lambda)}{<Z>\ N_e} \qquad (2)$$

where $<Z> = \Sigma Z_iN_i/\Sigma N_i$ and $N_e = \Sigma Z_iN_i$. With this definition an inversion of κ to $<Ln(\Lambda)>$ is very simple. Using the experimental values for κ, N_e, and T_e along with a Saha calculation of $<Z>$ (Figure 7) the experimental values for $<Ln(\Lambda)>$ are given in Figure 9. The curve alongside the data is that of the classical Coulomb logarithm[16] $Ln(\Lambda) = \ln[\lambda_D 3kT/<Z>e^2]$ where $\lambda_D = [kT/4\Pi(N_e+\Sigma Z^2N_i)]^{1/2}$ is the Debye length and Z is the average ionic charge. A calculation of the average Coulomb logarithm $<Ln(\Lambda)>$ is also plotted with the data as the dashed curve. The classical prediction fits the data well and the difference between $<Ln(\Lambda)>$ and $Ln(\Lambda)$ is not significant. For $Z \sim 3$ where all of the Al ions are in the same ionization stage $<Z>$ and Z are the same, thus the two calculations agree identically. The results of this experiment show that the classical Coulomb logarithm is still very accurate for Γ values as large as 1.4 even though the classical theory is not valid in this regime. It is interesting to note that the classical prediction fits the data very well for regions where the ions are strongly coupled but the electrons are not, whereas around 2eV where both the electrons and ions are strongly coupled and the data seems to deviate from the prediction. Clearly more work is needed to determine if this deviation is real.

Conclusion

Well characterized optically-thin strongly coupled plasmas with $\Gamma \approx .5 - 1.5$ have been produced by laser vaporization and heating of Al metal-film targets. To our knowledge, these plasmas have the highest Γ achieved in a two-component optically thin plasma. Independent diagnostics of the plasma temperature and density make these plasmas ideal for various studies of strong coupling in plasmas.

Experiments measuring the plasma opacity in the visible show that inverse bremsstrahlung is the dominant absorption mechanism in Al plasmas with $n_e \approx 10^{20}$ cm^{-3} and $T_e \approx 2-6$ eV and that the optical depth of these plasmas is on the order of 100µm. The Coulomb logarithm is determined from the measured absorption coefficient and we find that our measured value for $Ln(\Lambda)$ agrees almost exactly with that of the classical result based on Debye shielding even though the classical theory is not considered to be valid for $\Gamma > 1$.

References

1) Good reviews are found in: S. Ichimaru, H. Iyetomi and S. Tanaka, Physics Reports 149, 91 (1987); M. Baus and J. P. Hansen, Physics Reports 59, 1 (1980); H. E. DeWitt, Equilibrium Statistical Mechanics of Strongly Coupled Plasmas by Numerical Simulation, in: "Strongly Coupled Plasmas", eds. G. Kalman and P. Carini (Plenum Press, New York, 1978) pp. 81-115; "Strongly Coupled Plasma Physics", eds. F. J. Rogers and H. E. DeWitt (Plenum Press, New York, 1978); and references therein.

2) C.F. Driscoll and J.H. Malmberg, Phys. Rev. Lett. 50, 167 (1983). J.H. Malmberg and T.M. O'Neil, Phys. Rev. Lett. 39, 1333 (1977).

3) J.J. Bollinger and D.J. Wineland, Phys. Rev. Lett. 53, 348 (1984).

4) R.L. Shepherd, D.R. Kania, and L.A. Jones, Phys. Rev Lett. 61, 1278 (1988).

5) Y.V. Ivanov, V.B. Mintsev, V.E. Fortov, and A.N. Dremin, Sov. Phys. JETP 44, 112 (1976).

6) T.A. Hall et al., Phys. Rev. Lett. 60, 2034 (1988).

7) A. Ng et al., Phys. Rev. Lett. 57, 1595 (1986).

8) F.J. Mayer and G.E. Bush, J. Appl. Phys. 57, 827 (1985).

9) Ya. Zel'dovich and Yu. Raizer,"Physics of Shock Waves and High Temperature Hydrodynamic Phenomena" (Academic Press, New York, 1966), Vol. I, pp. 101-106.

10) R.A. Alpher and D.R. White, Optical Interferometry, in: "Plasma Diagnostic Techniques", eds. R.H. Huddlestone and S.L. Leonard (Academic Press, New York, 1965), pp. 431-476.

11) J.A. Stamper et al., Concepts and Illustrations of Optical Probing Diagnostics For Laser-Produced Plasmas, in: "Fast Electrical and Optical Measurements", eds. J.E. Thompson and L.H. Luessen (Martinus Nijhoff Publishers, Dordrecht, 1986), pp. 691-728.

12) H.R. Griem, "Plasma Spectroscopy", (McGraw-Hill, New York, 1964).

13) V.L. Ginzberg, "The Propagation of Electromagnetic Waves in Plasmas", (Pergamon Press, Oxford, 1970), pp. 224-303.

14) The bound-free absorption coefficient is calculated by summing over all of the individual absorption contributions of the bound states that fall within the energy range of the incident photons.

15) G.B. Rybicki and A.P. Lightman, "Radiative Processes in Astrophysics", (John Wiley, New York, 1979), pp. 155-165; and T.W. Johnston and J.M. Dawson, Phys. of Fluids 16, 722 (1973).

16) S. Ichimaru, "Basic Principles of Plasma Physics", (Benjamin/Cumming, Reading, 1973), pp. 1-8.

ION BEAM-PLASMA INTERACTION: A STANDARD MODEL APPROACH

C. DEUTSCH

L.P.G.P.[+], Université Paris Sud, 91405 ORSAY Cedex, France

The interaction of energetic multicharged ion beams with separately produced target plasmas is investigated within a projectile ion-target electron binary pattern. The validity conditions of the relevant standard model are asserted. A few extensions are suggested.
The corresponding stopping results are contrasted to recent experimental results obtained by producing a fully ionized plasma on accelerator beam lines with the SPQR2 setup.

1. INTRODUCTION

As already advocated several times[1], the stopping of intense and energetic ion beams in fully ionized plasmas may be correctly understood at zero order of the beam-plasma interaction, through the behaviour of a single projectile within the target. Simple order of magnitude estimates demonstrate that in the most intense beams of ICF interest, the average ion-ion distance remains two orders of magnitude larger than the mean electron screening length in cold target. This fortunate occurrence, often referred to as a reduction principle, provides us with two fundamental consequences.

First, it allows us to investigate plasma stopping properties within the theoretical framework already worked out for cold matter.

This latter is essentially patterned on the ion projectile-target electron encounter. In this approach, the excitation of one-electron orbitals bound to the target, plays a crucial role. So, it should be carefully evaluated.

Next, the reduction principle opens the way to a novel stopping metrology of ICF interest[1], which makes use of already existing accelerating facilities, provided the target plasma is separately ignited. The latter simulates a coronal plasma arising in a compression process through intense beams.

We are thus led to investigate at length the stopping of energetic ions in dense and hot matter. The resulting energy losses and charge exchange cross-sections will then be of a great help in assessing the performances of ICF driven by intense ion beams compressing and heating the pellet.

[+] associé au C.N.R.S.

The present approach relies on the neglect of any significant collective beam-plasma effect[3], as already suggested several times[1,2,4], by analytical estimates as well as numerical simulations demonstrating that intense ion beams (E/A 50 MEV) are unlikely to be significantly deflected by beam-target coronal instabilities.

In addition, even in a rather turbulent situation, the binary projectile-target electron pattern is expected to build up the beginning of a more complete understanding.

Sec. 2 is devoted to a systematic extrapolation of plasma stopping in the cold matter approach.

We thus explain the content of the various assumptions underlying the so-called Stopping Standard Model (SSM) implicitly used, up to now, in computing the stopping of intense ion beams within multilayered ICF targets[1,2,4].

Basic characteristics of plasma stopping and projectile effective charge are discussed within the SSM.

A few obvious extensions beyond the SMM are examined in Sec. 3.

The experimental setups currently developed to check out the present assertions are briefly displayed in Sec. 4. An emphasis is laid on the SPQR2 program which makes use of a dense plasma column inserted on the linear-accelerator beam line. Preliminary results are discussed. Future programs are sketched out.

2. STOPPING STANDARD MODEL (SSM)

The basic content of our standard knowledge about cold matter stopping is validated by a few assumptions and scaling laws underlying the use of the well-known 3B (Bohr-Bethe-Bloch) formula for swift particle stopping (see Table 1 and eq. (1)). The very content of the Stopping Standard Model (SSM) for hot matter, essentially consists in applying the same guidelines for ion-plasma interaction.

This is not an arbitrary assumption. It simply amounts to the most economical formalization of our present understanding of this topic[1,5], while allowing us to focus on the specific plasma effects. These latter, as detailed below, points out to a enhanced stopping capability of a plasma with a given linear electron density $n_e \ell$ (ℓ = range of projectiles into the medium) when compared to its cold gas equivalent (same $n_e \ell$). The various items listed in Table 1 may be systematically questioned. However, except very near the end of range, they can be checked out as highly robust statements.

Table 1 - Stopping Standard Model: Basic facts and assumptions

- Intense ion beams appear **dilute** in target
- rectilinear trajectoires
- pointlike projectiles
- Nonrelativistic regime ($\beta \leq 0.35$)
- $n_e \ell$-scaling $-\frac{\Delta E}{E} \sim \frac{n_e \ell}{E^2}$

- prefactor $\frac{4\pi Z_1^2 e^4 n}{m_e v_1^2}$ dominant for E/A between 1 Mev/a.m.u. and 1 Gev/a.m.u.

- log terms dominant at very small (end of range) and very large projectile velocity.

As mentioned above, the assumptions listed in Table 1 essentially sustain the validity range of the 3B formula for the stopping of nonrelativistic point charges by isolated atom or cold matter which may be straightforwardly extended to hot and partially ionized material. Then, bound and free electrons, and also to a very limited extent, plasma ions, contribute to the projectile stopping. Restricting our attention to the overwhelming electron contribution, one thus writes

$$-\frac{dE}{dx} = \frac{4\pi(Z_1 e^2)^2 n}{m_e v_1^2} \times [(Z_T - \overline{Z}) \ln \Lambda_B + \overline{Z} \ln \Lambda_F] \quad , \tag{1}$$

$\overline{Z}$ denotes target average ionization (number of free electrons/nucleus). For high (with respect to target electron velocities but not yet relativistic) velocities,

$$\Lambda_B = \frac{2m_e v_1^2}{I_{av}} \tag{2}$$

with I_{av} = mean excitation energy, while

$$\Lambda_F = \frac{m_e v_1^2}{\hbar \omega_p} \tag{3}$$

where ω_p = target plasma frequency.

Actually, Λ_F is a high T limit of a more general expression derived from a complete RPA dielectric function for the electron jellium.

A/ Enhanced Stopping (Fig. 1)

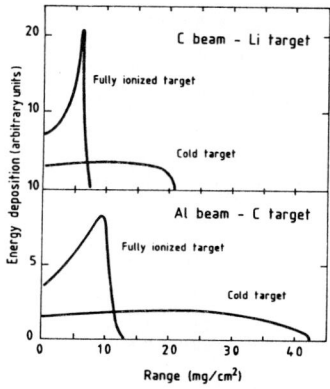

Figure 1— Energy deposition profiles. Top, 12 MeV carbon beam in a fully ionized and in cold lithium target. Bottom, 54 MeV Al beam in a fully ionized and in a cold C target (Nardi and Zinamon[5]).

The first basic discrepancy between stopping in a heated targed and in the cold equivalent with the same $n_e \ell$, arises from the enhanced stopping capabilities of the corresponding plasma. As long as the Born RPA approximation works, the small relative projectile ion-target electron velocity allows for an efficient energy exchange. Therefore, the highest electron orbitals are expected to respond in the most versatile fashion to the incoming electrostatic field. One thus gets the approximate rule of thumb, according to which free electron orbitals show up the most efficient stopping, while highly excited bound orbitals are more effective than deeply bound ones.

This latter statement may be illustrated by the range shortening arising from the bound electrons contribution.

For this purpose, let us remark that the isolated ion $\ell n I_B$ may still be used up to a 10 ev target temperature[8] (eq. (2)). Above 10 eV, plasma corrections systematically lower I_{av} values, thus demonstrating the above statement.

B/ Enhanced projectile effective charge ($Z_1(v_1)$)

The basic physics of the projectile effective charge is driven by the observation that it is much more difficult for the travelling ion to pick up bound orbitals out of plane waves than from target bound states[5].

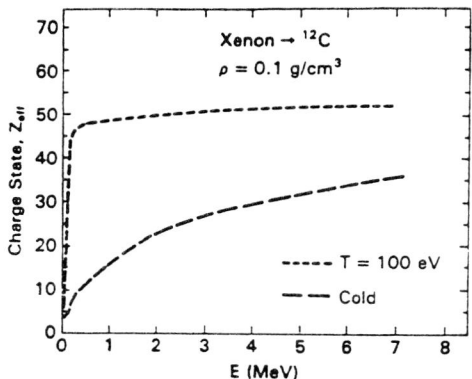

Figure 2 - The charge Z_{eff} of a xenon ion as a function of its energy as slowed in a fully ionized carbon target (Nardi and Zinamon[5]).

This specific behaviour explains that once the ionization fate of the projectile is settled at the target entrance (Fig. 2), it then remains nearly unchanged excepted very close to the end range, where it does recombine. Such a feature is at variance with projectile charge state in cold target which provides easily a lot of prepared bound orbitals, ready for pick up, so to speak. As a final result, we expect from the enhanced $Z_1(V_1)$ an additional contribution to the range shortening already considered in A/.

It is often a rather substantial one (see (Fig. 1).

3. BEYOND THE S.S.M.

The modelling developed up to now, seems quite adequate in investigating quantitatively ion stopping on most part of its range. However, the SSM is likely to break down near the end of range, where the beam-driver energy conversion takes place. Actually, the Coulomb logarithms (3) and (4), thus fall below unity, and the projectile behaves as a brownian particle assaulted by the target electrons. The stopping is then proportional to V_1. On the other hand, when the projectile kinetic energy is smaller than 100 kev/a.m.u., Coulomb ion-ion interactions are no longer negligible.

The very basic SSM assumption, according to which the beam appears dilute in target, might also be questioned in the presence of beam filamentation instabilities[1]. Then, clusters of beamlets are strongly compressed transversally. So, when the given projectiles flow with parallel velocities, they experience strong correlations. As a consequence, temperature extensions of cluster stopping currently used to determine the vicinage effect in solids, might be useful in addressing such questions.

4. BEAM-PLASMA EXPERIMENTS

A/ Target plasma

The theoretical predictions of the SSM for ion energy losses and effective charge in target can be checked out by firing a plasma located on the beam line of a conventional linear accelerating structure[1].

These beam/target interaction experiments can be easily scaled under the form (see Table 1)

$$-\frac{\Delta E}{E} \sim \frac{n_e \ell}{E^2} \quad , \tag{4}$$

where $n_e \ell$ is the product of the target electron number density with the projectile range in the target given by the linear part of

$$\int_{E_0}^{E_0/10} \frac{dE}{dE/dx}$$

Table 2 - Target plasma currently used in probing heavy ionbeam stopping

Set-up (location)	electron linear density $n_e \ell$ (cm^{-2}) in target	target electron temperature ev	projectile (energy in Mev/a.m.u.)
Laser ablated plate synchronized with a tandem accelerator (SPQR1) (Bruyère le Chatel)	10^{18}	80-200	Cu^{5+}, Si^{12+} at 0.5 Mev/a.m.u.
Dense hydrogen discharge on a beam line (SPQR2) (Orsay, IPN)	a few 10^{19}	2-5	C^{4+}, S^{7+} at 2 Mev/a.m.u.
GSI (Darmstadt)			Ar^{18+}, Ge^{20+} Ca^{13+}, U^{33+} at 1.4 Mev/a.m.u.
Z-Pinch (Aachen and GSI)	10^{21}	15	heavy ions available on UNILAC line ($\frac{E}{A} \geq 1.4$ Mev/a.m.u.)

The content of eq. (4) essentially amounts to saying that an optimum energy loss corresponds to a given line density $n_e \ell$. For instance, it is of little interest to fire too energetic ions in a moderately dense plasma. The corresponding E is likely to be too small. In these respects the projectile energy envisoned for SPQR3 lies near the ''real game'' value, i.e., 10 to 50 MeV/amu.

These considerations motivate the SPQR2 program with projectile energies in the 1 to 4 MeV/amu range and a target $n_e \ell \geq 10^{19}/cm^2$, described below.

Table 2 displays several projectile-target plasma pairs listed in order of increasing $n_e \ell$.

B/ SPQR2 SETUPS[11]

We detail a little bit more the SPQR2 setup which has been extensively used, very recently.

Our basic goal is to simulate an ablation corona resulting from an implosion conducted through intense ion beams. This is achieved with the discharge tube pictured on Fig. 3. Low pressure hydrogen (10 Torr) is introduced in a quartz capsule (neck diameter = 25 cm, length = 40 cm). High power electric discharge (20 kA) ignites a plasma. A free electron density ~ 4-5×10^{17} e-cm^{-3} is then established for 100 μsec. An axial and steady 2 kG magnetic field prevents the onset of Rayleigh-Taylor and other instabilities. The interelectrode distance is 40 cm.

Plasma parameters (n_e, T) are checked out through laser interferometry and line broadening (Stark effect) of H_α, H_β lines emitted by excited neutral species. Along the beam line, we can still rely on a neutral density $\sim 10^{15}$ cm^{-3}, sufficient for diagnostic purposes.

Electrodes are hollowed in their center to allow the ion beam through. Moreover, there are no solid plasma boundaries. An efficient differential pumping (Fig. 3) secures a swift circulation of helium or hydrogen.

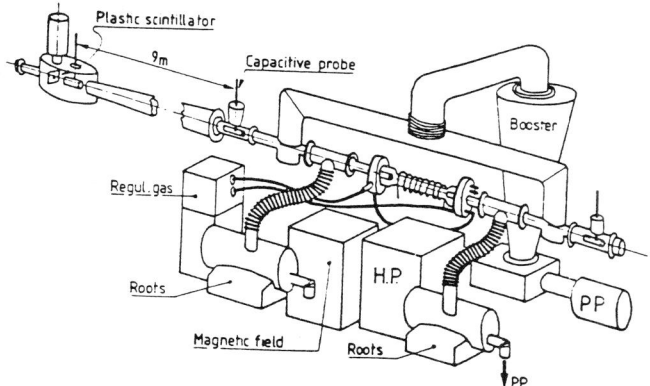

Figure 3- General set-up including discharge, differential pumping and time-of-flight. (Bimbot et al.,[11]).

The primary pump works at 17 ℓ/sec. The roots at 140 ℓ/sec, while the booster operates at 2000 ℓ/sec in air or 6000 ℓ/sec in H and He.

The whole set-up is located, altogether with the capacitor bank in a Faraday cage. In Figure 4, we display theoretical results closely related to the SPQR2 interaction experiment. One can see how the effective charge results from a dynamic balance between ionization and recombination. However, the latter is much less likely to occur in a plasma.

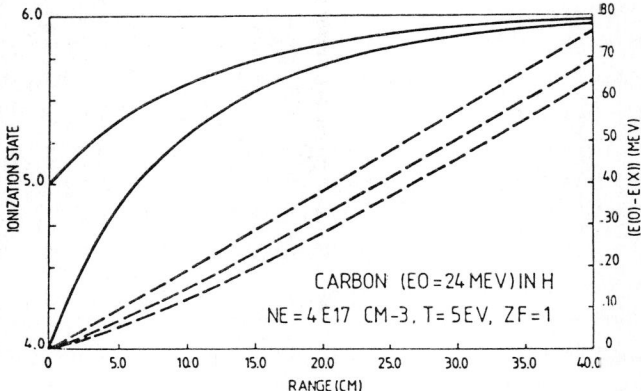

Figure 4 - Effective charges and energy losses of C^{4+} ions with 2 MeV/a.m.u. in fully ionized hydrogen

The important point explains a marked discrepancy between the effective charge in plasmas and in cold matter, where the usual expression

$$Z_{eff} = Z[1-1.034 \times \exp[-(v/2.19 \times 10^8 \text{ cm/s})]Z^{-0.688}]$$

applies. On the beam line (see Fig. 3) the three capicitive phase probes pick up the ion velocity before and after interaction with the plasma. The last two are disposed on both sides of a 9 m long time-of-flight which allows for a 4% accurate energy loss. When the beam current is too low, multichannel plates disposed at the end of the beam line can provide an alternative beam probe, provided the corresponding time structure is stable enough.

5. MEASURED ENERGY LOSS

The SPQR2 setup is particularly instructive, because it afford to contrast the behaviour of light ions to the heaviest ones, when flowing through the same plasma. As a rule of thumb, ions with the smallest ionicity/mass ratio are the easiest to detect after interaction with the plasma. They are less likely to be suddenly deflected by bursts of instabilities or other collective effects.

A/ Light ion stopping[12]

The energy losses of ^{12}C and ^{32}S beams in 9 torr H_2, deduced from time of flight measurements with and without gas, are found to be 0.61 ± 0.05 MeV and 3.11 ± 0.15 MeV respectively. These results are in fair agreement with semi-empirical tabulations for cold matter stopping.

The energy losses in plasma have been measured for ^{12}C and ^{32}S projectiles, and for two time windows corresponding to the beam intensity maximum mentioned above. The results are displayed in Fig. 5 for ^{32}S ions, and for a 13 kV discharge in 9 torr H_2. (A similar behaviour is observed for ^{12}C projectiles, the energy losses being reduced by a factor 5, consistent with the ratio of effective charges of C and S ions). The experimental energy loss ΔE (open symbols) is compared in Fig. 5 to the calculated value.

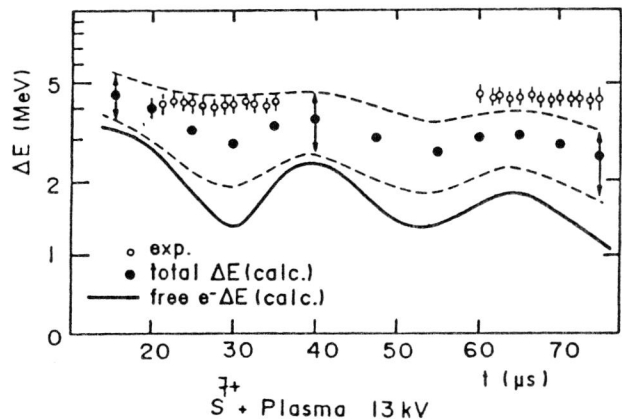

Figure 5 – Experimental and calculated energy loss in a fully hydrogen plasma (see text) for a S^{7+} ion beam (2 MeV/a.m.u.) (Gardès et al.[12])

B/ Heavy ion stopping

Figure 6 shows the energy loss of 333 MeV ^{238}U ions in neutral hydrogen gas at 6.5 mbar (dashed line) and the energy loss during the discharge time of the capacitor bank. Due to the oscillating current in the plasma discharge circuit we find a corresponding oscillation in the energy loss of the ions. The peaks in the energy loss occur when the plasma parameters temperature and density reach maximum values. After about 70 μs the energy loss drops below the value measured in neutral hydrogen. At this time a rarefaction wave reaches the axis of the plasma tube and the density of the target gas is reduced.

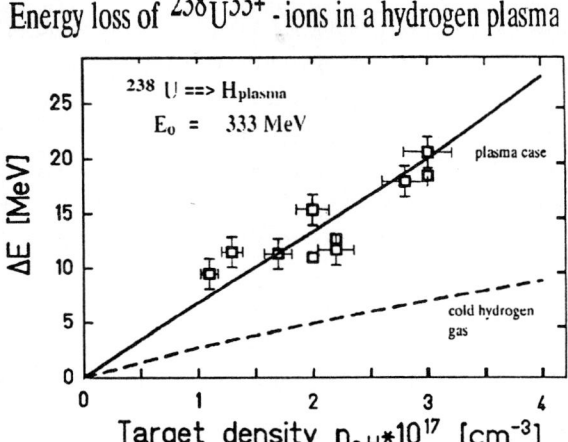

Figure 6 - Energy loss of U^{337} ions in a fully ionized hydrogen plasma and in its cold gas equivalent (Hoffmann et al.[13]) at the same pressure.

We compare the experimental results with theoretical precdictions for both the well known case of ions penetrating cold hydrogen gas and the plasma case. Figure 7 shows the evaluation of the SSM taking into account the charge state development of the ions passing through the target. The main difference between the cold gas case and the plasma stems however from the difference in the Coulomb logarithms in eq. (1). The enhanced energy loss of heavy ions in a plasma environment predicted by this theory is thus well documented by the experimental data.

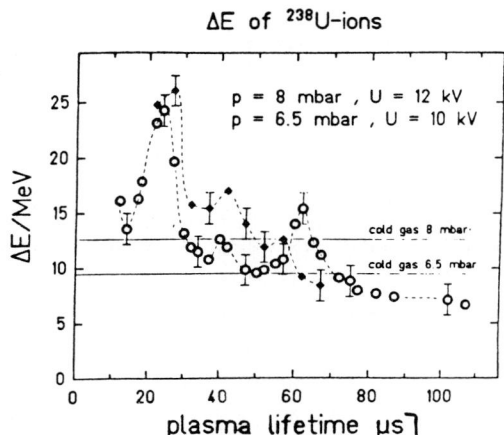

Figure 7 - dE/dx of Uranium ions in neutral hydrogen and in a plasma (Hoffmann et al.[13]).

6. FUTURE PROSPECTS

The success of the above SPQR2 program leads us to consider more dense plasma targets with higher $n_e \ell$ values. This is the purpose of our next step making use of a Z-pinch target, presently under development in Aachen, and displayed on the UNILAC beam line at GSI. One thus expects a two order of magnitude increase in $n_e \ell$, with a temperature in the 15-20 ev range for a highly ionized hydrogen discharge.

REFERENCES

1) C. Deutsch, Ann. Phys. Fr **11** (1986) 1.

2) R.C. Arnold, J. Meyer-Ter-Vehn, Rep. Prog. Phys. **50** (1987) 559.

3) C. Deutsch, P. Fromy, X. Garbet and G. Maynard, Fus. Techn. **13** (1988) 362.

4) R.O. Bangerter, Fus. Tech. **13** (1988) 348.

5) E. Nardi and Z. Zinamon, J. Physique **C8** (1983) 93.

6) R. Dei-Cas, Report CEA-R-5119 (1981).

7) G. Maynard and C. Deutsch, J. Physique **46** (1985) 1113.

8) X. Garbet and C. Deutsch, Europhys. Lett. **2** (1986) 761.

9) C. Deutsch and G. Maynard, Euro. Phys. Lett. **7** (1988) 31, and and Phys. Rev. A (to be published).

10) G. Maynard, Thesis, Orsay, 1987.

11) R. Bimbot, S. Della-Negra, D. Gardès, M.F. Rivet, C. Fleurier, B. Dumax, D.H.H. Hoffmann, K. Weyrich, C. Deutsch, G. Maynard, to be published IN Laser Interactions and Related Plasma Phenomena, ed. H. Hora, Vol. 8 (1988) 665.

12) D. Gardès, R. Bimbot, S. Della-Negra, M. Dumail, B. Kubica, A. Richard, M.F. Rivet, A. Servajean, C. Fleurier, A. Sanba, C. Deutsch, G. Maynard, D.H.H. Hoffmann, K. Weyrich, H. Wahl, Europhys. Lett. **8** (1988) 701.

13) D.H.H. Hoffmann, K. Wevrich, H. Wahl, T. Peter, J. Jacoby, R. Bimbot, D. Gardès, S. Della-Negra, M.F. Rivet, M. Dumail, C. Fleurier, A. Sanba, C. Deutsch, G. Maynard, R. Noll, R. Haas, R. Arnold, and S. Maurmann, Z. Phys. A **330** (1988) 339.

14) D.H.H. Hoffmann et al., letter of intend, GSI (1987) and C. Fleurier et al., J. Phys. (Paris) **C7** (1988) 141.

PARTICLE SIMULATIONS ON STATIC AND DYNAMIC PROPERTIES OF TWO COMPONENT HOT DENSE PLASMAS

H. FURUKAWA, K. NISHIHARA, M. KAWAGUCHI*, H. SAKAGAMI**,
T. HIRAMATSU and H. YASUI

Institute of Laser Engineering, Osaka University, Yamadaoka 2-6, Suita, Osaka 565
*NEC Software Kansai, Ltd. 1-4-24 Shiromi, Chuoku, Osaka 540
**Institute for Supercomputing Research, 2-11 Kachidoki, Chuoku, Tokyo 104

In laser produced dense plasmas, the Coulomb coupling constant for ions and the electron degeneracy parameter take on values on the order of unity. In order to simulate such a dense plasma, three dimensional two-component (ions and electrons) Particle-Particle Particle-Mesh (PPPM) code, which includes Coulomb interactions between particles and certain quantum effects, has been developed. Simulations are performed with PPPM code to investigate reduction in the bremsstrahlung emission from laser produced dense binary mixture plasmas, to estimate a contact potential and surface tension at fuel-pusher contact surface, and to investigate interaction between high intensity, ultra short laser and such dense plasmas.

1. Introduction

In a laser produced dense plasma, the Coulomb coupling constant for ions and the electron degeneracy parameter take on values on the order of unity. In order to simulate such a dense plasma, we have developed a three dimensional two-component (ions and electrons) Particle-Particle Particle-Mesh (PPPM) code in which short range forces are computed by using a direct particle-particle summation over spatially localized forces and long-range forces by particle-in-cell method.

The effective potential between particles is the same as that introduced by C.Deutsch et al[1], which includes quantum diffraction and symmetry effects. By using the pair distribution functions observed in the simulations, we estimate reduction in bremsstrahlung emission due to ion-ion correlation and electron shielding in a dense binary mixture (Carbon and Deuterium) plasma. The mass density more than 600 g/cm^3 has been recently obtained in a laser fusion experiment using a CD shell target[2]. Reduction in bremsstrahlung emission is mentioned in Sec. 2.

The contact surface between fuel and pusher may be subject to the Rayleigh-Taylor instability during the deceleration phase of the fuel compression. A surface tension, would however, act to reduces the growth rate of such an instability. We show a formation of a contact potential and existence of a surface tension between different plasmas in Sec. 3.

Interaction between high intensity, ultra short laser pulse and hydrogen plasma at a solid density is one of the current topics. By applying an external electric field oscillating at the frequency of 0.35 μm laser to such a plasma, we measure the complex electronic conductivity as a function of the quivering distance. We show the complex electronic conductivity in Sec. 4.

2. Reduction in bremsstrahlung emission from binary ionic mixture plasma

The bremsstrahlung emission coefficient $E(\omega)d\omega$ (energy emitted per time, volume, solid-angle and polarization) for binary ionic mixture plasma is given by[3]

$$E(\omega)d\omega = G(\omega)d\omega \sum_\alpha \sum_\beta Z_\alpha Z_\beta (n_\alpha n_\beta)^{1/2} \int dq F(q) S_{\alpha\beta}(q) P_\alpha(q) P_\beta(q) / q \quad (1)$$

$$F(q) = \ln\left[\frac{1 + \exp\{\mu/T - h^2(q/2 - 2\pi m_e \omega / hq)^2 / 8\pi^2 m_e T\}}{1 + \exp\{\mu/T - h^2(q/2 + 2\pi m_e \omega / hq)^2 / 8\pi^2 m_e T\}}\right] \quad (2)$$

where $S_{\alpha\beta}(q)$ is ion structure factor, while $P_\alpha(q)$ is electron shielding factor[3].

We estimate reduction of bremsstrahlung emission coefficients as a function of the frequency in a binary ionic mixture plasma, $Z_1=6$, $Z_2=1$, $n_1:n_2=1:1$, $\Gamma_{eff}=0.553$, $Z_{eff}=3.97$, $T=1keV$, by using pair distribution functions obtained by simulation with quantum diffraction and symmetry effects. Here $\Gamma_{eff} = Z_{eff}^2 e^2 / aT$, $a = (3/4\pi (n_1+n_2))^{1/3}$, $Z_{eff}^2 = <Z^{5/3}><Z>^{1/3}$. In Fig. 1, the dashed line represents bremsstrahlung emission coefficients included only ion-ion correlation effects, the solid-dashed line represents ion-ion correlation and free electron shielding effects, and solid line represents ion-ion correlation and total electron shielding effects. Bound electron shielding effects

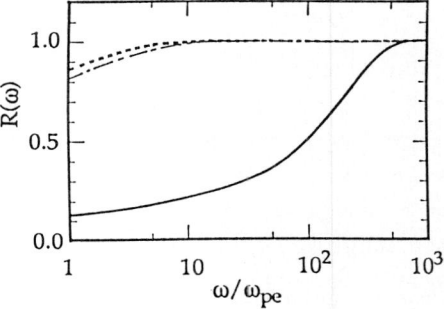

Figure 1
The reduction of bremsstrahlung emission coefficients for the case of $Z_1=6$, $Z_2=1$, $n_1:n_2=1:1$, $\Gamma_{eff}=0.553$, $Z_{eff}=3.97$, $T=1keV$

are dominant for reduction in bremsstrahlung emission.

We perform the simulation in a fictitious plasma, Z=3.97, Γ=0.553, T=1keV, to investigate the effective ionization state for bremsstrahlung emission[4]. Fig. 2 shows the reduction factor in the fictitious plasma. The three type of lines represent the same as mentioned above. We conclude if plasma is free ionized, the effective ionization state for bremsstrahlung emission can approximated by the plasma with Γ_{eff} and Z_{eff}, for two component plasma.

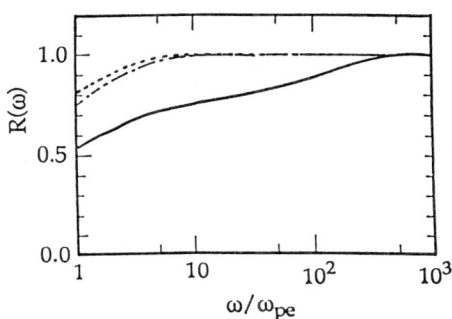

Figure 2
The reduction of bremsstrahlung emission coefficients for the case of Z=3.97, Γ=0.553, T=1keV

3. Contact potential and Surface Tension

We consider a pusher plasma with Z=5, and a fuel plasma with Z=1. The Coulomb coupling constant for ions and temperature of the pusher plasma are chosen to be Γ=3 and T=200eV. The plasma parameters of the fuel plasma are given so that the kinetic pressures and temperatures of the two plasmas are the same. The difference of the electron number densities in the both regions gives rise to a contact potential in the interface. By assuming the kinetic pressure balance and the Boltzmann distribution for electrons, we can estimate the contact potential for this case, Z=5, $e\phi/T = \ln(n_{ep}/n_{ef}) = \ln(2Z/(Z+1)) \sim 0.5$.

Simulation is performed to investigate contact potential and surface tension in the plasma mentioned above. The fuel plasma is initially located in the center in the x-direction and the pusher plasmas in the both sides of the fuel plasma.

Fig. 3 shows the contact potential, surface tension and charge densities at the time $\omega_{pi}t = 12$. Because of excess pressure due to Coulomb

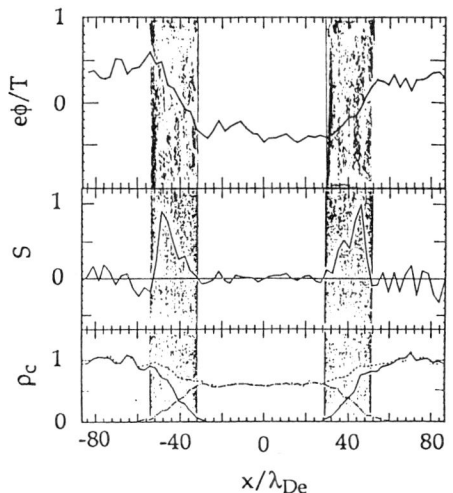

Figure 3
Contact potential, surface tension and charge densities calculated at $\omega_{pi}t = 12$

interaction, simulation result gives larger potential than that obtained from the simple estimation.

Surface tension is localized in the interface and has large values near the pusher plasma as expected. The calculated value of the surface tension is approximately takes on value on $S/\rho^3 \sim 3\times 10^8 \text{ cm}^3/\text{s}^2$.

4. Interaction between High Intensity, Ultra Short Laser and Hydrogen Plasma at a Solid Density

To simulate interaction between high intensity, ultra short laser pulse and dense plasma, we apply an external electric field oscillating at a frequency of 0.35µm laser to hydrogen plasma at a solid density. Density and temperature of the hydrogen plasma are assumed to be 5×10^{22} cm^{-3} and 13.6 eV. We observe the current density for the time duration of $6\times 2\pi/\omega_0$, where ω_0 is frequency of the external electic field. Fig. 4 shows the normalized electronic conductivity as a function of quivering distance normalized by mean particle distance, $\delta x/a$. When $\delta x/a$ becomes greater than 0.5, strong nonlinear effects are observed. Real part decreases with increase of the amplitude, approximately proportional to square root of the amplitude, for $\delta x/a > 0.5$.

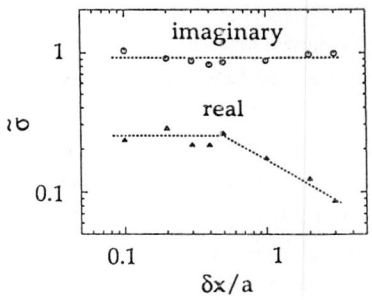

Figure 4
The normalized electronic conductivity as a function of quivering distance for the case of Z=1, n_i=5 x 10^{22} cm^{-3} and T=13.6 eV

REFERENCES

1) C.Deutsch, Phys. Lett. **60A**, (1977) 317.

2) S.Nakai , to be published.

3) R.Kawakami, K.Mima, H.Totsuji, and Y.Yokoyama, Phys. Rev. A **38**, (1988) 3618.

4) H.Totsuji, Memoirs of the School of Engineering, Okayama University, **21**, (1986) 45.

OPTICAL OBSERVATION OF LASER-COMPRESSED MATERIAL

Y.SAKAGAMI, T.NOMURA and H.YOSHIDA

Department of Electronics and Computer Engineering, Faculty of Engineering, Gifu University, Yanagido, Gifu 501-11, Japan

The properties of laser ablation driven shock waves in plane transparent target are investigated using photoelastic technique. The pressure distribution and shock front velocity are obtained. Five parameters of the equation of state(EOS) are determined.

1. INTRODUCTION

An advantage of using laser for compression of materials is the fact that high pressure can be achieved in light. Along the shock the three conservation laws can be applied to determine the state behind the shock front. These laws connect five unknown variables. If two of them are obtained experimentally, EOS can be clarified. Usually shock and particle velocities are measured. In this study, the pressure distribution and the shock velocity are obtained by the photoelastic technique.

2. EXPERIMENTS

The experiments were performed with the YAG(Oscillator)-Glass(Amplifiers) laser system with output energy of 100mJ in 30ps FWHM at 1,064nm wavelength.

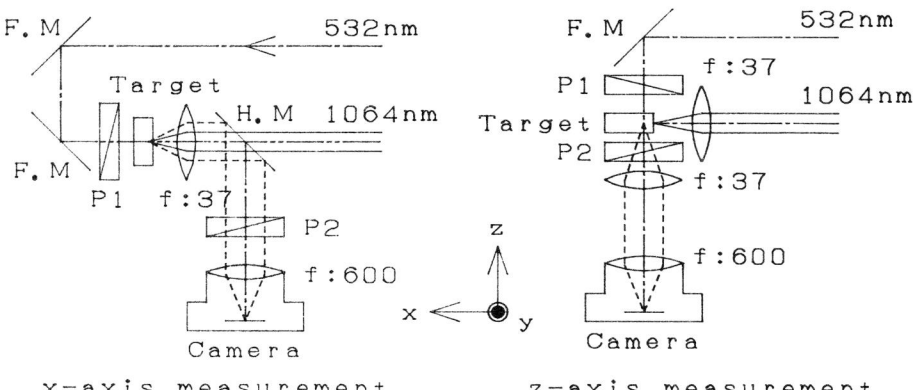

FIGURE 1
Configuration of experiment

It was focused on the target surface with the spot diameter of 20μm. The phenomena occurring inside the transparent epoxy target were observed by the second harmonics of the laser. The photoelastic system consisted of crossed polarizers. In figure 1, the configuration of this experiment is drawn. The compressed condition was observed from two directions. The photoelastic sensitivity of the target was decided 1.0×10^{-4} [$Pa^{-1} m^{-1}$] experimentally beforehand.

3. EXPERIMENTAL RESULTS AND ANALYSIS

In figures 2(a) and (b) are shown the examples of photoelastic pictures observed from the x and z axis respectively. The coaxial circular distribution of the isocolor fringes and the radial isoclinic lines show the hemispherical distribution of pressure in uniaxial stress condition.

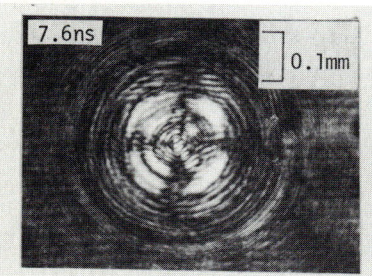

FIGURE 2(a)
Photoelastic picture
(x-axis observation)

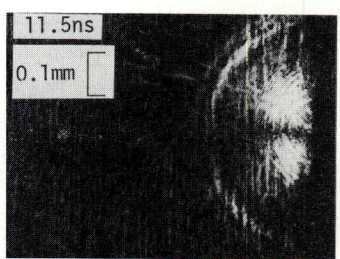

FIGURE 2(b)
Photoelastic picture
(z-axis observation)

In this case the pressure is analyzed by three dimensional photoelasticity. In figure 3 is shown the hemispherical model to analyze the pressure distribution in the x-axis observation. The principal stress is the radial pressure $p(r)$. The observed number of isocolor fringes are in proportion to the second principal stress $p_z(r,z)$, that is the z-axis component of $p(r)$. Therefore, the relationship between the isocolor fringe number $n(z)$ and pressure $p(r)$ is

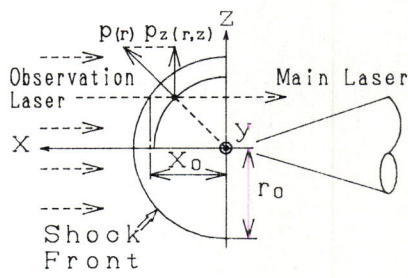

FIGURE 3
Hemispherical model of
pressure distribution

$$n(z) = e \int_0^{x_0} p(r) \cdot z/r \, dx, \tag{1}$$

where e is the sensitivity of photoelasticity. This can be rewritten as

$$N(z) = e \int_0^{x_0} P(r) \, dx, \tag{2}$$

where $N(z) = n(z)/z$, (3)
$P(r) = p(r)/r$. (4)

By the inverse Abel transformation we obtain

$$P(r) = - \frac{2}{e\pi} \int_r^{r_0} \frac{N'(z)}{(z^2-r^2)^{1/2}} dz. \tag{5}$$

The distribution $P(r)$ is calculated by numerical integration. Using equation (4), the pressure distribution $p(r)$ is determined.

In figure 4, is shown how $n(z)$ was determined. Experimental data at 1.8ns(upper) was fitted to a curve (lower) predicted by a self-similar theory[1] of a strong shock wave, in which z was normalized by a shock radius. The data obtained at other delay times are also fitted to the curve. Equation (4) and numerical calculation of (5) yielded the pressure distribution as shown in figure 5.

The self-similar theory was also applied to determine how the shock velocity depends on time. Fitting of the experimental data to the theory gave the next equation, $v_s[m/s] = 1.7 \times 10^{-1} t^{-3/5}[s]$.

From three conservation laws the parameters of EOS behind the shock front were determined. As an example, the case at 1.8ns is listed in table 1.

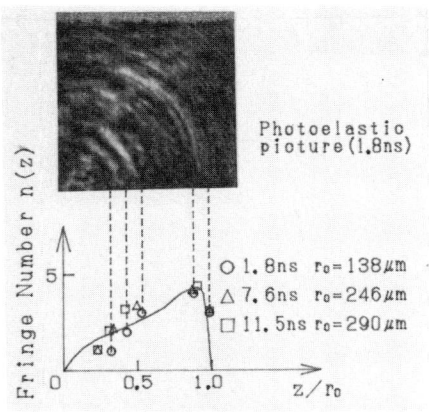

FIGURE 4
Fringe number distribution

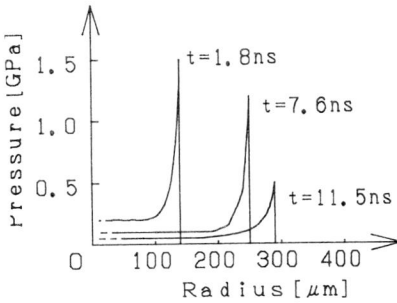

FIGURE 5
Pressure distribution

TABLE 1
Parameters of EOS

Pressure [GPa]	Shock Wave Velocity [km/s]	Particle Velocity [m/s]	Compression Ratio	Internal Energy per Unit Mass [mJ/mg]
1.5	3.1×10	4.1×10	$1 + 1.3 \times 10^{-3}$	8.3×10^{-1}

4. DISCUSSION

The energy absorbed in the solid target by laser irradiation, E_L, was estimated. In this experiment the laser energy was 100mJ. The absorption efficiency was about 40%. The fractional energy transfer around 10% to the compressed solid target gives the estimated energy, E_L, 4mJ. In the self-similar theory, the total energy involved should be conserved during the process. The pressure and the inferred kinetic energy distribution were integrated within this hemisphere, to give the total shock wave energy. This was found almost equal to the absorbed energy E_L.

5. CONCLUSION

In conclusion, by analyzing the simple hemispherical model of the photoelastic technique, the pressure distribution of laser ablation driven shock wave was obtained. Also, the shock velocity was obtained by application of self-similar theory. From them, parameters of EOS were determined.

REFERENCE
1) Ya.B.Zel'dovich and Yu.P.Raizer, Physics of Shock Waves and High-Temperature Hydrodynamic Phenomena (Academic Press, NY, 1966).

MECHANISM OF FUEL COMPRESSION IN ICF AND PROPERTY OF COMPPRESSED FUEL PLASMA

Keishiro NIU

Department of Energy Sciences, the Graduate School at Nagatsuta
Tokyo Institute of Technology, Nagatsuta, Midori-ku, Yokohama 227, Japan

A spherical shell target plays a role of supersonic converging nozzle in which the deuterium-tritium fuel is adiabatically compressed. The electron degenereracy in the fuel increases the fuel pressure from the value of ideal gas. In the final stage of compression, the fuel becomes a weakly coupled plasma whose shielding potential decreases a little due to its high density.

1. INTRODUCTION

In order to extract the fusion energy of 3GJ from a target, the deterium tritium (DT) fuel of 23mg must be compressed to more than 1000 times the solid density and heated to a temperature higher than 4keV. The final fuel pressure reaches a value higher than 10^{18}Pa. A sphercial shell target forms a supersonic converging nozzle. In the nozzle, fuel is accelerated by the pusher and its velocity reaches 3×10^5m/s after 40ns. This velocity corresponds to the Mach number 385 if the fuel is in the state of ideal gas. In reality, however, the electrons are degenerated and the real Mach number is about 10. In the supersonic nozzle, the fuel is adiabatically compressed to 269 times the solid density (5.12×10^4kg/m^3). The fuel is again compressed in the subsonic region and heated through a shock wave.

In the final stage, the fuel becomes a weakly coupled plasma whose coupling coefficient is 0.03. The shielding potential around ion decreases by the high dense electron, and fusion cross section will increase by 4%.

2. FUEL COMPRESSION

The structure of target for inertial confinement fusion (ICF) is schematically shown in Fig.1.[1] The target is irradiated by proton beam whose total energy is E_b=12MJ, pulse width is τ_b=30ns and particle energy is e_b=4MeV. Then 80% of beam energy is deposited in the radiater layer (lead, the thickness δ_{ra}=2757µm, the mass M_{ra}=149mg, and the density ρ_{ra}= 0.565g/cm^3), through the tamper layer (lead, the thickness δ_{pb}=23.4µm, the mass M_{pb}=120mg and the dentsity ρ_{pb}=11.3g/cm^3). The temperature of radiator layer reaches

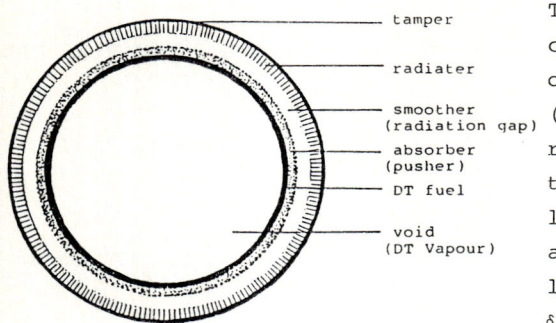

Fig. 1. Target structure. Target radius is r_t=9.056mm.

T_{ra}=1.76keV (expanding) and 60% of the deposited beam energy is changed to the radiative energy (soft x-ray). About 10% of radiative energy is lost outside the target through the tamper layer (T_{pb}=500eV) and the radiative energy flux in the smoother layer (radiation gap, thickness δ_{sm}=2mm) toward the absorber (pusher) layer (aluminum, the thickness δ_{Al}=217μm, the mass M_{Al}=264 mg, and the density ρ_{Al}=2.7g/cm^3) from the radiator layer will be 3.23x10^{13}W/cm^2. Initially, the temperature of absorber layer is very low (8K), but is heated to 200eV absorbing the intense radiative energy (expanding in the radiation gap). Thus the pressure of pusher (absorber) layer is elevated to p_{Al}=10^{12}Pa. The mass of DT fuel (thickness δ_{DT}=286μm and the density ρ_{DT}=0.19g/cm^3) is M_{DT}=23mg. The fuel is accelerated by the pusher toward the target centre with the acceleration α=7.2x10^{12}m/s^2. The implosion velocity of the fuel arrives at u=3x10^{5}m/s 40ns after the start of beam irradiation.

If we assume that the fuel is an ideal gas, the sound velocity is s=7.8x10^{2}m/s and the Mach number of the fuel is M_o=385. However, in the cold and high dense fuel, the electron is degenerated. According to SESAME Library,[2] the Mach number of the fuel is about M_o=10.

A spherical shell target forms a kind of supersonic converging nozzle for the fuel.[3] The sonic density ρ_s of the fuel is given by

$$\rho_s/\rho_o = [2/(\gamma+1)\{1+(\gamma-1)M_o^2/2\}]^{1/(\gamma-1)} = 269 \quad (1)$$

where ρ_o is the initial density and γ is the adiabatic exponent. The fuel is adiabatically compressed in the supersonic converging nozzle.

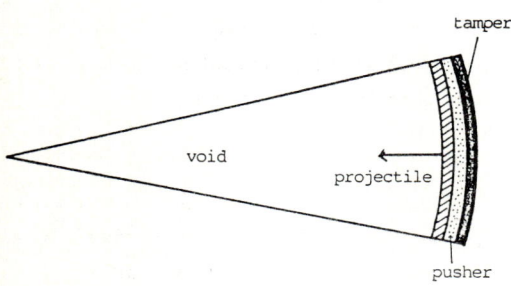

Fig. 2. Spherical target as a supersonic converging nozzle.

3. DEGENERCY OF ELLECTRON IN FUEL

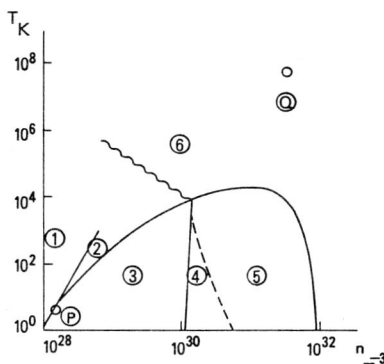

Fig. 3. Phase diagram for DT plasma.

In Fig.3, the phase diagram of DT fuel is shown.[3] The point ⓟ gives the fuel before implosion and the point ⓠ gives the fuel after implosion. The equation of state for plasma with a high temperature and a low density is, of course, that of the ideal gas and is writte by

$$p = 2nkT, \qquad (2)$$

where k is the Boltzmann constant, p is the pressure, n is the number density and T is the temperature. When the density of the plasma is high and its temperature is low, the electron in the plasma is weakly degenerated and the equation of state becomes

$$p = n_i kT_i + n_e kT_e \{1 + n_e (h^2/2\pi m_e kT_e)^{3/2} 2^{7/2} + \ldots\}, \qquad (3)$$

where h is the Planck constant, m is the particle mass, and suffices i and e refer to the ion and the electron, respectively. When the electron degenerates strongly, the equation of state becomes

$$p = n_i kT_i + 1/20 \cdot (3/\pi)^{2/3} h^2 n_e^{5/3}/5 n_e \cdot (1 + 5^2 m_e^2 k^2 T_e^2 n_e^{5/3}/3h^4 + \ldots). \qquad (4)$$

At the limit $T_i = T_e \to 0$, eq.(4) reduces to

$$p = 1/20 \cdot (3/\pi)^{2/3} h^2 n_e^{5/3}/5 m_e. \qquad (5)$$

Figure 4 shows the pressure p of the plasma versus the temperature T for the solid number density $n_s = 4.5 \times 10^{28}/m^3$ of the DT fuel. Below the temperature of $T = 10^4 K$, the electron is perfectly degenerated and the pressure is a constant of $p = 2.66 \times 10^{10}$ Pa regardless of the temperature. The electron degeneracy increases the pressure (decreases the Mach number) and suppresses the fuel compression during the implosion.

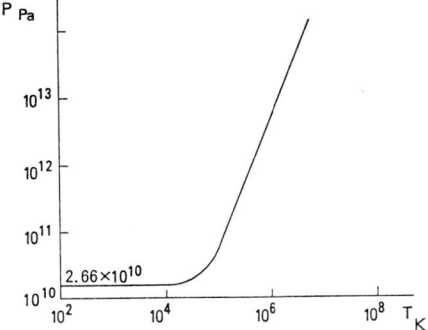

Fig. 4. Plasma pressure versus temperature for $n = 4.5E28/m^3$.

4. COUPLING OF FUEL PLASMA

The Coulomb coupling coefficient of the plasma is defined by

$$\eta = \text{(Coulomb potential energy)/(particle kinetic energy)}$$
$$= (e^2/4\pi\epsilon a)1/2.mv^2 = 7.2\times 10^{-5} n^{1/3}/T, \tag{6}$$

where ϵ is the dielectric constant in the vacuum and a is the mean distance between ions. In Fig.5, the Coulomb coupling coefficient of plasma is shown as functions of n and T. If $n=10^{32}/m^3$ and $T=10^8 K$ are substituted into eq.(6), η becomes 0.03. Thus the ICF plasma is weakly coupled. The Debye radius r_D is given by

$$r_D = (2kT_e/ne^2)^{1/2} = 6.90\times 10^3 (T_e/n)^{1/2}$$
$$= 6.9\times 10^{-9} m,$$

for $n=10^{32}/m^3$ and $T=10^8 K$, while the mean distance a between ions is

$$a = (3/4.n)^{1/2} = 0.62\times(1/n)^{1/3}$$
$$= 1.3\times 10^{-11} m,$$

for $n=10^{32}/m^{-3}$. A simple calculation suggests us that the fusion cross section $\langle\sigma v\rangle$ increases by 6% in this kind of high dense DT plasma.

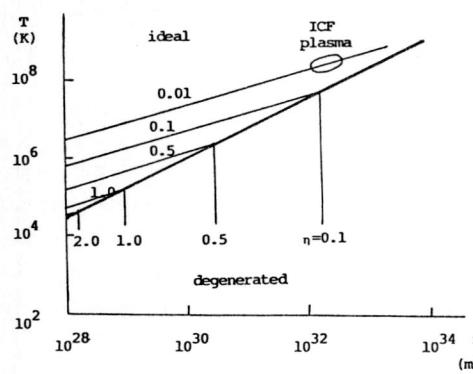

Fig. 5. Coulomb coupling coefficient of fuel plasma.

REFERENCES

1) K. Niu, Laser and Particle Beams 7 (1989) ooo.

2) SESAME-Library (Los Alamos Scientific Laboratory, 1981).

3) K. Niu and T. Aoki, Fluid Dyn. Res. 4 (1988) 195.

4) K. Niu, Laboratory and Space Plasma (Editor H. Kikuchi, Springer-Verlag, New York, 1986).

CHARGE NEUTRALIZATION DURING PROPAGATION OF INTENSE LIGHT ION BEAM FOR ICF DRIVER

Takeshi KANEDA and Keishiro NIU

Department of Energy Sciences, the Graduate School at Nagatsuta, Tokyo Institute of Technology, Midori-ku, Yokohama 227, Japan

The effects of electrostatic fields induced at the leading and tailing edges during the ion beam propagation are discussed. In this paper, analyses are given to derive the steady solution of the beam propagation, and then to show the mechanism to neutralize the charge of beam ions propagating in the background plasma.

1. INTRODUCTION

Although the charge and current of an intense light ion beam (LIB), required for LIB as a energy driver in inertial confinement fusion (ICF)[1], are partially neutralized by the motion of electrons in the background plasma, strong electrostatic fields remain at the leading and tailing edges because of the delay of neutralization by electron inertia. This strong electrostatic field causes the divergence of propagating beam. Charge-neutralization at the two edges is inevitably required [2,3]. In order to avoid the beam from divergence, a following method is proposed here to neutralize the beam charge; The beam passes through a high density and high temperature plasma. Then the electrostatic potentials of the beam trap the background electrons to neutralize the beam edge.

2. FUNDAMENTAL EQUATIONS

The governing equation for ion beam is the Vlasov equation, and for the background electron is the Krook equation,

$$\frac{\partial f_b}{\partial t} + \mathbf{v}\cdot\frac{\partial f_b}{\partial \mathbf{r}} + \frac{e_b}{m_b}(\mathbf{E} + \mathbf{v}\times\mathbf{B})\cdot\frac{\partial f_b}{\partial \mathbf{v}} = 0 \ , \tag{1}$$

$$\frac{\partial f_e}{\partial t} + \mathbf{v}\cdot\frac{\partial f_e}{\partial \mathbf{r}} + \frac{e_e}{m_e}(\mathbf{E} + \mathbf{v}\times\mathbf{B})\cdot\frac{\partial f_e}{\partial \mathbf{v}} = -(\nu_{ee} + \nu_{ei})(f_e - f_{eM}) \ . \tag{2}$$

The electromagnetic fields appearing in eqs.(1) and (2) satisfy the Maxwell equations.

3. STEADY SOLUTION OF BEAM

The solution to the steady (and one-dimensional in space) Vlasov equation for beam particles is

$$f_{bo} = f_{bo}(H, P_\theta, P_z) \ . \tag{3}$$

Here, H, P_z and P_θ are given by $H = \frac{1}{2}m_b v^2 + e_b \Phi$, $P_z = m_b v_z + e_b A_z$ and $P_\theta = m_b r v_\theta + e_b r A_\theta$, respectively.

4. SOLUTION OF BEAM WITH TWO EDGES

4.1 Velocity distribution function

The unsteady solution for the beam which has two edges is chosen in the form as

$$f_b = f_{b0}(H, P_\theta, P_z)\, g(Z) \tag{4}$$

where $g_A(Z) = \frac{1}{z_1}Z + 1 + \frac{l}{2z_1}$, $g_B(Z) = 1$, $g_C(Z) = -\frac{1}{z_1}Z + 1 + \frac{l}{2z_1}$ and $Z = z - V_{bz}t$. Here Z is a coordinate in the frame moving with beam velocity V_{bz}. The subscript A refers to the tailing edge of the beam, B the main part and C the leading edge. The solution (4) expresses that the gradients of initial number density are linear at the two edges.

4.2 Magnetic fields, induced electric fields and electrostatic fields

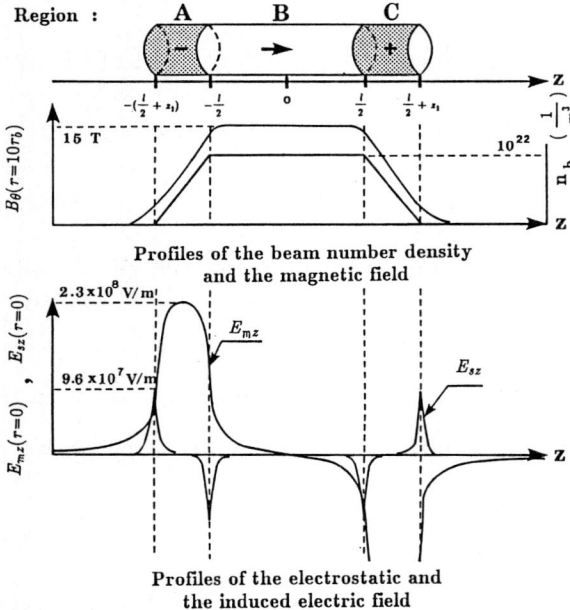

Profiles of the beam number density and the magnetic field

Profiles of the electrostatic and the induced electric field

Fig.1 Profiles of the Electromagnetic Fields

For the proton beam, the number density is very high ($n_b \sim 10^{22}/m^3$). Since the propagation velocity of the beam is very high, the delay of neutralization of beam charge due to the motion of electron in the background plasma is not so small. Hence the remaining charges at the two edges of the beam are rather significant under such a high density of the beam. The number density of nonneutralized beam particles at the end can be expressed approximately as $\bar{n}_b = \frac{1}{z_1} V_{bz}\tau n_{b0}$. The time delay τ of charge neutralization is inversely proportional to the electron plasma frequency ν_{pe} of the background plasma.

Thus the maximum electrostatic fields E_z and E_r are of order of 10^8V/m. Even in the main part of the beam, the beam current is not neutralized[1] and induces the strong magnetic field in the azimuthal direction. The typical profile of the magnetic field B_θ, and induced electric field E_{mz} are shown in Fig.1 as well as the electrostatic field E_{sz}.

5. MOTION OF ELECTRON IN BACKGROUND PLASMA

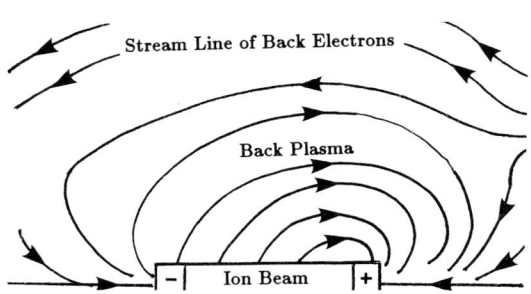

Fig.2 Flow of the Back Electrons

Motion of electrons in the background plasma plays an important role of determining the neutralized charge and current of the beam. For the velocity distribution function (4) of the beam at $t=0$, the velocity distribution function of the electron in the background plasma is chosen in the form as

$$f_e = f_{e0} + f_{e1} + f_{e2} \quad , \tag{5}$$

where

$$f_{e0} = n_{e0} \left(\frac{m_e}{2\pi k_B T_e} \right)^{3/2} exp \left(-\frac{\frac{1}{2}m_e(\mathbf{v}_e - \mathbf{v}_b - \mathbf{v}_{ed})^2 - e\Phi_0}{k_B T_e} \right) \quad . \tag{6}$$

Here $\mathbf{v}_e = (v_{er}, v_{e\theta}, v_{ez})$, $\mathbf{v}_b = (0, 0, v_{bz})$ and $\mathbf{v}_{ed} = (-\frac{E_z}{B_\theta}, 0, \frac{E_r}{E_\theta})$. In eq.(5), f_{e1} expresses the electron flow in the r-direction as follows,

$$f_{e1} = \frac{v_r(Z)}{r} S(r > r_b) \quad . \tag{7}$$

Here $S(r>r_b)$ stands for a step function. The last term of eq.(5) gives the dipole flow expressed by

$$\Psi(r,\theta,Z) = \frac{-m}{r + Z_1} + \frac{m}{r + Z_2} \quad , \tag{8}$$

where Z_1 and Z_2 are the coordinates of the sink and source, and the velocity potential Ψ is related with the velocity distribution function f_{e2} through $\int \mathbf{v} f_{e2} \, d\mathbf{v} = n_e \nabla \Psi$. By using (4) and (5), the Maxwell equations derive $\mathbf{E}$ and $\mathbf{B}$.

6. DETAILED ANALYSIS OF ELECTRON MOTION

In this section, the velocity distribution function of the background electrons is analyzed accompanied with the temporal development of the velocity distribution function of the beam. The function is chosen here as follows,

$$f_e(\mathbf{r}, \mathbf{v}, t) = f_{eM}(r, z, v_r, v_\theta, v_z) \, g(z, v_z, t) \quad . \tag{9}$$

Initially, distribution function should be the Maxwellian.

$$f_{eM} = f_e(t=0) = n_{e0} \left(\frac{m_e}{2\pi k_B T_e}\right)^{3/2} exp\left(-\frac{\frac{1}{2}m_e v_e^2 - e\Phi_0}{k_B T_e}\right) \quad . \tag{10}$$

If (9) is substituted into eq.(2), the equation for g is derived as

$$\frac{\partial g}{\partial t} + \dot{z}\frac{\partial g}{\partial z} + \frac{e_e}{m_e}(E_{mz} + E_{s0z} + E_{s1z} + v_r B_\theta)\frac{\partial g}{\partial \dot{z}}$$

$$= \left[\frac{e_e}{k_B T_e}\left\{E_{s1r}\dot{r} + (E_{mz} + E_{s1z})\dot{z}\right\} - (\nu_{ee} + \nu_{ei})\right]g + (\nu_{ee} + \nu_{ei}) \quad , \tag{11}$$

which leads to

$$g = \left[\frac{m_e}{F_z}(\nu_{ee} + \nu_{ei})\int exp\left[-\frac{m_e}{F_z}\int f(\dot{z})\,d\dot{z}\right]d\dot{z} + C\right]exp\left[\frac{m_e}{F_z}\int f(\dot{z})\,d\dot{z}\right] \quad , \tag{12}$$

where $f(\dot{z}) = \frac{e_e}{k_B T_e}\left\{E_{s1r}\dot{r} + (E_{mz} + E_{s1z})\dot{z}\right\} - (\nu_{ee} + \nu_{ei})$ and $F_z = e_e(E_{mz} + E_{s0z} + E_{s1z} + v_r B_\theta)$.

n_{e0}	$k_B T_e$
10^{22} /m^3	231 eV
10^{23}	186 eV
10^{24}	156 eV
10^{25}	134 eV

From (12), the peak of Maxwellian shifts time-dependently as a function of z, and the number density of electrons captured by the beam edge becomes clear. If the region of high density of the background plasma is located in the beam path, leading edge of the beam traps the electron. The required number density in the high density region is shown in Table 1 as a function of electron temperature.

Table 1

REFERENCES

1)   K.Niu and S.Kawata , Fusion Tech. **11** (1987) p.365.
2)   T.Kaneda and K.Niu , Laser & Particle Beams **7** (1989), part2, p.207.
3)   T.Kaneda and K.Niu , IPPJ-900, Mar.(1989) p.154.

Chapter XI:
Dense Multi-Ionic Systems

DYNAMICS AND MECHANISM OF DIFFUSION IN SUPERIONIC CONDUCTORS

Yutaka KANEKO and Akira UEDA

Department of Applied Mathematics and Physics,
Faculty of Engineering, Kyoto University, Kyoto 606, Japan

Dynamic correlations and the mechanism of diffusion in superionic conductors α-AgI and CaF_2 are studied with use of the molecular dynamics method. The characteristics of the collective motions in α-AgI are investigated by calculating dynamical structure factors. The diffusion mechanism of Ag^+ ions is presented from a dynamical view point. In the case of CaF_2, we closely examine the correlated motions of F^- ions. A new mechanism of the correlated jumps is suggested.

1. INTRODUCTION

Superionic conductors are solid state systems which exhibit ionic conductivities as high as those found in molten salts or in ionic solutions. α-AgI is a prototype of the superionic conductors in which the behavior of mobile ions is typically liquid-like. Iodine ions form a loosely-packed bcc lattice and Ag^+ ions occupy 12(d) sites. The number of the sites is six times as large as that of Ag^+ ions. In CaF_2, on the other hand, Ca^{2+} ions form a closely-packed fcc lattice and the ionic conduction is due to the disorder in the anion sublattice. According to the recent experimental studies[1,2], F^- ions do not occupy octahedral voids and diffuse among tetrahedral sites(t-sites). In this case there is no empty site for diffusing F^- ions.

In this work we study the diffusion mechanism in α-AgI and CaF_2 with use of the molecular dynamics (MD) method. Dynamical structure factors are calculated to examine the collective motions of Ag^+ ions and I^- ions. Summarizing the results, we propose the diffusion mechanism of Ag^+ ions from a dynamical view point. In contrast with the liquid-like behavior of Ag^+ ions, the diffusion of F^- ions in CaF_2 occurs by discrete hops between t-sites and the jumps of neighboring F^- ions strongly correlate to each other. We examine the mechanism of correlated jumps of F^- ions.

2. MODEL

We assume the simple interionic potential of the form[3]

$$\phi_{ij}(r) = \varepsilon(\frac{\sigma_i + \sigma_j}{r})^n + \frac{Z_i Z_j (fe)^2}{r}, \tag{1}$$

where σ_i and Z_i are the ionic radius and the valence of ion i, respectively, and f means the ionicity. For α-AgI, the potential parameters are chosen as : $n=7$, $\varepsilon=0.0851$eV, $\sigma_{Ag}=0.63$A,

σ_I=2.2A, f=0.6 and the lattice constant a=5.08A. For CaF_2, the parameters used are : n=7, ε=0.28eV, σ_{Ca}=1.28A, σ_F=1.28A, f=1.0 and a=5.9A. It should be noted that the ratios of core radii are σ_{Ag}/σ_I=0.286 and σ_F/σ_{Ca}=1.0. This suggests that the repulsive force between mobile ions is important in CaF_2.

The MD simulations are performed for the 500-ion system for α-AgI. The average temperature and the diffusion constant are 572K and 3.6×10^{-5} cm^2/sec, respectively. For CaF_2, the calculations are performed for the 324-ion and the 768-ion systems. The average temperatures of the 324-ion and the 768-ion systems are 1681K and 1752K, respectively, and the diffusion constants are 2.0×10^{-5} and 3.4×10^{-5} cm^2/sec, respectively.

3. RESULTS

3.1 α-AgI

The correlated motions of Ag^+ and I^- are investigated by calculating dynamical structure factors, current correlation functions and velocity autocorrelation functions.[4] The main results obtained are summarized as follows.

(1) For small wave vectors, the partial dynamical structure factor of Ag^+, $S_{++}(\mathbf{k},\omega)$, has almost the same feature as that of I^-, $S_{--}(\mathbf{k},\omega)$, and consists of a sharp peak of the longitudinal acoustic(LA) mode and the small broad peak of the longitudinal optic(LO) mode. (Fig.1(a)) This implies that the vibrational motion of Ag^+ ions couples with the low frequency LA mode of the I^--sublattice at long wavelengths.

(2) For large wave vectors, although the LO peak is clearly observed in $S_{--}(\mathbf{k},\omega)$, $S_{++}(\mathbf{k},\omega)$ consists of only a broad quasielastic peak and has no peak at high frequencies (Fig.1(b)) Therefore the motion of Ag^+ has no correlation with the high frequency LO mode of the I^--sublattice.

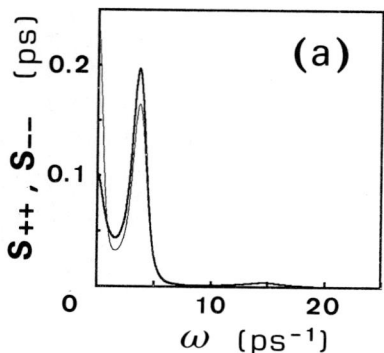 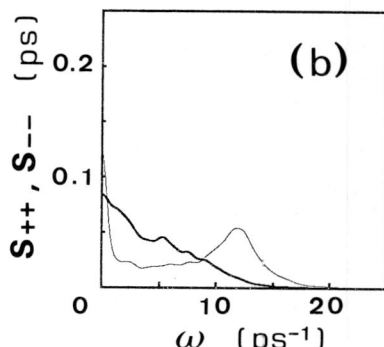

FIGURE 1

Examples of $S_{--}(\mathbf{k},\omega)$ (thin line) and $S_{++}(\mathbf{k},\omega)$ (bold line) normalized by static structure factors $S_{--}(\mathbf{k})$ and $S_{++}(\mathbf{k})$, respectively. (a) $\mathbf{k}=(2\pi/a)(0.2,0,0)$ (b) $\mathbf{k}=(2\pi/a)(0.8,0,0)$

Dynamic correlations between Ag^+ and a tetrahedron(TH) in the I^--sublattice were closely examined by Hokazono et al.[5] They showed that the mean volume of TH's which contain Ag^+ is somewhat smaller than that of empty TH's. Thus the movement of Ag^+ between TH's results in the local transportation of the strain energy. Moreover, as shown in ref.6, the potential barrier along the diffusion path becomes lower and lower as Ag^+ moves between TH's.

Summarizing all the results, the diffusion mechanism of Ag^+ ions is presented as follows.

(1) Ag^+ ions oscillate in TH's coupled with the low frequency LA mode.

(2) TH's also oscillate in the high frequency LO mode. When TH's deform so as to lower the potential barrier, Ag^+ moves to an adjacent TH.

(3) When Ag^+ moves between TH's, the relaxation of the local distortion of the I^--sublattice occurs and the strain energy is released to excite the lattice. Then I^- ions work to swing Ag^+ ions.

3.2 CaF_2

As is discussed in our previous paper,[7] the diffusion of F^- occurs by strongly correlated hops between t-sites. A jump of F^- always accompany jumps of neighboring F^- ions. Gillan and Dixon exhibited the correlated motions in CaF_2[8] and $SrCl_2$[9] diagrammatically.

In order to investigate the characteristics of the correlated jumps in more detail, we pick out jumps of F^- ions in the following manner. A jump between sites is identified as the movement of F^- between spheres centered at t-sites. Here we take the radius of the sphere as $a/10$. When we get a jump from the site n to the site m, we search for a jump of other F^- from the site m. In this way we follow the sequence of successive jumps. We find that some sequences of jumps form a closed loop. Most of the loops are made up of 4~8 jumps.

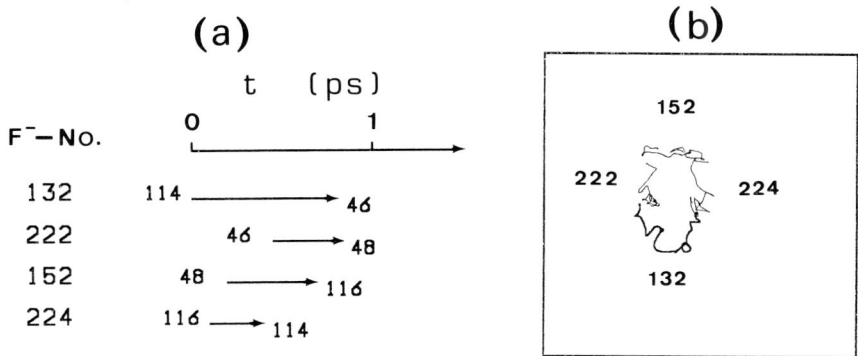

FIGURE 2

(a) The flight times of four F^- ions. Figures at both ends of the arrow show the numbers of t-sites. (b) Trajectories of F^- ions shown in (a), where F^- 132 is distinguished by a bold line.

Figure 2(a) shows the flight times of four F^- ions which make a loop. The lateral axis shows time and F^- ions are numbered as 132, 222, 152 and 224. The length of the arrow shows the flight time between spheres. The figures at both ends of the arrow stand for the numbers of the t-sites at which F^- stays before and after the movement. It should be noted that the flight time of F^- 132 is longer than that of other F^- ions. After the movement of F^- 132, F^- ions 224, 152 and 222 move between t-sites in a rather short time. The trajectories of these F^- ions are shown in Fig.2(b). The motion of F^- 132 largely deviates from the shortest path between TH's. These data suggest that the motion of F^- 132 induces the movements of other F^- ions. In other sequences of jumps, we have also observed the same kind of motions. That is, a large motion of F^- is always followed by jumps of neighboring F^- ions with a short flight time.

4. CONCLUDING REMARKS

Within the framework of the rigid ion model, the characteristics of the correlated motions in α-AgI are clarified. The diffusion mechanism of Ag^+ presented here is different from the jump diffusion mechanism usually applied to the diffusion in solids. In the case of CaF_2, a new mechanism of correlated jumps is suggested. As shown in Fig.2, some F^- moves largely out of t-site and seems to induce jumps of neighboring F^- ions. The large motion of F^- is considered to arise from the correlation with the motion of the surrounding Ca^{2+} ions or with that of other F^- ions.

REFERENCES

1) M.H.Dickens, W.Hayes, P.Schnabel, M.T.Hutchings, R.E.Lechner and B.Renker, J.Phys.C 16 (1983) L1.

2) M.T.Hutchings, K.Clausen, M.H.Dickens, W.Hayes, J.K.Kjems, P.G.Schnabel and C.Smith, J.Phys.C 17 (1984) 3903.

3) A.Fukumoto, A.Ueda and Y.Hiwatari, J.Phys.Soc.Jpn. 51 (1982) 3966.

4) Y.Kaneko and A.Ueda, Phys.Rev.B 39 (1989) 10281.

5) M.Hokazono, A.Ueda and Y.Hiwatari, Solid State Ionics 13 (1984) 151.

6) Y.Kaneko and A.Ueda, J.Phys.Soc.Jpn. 55 (1986) 3924.

7) Y.Kaneko and A.Ueda, J.Phys.Soc.Jpn. 57 (1988) 3064.

8) M.Dixon and M.J.Gillan, J.Phys.C 11 (1978) L165.

9) M.J.Gillan and M.Dixon, J.Phys.C 13 (1980) 1901, 1919.

PROPERTIES OF STRONGLY COUPLED MULTI-IONIC PLASMAS

Hugh E. DEWITT,* W.L. SLATTERY,** Guy S. STRINGFELLOW***

*Institute for Theoretical Physics, University of California, Santa Barbara, CA 93106 and Lawrence Livermore National Laboratory, University of California, P.O. Box 808, L-295, Livermore, CA 94550
**Los Alamos National Laboratory, Los Alamos, NM 87545
***University of California Observatories/Lick Observatory, Board of Studies in Astronomy and Astrophysics, University of California, Santa Cruz, CA 95064

A reexamination of the OCP fluid equation of state using new Monte Carlo results gives the EOS as $-\left(\frac{9}{10}\right)\Gamma + b\Gamma^s$ with $s \sim \frac{1}{3}$ rather than the $\frac{1}{4}$ value obtained previously. The EOS for the bcc crystalline phase indicates that first order anharmonic corrections are present. A recalculation of the freezing transition leaves the bcc transition unchanged at about $\Gamma = 178$, but the transition to fcc is lower, about 192. For ionic mixtures the linear mixture rule has near universal validity for fluid and solid mixtures. An extreme charge asymmetry test case using HNC equations shows near perfect agreement with the linear mixing rule.

1. INTRODUCTION

The possibility of separation of elements in strongly coupled multi-ionic plasmas requires a very accurate calculation of the equation of state of the ionic mixtures. In white dwarf stars consisting mainly of C and O nuclei in a nearly uniform background of degenerate electrons Coulomb interactions among the ions determine the possibility of separation, particularly in the freezing cores of these stars. In Jovian planets and brown dwarf stars the possible separation of elements is enhanced by electron screening and the calculations are considerably more involved. In both cases the thermodynamic properties of binary ionic mixtures require a detailed understanding of the one species case, the one component classical plasma or OCP when the point ions move in a uniform background. The properties of the OCP in the fluid and solid phases and of the binary mixtures in the fluid phase have been obtained with Monte Carlo numerical simulations and by coupled hypernetted chain (HNC) integral equations. These methods and results are well summarized in a recent review by Ichimaru, Iyetomi and Tanaka.[1]

In the crystalline phase of the OCP or the binary ionic mixture (BIM) there is a clear and rigorous distinction between Madelung energy of the crystal and the thermal energy due to lattice vibrations. The numerical simulations of the liquid phase indicate a similar separation of the internal energy into static energy with a 'fluid Madelung constant' and a thermal energy. This division persists from the melting value of the classical coupling parameter, $\Gamma = (ze)^2/a_{ws}kT$, a_{ws} =Wigner-Seitz or ion sphere radius, at about 180 down to $\Gamma \sim 1$, the limit of the strongly coupled region. For $\Gamma < 1$ the distinction between the static and fluid thermal energy breaks down, and in the low density or weak coupling limit the plasma is well described by the Debye-Huckel theory. Rosenfeld[2] has shown that the HNC equation applied to the OCP has an exact limiting form as $\Gamma \to \infty$ for which the internal energy is:

$$U/NkT = -\frac{9}{10}\Gamma + b\Gamma^{\frac{1}{2}} + \ldots \qquad (1)$$

with the coefficient, $-\frac{9}{10}$, and the exponent, $\frac{1}{2}$, being exact. Fits to the available MC data for the OCP[3] indicate a similar form, $a\Gamma + b\Gamma^s$, with s considerably smaller[4] than the HNC value of $\frac{1}{2}$, and usually taken as $s = \frac{1}{4}$. The difference between MC and HNC results for the fluid thermal energy is due to HNC approximation that puts the bridge function to zero. For ionic mixtures the corresponding static fluid energy is given by the ion sphere model which weights each ion by $Z_i^{\frac{5}{3}}$. Also it has been found that to a remarkable extent the internal energy of the mixture is given by the linear mixing rule[5,6] which relates the mixture energy to the OCP internal energy function:

$$U(x_1, Z_1, x_2, Z_2)/NkT = x_1 f(\Gamma_1) + x_2 f(\Gamma_2)$$
$$\Gamma_1 = Z_1^{\frac{5}{3}}\Gamma_e \quad , \quad \Gamma_2 = Z_2^{\frac{5}{3}}\Gamma_e$$
$$x_1 = \frac{N_1}{N_1 + N_2} \quad , \quad x_2 = 1 - x_1$$
$$\Gamma_e = e^2/a_e kT \quad , \quad a_e = \left(\frac{4\pi}{3} \cdot \frac{\bar{z}(N_1 + N_2)}{V}\right)^{-\frac{1}{3}} = \frac{a_{ws}}{\bar{z}^{\frac{1}{3}}}$$
$$\bar{z} = x_1 Z_1 + x_2 Z_2. \qquad (2)$$

The linear mixing rule gives the ion sphere mixture results for the static term, $\langle Z^{\frac{5}{3}}\rangle \Gamma_e$, which gives no change in free energy and hence no contribution to phase separation. Only the thermal energy of the mixture, which is approximately 2% of the static term at freezing, can result in phase separation. Thus it is quite important to have reliable and well understood results for the OCP and BIM fluid thermal energy. For the OCP the only theoretical model for the $\Gamma^{\frac{1}{4}}$ form of the thermal energy comes from a variational hard sphere minimization using the Percus-Yevick equation for hard spheres.[7] At the present time there is no generalization of this or any other theory for the BIM thermal energy, though the success of the linear mixing rule for the HNC strongly suggests a form for the BIM thermal energy.

2. N DEPENDENCE IN MC SIMULATIONS OF THE OCP

Accurate results for the fluid thermal energy of the OCP with MC simulation require long computer runs to reduce the MC noise *and* a reasonable understanding of the N dependence particularly when $N \approx 100 - 200$. Most of the MC data presented by

Slattery, Doolen, and DeWitt[3] (SDD) used adequately long runs (several million to a few tens of millions of configurations) and results were presented for bcc numbers, *i.e.*, $N = 2n^3 = 128, 250, 432, 686$, and 1024. This data showed a peculiar number dependence of the form $+0.035\Gamma/N$ in the region $80 < \Gamma < 170$. To obtain the thermodynamic limit ($N \to \infty$) the assumed N dependent correction and the data for all N were used to obtain the $N = \infty$ limit. The SDD data also included the center of mass correction for the thermal energy for both fluid and solid phases $[(U_{MC} - U_{bcc})N/(N-1)]$, *i.e.*, $O(1/N)$. Ogata and Ichimaru[8] provided new MC data for the bcc values of N, and concluded that the center of mass correction was in the wrong direction for the fluid state, hence questionable, but was justified for the solid phase.

It was recently noted that Ref. 3 also had MC data for fcc values of N, namely $N = 4n^3 = 108, 256, 500, \ldots$ The $N = 108$ from Ref. 3 and from Helfer, McCrory, and Van Horn[9] was all well *below* the thermodynamic limit. In particular the difference between $N = 108$ and 128 was far greater than any possible MC noise. Since these simulations were for the OCP fluid, it was hard to see why bcc numbers should be so different from fcc numbers. There is a simple explanation that has nothing to do with bcc and fcc numbers, but instead with the finite size of the basic cell in the MC simulation which is $L = N^{\frac{1}{3}}$. For large Γ the pair distribution function, $g(r)$ has large peaks and valleys. The MC value of the energy is influenced by these peaks and valleys of $g(r)$, since:

$$U(\Gamma,N)/NkT = \frac{1}{2}\rho \int_0^{\alpha L} 4\pi r^2 dr \beta \frac{(Ze)^2}{r} (g(r) - 1) \qquad (3)$$

where the upper limit of the integration limit αL, $L = N^{\frac{1}{3}}$, is some measure of size of the cubic cell. The factor, $g(r) - 1$, is important in this discussion. When the edge of the cell can accommodate $g(r)$ at a peak, the energy computed from Eq. (3) will be above the thermodynamic limit, but for larger N the maximum $g(r)$ will move into a valley and the computed energy will be below the thermodynamic limit.[10] Table 1 gives a few results that illustrate the effect. The final entry of Table 1 gives the thermodynamic limit. The only safe way to escape this oscillatory behavior is to use N large enough that the N dependence is approximately the MC noise. $N = 500$ and 686 is large enough. Consequently only $N = 686$ data was used in a recent study of the fluid OCP by Stringfellow, DeWitt, and Slattery.[11]

TABLE 1

N	$U(N)_{\mathrm{MC}}/NkT$	$U_{th}(N)/NkT$	$(U_{\mathrm{MC}}(N) - U(\infty))/NkT$
108	−132.1737	2.2156	−.0651
118	−132.1346	2.2548	−.0259
128	−132.0961	2.2933	+.0126
138	−132.1015	2.2879	+.0072
162	−132.1143	2.2751	−.0056
174	132.1228	2.2666	−.0141
∞	−132.1087	2.2807	.0000

OCP Fluid Energies vs N for $\Gamma = 150$

3. OCP FLUID AND SOLID

Until recently the MC OCP fluid date[3,8,9] for the internal energy has been fitted to the form:

$$U/NkT = f_{\mathrm{OCP}}(\Gamma) = a\Gamma + b\Gamma^{\frac{1}{4}} + c + d\Gamma^{-\frac{1}{4}} + \ldots \qquad (4)$$

as suggested by the variational hard sphere method of Ref. 6. With four parameters, $a, b, c,$ and d, it is relatively easy to obtain an adequate fit to the MC data, particularly in the lower range of Γ. The exponent $s = \frac{1}{4}$ had first been obtained by DeWitt[4] from Hansen's 1973 data[12] for $N = 128$ in the range $1 < \Gamma < 40$. Because of insufficiently long runs and lack of understanding at that time of the N dependence, it was not possible to obtain the exponent s to good accuracy in the larger Γ range, $40 < \Gamma < 180$. It was also found that the 'fluid Madelung constant', a, came out to be intermediate between $a_{\mathrm{bcc}} = -0.895929$ and $A_{\mathrm{HNC}} = -0.9$. The difference between the MC result for a and a_{bcc} is important. With the transition Γ at about 180 one has $a_{\mathrm{bcc}} - (-.9) \times 180 = 0.73$. This is a significant fraction of the OCP fluid thermal energy which is 2.370 at $\Gamma = 180$.

The N dependence problem can be avoided by fitting only data for a single large value of N. We chose $N = 686$, and then generated very long run MC values for fluid and solid phases for a number of large values of Γ from 150 to 200, as shown in Table 2.

TABLE 2

Q	$\frac{U_{MC}}{NkT}$	$\pm\sigma$	N	Thermalizations (10^6)	Configurations (10^6)	IC
			BCC			
150	−132.1070	0.0016	686	0.1	31.9	F
160	−141.7254	0.0010	686	2.0	77.2	L
180	−158.8972	0.0008	686	1.0	143.0	F
180	−159.6684	0.0009	686	32.0	80.0	L
200	−176.7739	0.0011	686	11.0	83.0	F
200	−177.6060	0.0010	686	0.7	46.5	L
			FCC			
160	−141.0381	0.0012	500	2.0	78.0	F
160	−141.7006	0.0010	500	0.5	111.5	L
180	−158.8943	0.0012	500	6.0	90.0	F
180	−159.6489	0.0010	500	0.1	79.9	L
200	−176.7697	0.0010	500	2.0	142.0	F
200	−177.5860	0.0010	500	0.1	79.9	L

New Monte Carlo internal energies, uncorrected for center-of-mass motion, computed for selected values of N. Initial conditions (IC) are either fluid (F) or lattice (L).

Some of the fluid runs were taken up to as much as 143 million configurations, thus several times the length of the runs used in 1980 and 1982.[3] The 22 OCP fluid data values from $\Gamma = 1$ to 200 at $N = 686$ were fitted by a non-linear least squares routine to the form:

$$U/NkT = a\Gamma + b\Gamma^s + c \tag{5}$$

with the exponent s also a variable to be fitted. The result is:

$$U/NkT = -.8992\Gamma + .596\Gamma^{.3253} - .268 \tag{6}$$

with a standard deviation of $\sigma = \pm.0016$. Thus the fluid Madelung constant is close to the HNC value of $-\frac{9}{10}$ and the exponent is significantly larger than .25. The inclusion of a fourth term, d/Γ^s, brings the fitted values of a and s closer to $-.9$ and $\frac{1}{3}$ respectively,[11] however the three term form of Eq. (5) is useful for astrophysical purposes.

The MC fluid data, the fitted values, the residuals, and the heat capacity are shown in Table 3.

TABLE 3

Γ	U_{MC}/NkT	$f(\Gamma)_{\mathrm{fit}}$	$f_{\mathrm{fit}} - U_{\mathrm{MC}}/NkT$	C_v/nK
1.000	$-5.72000e-01$	$-5.71561e-01$	$4.39237e-04$	$1.33397e-01$
2.000	$-1.32100e+00$	$-1.32008e+00$	$9.20713e-04$	$2.34589e-01$
3.000	$-2.11400e+00$	$-2.11404e+00$	$-4.20317e-05$	$3.06566e-01$
4.000	$-2.92900e+00$	$-2.92970e+00$	$-6.97521e-04$	$3.63600e-01$
5.000	$-3.75800e+00$	$-3.75849e+00$	$-4.94150e-04$	$4.10910e-01$
6.000	$-4.59500e+00$	$-4.59626e+00$	$-1.25564e-03$	$4.53136e-01$
10.000	$-7.99900e+00$	$-8.00020e+00$	$-1.19759e-03$	$5.83250e-01$
15.000	$-1.23200e+01$	$-1.23186e+01$	$1.38340e-03$	$7.00549e-01$
20.000	$-1.66750e+01$	$-1.66737e+01$	$1.34432e-03$	$7.95763e-01$
30.000	$-2.54430e+01$	$-2.54432e+01$	$-2.35445e-04$	$9.47548e-01$
40.000	$-3.42570e+01$	$-3.42587e+01$	$-1.69878e-03$	$1.06826e+00$
50.000	$-4.31040e+01$	$-4.31020e+01$	$2.04230e-03$	$1.16502e+00$
60.000	$-5.19640e+01$	$-5.19642e+01$	$-1.78706e-04$	$1.25494e+00$
80.000	$-6.97280e+01$	$-6.97271e+01$	$8.50998e-04$	$1.40333e+00$
100.000	$-8.75270e+01$	$-8.75249e+01$	$2.05390e-03$	$1.52805e+00$
120.000	$-1.05346e+02$	$-1.05347e+02$	$-5.01982e-04$	$1.64049e+00$
140.000	$-1.23182e+02$	$-1.23185e+02$	$-3.48347e-03$	$1.74161e+00$
150.000	$-1.32107e+02$	$-1.32110e+02$	$-3.19999e-03$	$1.78687e+00$
160.000	$-1.41039e+02$	$-1.41038e+02$	$1.11178e-03$	$1.82609e+00$
170.000	$-1.49970e+02$	$-1.49968e+02$	$1.75681e-03$	$1.86718e+00$
180.000	$-1.58900e+02$	$-1.58901e+02$	$-1.00704e-03$	$1.91006e+00$
200.000	$-1.76775e+02$	$-1.76773e+02$	$2.08888e-03$	$1.98289e+00$

Fit to OCP Fluid Data

The 'fluid Madelung constant' in Eq. (6) is remarkably close to the HNC and ion sphere result of $-\frac{9}{10}$, and significantly below the a_{bcc} value of $-.895929$. This result can only be valid for the strongly coupled fluid regime and does not indicate a more stable disordered state at some large value of Γ. The exponent of the thermal energy, $s = .3253$, in this fit to the data is well above the $\frac{1}{4}$ value used in Refs. 3, 8 and 9. This exponent is closely tied to the behavior of the hard sphere fluid entropy in the limit for which the packing faction of hard spheres, $\eta = \left(\frac{1}{6}\right)\pi\left(\frac{N}{V}\right)\sigma^3$ goes to 1. DeWitt and Rosenfeld[7] obtained the form of the OCP fluid energy, Eq. (4), by expanding in powers of $x = 1 - \eta$ around the limit of $\eta = 1$ using the hard sphere system with the Percus-Yevick solution as a basis system.

$$\beta F/N \leq -S(\eta)/Nk + \beta U_{PY}(\eta)/N$$

where

$$\beta U_{PY}/N = -3\Gamma\eta^{\frac{2}{3}}\left(1 - \frac{1}{5}\eta + \frac{1}{10}\eta^2\right)\Big/(1+2\eta)$$
$$= -\Gamma\left\{\frac{9}{10} - \frac{1}{9}x^3 - \frac{1}{27}x^4 + \ldots\right\}. \tag{7}$$

They used the PY virial entropy which has a first order singularity at $\eta = 1$:

$$-S(\eta)_{N/Nk} = \frac{6}{1-\eta} + 2ln(1-\eta) - 6$$

to obtain

$$(U/NkT)_{PY,V} = -\frac{9}{10}\Gamma + \left(\frac{8}{9}\right)^{\frac{1}{4}}\Gamma^{\frac{1}{4}} - \frac{1}{2}. \tag{8}$$

The PY equation, however, is seriously thermodynamically inconsistent in that the compressibility entropy has a second order singularity:

$$-S(\eta)_C/Nk = \frac{\frac{3}{2}}{(1-\eta)^2} - ln(1-\eta) - \frac{3}{2}.$$

This gives

$$(U/NkT)_{PY,C} = -\frac{9}{10}\Gamma + \frac{1}{3^{\frac{4}{5}}}\Gamma^{\frac{2}{5}} + \ldots \tag{9}$$

with an exponent, $s = \frac{2}{5}$, that is too large. It is well known that the PY pressure obtained from the virial entropy is too small, and the PY pressure from the compressibility entropy is too large. An arithmetic combination of the two forms of the entropy,[13] the Carnahan-Starling form:

$$-(S/Nk)_{CS} = \frac{2}{3}(S/Nk)_C + \frac{1}{3}(S/Nk)_V$$
$$= \frac{1}{(1-\eta)^2} + \frac{2}{(1-\eta)} - 3 \tag{10}$$

provides a remarkably good result for the hard sphere pressure for $\eta < .5$, but is no improvement on the OCP thermal energy which still comes out as $\Gamma^{\frac{2}{5}}$ because of the second order singularity.[14] An alternative combination of the two forms of the PY entropy is the geometric mean (GM) which gives an intermediate singularity:

$$(S/Nk)_{PY,GM} = \{(S/Nk)_V (S/Nk)_C\}^{\frac{1}{2}}$$
$$= \frac{3}{(1-\eta)^{\frac{3}{2}}} + \cdots \quad (11)$$

This yields

$$(U/NkT)_{PY,GM} = -\frac{9}{10}\Gamma + \frac{1}{2^{\frac{3}{2}}}\Gamma^{\frac{1}{3}} + \cdots \quad (12)$$

which is in excellent agreement with the fit to the MC data, Eq. (6).

Using the three term fit to the fluid internal energy, Eqs. (5) and (6), gives the excess Helmholtz free energy as:[3]

$$\beta F/N = a\Gamma + \frac{b}{s}\Gamma^s + (3+c)ln\Gamma - \left(a + \frac{b}{s} + 1.1516\right). \quad (13)$$

The bcc lattice MC data for $N = 686$ gives the internal energy as:

$$(\beta U/N)_{bcc} = a_{bcc}\Gamma + \frac{3}{2} + \frac{g}{\Gamma} + \frac{h}{\Gamma^2} + \cdots \quad (14)$$

where $g = 3.944$ and $h = 2490$. The corresponding bcc Helmholtz free energy is:

$$(\beta F/N)_{bcc} = a_{bcc}\Gamma + \frac{9}{2}ln\Gamma - \frac{g}{\Gamma} - \frac{h}{2\Gamma^2} - 1.8856. \quad (15)$$

Our bcc lattice data clearly indicate that there is a first order anharmonic term ($g \neq 0$ in Eqs. (14) and (15)). It has been believed for many years that the anharmonic correction began with the second order term, h/Γ^2, because the cell model calculation of Pollock and Hansen[15] gave $g = 0$. A lattice dynamics calculation of g is needed since the cell model result is incorrect.

The crossing of the fluid and bcc free energies gives the phase transition at $\Gamma_m = 178$, in good agreement with earlier calculations.[3,8] A similar calculation of the free energy for the fcc free energy ($a_{fcc} = -.895873$) using $N = 500$ solid MC data gives the transition to the fcc lattice as $\Gamma_m = 192$, slightly lower than the 196 result obtained by Helfer, McCrory, and Van Horn,[9] which again confirms that the bcc lattice is the most stable lattice.

In Table 2 we give OCP internal energy results for $\Gamma = 200$ for $N = 686$ and 500. Since this value of Γ is well above (i.e., lower temperature) the transition values of 178 and 192 respectively for bcc and fcc, the OCP fluid should be in a supercooled state. We made very long runs (83 million configurations for $N = 686$ and 142 million for $N = 500$), and found the supercooled state to be very stable, *i.e.*, no sign of dropping into a crystal or disordered solid (glass), though this is expected to happen eventually. At $\Gamma = 250$ the simulation of the supercooled fluid was not stable as shown in Fig. 1. Here we ran for 224 million configurations (each + mark on the figure represents 100,000 configurations). The system remained in a supercooled state for about 135 million configurations, and then dropped into a disordered state with an energy of $U/NkT = -222.348$ which is slightly above the bcc value of -222.427.

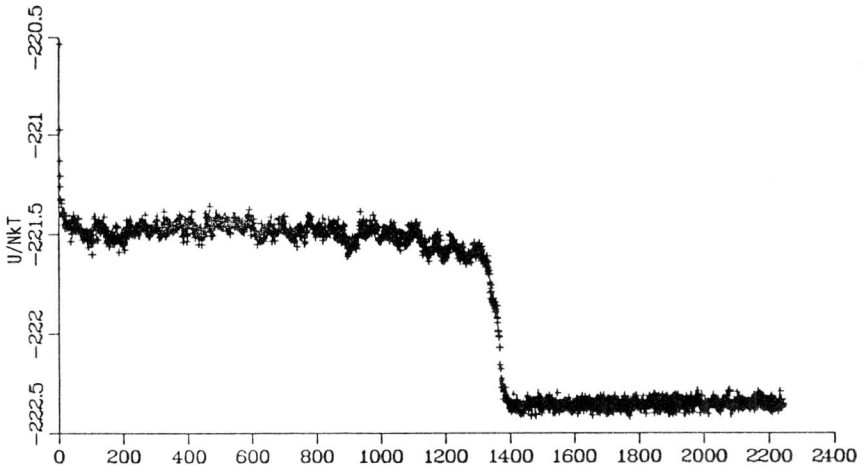

FIGURE 1

The nature of the disordered state has been studied thoroughly in a series of recent papers by the Tokyo group led by Prof. Ichimaru.[16,17,18] They have done a very impressive study of the microcrystalline structure of the "glassy" state with random polycrystalline structures for values of $\Gamma \geq 200$ and with N up to 1458.

It is apparent from Eq. (1) that the HNC thermal energy is too large for large Γ ($\Gamma^{\frac{1}{2}}$ vs. approximately $\Gamma^{\frac{1}{3}}$ for the OCP). In the past ten years there has been considerable effort to introduce a reasonable bridge function expression into the HNC equation to bring the modified HNC results for U/NkT into closer agreement with the MC results. Rosenfeld and Ashcroft[19] introduced the PY bridge function into the HNC and imposed a thermodynamic consistency requirement. Rogers, et al.[20] solved this modified HNC equation numerically for the OCP fluid and obtained agreement with the MC data of Ref. 3 to the 4th and 5th decimal places; thus to within a few percent of the thermal energy. The MC data for $\Gamma = 100$ has also been used as a means of deducing the 'experimental' form of OCP bridge function.[21] Other approaches to the bridge function problem from Prof. Ichimaru's group in Tokyo are discussed in Ref. 1. Rosenfeld[20] recently reported the 'Onsager-molecule' approach to liquid structure which yields another useful and fairly accurate means of modifying the HNC equation to incorporate a reasonable bridge function and obtained results in good agreement with MC results for the fluid OCP.

4. BINARY IONIC MIXTURES

Numerous studies in recent years of strongly coupled binary ionic mixtures have shown that the linear mixing rule, Eq. (2) is remarkably accurate for a wide variety of mixtures of charges of z_1 and z_2. Thus a knowledge of the OCP fluid and solid equations of state, Eqs. (6) and (14), allows a convenient and reasonably accurate result for the mixture equation of state. Ichimaru, et al. (Ref. 6) in their Table 1 demonstrate the level of accuracy for the linear mixing rule for C and O fluid and solid mixtures near the phase transition. Thus the deviation from linear mixing they find for fluid mixture is roughly .1% of the total energy, and consequently a few percent of the thermal energy. Their results for the solid mixture show even better agreement with the linear mixing rule.

The coupled HNC equations for ionic mixtures are inexact because of the neglect of bridge functions. However, they are useful to study because some exact results are known,[2] notably the form of the fluid Madelung term and the $\Gamma^{\frac{1}{2}}$ dependence of the thermal energy. The $\overline{Z^{\frac{5}{3}}}$ charge average cannot result in change of the free energy with fixed electron density. Consequently only a change in the thermal energy can contribute to phase separation of highly asymmetric mixtures in the HNC approximation. Brami, et al. (Ref. 5) verified the validity of the linear mixing rule for the HNC equations for z_2/z_1 as large as 8 (H and O). The ion sphere result for the Madelung term (the $\overline{Z^{\frac{5}{3}}}$) makes it possible to separately examine the thermal energy in the HNC approximation. Thus knowing the form of HNC internal energy, Eq. (1), the mixing rule would say that the mixture internal energy in a uniform background is:

$$U_{\text{HNC}}(x_1Z_1x_2Z_2,\Gamma_e)/NkT = -\frac{9}{10}\overline{Z^{\frac{5}{3}}}\Gamma_e + b\overline{Z^{\frac{5}{6}}}\Gamma_e^{\frac{1}{2}} + c. \qquad (16)$$

We tested the HNC mixture result, Eq. (16), with a number of mixtures up to $z_2 = 24$. One can see the behavior of the thermal energy (the $O(\Gamma^{\frac{1}{2}})$ term) by doing mixture runs that hold the effective interaction of the mixture:

$$\Gamma_{\text{eff}} = \overline{Z^{\frac{5}{3}}} \, \Gamma_e$$

equal to a constant. Equation (16) can then be written as:

$$U_{\text{HNC}}/NkT = -\frac{9}{10}\Gamma_{\text{eff}} + b\Gamma_{\text{eff}}^{\frac{1}{2}} g(x,Z) + c \qquad (17)$$

where

$$g(x,Z) = \frac{\overline{Z^{\frac{5}{6}}}}{(\overline{Z^{\frac{5}{3}}})^{\frac{1}{2}}} = \frac{((1-x) + xZ^{\frac{5}{6}})}{((1-x) + xZ^{\frac{5}{3}})^{\frac{1}{2}}} \qquad (18)$$

with $Z = Z_2$ and $x = x_2 = N_2/(N_1 + N_2)$. The function $g(x,Z)$ is clearly equal to 1 at $x = 0$ and $x = 1$, and has a minimum between. This occurs at:

$$x_m = \frac{1}{Z^{\frac{5}{6}} + 1} \qquad (19)$$

for which

$$g(x_m) = \frac{2Z^{\frac{5}{12}}}{Z^{\frac{5}{6}} + 1}. \qquad (20)$$

$g(x_m)$ is a measure of how much the mixture thermal energy is reduced compared with the OCP result. To check this result for the HNC we[24] used $\Gamma_{\text{eff}} = 100$, and $Z = Z_2 = 2, 4, 8, 16,$ and 24. In Table 4 some results are shown with the mixture evaluated at the

TABLE 4

Z	x_m	$g(x_m)$	U/NkT	U_{th}/NkT	F/NkT	F_{th}/NkT
1		1.	-86.96811	3.0316	-83.7586	6.2414
2	.35948	.9597	-87.0901	2.9100	-83.9671	6.0329
4	.23953	.8536	-87.3955	2.6045	-84.5883	5.4007
8	.15022	.7146	-87.7781	2.2219	-85.4410	4.5596
16	.090258	.5731	-88.1985	1.8015	86.2718	3.7282
24	.066089	.4969	-88.4227	1.5773	-86.7063	3.2937

HNC Minimum Mixture Thermal Energy

minimum defined by Eq. (20). The actual thermal energy is obtained by subtracting out the $O(\Gamma_{\text{eff}})$ term which is held constant. $g(x_m)$ gives the reduction in the $O(\Gamma_e^{\frac{1}{2}})$ term.

U_{th}/NkT is the actual computed value of the HNC mixture thermal energy. The check on the linear mixture rule here is the extent that U_{th}/NkT agrees with $b\Gamma_{\text{eff}}^{\frac{1}{2}}g(x_m) + c$ with $b = .272$ and $c = .302$. Even with $z_2 = 24$ the agreement with the linear mixing rule is good to about 1% of the thermal energy. This is a much more stringent test and confirmation of the linear mixing rule than the results given in Ref. 5. So far there is no known mathematical proof for the form of the mixture thermal energy in Eqs. (16) and (17), but for any contemplated astrophysical ionic mixtures with both components strong coupled the linear mixture rule is nearly perfect in the HNC approximation. The HNC equations give also the Helmholtz free energy. Its thermal component in Table 5 is obtained by subtracting off the static term.

For more accurate mixture results as obtained from the IHNC equation[23] and Monte Carlo[6] the linear mixing rule using the OCP fluid equation of state, Eq. (6), gives:

$$U_{\text{MC}}/NkT \cong -\frac{9}{10} \overline{Z^{\frac{5}{3}}} \Gamma_e + b\overline{Z^{\frac{5}{9}}} \Gamma_e^{\frac{1}{3}} + c + \ldots \qquad (21)$$

Again there is no mathematical proof of the correctness of the $\overline{Z^{\frac{5}{9}}}$ charge average in the mixture thermal energy, and the results of Ref. 6 suggest small deviations.

Phase separation cannot occur unless there is a sizeable redistribution of energy in the two different phases. This can in principle occur for extremely asymmetric mixtures, *i.e.*, $z_2 \gg z_1$. It is also much more likely when the electrons can strongly screen the ions as in Jovian planetary interiors and in low mass stellar interiors. Thus the phase separation criterion for fully ionized H and He mixtures found by Hubbard and DeWitt[21] could happen mainly because of screening effect on the static term of the internal energy rather than the thermal energy itself. A similar statement is true for the criterion of separation of Fe^{+24} from H in the interior of the sun as studied by Iyetomi and Ichimaru.[22] By contrast the C and O nuclei in a white dwarf cannot separate since the degenerate electrons screen the nuclei hardly at all, and the static term in the internal energy remains unchanged.

ACKNOWLEDGEMENTS

Work of HED performed under the auspices of the U.S. Department of Energy by the Lawrence Livermore National Laboratory under contract number W-7405-ENG-48, and by Institute of Nonlinear Sciences, University of California, Santa Cruz, CA 95064. This research was supported in part by the National Science Foundation under Grant No. PHY82-17853, supplemented by funds from the National Aeronautics and Space Administration, at the University of California at Santa Barbara.

REFERENCES

1). S. Ichimaru, H. Iyetomi, and S. Tanaka, Physics Reports **149** (1987) 91.
2). Y. Rosenfeld, Phys. Rev. **A33** (1986) 2025.
3). W.L. Slattery, G.D. Doolen, and H.E. DeWitt, Phys. Rev. **A21** (1980) 2087; Phys. Rev. **A26** (1982) 1982.
4). H.E. DeWitt, Phys. Rev. **A14** (1976) 1290.
5). B. Brami, J.P. Hansen, and F. Joly, Physica **905A** (1979) 505.
6). S. Ichimaru, H. Iyetomi, and S. Ogata, Astrophys. J. **334** (1988) L17.
7). H.E. DeWitt and Y. Rosenfeld, Physics Lett. **75A** (1979) 79.
8). S. Ogata and S. Ichimaru, Phys. Rev. **A36** (1987) 5451.
9). H.L. Helfer, R.L. McCrory, and H.M. Van Horn, J. Stat. Phys. **37** (1984) 577.
10). This suggestion is due to Prof. Karl Kratky.
11). G.S. Stringfellow, H.E. DeWitt, and W.L. Slattery, submitted to Phys. Rev. A.
12). J.P. Hansen, Phys. Rev. **A8** (1973) 3096.
13). J.P. Hansen, Theory of Simple Fluids (Academic Press, 1976) pp. 116–119.
14). H. Gould, R.G. Palmer, and G.A. Estevez, J. Stat. Phys. **21** (1979) 55.
15). E.L. Pollock and J.P. Hansen, Phys. Rev. **A8** (1973) 3110.
16). S. Ogata and S. Ichimaru, Phys. Rev. **A39** (1989) 1333.
17). S. Ogata and S. Ichimaru, J. Phys. Soc. of Japan **58** (1989) 356.
18). S. Ogata and S. Ichimaru, Phys. Rev. Lett. **62** (1989) 2293.
19). Y. Rosenfeld and N.W. Ashcroft, Phys. Rev. **A20** (1979) 1208.
20). F.J. Rogers, D.A. Young, H.E. DeWitt, and M. Ross, Phys. Rev. **A28** (1983) 1208.
21). P.D. Poll, N.W. Ashcroft, and H.E. DeWitt, Phys. Rev. **A37** (1988) 1672.
22). Y. Rosenfeld, D. Levesque, and O.O. Weis, Phys. Rev. **A39** (1989) 3079.
23). H. Iyetomi and S. Ichimaru, Phys. Rev. **A25** (1982) 2434; Phys. Rev. **A27** (1983) 3241.
24). These calculations were done with the HNC code of F. Rogers at LLNL.
25). W.B. Hubbard and H.E. DeWitt, Astrophys. J. **290** (1985) 385.
26). H. Iyetomi and S. Ichimaru, Phys. Rev. **A34** (1986) 3203.

LINEAR AND ELECTRONIC TRANSPORT IN STRONGLY COUPLED BINARY IONIC MIXTURES

D. LEGER and C. DEUTSCH

Laboratoire de Physique des Gaz et des Plasmas*, Bât. 212
Université Paris XI, 91405 ORSAY Cedex, France.

The paper is devoted to a systematic investigation of linear transport properties in strongly coupled binary ionic mixtures of pointlike ions interacting solely through Coulomb interactions. The basic formation rests upon suitable extensions of the Boltzmann-Ziman equation. High temperature and inelastic contributions to electron transport are emphasized out.

Let us consider a nearly homogeneous and three components system built on degenerate electrons and two pointlike ion species[1-4] $\alpha = 1$ and 2, endowed with electric charge $-Z_\alpha e$ (e = electron charge), mass M_α and particle number N_α. Mean ion sphere radius $a_i = 3/(4\pi n_i)^{1/3}$ defines a convenient length unit.

As usual the classical ion plasma parameter reads

$$\Gamma = \frac{\beta e^2}{a_i} \quad , \quad \beta = (k_B T)^{-1} \tag{1}$$

The strongly coupled BIM is thus conveniently parametrized by the triplet (Γ, r_s, c_2) of dimensionless and independent variables[1-4].

Turning to the ion fluid properties, we consider the significant ratio of ion thermal wavelength $\Lambda = [\beta h^2/(2\pi \overline{M})]^{1/2}$ to relative ion distance a_i, as a criterion for estimating the classical ion behavior

$$\delta = \frac{\Lambda}{a_i} \ll 1 \tag{2}$$

with

$$\delta^2 = \frac{2\pi}{\overline{Z}^{1/3}} \frac{\Gamma}{r_s} \frac{m_e}{\overline{M}} \tag{2a}$$

---

*Associé au C.N.R.S.

Departure from elastic scattering at the Fermi surface is conveniently measured by

$$y = \frac{\beta \hbar^2 k_F^2}{2 \overline{M}} \equiv \alpha^{-1} \frac{m_e}{\overline{M}} \qquad (3)$$

where $k_F = (2m_e E_F)^{1/2} \hbar^{-1}$ denotes the Fermi wave vector. So one gets

$$\delta^2 = \frac{4\pi}{\lambda^2} \frac{y}{Z^{2/3}} \equiv 3.41 \frac{y}{Z^{2/3}} \qquad (4)$$

A detailed analysis shows that inelastic contributions to the BZ formalism are to be measured in $y/10$ units. For our purposes, the classical ion range is thus well characterized by $y \ll 1$ for $\Gamma < 10$ and $y/10 \ll 1$ for $\Gamma \geq 10$. The various $\Gamma - r_s$ domains are pictured in Fig. 1. Inelastic contributions are expected to act significantly for $0.1 \leq y/10 \leq 1$, i.e., in domain B. Hopefully, it is well-decoupled from domain A ($\Gamma \leq 10$) where temperature effects due to finite degeneracy (α-dependent) corrections to the $T = 0$ jellium play an important role.

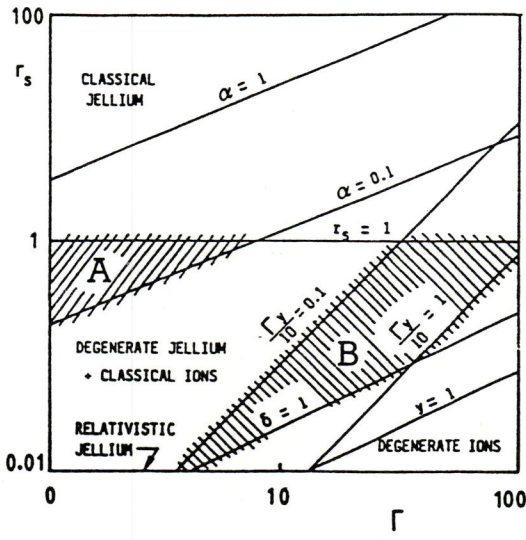

Figure 1- : $\Gamma - r_s$ plane for the $H^+ - He^{2+}$ mixture. $c_2 = 0.1$ (10% He)
Domain A : important T corrections
Domain B : important inelastic corrections.

Basic transport quantities are expressed under a reduced form which allows to an easy analytical treatment of temperature and inelastic corrections, parametrized with $\alpha = T/T_F$ and $\bar{\omega} = \beta\hbar\omega$ respectively. The former are derived from exact solutions of transport equation through various jellium dielectric functions. The Thomas-Fermi one appears as a paradigm of regular behavior at $q = 2k_F$, while the Lindhard's and its T-dependent extensions head the singular class: there the given electron-ion interactions displays diverging derivatives at $q = 2k_F$. Calculations are performed within binary ionic mixture (BIM) and polarized BIM (PBIM) frameworks respectively[1]. Ion-ion structure factors are computed through the hypernetted-chain (HNC) scheme, by neglecting bridge diagrams. Most of our results pertain to homogeneous phases. First, we focus on BIM and perform a systematic investigation of the $\Gamma - r_s$ and c_2 dependence of the transport quantities. We also pay attention to the influence of the jellium dielectric function. We consider first the T = 0 elastic limit. Next, we turn to inelastic and finite temperature corrections, and we compare BIM to PBIM results.

Typical elastic results for the H^+-He^+ mixtures are illustrated on Fig. 2.

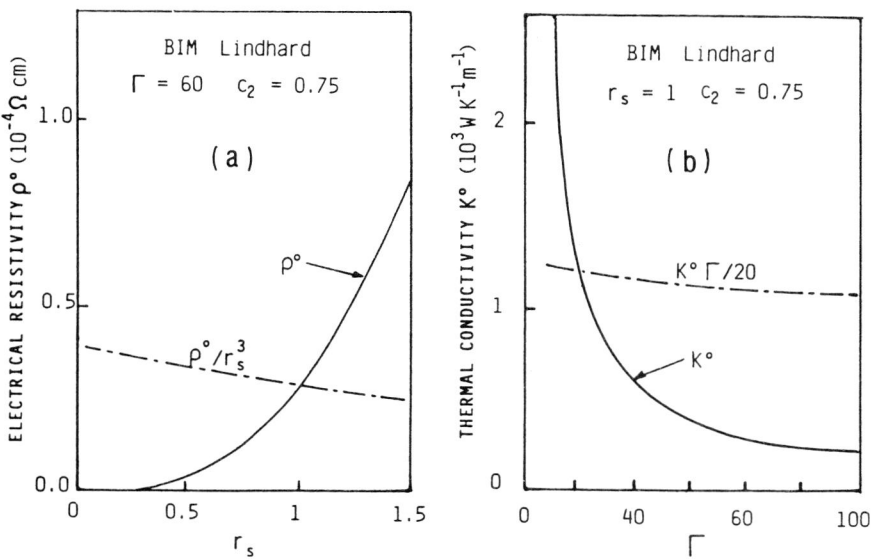

Figure 2 : Variations of (a) electrical resistivity ρ^o in terms of r_s and (b) thermal conductivity K^o in terms of Γ. ρ^o and K^o are computed in the elastic limit at T = 0 within the H^+-He^{2+} mixture. Those results pertain to BIM model with HNC structure factors and lindhard $\varepsilon(q)$ within electron-ion interaction.

REFERENCES

1) D. Léger and C. Deutsch, Phys. Rev. **A37** (1988) 4916 and **37** (1988) 4930.

2) W.B. Hubbard, Astrophys. J. **148** (1966) 858.

3) W.B. Hubbard and H.E. DeWitt, Astrophys. J. **290** (1985) 388.

4) J.P. Hansen, G.M. Torrie and P. Vieillefosse, Phys. Rev. A **16** (1977) 2153.

5) J.M. Ziman, Electrons and phonons (Oxford at the Clarendon Press, 1960).

6) (a) C. Deutsch, J. Non-Cryst. Solids **8–10** (1972) 713.
   (b) H. Minoo, C. Deutsch and J.P. Hansen, Phys. Rev. A **14** (1976) 840.

7) S. Ichimaru, H. Iyetomi and S. Tanaka, Phys. Rep. **149** (1987) 91.

STATISTICAL-MECHANICAL EFFECTS ON COLD NUCLEAR FUSION IN METAL HYDRIDES

Setsuo ICHIMARU, Shuji OGATA, Aiichiro NAKANO, Hiroshi IYETOMI, †Toshiki TAJIMA

Department of Physics, University of Tokyo, Bunkyo-ku, Tokyo 113, Japan
†Department of Physics and Institute for Fusion Studies, University of Texas, Austin TX78712, U.S.A.

We perform Monte Carlo simulation study for short-range correlations between itinerant hydrogen, interacting mutually via electron-screened repulsive forces, in periodic and aperiodic (due to defects) lattice fields of metal hydrides. We find that the screening potentials and the resultant fusion rates depend extremely sensitively on microscopic details in the lattice fields, corroborating qualitatively the varied results in recent "cold fusion" experiments.

Observation of nuclear fusion reaction between itinerant hydrogen in metal hydrides claimed in recent experiments[1] has created a challenge to condensed matter physics, calling for a theoretical account of how two hydrogen nuclei can come to fuse by overcoming the Coulombic repulsive forces in such a metallic environment. Other experiments[2] performed under analogous settings, however, have not shown significant observation of nuclear reaction. One therefore speculates that the rate of nuclear reaction should depend extremely delicately on the states of reacting pairs at short distances.

The rate of fusion between hydrogen isotopes, a and b (= p, d, t), is proportional to $g_{ab}(r)$, the value of the joint probability density $g_{ab}(r)$ at $r = 0$ for the reacting pairs averaged over the motion of their center-of-mass and the remainder of the metal-hydride environment. The probability density can be factored as

$$g_{ab}(r) = g_{ab}^0(r)\exp[\beta H_{ab}(r)] \tag{1}$$

where $g_{ab}^0(r)$ represents the joint probability in the partial system consisting only of the reacting pair; β is the inverse of temperature in energy units. Equation (1) defines the screening potential, $H_{ab}(r)$. Since the contribution of the direct interaction between a and b has been factored in (1), $H_{ab}(r)$ results from many-body effects in scattering with the rest of particles. It follows from (1) that the rate of fusion is enhanced by a factor, $\exp[\beta H_{ab}(0)]$, over the value due to the direct binary-interaction between hydrogen.[3] We here present a Monte Carlo (MC) simulation study on the screening potential between itinerant hydrogen in PdH and TiH_2, two of the typical metal hydrides used in "cold fusion" experiments; the detail will be published elsewhere.[4]

The potential $V(r)$ of binary interaction between hydrogen was calculated in terms of the charge-form factors derived from the s-d hybridized electrons and by taking account of the dielectric screening due to valence electrons.[4] The rate of fusion per a pair of a and b with the binary interaction $V(r)$ is then calculated in terms of a standard penetration integral.[3,4] Assuming n_a and n_b to be $6\times10^{22} cm^{-3}$ and T = 300 K, we obtained $\lambda_{dd}^0 = 2\times10^{-53} s^{-1}$ in PdH and $3\times10^{-58} s^{-1}$ in TiH_2. Those values of λ_{dd}^0 are far below an observable level, which we arbitrarily take as $\lambda_{ab} > 10^{-25} s^{-1}$.

To analyze the statistical-mechanical enhancement factor, $\exp[\beta H_{ab}(0)]$, we split the screening potential into the attractive and repulsive parts: $H_{ab}(r) = H_A(r) + H_R(r)$. A quantum-mechanical treatment of the attractive part through a consideration of quasi-bound pairs at the potential trough showed,[4] however, $\exp[\beta H_A(0)] \approx 1$.

The major contribution to $H_R(r)$ comes from cumulative scattering through the long-range repulsive part[4] $V_R(r)$ of $V(r)$, for which the classical approximation,[3]

$$H_R(r) = V_R(r) + \ln[\,g_R(r)\,]/\beta \qquad (2)$$

can be used. We thus perform MC simulation for $g_R(r)$ in a *fictitious* system of particles interacting via $V_R(r)$.

FIGURE 1
Equipotential contours for hydrogen in eV on the {110} plane of fcc Pd lattices. Zero level is at the O-sites (with dark shadow) in a periodic lattice (left); potential in a lattice with defect is on the right

FIGURE 2
Same as Fig. 1, but for Ti lattices. Zero level is at the T-sites

The lattice fields for itinerant hydrogen are constructed so that the following observed features may be met: In densely hydrated phases, both Pd and Ti assume a *fcc* structure with lattice constants, d = 4 Å and 4.4 Å, respectively. In Pd lattice, hydrogen sits around the octahedral (O) sites where the potential assumes local minima with curvature ~ 1.1 eV Å^{-2}; barrier height between the minima is ~ 0.23 eV inferred from diffusivity. In Ti lattice, the tetrahedral (T) sites are the local minima with curvature ~ 5.1 eV Å^{-2} and barrier height ~ 0.51 eV. The equipotential contours on the {110} plane so calculated for Pd and Ti fcc lattices with and without missing atoms (defects) are portrayed in Figs. 1 and 2. As expected, broad potential minima appear around a defect, which may trap one or more hydrogen atoms.

Periodic lattice fields for the MC simulation are obtained by placing 500 metal atoms at fcc sites in a MC cell with the periodic boundary conditions. Fields with defects are produced by removing 8 metal atoms randomly in such a way that no pairs of defects occupy nearest-neighbor positions. The corresponding numbers, 500 for PdH and 1000 for TiH$_2$, of hydrogen atoms are placed in the cell at random. A sequence of MC configurations are then generated through the random displacements of hydrogen positions in the Metropolis algorithm.[3] Several runs of such a simulation have been performed to cover various cases of metal hydrides at temperatures: 1200, 600, 300 and 200 K. Each run consisted typically of $(1-3)\times 10^4$ configurations per hydrogen atom to assure thermalization.

The joint probability density $g_R(r)$ is calculated in the statistical ensemble of particle configurations generated by the simulation. As Fig. 3 illustrates, the resulting values of $H_R(r)$ in the "visible" short-range part can be fitted quite accurately by a functional form, $H_R(r) = A + B \exp(-Cr)$. Extrapolation of this fit to r = 0 would yield $H_R(0) = A + B$. Through a comparison between a control experiment by MC simulation, where the particles interacting via $V_R(r)$ were distributed in the MC cell *without* the lattice fields, and the results for $\beta H_R(0)$ from the solution to integral equations,[3,4] we have concluded that no corrections are necessary for the extrapolated values of simulation data in the PdH potential while a reduction by 14% is required in the cases of TiH$_2$. TABLE I summarizes the values of $\beta H_R(0)$ so obtained in various cases of metal hydrides.

The total rates of fusion, $\lambda_{ab} = \lambda_{ab}^0 \exp[\beta H_{ab}(0)] \approx \lambda_{ab}^0 \exp[\beta H_R(0)]$, now exhibit a steep increase with β, a usual characteristic of a

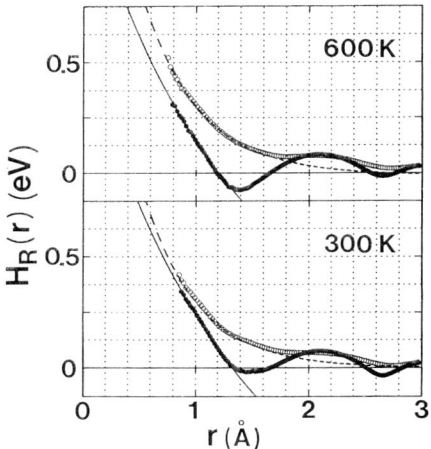

FIGURE 3
Repulsive screening potentials between hydrogen in Ti and their short-range fitting by the formula: A + B exp(–Cr). Solid circles and lines, periodic lattice; open circles and dashed lines, lattice with defect

statistical-mechanical effect. At T = 300 K, the periodic lattice fields are found to raise the fusion rate over the homogeneous cases by a factor 3×10^{10} in PdH and 7×10^{20} in TiH$_2$, resulting in $\lambda_{dd} \approx 10^{-41}$s^{-1} (PdH) and 10^{-36}s^{-1} (TiH$_2$).

When the periodic lattice fields are perturbed by defects, the enhancement factor from TABLE I increases drastically in TiH$_2$ while it rather decreases in PdH. Since the adopted potential for Ti-H has an extra term of long-range repulsion, which is absent in the adopted Pd-H potential, the fields around defects in TiH$_2$ have much finer structures with microscopic undulation than those in PdH. As seen in Fig. 3, the resulting short-range modulation of density is the cause of enhancement. The value in TABLE I implies $\lambda_{dd} \approx 7\times 10^{-23}$s^{-1} for TiH$_2$ with defects at T = 300 K. Possibility of such an "observable" fusion rate thus depends extremely sensitively on microscopic details in lattice fields, corroborating quantitatively the varied results in recent experiments.[1,2]

TABLE I
Extrapolated MC values of $\beta H_R(0)$ with corrections as described in the text.

T(K)		1200	600	300	200
PdH	[L]	5.8	12.6	27.1	40.6
	[D]	4.4	8.7	11.6	-----
TiH$_2$	[L]	9.6	21.6	48.2	84.8
	[D]	16.6	38.1	81.5	129.5

[L] periodic lattice fields, [D] aperiodic lattice fields with defects. (Simulation was not performed for PdH with defects at T = 200 K.)

The work was supported in part through Grants-in-Aid for Scientific Research by Japanese Ministry of Education, and by U. S. Department of Energy.

REFERENCES

1) M. Fleischmann, S. Pons and M. Hawkins, J. Electroanal. Chem. 262 (1989) 301; E.S. Jones, E.P. Palmer, J.B. Czirr, D.L. Decker, G.L. Jensen, J.M. Thorne, S.F. Taylor and J. Rafelski, Nature **338** (1989) 737; A. DeNinno, A. Frsttolillo, G. Lollobattista, L. Martins, L. Mori, S. Podda and F. Scaramuzzi, submitted to Europhys. Lett.

2) J.F. Ziegler, T.H. Zabel, J.J. Cuomo, V.A. Brusic, G.S. Cargill, III, E.J. O'Sullivan and A.D. Marwick, Phys. Rev. Lett. **62** (1989) 2929; M. Gai, S.L. Rugari, R.H. France, B.J. Lund, Z. Zhan, A.J. Davenport, H.S. Isaacs and K.J. Lynn, submitted to Nature.

3) For a review, S. Ichimaru, Rev. Mod. Phys. **54** (1982) 1017.

4) S. Ichimaru, A. Nakano, S. Ogata, S. Tanaka, H. Iyetomi and T. Tajima, to be published.

Chapter XII:
Strong-Coupling Theories and Experiments in General

CRITICAL COMPRESSIBILITY FACTOR OF LATTICE GAS

Ryuzo ABE

Department of Pure and Applied Sciences, University of Tokyo, Komaba, Meguro-Ku, Tokyo 153

The critical compressibility factor Z_c is studied for the lattice gas, by using its equivalence with the Ising model. The Z_c is calculated exactly for the case of two-dimensional lattice gas with nearest neighbor interaction. Also, by means of high temperature expansion and proper extrapolation, the Z_c's of three-dimensional system, the Penrose and its dual lattices are studied numerically.

1. INTRODUCTION

The critical compressibility factor Z_c at the gas-liquid critical point defined by

$$Z_c = \frac{p_c V_c}{N k_B T_c}$$

(p_c: critical pressure, V_c: critical volume, T_c: critical temperature, k_B: Boltzmann's constant, N: number of molecules) is an important quantity which characterizes the property of critical point. In contrast to the critical exponents which are universal, Z_c depends on the details of the system, i.e., it is nonuniversal. However, all the fluids in which the molecules interact through the Lennard-Jones potential $\phi(r) = 4\epsilon[(\sigma/r)^{12} - (\sigma/r)^6]$ should have the same Z_c independent of ϵ and σ. For this reason, Z_c is expected to describe some category of substance. If we consider an isotherm at T_c on pV plane and compare it with the one for ideal gas, it may be seen that in general $Z_c < 1$. Typical experimental values[1] for Z_c of nonmetallic and metallic fluids are: Ar: 0.291, Xe: 0.293, Cs: 0.22, Rb: 0.22 and Hg: 0.385. The purpose of this paper is to study Z_c of lattice gas.

2. LATTICE GAS AND THE ISING MODEL

Lee and Yang[2] have shown that the lattice gas is equivalent to the Ising model from statistical mechanical point of view. By using this equivalence, Z_c for the case of nearest neighbor interaction is shown to be expressed as[3]

$$Z_c = (2/L)\ln Z(K_c) - zK_c.$$

Here K is defined by $K = J/k_B T$ with J the strength of exchange interaction and K_c implies its value at the transition point. The L is the total number of lattice points and Z(K) is the partition function of the Ising model in the absence of magnetic field. The z stands for the coordination number of the lattice under consideration.

3. EXACT Z_c FOR TWO-DIMENSIONAL SYSTEM

The two-dimensional Ising model with nearest neighbor interaction is solved exactly so that the exact Z_c of corresponding two-dimensional lattice gas can be derived. Since the details of calculation were published elsewhere[3], we quote only the results in the following:

$$Z_c = 0.09664362264... \quad \text{(square lattice)},$$
$$Z_c = 0.0746813475... \quad \text{(honeycomb lattice)},$$
$$Z_c = 0.1112523394... \quad \text{(triangular lattice)}.$$

A tendency is observed that Z_c increases as z increases.

4. APPLICATION OF HIGH TEMPERATURE EXPANSION

In the case where the exact solution for $(1/L)\ln Z$ is impossible, high temperature expansion is a useful tool to study Z_c. As in usual, we introduce an expansion parameter w defined by $w = \text{th } K$ and expand $(1/L)\ln Z$ as

$$\frac{1}{L}\ln Z = \ln 2 + \frac{z}{2}\ln(\text{ch } K) + \sum_{r=3}^{\infty} a_r w^r .$$

Then, Z_c is expressed as

$$Z_c = 2\ln 2 + z\ln(\text{ch } K_c) - zK_c + \sum_{r=3}^{\infty} b_r w_c^r$$

with $b_r = 2a_r$. If we note that $(1/L)\ln Z$ is essentially the Helmholtz free energy, its critical behavior near T_c is written as

$$(1/L)\ln Z \simeq A(1-x)^\phi + B$$

with $x = w/w_c$ and $\phi = 2-\alpha$ (α: the critical exponent for specific heat). If we truncate the infinite series for Z_c at $r = R$, it is shown that

$$Z_c(R) = a + b/R^\phi , \quad (R \to \infty).$$

Thus, we expect that $Z_c(R)$ vs $1/R^\phi$ plot yields a linear relation between them and an extrapolation to $1/R^\phi \to 0$ leads to a or Z_c.

4.1. Two-dimensional square lattice

As a test case of the above procedure, let us consider the two-dimensional square lattice. In this case we have $\alpha = 0$ so that $\phi = 2$. Also, terms of even powers appear in the expansion. With the exact value $w_c = \sqrt{2} - 1$, we find

$$Z_c(R) = \sum_{r=2}^{R} b_{2r} w_c^{2r} .$$

We have studied the expansion up to w^{12} and $Z_c(R)$ vs $1/R^2$ plot is demonstrated in figure 1. We see that the linear relation between them is realized very well. In this figure, the same plot for the Penrose lattice discussed in 5. is also shown. If we take a pair of R = 3 and 4 and determine a in the above equation, as an extrapolation to $R \to \infty$ limit, we have Z_c = 0.096903. As compared to exact value discussed in 3., the relative error is about 0.3%. This indicates that a relatively few terms in high temperature expansion can give rise to rather a good result for Z_c.

4.2. Three-dimensional lattices

The most reliable critical data for the three-dimensional simple cubic (sc) lattice are probably due to Monte Carlo renormalizaion group method[4]:

$$K_c = 0.221654, \quad \alpha = 0.113 \quad \text{and} \quad \phi = 1.887.$$

In this case, $Z_c(R)$ is expressed as

$$Z_c(R) = 0.20570 + \sum_{r=2}^{R} b_{2r} w_c^{2r}.$$

Expansion coefficients b_{2r} are known[5] up to b_{18}. Figure 2 shows a plot similar to figure 1. As a result, we have

$$Z_c = 0.2257 \quad (sc).$$

The same analyses for the body-centered cubic (bcc) and face-centered cubic (fcc) lattices lead to

$$Z_c = 0.2488 \quad (bcc), \quad Z_c = 0.2584 \quad (fcc).$$

As in the two-dimensional system Z_c increases as z increases.

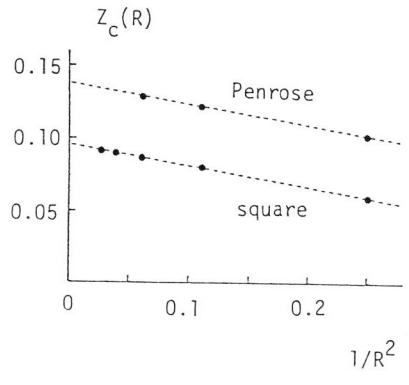

FIGURE 1

The $Z_c(R)$ vs $1/R^2$ plot for the square and Penrose lattices.

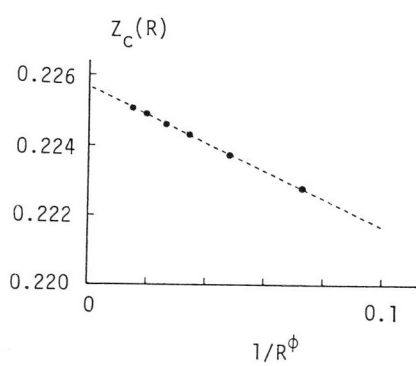

FIGURE 2

The $Z_c(R)$ vs $1/R^\phi$ plot (ϕ = 1.887) for the simple cubic lattice.

5. PENROSE AND ITS DUAL LATTICES

The Penrose or its dual lattice is a kind of quasicrystal (figure 3). Monte Carlo simulations[6] of the Ising models on these lattices reveal that they belong to the same universality class as the usual one. Thus, the method used in 4.1. is also applicable to these systems. Since both lattices have the same coordination number $z=4$, study of these systems is expected to yield some information as to how Z_c depends on the structure of the system apart from z.

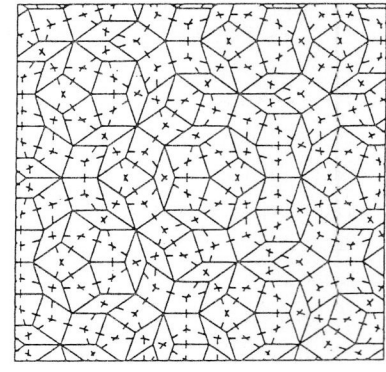

FIGURE 3
Full or dashed lines represent the Penrose or its dual lattice.

We have carried out high temperature expansion up to w^8 for the Penrose lattice[7]. Also, by means of Monte Carlo simulation $1/K_c$ is found to be[6] $1/K_c = 2.392 \pm 0.004$. The result for this case is shown in figure 1 and extrapolation leads to $Z_c = 0.138 \pm 0.002$. For the dual Penrose lattice, we have made[8] expansion up to w^9. In this case, the linearity between $Z_c(R)$ and $1/R^2$ is not so good. However, a simple-minded extrapolation procedure yields $Z_c = 0.0858 \pm 0.0015$. A more reliable method is to rely on duality relation which enables us to derive Z_c of the dual lattice from the data for the Penrose lattice. This procedure gives rise to $Z_c = 0.0846 \pm 0.0007$.

REFERENCES

1) F. Hensel, High Pressure Chemistry and Biochemistry, ed. R. van Eldik and J. Jonas (Reidel, Dordrecht, 1987) p. 244.

2) T. D. Lee and C. N. Yang, Phys. Rev. 87 (1952) 410.

3) R. Abe, Prog. Theor. Phys. 81 (1989) 990.

4) G. S. Pawley, R. H. Swendsen, D. J. Wallace and K. G. Wilson, Phys. Rev. B29 (1984) 4030.

5) C. Domb, Phase Transitions and Critical Phenomena, ed. C. Domb and M. S. Green (Academic Press, London, 1974) Vol. 3, p. 357.

6) Y. Okabe and K. Niizeki, J. Phys. Soc. Jpn 57 (1988) 16, 1536.

7) R. Abe and T. Dotera, J. Phys. Soc. Jpn. 58 (1989) No. 9.

8) T. Dotera and R. Abe, preprint.

STRUCTURAL PHASE TRANSITIONS IN DENSE HYDROGEN

Hitose NAGARA

Department of Material Physics, Osaka University,
Toyonaka, Osaka 560, Japan

The ground-state configurations of compressed hydrogen are studied, by the band theoretical calculation using the local-density approximation and plane-wave basis. Results for the total energies are shown for several structures in molecular and metallic phases and compared with previous results by the structural expansion, the FLAPW method and the plane-wave method. Possibilities for the appearance of "filamentary" and "planar" structures are pointed out, in metallic hydrogen. Density dependence of the energy band gap is examined for Pa3 and HCP structures of molecular phase, in comparison with previous results. The energy corrections from the omitted plane-wave basis are evaluated by a perturbative approach. The intramolecular vibron frequency and the equilibrium bond length are calculated and compared with the experimental results.

1. INTRODUCTION

A great progress has been achieved in the experimental study of highly compressed hydrogen[1-3], with pressure approaching to $\sim$ 300 GPa. Challenging experiments[4] are under way aiming at producing metallic hydrogen. The experimental determinations of the equation of state[1-2] have been carried out up to 50-GPa pressure range. Detailed structural data[2] for the compressed solid hydrogen are obtained up to 30 GPa. Recently, a yet unspecified structural phase transition has been observed[3] at $\sim$ 150 GPa, and the metallization have been reported[4] to occur at higher pressure than 250 GPa.

Extensive theoretical studies were carried out in the last decades using a perturbative calculation.[5] And possibility of highly anisotropic structures were pointed out[6] for a low pressure phase of metallic hydrogen, though the zero-point motion energy (ZME) effect may be important for these structures.[7] Recently total energy calculations using the full-potential linearized-augmented-plane-wave (FLAPW) method in the local density approximation(LDA) were carried out by Min, Jansen and Freeman (MJF)[8], who discussed the phase transition from a molecular phase to an atomic one in the same footing of approach for both phases. Meanwhile quantum Monte Carlo calculations[9] of the properties of hydrogen at zero temperature were also performed. This method

---

*Supported by the Grant-in-Aid for Scientific Research on Priority Areas (Origin of the Solar System) of Japanese Ministry of Education, Science and Culture (No. 63611002 and No. 0161002).

provides us with a promising technique in which the quantum many-body problem is solved simultaneously both for the electron and proton degrees of freedom. And advances in the computing power could, in the near future, improve the numerical accuracy.

Very recently there appeared the total energy calculation using the plane-wave (PW) basis in the local density approximation by Barbee, Garcia, Cohen and Martins (BGCM)[10], who claimed that the primitive hexagonal (PH) structure to appear as the first phase of metallic hydrogen. This is the first study which treated the anisotropic phase of metallic hydrogen in the band-theoretical approach. The same PW approach was also taken up by the present author[11], in studying the lowest configurations of compressed hydrogen. In the present paper we summarize our results in comparison with BGCM, MJF and our previous results based on the structural expansion of Brovman et al, Hammerberg and Ashcroft , and others.[5]

In section 2 we give the scheme of the present calculation using the plane-wave basis in the local density approximation, with a perturbative approach to the energy correction from the omitted plane-wave basis. In section 3 we give the total energy for several structures in metallic and molecular phases and discuss the phase transitions in compressed hydrogen. Effects of the energy correction are given in section 4.

2. TOTAL ENERGY CALCULATION USING PLANE-WAVE BASIS FUNCTION

2.1. Outline of the calculation

In the PW method, we use the one-electron wave function expanded in plane-waves,

$$\psi_k^{(\alpha)}(r) = \Sigma_g \, A_g^{(\alpha)} \exp\{i(k+g)\cdot r\} / \sqrt{V} \tag{1}$$

and calculate the electron density by

$$\rho(r) = \Sigma_{k\alpha} |\psi_k^{(\alpha)}(r)|^2 , \tag{2}$$

where the summation is taken over occupied states. The coefficient A_g is determined by the eigen-value equation

$$\Sigma_{g'} H_{g,g'} \, A_{g'}^{(\alpha)} = \varepsilon_k^{(\alpha)} A_g^{(\alpha)} \tag{3}$$

The one-electron Hamiltonian H includes the exchange-correlation term as well as the Coulomb contribution, of which both depend upon the electron density ρ. In the above expressions, k denotes a wave-vector, g a reciprocal lattice vector, α the band index and V the volume of the system. We solve the equations above by the iterative method and then evaluate the total energy by

$$E_{tot} = E_m + \Sigma_{k\alpha} \varepsilon_k^{(\alpha)} - (N/2) \Sigma_g |\Omega_a \rho(g)|^2 v(g)$$
$$- N \Sigma_g \Omega_a \rho(g) [v_{xc}(g) - \varepsilon_{xc}(g)] \quad (4)$$

where E_m is the Madelung energy, N the total number of electrons, Ω_a atomic volume, and $v(g)$, $\rho(g)$, $v_{xc}(g)$ and $\varepsilon_{xc}(g)$ denote the Fourier-transforms of the Coulomb potential, the electron density, the exchange-correlation potential and the exchange-correlation energy density, respectively. In the iteration procedure we can efficiently use the initial potential which are obtained from the Coulomb potential screened by the LDA dielectric function. To achieve the convergence of six significant figures of the total energy, two or three iterations were sufficient for most structures, while one more iteration was needed at the worst convergence.

Advantageous points of using the plane-wave basis are (i) no need to consider Muffin-tin spheres, with simplification of the the treatment and accordingly with smaller resultant error, (ii) simplicity in the evaluation of matrix elements of the one-electron Hamiltonian, requiring only computation of the Fourier-Transforms such as $\rho(g)$, $v_{xc}(g)$ and $\varepsilon_{xc}(g)$, (iii) the increasing accuracy in the region of higher density where the plane-wave description works better.

The disadvantage lies in the use of increasing number of PW for the description of the spatially localized electron states with lowering density, but computer memories and machine times are limited. However we can employ a perturbative approach[12] to evaluate the energy correction from the omitted PW basis.

For the exchange-correlation energy, we use the Barth-Heddin's expression[13]

$$\varepsilon_{xc} = - A/x - B \{ (1+x^3) \ln(1+1/x) + x/2 - x^3 - 1/3 \} \quad (5)$$

where A = 0.0436348, B = 0.0449962 and $x = r_s/21$. The density parameter r_s is defined by $r_s = (3\Omega_a/4\pi)^{1/3}$ (in units of bohr). The parametrized expression for Ceperley and Alder's correlation energy[14] gives slightly higher total energies but it brings no essential differences in our results.

2.2. Perturbative approach to the energy correction

In the actual calculations, the number of the PW basis, N_{pw}, is set about 2000 at most, owing to the limitation of machine times and the memory capacity of computers. So the total energies depend on the number of PW considerably and the energy corrections from the omitted PW basis are needed to get better results of the total energy or to reduce the machine times with the use of a small number of plane waves, N_{pw}.

To obtain the energy correction we utilize a perturbative approach with the use of a standard formula

$$\Delta \varepsilon_k^{(\alpha)} = \Sigma_{g > g_c} \frac{\Sigma_{g',g''} A_{g'}^{(\alpha)} U(g',g) U(g,g'') A_{g''}^{(\alpha)}}{\varepsilon_k^{(\alpha)} - \varepsilon_{k+g}} \quad (6)$$

for the one-electron energy correction, where $U(g,g')$ is the matrix element of the potential term of the one-electron Hamiltonian, and g_c is the cut off length of the reciprocal lattice vectors defined by $g_c = [N_{pw}/n]^{-1/3}$ using N_{pw} and the number, n, of atoms in a unit cell.

In evaluating the energy correction, we neglect terms due to the electron-electron interaction. This procedure is relevant because the correction comes mainly from modification of the electron density in the vicinity of protons. Thus the correction to the total energy is obtained merely by summing up the above $\Delta \varepsilon_k$ over the occupied states obtained earlier.

Since $g \gg k$, the denominator is approximated by $-g^2$ in the relevant unit of energy and $U(g,g')$ by the Fourier-transform of the bare Coulomb interaction. Then the quantity we need to calculate is

$$\Sigma_{g > g_c} \frac{1}{g^2} \frac{1}{|g'-g|^2} \frac{1}{|g-g''|^2} e^{ig \cdot (\rho_t - \rho_{t'})} \quad (7)$$

where ρ_t is the position vector of an atom in a unit cell. We replace the summation by the corresponding integration and use the following series expansions,

$$\frac{1}{|g'-g|^2} = \frac{1}{g^2} \times \{(1 + \frac{u^2}{3}) + 2u P_1(\cos\Theta_{gg'}) + \frac{8}{3} u^2 P_2(\cos\Theta_{gg'}) + \cdots \} \quad (8)$$

and

$$e^{ig \cdot \rho} = \Sigma_\ell (2\ell + 1) i^\ell j_\ell(g\rho) P_\ell(\cos\Theta_{g\rho}) \quad (9)$$

where $P_\ell(x)$ and $j_\ell(x)$ are the Legendre function and the spherical Bessel function respectively and $\Theta_{g,g'}$ denotes the angle of g making with g', and we put $u = g'/g$. In this way we evaluate the sum (7). A few terms of the series are sufficient for the energy calculation with relevant accuracy.

3. PHASE TRANSITIONS IN THE METALLIC AND MOLECULAR PHASE

We calculated the total energy for several structures in metallic and molecular phases at several values of r_s.

3.1. Results

First we show our results in comparison with that of BGCM.[10] The results below does not include the energy correction, whose effects will be shown in the next section.

In the atomic phase, we consider families of primitive tetragonal (PT), of rhombohedral (RH) and of primitive hexagonal (PH) lattices. BGCM looked for the lowest structure mainly in a family of primitive hexagonal lattices. In Figure 1 the ground-state energies of PT, PH and RH are plotted as functions of c/a, for several values of the density parameter r_s. For RH lattice, a denotes the atomic distance in a plane perpendicular to the trigonal

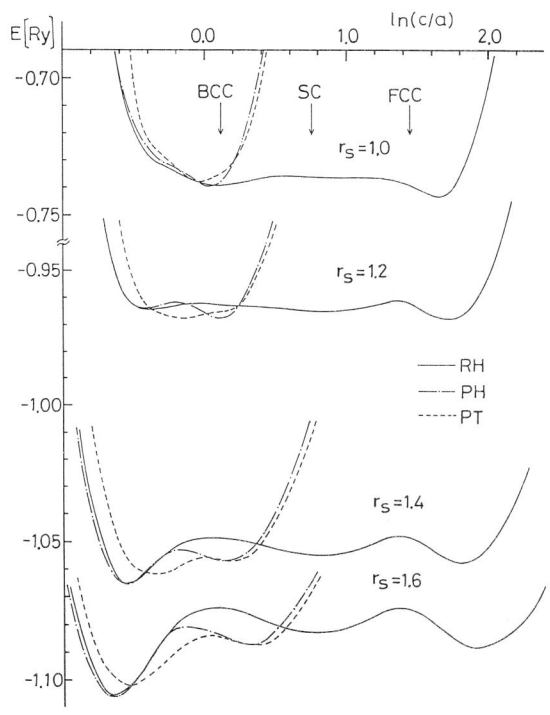

FIGURE 1. Total energy as a function of c/a

axis and c denotes treble the distance between adjacent planes. For PH and PT, the structures for c/a < 1 are called "filamentary" and those for c/a > 1 "planar", which are designated by (I) and (II) respectively. Particularly for RH, the planar structure (RH(II)) appears in the region of c/a > 5 and the filamentary one (RH(I)) in c/a < 1, according to our definition of c/a. At lower density, the curves have several local minima, of which some disappear at higher density. In PH the filamentary structure disappears at $r_s \sim 1.2$, but the planar one remains at higher density. Similar behavior can be observed also in RH, where the filamentary structure disappears at $r_s \sim 1.2$, but the planar one remains until $r_s \sim 1.0$. We note here that the BCC and FCC structures are unstable at lower density. On the other hand, for PT the stable planar structure disappears at $r_s \sim 1.4$ while the filamentary one tends to SC at $r_s \sim 1.0$.

Here we compare our results for FCC and BCC with those by MJF[8] for the purpose of examining accuracy of the results. Our results show that the total energy for FCC is lower than that for BCC in the region $r_s > 1.1$, which is in accordance with the earlier results obtained by using the perturbative theory,[5] but in contradistinction with MJF. We have studied this discrepancy earlier[15], as will be shown later (Figure 5).

For the Gibbs-free-energy G for various structures, BGCM plotted the difference, $\Delta G = G - G_{PH}$, as a function of pressure, where G_{PH} is the free energy for PH. Their curves must have a kink in

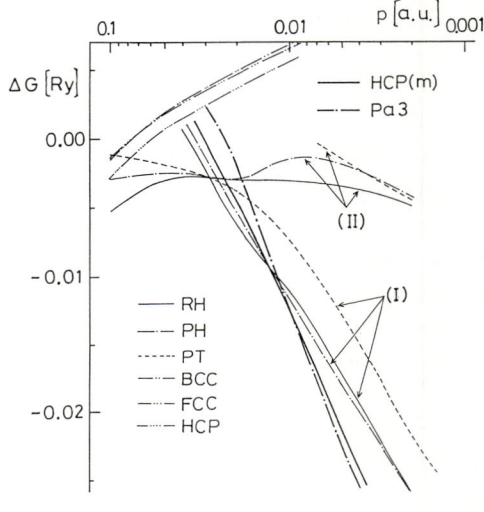

FIGURE 2. Gibbs free-energy differences

accordance with disappearance of the filamentary structure for PH. However no relevant kink is observed in BGCM's free-energy curves.

In the molecular phase we consider Pa3 and HCP structures, of which the latter is designated by HCP(m). In HCP(m) we assume molecules to direct along c-axis. For both structures, we fixed the bond-length at 1.4 bohr, for a moment. Later in section 4, we shall show effects of the relaxation.

In Figure 2 we plot the Gibbs-free-energy differences, $\Delta G = G - G_{SC}$, as a function of the pressure, where G_{SC} refers to SC lattice. For the interpolation of the energy and pressure, we used the cubic splines. Our EOS is very close to BGCM's at pressure higher than 100 GPa. In the pressure region $p < 140$ GPa, the lowest structure is Pa3, which is replaced by HCP(m) at 140 GPa. In the region of 190 GPa $< p <$ 420 GPa, RH(I) is the lowest and then replaced by PT(I) at $p = 420$ GPa. Beyond a very small region of 420 GPa $< p <$ 450 GPa with stable PT(I), there exists a region of 450 GPa $< p <$ 2 TPa with stable RH(II).

At p∼ 2 TPa, HCP becomes the lowest, and finally the BCC becomes lowest at∼ 4TPa. Thus there exists no stable region of PH(I). We note here that there is no stable region of SC, in agreement with MJF. The overall feature is similar to the third-order perturbation results[5,6], though the pressure is somewhat shifted. The appearance of anisotropic phase reflects tendency of covalent bonding of hydrogen atoms.

3.2. Metal-insulator transition

We first discuss a structural transition from HCP(m) to RH(I), which occurs at 190 GPa according to our result. This transition pressure is lower than the value expected from the experiment, which has proved to be higher than 250 GPa.[3,4] On the other hand, BGCM predicted the transition from HCP(m) to PH to occur at 380 GPa. This estimate is quite high compared with ours, ∼ 200 GPa. We barely compare BGCM's estimates with ours since they included ZME by an approximate procedure. However BGCM took into account PW whose kinetic energy is less than a certain value (36Ry). The number of PW thus chosen is comparable to ours at low density, but fairly smaller than ours at the density of interest. It is recalled here that the free-energy curves of HCP(m), RH(I) and PH(I) are close to each other in a fairly broad region of pressure (Figure 2). Thus the position of the free-energy crossing depends very sensitively upon the accuracy of energies. This point will be studied further in section 4.

The metallization of hydrogen may also occur when the energy gap vanishes between the valence and conduction bands, in the molecular phase. In Figure 3 our results for the density dependence of the energy gaps are shown. For HCP(m), the pressure at which the gap vanishes is calculated to be ∼ 55 GPa, to which BGCM's value, 40 GPa, is fairly close. Both values are much lower than the value for Pa3, 180 GPa (MJF's 170GPa). Large difference shows that the metallization through band closing depends very sensitively upon the crystal structure of molecular solid hydrogen. It is however known that the LDA underestimates

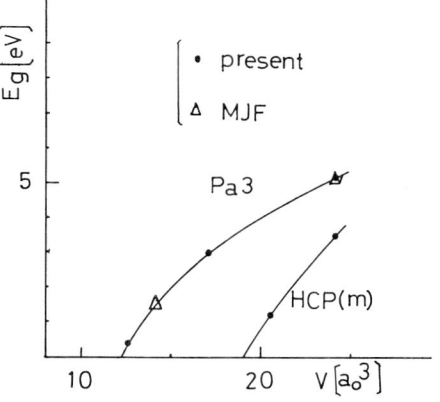

FIGURE 3. Band gap for Pa3 and HCP

considerably the energy gap; for solid hydrogen at normal density a LDA estimate, 9.3 eV,[8] of the energy gap is quite lower than the experimental value, 17 eV. Thus, in the molecular phase the metallization pressure may be much higher than the LDA estimation.

4. BOND LENGTH AND VIBRON FREQUENCY WITH ENERGY CORRECTION

In the above results, the bond length are fixed at 1.40 bohr for the structures in the molecular phase. Ramaker et al.[16] are the first who studied the bond length in molecular phase as a function of density. Similar studies have also been done by MJF. We shall below give our preliminary results for the total energy as a function of internuclear distance, where the energy correction from the omitted plane-waves plays some important role, particularly in the low density region.

In Figure 4, we plot the curves for the total energy as functions of the internuclear distance in a molecule for HCP structure with inclusion of the energy correction. The bond length can be determined from the lowest position of the energy curves. In this way we estimate the bond length to be $R_b=1.45a_0$ for $r_s=1.6$ and 1.8 for both HCP(m) and Pa3, where a_0 denotes the Bohr radius.

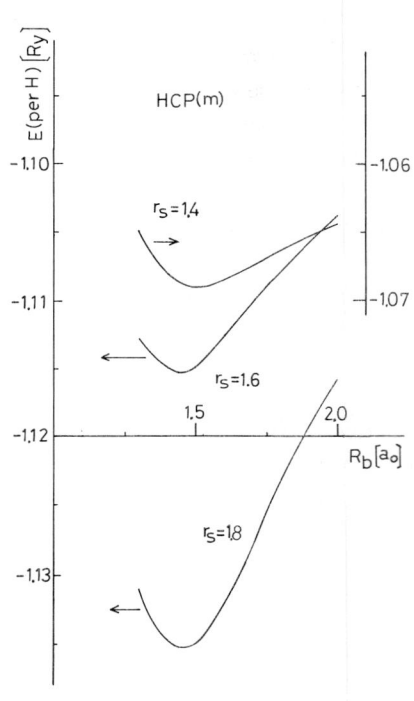

FIGURE 4. Total energy as a function of R_b.

Without the energy correction we estimate $R_b=1.47\ a_0$.

To the energy curve of interest we fit cubic polynomials, whence the vibration frequencies for the stretching mode of a molecule can be obtained by estimating the second derivative at the equilibrium internuclear distance, in the Born-Oppenheimer approximation. In Table 1, we give a few results of our calculation. We note that this is the first study of vibron frequencies by the

TABLE 1. Bond length and vibron frequency

	$r_s=1.8$		$r_s=1.6$	
	R_b [a_0]	ω [cm^{-1}]	R_b [a_0]	ω [cm^{-1}]
Pa3	1.45	4130	1.43	4130
HCP(m)	1.45	4000	1.44	3570

band theoretical calculation, though our study is preliminary. The calculated frequencies at $r_s=1.8$ are $\omega = 4000$ cm^{-1} for HCP(m) and 4130 cm^{-1} for Pa3. The bond lengths and the vibron frequencies at $r_s=1.8$ are very close to the results for a single molecule using the local density functional theory[17]: the experimental values[18] of R_b and ω of isolated molecules are 1.40 a_0 and 4400 cm^{-1}, respectively. However our vibron softening seems much steeper for HCP(m) in comparison with the experiment.[3] We mention the vibron frequencies which are rather insensitive to the energy correction.

Energy corrections are evaluated respectively for the filamentary and the molecular structures, for which the total energies are very close to each other.

With inclusion of the energy correction thus obtained, we study the bond relaxation effects. The overall feature described before (section 3) remains unchanged. Contribution of the energy correction to EOS is rather small at higher density ($r_s > 1.6$), though the pressure increases by 10% at $r_s=1.7$. However the pres-

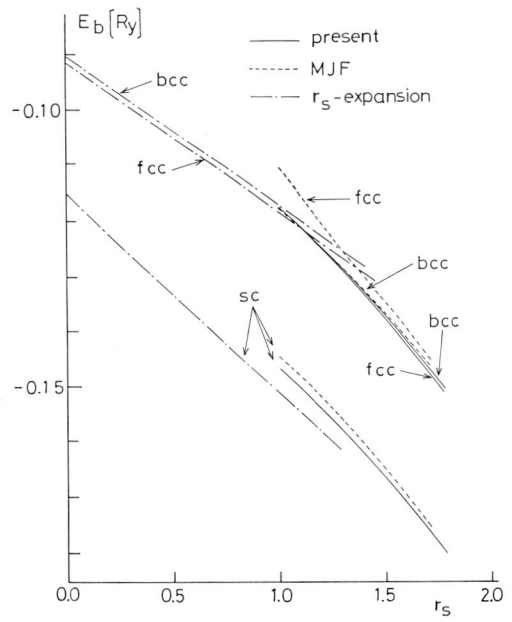

FIGURE 5. Band structural energy

sure for HCP(m)-RH(I) transition increased remarkably from 190 GPa to 360 GPa. This transition pressure is comparable to BGCM's estimate for molecular HCP to PH transition. The increase of the transition pressure comes mainly from the increase of the relative stability of HCP(m) by the lowering of the total energy due to bond relaxation. As a result, the stable region of the filamentary structure, RH(I), becomes narrow around 400 Gpa. We note here that the energy correction works to lower the total energy more appreciably for anisotropic and molecular phases than for cubic atomic phases.

5. CONCLUDING REMARKS

We have studied the structural phase transitions in highly compressed hydrogen. The hydrogen undergoes a series of structural phase transitions, HCP(m) --> RH(I) --> RH(II) --> HCP --> BCC as the density increases. The region of the filamentary structure, RH(I), is very small. In the present study the effect of ZME is not taken into account though it would play some role in the phase transition among anisotropic structures. Possibility of metallization due to the band overlapping in molecular phases is not excluded, in view of a wide stable region of molecular phase, HCP(m). It is however noted that there may exist other structures more stable than HCP(m). We have not clarified a phase transitions at 150 GPa, as evidenced in a jump observed for the vibron frequency.[3]

We finally mention the band structural energy which is defined by the total energy subtracted by the sum of the Madelung energy and the energy of electron gas. The band-structural energy thus defined are plotted in Figure 5 for cubic metallic hydrogen by various methods. As can be seen in the figure, at high densities the results from the LDA approaches those from the r_s-expansion in which electron-electron interaction is properly taken into account. The above comparison may serve as another demonstration for the effectiveness of the LDA.

The accuracy of the present calculation has been checked by varying the number of mesh points in the Fourier-transform and that of the k-point sampling. In the above results, the number of plane-wave basis has been taken to be 250 times n, the number of atoms in a unit cell, roughly. Examinations on the accuracy of the present results will be given elsewhere.

ACKNOWLEDGEMENT

The author is grateful to Prof.(emeritus) T. Nakamura, Prof. H. Miyagi and Mr. T. Hatano for their cooperation in the course of the present study.

REFERENCES

1) H. Shimizu, E. M. Brody, H. K. Mao and P. M. Bell, Phys. Rev. Lett. 47 (1981) 128; J. van Straaten and I. F. Silvera, Phys. Rev. B37 (1988) 1989.
2) H. K. Mao, A. P. Jephcoat, R. J. Hemley, L. W. Finger, C. S. Zha, R. M. Hazen, and D. E. Cox, Science 239 (1988) 1132.
3) R. J. Hemley and H. K. Mao, Phys. Rev. Lett. 61 (1988) 857.
4) H. K. Mao and R. J. Hemley, Science 244 (1989) 1462.
5) E. G. Brovman, Yu. Kagan, and A. Kholas, Zh. Eksp. Teor. Fiz. 61 (1971) 2429 [Sov. Phys.-JETP 34 (1971) 1300]; J. Hammerberge and N. W. Ashcroft, Phys. Rev,B9 (1974) 409; T. Nakamura, H. Nagara and H. Miyagi, Prog. Theor. Phys. 63 (1980) 368; H. Nagara, H. Miyagi and T. Nakamura, Prog. Theor. Phys. 56 (1976) 396; 64 (1980) 731.
6) E. G. Brovman, Yu. Kagan, and A. Kholas, Zh. Eksp. Teor. Fiz. 62 (1972) 1492 [Sov. Phys.-JETP 35 (1972) 783]; H. Miyagi, T. Nakamura and H. Nagara, Phys. Lett. 62A (1977) 171.
7) D. M. Strause and N. W. Ashcroft, Phys. Rev. Lett. 38 (1977) 415.
8) B. I. Min, H. J. F. Jansen and R. J. Freeman, Phys. Rev. B30 (1984) 5076; B. I. Min, H. J. F. Jansen and R. J. Freeman, Phys. Rev. B33 (1986) 6383.
9) P. M. Ceperley and B. J. Alder, Phys. Rev. B36 (1987) 2092.
10) T. W. Barbee,III, A. Garcia and M. L. Cohen, Phys. Rev. Lett. 62 (1989) 1150.
11) H. Nagara, to appear in J. Phys. Soc. JPN. 58, No. 10.
12) H. Nagara and T. Nakamura, to be published.
13) U. von Barth and L. Hedin, J. Phys. C5 (1972) 1629.
14) D. M. Ceperley and B. J. Alder, Phys. Rev. Lett. 45 (1980) 566; J. P. Perdew and A. Zunger, Phys. Rev. B23 (1981) 5048.
15) H. Miyagi, T. Hatano and H. Nagara, J. Phys. Soc. JPN. 57 (1988) 2751.
16) D. E. Ramaker, L. Kumar, and F. E. Harris, Phys. Rev. Let. 34 (1975) 812.
17) Sea, for example, F. W. Averill and G. S. Painter, Rev. B24 (1981) 6795.
18) K. P. Huber, Constants of diatomic molecules, in: American Institute of Physics Handbook (McGraw-Hill, New York,1972) 7g.

PLASMA CONTRIBUTIONS TO THE COHESIVE ENERGY OF CHARGE STABILISED COLLOIDAL SYSTEMS

E. CANESSA+, M.J. GRIMSON+ and M. SILBERT*

+ Theory and Computation Science Group, AFRC-IFRC, Norwich NR4 7UA, UK.
* School of Physics, University of East Anglia, Norwich NR4 7TJ, UK.

Purely volume dependent forces in charge stabilised colloidal suspensions are briefly discussed within the one component macroparticle description, assuming the macroions to be adiabatically decoupled from the small ions. Such forces are postulated to be due to chemical equilibrium and counterion-macroion coupling and characterized by the one component plasma and Debye-Hückel theory, respectively.

Charge stabilised colloidal suspensions are usually viewed as one component macrofluids. The colloidal macroparticles are assumed to interact through a specified DLVO pair potential[1] and to be the only objects present in the suspension. The solvent is taken to be structureless and have no correlations with the colloids. It is also assumed that any contributions from the dissolved ionic species, such as counterions and added electrolyte, toward the correlations between the colloidal macroparticles appear only in the screening of the DLVO pairwise potential.

The model of colloidal macroparticles as electrically neutral objects is obviously only an approximation. One must bear in mind that each colloid in solution is susceptible to charge dissociation in solvents such as water. Some fraction of these provide an effective charge of Ze per macroparticle, as well as an equivalent density Zn of monovalent counterions in solution. Independently of such counterions the dispersion may also contain some extra concentration n_s of added salt. This system is globally neutral and all of the ions it contains interact via electrostatic forces. Nevertheless, in constructing a one component model for a system of charged stabilised macroparticles special care is needed in treating the mobile screening ions present in the solution[2].

In the pseudoatom description of metallic systems the screening electrons provide a contribution to the system free energy that is independent of the ionic structure as well as producing a density dependent effective pair potential between the pseudoatoms.

In charge stabilised colloids the screening ions present from the addition of salt cannot be considered as localized to the colloid phase volume, which may be a fraction of the system volume, from the simple consideration of the Donnan equilibrium[3]. Thus there will be also a system free energy contribution arising from added electrolyte that is independent of the macroion configuration. This is in addition to the screening of the effective two-body interaction of the macroparticles and to counterion-macroion coupling which is assumed to be small but still important.

Here, it is assumed that the macroions are adiabatically decoupled from the small ions. Whence purely volume dependent forces due to the chemical equilibrium and counterion-macroion coupling contribute to the cohesive energy of charge stabilised colloidal systems. We suggest that the total energy E per unit volume of a charge stabilised colloidal crystal has the general approximate form

$$E = U_o + (\rho/2) \sum_{i \neq j} v(r_{ij}; V) \quad , \qquad (1)$$

where $\rho = N/V$ is the macroparticle number density and the sum is over all the macroparticle positions. The energy contribution from the screening ions, U_o, can be written as the sum of two terms, i.e.

$$U_o = U_\kappa + U_\Gamma \quad . \qquad (2)$$

The plasma contribution U_Γ arises from the self energy of the counterions in solution - part of the macroion charging process - and is treated as a one component plasma to a first approximation[4]. This plasma is considered as a statistical system of mobile charged particles embedded in a uniform background of neutralizing charged macroparticles. The second plasma contribution, U_κ, is due to the addition of salt and can be handled within the Debye-Hückel theory of electrolytes.

The one component plasma description for counterions, where the macroions form a jellium, is adopted within the Debye-Hückel plus hole approach, as proposed in Ref.5. Such an approach extends the Debye-Hückel theory at high coupling strength. Then

$$U_\Gamma/Nk_BT = -(1/4)\{[1 + (3\Gamma)^{3/2}]^{2/3} - 1\} \quad , \qquad (3)$$

where

$$\Gamma = (Ze)^2/4\pi\varepsilon_o\varepsilon k_BT\, a \quad , \qquad (4)$$

corresponds to the potential energy of two macroparticles at the mean nearest neighbour separation divided by the thermal energy and

$$a = (3/4\pi\rho)^{1/3} \qquad (5)$$

is the ion-sphere radius. Note that U is non-linear in V for a fixed number N of macroparticles.

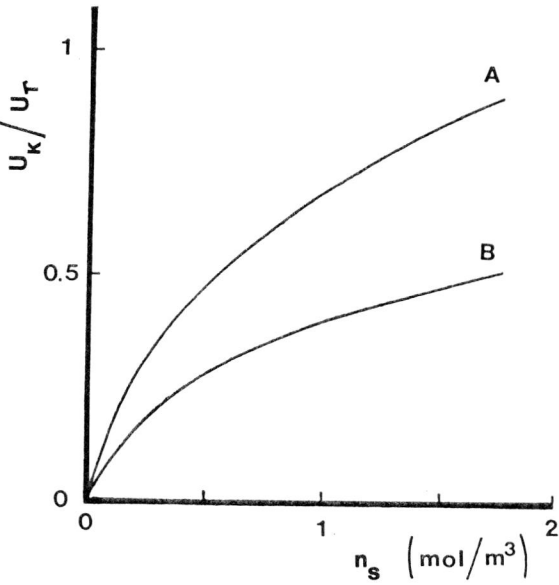

FIGURE 1

The ratio U_κ/U_Γ as a function of added electrolyte. Here $Z = 200$:
(A) $\rho = 2.8 \; 10^{13}$ (particles/cm^3), $a = 0.204$ μm, $\Gamma = 136$; (B) $\rho = 1.4 \; 10^{14}$ (particles/cm^3), $a = 0.120$ μm, $\Gamma = 233$.

Within Debye-Hückel theory, where the continuous phase now forms a jellium, U_κ is expressed as[6]

$$U_\kappa/k_B T = -\kappa^3 V_2/8\pi \quad , \qquad (6)$$

where $V_2 \leqslant V$, i.e. the volume occupied by the added electrolyte V_2 is greater than or equal to the volume V occupied by counterions (c.f. the Donnan equilibrium). Thus

$$\kappa^2 = 2n_s(z_2 e)^2/\varepsilon_o \varepsilon k_B T \quad , \qquad (7)$$

where $n_s = N_2/V_2$ is the added salt concentration. For NaCl $z_2 = 1$.

Figure 1 shows the ratio U_κ/U_Γ for the extra plasma contributions to the total energy as a function of salt concentration. It can be seen that for typical parameter values of macroparticle diameter, number of charges and Γ the contributions U_κ due to the addition of salt is rapidly increasing for the case of highly concentrated colloidal dispersions. This simple result serves to outline the importance of the extra volume dependent (cohesive) terms in charge stabilised colloids.

Summing up, in this work, a one component plasma model for counterions has been adopted in which the macroions form the jellium. Added electrolyte have been treated as an ionized gas within the Debye-Hückel theory.

ACKNOWLEDGEMENT

We are grateful to Gary C. Barker and Malcolm J. Stott for many helpful discussions. One of us (MS) also gratefully acknowledges financial support by SERC, UK (Grant GR/E63698).

REFERENCES

1) E.J.W. Verwey and J.Th. Overbeek, Theory of Stability of Lyophobic Colloids (Elsevier, Amsterdam, 1948).

2) E. Canessa, M.J. Grimson and M. Silbert, Mol. Phys. 64 (1988) 1195.

3) G. Stell and C.G. Joslin, Biophys. J. 50 (1986) 855.

4) S. Ichimaru, Rev. Mod. Phys. 54 (1982) 1017.

5) S. Nordholm, Chem. Phys. Lett. 105 (1984) 302.

6) E.M. Lifschitz and L.P. Pitaevskii, Statistical Physics (Pergamon Press, London, 1980).

A TWO-DIMENSIONAL POLYMER CHAIN WITH SHORT-RANGE INTERACTIONS

Masako TAKASU, Jun TAKASHIMA, Yasuaki HIWATARI

Department of Physics, Faculty of Science, Kanazawa University, 1-1 Marunouchi, Kanazawa, Ishikawa 920 Japan

The effects of short-range interactions on a two-dimensional polymer chain are studied with Monte Carlo simulations. The quantities such as the average length of turns along the chain, the persistence length, and the winding angle are investigated, for different interaction ranges, as functions of temperature. The scaling properties of the winding angle are also studied.

1. INTRODUCTION

Charged polymers have recently been studied by many people[1-4]. In our previous paper[4], in order to study the conformation characteristics of the polymer chain in the presence of the short-range Coulomb interactions, we calculated the gyration radius, and the average length of turns. We studied the conformations of the chain in the case of nearest neighbor interactions ($r_c=1$) and the third-nearest neighbor interactions ($r_c=2$). We found that the polymer has more rod-like structures at low-temperatures for $r_c=2$ than for $r_c=1$.

In this paper, we study the winding angle θ which is the total angle of rotation along the chain. For the neutral polymer case, Duplantier et al[5] proved the following asymptotic form of the distribution of the winding angle.

$$P(x) \sim \pi^{-1/2} \exp(-x^2) \qquad N \to \infty \qquad (1.1)$$

where $x=(4\ln N)^{-1/2} \cdot \theta$, and N is the length of the chain. We calculated the winding angle for the charged polymer with the model described in the next section.

We also report the end effects of the polymer and discuss the polymer conformations.

2. MODEL

The details of our model for the two-dimensional polymer are given in our previous paper[4]. The monomers and counter-ions of opposite sign move on two separate square lattices. They interact through the following logarithmic potential.

$$v_{ij}(r) = - q_i q_j C \ln(r/l) \qquad r \le r_c$$
$$= 0 \qquad r > r_c \qquad (2.1)$$

where $q_i=+1$ for monomers and $q_i=-1$ for counter-ions. r is the distance between i-th ion and j-th ion. l and C are positive constants. The interaction range r_c is taken to be 1 or 2 in the unit of lattice spacing.

We impose the periodic boundary condition, and configurations of the polymer and ions are generated with Metropolis method[6]. The polymer chain is moved by the slithering-snake method.[1,7]

3. RESULTS

First, we investigated the scaling property of the winding angle. Figure 1 shows the second moment of the winding angle as a function of ln N for the neutral polymer case. It is found that there is a very good linear dependence between θ^2 and ln N, even at small N, like Eq. (1.1). The fitted line deviates from the origin, but this could be considered to be due to a finite-size effect.

Next, various quantities are studied by dividing the polymer chain into a number of segments of length 8. Figure 2(a) shows the values of θ^2 on each segment. At low-temperatures, the values of θ^2 become large on the end-segments, indicating that the polymer tends to curl up at the end. This tendency appears for both cases of interaction lengths, $r_c=1$ and 2.

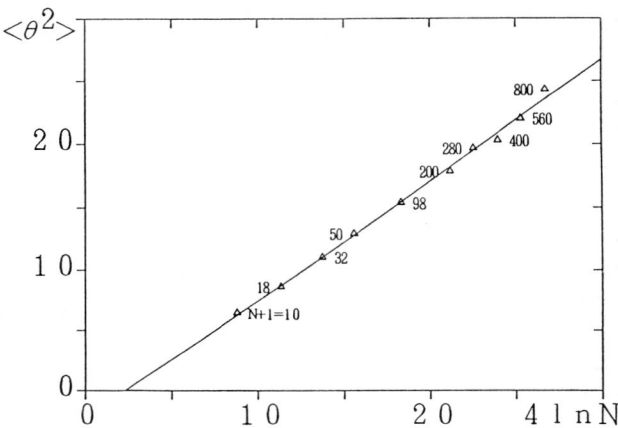

FIGURE 1
Squared winding angle of the neutral polymer as a function of 4ln N

In order to investigate the detailed chain conformation characteristics, we study the temperature-dependence of the persistence length, which is the length of successive links (monomer-monomer distances) of the same direction. In Fig. 2(b), the values of persistence length are shown for each segment. As the temperature is lowered, the persistence length of the end-segments decreases. This is observed for both $r_c=1$ and $r_c=2$.

We also study the average length of turns of the chain[4], which is the mean distance between subsequent inflection points along the chain.

$$S = \left\langle \frac{1}{K-1} \sum_{i=1}^{K-1} l_i \right\rangle \quad (3.1)$$

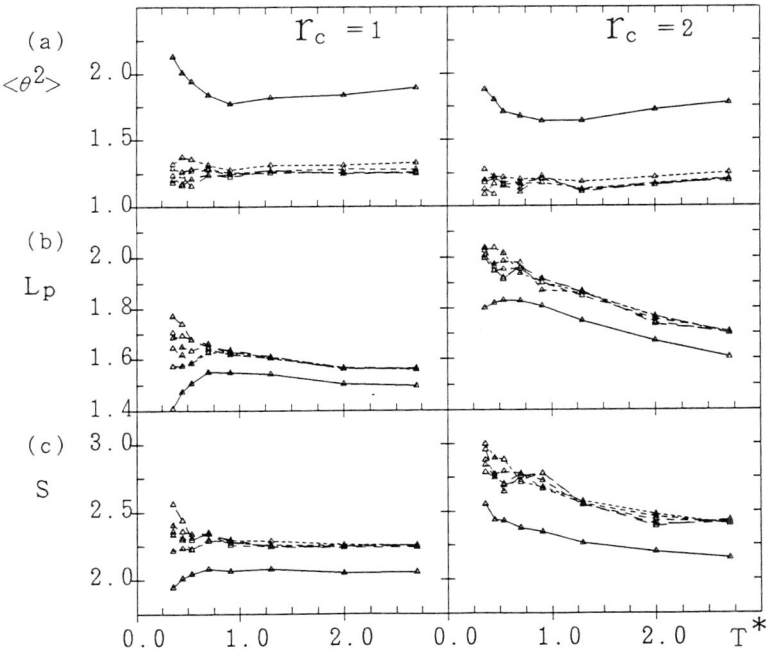

FIGURE 2
The temperature dependence of (a) squared winding angle in the unit of $(\pi/2)^2$, (b) persistence length and (c) average length of turns, for $r_c=1$ (left) and for $r_c=2$ (right). The length of the polymer is $N=97$. The solid lines are for the end-segments, and dotted lines are for the middle-segments.

where K is the total number of inflection points. This quantity is related to the persistence length, but it is more susceptible to the variation of the interaction range. Figure 2 (c) shows the temperature-dependence of S for each segment. For $r_c=1$, the values of S for the end-segments become small at low-temperatures. For $r_c=2$, on the contrary, the values of S for the end-segments increase as the temperature is lowered. Thus, for $r_c=2$, the polymer tends to have rod-like structure even at the end-segments, at low-temperatures.

4. DISCUSSIONS AND CONCLUSIONS

In this paper, we studied the detailed structure of the conformations of the polymer. At high temperatures, the system has the property of the self-avoiding walk. As the temperature is lowered, the charged polymer starts to have different conformations depending on the range of interactions. For the neutral polymer case, the scaling property (1.1) of the winding angle was observed even for short chains. For the charged polymer case, there are too much fluctuations of the data to confirm the scaling relation. These questions as well as three-dimensional polymers are interesting future problems.

ACKNOWLEDGMENTS

This work is partially supported by the Grant-in-Aid from the Ministry of Education, Science and Culture, and also by the Nissan Science Foundation. Computations were made at the Information Process Center of Kanazawa University and the Plasma Institute of Nagoya University.

REFERENCES

1) J. M. Victor, Europhys. Letter 3 (1987) 1161.

2) G. S. Manning, J. Chem. Phys. 51 (1969) 924.

3) K. Kaji, H. Urakawa, T. Kanaya, and R. Kitamura, J. Phys. (France) 49 (1988) 993.

4) J. Takashima, M. Takasu, Y. Hiwatari, Phys. Rev. A. 40 (1989), in press.

5) B. Duplantier, H. Saleur, Phys. Rev. Lett. 60 (1988) 2343.

6) K. Binder, Monte Carlo Methods in Statistical Physics (Springer, New York, 1981).

7) F. T. Wall, F. Mandel, J. Chem. Phys. 63 (1975) 4592.

NEW EMPIRICAL BRIDGE FUNCTIONS OF INTEGRAL EQUATION: APPLICATION TO THE BINARY SUPERCOOLED LIQUIDS OF THE TWELFTH INVERSE POWER POTENTIAL

Shaw KAMBAYASHI and Yasuaki HIWATARI*

Computing Center, Japan Atomic Energy Research Institute, Naka, Tokai, Ibaraki 319-11, Japan
*Department of Physics, Faculty of Science, Kanazawa University, Kanazawa, Ishikawa 920, Japan

The MHNCS equation, proposed very recently by us, based on the improved hypernetted-chain approximation for the classical one-component plasma has been found to work well for one-component soft-sphere fluids above and below the freezing point. For a more crucial test, the MHNCS equation is extensively studied for the binary soft-sphere liquids. Thermodynamic properties depending on the core size ratio and the concentration will be discussed.

1. INTRODUCTION

The pair structure of a binary mixture of simple liquids can be described by pair distribution functions (PDF) $g_{ij}(r)$, where subscripts i and j (=1 or 2) denote the species indices. It is well known that $g_{ij}(r)$ for prescribed potential $u_{ij}(r)$ are uniquely determined from the simultaneous solution of the Ornstein-Zernike relation

$$h_{ij}(r) = c_{ij}(r) + \rho \sum_{k=1}^{2} x_k \int d\vec{r}' h_{ik}(r') c_{kj}(|\vec{r}-\vec{r}'|) \qquad (1)$$

and the closure relation

$$g_{ij}(r) = \exp[-\beta u_{ij}(r) + h_{ij}(r) - c_{ij}(r) + B_{ij}(r)], \qquad (2)$$

where $h_{ij}(r) = g_{ij}(r) - 1$ are the pair correlation functions, $c_{ij}(r)$ the direct correlation functions, ρ the number density, β the inverse temperature $1/k_BT$, and x_k the number concentration of the k-th species[1]. Equation (2) contains the bridge function $B_{ij}(r)$. It is well known that $B_{ij}(r)$ can be given by an expansion of highly connected h-bond elementary diagrams,

$$B_{ij}(r) = \sum_{n=4}^{\infty} \{\varepsilon_{n;ij}(r)\}, \qquad (3)$$

where $\{\varepsilon_{n;ij}(r)\}$ represents a set of n-points elementary diagrams[1]. However, the convergence of Eq. (3) is generally too slow to be applicable to practical calculations for highly densed liquid states. For such a difficulty with the calculation of $B_{ij}(r)$, various approximations for the integral equation have been proposed. For example, well known classical hypernetted-chain (HNC) and

Percus-Yevick (PY) approximations are equivalent to substituting in Eq. (2) $B_{ij}(r)=0$ and $-h_{ij}(r)+c_{ij}(r)+\log[1+h_{ij}(r)-c_{ij}(r)]$, respectively[1].

The reliability of approximated integral equations can be tested by comparing their solutions with the "exact" results obtained by computer simulations for a wide density range. It has been shown that the HNC and PY approximations break down, when a liquid approaches to the freezing point. Since the properties of approximations involved in the integral equation are directly related to the approximation for $B_{ij}(r)$, a more reliable approximation could be obtained by modifying $B_{ij}(r)$ in some manner.

2. The MHNCS approximation

Very recently, we have proposed a new modified HNC approximation for one-component soft-sphere fluids, which reproduce the correct behavior of the PDF in both stable and supercooled liquids[2]. The PDF obtained yields a clear splitting of the second peak near and below the glass transition temperature, compatible to that of computer simulations. Our approximation is based on the universality of the short range part of $B(r)$, suggested by Rosenfeld and Ashcroft[3], and the relevant work for the one-component plasmas by Iyetomi and Ichimaru[4]. According to the latter, it has been shown that an approximated $B(r)$ based on the leading term of Eq. (3), i.e.,

$$\varepsilon_4(r) = \frac{1}{2}\rho^2 \iint d\vec{r}' d\vec{r}'' h(r')h(r'') \cdot h(|\vec{r}'-\vec{r}''|)h(|\vec{r}-\vec{r}'|)h(|\vec{r}-\vec{r}''|)$$

together with a rescaling assumption could be in a good agreement with computer simulations, resulting in the splitting of the second peak of the PDF in a highly supercooled state[4]. Similar results have successfully been obtained for two-component plasmas by Ballone, Pastore, and Tosi[5]. In our approximation, the short range part of the bridge function is approximated by that of the PY hard-sphere fluid $B_H^{PY}(r,d)$ with core diameter d, while the intermediate and long range part of it by $\varepsilon_4(r)$ in Eq. (3) with the pair correlation function $h(r)$ obtained from the HNC approximation;

$$B(r) = (1-f(r,d))B_H^{PY}(r,d) + f(r,d)\varepsilon_4(r), \qquad (4)$$

where d is an adjustale hard-sphere diameter and $0 \le f(r,d) \le 1$ a continuous mixing function. The mixing function is conveniently assumed to be of the form[6],

$$f(r,d) = \frac{1}{2}[1+\tanh(\frac{r-d}{W})], \qquad (5)$$

where W is a dumping parameter, which may be determined from magnitude of the thermal vibration of particles (root mean square amplitude) or the width of the first peak of the PDF. We call our approximation MHNCS. The adjustable parameter d in Eq. (4) has been determined by using the property of the screening potential $H(r)=-h(r)+c(r)-B(r)$ at $r=0$;[3]

$$H(0)=\beta[F^{ex}(0,N)-F^{ex}(1,N-2)], \qquad (6)$$

where $F^{ex}(n,m)$ is the excess free energy of the system with n coupled particles and m single particles. We also assume that the entropy difference between two terms in Eq. (6) is negligible, so that H(0) can be evaluated by the internal energy.

3. MHNCS RESULTS FOR THE BINARY SOFT-SPHERE MIXTURES

The MHNCS approximation can be applied to binary mixtures without any difficulties. We consider binary mixtures composed of two species with diameters σ_1 and σ_2, interacting through the purely repulsive inverse power potential;

$$u_{ij}(r)=\varepsilon(\sigma_{ij}/r)^{12} \quad ; \quad \sigma_{ij}=\frac{1}{2}(\sigma_i+\sigma_j).$$

The advantage of the inverse power potential is due to its scaling property, according to which all excess thermodynamic quantities depend on two independent variables, which are conveniently chosen to be the number concentration of species 1, i.e., x_1, and the dimensionless coupling constant $\Gamma=\rho\sigma_1^3(\varepsilon\beta)^{1/4}$. An "equivalent" one-component soft-sphere fluids can be approximated by using an effective diameter

$$\sigma_{eff}^3 = \Sigma\Sigma_{ij} x_i x_j \sigma_{ij}^3$$

and corresponding effective coupling constant, $\Gamma_{eff}=\Gamma(\sigma_{eff}/\sigma_1)^3$. We use $\{\Gamma_{eff}, x_1, \sigma_2/\sigma_1\}$ to assign each thermodynamic state of binary soft-sphere mixtures. The freezing and glass transition temperatures of this model are at $\Gamma_{eff}\simeq 1.15$ and 1.56, respectively. The following thermodynamic states were investigated: $\Gamma_{eff}=0.8$, 1.2 and 1.5; $x_1=0.1$, 0.5 and 0.9; $\sigma_2/\sigma_1=1.1$, 1.2, 1.3 and 1.4.

The equation of states obtained for each $\{\Gamma_{eff}, x_1, \sigma_2/\sigma_1\}$ yields almost the same curve irrespective with the concentration and the size ratio. The partial PDF's yield a clear splitting of the second peak near and below the glass transition temperature, similar to the results for one-component soft-sphere fluids. Deatiled structures, however, significantly depend on the concentration and the size ratio, which are the most effective to lead to the phase separation: Figure 1 shows the partial PDF's for thermodynamic state {1.5, 0.9, 1.3}. It is clearly seen that the first peak of $g_{22}(r)$ becomes very sharp, suggesting a phase separation to be predicted. We also calculated concentration-concentration structure factor defined as[7]

$$S_{cc}(q) = x_1 x_2 \{1 + x_1 x_2 \rho (\hat{h}_{11}(q) + \hat{h}_{22}(q) - 2\hat{h}_{12}(q))\},$$

where $\hat{h}_{ij}(q)$ are the Fourier transform of $h_{ij}(r)$. The ratio $x_1 x_2 / S_{cc}(0)$ equal to unity for the random mixing (ideal mixing), otherwise it deviates from unity. It is clearly seen from figure 2 that the demixing tendency becomes remarkable as the size ratio exceeds the value $\sigma_2/\sigma_1 \approx 1.25$ for $\Gamma_{eff} = 1.5$.

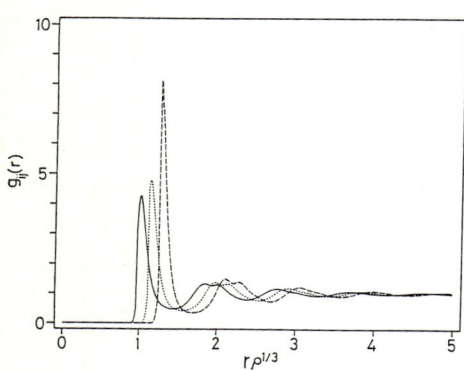

FIGURE 1
The partial PDF's of for thermodynamic state (1.5, 0.9, 1.3) (g_{11}; full curve, g_{12}; dotted curve, g_{22}; dashed curve).

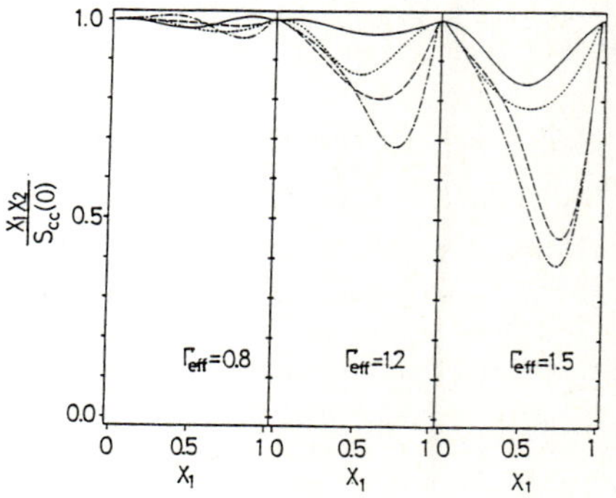

FIGURE 2
$x_1 x_2 / S_{cc}(0)$ versus concentration for $\sigma_2/\sigma_1 = 1.1$ (full curve), 1.2 (dotted curve), 1.3 (dashed curve) and 1.4 (dott-dashed curve).

ACKNOWLEDGEMENTS

This work was partly supported by a Grant-in-Aid from the Ministry of Education, Science and Culture. Computations were carried out on FACOM M-780 and VP-100 at Computing Center of Japan Atomic Energy Research Institute.

REFERENCES
1. J.P. Hansen and I.R. McDonald, Theory of Simple liquids, (Academic, London, 1987).

2. S. Kambayashi and Y. Hiwatari, "A modified hypernetted-chain integral equation for the supercooled liquids of inverse power potentials", Proceedings of the 7th international Conference on Liquid and Amorphous Metals, ed. H. Endo, J. Non-Cryst. Solids (1989), to be published.

3. Y. Rosenfeld and N.W. Ashcroft, Phys. Rev. A 20 (1979) 1208.

4. H. Iyetomi and S. Ichimaru, Phys. Rev. A 27 (1983) 3241.

5. P. Ballone, G. Pastore, and M.P. Tosi, J. Chem. Phys. 81 (1984) 3174.

6. S.M. Foiles, N.W. Ashcroft, and L. Reatto, J. Chem. Phys. 80 (1984) 4441.

7. A.B. Bhatia and D.E. Thornton, Phys. Rev. B 2 (1970) 3004.

EXTENDED MEAN DENSITY APPROXIMATION FOR STRUCTURE FACTORS OF FLUIDS

M. ITOH, O. HONDA and K. NAKAYAMA

Department of Physics, Faculty of Science, Shimane University, Matsue 690, Japan

A self-consistent perturbation theory is developed to calculate the structure factor and other thermodynamic properties of fluids. The theory provides an accurate description of the long-wavelength density response, so that it is suitable to deal with low-density metallic fluids away from their triple points in the small-q region, where the attractive and oscillatory part of the interatomic potential plays essential roles. Results of preliminary model calculations for the low-angle structure factor are reported.

1. INTRODUCTION

The thermodynamic perturbation theory for fluids, which relies on a separation of the interatomic potential into the harsh repulsive part and the long-range attractive tail, is particularly useful for studying low-angle structure factors at low densities. Recently one of the present authors[1] has emphasized the importance of the fluctuation-dissipation theorem for this problem and developed a self-consistent theory by extending the mean density approximation (MDA) due to Henderson and Ashcroft.[2] In this formalism the two equations of state (the energy and the compressibility equations of state) are bound to coincide. In other words the small-q limit of the structure factor is given correctly, so that it is expected to provide a reliable data for the low-angle structure factors. We present here the result of our numerical test using a simplified model system.

2. REVIEW OF THE FORMALISM

For the correct small-wavelength limit of the structure factor, we impose the condition that the compressibility sum rule

$$\frac{\beta}{V}\left(\frac{\partial^2 F}{\partial \rho^2}\right)_{T,V} = \rho^{-1} - C(\mathbf{q}=0) , \qquad (2.1)$$

where $C(\mathbf{q})$ is the direct correlation function, is to be satisfied. In other words the free energy and the structure factor must be calculated self-consistently. For this purpose we start from the generalization of (2.1):

$$\left(\frac{\beta}{V}\right)\frac{\delta^2 F[\rho]}{\delta\rho(1)\delta\rho(2)} = \rho(1)^{-1}\delta(1-2) - C(1,2) \qquad (2.2)$$

In the above equation $F[\rho]$ is the free energy functional of the *inhomogeneous* system, with ρ being the one-particle density as a function of the location, and the number 1 or 2 denotes the space variable. If the explicit form of the functional $F[\rho]$ is known, the direct correlation function can be obtained by calculating the l.h.s. of Eq.(2.2) and then taking its homogeneous limit. Itoh[1] has actually performed this procedure by introducing the following approximations:

(1) The ring diagram expansion for $F[\rho]$, and

(2) the mean density expansion[2] for the reference system.

The calculation involves a complicated diagram analysis. The result is expressed in terms of the direct correlation function:

$$\begin{aligned}
C(\mathbf{q}) = {}& C_0(\mathbf{q}) - \beta V_1(\mathbf{q}) \\
& - \frac{\beta}{2}\int\frac{d\mathbf{k}}{(2\pi)^3}\frac{\partial^2 \Lambda_0(\mathbf{q}-\mathbf{k})}{\partial\rho^2}\cdot\frac{V_1(\mathbf{k})}{1+\beta V_1(\mathbf{k})\Lambda_0(\mathbf{k})} \\
& + \frac{\beta^2}{8}\int\frac{d\mathbf{k}}{(2\pi)^3}\left(\frac{\partial\Lambda_0(\mathbf{k})}{\partial\rho} + \frac{\partial\Lambda_0(\mathbf{q}-\mathbf{k})}{\partial\rho}\right)^2 \\
& \qquad \times \frac{V_1(\mathbf{k})}{1+\beta V_1(\mathbf{k})\Lambda_0(\mathbf{k})}\frac{V_1(\mathbf{k}-\mathbf{q})}{1+\beta V_1(\mathbf{q}-\mathbf{k})\Lambda_0(\mathbf{q}-\mathbf{k})}
\end{aligned}$$

$$(2.3)$$

where $V_1(\mathbf{q})$ is the Fourier transformation of the perturbation potential. We have also introduced the function $\Lambda_0(\mathbf{q}) \equiv \rho S_0(\mathbf{q})$, with $S_0(\mathbf{q}) \equiv [1 - C_0(\mathbf{q})]^{-1}$ being the structure factor of the reference system. In the small-q limit, Eq (2.3) reduces to the the sum rule (2.1), with F being given by the free energy in the in the ring diagram approximation;

$$C(\mathbf{o}) = C_0(\mathbf{o}) - V_1(\mathbf{o}) - \frac{1}{2}\frac{\partial^2}{\partial\rho^2}\int\frac{d\mathbf{k}}{(2\pi)^3} \ln\left[1 + \beta V_1(\mathbf{k})\Lambda_0(\mathbf{k})\right].$$

(2.4)

The first two terms of Eq (2.3) represents the random phase approximation (RPA), and the first three terms include the MDA.

3. RESULTS AND DISCUSSIONS

We study a hard sphere system with an attractive Lennard-Jones tail as the perturbation;

$$V_1(r) = \begin{cases} -\epsilon & : r < 2^{1/6}d \\ 4\epsilon\left[(d/r)^{12} - (d/r)^6\right] & : r > 2^{1/6}d \end{cases}$$

(3.1)

where d is the hard sphere diametre. The reason for this choice is that the results of the Monte-Carlo (MC) simulation study[3] are available for this system. Numerical tests of the MDA have also been performed[4] for the same system. We take ϵ and d appearing in (3.1) to be the units of the energy and the length respectively. The perturbation is cut off at the hard sphere diametre in the present study. We use for $S_0(\mathbf{q})$ the analytic solution of the Percus-Yevic equation for hard spheres in this preliminary report. In this paper we shall study only the low-density case of $\rho = 0.6$ and $kT = 1.6$, for which the packing fraction is $\eta = 0.314$, so we expect that the above structure factor will do for the purpose of our numerical test.

In Fig 1 we plot the small angle structure factor for the above set of parametres. Three different approximations are shown — the RPA, the MDA and the present theory. In the figure we have also plotted the result of the MC simulations by Stell and Weis for a tentative comparison, although we need more accurate structure factor of the reference system for a quantitative argument. It is tempting to conclude that the present theory is a considerable improvement over the previous ones. In particular, the values in the vicinity of q = 0 are pushed up by the corrections to the MDA, improving the insufficiency common to the previous theories (see reference 4). This is due to the consistency restriction imposed on the formalism; the correct small-q limit is fixed as is seen from (2.4). The MDA curve calculated by us is slightly lower than that by Bowles and Silbert[4], which lies in between our MDA and the present theory. This is because we have used a different $S_0(q)$, and it suggests that the theory would predict even higher values of $S(q)$ in small-q region. It is not clear whether our theory will exceed the MC simulation.

A full report of the present study, using a more accurate $S_0(q)$, will be reported elsewhere.

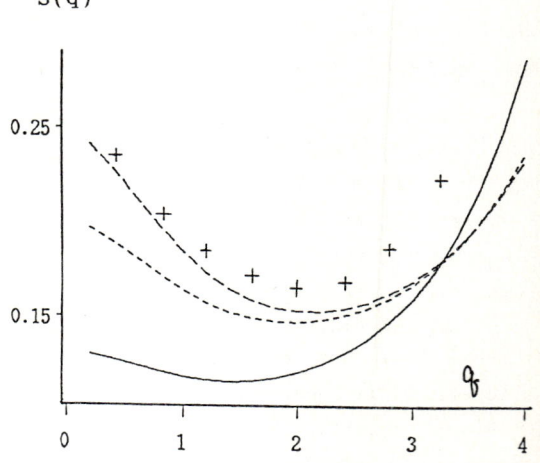

The structure factors calculated by various perturbation theories.

—————— RPA   ············ MDA
– – – – Present theory,   +   MC

REFERENCES

1. M. Itoh, J. Phys. C 21 (1988) 3387
2. R. C. Henderson and N. W. Ashcroft, Phys. Rev. A 13 (1976) 859
3. G. Stell and J. J. Weis, Phys. Rev. A 21 (1980) 645
4. R. J. Bowles and M. Silbert, J. Phys. C 17 (1984) 207

INTEGRAL EQUATION APPROACH FOR CHARGED COLLOIDAL DISPERSIONS

Makoto FUSHIKI*

Department of Physics, Faculty of Science, Kyoto University, Kyoto 606, Japan

The mean force for charged colloidal particles is calculated by using the BGY hierarchy equation. Two kinds of triplet correlation functions which appear in the equation are approximated with the second order h-bond expansion and truncated density-functional expansion.

1. INTRODUCTION

A charged colloidal dispersion is one example of an asymmetric multi-component plasma in chemistry. It consists of large colloid particles and small salt ions.

We have studied this system with the multi-component HNC equation, based on the primitive model(PM).[1] According to this theory, the height of the first peak of the structure factor, which is a measure of the correlation between colloids, has a maximum as a function of colloidal charge. This maximum may, however, be an artifact because thermodynamic consistency between the virial and compressibility equations is violated for the corresponding colloidal charge.

A similar problem is encountered in the system of ions between two parallel plates.[2] The potential of mean force between the plates, deduced by the HNC equation, does not tend to the Verwey Overweek (VO) potential in the point ion limit. On the other hand, the mean force, deduced by the BGY hierarchy equation, correctly tends to the VO potential.[2] Thus, in this work, we use the BGY equation to obtain the mean force between colloids.

2. METHOD

In this work, we are only concerned with a salt free system for simplicity, and assume counterions as point charges. By using the BGY equation, we find the mean force between colloids as

$$\nabla \ln[g_{00}(r)] = -\nabla \beta u_{00}(r) - \sum_i \int \nabla \beta u_{i0}(r_1) \, n_i(\mathbf{r}';r) \, d\mathbf{r}' \quad , \tag{1}$$

---

*Present address: Physical Chemistry 2, Chemical Center P.O.Box 124, S-221 00, Lund, Sweden where $r_1 = r - r'$, $u_{ij}(r)$ is a PM interaction potential between i,j pair (the subscripts 0, 1 refer to colloids and counterions, respectively), and $n_i(r';r)$ = $n_i g_{00i}(r,r',r_1)/g_{00}(r)$ is the density distribution of the i-th particle around two fixed colloid particles separated by r. The density distributions $n_i(r';r)$ are approximated as follows.

For $n_0(r';r)$, we use the second order h-bond expansion introduced by Haymet, Rice, and Madden (HRM).[3] This theory considerably improved the Kirkwood superposition approximation (SA) in obtaining the mean force of the Lennard-Jones system.[3]

Once $n_0(r';r)$ is obtained, the ion density $n_1(r';r)$ is deduced from the truncated density functional (TDF) expansion

$$\ln[\bar{n}_1(r';r)] = -\beta u(r';r) + \sum_i \int c_{i1}(r_1) \delta n_i(r';r) dr', \qquad (2)$$

where $\bar{n}_i(r',r) = n_i(r';r)/n_i$, $\delta n_i(r',r) = n_i(r';r) - n_i$ and $u(r';r)$ is the field created by the two colloid particles. The function $c_{i1}(r)$ is a two particle direct correlation function, which is obtained from the colloid-ion and ion-ion part of the HNC equation.[1] We further approximate $c_{11}(r)$ with the Debye-Huckel approximation. Equation (2) then becomes a self-consistent equation for $n_1(r';r)$.

By putting these $n_i(r';r)$ into eq.(2), and integrating it, we obtain $g_{00}(r)$, which is again used to get $n_0(r';r)$ in the next iteration. We continue this process until a convergence criterion for $g_{00}(r)$ is satisfied.

3. RESULTS

So far we have found only the mean force (2) from the initial $g_{00}(r)$ which

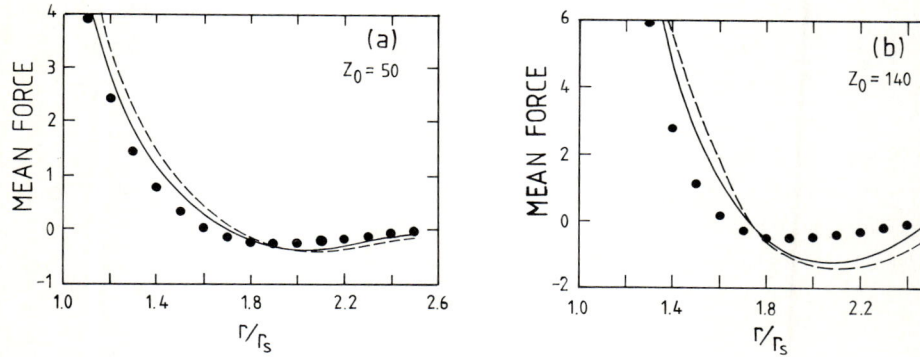

FIGURE 1
The mean force calculated with BGY/HRM/TDF (——), BGY/SA/TDF (---), and HNC (●), for $\sigma_0 = 320$ Å, $\eta_0 = 0.04$, and $Z_0 = 50$ (a) and 140 (b).[r_S is the ionsphere radius, $(3\eta_0/4\pi)^{-1/3}$.]

is obtained with the HNC equation. It is plotted in Fig.1. There we set the particle diameter σ_0 = 320 Å, volume fraction η_0 = 0.04, and charge Z_0 = 50 and 140, (T = 298 K, ε = 78.4). We refer to the results using HRM and SA for $n_0(r';r)$ as BGY/HRM/TDF and BGY/SA/TDF, respectively. The derivative of the initial $g_{00}(r)$ is also plotted for comparison.

For lower charge, all three curves are close to each other, though BGY/HRM/TDF is a little closer to HNC than BGY/SA/TDF.

For larger charge, although BGY/HRM/TDF is still closer to HNC than BGY/SA/TDF, the difference between BGY/HRM/TDF and HNC is considerable. It suggests that the final $g_{00}(r)$ will be quite different from HNC one.

REFERENCES

1) M. Fushiki, J. Chem. Phys. 89 (1988) 7445.

2) M. Lozada-Cassou and E. Diaz-Herrera, The Interaction of Electrical Double Layers, in: Ordering and Organization in Ionic Solutions, eds. N. Ise and I. Sogami (World Scientific, Singapore, 1988) pp. 555-564.

3) A. D. J. Haymet, S. Rice, and W. G. Madden, J. Chem. Phys. 74 (1981) 3033, and 75 (1981) 4696.

DENSITY FUNCTIONAL THEORY AND LANGEVIN-DIFFUSION EQUATION

Toyonori MUNAKATA

Department of Applied Mathematics and Physics,
Faculty of Engineering, Kyoto University, Kyoto 606 Japan

We present a dynamical extension of the density functional theory(DFT) for equilibrium states by deriving a Langevin-diffusion equation for the density field $\rho(\mathbf{r},t)$, which is shown to have an appropriate stationary distribution functional expressed in terms of the free energy functional used in the DFT. We apply the Langevin equation to study dynamics of liquid-crystal interface.

1. INRODUCTION

The density functional theory(DFT) plays an important role in studies on freezing and melting transitions[1,2]. The liquid-crystal interface[3], nucleation[4], glass transition[5] and quasi-crystals[6] are also treated within the framework of the DFT.

In view of the wide applicability of the theory we try to obtain a dynamic version of the DFT, which might enable us to investigate some dynamic aspects of the processes mentioned above. In this report we first derive a nonlinear Langevin equation for the density $\rho(\mathbf{r},t)$ by combining the DFT with hydrodynamics under the assumption that $\rho(\mathbf{r},t)$ is the only relevant variable that changes slowly in time. The corresponding Fokker-Planck (FP) equation is shown to have the stationary solution of the form $\exp\{-\beta F[\rho]\}$ with $F[\rho]$ denoting the free energy functional used in the DFT and $\beta=1/k_BT$ (Sec.2). After reviewing statics of the freezing and liquid-crystalinterface problems, we derive a time-dependent Ginzburg-Landau type equation for the order parameters of the problem as an application of our dynamic DFT(Sec.3).

2. LANGEVIN-DIFFUSION EQUATION

In the derivation of the Langevin equation for the density field $\rho(\mathbf{r},t)$ special attention is paid to satisfying (a) the conservation law for the number of the particles and (b) the requirement that the stationary distribution functional $f_{eq}[\rho]$ should be given by $\exp\{-\beta F[\rho]\}$ with $F[\rho]$ denoting the free energy functional to be used in the DFT.[7]

We first consider a single-component system with the density $\rho(\mathbf{r},t)$ and start from the hydrodynamic equations

$$\partial \rho(\mathbf{r},t)/\partial t = -\nabla \cdot \mathbf{g}(\mathbf{r},t)/m. \tag{1}$$

$$\partial \mathbf{g}(\mathbf{r},t)/\partial t = -\rho(\mathbf{r},t)\nabla \delta F/\delta \rho(\mathbf{r},t) - \int d\mathbf{r}' \int dt' \Gamma_m(\mathbf{r},\mathbf{r}';t-t') \cdot \mathbf{g}(\mathbf{r}',t') + \mathbf{f}(\mathbf{r},t), \tag{2}$$

where $g(r,t)$ denotes the momentum density and the first term on the r.h.s of (2) represents a generalized pressure gradient.[8] Usually the dissipative matrix Γ_m with the element Γ_{ij} (i,j=1, 2,3) is related to the random force $f(r,t)$ with zero mean through

$$\Gamma_{ij}(r,r';t-t') = \Sigma_k \int dr'' <f_i(r,t)f_k(r'',t')> <g(r'')g(r')>^{-1}{}_{kj}, \qquad (3)$$

where

$$<g_i(r)g_j(r')> = mk_BT\delta(r-r')\rho_{eq}(r)\delta_{ij}. \qquad (4)$$

Now we assume that the density field $\rho(r,t)$ is the only relevant dynamical variable which changes slowly in time. Thus we replace the quilibrium density $\rho_{eq}(r)$ in (4) by the (non-equilibrium) density $\rho(r,t)$ to obtain from (3)

$$<f_i(r,t)f_j(r',t')> = mk_BT\rho(r',t')\Gamma(r,r';t-t')\delta_{ij}$$
$$= 2mk_BT\Gamma\rho(r,t)\delta(r-r')\delta(t-t')\delta_{ij}, \qquad (5)$$

where $\Gamma(r,r';t)$ denotes the element of the isotropic diagonal matrix Γ_m and it is put equal to $2\Gamma\delta(r-r')\delta(t-t')$ for the sake of simplicity. From the above the equation of motion (2) becomes

$$\partial g(r,t)/\partial t = -\rho(r,t)\nabla\delta F/\delta\rho(r,t) - \Gamma g(r,t) + f(r,t). \qquad (6)$$

Under the assumption that Γ^{-1} is much smaller than the characteristic time for the variation of the density field, (6) reduces to

$$g(r,t) = \{-\rho(r,t)\nabla\delta F/\delta\rho(r,t) + f(r,t)\}/\Gamma. \qquad (7)$$

From (7) and (1) we finally obtain the Langevin equation with $D = k_BT/m\Gamma$,

$$\partial\rho(r,t)/\partial t = -\nabla\cdot j(r|\rho) + \xi(r,t), \qquad (8)$$

$$j(r|\rho) = -(m\Gamma)^{-1}\rho(r,t)\nabla\delta F/\delta\rho(r,t) \qquad (9)$$

$$<\xi(r,t)\xi(r',t')> = 2D(\nabla\cdot\nabla')\rho(r,t)\delta(r-r')\delta(t-t') \qquad (10)$$

Among the various free energy functionals used for a one-component system[2,3] let us take here

$$F[\rho]/k_BT = \int dr\rho\{\ln(\rho\lambda^3) - 1\} - (1/2)\int dr\int dr'\delta\rho(r)c(|r-r'|)\delta\rho(r'), \qquad (11)$$

where $\lambda = (h^2/2\pi mk_BT)^{1/2}$ and $\delta\rho(r) = \rho(r) - \rho_L$ with ρ_L and $c(r)$ denoting the density and the direct correlation function of a reference liquid. In (11) we neglected terms which play no role in (9). From (8),(9) and (11) we readily obtain, omitting the noise part,

$$\partial\rho(r,t)/\partial t = D\{\nabla^2\rho - \nabla\cdot\rho\nabla\int c(|r-r'|)\delta\rho(r',t)\}. \qquad (12)$$

Equation (12) is the nonlinear diffusion equation which has been used in discussing instability of a supercooled liquid.[9] At this point we note that in (11) we neglected terms higher than second in $\delta\rho$. Inclusion of these terms in (11) results in a three(and more)-body force term in (12) and requires knowledge of many-body correlations.

For a multi-component system with the density $\{\rho_i(r,t)\}, i=1,\ldots,\nu$ we can go through in the same way. The Langevin equation for $\rho_i(r,t)$ takes the form

$$\partial\rho_i(r,t)/\partial t = -\nabla\cdot j_i(r|\rho) + \xi_i(r,t), \qquad (13)$$

$$\mathbf{j}_i(\mathbf{r}|\rho) = -(m_i\Gamma_i)^{-1}\rho_i(\mathbf{r},t)\nabla\delta F\nu/\delta\rho_i(\mathbf{r},t) \quad (14)$$

$$\langle\xi_i(\mathbf{r},t)\xi_j(\mathbf{r}',t')\rangle = 2D_i(\nabla\cdot\nabla')\rho_i(\mathbf{r},t)\delta(\mathbf{r}-\mathbf{r}')\delta(t-t') \quad (15)$$

where $D_i = k_B T/(m_i\Gamma_i)$. If we take, as the free energy functional of the ν-conponent system,

$$F\nu[\rho]/k_B T = \Sigma\int d\mathbf{r}\rho_i\{\ln(\rho_i\lambda^3)-1\} - (1/2)\Sigma\int d\mathbf{r}\int d\mathbf{r}'\delta\rho_i(\mathbf{r})c_{ij}(|\mathbf{r}-\mathbf{r}'|)\delta\rho_j(\mathbf{r}'), \quad (16)$$

where $c_{ij}(r)$ denotes the direct correlation function between the species i and j and $\delta\rho_i(\mathbf{r}) = \rho_i(\mathbf{r}) - \rho_{i,L}$, we have a nonlinear diffusion equation similar to (12),

$$\partial\rho_i(\mathbf{r},t)/\partial t = D_i[\nabla^2\rho_i - \nabla\cdot\rho_i\nabla\{\Sigma_j\int c_{ij}(|\mathbf{r}-\mathbf{r}'|)\delta\rho_j(\mathbf{r}',t)\}]. \quad (17)$$

Single-particle motion can also be treated in a similar way.[7]

We now turn to the Fokker-Planck equation which governs time evolution of the istribution functional $f(t|\rho)$. Interpreting the stochastic integral, associated with the multiplicative noise $\xi(\mathbf{r},t)$ in (8) as an Ito-type one[18] we have from (10) and (8)

$$\langle\Delta\rho(\mathbf{r},t)\rangle/\Delta t = -\nabla\cdot\mathbf{j}(\mathbf{r},t), \quad (18)$$

$$\langle\Delta\rho(\mathbf{r},t)\Delta\rho(\mathbf{r}',t)\rangle = 2D(\nabla\cdot\nabla')\rho(\mathbf{r},t)\delta(\mathbf{r}-\mathbf{r}'). \quad (19)$$

where $\Delta\rho(\mathbf{r},t)$ denotes the increment of $\rho(\mathbf{r},t)$ in a small time interval Δt. Under the Gaussian noise we need not go beyond the second order in $\Delta\rho$ and in a continuum limit, we obtain from (18) and (19) the FP equation

$$\partial f(t|\rho) = -\int d\mathbf{r}\delta/\delta\rho(\mathbf{r})J[\rho,f], \quad (20)$$

where

$$J[\rho,f] = -f\nabla\cdot\mathbf{j} + D\nabla\cdot\rho(\mathbf{r})\nabla\delta f/\delta\rho(\mathbf{r}). \quad (21)$$

It is readily confirmed from (21) and (9) that for $f_{eq}[\rho]\propto \exp\{-\beta F[\rho]\}$ J,(21), vanishes, ensuring that it is a stationary solution to (20).

3. LIQUID-CRYSTAL INTERFACE

Equilibrium aspects of freezing and the liquid-crystal interface [1,3] are obtained by studying the solution to $\mathbf{j}=0$ in (8,9). This is equivalent to the integral equation

$$\rho_{eq}(\mathbf{r}) = \rho_L\exp\{\int d\mathbf{r}'c(|\mathbf{r}-\mathbf{r}'|)\delta\rho_{eq}(\mathbf{r}')\}, \quad (22)$$

which has a solution $\rho_{eq}(\mathbf{r}) = \rho_L$, representing a liquid state. A uniform crystal is expressed by

$$\rho_{eq}(\mathbf{r}) = \rho_L[1+\eta_{eq}+\Sigma_n\mu_{n,eq}\exp\{i\mathbf{G}_n\cdot\mathbf{r}\}], \quad (23)$$

where $\{\mathbf{G}_n\}$ denotes reciprocal lattice vectors of the assumed crystal. The transition point is found by putting the difference in the grand-potential $\Delta\Omega = \Omega_C - \Omega_L = k_B T\rho_L V\omega$ zero, i.e.

$$\omega(\eta,\mu_n) = (c_0-1)\eta + c_0\eta^2/2 + \Sigma_n c_n\mu_n^2/2 = 0. \quad (24)$$

If we are interested in a flat interface we generalize (18) by assuming the order parameters η_{eq} and μ_{eq} to be slowly varying functions of z, the coordinate normal to the interface.³ In this case $\eta(z)$ and $\mu_n(z)$ should satisfy a coupled set of differential equations, which is the Euler-Lagrange equation for the interface free energy functional

$$\Delta\Omega = \rho_L k_B T \int d\mathbf{r} \{f(z) - c_\theta''(\eta')^2/4 - \Sigma_n c_n'' \alpha_n^2 (\mu_n')^2\} \tag{25}$$

with $f(z) = \omega(z) - \beta U_\theta(z)\{1+\eta\} - \Sigma_n \beta U_n(z)\mu_n$ where $c' = dc/dk$, $\eta'(\mu') = d\eta(\mu)/dz$ and $c_n = c_{G_n}$. α_n denotes the direction cosine of $\mathbf{G}_n$ and z-axis. U_θ and U_n, each function of $\{\eta, \mu_n\}$, are the Fourier components of the external force necessary to stabilize the uniform crystal, thus for $\{\eta_{eq}, \mu_{eq}\}$ and $\{\eta=0, \mu_n=0\}$, $U_\theta = U_n = 0$ and $f(z)$ is known to have the double-well form³.

Now we turn to dynamics and derive equations for $\eta(z,t)$ and $\mu_n(z,t)$. First we linearize (9) as $\mathbf{j}(\mathbf{r}|\rho) = -(m\Gamma)^{-1} \rho_L \nabla \delta F / \delta \rho(\mathbf{r})$ and use the functional (24) as F above to obtain

$$d/dt\{\eta + \Sigma\mu_n \exp(i\mathbf{G}_n \cdot \mathbf{r})\} = D\nabla^2 \{A_\theta + \Sigma A_n \exp(i\mathbf{G}_n \cdot \mathbf{r})\}, \tag{26}$$

where

$$A_\theta = \partial f/\partial \eta - |c_\theta''|\eta''/2, \quad A_n = \partial f/\partial \mu_n - |c_n''|\mu_n''/2 \tag{27}$$

and use is made of the relation that

$$\delta F/\delta\rho(\mathbf{r}) = \{\delta F/\delta\eta(\mathbf{r}) + \Sigma_n \delta F/\delta\mu_n(\mathbf{r}) \exp(-i\mathbf{G}_n \cdot \mathbf{r})\}. \tag{28}$$

If we further neglect ∇A compared with $\nabla \exp\{i\mathbf{G} \cdot \mathbf{r}\}$ we arrive at

$$d\eta/dt = DA_\theta'', \quad d\mu_n/dt = -DG_n^2 A_n. \tag{29a,b}$$

Equations (29a) and (29b) are the TDGL equations for the conserved (η) and the non-conserved (μ) variables, respectively. Similar equation has been used to analyze dynamics of liquid-crystal interface for the b.c.c lattice by Harrowell and Oxtoby.[11]

1. T.V. Ramakrishnan and M. Yussouff, Phys. Rev. **B32**(1979)2775.
2. For reviews see R. Evans, Adv. Phys. **28**(1979)143, A.D.J. Haymet, Ann. Rev. Phys. Chem. **38**(1987)89.
3. A.D.J. Haymet and D.W. Oxtoby, J. Chem. Phys. **74** (1981)2559, D.W. Oxtoby and A.D.J. Haymet, ibid. **76**(1982)6262.
4. P. Harrowell and D.W. Oxtoby, J. Chem. Phys. **80** (1984)1639.
5. Y. Singh, J.P. Stoessel and P.G. Wolynes, Phys. Rev. Lett. **54**(1985)1059.
6. S. Sachdev and D.R. Nelson, Phys. Rev. B32(1985)1480.
7. T. Munakata, J. Phys. Soc. Japan **58**(1989)2434.
8. T.R. Kirkpatrick and P.G. Wolynes, Phys. Rev. A **35**(1987)3072.
9. T. Munakata, J. Phys. Soc. Japan **43**(1977)1723, B. Bagchi, Physica **145A**(1987)273.
10. C.W. Gardiner, Handbook of Stochastic Method, (Springer, 1982)
11. P. Harrowell and D.W. Oxtoby, J. Chem. Phys. **86** (1987)2392.
    P. Harrowell and D.W. Oxtoby, Phys. Rev. **B33**(1986)6293.

AUTHOR INDEX

Abe, R., 659
Alastuey, A., 377
Alder, B.J., 65
Ando, T., 263
Beck, B., 313
Benage, Jr., J.F., 429
Bespalov, V.E., 571
Bollinger, J.J., 177
Brochot, S., 297
Canessa, E., 675
Cauble, R., 439
Charbier, G., 495
Choquard, Ph., 225
Degani, M.H., 385
DeSilva, A.W., 449
Deutsch, C., 601, 649
DeWitt, H.E., 635
Dharma-wardana, C., 409, 421
Dubin, D.H.E., 189
Dufty, J.W., 533
Ebina, K., 397
Fajans, J., 313
Fesser, K., 255
Forrest, W.J., 33
Fortov, V.E., 571
Fu, R., 259
Furukawa, H., 613
Furutani, Y., 421
Fushiki, M., 691
Garnett, J.D., 33
George, T.F., 259
Gilbert, S.L., 177
Girardeau, M.D., 521
Grimson, M.J., 675
Harigaya, K., 255
Harris, J., 93

Hasegawa, M., 129, 201, 401
Heinzen, D.J., 177
Hensel, F., 455
Hernandez, J.P., 479
Hess, H., 483
Hiramatsu, T., 613
Hitawari, Y., 163
Hiura, J., 55
Hiwatari, Y., 167, 679, 683
Hjorth, P.G., 313
Honda, O., 687
Hubbard, W.B., 21
Ichimaru, S., 59, 101, 337, 405, 545, 653
Iglesias, C.A., 549
Imada, M., 81
Ishida, A., 129
Isihara, A., 301
Itano, W.M., 177
Itoh, M., 687
Iyetomi, H., 59, 653
Jancovici, B., 285
Jones, L.A., 429, 433
Kaburagi, M., 397
Kajita, K., 275
Kalia, R.K., 93, 385
Kambayashi, S., 683
Kaneda, T., 625
Kaneko, Y., 631
Kania, D.R., 433
Kato, K., 445
Kawaguchi, M., 613
Kawakatsu, T., 237
Kawasaki, K., 237
Kearney, K.J., 589
Kilimann, M.K., 507

Kohn, W., 331
Kondo, I., 401
Kraeft, W.-D., 507
Kremp, D., 507
Kulish, M.I., 571
Kuz, S.I., 571
Lavaud, M., 297
Leeuw, S.W. de, 93
Leger, D., 649
Lin, D.L., 259
Makishima, K., 43
Malmberg, J.H., 313
Martin, Ph.A., 377
Mihajlov, A.A., 561
Minami, K., 445
Miyagawa, H., 167
Mostovych, A.N., 589
Munakata, T., 695
Nagano, S., 171
Nagara, H., 663
Nakano, A., 337, 653
Nakayama, K., 687
Ninkov, Z., 33
Nishihara, K., 613
Niu, K., 621, 625
Nomura, T., 445
Nomura, T., 617
Nothwang, D.P., 429
O'Neil, T.M., 313
O'Neil, T.M., 189
Odagaki, T., 163
Ogata, S., 59, 101, 653
Ohnishi, S., 171
Ousaka, Y., 369
Pastore, G., 145
Perrot, F., 421
Popovic, M.M., 561
Rogers, F.J., 439, 549
Rozmus, W., 439
Sakagami, H., 613
Sakagami, Y., 617
Sakamoto, S., 467
Saumon, D., 495
Schiffer, J.P., 113
Schneider, D.H., 433
Senatore, G., 145
Shanahan, W.R., 429

Shepherd, R.L., 429, 433
Shirakawa, T., 129
Shure, M., 33
Silbert, M., 675
Singwi, K.S., 349
Skrutskie, M.F., 33
Slattery, W.L., 635
Stamper, J.A., 589
Stewart, R.E., 433
Stringfellow, G.S., 635
Sugawara, A., 445
Sun, X., 259
Suttorp, L.G., 325
Tajima, T., 653
Takada, Y., 357
Takashima, J., 679
Takasu, M., 679
Takatsuka, T., 55
Tanaka, M., 125
Tanaka, S., 405, 545
Theilhaber, J., 65
Tosi, M.P., 135
Totsuji, H., 203
Trainor, Jr., R.J., 429
Ueda, A., 631
Van Horn, H.M., 3
Vashishta, P., 93, 385
Vitel, Y., 561
Wada, Y., 243, 255
Watabe, M., 201
Watabe, M., 401
Watabe, M., 129, 201, 401
Wineland, D.J., 177
Wu, C., 259
Yan, X-Z., 405
Yasuhara, H., 369
Yasui, H., 613
Yip, S., 149
Yonemitsu, K., 373
Yonezawa, F., 467
Yoshida, F., 425
Yoshida, H., 617
Young, W.H., 401
Younger, S.M., 583
Zhang, Y.Q., 449
Zogaib, L., 533

SUBJECT INDEX

Activity expansion, 549
Adiabatic invariant, 313
Amorphization, 149
Anharmonic corrections, 635
Anyon statistics, 87
APEX approximation, 538
Argon plasmas, 449
Atomic structures, 467
$Ba_{1-x}K_xBiO_3$, 385
Ballistic compressor, 486
Balmer alpha line, 571
Band-gap renormalization, 263
Band
— structural energy, 672
— structures, 255
Baranger-Mozer expansion, 538
BCS, 81
Beam propagation, 625
Betatron oscillation, 117
Bethe Salpeter equations, 510, 521
Bijl-Dingle-Jastrow form, 358
Binary
— ionic mixtures, 614, 649
— mixtures, 237
— soft-sphere liquids, 683
Birth-rate function, 28
Blackbody radiation, 564
Bogoliubov de Gennes equations, 331
Bohr-Bethe-Bloch formula, 602
Boltzmann collision operation, 318

Boltzmann-Ziman equation, 649
Bond charge model, 135
Bound-state poles, 521
Bounded ion crystal, 190
Bremsstrahlung, 46, 613
Bridge functions, 683
Brillouin density, 181
Brown dwarfs, 3, 21, 33, 495
Capillary discharge, 429
Cell dynamics method, 237
Cepheid variable stars, 549
Chapman-Enskog polynomial expansion, 440
Charge
— neutralization, 625
— stabilised colloidal suspensions, 675
Charge-charge correlation function, 228, 378
Charged hard discs, 287
Clausius-Clapeyron equation, 487
Clausius-Mossotti relation, 225
Cluster, 65
CO_2 laser, 449
Coherent
— medium approximation, 163
— potential approximation, 255
Cohesive energy, 675
Cold
— confined plasmas, 113
— nuclear fusion, 653

Collision
— integral, 513
— operator, 440
— relaxation, 313
Concentration-concentration structure factor, 685
Conducting polymers, 243, 259
Configuration integral, 297
Contact potential, 613
Convolution approximation, 339
Cooling transition, 179
Coordination number, 467
Correlated basis functions (CBF) theory, 260
Correlation
— energy, 357
— operator, 360
Coulomb
— coupling constant, (see *Coulomb coupling parameter*)
— coupling parameter, 3, 101, 113, 178, 589, 635
— logarithm, 324, 589
Critical compressiblity factor, 659
Critical point, 483
Cryogenic
— helium gases, 445
— temperature range, 314
Crystallization, 3, 281 (see also *Freezing*)
Current-current correlation function, 98
Current correlation functions, 167
Cyclotron
— frequency, 313
— resonance, 43
DC electrical conductivity, 561
Debye-Hückel
— approximation, 232

— theory, 675
Dense
— helium, 93, 421, 583
— hydrogen plasmas, 405, 421, 545
— nonideal plasmas, 561
— plasmas, 3, 101, 409, 483, 549
Density
— correlation function, 152
— functional theory (DFT), 135, 145, 171, 201, 265, 331, 409, 421, 695
— gradient expansion, 203
— matrix, 66
— profiles, 203, 290, 565, 594
dHvA oscillation, 301
Dielectric function, 270, 301, 337, 459
Diffusion, 631
— coefficient, 152
Disconnected approximation, 440
Discreteness of charges, 213
Dissociation, 65
DLVO pair potential, 675
Doping disorder, 255
Dynamic
— form factor (see *dynamic structure factor*)
— screening, 507, 584
— structure factor, 337, 425, 449, 631
Dynamical
— conductivity, 301
— transition, 149
Dyson equations, 337, 508
Effective
— interaction, 55, 455
— mass, 369
— potential expansion, 357
— temperature, 21, 33
Elastic constants, 158

Electric
— conductivity, 109, 405, 456, 516
— fields, 533
— levels, 545
Electrical
— double layers, 289
— resistivity, 429, 439, 651
Electron
— bubble, 93
— collisions, 567
— correlations, 243, 259, 275, 301, 349
— crystal, 275
— degeneracy, 621
— liquid, 357, 369
— mean free paths, 337
— mobility, 99, 277
— resistivity, 433
— tunneling, 385
— velocity distribution, 313
Electron-electron interactions, 263
Electron-hole plasmas, 263
Electron-ion correlation potentials, 421
Electronic
— conductivity, 616
— structures, 409, 468
Elemental semiconductors, 135
Eliashberg gap equations, 385
Emission profiles, 571
Energy
— band gap, 663
— loss, 425, 608
— transport, 433
Entropy, 21
Equation of state, 405, 495, 549, 617, 635
Equipartition rate, 313
Escape rate, 278
Excess electron, 93
Exchange-correlation
— functional, 332

— hole, 349
— potential, 352
Exchange specific heat, 301 349
Expanded fluid metals, 455
Exponential screening, 377
Extended mean density approximation, 687
Fast Fourier transform, 94
Fermi statistics, 65
Fermi-edge singularities, 263
Fermi-liquid parameters, 365
Fermion simulation, 81
Fluid hydrogen, 495
Fokker-Planck equation, 536
Fractional quantum Hall effect, 306, 409
Free energy functional, 136
Freezing, 101, 135, 635
Frequency-dependent diffusion constant, 163
Frequency-dependent local-field factor, 349
Friedel oscillations, 462
Full-potential linearized-augmented-plane-wave method, 663
g-factor, 301
Galaxy, 21, 33
Gaussian limit, 533
Generalized Schrodinger equation, 521
Giant planets, 3, 21, 495
Gibbs-Bogoliubov method, 401
GINGA, 43
Ginsburg-Landau theory, 331
Glass transition, 163, 167
Glassy states, 3, 125, 149
Grand-partition function, 549
Green's function, 270, 337, 507, 521
Guiding center
— approximation, 315
— center equations of motion, 193

GW approximation, 341
Hartree-Fock
— approximation, 350
— equations, 55
Hartree-Fock-Roothaan
  method, 469
Helium film, 281
Herzfeld criterion, 484
High field limit, 533
High temperature expansion, 659
High-T_c superconductivity, 81, 335, 373
Hubbard model, 81, 373
Hyades stellar cluster, 549
Hydrodynamic heating, 571
Hydrogen
— phase diagram, 21
— plasma, 65, 171
Hypernetted-chain (HNC)
— approximation, 337, 405, 683
— integral equations, 635
Impurity ions, 545
Independent particle model, 533
Inelastic neutron scattering, 385
Inertial confinement fusion (ICF), 601, 621, 625
Infrared observations, 34
Insulator-metal transition, 479, 483
Interferogram, 593
Intra-layer and inter-layer Coulomb interactions, 213
Intramolecular vibron frequency, 663
Ion
— beam-plasma interaction, 601
— diffusion, 185
— microfield, 409
— plasmas, 177, 189

Ion-electron two-component plasma, 397
Ion-sphere radius, 102, 635
Ionic mixture, 325
Ionization, 507
— energy, 467, 483
— equilibrium, 464, 479
Isentropes, 23
Isentropic compression, 486
Ising model, 659
Isotope effect, 385
Joint probability density, 533
Jupiter, 21
Kinetic equation method, 350
Kirkwood decomposition, 416
Kohn-Sham (KS) equations, 332, 413
Korteweg-de Vries (KdV) equation, 243
Landau-Ginzburg free energy, 397
Landau interaction function, 369
Landau-level, 266
Langevin-diffusion equation, 695
Larmor formula, 323
Laser
— ablation driven shock waves, 617
— cooling, 179
— induced fluorescence, 182
— ionization and heating, 589
Lattice
— gas, 659
— instability, 259
— model, 286
Laughlin's state, 301
Law of rectilinear diameter, 464
Layered structures, 105, 123, 213

Level
— shift, 531
— spacing, 467
— width, 523
Light
— ion beam, 625
— scattering, 449
Linear mixture rule, 635
Liouville-space methods, 521
Liquid-and amorphous metals, 425
Liquid-crystal interface, 695
Liquid
— helium, 279
— metal, 401
— metal surfaces, 201
Liquid-vapor critical point, 455
Local
— density approximation, 264, 413, 663
— field correction, 138, 145, 337, 426
— field factor (see local field correction)
— thermodynamic equilibrium, 557
Logarithmic interaction, 285
Long-time tails, 167, 325
Low-mass objects, 38
Luminosities, 33
Luminosity, 4, 22
— function, 6
Madelung energies, 103, 215
Magnetohydrodynamics, 325
Many-body effects, 263, 334
Mass
— accretion, 43
— action law, 507
Metal hydrides, 653
Metal-insulator
— boundaries, 405
— transition, 455
Metallic hydrogen, 663
Metallization, 3

Microclusters, 467
Microfield method, 521
Missing mass, 33
Mode-coupling theory, 325
Modified convolution approxmation (MCA), 405–406
Molar refractivity, 484
Molecular dynamics (MD), 113, 125, 149, 171, 385, 631
— simulations, 167, 192, 313
Molecular-orbital method, 467
Momentum distribution function, 357
Monte Carlo (MC), 192
— simulation, 59, 81, 101, 129, 192, 201, 636, 653, 679
Mott criterion, 484
Multi-ionic plasmas, 635
Neutron stars, 3, 43, 55, 109
Nonideal plasmas, 571
Nonlinear transport, 282
Nuclear reaction rate, 111
Nuclear reactions, 59
Occupation numbers, 549
One-component
— plasma (OCP), 101, 125, 129, 135, 178, 189, 201, 213, 293, 635, 675
— soft-sphere fluids, 683
Opacity, 21, 549
Optical absorption spectrum, 98
Ornstein-Zernike
— correlation function, 137
— relation, 233, 416
Pair correlation function, 357, 457
Pair-function, 331
Partially ionized hydrogen plasma, 521
Partially-ionized plasmas, 507, 521, 549

Participation ratio, 96
Particle correlations, 409
Particle-particle particle-mesh (PPPM) code, 613
Partition function, 66
Path-integral, 65
Pattern formation, 237
Pauli blocking, 511
Penetration effects, 545
Penning trap, 122, 177, 189, 213
Penrose lattice, 662
Percus-Yevick (PY) approximations, 684
Perfect screening, 229
Phase
— diagram, 12
— separation, 3
— transitions, 3, 21
Phonon density of states, 385
Phonons, 243, 385
Photo-induced infrared absorption, 243
Photo-induced Raman scattering, 251
Photoelastic technique, 617
Photoemission bandwidths, 337
Photoluminescence, 263
Plane-wave basis, 663
Plasma
— frequency, 459
— opacities, 589
— phase transition, 483, 495,
— stopping, 602
— turbulence, 43
Point defects, 149
Polarization
— catastrophe, 484
— potential, 499
Polaron, 243
Polyacetylene, 243, 255
Polymer chain, 679

Pressure
— dissociation, 495
— ionization, 495
Probe laser, 181
Projectile effective charge, 602
Proton abundance, 55
Pseudopotentials, 201, 409
Pulsed
— arc, 561
— discharges, 445
Pycnonuclear reaction, 111
Quantum
— electron liquids, 337
— mechanical plasmas, 377
— molecular dynamics, 93
— Monte Carlo, 65
— wells, 263
— Wigner crystallization, 145
Quasi-molecular effects, 583
Quenches, 101
Radial distribution functions, 67, 157
Radiation, 571
Random-phase approximation (RPA), 270, 301, 349, 358, 373, 689
Reaction diffusion equation, 515
Reciprocal lattice vectors, 136
Recombination, 507
Reduced density matrix, 521
Renormalization factor, 364
Ring energy, 301
r_s parameter, 101
Saha equation of state, 589
Saturn, 21
Scaling properties, 679
Schrödinger equation, 94
Screening potentials, 59, 109, 653
Self-consistent field
— method, 521
— molecular dynamics, 583

Self-energy, 270, 337, 507
Self-similar time variability, 43
Semiconductor heterostructures, 263
Shape-dependent thermodynamic limits, 225
Shape spectroscopy, 104
Shear stress autocorrelation function, 167
Shell structures, 182
Shock
— front, 571, 617
— wave, 621
— wave experiments, 489
Short time limit, 533
Si inversion layers, 303
Size dependences, 467
Slab geometry, 190
Soft-sphere fluids, 401
Solar neutrino, 3, 549
Solid neon, 279
Soliton, 243
Solvable models, 285
Spectral
— line shape, 561
— lines, 571
Spin Zeeman splitting, 268
Spitzer value, 561
Stark shift, 573
Static structure factors, 101, 364
Stillinger-Lovett sum rule, 225
Stochastic dynamics, 163
Stopping
— power, 433
— standard model, 602
Storage ring, 113
Strong coupling effects, 545
Strongly
— coupled plasmas, 429, 433, 439, 445, 533, 571, 589

— magnetized pure electron plasma, 313
Structural
— expansion, 397, 663
— phase transitions, 663
— transitions, 213
Structure factors, 461, 687
Subpicosecond laser pulses, 439
Sum rules, 337, 352
Superconductivity, 81, 331, 385
Superionic conductors, 631
Supernova SN1987A, 44
Supersonic nozzle, 621
Surface
— correlations, 225
— electrons, 275
— energy, 204
— properties, 129, 201
— tension, 613
Surfactants, 237
Susceptibility, 229
Taurus, 33
Taurus-Auriga dark clouds, 33
Temperature profiles, 561
Thermal
— conductivities, 109, 651
— de Broglie wavelength, 102
— evolution, 22
Thermodynamic
— functions, 297, 421
— properties, 401
Time correlation functions, 325
Time-dependent Ginzburg-Landau (TDGL) model, 237
Total energy, 467
Transport cross section, 557
Trap parameter, 192
Triplet distribution functions, 125

Two-component plasmas, 292, 331
Two-dimensional
— Coulomb systems, 285
— electron gas, 301, 349
— two-component Coulomb gas, 297
Two-fluid model, 331
Two particle bound state, 507
Vapor-liquid phase transition, 479
Variational theory, 357
Velocity autocorrelation functions, 586
Vertex correction, 337, 373
Vitrification, 149

Vlasov equation, 625
Voronoi polyhedron, 125
WCA fluid perturbation expansion, 496
White dwarfs, 3, 59
Wigner
— crystallization, 350
— transition, 102
Wigner-Kirkwood (WK) expansion, 378, 498
Wigner-Seitz radius, 178
Winding angle, 679
X-ray
— observations, 43
— emission, 43
Ziman formula, 458